The First Sourcebook on Asian Research in Mathematics Education: China, Korea, Singapore, Japan, Malaysia, and India

International Sourcebooks
in Mathematics and Science Education

Series Editor:
Bharath Sriraman, *The University of Montana*

International Sourcebooks in Mathematics and Science Education

Bharath Sriraman, Series Editor

The First Sourcebook on Nordic Research in Mathematics Education: Norway, Sweden, Iceland, Denmark, and Contributions From Finland (2010)
edited by Bharath Sriraman, Christer Bergsten, Simon Goodchild, Gudbjorg Palsdottir, Bettina Dahl Søndergaard, and Lenni Haapasalo

The First Sourcebook on Asian Research in Mathematics Education: China, Korea, Singapore, Japan, Malaysia, and India (2014)
Edited by Bharath Sriraman, Jinfa Cai, Kyeong-Hwa Lee, Lianghuo Fan, Yoshinori Shimizu, Chap Sam Lim, and K. Subramaniam

The First Sourcebook on Asian Research in Mathematics Education: China, Korea, Singapore, Japan, Malaysia, and India

Edited by

Bharath Sriraman
The University of Montana

Jinfa Cai
University of Delaware

Kyeong-Hwa Lee
Seoul National University

Lianghuo Fan
University of Southampton

Yoshinori Shimizu
University of Tsukuba

Chap Sam Lim
Universiti Sains Malaysia

K. Subramaniam
Homi Bhabha Centre for Science Education (TIFR), India

Information Age Publishing, Inc.
Charlotte, North Carolina • www.infoagepub.com

ISBNs:

China and Korea Sections

Paperback: 978-1-62396-028-5
Hardcover: 978-1-62396-029-2
eBook: 978-1-62396-030-8

Singapore, Japan, Malaysia, and India Sections

Paperback: 978-1-68123-236-2
Hardcover: 978-1-68123-237-9
eBook: 978-1-68123-238-6

Printed in the United States of America

To the present and future generations of children in China, Korea, Singapore, Japan, Malaysia and India.

CONTENTS

CHINA

PART VI. TECHNOLOGY

KOREA

SINGAPORE

MALAYSIA

JAPAN

INDIA

SINGAPORE

SECTION EDITOR
Lianghuo Fan
The University of Southampton, UK

Editorial Board

Authors

Shaljan Areepattamannil, *Nanyang Technological University, Singapore*
Chun Ming Eric Chan, *Nanyang Technological University, Singapore*
Kok Leong Boey, *Nanyang Technological University, Singapore*
Lianghuo Fan, *The University of Southampton, UK*
Berinderjeet Kaur, *Nanyang Technological University, Singapore*
Kim Hong Koh, *The University of Calgary, Canada*
Lionel Pereira-Menzoda, *Educational Consultant, Canada*
Swee Fong Ng, *Nanyang Technological University, Singapore*
Yan Zhu, *East China Normal University, China*

CHAPTER 41

RESEARCH IN SINGAPORE MATHEMATICS EDUCATION

An Introduction

Lionel Pereira-Mendoza, Swee Fong Ng, and Lianghuo Fan

As the recent large scale international studies have shown, Singapore students' performance in mathematics is amongst the best in the world. When discussing the latest results of Trends in International Mathematics and Science Study (TIMSS) and Progress in International Reading Literacy Study (PIRLS) in a 2012 press release, Ho Peng, Director-General of Education of the Singapore Ministry of Education, by saying, "Overall, Singapore students have done well. In particular, students across all ability groups have improved. We are especially encouraged by the progress made by our academically-weaker students" (Ministry of Education, 2012).

Singapore students' outstanding performance is the result of many factors. These include the commitment of all levels, government, teachers, parents and students to the importance of education, particularly mathematics. This has developed a culture which places the need and desire for, and commitment to, education high on everyone's agenda. In addition to

The First Sourcebook on Asian Research in Mathematics Education: China, Korea, Singapore, Japan, Malaysia, and India, pp. 927–932

the education culture, teachers are well paid, schools well resourced, and an extensive network of tuition teachers and centres available to students. Finally, the centralised assessment system, which ensures that the curriculum is implemented, and a system which rewards meritocracy, mean that success is recognised within the society.

This does not mean that Singapore is not taking major steps to review and reform its curriculum and policies. It is continually implementing changes to enhance the system and ensure that performance in mathematics and other subjects will continue to be among the best in the world and support Singapore's economic development.

The articles in this section give a snapshot of some of the research that has been undertaken in Singapore. It is worth noting that the Ministry of Education is aware of the research and takes the information into account when making decisions.

The remainder of this introduction will look at the specific articles. It should be emphasised that our intention is to give the reader a sense of focus in each chapter, as a starting point prior to reading it in detail. Needless to say, the introduction for each chapter is brief and should not be considered to show a complete picture of the chapter. This introductory chapter concludes with a general discussion that might help the reader contextualise the research and place it into a more global perspective.

Chan (Chapter 42) provided an overview on recent research conducted in Singapore on mathematics problem-solving, which is at the core of the Singapore Mathematics Curriculum. As with research in many other countries, there are many topics covered under this broad heading and many questions still to be researched. However, there are a few points worth noting in this introduction. The Model-Drawing method has been central to helping students solve word problems in primary school through providing affordances for the establishing of relationships between variables by way of text, pictorial and symbolic representations. The research shows that it is an effective method in meeting this goal, but does note that, as with any other approach, it has limitations. Another area of research looked at solving open-ended problems and concluded that this was effective in enhancing performance as well as developing students' positive attitude towards the learning of mathematics. Chan also looked at other aspects of problem-solving research in Singapore including the model method, solving non-routine problems, sense making, metacognition, affect and the use of ICT. Finally, a brief discussion on future research directions in mathematical problem solving is provided to conclude the review. This chapter provides the reader with an excellent overview of recent problem-solving research in Singapore.

Zhu and Fan (Chapter 43) reviewed research on the Singapore mathematics curriculum and Singapore textbooks. There is a general consensus that while textbooks do not equate to the curriculum, they are an important component of the implemented curriculum, or as the TIMSS researcher group defined, the potentially implemented curriculum (e.g., see Schmidt et al., 2001). Zhu and Fan concluded that the Singapore mathematics curriculum has been overall well planned and effectively implemented. They also found that, while there are gaps between the national curriculum and Singapore textbooks, for example in the area of presenting problem solving, in general there is good alignment between the curriculum and the textbooks. They pointed out that while textbooks were a major resource for teachers, many other teaching resources were also used, "including worksheets, assessment books, and IT resources". Zhu and Fan also reviewed research involving textbooks in Singapore and other countries such as the USA which provides the reader with an international comparison on components of textbook use in other countries.

Koh's contribution, Chapter 44, began with a discourse on the importance for realigning mathematics classroom assessments to the desired learning outcomes of the 21st century as well as the need for improving teachers' assessment literacy in designing and implementing authentic assessments in mathematics using the criteria for authentic intellectual quality. Koh also reviewed the literature on mathematics assessments and used two of her own empirical studies in Singapore mathematics classrooms to argue for the importance of improving mathematics teachers' assessment literacy, especially for quality task design. In her first study, she found that the majority of mathematics assessment tasks in Singapore classrooms tended to focus on the drill-and-practice of mathematical facts and procedures, which contradicts the teaching and learning of the 21st century competencies. Her second study demonstrated the benefits of providing mathematics teachers with on-going, sustained professional development in designing and implementing authentic assessments, which in turn led to improved student learning. The implications of her evidence-based findings on the planning and implementation of teacher training courses and professional development programmes in mathematics assessment were discussed at the end of the chapter.

Ng (Chapter 45) discussed the various issues students have when they used the model method, an integral part of the Singapore curriculum's approach in primary school, to solve structurally complex word problems. Although primary students found the model method useful, their knowledge of the method could be an obstacle to their learning of letter-symbolic algebra, the method of choice of secondary mathematics teachers. Studies applying neuroimaging methodologies were conducted, better to understand how the brain responds to the two different methods. The findings suggest that, of the two methods, letter-symbolic algebra is more attention resource intensive. This is an important finding since other research shows that proficiency with the concrete and more visual model method could inhibit a move towards abstract letter-symbolic algebra in secondary school. An important component of this chapter is the development of a theoretical model to explain the greater resource demands on an individual using the model method then using the letter-symbolic algebra. This is of particular importance for readers who might be interested in using the model approach, since the transition to letter-symbolic algebra is a key component of the overall development of a mathematics programme.

Boey, Areepattamannil, and Kaur (Chapter 46) looked at grade 8 students' understanding of mathematics, which comprised four dimensions based on the SPUR approach to the learning of mathematics. These dimensions include skills (knowing procedures to obtain and check solutions), properties (knowing properties that identify and justify reasoning), uses (recognizing situations in which mathematics is applied), and representations (drawing a picture or otherwise to represent the mathematical idea). This final chapter differs from the others since it focuses on one study undertaken by the authors. They noticed that students' self-concepts such as self-confidence in learning mathematics, and school climate were positively correlated with each of the four dimensions of understanding. They also indicated that the amount of homework has a positive association with the students' performance in every dimension of understanding. They concluded that the mathematics curriculum should include some of the SPUR approaches since they identify different aspects of students' weaknesses in their understanding of mathematics.

To conclude, we wish to offer a few general comments that are relevant to the body of research reviewed in this section. Singapore has a centralised education system with a single curriculum for a particular level or stream (for example, the normal academic and express streams in secondary school). It also has a series of high stakes examination. This structure is directed and overseen by a single Ministry of Education. While schools have considerable autonomy in their pedagogy, it operates within a very structured environment. Thus teachers are very focused in what they do and the expectations they have for their students.

It is also worth noting that the studies reported in this section have been largely undertaken by researchers at the National Institute of Education, which is part of Nanyang Technological University and the sole teacher training institution in Singapore. These researchers are cognizant of the key issues related to the teaching and learning of mathematics. It is therefore not surprising that much of the research reported focus on critical issues related to the Singapore mathematics curriculum. In particular, problem solving is at the core of the national curriculum, and to enable students to succeed in problem solving, it is necessary to provide them with the appropriate tools. Hence it is important for teachers to teach students problem solving heuristics. Because the so-called model method could be used to solve arithmetic and algebraic-type word problems and could be used as a bridge to learn letter-symbolic algebra, it explains why the model method is treated in teaching and research more prominently than other problem solving heuristics. Furthermore, students must value the mathematics they learn and be confident that the knowledge they have acquired will serve them well in the 21st century. Assessment largely drives teaching and learning. Hence it is important to improve assessment literacy of teachers. Schools must also be supportive of what students learn and how they are assessed. Hence professional development of teachers in the area of assessment is also an important factor to improve the teaching and learning of mathematics. In a sense, although the research conducted and reported about Singapore mathematics education may not be as extensive as that carried out in some other larger countries, it is, nevertheless, very focused on significant local issues. As such, the research is likely to have a greater impact on practice within Singapore than a larger body or more diverse research. This type of focus might be of particular interest to colleagues who are particularly interested in translating research into classroom practice.

Finally, it should be pointed out that this section is not intended to present a comprehensive review of research about Singapore mathematics education. For those readers who are interested in such a more comprehensive view, we suggest they refer to the book *Mathematics Education: The Singapore Journey* (Wong, Lee, Kaur, Foong, & Ng, 2009). However, it is our hope that the chapters in this section will provide the readers with a meaningful window into the scope of the research about Singapore mathematics education and generate more informed discussion on Singapore's achievement in mathematics.

ACKNOWLEDGMENT

In addition to the three editors, Drs. Eirini Geraniou, Kim Hong Koh, Charis Voutsina and Yan Zhu also served as reviewers for different chapters in their section. We wish to record here our sincere appreciation to their valuable help.

REFERENCES

Ministry of Education (2012, December 12). International studies affirm Singapore students' strengths in reading, mathematics & science. Retrieved from http://www.moe.gov.sg/media/press/2012/12/international-studies-affirm-s.php

Schmidt, W. H., C., M. C., Houang, R. T., Wang, H. C., Wiley, D. E., Cogen, L. S., & Wolfe, R. G. (2001). *Why schools matter: A cross-national comparison of curriculum and learning*. San Fransisco: Jossey-Bass.

Wong, K. Y., Lee, P. Y., Kaur, B., Foong, P. Y., & Ng, S. F. (Eds.). (2009). *Mathematics education: The Singapore journey.* Singapore: World Scientific.

CHAPTER 42

MATHEMATICAL PROBLEM SOLVING RESEARCH INVOLVING STUDENTS IN SINGAPORE MATHEMATICS CLASSROOMS (2001 TO 2011)

What's Done and What More Can Be Done

Chun Ming Eric Chan
National Institute of Education, Nanyang Technological University, Singapore

INTRODUCTION

Singapore gained independence in 1965. In its relatively short history of nationhood, mathematics education in Singapore has made significant progress as affirmed by the outstanding results Singapore has attained in international comparative studies such as the Trends in Mathematics and Science Studies (TIMSS) since 1995 and the recent Programme for International Student Assessment (PISA) in 2010. While numerous factors con-

The First Sourcebook on Asian Research in Mathematics Education: China, Korea, Singapore, Japan, Malaysia, and India, pp. 933–957

tribute to the apparently successful education system, it cannot be denied that it is the mathematics curriculum that drives instruction and assessment.

Singapore's mathematics curriculum is popularly known to teacher-practitioners and educators as the problem-solving curriculum. Because of the problem-solving emphasis, research, particularly in the area of mathematical problem solving has grown since the 1990s (Fan & Zhu, 2007; Foong, 2007) and has been the dominant area of interest in research.

In this chapter, a brief history on the evolution of the Singapore Mathematics Curriculum is presented as overview and background information towards understanding the core components essential to developing good problem solvers. This is followed by a review of research on problem solving involving students in the last decade reflective of the growing interest in this field of study. Various strands have been identified and these include the Model-Drawing Method, heuristics for solving non-routine problems, the solving of open-ended and real-world problems, the relation of metacognition, affect, sense-making, and the innovative use of technology in problem solving. Finally, some implications and challenges are discussed in particular the classroom pedagogy in promoting problem solving, and some recommendations posed as the way forward to meet the goals of the curriculum.

TOWARDS A PROBLEM-SOLVING MATHEMATICS CURRICULUM: A BRIEF BACKGROUND

Unlike other big countries, Singapore is a small country with an estimated area of 700 m^2 and has a population of about 5 million. Singapore has no natural resources and much of the economic progress made towards achieving a developed country status has been dependent on human capital.

Singapore has a highly centralized education system under the purview of the Ministry of Education (MOE). A small country coupled with a centralized system "*allows for a high degree of homogeneity and coherence in curriculum coverage*" (Kaur, 2009, p. 456) and makes it easier to implement the curriculum compared to countries with many diverse curriculums. Through the years after gaining independence, Singapore's mathematics education system has evolved based on the nation's development as well as the influences from reform movements in the United States (e.g. the *Agenda for Action* (NCTM, 1980)) (Kho, Yeo & Lim, 2009). In 1990, mathematical problem solving became the central focus of the mathematics

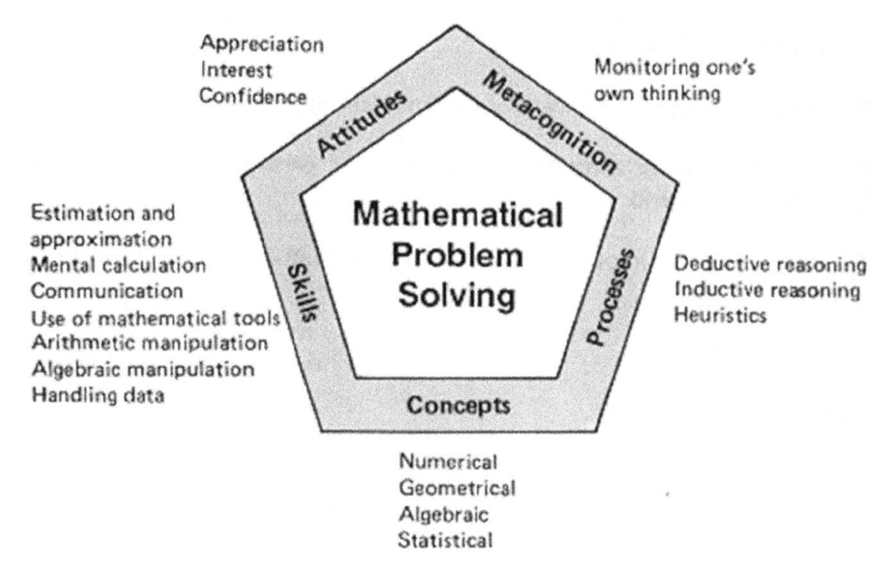

Figure 42.1. The Singapore Mathematics Curriculum Framework (1990).

curriculum and with importance placed on processes in mathematics learning (Ministry of Education, 1990a; 1990b).

Figure 42.1 shows the Singapore Mathematics Curriculum Framework developed in 1990. This framework has problem solving at its centre and has five inter-related core components that are deemed to be essential for successful problem solving: *concepts*, *skills*, *processes*, *metacognition*, and *attitude*.

The other key aspect about this curriculum is that it adopts a spiral structure in the development of concepts, skills and processes where similar mathematics topics are found across each level but the content of the topic expands in increasing complexity as students move up the levels. Through such an approach, correspondingly, students learn to solve problems of increasing difficulty within specific topics as they move up the levels as well.

The next significant revision made to the mathematics curriculum framework was in 2000. The move was towards the attainment of a more comprehensive framework that befitted the relevance of the mathematics education landscape while maintaining mathematical problem solving to be at the heart of the curriculum. For instance, the revised version in 2000 included "Perseverance" under *Attitudes* to take into account the need to cultivate the spirit of not giving up as students work on non-routine and open-ended problems (Kho, Yeo & Lim, 2009). The emergence of non-

Thinking Skills	Heuristics
• Classifying • Comparing • Sequencing • Analyzing parts and Whole • Identifying patterns & Relationships • Induction • Deduction • Spatial Visualization	• Act it out • Use a diagram / model • Make a systematic list • Look for pattern(s) • Work backwards • Use before-after concept • Use guess-and-check • Make suppositions • Restate the problem in another way • Simplify the problem • Solve part of the problem

Figure 42.2. Thinking skills and heuristics in the mathematics curriculum (2000).

routine and open-ended problems would require students to use a wide range of heuristics and thinking skills (Ministry of Education, 2000) as shown in Figure 42.2.

Subsequently, in 2006, the framework was revised to cover more components. The 2006 version of the Singapore Mathematics Curriculum Framework (Ministry of Education, 2006a; 2006b) is shown in Figure 42.3.

According to ministry officials, Kho, Yeo and Lim (2009), "Beliefs" was introduced as they were deemed to influence student attitudes in mathematics and problem solving as well as to inculcate a sense of ownership in learning. "Self-Regulation" was included under *Metacognition* to encompass a wider interpretation of metacognition involving both monitoring of thinking during problem solving and regulation of learning behaviours by the students themselves. Under the *Processes* core component, "Reasoning, communication and connections" and "Applications and modelling" were included to highlight the importance of these processes to meet the challenges of the 21st century. This is a significant change as these processes now require students to make their mathematical thinking visible (Chan, 2008) as they search for patterns, structures, relationships, and question why certain things work and how they can validate responses. The review of the curriculum also resulted in taking an integrated approach to using the problem-solving heuristics. In other words, the listed heuristics of the 2000 version should not be viewed as stand-alone strategies since some problems require a mix of different heuristics to solve. The heuristics were thereby categorised as "giving a representation" (e.g. drawing a diagram,

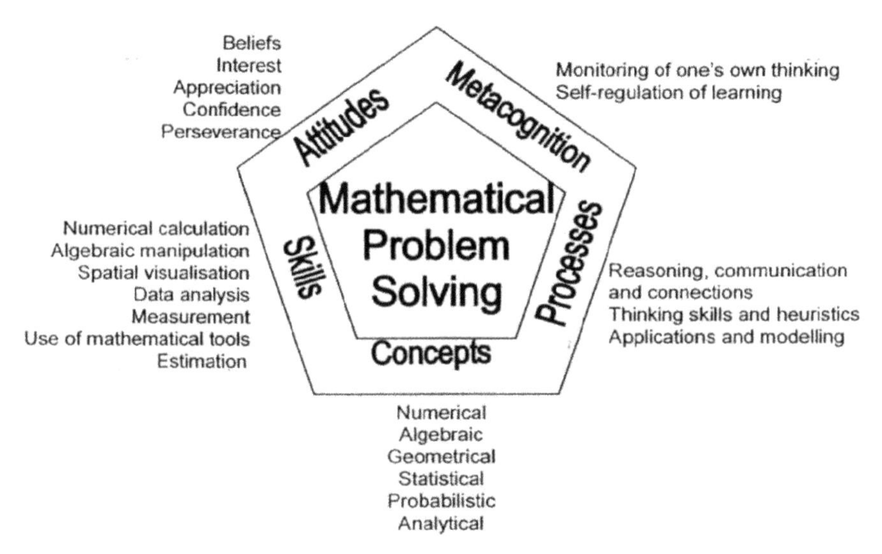

Figure 42.3. The Singapore Mathematics Curriculum Framework (2006).

making a list, using equations), "making a calculated guess" (e.g. guessing-and checking, looking for patterns, making suppositions), "going through the process" (e.g. working backwards, acting it out, using before-after concepts) and "changing the problem" (e.g. simplifying the problem, restating the problem, solving part of the problem).

DESCRIPTION OF PROBLEM-SOLVING IN BROAD STOKES

The Singapore Mathematics Curriculum Framework does not define what problem solving is but puts in broad strokes what problem solving encompasses.

> Mathematical problem solving includes using and applying mathematics in practical tasks, in real life problems and within mathematics itself. In this context, a problem covers a wide range of situations from routine mathematical problems to problems in unfamiliar contexts and open-ended investigations that make use of the relevant mathematics and thinking processes. (Ministry of Education, 1990a, p. 3)

> Mathematical problem solving is central to mathematics learning. It involves the acquisition and application of mathematics concepts and skills in a wide range of situations, including non-routine, open-ended and real-world problems. (Ministry of Education, 2006a, p. 6)

Both the descriptions offer a range of mathematical problems that students could experience. One notes that the 1990 version includes "routine mathematical problems" but it is the 2006 version that captures what a problem-solving situation should entail if one is to think about the problem-solving actions exercised using non-routine, open-ended and real-world problems. The omission of "routine mathematical problems" aids the cause in distinguishing solving problems and doing routine exercises.

A REVIEW OF MATHEMATICAL PROBLEM-SOLVING RESEARCH INVOLVING STUDENTS IN SINGAPORE MATHEMATICS CLASSROOMS (2001 TO 2011)

Several local researchers have provided reports on the state of research in mathematics education and problem solving in Singapore. In 1991, Chong et al. (1991) published the so-called "state-of-the-art" review of the mathematics education scene from the early 1980s to the early 1990s. This was followed by Foong (2009) whose review on research in problem solving from the 1990s to 2007 was seen through a categorization of specific themes related to problem solving. Fan and Zhu's (2007) review offers an overview of the development of mathematical problem solving research that relates to the curriculum and classroom teaching. Although the state of research in mathematics education is growing, it must be noted that compared to the practice of mathematics education, the history of research in mathematics education is relatively short (Foong, 2007).

This section provides a review of problem solving research that involves students in the last decade (from 2001 to mid-2011). The scope of the review covers research areas based on broad strands that align with the emphases of the Singapore mathematics curriculum framework. Most of the studies reported in this chapter were found to have Primary and Secondary students as the participants. Indeed as Foong (2007) found out, most of the research studies were from teacher practitioners taking post-graduate studies who carried out the research in their Secondary or Primary schools. Individual studies that are topic based (e.g. Fractions or Speed) and where only one or two of such are located are not presented in this chapter. Additionally, most of the studies did not adopt a longitudinal design.

Research Involving the Model-Drawing Method

All students in Singapore primary schools are taught to use the Model-Drawing Method (MDM) as a heuristic to solve structured translational word problems from Primary 1 through Primary 6 (Ng & Lee, 2009). The

MDM essentially is a heuristic that involves drawing a diagram comprising a bar or bars that can be sub-divided into parts or units. At the lower-primary levels, students solve one-step or two steps word problems that usually have an additive, multiplicative or comparative structure. At the upper-primary levels, the word problems usually involve more than two steps (also known as multi-step problems) that may incorporate different topics like fraction, ratio and percentage. According to Kho, Yeo and Lim (2009), this method was an innovation developed by a project team under the Curriculum Development Institute of Singapore (CDIS) in the 1980s to address the difficulties of students solving word problems in the early years of primary school. Kho (1987), a member of the pioneering project team, asserted that using the MDM can help students gain better insights into mathematical concepts such as fractions, ratio and percentage, plan for solution steps, stimulate students to solve challenging problems and it is a method that is comparable to but less abstract than the algebraic method. Figure 42.4 shows an example of a solution to a moderately sophisticated word problem solved at the Primary 5 level through model-drawing. The MDM has since become a distinguishing feature of the Singapore mathematics curriculum and approved textbooks at the primary levels.

Although this method was introduced in the 80s, it is only quite recent that research was carried out to determine its impact on solving structured word problems. Researchers are keen to know what makes the MDM an important heuristic in the curriculum and how students use it. Ng and Lee's (2005) research had 151 Primary 5 students that used the MDM to solve arithmetic and algebraic word problems. They found that the students were more successful when solving arithmetic word problems compared to algebraic word problems and that the method cannot solve all algebraic word problems. With the algebraic word problems, the rate of success decreased when the problems involved two additive non-homogenous relationships with errors clustered around relational phrases such as 'more than', 'less than' and 'n times as many'. When students used the MDM specifically in solving problems that involved fractions, the pupils showed that they had a superior command of how they visualised the problem and were able to alternate between the symbolic representations of the problem and its visual analogue. Their later study posited that the MDM engaged students through three modes of representations during problem solving: text, pictorial and symbolic (Ng & Lee, 2009). In particular, this method provides students with "visual analogues" in capturing all the information in a word problem and in its ability to show relationships between variables thus enabling students to construct appropriate steps towards solving the problems.

Ali bought 5 buns and 2 sandwiches. He paid a total of $4.05 for them. If a bun costs half the price of a sandwich, how much does a sandwich cost?

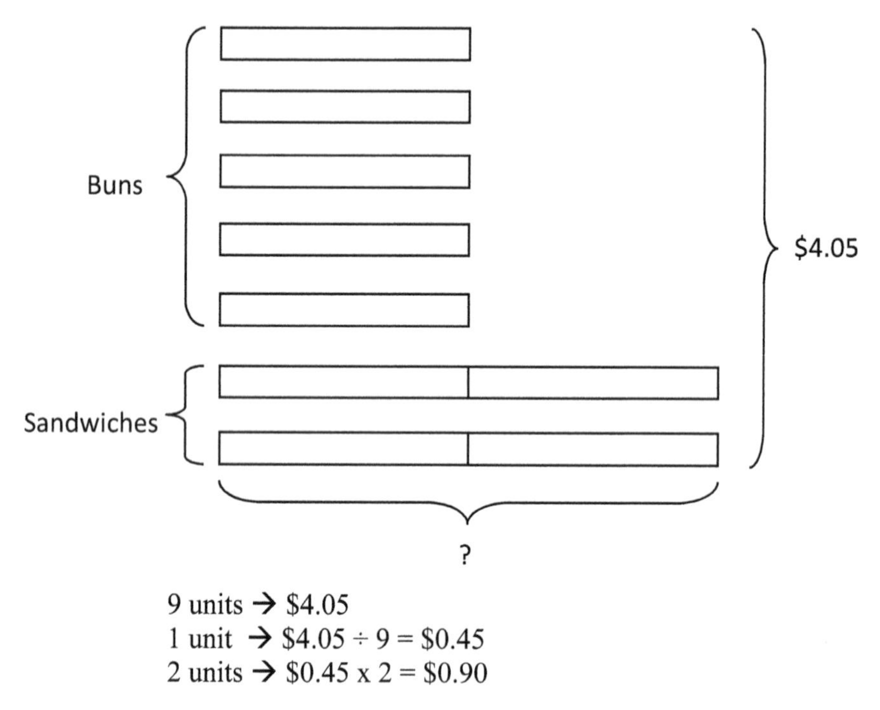

Figure 42.4. Solution to a primary 5 problem solved using the MDM.

Learning to use the MDM is not without difficulties as well. A study by Goh (2009) revealed several difficulties Primary 5 students faced in using the MDM for solving before-after concept word problems. She found that students tended to draw inaccurate or incorrect models for one-step translational word problems which probably were due to their inability to comprehend the language in before and change situations used in the problems or their lack of basic skills in transforming models. Where multi-step problems were concerned, students tended to face difficulties when the problems had elements of comparative relationships that involved the value of change.

The MDM has been asserted to be a bridge to help students transit towards symbolic algebra when they go to secondary schools. A concern was whether Primary school students who go to Secondary schools would still draw models when the preferred method in Secondary schools was to use symbolic algebra. Ng's (2003) research found that, from a sample of 145 Express Stream Secondary Two students, symbolic algebra was the preferred method for solving problems instead of the MDM after having learned algebra. With the MDM, the limitation was mainly with the uncertainty of determining where the unknowns were, which had implications for Primary school teachers to be more explicit in making the visual connections of the rectangle bars to representing the unknowns as presented in the word problems.

In an effort to ease the transition from use of the MDM to the use of letter-symbolic algebraic methods, Looi and Lim (2009) developed a computer software tool "AlgeBAR" and compared the performances of two classes of Secondary 1 students involved in the research. One class was the experimental class that used AlgeBAR where the students used the model-drawing software tools for solving word problems while the other used the MDM as bridging lessons towards learning algebra. The intervention was a four week period held in between a pre- and post- algebra test. The post test scores revealed that the experimental class has performed significantly better than the control class after the intervention, and this outcome was significant as well even after adjusting for the students' PSLE (Primary School Leaving Examination) mathematics grades. As this was a preliminary study and part of a bigger study, Looi and Lim (2009) believed that the software tool was an important enabler of the bridging process and sought to research further the efficacy of using the tool.

In another study by Poh (2007) towards understanding the inclination of Primary 3 students in using the MDM to solve word problems, it was found that more than half the class did not use models to solve the problems. Poh (2007) asserted that the students, when they have other means, do not see the need to use models to solve the word problems.

Research Involving Other Heuristics to Solve Non-Routine Problems

Not all problems can be solved by or require the use of the Model-Drawing Method because there are problems with structures that do not conform to the additive, multiplicative, comparative or algebraic structure of the standard translational word problems. In this respect, teachers are encouraged to use a variety of non-routine mathematics problems in their teaching to de-emphasize the use of routine problems that promote basic algorithmic skills (Foong, 2002) and students would thus need to learn a repertoire of different heuristics (see Figure 2) to engage such problems. The non-routine problems that students typically solve are not ill-structured problems but are well-structured tasks with a "*non-standard nature*" that requires "*the generation of procedures*" to solve them (Wong, 2002, p. 11).

A research project on "Developing the Repertoire of Heuristics for Mathematical Problem Solving" was started in 2004 by a team from the Centre for Research in Pedagogy and Practice in the National Institute of Education (Hedberg et al., 2005). Although the project primarily involved examining teachers' classroom practices, it also involved 218 Primary 5 and 145 Secondary 1 students who sat for a pre- and post-test and their solution scripts were analysed according to the types of heuristics used. The analysis revealed a limited range of heuristics students had used to solve problems. The students did better in the later set of tasks though, suggesting some improvement in their performance towards the end of the project. According to Wong (2008) who was part of the project team, the results provided some evidence that upper primary and lower secondary students had different success in using five of the common heuristics (systematic listing, guess-and-check, equations, logical argument and diagrams) to solve mathematics problems. Of interest to the researchers was the outcome where a sizable number of students did not use the same heuristics to solve parallel problems. The implications were for teachers to improve their pedagogy with a more accurate delineation of the interaction between heuristics, task and students.

Other research that involves students solving non-routine problems using heuristics mainly comprises smaller samples such as intact classes. Ho's (2007) study had Secondary 3 students working in small cooperative groups to solve non-routine problems. She found that the students could apply most of the heuristics learned and were less reliant on strategies such as guess-and-check, logical reasoning and unsystematic listing in the post test. In another study, Wong (2002) revealed that Primary 5 students who were taught heuristics to solve non-routine problems were more proficient in selecting and implementing appropriate heuristics and they showed more flexibility in getting diverse solutions.

Research Involving Solving Open-Ended and Real World Problems

Moving forward, the Ministry of Education sought to make progress in the areas of fostering in students greater motivation and stronger habits of independence and reflective learning. They published a monograph, SAIL (Strategies for Active and Independent Learning), in 2004 (Ministry of Education, 2004). Under SAIL, the instructional strategies are to be part of the repertoire of strategies to promote student-centred learning.

Rubrics are introduced in that document as a means to assess students in mathematical problem solving and teachers are to provide qualitative feedback so that students would know their strengths and weaknesses. In this respect, teachers are to select more "*open-ended tasks that encourage divergent thinking and the generation of alternative responses*" (p. 9).

Subsequently, under the *Teach Less, Learn More* initiative mooted in 2005 towards engaged learning, the Ministry of Education produced a document they termed as *Principles of Engaged Learning* (Ministry of Education, 2005). This document serves as a toolkit in providing teachers with examples of various types of pedagogies such as problem-based learning (PBL) and cooperative learning (CL) to support instruction. The productions of these supplementary materials paved the way in making the intent known in the latest 2006 curriculum revision. As mentioned earlier, *Reasoning, Communication and Connections* as well as *Applications and Modelling* were included under the *Processes* component in the 2006 revision. Although Singapore schools have a strong culture of teacher-centred pedagogy where seat work and practice overwhelmingly make up the duration during problem solving (Yeo & Zhu, 2005; Ho and Hedberg, 2005), there appears to be increased research interest in open-ended and real-world problem solving in the last decade. The findings of related studies are reported below.

Chow (2004) studied the impact of open-ended problem solving as an alternative assessment method on Secondary One students. Her findings showed a significant difference in the experimental groups' overall performance in open-ended problems with improvement in cognitive areas like mathematical knowledge, strategic knowledge and mathematical communication. Seoh (2002) engaged Secondary Five students through an *Open-Ended Mathematics Programme* for eight sessions and found that the paired students working on this programme enhanced their critical thinking skill in mathematics. Chang (2005) investigated how the use of an open-ended approach could enhance higher-order thinking in mathematics with her Secondary Two class. She found that her students demonstrated the skills of analysing, synthesizing and evaluating. In both Seoh's and Chang's research, students gave positive feedback and were receptive towards solving such problems.

In an exploratory study where secondary school students solved authentic and open-ended problems termed as performance tasks, Fan and Zhu (2008) found that the students performed significantly better than their counterparts in solving conventional exam problems, and in general they also showed more positive changes in attitudes towards mathematics and mathematics learning. The findings gave credence to the use of contextualised problems in real-life situations and open-ended investigations in students' learning of mathematics.

Ng's (2010) case study of two groups of Secondary Two students that participated in a real-world design-based mathematical project showed that the students had contrasting reasoning and sense making processes related to the reasonableness of measurements within their interpretations of real-world constraints of the project mainly due to the group dynamics and their ability to communicate mathematical arguments effectively during the discourse.

Chan (2009, 2010) investigated Primary 6 students' mathematical modelling process in a problem-based learning setting. In the study, mathematical modelling was seen as a problem-solving activity where students worked collaboratively towards problem resolution. He found that the students were able to model problem situations in varying degrees of sophistication through conceptualizing representations and giving mathematical meanings to them. The students exercised cycles of expressing, testing and revising their models as they develop the models. In Chan's (2011a) dissertation, he compared and found the students' engagement in problem solving (of modelling tasks) to be much higher than the problem-solving sessions that teachers do with the students in typical lessons (Ho and Hedberg's (2005) research).

Research Involving Metacognition

Metacognition is asserted to play a critical role in successful learning when problem solvers monitor and regulate their own thinking (Schoenfeld, 1985). As metacognition is one of the five core process components in the Singapore Mathematics Curriculum Framework, it is not surprising that the field of metacognitive thinking and its relevance to mathematical problem solving is an area of interest among researchers to determine how students can be taught to better apply their cognitive resources through metacognitive control.

In a study to examine the metacognitive behaviours of Secondary 2 students as they solve open-ended problems in pairs, Lioe, Ho and Hedberg (2006) analyzed the students' protocols based on an adaptation of Artzt and Armour-Thomas' (1992) framework towards distinguishing the students' cognitive and metacognitive behaviours. The analysis revealed that regulation of behaviours ranges from *well-regulated* to *not-so regulated*. Well-regulated behaviours were characterized as being exercised in an orderly manner like reading, analyzing/planning, implementing and verifying. Well-regulated behaviours were found to be highly metacognitive. Patterns that pertained to not-so regulated behaviours saw the students engaged in a "wild-goose chase" (Schoenfeld, 1985) as well as the inability to achieve agreement on the solution paths.

To find out how students engage in problem solving, Wong (2007) administered a questionnaire on metacognition asking about what they did during problem solving. The instrument was administered as a pre-test and post-test to samples of Primary 5 and Secondary 1 students. It was found that there were not many changes in students' metacognitive awareness between the two tests. Their responses were fairly general, lacking in deep awareness of personal metacognition. While many students wrote about trying to understand the problem, very few mentioned about checking their work. Most students expected the teachers to teach and explain more, but few could give specific suggestions. The researcher suggested that some form of structured programme to inculcate metacognitive thinking might help.

Teo (2006) investigated the effects of metacognition on students in A-level Sequences and Series problem-solving. Although the study looked into the confidence level of the students as well, findings based on questionnaire analysis and interviews that dealt with the students' perception on how to solve mathematical problems and control decisions respectively was analysed. She found that an individual's control decision affected their test performance. Inefficient control behaviour often caused the students to rush into solving the problem without much analysis. On the other hand, students who were good in accessing their resources and had good control were more successful. The implications were that students with good control demonstrated how they approach the problems with ease. More control decisions were made when working with familiar problems. The issue of control becomes crucial to success in their attempts with unfamiliar problems.

Lee's (2008) PhD study was to explore the use of metacognitive instructional strategies as an intervention programme to address the needs of mathematically weaker learners in Secondary 1 classes based on a quasi-experimental approach. There were several positive outcomes that emerged in that the study had contributed to a positive change in the self-regulated learning strategies of the students. Besides the aspect of self-regulation, Lee also found the intervention to result in a positive change in the intellectual self-concept and mathematics self-efficacy as well as an increased level of perseverance in problem solving. He found that the positive outcomes were more pronounced among the Malay students. Another study (Teong, 2003) on the use of metacognitive instruction was facilitated through a computer environment with four 11 and 12 year olds from a Primary school over a period of eight weeks. Students worked collaboratively in a WordMath (Looi & Tan, 1998) environment where a metacognitive intervention strategy was introduced to promote students' awareness of their cognition. Using a case study approach and the analysis of two pairs of students' think aloud protocol data, she found that the students had distinctive progressions of word problem solving activity which could be represented by cognitive-metacognitive word problem solving models that appeared to be related to students' success in word problem solving.

The review of literature on metacognition shows that metacognition is a multi-faceted strand of study and while the various studies are different in their respective research objectives, the common stance lies in the creation of learning environments to support and develop metacognitive behaviours and skills.

Research Involving Sense-Making in Problem Solving

There have been some interests in researching students' sense making in the solving of structured word problems. These studies were taken up mainly due to the realization that students have the tendency to disregard the situations described in the context of the problems and accept illogical answers thus resulting in the suspension of sense-making. Three of such studies were located.

Chang's (2004) study of 307 Primary 5 students solving arithmetic word problems adapted Verschaffel, De Corte and Lasure's (1994) instrument to identify the students' sense-making ability. The results of the study indicated that about 60% of the students solved the problems by applying the standard straightforward arithmetic operations without considering the realities of the context of the problem situation. Heng's (2007) study was concerned with whether Primary 4 students' suspension of sense making was related to gender, achievement and task familiarity through their consideration of contextual factors when solving word problems. The study used a Word Problem Test comprising familiar and unfamiliar tasks which might or might not require sense making. The findings showed sense making was not gender related. Both the males and females had the tendency to solve familiar tasks better than unfamiliar tasks. Findings also indicated that the high achievers had the tendency to consider the context of familiar non-standard problems and they made more sense in familiar non-standard tasks than unfamiliar non-standard tasks.

Teo's (2005) research was to determine the effects of an intervention programme on the sense-making ability of three Primary 3 classes. The intervention programme required the students to be engaged in solving variants of typical textbook word problems where they had to extend the textbook problems through posing "what-if" questions. The findings showed that the intervention programme did improve the students' performance in standard items and non-standard items. The students' performance in standard items was better than non-standard items in both the pre-test and post-test. Teo reported that the students performed better when the items were of the near-transfer type (similar items as in the intervention programme) but not so well for the far-transfer type (items not found in the intervention programme).

Research Involving the Affective Domain

Mathematics education and problem solving tends to focus on the cognitive. In this regard, it is reasonable to suggest why there is a lack of research locally in the area of the affective domain in mathematical problem solving. According to Goldin (2002), the affective domain should not be considered unimportant. He argued that affect is not merely auxiliary to cognition but is indeed central, that is, affect as a representational system is intertwined with cognitive representation, and it affects student actions. The importance of the affective domain has recently been given greater emphasis when the Ministry of Education promoted holistic assessment (Ministry of Education, 2010) for Primary schools towards developing students to become more confident and communicative learners. This has implications in that it will impact pedagogies in the ways lessons are designed and how problem-solving is carried out. There are three studies that investigated the role of the affective domain on mathematical problem solving Their findings are discussed below.

Yeo (2005) research was to determine the interrelationship between mathematics anxious Secondary 2 students with solving non-routine problems. He found that mathematics anxiety and performance on a non-routine mathematical problem solving test showed a marginal linear relationship whereas test anxiety had almost no relationship with the performance on the non-routine mathematical problem solving test. It appears that the students at the low mathematics-anxiety level performed better on a non-routine mathematical problem-solving test than the high mathematics-anxiety students.

In investigating the effects of the use of alternative assessment, Foo (2007) had a Secondary 1 class to work on mathematics performance tasks. She found that the authentic context of the performance tasks helped students to relate mathematics to real-life situations and the non-routine and open-ended nature of the problems motivated some students to persevere in the face of challenges and fostered the positive perception that they were able to solve difficult problems.

In another study, Chan (2011b) investigated 80 Primary 5 students' Attitudes towards solving a series of five modelling problems in a problem-based learning setting. The students were administered the Attitude Questionnaire (AQ) at the end of the final task. The AQ had items that were related to aspects of *interest, confidence* and *perseverance*. The findings revealed high positive responses in the attributes of *Interest, Perseverance,* and *Confidence* and that solving problems in a PBL setting can impact their attitudes towards mathematics learning. It was found that the mixed-ability students registered higher, but statistically not significant, mean scores in these three areas than high-ability students. This study offers a different outcome to a study by Fan et al. (2005) who had 1215 Secondary 1 students completed a perception questionnaire on mathematics learning. They found that the students had generally positive attitudes towards mathematics and mathematics learning. However, they held relatively negative attitudes about working on challenging mathematics problems (31% disliked these problems) and that 21% of the students indicated that they would rather have someone give them the answers rather than having to solve the challenging problems themselves. Based on this rather negative response from students, it was recommended that school teachers provide students with opportunities to work on non-routine and challenging problems in an authentic way to maximize their higher-order thinking skills.

Research Involving the Innovative Use of Technology

Technology is an area that keeps advancing with increasing pace. To date, Singapore has launched her third instalment in the implementation of her Masterplan for ICT in Education since it was first started in 1997. This third Masterplan serves as a follow-up to the first two but with the emphasis to "*enrich and transform the learning environments of our students and equip them with the critical competencies and dispositions to succeed in a knowledge economy*" (Ministry of Education, 2008). However, research in the use of ICT in innovative ways in mathematical problem solving has been limited. Chan (2002) who did a review of effective use of ICT for primary school mathematics education puts it that the primary school teachers' limited use of ICT was due to the lack of understanding of the correct approach to using ICT for effective mathematics learning.

The following are two studies related to ICT and mathematical problem solving with students: Lee's (2002) research had Primary 4 students use LOGO to solve open-ended problems as an intervention programme to teach geometry. One of the tasks involved manoeuvring through different obstacles to get to the treasure shown on a map. In the post test, it was found that for items assessing students' conceptual understanding of angles, their performance improved in terms of both performance (correct answers) and quality of their explanations. The responses from students also seemed to suggest that the teaching process incorporating both

IT and thinking has helped them in their learning process of the concept of angles. For the thinking items, students were able to sequence and rank better in the post-test as compared to the pre-test.

In another study, Ahmed Ibrahim (2004) involved 46 Primary 5 students viewing a web-stream video to plan a budget for an outing. The goal was to develop the students' problem solving in an authentic context through negotiating meanings. The students made use of the Knowledge Forum platform which is a computer-supported collaborative learning platform for discussing mathematical ideas (costing, quotations, etc.) as well as selecting relevant and irrelevant information towards solving the problem. The students who obtained scaffolding support provided by the teacher were found to have performed better than the control group.

FUTURE CHALLENGES

The positioning of mathematical problem solving as the core of the curriculum has influenced research in student problem solving in the Singapore mathematics classroom. Although locally there is not a large body of research, the domain of mathematical problem solving has continued to generate interest (Foong, 2007). What does the review tell us then about student problem solving in Singapore? The following segments summarize the findings of the problem-solving research into three categories that align with the broad goals of the mathematics curriculum: (i) using effective tools and applying mathematical concepts and skills, (ii) developing mathematical thinking, positive attitudes and making connections, and (iii) reasoning logically and communicating mathematically (Ministry of Education, 2006, p.5).

Using Effective Tools and Applying Mathematical Concepts and Skills

From the review, what is known about the Model Drawing Method (MDM) now is that it is a powerful tool that students use to represent problem situations visually and aid concept development. The MDM is taught in class but it is not one that students would insist on using if they have alternative methods to solving problems (e.g. Poh, 2007) or when they have learned symbolic algebra (e.g. Ng, 2003). It can be quite challenging to use when the problems are complex, especially with two additive non-homogenous comparative relationships. To use it well, students must master the skill of drawing it out accurately (e.g. Goh, 2009; Ng & Lee, 2009). Erroneous representations tend to lead to incorrect or incomplete solutions. According to Yan (2002), the models drawn need to look "*realistic, with the units and known parts to be sensibly proportionate to enable any meaningful relationship to be deduced*" (p. 62). In this light, it will be helpful for teachers to model the drawing well for students to learn to pick up the skill of drawing proportionately. The challenge is also for teachers to be proficient in employing the method towards teaching students how to solve the more complex types of word problems. For students to be successful in using the MDM especially in solving word problems involving fractions or ratio, they will also need to have good conceptual knowledge of the topics to know how to transform part-whole relationships (Ng & Lee, 2009). Further research can be carried out to determine how instruction of this method can impact students in successfully solving the more complex types of problems, and even easing the transition towards using learning symbolic algebra through the AlgeBAR technology that Looi and Lim (2009) are researching.

Where the solving of non-routine problems in Primary and Secondary schools is concerned, the explicit teaching of heuristics is espoused (Tiong, 2005). From the review, generally most of the research shows that students are able to use a limited variety of common heuristics to solve certain non-routine problems (e.g. Hedberg et al., 2005; Ho, 2007; Wong, 2002; Wong, 2008). However, the findings do not provide an answer to knowing the students' problem-solving capabilities when they confront problems in new and unfamiliar situations for the application of such skills and strategies in real-world contexts. Currently, explicit teaching has paved the way for students to apply those learned heuristics to solve similar types of non-routine problems. This practice of problem solving can be attributed to the textbooks that students use, which comprise problem items that are primarily of a routine or well-structured nature (Fan & Zhu, 2000; Ng, 2002; Yeap, 2005). It is not surprising then that teachers take on problem-solving roles referred to as teaching *for* problem solving (where the emphasis is on learning the mathematics to apply towards solving the problem) and teaching *about* problem solving (where the teaching is about what the specific heuristics are and how to use them) rather than teaching *via* problem solving (to surface mathematical ideas from the problem solving process itself) (Foong, 2009). It is thus highly probable that teachers see themselves as indeed helping students to be engaged in problem solving if they regard problem solving to be merely the process of achieving a solution and it is their job to teach it explicitly.

In looking towards the future of problem-solving research, English and Sriraman (2010), from their review of problem-solving research, noted that using heuristics based on the standpoint of "What should I do when I get stuck?" has not resulted in improving students' problem solving. They contended that research can be more expansive to investigate the when, why, and how of problem solving, that is, knowing which tools to apply, when to apply, and how to apply as students systematically reason as they identify the underlying structure of a problem that lends itself to the use of the tools. In this light, there is certainly scope to research apart from the use of heuristics and student problem-solving abilities, how and in what ways the understanding of heuristics develop with problem solving.

Developing Mathematical Thinking, Positive Attitudes and Making Connections

It is encouraging to note that in the last decade, pockets of research concerning solving open-ended and real-world problems have surfaced. These studies (e.g. Chan, 2009, 2010; Ng, 2010; Seoh, 2002), although usually carried out in small sample sizes or intact classes, have informed about what the students are capable of when solving such problems and also point to the importance of developing their social and affective dimensions (e.g. Chan 2011; Foo, 2007). Notably, the review suggests that students indicated that they enjoyed solving investigative and performance-based tasks, and the findings also presented features of a conceptually enriched shared learning environment where students are held accountable by peers to explore and explain. Importantly, such research provides students with opportunities to engage in discourse towards generating and uncovering meanings and making connections, thereby seeing mathematics as a process of enquiry. As well, in the field of ICT research, the use of ICT has taken on a more interactive approach and students have used ICT more communicatively instead of merely computer-aided instructions as in the 90s. In particular, the incorporation of knowledge building and collaboration theories account for intentional problem-solving resolution when students are engaged in online platforms (e.g. Ahmed Ibrahim, 2004).

The newer interest in research that focuses on communication, affect and collaboration with collaborative open-ended tasks has implications seen from two sides of the coin. On the one hand, there seems to be a sense of greater readiness to embrace problem solving that takes on a contemporary stance where the focus is more on the process and the mathematical reasoning involved in the problem solving. On the other hand, it must be noted that the researchers involved conducted the research within a research time frame for a particular research purpose and after the research was done, such problem-solving experiences would likely have ceased or at best be carried out as enrichment lessons after examinations. The benefits of research need to reach a larger population of the students concerned. To maximize research and work towards creating better problem-solving learning environments for students, it is desirable that research findings be promoted in the schools so that teachers and researchers can work in partnership to advance the knowledge development of all participants (students, teachers and researchers) in understanding the role of student discourse and negotiations in problem solving.

Reasoning Logically and Communicating Mathematically

From the review, student problem solving that encompasses metacognitive and sense-making aspects were meaningful because they point to the importance of having good control of one's thinking to be successful in problem solving and the recognition of metacognitive instruction to scaffold the thinking process for positive outcomes. In a sense, the review findings of the metacognitive and sense-making studies tend to validate the findings of other research done in international circles (e.g. Lioe, Ho & Hedberg, 2006; Teo 2006 – having control is crucial to success; Lee, 2008; Teo, 2005 – metacognitive instructions/scaffolding had positive outcomes; Wong, 2007 – students' tendency to exercise routine monitoring). Future research could determine the role of the teacher in acquiring sufficient knowledge about metacognition towards impacting student problem solving. It will be of interest to know if teachers have the "tools" for implementing metacognition and making students aware of their metacognitive activities and the utilities of those activities (Veenman, Van Hout-Wolters & Afferback, 2006). Apart from researching on behaviours observed for what constitute expert behaviours, research can also focus on how students interpret and mathematize problem situations in a more holistic sense (problematizing, mathematizing, communicating, collaborating and affect). Since there is greater interest in collaborative problem solving of open-ended and investigative tasks, these can be areas of research to understand more about social metacognition and metacognitive questioning by peers as well.

CLOSING THOUGHTS

Although research in mathematical problem solving that involves students is growing, there are still issues to address and follow-up research to be carried out as well as newer areas to explore. What constitutes the future of mathematical problem-solving involving students in Singapore? As classroom pedagogy seeks to embrace reform efforts, the future of mathematical problem solving research looks promising and yet challenging. Research in problem solving will become richer when learning environments for problem solving becomes one that provides creative space for students to unpack their cognitive processes and where teachers become the triggers to advance student thinking.

REFERENCES

Ahmad Ibrahim, E. (2004). Computer-supported collaborative problem solving and anchored instruction in a mathematics classroom: an exploratory study. *International Journal of Learning and Technology, 1*(1), 16-39.

Artzt, A. F., & Armour-Thomas, E. (1992). Development of a cognitive-metacognitive framework for protocol analysis of mathematical problem solving in small groups. *Cognition and Instruction, 9*, 137-175.

Chan, C. M. E. (2008). Using model-eliciting activities for primary mathematics classrooms. *The Mathematics Educator, 11*(1/2), 47-66

Chan, C. M. E. (2009). Mathematical modelling as problem solving for children in the Singapore Mathematics Classroom. *Journal of Science and Mathematics Education in Southeast Asia, 32*(1), 36-61.

Chan, C. M. E. (2010). Tracing primary 6 pupils' model development within the mathematical modelling process. *Journal of Mathematical Modelling and Application, 1*(3), 40-57.

Chan, C. M. E. (2011a). *Investigating primary 6 pupils' mathematical modellingprocess in a PBL setting* (Unpublished doctoral dissertation). National Institute of Education, Nanyang Technological University, Singapore.

Chan, C. M. E. (2011b). Primary 6 students' attitudes towards mathematical problem-solving in a problem-based learning setting. *The Mathematics Educator, 13*(1), 15-31.

Chan, L. K. (2002). *Beyond drill-and-practice: the use of ICT in enhancing mathematics concept learning* (Unpublished master's thesis). National Institute of Education, Nanyang Technological University, Singapore.

Chang, C. Y. (2005). *An open-ended approach to promote higher order thinking in mathematics among Secondary Two Express students* (Unpublished master's thesis). National Institute of Education, Nanyang Technological University, Singapore.

Chang, S. H. (2004). *Sense-making in solving arithmetic word problems among Singapore primary school students* (Unpublished master's thesis). National Institute of Education, Nanyang Technological University, Singapore.

Chong, T. H., Khoo, P. S., Foong, P. Y., Kaur, B., & Lim-Teo, S. K. (1991). *A state-of-the art review on mathematics education in Singapore*. Southeast Asian Review and Advisory Group (SEARRAG).

Chow, I. V. P. (2004). *Impact of open-ended problem solving as an alternative assessment on Secondary One Mathematics students* (Unpublished master's thesis). National Institute of Education, Nanyang Technological university, Singapore.

Cockcroft, W. H. (1982). *Mathematics Counts.* London: Her Majesty's Stationery Office.

English, L. D., & Sriraman, B. (2010). Problem solving for the 21st century. In B. Sriraman & L. D. English (Eds.), *Theories of mathematics education: Seeking new frontiers* (pp. 263-285). New York: Springer.

Fan, L., Quek, K. S., Zhu, Y., Yeo, S. M., Pereira-Mendoza, L., & Lee, P. Y. (2005, August). *Assessing Singapore students' attitudes toward mathematics and mathematics learning: Findings from a survey of lower secondary students.* Paper presented at the Third East Asia Regional Conference on Mathematics Education, Shanghai, China.

Fan, L., & Zhu, Y. (2000). Problem solving in Singaporean secondary mathematics textbooks. *The Mathematics Educator, 5*(1/2), 117–141.

Fan, L., & Zhu, Y. (2007). From convergence to divergence: The development of mathematical problem solving research, curriculum, and classroom practice in Singapore. *ZDM-International Journal on Mathematics Education, 39*(5-6), 491-501.

Fan, L., & Zhu, Y. (2008). Using assessment performance in secondary school mathematics: An empirical study in a Singapore classroom. *Journal of Mathematics Education, 1*(1), 132-152.

Foo, K. F. (2007). *Integrating performance tasks in the secondary mathematics classroom: An empirical study* (Unpublished master's thesis). National Institute of Education, Nanyang Technological University, Singapore.

Foong, P. Y. (2002). The role of problems to enhance pedagogical practices in the Singapore mathematics classroom. *The Mathematics Educator, 6*(2), 15-31.

Foong, P. Y. (2007). Teacher as researcher: A review on mathematics education research of Singapore teachers. *The Mathematics Educator, 10*(1), 3-20.

Foong, P. Y. (2009). Review of research on mathematical problem solving in Singapore. In K. Y. Wong, P. Y. Lee, B. Kaur, P. Y. Foong, & S. F. Ng (Eds.), *Mathematics education: The Singapore journey* (pp. 263-300). Singapore: World Scientific.

Goh, S. P (2009). *Primary 5 students' difficulties in using the model method for solving complex relational word problems* (Unpublished master's thesis). National Institute of Education, Nanyang Technological University, Singapore.

Goldin, G. A. (2002). Affect, meta-affect, and mathematical belief structures. In C. G. Leder, E. Pehkonen, & G. Torner (Eds.). *Beliefs: A hidden variable in mathematics education?* (pp. 59-72). Dordrecht, The Netherlands: Kluwer Academic Publishers.

Hedberj, J. G., Wong, K. Y., Ho, K. F., Lioe, L., & Tiong, J. (2005). *Developing the repertoire of heuristics for mathematical problem solving.* Executive Summary Report for Project No. CRP 38/03 TSK. Retrieve from http://repository.nie.edu.sg/jspui/bitstream/10497/258/4/CRP38_03TSK_Summary.pdf.

Heng, C. H. J. (2007). *Primary pupils' ability to engage in sense making when solving word problems* (Unpublished master's thesis). National Institute of Education, Nanyang Technological University, Singapore.

Ho, G. L. (2007). A cooperative learning programme to enhance mathematical problem solving. *The Mathematics Educator, 10*(1), 59-80.

Ho, K. F., & Hedberg, J. G. (2005). Teachers' pedagogies and their impact on students' mathematical problem solving. *Journal of Mathematical Behavior, 24 (3 & 4)*, 238-252.

Kaur, B. (2009). Performance of Singapore students in Trends in International Mathematics and Science Studies (TIMSS). In K. Y. Wong, P. Y. Lee, B. Kaur, P. Y. Foong, & S. F. Ng (Eds.), *Mathematics Education: The Singapore Journey* (pp. 439-463). Singapore: World Scientific.

Kho, T. H. (1987). Mathematical models for solving arithmetic problems. In *Proceedings of the Fourth Southeast Asian Conference on Mathematical Education (ICMI-SEAMS)* (pp. 345-352). Singapore: Institute of Education.

Kho, T. H., Yeo, S. M., & Lim, J. (2009). *The Singapore model method for learning mathematics*. Singapore: EPB Pan Pacific.

Lee, C. M. (2002). *Integrating the computer and thinking into the primary mathematics classroom* (Unpublished master's thesis). National Institute of Education, Nanyang Technological University, Singapore.

Lee, N. H. (2008). *Enhancing mathematical learning and achievement of secondary one Normal (Academic) students using metacognitive strategies* (Unpublished doctoral dissertation). National Institute of Education, Nanyang Technological University, Singapore.

Lioe, L. T., Ho, K. F., & Hedberg, J. (2006). Students' metacognitive problem solving strategies in solving open-ended problems in pairs. In W. D. Bokhorst-Heng, M. D. Osborne, & K. Lee (Eds.), *Redesigning pedagogy: Reflection on theory and praxis* (pp. 243-260). Rotterdam, The Netherlands: Sense Publishers.

Looi, C. K. & Tan, B. T. (1998). A cognitive -apprenticeship-based environment for learning word problem solving. *Journal of Computers in Mathematics and Science Teaching, 17*(4), 339-354.

Looi, C. K., & Lim, K. S. (2009). From bar diagrams to letter-symbolic algebra: A technology-enabled bridging. *Journal of Computer Assisted Learning, 25*(4), 358–374.

Ministry of Education. (1990a). *Mathematics syllabus (Primary).* Singapore: Curriculum Planning and Development Division.

Ministry of Education. (1990b). *Mathematics syllabus (Lower Secondary).* Singapore: Curriculum Planning and Development Division.

Ministry of Education. (2000). *Mathematics syllabus (Lower Secondary).* Singapore: Curriculum Planning and Development Division.

Ministry of Education. (2004). *Strategies for active and independent learning.* Singapore: Curriculum Planning and Development Division.

Ministry of Education. (2005). *Principles of engaged learning: PETALS.* Singapore: Curriculum Planning and Development Division.

Ministry of Education. (2006a). *Mathematics syllabus: Primary.* Singapore: Curriculum Planning and Development Division.

Ministry of Education. (2006b). *Mathematics syllabus: Secondary.* Singapore: Curriculum Planning and Development Division.

Ministry of Education. (2008). *MOE launches third masterplan for ICT in education.* Retrieved from http://www.moe.gov.sg/media/press/2008/08/moe-launches-third-masterplan.php.

Ministry of Education (2010). *PERI holistic assessment seminar 2010 -16: Schools share holistic assessment practices and resources.* Retrieved from http://www.moe.gov.sg/media/press/2010/07/peri-holistic-assessment-seminar.php.

NCTM. (1980). *An agenda for action: Recommendations for school mathematics of the 1980s*. Reston, VA: Author.

Ng, D. J., Liew, E., Menon, L., Lim, P. L., & Wang, A. (2007, June). Learning support for mathematics (LSM): Using the 4-pronged intervention approach. Paper presented in the Redesigning Pedagogy: Culture, Knowledge and Understanding Conference, Singapore

Ng, L. E. (2002). *Representation of problem solving in Singaporean primary mathematics textbooks with respect to types, Polya's model and heuristics* (Unpublished master's thesis). National Institute of Education, Nanyang Technological University, Singapore.

Ng, K. E. D. (2010). Collective reasoning and sense making processes during a real-world mathematical project. In Y. Shimizu, Y. Sekiguchi & K. Hino (Eds.), *In search of excellence in mathematics education: Proceedings of the 5th East Asia Regional Conference on Mathematics Education* (Vol. 2, pp. 771-778). Tokyo, Japan: Japan Society of Mathematical Education.

Ng, S. F. (2003). How secondary two express stream students used algebra and the model method to solve problems. *The Mathematics Educator, 7*(1), 1-17.

Ng, S. F. & Lee, K. (2005) How primary five pupils use the model method to solve word problems. *The Mathematics Educator, 9*(1), pp. 60–83.

Ng, S. F., & Lee, K. (2009). The model method: Singapore children's tool for representing and solving algebraic word problems. *Journal for Research in Mathematics Education, 40*(3), 282-313.

Poh, B. K. (2007). *Model method: Primary three pupils' ability to use models for representing and solving word problems* (Unpublished master's thesis). National Institute of Education, Nanyang Technological University, Singapore.

Schoenfeld, A. (1985). *Mathematical problem solving*. Orlando, FL: Academic.

Seoh, B. H. (2002). *An open-ended approach to enhance critical thinking skill in mathematics among secondary five normal (Academic) pupils* (Unpublished master's thesis). National Institute of Education, Nanyang Technological University, Singapore.

Teo, A. L. (2005). *Effects of an intervention programme on the sense-making ability of primary three pupils* (Unpublished master's thesis). National Institute of Education, Nanyang Technological University, Singapore.

Teo, O. M. (2006). *A small-scale study on the effects of metacognition and beliefs on students in A-level sequences and series problem-solving* (Unpublished master's thesis). National Institute of Education, Nanyang Technological University, Singapore.

Teong, S. K. (2003, December). Metacognitive intervention strategy and word problem solving in a cognitive-apprenticeship-computer-based environment. Paper presented at the Australian Association for Research in Education Conference, Auckland, New Zealand.

Tiong, Y. S. J. (2005). Top-down approach to teaching problem solving heuristics in mathematics. Paper presented at the Mathematics Education Symposium, Srinakharinwirot University, Bangkok, Thailand.

Veenman, M.V.J., Van Hout-Wolters, H.A.M., & Afflerbach, P. (2006). Metacognition and learning: Conceptual and methodological considerations. *Metacognition and Learning, 1*, 3-14.

Verschaffel, L., De Corte, E., & Lasure, S. (1994). Realistic considerations in mathematical modeling of school arithmetic word problems. *Learning and Instruction, 4*, 272-294.
Wong, K. Y. (2007, June). Metacognitive awareness of problem solving among primary and secondary school students. Paper presented in the Redesigning Pedagogy: Culture, Knowledge and Understanding Conference, Singapore
Wong, K. Y. (2008). Developing the Repertoire of Heuristics for Mathematical Problem Solving. In M. Goos, R. Brown, & K. Makar (Eds.), *Proceedings of the 31st Annual Conference of the Mathematics Education Research Group of Australasia* (pp. 589-595). Brisbane: MERGA.
Wong, S. O. (2002). *Effects of heuristics instruction on pupils' achievement in solving non-routine problems* (Unpublished master's dissertation). National Institute of Education, Nanyang Technological University, Singapore.
Yan, K. C. (2002). The model method in Singapore. *The Mathematics Educator, 6*(2), 47-64.
Yeap, B. H. (2005, August). Building foundations & developing creativity: An analysis of Singapore mathematics textbooks. Paper presented at the Third East Asia Regional Conference on Mathematics Education, Shanghai, China.
Yeo, K. K. J. (2005). Anxiety and performance on mathematical problem solving of secondary two students in Singapore. *The Mathematics Educator, 8*(2), 71-83.
Yeo, S. M., & Zhu, Y. (2005, June). Higher-order thinking in Singapore mathematics classrooms. Paper presented in the Redesigning Pedagogy: Research, Policy and Practice Conference, Singapore.

CHAPTER 43

RESEARCH ON SINGAPORE MATHEMATICS CURRICULUM AND TEXTBOOKS

Searching for Reasons Behind Students' Outstanding Performance

Yan Zhu
East China Normal University, China

Lianghuo Fan
The University of Southampton, UK

Over the last 15 or so years, the Singapore mathematics curriculum and textbooks have attracted a considerable amount of attention internationally, particularly from educational researchers, reformers and policy makers. This phenomenon is closely related to people's interest in searching for the reasons for Singapore students' outstanding performances in well referred international comparative studies, such as The Trends in International Mathematics and Science Study (TIMSS) and The Programme for International Student Assessment (PISA). This chapter is intended to

The First Sourcebook on Asian Research in Mathematics Education: China, Korea, Singapore, Japan, Malaysia, and India, pp. 959–979

provide readers with an overall look at the development of the Singapore mathematics curriculum and textbooks and research that has been carried out over the last decades on these two related areas, as well as discuss relevant issues concerning mathematics curriculum and textbook development.

HISTORICAL BACKGROUND

Curriculum is a social and cultural existence. To understand issues related to the Singapore mathematics curriculum, or more generally, school curriculum, we shall start with its historical background, followed with a brief description of the general education system in Singapore.

It has been said that the modern history of Singapore began in 1819 when it came under British colonial rule and it remained a British colony until 1959 (except for the period of Japanese occupation from 1942 to 1945). At the beginning, the British administers' interest in Singapore was primarily commercial, with scant attention being paid to education. Gopinathan (1997) characterized the colonial education in Singapore as "benign neglect, ad hoc policy making and indifference to consequences" (p. 593). In fact, the then education was left to the different communities. By 1939, of the 72,000 children in Singapore schools, there were 38,000 in Chinese schools, 27,000 in English schools, 6,000 in Malay schools, and 1,000 in Tamil schools. These schools not only used different languages as the medium of instruction, but also adopted different syllabi and textbooks. For mathematics, Chinese schools used American textbooks such as *College Algebra* by Fine (in a Chinese translation) at the senior secondary two level (i.e., Grade 11), while English schools used British mathematics textbooks by Durell, such as *New Algebra for Schools*, as their teaching resources (also see Ho, 2008).

In 1959, Singapore attained self-government and the government placed emphasis on educating the masses, particularly on the study of mathematics, science and technical subjects. In order to meet the needs of a multi-racial, multi-religious population in Singapore, the first localized syllabus in mathematics, covering all school stages from Year 1 to Year 13, was drafted in 1957 and started to be implemented in 1959. The syllabus made no distinction except it was published in four different languages, i.e., English, Chinese, Malay, and Tamil, for different schools. Another important feature of this syllabus was that it adopted a spiral approach and treated mathematics as a unified subject rather than a "many-branched" discipline. Further, it highlighted the close relationship among the various branches of mathematics as well as between mathematics and other science subjects. The part of this syllabus for the secondary level was commonly known in Singapore as Syllabus B, which was used across all schools to prepare students for the mathematics examinations of the Cambridge Certificate of Education conducted by the University of Cambridge Local Examination Syndicate (Lee & Fan, 2004).

According to Lee (2008), the most important event in the history of mathematics education in modern Singapore was the Math Reforms in the 1970s, which were apparently influenced by the new math movement in the US or the modern mathematics movement in the UK. The reforms hastened the localization of the syllabus and textbooks. In particular, the primary mathematics syllabus was revised in 1971 and again in 1976, resulting in a reduction of the content. There were emphases on numerical skills including approximation, and the concept of algebra was introduced at the upper primary level (Grades 5–6). At the secondary level, the school mathematics syllabus (i.e., Syllabus C) was introduced in 1968 and revised in 1973. Modern mathematics, such as sets, was introduced at Secondary One, while Euclidean geometry was drastically reduced. In addition, these documents first presented the "General Aim and Objectives of the Singapore Mathematics Curriculum", which were absent from the earlier syllabi. In 1969, the first set of locally produced secondary school textbooks, *Modern Mathematics for Secondary Schools*, were published (Teh, 1969) and others followed shortly. For examination, elementary mathematics at the O level (Grade 7–10) was made compulsory as an examination subject in 1974. Furthermore, mathematics was taught and examined in English starting from 1976 but without being fully implemented until 1984.

Internationally, the math reforms eventually led to the falling of mathematics standards across the world. Therefore, the call for back-to-basics gained momentum in the late 1970s. Singapore also followed this worldwide trend. In particular, for the secondary mathematics, Syllabus C was phased out progressively and replaced by Syllabus D and the later secondary school syllabus is in a sense Syllabus D plus, a modified version. Syllabus D retains traditional topics at the lower secondary level but moved out those so-called modern topics with some topics either disappearing or being kept at Secondary Four (Grade 10). As Lee (2008) commented, one rationale was to make sure that students completing lower secondary learning would have enough trigonometry to learn a trade and to work in the maritime industry, for which modern mathematics was considered to be not so helpful.

Along with this backdrop, Singapore started to implement a New Education System (NES) in 1979, under which students were streamed on the basis of their academic ability. Each stream had its own syllabus which is to ensure students could learn at their own paces. As a result, for the first time in the history of Singapore, there existed different syllabi at the primary levels, though the objectives remained largely unchanged. To better communicate the subtle yet different expectations of the various streams, the column "Sample Exercises" was included in the outcome charts for each topic to be taught. In addition, under the NES, schools using various languages as their medium of instruction gradually evolved into a single system using only English. Further, with the establishment of the Curriculum Development Institute of Singapore (CDIS) in 1980, it became the only textbook developer as well as the sole publisher responsible for primary school textbooks until 2001. Meanwhile, popular secondary school textbooks were still published commercially. By the early 1990s, Singapore had reached a kind of equilibrium state for syllabi and textbooks with both being designed locally.

Starting from around 1990, problem solving became the central theme in both the primary and secondary mathematics curriculum, which was explicitly stated in the syllabus. In particular, it was the first time that developing students' ability in mathematical problem solving was set as the primary aim of the mathematics curriculum and problem solving was also centered in the well-known pentagonal framework surrounded by five inter-related important qualities – concepts, skills, attitudes, metacognition, and progresses. Figure 43.1 shows the latest pentagonal framework (also see Chan or Koh, this volume). Further, as part of processes, heuristics for problem solving were explicitly listed in the syllabus.

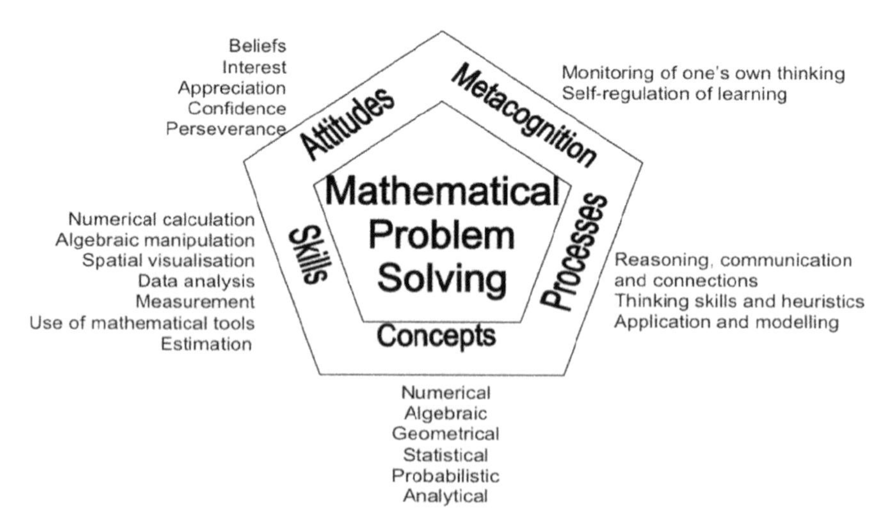

Source: MOE (2006a, 2006b).

Figure 43.1. Singapore Mathematics Curriculum Framework.

In 1997, the Ministry of Education (MOE) put forward the idea of "Thinking Schools, Learning Nation" (TSLN), which started a new era of education reform, with three initiatives being National Education (NE: good citizenship and nation building), Information Technology (IT Master Plan I: 1997 – 2002 to bring computers into schools and train teachers to use them; IT Master Plan II: 2002 – 2008 to integrate IT into the curriculum) and Thinking. In particular, 'Thinking' refers to *more learning and less teaching*, which led to a content reduction by about 10%, with more reduction at the secondary school level and less at the primary level. For mathematics, it encourages students to spend more time with problem solving rather than route learning. For adult education outside the school environment, learning nation advocates instilling a culture of life-long learning.

In 2006, the mathematics curriculum was revised again, with greater emphasis placed on developing mathematical concepts and fostering the ability to apply them in mathematical problem solving situations. There were some new guidelines, including emphasizing computational skills along with more conceptual and strategic thinking, covering fewer topics in-depth in a careful grade-by-grade sequence, covering concepts in one grade and in later grades at a more advanced level, ensuring students master prior content without repetition, and encouraging mathematical representation, reasoning and mathematics communication.

Starting from 2001, the development of primary school textbooks has been privatized in the hope that it could not only provide students with more choices of textbooks, but also respond to changes more efficiently. Furthermore, it was expected that market force could drive commercial publishers to produce quality textbooks at more affordable prices. Meanwhile, the secondary mathematics textbooks have been all along dominated by commercial publishers. There are no standard textbooks at the Junior College (JC) level, as JC teachers normally do not use textbooks.

Lee once divided the past few decades of mathematics education in Singapore into five periods: (1) Early days from 1945-1960; (2) First local syllabus from 1960–1970; (3) Math reforms from 1970–1980; (4) Back to basics from 1980–1995; and (5) New initiatives from 1995–present. Similarly, in terms of general education in Singapore, researchers often categorize the post-1945 period into three phases of development, that is, survival-driven phase (1959–1978), efficiency-driven phase (1979–1996), and ability-driven phase (1997–present) (for more details, see Dimmock & Goh, 2011). Below we shall give a brief description of the current educational system in Singapore.

CURRENT EDUCATIONAL SYSTEM IN SINGAPORE

Like many other Asian countries, Singapore has a centralized school system. Under this system, students usually receive six years of compulsory primary education, four or five years of secondary education, two or three years of pre-university education at the Junior Colleges/Centralized Institutes or one or two years of studies at the Institutes of Technical Education (ITE) or three years of practice-oriented studies at the Polytechnics, followed by higher education at universities, polytechnics or the ITE.

A fundamental aim of primary education is to give students a good grasp of English, mother tongue and mathematics, which, normally starting at age seven, consists of a 4-year foundation stage (Primary 1 to 4) and a 2-year orientation stage (Primary 5 to 6). The foundation stage is the first stage of formal schooling in Singapore and all students follow a common curriculum that emphasizes English, mother tongue, science, and mathematics. Other subjects include civics and moral education, arts and crafts, music, health education, social studies, and physical education. Science is taught from Primary 3 and onwards.

Streaming is a key feature of the Singapore education system. It was designed to allow students to progress at their own pace at the orientation stage, i.e., the final primary stage starting from Primary 5. Previously, students were divided at Primary 5 to the EM1 (more capable), EM2 and EM3 (weakest) streams (E stands for English and M stands for Mother Tongue). In 2004, EM1 and EM2 were merged into one single stream. Since 2008, streaming was replaced with subject-based banding. Under the new system, students are given the choice to take a mix of standard or foundation subjects depending on their strengths. For instance, students who are strong in mathematics but very weak in language may choose to take mathematics at the standard level and both English and Mother Tongue at the foundation level. By doing so, students could focus and improve on their strong subjects, while building up the fundamentals in their weak subjects. It is believed that the new system encourages students to stretch their potential in the areas they naturally excel in.

At the end of Primary 6, all students are required to take the Primary School Leaving Examination (PSLE) in English, mathematics, Mother Tongue, and science. Except for the students whose performance is so poor to be required to repeat P6, based on their performance on the PSLE all the other students are allocated places in one of the four different education tracks or streams at the secondary school level: Special, Express, Normal Academic, and Normal Technical. The different curricular emphases are designed to match students' learning abilities and interests. In particular, students in the Special or Express course follow a 4-year programme culminating in the Singapore-Cambridge General Certificate of Education

'Ordinary' Level Examination (GCE 'O' level). The difference between the two courses is that students in the Special course take "Higher Mother Tongue", whereas those in Express course take "Mother Tongue".

The Normal course is a 4-year programme leading up to GCE 'Normal' Level examination ('N' level), with the possibility of a fifth year followed by an 'O' level. The Normal course is further spilt into Normal Academic and Normal Technical. In the Normal Technical course, students take subjects of a more technical nature, such as Design and Technology, which prepares them for technical higher education, job or postsecondary ITE, while the Normal Academic course prepares students to take the 'O' level examination and they normally take subjects such as Principles of Accounting. In 2004, the MOE announced that selected students in the Normal course would have opportunity to sit for the 'O' level examination directly without first taking the 'N' level examination.

In recent years, more choices have been offered to students at the secondary school level, with a wider range of subjects at 'O' level and elective modules. Besides the four traditional routes, students also have other options such as the Integrated Programme which allows students to leave secondary schools to enter junior college education without taking the 'O' level examinations, the International Baccalaureate (IB) programme, Specialized Programme which aims to develop the students' talents in specific areas, Privately-funded School programmes that determine their own curriculum and provide more options for students, or the Special Education School programmes which provide either mainstream curriculum with programmes catering to students' special needs or customized special education curriculum. All these alternative courses run for 4 to 6 years.

With 'O' level certificate, students can progress to their postsecondary education in the 2-year junior colleges or a 3-year centralized institute taking pre-university course, which leads to the GCE 'Advanced' Level Examination ('A' level), or the 3-year polytechnics, which offer a variety of diploma courses such as Banking and Financial Services, Mass Communications, Engineering, etc. to prepare for practice-oriented and knowledgeable middle-level professionals, or the 1- or 2-year Institutes of Technical Education (ITE) provided mainly for skilled workforce. Depending on postsecondary assessment examinations, the graduates may elect to continue to the 3- or 4-year undergraduate university level studies or join the workforce.

RESEARCH ON SINGAPORE MATHEMATICS CURRICULUM AND ITS DEVELOPMENT

As mentioned earlier, Singapore students' outstanding performance as consistently demonstrated in international comparison studies has aroused both local and international researchers' attentions. Many efforts have been made to search for the reasons behind their success. Singapore's national mathematics curriculum is believed to be one important component that contributes to Singapore's world-class mathematics system (Ginsburg, Leinwarnd, Anstrom, & Pollock, 2005). In this connection, researchers have conducted rather detailed investigations on the mathematics curriculum with some notably being from comparative perspectives.

Dindyal (2005) compared the Singapore mathematics curriculum and the National Council of Teachers of Mathematics (NCTM) *Standards* at the level of the intended, implemented, and attained curriculum. He found that with the aims of improving the mathematics education of students in their own socio-cultural contexts, both documents put emphases not only on the content at various levels but also the process of learning mathematics. Compared to the similarities, more differences were revealed between the two documents in several aspects. In particular, the *Standards* were produced by an autonomous professional organization (i.e., NCTM) and therefore do not carry legal weight (also see Usiskin & Dossey, 2004). Correspondingly, individual states, school districts, and schools in the US are free to opt out of the reforms as suggested by the *Standards*. In contrast, in a centralized education system, all the schools in Singapore have to follow the prescribed guidelines. Compared to the *Standards*, the Singapore curriculum included details on not only the desirable aspects of mathematics education at various levels, but also provided guidance on how to implement the vision as well as how to assess achievement in schools. Furthermore, besides the emphasis on problem solving, the Singapore curriculum also paid greater attention to affective issues in the learning of mathematics, which is in a sense less emphasized in the *Standards*.

Concerning students' different learning abilities, Singapore provides a differentiated curriculum for their mathematics learning. It is believed that such a curriculum could provide a more focused learning experience to students. In fact, in another paper, related to the results from the TIMSS, Dindyal (2006) argued that a differentiated curriculum could be a reason for students' good performance. In particular, there are four countries using differentiated curriculum (i.e., Belgium [Flemish], the Netherlands, the Russian Federation, and Singapore) in TIMSS 2003 whose students achieved good rankings at the assessed grade levels.

Focusing on the presentation of problem solving and process aspects of mathematics, Stacey (2005) compared the curriculum documents from Australia, UK, US and Singapore. She found that the Singapore curriculum presents a distinctly different structure, which puts mathematical problem solving at the center of its curriculum framework as a goal and purpose of mathematics education. In particular, the Singapore syllabus explicitly states that the primary aim of the mathematics programme is to develop students' ability in mathematical problem solving. In contrast, the curriculum structure of other countries treats problem solving as one of the components of a successful mathematics education rather than its overall goal.

While the local mathematics curriculum placed great emphasis on problem solving and mathematical investigation at both the primary and secondary levels, Fong (1994) argued that some bridging problems arise between the two school levels. According to Fong, when students move from the primary level to the secondary level, they tend to carry along with the model approach, or the so-called model method (see Ng, this volume), in solving problems which has been overemphasized at the primary level. Therefore, students feel lots of pressure when being asked to use abstract approaches for solving problems at the secondary level. As for secondary teachers, many of them are unaware of the repetition of some topics which have been taught at the primary level so that students may feel bored going through again the same types of curriculum materials. Fong, therefore, called for a reexamination for the secondary syllabus and/or revising the corresponding curriculum materials.

Ng's (2004) investigation focused on how the Singapore primary mathematics curriculum developed algebraic thinking in the early grade levels. The study analyzed the curricular materials, including the syllabus, teachers' guides, textbooks and workbooks produced by the MOE. The results showed that the then Singapore mathematics curriculum adopted three main approaches to cultivate students' algebraic thinking, namely, problem solving, generalization and functional, which were further supported by three thinking processes–analyzing parts and whole, generalizing and specializing, and doing and undoing. Ng concluded that the Singapore primary mathematics curriculum utilized a wide spectrum of activities, aiming at problem solving, identifying patterns, constructing rules and functions for number sequences and using letters as variables well engaged students in algebraic thinking. Furthermore, via a spiral approach, the curriculum presents a comprehensive content coverage; that is, content covered at a lower level is revisited with each visit ensuring that students engage with similar ideas and concepts, each at a higher and more complex level.

With the notion that teachers play key roles in implementing any suggested policy changes, Foong, Yap, and Koey (1996) investigated teachers' concerns and constraints regarding the changes brought by the revised mathematics curriculum. The researchers highlighted four important changes including shifts from an emphasis on rote memorization of facts and procedures to meaningful understanding of concepts and problem solving, from a dependence on paper-and-pencil manipulative computations and skills to mental calculations and thinking strategies, from teaching in lecture style to teaching through activities, group work and communication in mathematics, and from paper-and-pencil test for the sole purpose of assigning grades to alternative assessment methods. Their study found that though many teachers were aware of the new changes and had already started to teach the revised curriculum, they felt that they greatly needed to have adequate knowledge and understanding of the rationale for new topics and develop new teaching skills to cope with these changes. Strong constraints come from time limitation which is mainly related to the increasing content load, lack of knowledge of the new topics and teaching skills. Furthermore, the teachers reported that they were bounded by the syllabus and the stipulated requirements in terms of emphasis on problem solving and new teaching methods, and quantity of student output like workbook and worksheet. Support from change facilitators and school administrators is another important aspect in teachers' concerns.

Yeap (2005) analyzed primary school national examination released items during the period from 2000 to 2004. The examination was in the form of paper-and-pencil test consisting of 15 selected-response type items and 35 constructed-response type items, which takes about two hours and fifteen minutes to complete. The analysis identified all the 196 released items as either procedural items or challenging items with the former assessing knowledge, basic skills, routine procedures and familiar word problem solving, and the latter assessing students' ability beyond routine procedures. It was found that about a quarter of the items were classified as challenging ones. Yeap related this finding to the culture of challenges developed in a typical Singapore mathematics classroom, where every student, not just a high-achiever, is expected to do challenging mathematics (also see Yeap, 2008).

Seng's (2000) study examined the learning and teaching of mathematics at the primary school level in Singapore. It reported that while key mathematical topics and concepts are prescribed in a series of teachers' guides and worksheets developed by the Curriculum and Development Unit in the MOE, there is also a large element of freedom for teachers to choose other appropriate topics and concepts for teaching, which is usually carried out by a group of mathematics teachers under the guidance of the head of the mathematics department within a school. Furthermore, the study found that education polices influenced the school learning climate in Singapore. In particular, the official mathematics curriculum has been used as a guide for Singaporean teachers to plan their mathematics programmes, though the teachers are not bound by the official documents but encouraged to maintain the linkage among topics within the curriculum. The study also observed that recommended big changes have taken place at the primary school level, including project work, multiple modes of assessment, creative modes of learning, emphasis on basic numeracy skills, promotion of critical thinking, and creativity nurturance.

Kaur (2005) looked into the assessment issue in Singapore mathematics curriculum. She found that, with the main purpose being to improve the teaching and learning of mathematics, the syllabi stated that the mathematical assessment must continually provide accurate and useful information about students' learning and also support the curriculum in its aim to enable students to develop their problem solving ability, vis-à-vis concepts, skills, processes and attitudes. Furthermore, the syllabi recommended two types of assessment, which are continual assessment (CA: formative) and semestral assessment (SA: summative). Kaur noted that it is a common practice in many schools to carry out both types of assessment just in the form of paper-and-pencil tests and examinations, though a variety of assessment modes have been recommended and encouraged. Moreover, it is found that many schools conducted CA in the form of a common test for the entire year level at the same time and day, which functions as a mini SA. In this sense, the difference between the two types of assessment in the actual usage is just in the coverage of topics with CA covering fewer topics than SA. Besides the two main assessment modes, many school teachers also use topical tests, class quizzes, regular homework assignments, and revision worksheets to monitor students' learning progress.

When being asked for the reason why they used limited recommended modes of assessment in daily teaching, teachers often attributed it to lack of know-how and resources. To improve the situation, the Assessment Guides to both the primary and lower secondary mathematics were produced in 2004 and distributed to all schools. The guides present to teachers a holistic perspective on assessment and assist the heads-of-departments and classroom teachers in carrying out different types of assessment via resources, tools and ideas which could help teachers incorporate meaningful assessment, such as mathematical investigation, journal writing, classroom observation, self-assessment and portfolio assessment in their schemes of work other than written paper-and-pencil tests and examinations. Further, through conducting workshops, the Curriculum Planning and Development Division (CPDD) of MOE tend to help mathematics teachers be familiarized with those alternative modes of assessment proposed in the Assessment Guides (for a relatively comprehensive review about assessment, see Koh, this volume).

RESEARCH ON SINGAPORE MATHEMATICS TEXTBOOKS

Besides the national curriculum as reflected in the syllabus and other policy documents, the quality of Singapore textbooks is also often regarded as one important reason for Singapore students' excellent performance in the international comparisons. Schools in many countries, notably the US, directly imported Singapore mathematics textbooks and used them for teaching and learning (e.g., see Quek, 2002). However, researchers also found that a successful adoption of another nation's textbooks took much more than simply using the textbooks (e.g., Ginsburg, Leinwand, Anstrom, & Pollock, 2005). In this sense, it is important for the decision makers to have a good understanding about the features of the curriculum materials and situate them into the respective system contexts (Li, 2007). In the following, we shall review research studies with a focus on Singapore textbooks, mainly including textbook analysis, textbook comparison, and textbook use.

As mentioned earlier, there was only one national primary mathematics textbook series available during the period of 1981 to 2000, and its influence on the classroom teaching and learning is easy to see. The series became the research subject of Ng's study (2002) with a particular focus on the representation of problem solving in this series of textbooks. The results showed that the series presented a high portion of routine, close-ended problems, and problems with exactly sufficient information out of all the 2819 problems from Grade 1 to Grade 6 (see Table 43.1).

Furthermore, the results revealed that nearly 90% of the worked-out examples modeled Pólya's first two problem-solving phases, 64% with the first three, and none with the final phase, "looking back", which appears not aligning well with the syllabus prescription. Moreover, the analysis also found that not all the heuristics suggested in the syllabus were presented in the textbooks, which clearly revealed a gap between the national syllabus and the textbooks. This requires attention from both curriculum

Table 43.1. Ratio of Different Types of Problems in Singapore Primary Mathematics Textbooks

Types of Problems	*Ratio in Number*	*Ratio in Percent*
Routine prob. vs Non-routine prob.	2790 : 29	99.0 : 1.0
Close-ended prob. vs Open-ended prob.	2747 : 72	97.4 : 2.6
Prob. with exactly sufficient info. vs Prob. with insufficient info. vs. Problem with extra info.	2819 : 0 : 0	100 : 0 : 0

Source: Ng (2002).

developers and textbook developers. On the other hand, three heuristics were predominantly modeled in the problem solving procedures of the worked-out examples.

Similar findings were also reported in Zhu's (2003) examination of one main lower secondary mathematics textbook series, which was conducted as part of a larger research effort (see Fan & Zhu, 2007). The series was intended for Special/Express courses and used in most schools in Singapore (also see Fan & Zhu, 2000). Again the examination found that the majority of problems presented in the textbooks were routine, traditional, close-ended, and un-contextualized in real world situations. In most cases, the solutions to the worked-out examples only demonstrated "carrying out the plan". The four problem solving phases were usually modeled in an implicit way (i.e., no explicit labels). Compared to the primary series, the secondary textbooks explicitly illustrated all but one of the recommended heuristics. In addition, the study found that the lower secondary series devoted one whole section to the introduction of problem solving in 11 out of 27 chapters, which to a certain extent implies that the textbook authors treated "problem solving" as an independent topic as with many other mathematics topics.

Yeap's (2005) study investigated the extent to which Singapore textbook tasks engaged students in creative processes and the explicitness at the primary school level. The results showed that the Singapore textbooks help students acquire strong mathematics foundations through the extensive use of pictorial representations and pattern observations as well as variations in practice tasks. Creative thinking is encouraged in both explicit and implicit manners with some tasks having no fixed responses and others being organized to reveal a pattern/relationship which could be identified with or without explicit prompts. However, the study also revealed that the proportion of tasks encouraging creativity decreased as the grade level increased.

Seah and Bishop (2000) conducted a comparative study on the representative lower secondary mathematics textbooks between Singapore and Victoria, Australia. They reported that a variety of problem solving strategies were explicitly covered in the Singapore textbooks. In particular, problem posing as a focused strategy was introduced via worked-out examples, which was unique to Singapore textbooks. The analysis showed that while Victoria textbooks presented opportunities for different interpretations and possible solutions in mathematics as extension to practice questions, Singapore textbooks did not feature questions testing for mathematical reasoning and communication, though the Singapore mathematics curriculum advocated using mathematics as a means of communication. Furthermore, the comparison found that while the textbooks in both places presented the majority of questions to test for students' knowledge and drills, the proportion was smaller among Singapore questions than Victoria ones. The researchers related this finding to Singapore's more explicit official emphasis on problem solving and the wide availability of supplementary assessment materials in Singapore bookstores. Interestingly, the study showed that the presentation of problem solving questions in Singapore textbooks was in the form of drills with less involving subjective or multiple solutions. The authors argued that this may reflect the writers' internalization of some time-tested, more traditional mathematics educational values. They believe this could be one reason to explain Singaporean students' outstanding performance in the TIMSS but not in the International Mathematics Olympiad (IMO).

Among available comparative studies involving Singapore mathematics curriculum and textbooks, Ginsburg, Leinwand, Anstrom, and Pollock's (2005) study is particularly noteworthy. The study represented a most comprehensive and, in our view, thorough comparison between the school mathematics systems in the United States and Singapore. In their 192-page research report, an entire chapter was devoted to the comparison of textbooks in the two countries, which examined, in Singapore, mainly the series of Singapore CPPD primary mathematics textbooks as mentioned earlier, and in the United States, the series published by Scott-Foresman Addison-Wesley Mathematics (2004) for grades 1–6, which was regarded as one of the four major traditional basal textbooks in the US, and the *Everyday Mathematics* textbook series published by Everyday Learning Corporation (2001) for grades 1–6, a widely used non-traditional mathematics curriculum. With a focus on how different these textbooks are in structure, approach, and content, the researchers found that the Singapore textbooks were much thinner, containing less mathematical topics but each covered in greater depth. They were stronger in their treatment of mathematical concepts and provided rich problem sets that would potentially give students many and varied opportunities to apply the concepts they have learned. On the other hand, it was found that the Singapore textbooks were weaker, particularly compared to the non-traditional US mathematics textbook series, in that the problems provided in the textbooks were not often based on real-world situations, which, according to the researchers, would probably make the mathematics less applicable to real life than it should be.

Another important study comparing the Singapore mathematics curriculum, mainly referring to textbooks, and the US mathematics curriculum was conducted by Adams, Tung, Warfield, Knaub, Yong, and Mudavanhu (2002), with a focus on the middle school level, i.e., from Grade 4 (Primary 4) to Grade 8 (Secondary 2). The researchers compared the Singapore mathematics textbook series with two reform mathematics textbook series, namely Connect Mathematics Programme, and Mathematics in Contexts. The National Council of Teachers of Mathematics (NCTM) *Principles and Standards* (NCTM, 2000) was also used for comparison purpose. The study found that Singapore mathematics was more traditional in orientation and emphasised students' acquisition of proficiency in mathematical skills and teacher-directed learning. On the other hand, the mathematics problems presented in the Singapore textbooks were at a more advanced level. Moreover, it was found that Singapore textbooks scored the lowest against the NCTM *Principles and Standards*, which however seems quite understandable as Singapore textbooks were developed based on Singapore's national mathematics syllabus. The researchers argued that Singapore mathematics textbooks might not be suitable for American schools to use as the source of principle texts.

Specifically focusing on the lower secondary grade level, Fan and Zhu (2007) examined how selected school mathematics textbooks in Singapore, China, and US represent problem-solving procedures from a comparative perspective (also see Zhu, 2003). The then most widely used series of mathematics textbooks in Singapore and China, the former by Shing Lee Publishers, and the later by People's Education Press, and reform textbooks developed by the University of Chicago School Mathematics Projects were selected for that study. The analysis of problem-solving procedures was carried out in two layers including the general strategies, which adopted Pólya's four-stage problem-solving model, and specific strategies, which consisted of 17 different problem-solving heuristics. The study showed that concerning the general strategies, the two Asian series, in most cases, merely presented 'how to carry out the plan' in the solutions to problems, whereas more than two-thirds of the problem solutions in the US textbooks modelled at least two problem-solving stages in Pólya's model (also see Table 43.2).

Furthermore, all the three textbook series introduced a considerable number of problem-solving heuristics, as shown in Table 3. Nine of the heuristics were common across the three series and two were introduced only in the Singapore series. The distribution of different heuristics was found most concentrated in the Chinese series, and most widespread in Singapore textbooks. Singapore textbooks also presented specific heuristics in the most explicit way.

It is interesting that part of the TIMSS studies was an investigation of how the mathematics textbooks were used in mathematics classroom in different countries, including Singapore. According to TIMSS 2007 survey (Mullis, Martin, & Foy, 2008), there were about 75% of the fourth graders and 51% of the eighth graders whose mathematics teachers reported that they used textbooks as primary resources for lessons. Interestingly, the corresponding percentages in TIMSS 2003 were smaller at

Table 43.2. Numbers of Problems Whose Solutions Modeled Pólya's Problem Solving Stages in the Chinese, Singapore, and US Textbooks

	China	*Singapore*	*United States*
Understanding the problem	122 (15.6%)	182 (20.2%)	259 (27.8%)
Devising a plan	149 (19.0%)	78 (8.7%)	188 (20.2%)
Carrying out the plan	783 (100%)	899 (100%)	930 (100%)
Looking back	209 (26.7%)	132 (14.7%)	402 (43.2%)

Source: Fan and Zhu (2007).

Table 43.3. Numbers of Problems That Were Solved Using the Specific Problem Solving Heuristics in the Chinese, Singapore, and US Textbooks

	China	*Singapore*	*United States*
1. Act it out	1	1	0
2. Change your point of view	0	2	1
3. Draw a diagram	51	60	119
4. Guess and check	0	1	5
5. Logical reasoning	4	6	24
6. Look for a pattern	2	6	8
7. Make suppositions	1	6	7
8. Make a systemic list	3	9	9
9. Make a table	11	11	23
10. Restate the problem	24	25	49
11. Simplify the problem	5	9	3
12. Solve part of the problem	0	4	3
13. Think of a related problem	13	0	16
14. Use a model	0	3	0
15. Use an equation	40	44	39
16. Use before-after concept	0	2	0
17. Work backwards	0	3	2

Source: Fan and Zhu (2007).
Note: Only the problems that were solved using specific heuristics at the first three stages in Pólya's model are included. In addition, some problems were solved by using two or more different heuristics.

the primary level (66%) but greater at the secondary level (74%). Consistently, the survey also found that there were more secondary students taught by teachers who reported that they did not use textbooks for mathematics teaching in 2007 (9%) than 2003 (0%) and the difference at the primary level is negligible (2007: 1% vs. 2003: 0%).

Using a questionnaire survey, Zhu and Fan conducted a study which was specifically aimed to investigate how Singapore mathematics teachers used two widely used textbooks at the lower secondary level in their teaching practice (Zhu & Fan, 2002; Zhu, 2003). It was found that while textbooks were still the major written resource for the local mathematics teachers, many other teaching resources were also in use, including worksheets, assessment books, and IT resources (e.g., CD-ROMs and internet-based resources). The results also showed that while the majority of in-class example problems were taken from the textbooks, more than half of the example problems presented in the textbooks were not actually used in class. Moreover, more than one third of the problem solving heuristics, which were required in the syllabus and introduced in the textbooks, were not taught by the teachers in classroom instruction, which also indicated that there was a gap between the "intended" curriculum and "implemented" curriculum in Singapore mathematics classrooms. In addition, the study revealed that, in general, the types of schools and students, teachers' teaching experience, and particularly teachers' gender and textbook use experience did not have significant impact on the ways how teachers used their textbooks for teaching. From the results, the researchers also argued that although it was unlikely that textbooks would be eventually replaced by other instructional resources such as worksheets, assessment book, and IT materials, it is likely that the use of non-textbook materials would increase in the future.

CONCLUDING REMARKS

In general, the mathematics curriculum and its related textbooks have been regarded as one of the important factors related to students' academic performance. From the research literature reviewed earlier, we can see that the Singapore mathematics curriculum and textbooks have overall received considerable attention over the last decades, mainly due to Singapore students' outstanding performance in international mathematics comparisons. Researchers generally found that the Singapore curriculum, which has historical connection with both Western and Eastern traditions, has been carefully planned, coherently developed, and effectively implemented. It placed clear emphasis on problem solving by making it the centre of the curriculum framework, created more differentiated learning for different students by streaming, and provided effective guidance and support for teachers and students to achieve the curriculum goals. Based on the Singapore's national curriculum (syllabus), the Singapore mathematics textbooks emphasized fundamental knowledge and skills, provided clear instruction on specific problem solving heuristics and procedures (though not all that the syllabus required), and presented more challenging mathematics problems. Nevertheless, they are more traditional in orientation, or in other words, can hardly be called reformed textbooks.

As mentioned earlier, researchers have also found there were gaps between Singapore's national mathematics curriculum and its mathematics textbooks, which was particularly evident is textbooks' representation of problem types and problem solving procedures. It will be interesting as well as important to further investigate why there were such gaps and how the curriculum policy makers and textbook developers can work together so textbooks can be better aligned with the curriculum. As Fan (2010) point out, past experience has suggested that Singapore has overall achieved this alignment well, but it is clear that further research is needed in this direction.

Finally, although many researchers conducted research on the Singapore mathematics curriculum and textbooks because it is supposedly an important factor for its students' excellent achievement as revealed in international comparisons, there has been virtually no specific research that directly aims to address if there is direct or causal relationship between the mathematics curriculum/textbooks and students' achievement in Singapore, or more generally in any other country. In this connection, this category of curriculum and textbook research remains much needed in order to obtain more scientifically reliable conclusion for curriculum reform and textbook development, though the method to conduct such research is likely more challenging (also see Fan, 2013).

REFERENCES

Adams, L. M., Tung, K. K., Warfield, V. M., Knaub, K., Yong, D., & Mudavanhu, B. (2002). *Middle school mathematics comparisons for Singapore Mathematics, Connected Mathematics Program, and Mathematics in Context (including comparisons with the NCTM Principles and Standards 2000).* Retrieved from http://www.amath.washington.edu/~adams/full.pdf.

Dimmock, C., & Goh, J. (2011). Transforming Singapore schools: The economic imperative, government policy, and school principalship. In T. Townsend & J. MacBeath, (Eds.), *International handbook of leadership for learning* (pp. 225-244). Dordrecht, The Netherlands: Springer.

Dindyal, J. (2005, August). *An overview of the Singapore mathematics curriculum framework and the NCTM standards.* Paper presented at the third East Asian Regional Conference on Mathematics Education, Shanghai, China.

Dindyal, J. (2006). The Singapore mathematics curriculum: Connections to TIMSS. In Grootenboer, P., Zevenbergen, R., & Chainnappan, M. (Eds.), *Proceedings of the 29th annual conference of the mathematics education research group of Australia* (pp. 179-186). Adelaide: MERGA Inc.

Everyday Learning Corporation (2001). *Everyday mathematics*. Chicago: Author.

Fan, L. (2010, March). *Principles and processes for publishing textbooks and alignment with standards: A case in Singapore*. Paper presented at APEC Conference on Replicating Exemplary Practices in Mathematics Education (APEC Human Resources Development Working Group, APEC#210-HR-01.4). Koh Samui, Thailand. Retrieved from http://publications.apec.org/publication-detail.php?pub_id=1047

Fan, L. (2013). Textbook research as scientific research: Towards a common ground on issues and methods of research on mathematics textbook. *ZDM-International Journal on Mathematics Education, 45*(5), 765-777.

Fan, L., & Zhu, Y. (2000). Problem solving in Singaporean secondary mathematics textbooks. *The Mathematics Educator, 5*(1-2), 117-141.

Fan, L., & Zhu, Y. (2007). Representation of problem-solving procedures: A comparative look at China, Singapore, and US mathematics textbooks. *Educational Studies in Mathematics, 66*(1), 61-75.

Fong, H. K. (1994). Bridging the gap between secondary and primary mathematics. *Teaching and Learning, 14*(2), 73-84.

Foong, P. Y., Yap, S. F., & Koay, P. L. (1996). Teachers' concerns about the revised mathematics curriculum. *The Mathematics Educator, 1*(1), 99-110.

Ginsburg, A., Leinwand, S., Anstrom, T., & Pollock, E. (2005). What the United States can learn from Singapore's world-class mathematics system (and what Singapore can learn from the United States): An exploratory study. Washington, DC: American Institutes for Research.

Gopinathan, S. (1997). Singapore educational development in a strong developmentalist state: The Singapore experience. In W. Cummings & N. McGinn (Eds.), *International handbook of education & development: Preparing schools, students and nations for the twenty-first century*. Exeter: BPC Wheatons.

Ho, W. K. (2008, February). Using history of mathematics in teaching and learning of mathematics in Singapore. Paper presented at The First Raffles International Conference on Education, Singapore. Retrieved from http://math.nie.edu.sg/wkho/Research/My%20publications/Math%20Education/hom.pdf

Kaur, B. (2005, August). *Assessment of mathematics in Singapore schools – The present and future*. Paper presented at the third East Asian Regional Conference on Mathematics Education, Shanghai, China.

Lee, P. Y. (2008). Sixty years of mathematics syllabi and textbooks in Singapore. In Z. Usiskin & E. Willmore (Eds.), *Mathematics curriculum in Pacific rim countries – China, Japan, Korea and Singapore: Proceedings of a conference* (pp. 85-94). Charlotte, NC: Information Age Publishing.

Lee, P. Y., & Fan, L. (2004). The development of Singapore mathematics curriculum: Understanding the changes in syllabus, textbooks and approaches. In P. Y. Lee, D. Zhang, & N. Song (Eds.), *Proceedings of the ICM 2002 Satellite Confer-*

ence on the Reform of Mathematics Curriculum and its Education in the 21st Century (pp. 31–36). Chongqing, China: Chongqing Publishing House.
Li, Y. (2007). Curriculum and culture: An exploratory examination of mathematics curriculum materials in their system and cultural contexts. *The Mathematics Educator, 10*(1), 21-38.
Ministry of Education. (2006a). Mathematics syllabus (Primary). Singapore: Author.
Ministry of Education. (2006b). Mathematics syllabus (Secondary). Singapore: Author.
Mullis, I. V. S., Martin, M. O., & Foy, P. (2008). TIMSS 2007 international mathematics report: Findings from IEA's Trends in International Mathematics and Science Study at the fourth and eighth grades. Chestnut Hill, MA: TIMSS & PIRLS International Study Center.
National Council of Teachers of Mathematics. (2000). *Principles and standards for school mathematics*. Reston, VA: Author.
Ng, L. E. (2002). *Representation of problem solving in Singaporean primary mathematics textbooks with respect to types, Pólya's model and heuristics* (Unpublished master's thesis). National Institute of Education, Nanyang Technological University, Singapore.
Ng, S. F. (2004). Developing algebra thinking in early grades: Case study of the Singapore primary mathematics curriculum. *The Mathematics Educator, 8*(1), 39-59.
Quek, T. (2002). US, Malaysia, Thailand, Viet Nam, India, Pakistan, Bangladesh, Finland ... Now, Israel uses S'pore maths textbooks too. *The Straits Times*, Sept. 23, 2002, Singapore.
Scott Foresman-Addison Wesley (2004). *Mathematics*. US: Pearson Education.
Seah, W. T., & Bishop, A. J. (2000, April). *Values in mathematics textbooks: A view through two Australasian regions*. Paper presented at the 81st annual meeting of the American Educational Research Association, New Orleans, LA.
Seng, S. (2000, April). *Teaching and learning primary mathematics in Singapore*. Paper presented at the annual International Conference and Exhibition of the Association for Childhood Education International, Baltimore, MD. (ERIC Document Reproduction Service No. ED439812).
Stacey, K. (2005). The place of problem solving in contemporary mathematics curriculum documents. *Journal of Mathematical Behavior, 24*(3-4), 341-350.
Teh, H. H. (1969). *Modern mathematics for secondary schools*. Singapore: Pan Asian.
Usiskin, Z., & Dossey, J. (2004). *Mathematics in the United States 2004*. Reston, VA: National Council of Teachers of Mathematics.
Yeap, B. H. (2005, August). *Building foundations & developing creativity: An analysis of Singapore mathematics textbooks*. Paper presented at the third East Asian Regional Conference on Mathematics Education, Shanghai, China.
Yeap, B. H. (2008, June). *Challenging mathematics in primary school national examination in Singapore.* Paper presented at ICMI Study 16: Challenging mathematics in and beyond the classroom, Canberra, Australia. Retrieved from http://www.amt.canberra.edu.au/icmis16psinyeap.pdf
Zhu, Y. (2003). *Representations of problem solving in China, Singapore and US mathematics textbooks: A comparative study* (Unpublished doctoral dissertation).

National Institute of Education, Nanyang Technological University, Singapore.

Zhu, Y., & Fan, L. (2002). Textbook use by mathematics teachers at lower secondary school level in Singapore. In D. Edge & B. H. Yeap (Eds.), *Proceedings of EARCOME?2 & SEACME?9 Conference* (Vol. 2, pp. 194-201). Singapore: Association of Mathematics Eucators.

CHAPTER 44

TEACHERS' ASSESSMENT LITERACY AND STUDENT LEARNING IN SINGAPORE MATHEMATICS CLASSROOMS

Kim Hong Koh
University of Calgary, Canada

INTRODUCTION

To be productive in the workplace and to be informed citizens in the 21st century, our knowledge-based society requires individuals to possess mathematical literacy, which denotes the ability to grasp the implications of many mathematical concepts, to reason and communicate mathematically, and to solve non-routine, real-world problems effectively using a variety of mathematical methods. This set of competencies has been given more emphasis in the 21st century curricula and teacher education programs around the globe, including in Singapore.

This chapter begins with a discourse on the importance of realigning assessment and curriculum to the desired learning outcomes for the 21st century in the teaching and learning of mathematics. The success of a

The First Sourcebook on Asian Research in Mathematics Education: China, Korea, Singapore, Japan, Malaysia, and India, pp. 981–1010

nation in achieving the goals of the 21st century education hinges on the quality of teachers' assessments of student learning and performance during daily classroom instruction. Teachers' assessment literacy plays an important role in influencing the content and format of classroom assessments. As we enter the second decade of the 21st century, there is an increasing need for education systems around the world to improve teachers' assessment literacy, especially in designing and implementing new forms of assessments that allow for a tight coupling between the learning of mathematics and the attainment of the 21st century competencies. Therefore, the focus of this chapter is a review and discussion of the empirical studies that pertain to the efforts of Singapore educators and researchers to improve the quality of the teachers' assessments in the teaching and learning of mathematics. The chapter concludes with a list of recommendations for the preparation of mathematics teachers in teacher education and professional development programs.

REALIGNING ASSESSMENT AND CURRICULUM TO 21ST CENTURY COMPETENCIES

Since the first decade of the 21st century, many developed countries, including Singapore, have been implementing education reforms that center on aligning changes in assessment and curriculum with the vision of teaching and learning 21st century competencies. This new set of competencies requires students to go beyond the mastery of factual and procedural knowledge. Rather, students are expected to develop a deep conceptual understanding and demonstrate higher-order thinking, inquiry habits of mind, real-world problem-solving skills, effective communication, and collaborative skills. In the US, for example, President Barack Obama has called for the development of new standards and alternative forms of assessment that measure whether students possess 21st century competencies (Darling-Hammond & Adamson, 2010).

Such a reform movement in assessment and curriculum is not new to mathematics education. Since the late 20th century, many mathematics educators have advocated a shift of focus from the drill-and-practice of basic mathematical concepts and procedural skills to students' active learning and understanding of complex mathematical concepts through non-routine problem solving, mathematical thinking and reasoning, communication, and making connections to the real world (Hiebert & Carpenter, 1992; Putnam, Lampert, & Peterson, 1990; Schoenfeld, 2002). This focus on learning mathematics with understanding, or mathematical literacy, has also been endorsed by the standards of the National Council of Teachers of Mathematics in the US (NCTM, 1989) and the earlier Cockcroft Report

Mathematics Counts in the UK (Cockcroft, 1982). According to the NCTM standards (1989, 1991, 1995), the following five general goals are essential for students to become mathematically literate: becoming a mathematical problem solver, learning to reason mathematically, learning to communicate mathematically, learning to value mathematics, and becoming confident of one's own ability. The Principles and Standards for School Mathematics (NCTM, 2000) also call for teachers to use new forms of assessment that are well aligned with these higher-order learning goals and that yield formative information about students' mathematical literacy. In essence, classroom assessment tasks must be cognitively demanding and intellectually challenging to engage students in learning mathematics with understanding, and assessment information should help teachers document and support student learning (i.e., formative assessment).

A similar focus on mathematical literacy as the desired outcome of learning mathematics is noted in the Singapore Mathematics Curriculum Framework. Since its publication in 1990, the framework has undergone some revisions with the inclusion of key 21st century competencies, such as reasoning, communication, connections, application, and modeling. In response to the *Thinking Schools, Learning Nation* policy intiative (Goh, 1997), thinking skills have been added to the curriculum framework. The latest mathematics curriculum framework (i.e., pentagon model, MOE, 2006) is presented in Figure 44.1.

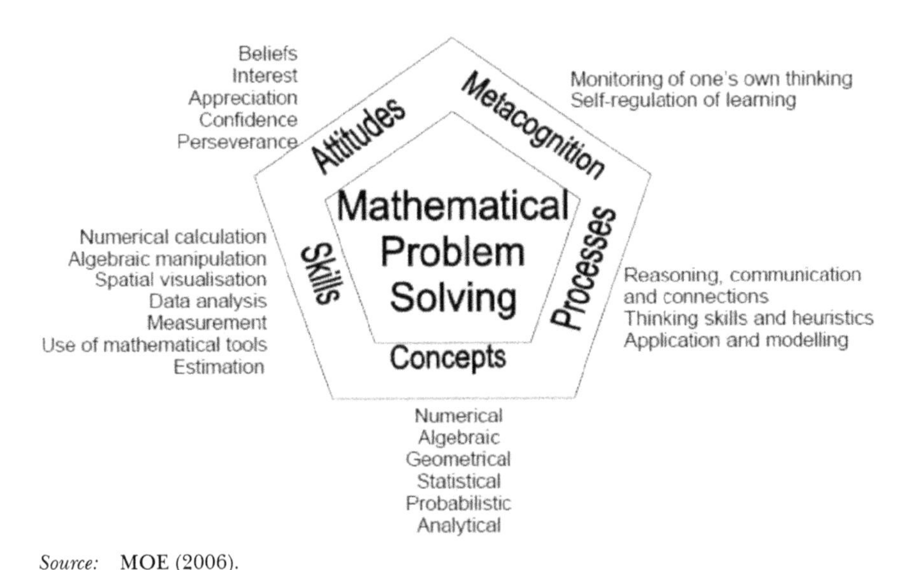

Source: MOE (2006).

Figure 44.1. Singapore Mathematics Curriculum Framework.

In the mathematics curriculum framework (see Figure 44.1), mathematical problem solving is at the core of mathematics learning, and its development is dependent on five interwoven components: concepts, skills, processes, attitudes, and metacognition. To become mathematically literate, students must not only learn the concepts and procedural skills of mathematics, but also learn to use these skills to solve non-routine problems through reasoning, communication, and connections between mathematical concepts and real-world applications. At the same time, students must be confident of their own abilities in mathematical thinking and problem solving as well as be able to value mathematics and self-regulate their own learning. This affective domain of learning outcomes has been deemed to be one of the key 21st century competencies. Although the Singapore Mathematics Curriculum Framework reiterates the importance of mathematical literacy, a greater emphasis has been placed on mathematical problem solving, which is believed to be influenced by the problem-solving reform movement in the US and the UK. The NCTM standards in the US place a greater emphasis on applications and connections to real-world problems, whereas the Cockcroft Report in the UK (Cockcroft, 1982) endorses real-world problem solving and the use of mathematics in daily life as a way to provide enjoyment when learning mathematics and to promote increased mathematical understanding.

The Singapore Mathematics Curriculum Framework serves as a guide for the improvement of mathematics instruction and assessment in Singapore schools. Similar to the teaching professionals in the US and the UK, mathematics teachers in Singapore have been urged to shift their assessment practices toward the use of alternative forms of assessment (i.e., open-ended questions, performance tasks, projects, journals, and portfolios) that are well aligned with higher-order curricular goals. Because the goal of mathematics is to enable students to learn mathematics with understanding, a good way to assess students' attainment of this important goal is through their flexible performances of understanding. Traditional assessment methods, such as summative tests, only allow for the assessment of factual and procedural knowledge due to the nature of the response format (i.e., choosing among pre-determined options or giving short answers) and the teachers' tendency to mimic high-stakes examinations. Performance assessment is touted as a viable assessment method that can capture students' demonstration of understanding (Darling-Hammond & Adamson, 2010; Perkins, 1998). In performance assessment, students must construct an extended response, produce a product, or perform an activity in real-world contexts or contexts that resemble the real world. As such, students are required to engage in higher-order thinking, reasoning, communication, and collaboration to solve realistic, non-routine problems. Arter (1999) has suggested that performance assessment requires teacher observation and judgment based on clear and explicit performance criteria. In the assessment literature, the terms *performance assessment* and *authentic assessment* are used interchangeably. According to Wiggins (1989, p. 703), authentic assessment is "a true test of intellectual ability [that] requires the performance of exemplary tasks." In other words, authentic assessments replicate the challenges and standards of performance in a profession.

The success of a nation in achieving the goals of the 21st century education hinges on the quality of the teachers' assessments of student learning and performance during daily classroom instruction. Teachers' assessment literacy plays an important role in influencing the contents and formats of classroom assessment. Although there have been various initiatives over the past two decades to adopt performance assessments in mathematics teaching and learning, empirical research in the US has shown that teachers' low level of assessment literacy impedes the implementation of high-quality performance assessments (Borko, Mayfield, Marion, Flexer, & Cumbo, 1997; Silver, Mesa, Morris, Star, & Benken, 2009). In Singapore, performance assessment has been introduced into mathematics classrooms through the research conducted by both teacher educators and mathematics teachers who are graduate students. At the systemic level, there are concerns with respect to the misalignment between the intended mathematics curriculum and the enacted assessment. This is partly due to the teachers' lack of assessment literacy in designing and implementing high-quality performance assessments. According to Schoenfeld (2002), "standards-based reform appears to work when it is implemented as part of a coherent systemic effort in which curriculum, assessment, and professional development are aligned." (p. 17). This chapter examines the empirical studies conducted in relation to teachers' assessment practices in Singapore's 21st century mathematics classrooms in general and the global mathematics reform movement in particular. This will allow us to reflect on how best to provide high-quality teacher training and professional development to improve the teachers' assessment literacy (e. g., contemporary knowledge and skills concerning authentic assessment task design and rubric development) at the systemic level, and it will eventually lead to the improvement of the quality of mathematics teaching, assessment, and learning in Singapore schools.

POLICY INITIATIVES, SINGAPORE'S MATHEMATICS CURRICULUM, AND ASSESSMENT

As one of the developed countries in the Asia Pacific region, the economic success of Singapore is based, in part, on the scientific and mathematical literacy of its people. Due to its lack of natural resources, the continuous improvement of the quality of education is an imperative for the Singapore government. In two international achievement assessments, that is, the Trends for International Mathematics and Science Study (TIMSS) and the Program for International Student Assessment (PISA), Singapore students in grades 4 and 8 performed among the top countries in both mathematics and science.

The PISA 2012 defines mathematical literacy as "an individual's capacity to formulate, employ, and interpret mathematics in a variety of contexts. It includes reasoning mathematically and using mathematical concepts, procedures, facts, and tools to describe, explain, and predict phenomena. It assists individuals to recognize the role that mathematics plays in the world and to make the well-founded judgments and decisions needed by constructive, engaged and reflective citizens" (OECD, 2010, p. 4). Clearly, the construct of *mathematical literacy* is well represented by the 21st century competencies and mathematical understandings that are central to students' preparedness for life and work in a knowledge-based economy. Again, the competencies required by the PISA mathematics assessment are similar to those that are outlined in the Singapore mathematics curriculum framework and the NCTM standards. According to Romberg (2001), the PISA framework, since its implementation in 2000, uses the mathematics reform epistemology, which espouses a shift from the mastery of factual and procedural knowledge toward mathematical literacy. From the TIMSS and PISA results, we can infer that the quality of Singapore mathematics education is one of the best in the world. To understand this phenomenon, we must consider the impact of the various policy initiatives by the Ministry of Education (MOE) on the improvement of the quality of education in general and mathematics education in particular.

To achieve the vision of providing all students with a quality education, the MOE has invested in a number of new policy initiatives in response to the global changes. Often, the new policy initiatives have direct implications for curriculum, assessment and teacher education. Prior to 2000, both deductive and inductive reasoning skills were incorporated into the processes domain of the mathematics curriculum framework. When the *Thinking Schools, Learning Nation* initiative was launched in 1997 by then Prime Minister Goh Chok Tong, the deductive and inductive reasoning skills were replaced with thinking skills in the revised mathematics curriculum framework. At the same time, project-based work was introduced as an alternative form of assessment to promote the integration of thinking skills and content knowledge. In the 2000 mathematics syllabus, thinking skills were further defined by the following sub-skills used in problem solving: classifying, comparing, sequencing, analyzing parts and wholes, identifying patterns and relationships, induction, deduction, and spatial visualization (MOE, 2000).

In 2004, the *Teach Less, Learn More* initiative was launched by Prime Minister Lee Hsien Loong at the national rally. This policy initiative resulted in more significant changes to curriculum and assessment. Since teacher professional development is deemed to be important for the improvement of the quality of assessment, teaching, and learning of all subject areas including mathematics, it was important that it incorporated this new focus. For example, the MOE created curriculum white space and a timetable for teachers to participate in professional development. Moreover, the school-based curriculum has been touted as one of the bottom-up initiatives for teachers to innovate through pedagogical methods (Shanmugaratnam, 2005). Many schools also participated in the TLLM Ignite! programs, which provide teachers with more research-based training in curriculum development and assessment. In 2009, in its official report that was recommended for the lower primary school teachers, the national Primary Education Review and Implementation Committee (PERI) emphasized the holistic assessment of students and the use of bite-sized formative assessments to improve student learning (PERI, 2009). The PERI initiative was timely because of the global attention on the role of assessment for supporting classroom teaching and learning (OECD, 2005). Furthermore, the paradigm shift from the 20th century traditional approach to teaching and psychometric testing to the 21st century social-constructivist approach to learning (i.e., student-centered learning) and assessment has played a significant role by promoting the practices of formative assessment, that is, the *assessment for learning* (Shepard, 2000), in contemporary classrooms around the globe. The increasing focus on formative assessment echoes the call for the use of assessment as an integral part of instruction by NCTM's Principles and Standards for School Mathematics (2000).

Although the 2006 Singapore mathematics curriculum framework includes self-regulation of learning as an important metacognitive skill in mathematics teaching and learning, there are no clear guidelines for mathematics teachers to consider with respect to the use of assessment for learning strategies, and most teachers may not know that self- and peer-assessment can be used to promote student self-regulated learning in the process of learning mathematics. A review of the articles published in *The Mathematics Educators Journal,* a publication by the Association of Mathematics Educators in Singapore from 1996 to 2010, revealed that none of the authors had focused on formative assessment (assessment for learning) as a component of mathematics teachers' classroom practices.

Although the intended mathematics curriculum has been written and revised by the curriculum developers according to world-class mathematics standards and new developments in 21st century competencies, the enactment of the intended curriculum depends on the teachers' capacity for curriculum design, content knowledge, pedagogical content knowledge, and assessment.

In the next section of the chapter, the empirical studies of Singapore teachers' assessment practices in mathematics teaching and learning will be used to examine and reflect on the extent of the changes in mathematics assessment since the implementation of the mathematics curriculum.

TEACHERS' ASSESSMENT PRACTICES IN SINGAPORE MATHEMATICS CLASSROOMS

Improving the quality of mathematics teaching and learning hinges, to a large degree, on access to sound empirical evidence regarding what happens in classrooms, that is, what teachers actually do with students to promote the development of students' mathematical literacy. A review of the empirical studies conducted in the context of Singapore classrooms from 1996 through 2010 in published journals, monographs, and conference proceedings indicates that almost all of the studies were related to the teachers' instructional practices or use of new teaching methods to improve students' content knowledge and problem-solving skills in mathematics. Five studies were conducted by mathematics teachers and their supervisors at the National Institute of Education. The purpose of these studies was to explore the use of journal writing, oral presentation, mathematics trails, and a mathematics dictionary as alternative assessment methods in primary and secondary mathematics classrooms (Amir & Fan, 2002; Ng, 2004; Seto, 2002; Soh, 2002; Fan & Yeo, 2007). Table 44.1 presents a summary of the empirical studies conducted in Singapore.

Only three empirical studies have been conducted by teacher educators who examined mathematics teachers' assessment strategies and practices during daily classroom instruction (Fan & Quek, 2005; Fan & Zhu, 2008; and Koh & Luke, 2009). Fan and his associates focused on the use of performance tasks as a classroom intervention and its effects on students' learning outcomes. The Koh and Luke study was a large-scale comparison of the quality of teachers' assessment tasks and of students' work across the core subject areas in Singapore.

Fan and Quek's (2005) two-year intervention study (2003-2004) examined the integration of new assessment strategies into the teachers' day-to-day mathematics teaching and the students' learning. The new assessment strategies included project, performance tasks, journal writing, oral presentation, and self-assessment. The sample consisted of 55 teachers and 2323 students from 8 primary and 8 secondary schools. Two grade

Table 44.1. Empirical Studies on Alternative Assessment Methods

Authors and Year	*Sample*	*Study Design*	*Assessment Methods*	*Key Findings*
Amir, Y., & Fan, L. (2002)	36 grade 5 students	Action research	Journal writing	Students provided feedback on school programs and expressed their needs in terms of mathematics teaching and learning. Teachers used the assessment information to monitor students' understanding of the topic (i.e., formative assessment).
Ng, L. S. (2004)	48 grade 6 students	Pre-test and Post-test control group	Use of mathematics dictionary in geometry	Students in the experiment group performed better than those in the control group on defining geometric concepts in writing and solving geometric problems
Seto, C. (2002)	37 grade 4 students	Action research	Oral presentation on the use of whole numbers, fractions, and decimals	Students learned how to re-evaluate their own thinking processes (i.e., metacognition) and became confident in mathematical communication
Soh, K. J. (2002)	42 grade 6 students	Pre-test and post-test	Use of mathematics trails	Students displayed more positive attitudes toward the learning of mathematics, were able to appreciate mathematics, and found enjoyment in learning mathematics
Fan, L. & Yeo, S. M. (2007)	200 grades 7-8 students	Pre- and post-intervention in two different types of schools (high- versus non-high performing)	Oral presentation	Students found oral presentation a challenging task in regular classroom teaching, and their presentation was not graded. However, they had positive feelings about the benefits of oral presentation in terms of helping them to learn and understand mathematics and in encouraging them to engage in deep thinking about mathematics. Teachers changed their beliefs and views about oral presentations after the initial challenges were overcome.

levels were involved: grade 3 and grade 7. The Fan and Quek study was a large-scale study and used a quasi-experimental pretest-posttest control group design, which allowed for the investigation of the impact of using various new assessment strategies on students' mathematics learning in both academic and affective domains. Fan and Quek (2005) reported that students from the experimental classes benefited from their exposure to new assessment strategies. However, in his paper, Fan concluded (Fan & Zhu, 2007) that developing the students' ability to solve challenging mathematical problems could take longer than expected. This is in agreement with Wright, Palmar, and Kavanaugh (1995) who contended that any particular innovation in education could take many years before it could produce significant results.

Fan and Zhu's (2008) intervention study consisted of 38 grade 7 students from a high-performing secondary school in Singapore. The researchers examined the effects of incorporating authentic performance tasks during mathematics instruction on students' academic achievements and attitudes toward mathematics. A parallel class of 40 students served as the comparison group. The Fan and Zhu study found that students who received the intervention (i.e., use of performance tasks in mathematics instruction for one and a half years) performed significantly better than their counterparts in solving conventional exam problems. They also showed more positive changes in their attitudes toward mathematics and mathematics learning. Although the students expressed positive views concerning the benefits of using performance tasks, their performance on solving unconventional tasks was not statistically significantly different from their peers in the comparison group. As cautioned by the authors, the limitations of the study are that only one experimental class and one parallel comparison class were involved, the students' motivation to work on the new assessment tasks was negatively affected by the fact that their performance was still assessed by conventional assessments at the school level, and the implementation of the new assessment tasks were interrupted by unexpected school activities (i.e., fidelity of intervention implementation).

The following empirical findings are derived from a large-scale study conducted by Koh and Luke (2009) regarding the quality of teachers' assessment tasks and student work in Singapore classrooms. The artifacts of both teachers' assessment tasks and related student work samples in grades 5 and 9 were collected by Koh and her research team from a representative sample of Singapore schools (30 primary and 29 secondary) in 2004 and 2005. The majority of the student work samples at both grades 5 and 9 consisted of class work and homework, which included mathematics worksheets and routine problem-solving tasks from the workbooks and textbooks (see Table 44.2). A closer analysis of the student work samples

Table 44.2. Types of Student Work in Singapore Mathematics Classrooms

	Type of Student Work				
	Worksheet	*Workbook*	*Textbook*	*Test*	*Others*
Grade 5	44%	26%	12.7%	9.4%	8.9%
Grade 9	31%	9%	44%	16%	-

revealed that students were seldom required to explain their mathematical reasoning, applications, or problem-solving strategies. Tests given to students were used predominantly for summative purposes.

The teachers' rationales for assessment tasks typically reflect their purposes of assessment. When the teachers in this study were asked to report their rationales for the assessment tasks assigned to their students, the mean scores shown in Table 44.3 were high for two of the statements: *This assessment task will prepare students for the examination* and *This assessment task is required by the syllabus*. It is evident that most of the assessment tasks were used to prepare students for the high-stakes national examination because grades 5 and 9 students are approaching their Primary School Leaving Examinations (PSLEs) and their O-level examinations, respectively, in the following year. Although many teachers reported that their assessment tasks were required by the syllabus, this does not necessarily indicate an alignment between the assessment tasks and the mathematics curriculum. This is because most of the tasks only assessed the students' reproduction of factual and procedural knowledge, which is not consistent with the objectives of the mathematics syllabus.

Koh (2006) developed two sets of criteria for assessing the authentic intellectual quality of both teachers' assessment tasks and student work based on her adoption of the authentic intellectual work (AIW) framework by Newmann and associates (1996), the revised Bloom's educational objectives by Anderson and Krathwohl (2001), and the different dimensions of learning by Marzano (1992). The two manuals, both written by Koh (2006, 2011), with mathematics indicators for each of the authentic intellectual quality criteria, were used in the training of experienced mathematics teachers to judge and moderate the quality of the assessment tasks and the related student work samples.

The teachers' judgment of each of the criteria for authentic intellectual quality was based on a 4-point rating scale. As shown in Table 44.4, the mean scores of the teachers' assessment tasks were high for factual knowledge, procedural knowledge, reproduction, and the presentation of knowledge as given. In contrast, the mean scores were much lower for

Table 44.3. Teachers' Rationales for Assessment Tasks at Grades 5 and 9

Statement	*Grade 5* $n = 45$		*Grade 9* $n = 26$	
This assessment task ...	*Mean*	*SD*	*Mean*	*SD*
• will prepare students for the examination	4.38	.75	4.65	.63
• is required by the syllabus	4.29	.84	3.92	1.09
• is required by my department head	3.42	1.60	2.81	1.27
• gives my students something to do	3.13	1.66	4.27	.96
• was suggested in a professional development session	2.47	1.50	1.85	.93
• is included in the class textbook	3.29	1.49	3.15	1.59
• is not really necessary at all	1.62	.58	1.62	.64

Note: Mean scores were based on a 5-point Likert scale, ranging from 1 (*strongly disagree*) to 5 (strongly agree); *SD* = standard deviation; *n* = number of teachers' assessment tasks.

conceptual understanding; organization, analysis, interpretation, evaluation, and synthesis of information; problem solving or application; generation of new knowledge; sustained writing or extended communication (e.g., mathematics argumentation and reasoning); and making connections to the real world. These results indicate that the mathematics classroom assessment tasks were of low authentic intellectual quality. Additionally, the mean scores for student control and the sharing of explicit performance standards were relatively low, implying a lack of formative assessment in mathematics classrooms.

These findings imply that the majority of classroom assessment tasks designed and implemented by the mathematics teachers tend to mirror the contents and formats of high-stakes examinations. These findings are also in agreement with the literature on teachers' instructional and assessment practices in mathematics in other developed countries (Silver, Mesa, Morris, Star, & Benken, 2009). *Teaching to the test* is a common classroom practice as a result of high-stakes accountability demands. Studies on teachers' instructional practices in mathematics classrooms have found that mathematics instruction and instructional tasks tend to emphasize low-level rather than high-level cognitive processes, that is, memorizing and recalling facts and procedures rather than reasoning about and connecting ideas or solving complex, non-routine problems (Silver et al., 2009).

Research in educational reform has consistently shown that there is a significant relationship between the quality of teachers' assessment tasks and the quality of student work. Students' learning and performance can

Table 44.4. The Quality of Teachers' Assessment Tasks in Mathematics

Criteria	Grade 5 n = 55 Mean (SD)	Grade 9 n = 41 Mean (SD)
Depth of Knowledge:		
• Factual Knowledge	3.85 (.41)	3.95 (.22)
• Procedural Knowledge	2.69 (1.10)	2.83 (1.07)
• Advanced Concepts	1.49 (.57)	1.63 (.77)
Knowledge Criticism:		
• Presentation of Knowledge as Given	3.85 (.45)	3.93 (.35)
• Compare and Contrast Knowledge	1.35 (.58)	1.59 (.59)
• Critique of Knowledge	1.05 (.23)	1.00 (.00)
Knowledge Manipulation:		
• Reproduction	3.67 (.67)	3.85 (.48)
• Organization, Analysis, Interpretation, Evaluation, or Synthesis of Information	2.09 (.78)	2.12 (.68)
• Application/Problem-Solving	1.76 (.64)	1.54 (.60)
• Generation/Construction of Knowledge New to Students	1.27 (.56)	1.10 (.30)
• Sustained Writing/Extended Communication	2.20 (1.11)	2.24 (.94)
• Connections to the Real World beyond the Classroom	1.33 (.55)	1.12 (.40)
• Clarity and Organization	3.73 (.68)	3.78 (.53)
Learner Support:		
• Student Control	1.29 (.60)	1.32 (.47)
• Explicit Performance Standards/Marking Criteria	1.33 (.51)	1.24 (.44)

Note: Mean scores were based on a rating scale of 1-4; *SD* = standard deviation; *n* = number of teachers' assessment tasks.

be restricted when the cognitive demands of assessment tasks are low. In the Chicago school reform project, Newmann and his associates (1996) found that if teachers made high intellectual demands in their assessment tasks, the students were more likely to demonstrate high authentic intellectual quality in their work. As shown in Table 44.5, the mean scores of student work were found to be high for factual knowledge, procedural

Table 44.5. The Quality of Student Work in Mathematics

Criteria	Grade 5 n = 637 Mean (SD)	Grade 9 n = 503 Mean (SD)
Depth of Knowledge:		
• Factual Knowledge	3.56 (.72)	3.37 (.77)
• Procedural Knowledge	3.42 (.85)	2.85 (1.15)
• Advanced Concepts	1.33 (.53)	1.36 (.59)
Knowledge Criticism:		
• Presentation of Knowledge as Given	3.89 (.46)	3.94 (.24)
• Compare and Contrast Knowledge	1.32 (.50)	1.38 (.50)
• Critique of Knowledge	1.00 (.00)	1.02 (.13)
Knowledge Manipulation:		
• Reproduction	3.72 (.61)	3.88 (.41)
• Organization, Interpretation, or Evaluation of Information	2.16 (.77)	1.82 (.66)
• Application/Problem-Solving	1.75 (.54)	1.49 (.55)
• Generation/Construction of Knowledge New to Students	1.16 (.45)	1.00 (.00)
• Sustained Writing/Extended Communication	2.29 (1.23)	2.23 (.85)
• Quality of Student Writing/Answers	3.23 (.77)	3.09 (.84)
• Connections to the Real World beyond the Classroom	1.28 (.45)	1.03 (.18)

Note: Mean scores were based on a rating scale of 1-4; *SD* = standard deviation; *n* = number of student work.

knowledge, presentation of knowledge as given, and reproduction. In contrast, their mean scores for the criteria related to mathematical literacy were relatively low. These results indicate that most students' work in Singapore mathematics classrooms has low authentic intellectual quality.

IMPROVING TEACHERS' ASSESSMENT LITERACY IN SINGAPORE MATHEMATICS CLASSROOMS

Efforts to improve teachers' assessment literacy in Singapore mathematics classrooms have been undertaken by both mathematics educators and researchers through in-service teacher training and professional development.

For in-service teacher training, Fan (2002) conducted a course using the action research approach with three groups of mathematics teachers from 2000 to 2001. In the 30-hour course, 59 primary mathematics teachers were provided with extensive training in the following aspects of mathematics assessment: (1) application of assessment theory to test construction, (2) application of alternative assessment methods to assess students in mathematics, (3) application of information technology in mathematics assessment, and (4) evaluation of mathematics programs for low- and high-ability students. The participating teachers were required to integrate one type of alternative assessment method they had learned from the training program into their daily instruction for the classes they were teaching. Fan (2002) found that most of the participating teachers felt most familiar with two types of alternative assessment, namely, projects and oral presentations. The teachers' familiarity with projects could be attributed to their earlier exposure to project work, which was introduced by MOE in conjunction with the *Thinking Schools, Learning Nation* policy initiative (Goh, 1997). However, many teachers reported that they were neither familiar with performance-based tasks and portfolios nor had they used them before. Fan's (2002) study also revealed that there was a misalignment between the mathematics curriculum and assessment in Singapore schools. He further suggested that systematic training in formative assessment was needed for school teachers to help them integrate this new form of assessment into the mathematics curriculum.

As a result of the large-scale empirical findings of the quality of teachers' assessment tasks and student work in Singapore mathematics classrooms, as reported in the earlier section, we concur with Fan (2002) that there is a need for more systematic in-service training or professional development for mathematics teachers to improve their assessment literacy in designing and implementing classroom assessments that are well aligned with the objectives of the mathematics curriculum. In 2006-2007, Koh (2011) conducted a 2-year longitudinal intervention study with a group of mathematics teachers who taught grade 5 in two neighborhood schools. Another group of mathematics teachers from two other neighborhood schools served as the comparison group. The intervention and comparison schools were matched based on two variables, namely, type of school and MOE ranking. The approach to this intervention study was different from those conducted by Fan (2002), Fan and Quek (2005), and Fan and Zhu (2008) because the focus of the current study was more on the improvement of the quality of assessment tasks set by mathematics teachers using the criteria of authentic intellectual quality and the changes in the quality of student work in response to the assessment tasks.

In addition, the study also aimed to examine the impact of professional development on teachers' assessment literacy through the use of a quasi-experimental design.

The teachers from the intervention group were involved in ongoing, sustained professional development in which the participating teachers collaborated with the author and her research assistants in authentic assessment task design and rubric development, whereas their counterparts in the comparison group received only one professional development workshop in authentic assessment at the end of each year of the study. The criteria for authentic intellectual quality were used as the guidelines for teachers to design authentic assessment tasks. The criteria and their elements were as follows: depth of knowledge (factual knowledge, procedural knowledge, and advanced concepts), knowledge criticism (presentation of knowledge as given, comparing and contrasting knowledge, and critiquing knowledge), knowledge manipulation (reproduction; the organization, interpretation, analysis, evaluation, synthesis of information; application/problem solving; and generation/construction of new knowledge), sustained writing/extended communication, and making connections to the real world beyond the classroom. These criteria are closely relevant to mathematical literacy and 21st century competencies. More importantly, they are useful for the teaching and assessment of mathematics to *increase student understanding*. To develop students' conceptual understanding of mathematics, it is important that mathematics tasks are drawn from a broad array of school mathematics content domains and are cognitively demanding. At the same time, the mathematics tasks should drive the instructional practices to support mathematical discourse and collaboration among students as well as their engagement with mathematical reasoning and explanation and consideration of real-world applications (Hiebert & Carpenter, 1992).

The artifacts of teachers' assessment tasks and related student work samples were collected from both groups of teachers prior to intervention (as baseline data) and over two phases of intervention. The quality of the teachers' assessment tasks and related student work samples was judged and moderated by the participating teachers who were trained to use the criteria of authentic intellectual quality in teacher moderation sessions.

Figures 44.2 and 44.3 present the comparison of the mean scores on the criteria of authentic intellectual quality for the teachers' assessment tasks and the students' work, respectively. The mean score differences were compared between the intervention and comparison schools across three time points: baseline intervention vs. baseline comparison, post-intervention Phase I vs. comparison Phase I, and post-intervention Phase II vs. comparison Phase II.

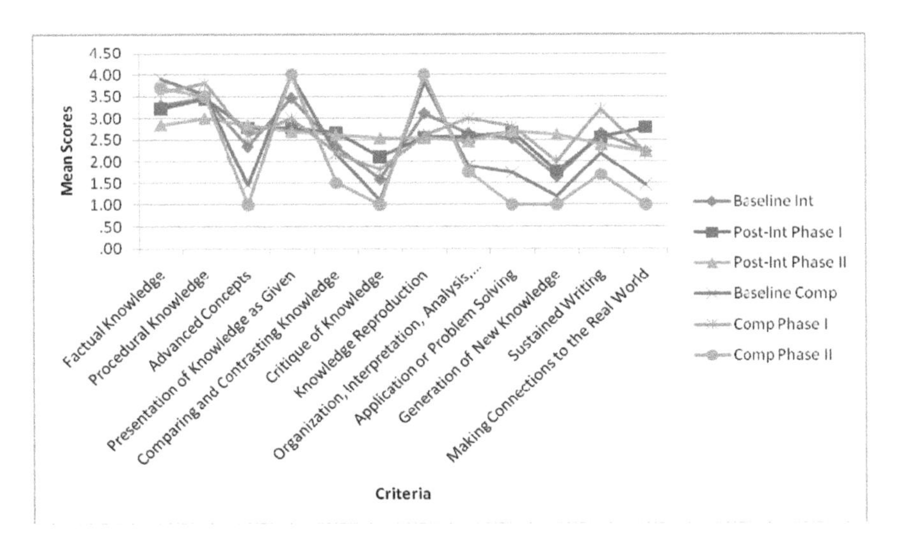

Note: *Baseline int = Baseline in intervention schools; Post-int Phase I = Phase I in intervention schools; Post-int Phase II = Phase II in intervention schools; Baseline Comp = Baseline in comparison schools; Comp Phase I = Phase I in comparison schools; Com Phase II = Phase II in comparison schools.

Figure 44.2. Comparisons of the quality of mathematics assessment tasks.

In Figure 44.2, the mean scores from the intervention schools were lower than those from the comparison schools during Phase II on *factual and procedural knowledge*, *presentation of knowledge as given*, and *knowledge reproduction*. In contrast, the mean scores were higher for the intervention schools than those for the comparison schools on the higher-order thinking criteria that included *understanding advanced concepts*; *comparing and contrasting knowledge*; *critique of knowledge*; *organization, interpretation, analysis, synthesis, and evaluation*; *problem solving*; *generation of new knowledge*; *sustained writing*; and *making connections to the real world*. In the comparison schools, the mean scores at Phase II were lower than those at baseline with respect to the higher-order thinking criteria such as *advanced concepts*; *comparing and contrasting knowledge*; *organization, interpretation, analysis, synthesis, and evaluation; problem solving; generation of new knowledge; sustained writing; and making connections to the real world*. Using the *t*-test, all of the above-mentioned mean score differences were found to be significant with $p < .05$. Due to space limitations, the mean score tables are not presented here.

Figure 44.3 shows that the mean scores on all of the authentic intellectual criteria at Phase II were higher in the intervention schools than in the

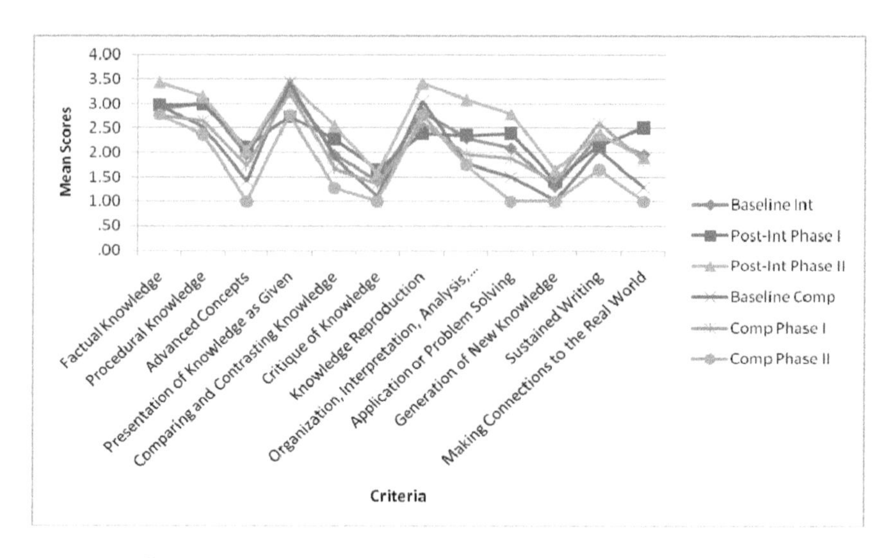

Note: *Baseline int = Baseline in intervention schools; Post-int Phase I = Phase I in intervention schools; Post-int Phase II = Phase II in intervention schools; Baseline Comp = Baseline in comparison schools; Comp Phase I = Phase I in comparison schools; Com Phase II = Phase II in comparison schools.

Figure 44.3. Comparisons of the quality of student work in mathematics.

comparison schools. The changes in mean scores in the intervention schools indicates that mathematics student work demonstrated less *presentation of knowledge as given* and *knowledge reproduction* at Phase I as compared to baseline. However, the mean scores on these two criteria at Phase II had increased, although there was also a significant increase on *comparing and contrasting knowledge*; *organization, interpretation, analysis, synthesis and evaluation*; and *problem solving*. For the comparison schools, the mean scores on all of the authentic intellectual criteria decreased from baseline to Phase II. Further, there was a significant decrease on *problem solving* and *sustained writing*. All of the mean score differences were found to be significant with $p < .05$.

The results indicate that teachers' assessment tasks in the comparison schools focused less on developing students' mathematical understanding and thinking, problem solving, and connections. This is not surprising given that the teachers did not undergo ongoing, sustained professional development. As such, we can conclude that providing teachers' with ongoing, sustained professional development in designing authentic assessment tasks and rubrics improved the teachers' assessment literacy in the redesign of their classroom assessment tasks. Through effective professional development, the teachers in the two intervention schools increased their competence to design assessment tasks that were of high authentic intellectual quality. As a result, their students were able to demonstrate better performance on the tasks assigned by the teachers.

The findings correspond to those in Newmann and associates' (1996) Chicago school reform study and Lingard et al.'s (2001) Queensland School Reform Longitudinal study, which showed that when teachers assigned more cognitively demanding classroom tasks, most students were able to demonstrate more complex intellectual performance. In other words, students' performance was strongly dependent upon what was asked of them in the assessment tasks.

EXAMPLES OF TEACHERS' ASSESSMENT TASKS BEFORE AND AFTER INTERVENTION

Two samples of mathematics assessment tasks before and after intervention from one of the intervention schools can be found in Appendices A and B, respectively. Before intervention, the assessment task given to students contained routine problem solving questions, which only required students to show the procedures for getting the correct answers. This kind of performance task is a"camouflage" because the inclusion of the real-world elements (e.g., local names and everyday problems) does not change the context of the problem or provide a degree of situatedness that facilitates a solution (Cumming & Maxwell, 1999, p. 188). Moreover, the task does not allow students to engage in high-level mathematical reasoning, communication, or connections. After the intervention, there was a significant improvement in the quality of the mathematics tasks. As shown in Appendix B, the performance task focused on non-routine problem solving in a real-world context where students were asked to compare and contrast the prices of different models of cameras from an advertisement and to show the differences of prices paid using two different methods (i.e., cash vs. installments). This performance task is clearly situated in a real-world context that is authentic. The task demands involve not only factual and procedural knowledge, but also most of the aspects of mathematical understanding, including organizing, analyzing, interpreting, evaluating, and synthesizing information; understanding the concepts of multiplication and subtraction of money with decimal notations; making connections to real-world events (i.e., differences between payment by cash and installments); and communication (i.e., explanation of the type of camera that is the best buy for installments). The authentic intellectual quality of this performance task was rated highly by the teachers in the moderation.

IMPLICATIONS FOR TEACHER PROFESSIONAL DEVELOPMENT IN MATHEMATICS EDUCATION

Despite the many waves of reform since the 1990s, the effort to develop teachers' assessment literacy and capabilities in their daily classroom practice remains enormous (Wyatt-Smith and Cumming, 2009). Many teachers adhere to the assessment formats and scoring practices found in high-stakes examinations, which are often in the form of traditional paper-and-pen tests. Without a rigorous and systematic training in classroom assessment that is linked to the assessment of higher-order learning outcomes, teachers' inclination to teach to the tests are forgivable for two reasons: lack of assessment literacy and accountability demands.

There is a consensus in the assessment community that teacher education and professional development programs should help to equip both pre-service and in-service teachers with the knowledge and skills required for designing richer, more intellectually challenging classroom assessment tasks that allow for the holistic assessment of students' knowledge, skills, and dispositions in the context of 21st century learning and teaching (Pellegrino, Chudowsky, & Glaser, 2001; Shepard et al., 2005; Wiliam & Thompson, 2008). The use of authentic or performance assessments is touted as a mechanism for assessment reform in the literature. According to Shepard et al. (2005, pp. 302-303), "assessment reform has been an integral part of educational reform because of the need to engage students in authentic tasks so as to develop, use, and extend their knowledge". Authentic tasks that require higher-order thinking and active problem solving are found to be intrinsically more interesting and motivating to students than memorizing or applying simple procedures. For example, research has shown that students' beliefs about the real-world significance of what they are learning are a strong predictor of their interest and enjoyment in a mathematics class.

Shepard et al. (2005) have further emphasized the importance of focusing on the four following pedagogical approaches in the teaching of assessment to pre-service teachers: (1) analysis of student work and learning, (2) engagement in assessment design, (3) examining motivation and learning theories and how they relate to assessment, and (4) working with standards to design and evaluate assessments for accountability. In the Singapore intervention study that sought to improve teachers' assessment literacy (Koh, 2011), the author included both assessment task design and analysis of student work and learning as the key activities in the professional development sessions. The task design and analysis of student work were based on the framework of authentic intellectual quality. In other words, the participating teachers were actively involved in learning the features of authentic assessment by using the criteria of authentic intellectual quality, which are also well aligned with the desired learning outcomes in the Singapore mathematics curriculum framework. The criteria of authentic intellectual quality also reflect the knowledge and skills of mathematical literacy and the 21st century competencies. As reported in an earlier section of this chapter, this study has shown promising results in improving the quality of teachers' assessment tasks and student work in mathematics, albeit the intervention took place only for two years. As such, the criteria of authentic intellectual quality can also be used in the teaching of assessment task design to pre-service teachers who enroll in the teacher education programs at the National Institute of Education and elsewhere.

As we enter the second decade of the 21st century, Curriculum 2015 in Singapore calls for a more holistic development of student learning in the contemporary classroom, which aims to prepare students to be productive workers, informed and responsible citizens, and lifelong learners. As such, all teachers, including mathematics teachers in Singapore, are expected to move toward a more holistic assessment approach to student learning and performance. This signifies that changes in teachers' assessment practices are imperative and that there is a need to improve teachers' assessment literacy in designing and implementing authentic assessments that enable a richer demonstration and a more holistic representation of what students know and can do in a real-world context. It is my view that to support teachers' professional growth and to effect positive changes in teachers' assessment practices, preparing them for the mastery of contemporary knowledge and skills in designing and implementing authentic assessments is far more important than passing them the assessment tools.

The success of assessment reform initiatives hinges, in large part, on the effectiveness of teachers. As such, to help move the mathematics assessment reform in Singapore and other countries beyond the rhetoric, the seven features of effective professional development from Garet, Porter, Desimone, Birman, and Yoon (2001) must be considered in the planning and implementation of teacher training courses and professional development programs for mathematics assessment.

1. *Type of activity.* The type of activity should be localized and contextualized in the teachers' mathematics classrooms rather than in workshops that occur outside of the teachers' classrooms. Reform types of activities, such as collaborative research between school teachers and university researchers, coaching and mentoring of teachers, and school-based teacher learning communities, have been found to be more effective than the traditional forms of professional development (i.e., workshops, courses, and conferences).

2. *Duration*. The literature on teacher learning and professional development has proven that ongoing, sustained professional development is more effective than ad-hoc, 1 to 2 day professional development sessions. This is because teachers need more time to engage in in-depth discussion of content, pedagogical and assessment strategies, and student learning. In addition, sufficient time is needed for teachers to experiment with new assessment practices in the classroom and to obtain feedback on their practices.
3. *Collective participation*. According to Garet et al. (2001), effective professional development should focus on a group of teachers from the same school, subject, and/or grade level so that the teachers can benefit from their collective participation. There are at least four advantages for involving teachers from the same school, subject, or grade level to work together as a group in developing their assessment literacy. First, they will have the opportunity to discuss knowledge and skills that they obtain and problems that arise during their professional development experiences. Second, the same group of teachers can work together on authentic assessment task design and rubric development based on the same topic and contents as they share common curriculum materials, course offerings, and assessment requirements. Third, teachers can share assessment information and discuss the needs of students across classes and grade levels. Lastly, teachers can contribute to and receive support from a shared professional culture. This will allow them to develop and sustain their professional growth over time.
4. *Focusing on content*. Effective professional development in mathematics assessment should focus on teachers' mastery of both content knowledge (i.e., knowledge of the subject matter content needed to select appropriate tasks) and pedagogical content knowledge (i.e., knowledge of how students learn specific content needed to select appropriate tasks) so that changes in teachers' classroom practices will result in positive student learning outcomes. Research in mathematics education has found that professional development that focuses on specific content and how students learn that content has greater positive effects on student learning outcomes than professional development that focuses on general pedagogy.
5. *Active learning*. There are four dimensions of active learning: observing and being observed; planning for classroom implementation; reviewing student work; and presenting, leading, and writing (Garet et al., 2001). These dimensions should be considered in the planning of mathematics assessment courses for pre-service

teachers and in the creation of professional development programs for in-service teachers. The opportunity to review or analyze student work enables both pre-service and in-service teachers to reflect on the quality of their own assessment tasks and the impacts of their assessment practices on student learning. Teachers will also develop competency in diagnosing the needs of individual students because certain information can serve as formative feedback for teachers to adjust their instructional plans and for students to improve the quality of their work.

6. *Fostering coherence and communication.* Effective professional development in mathematics assessment is characterized by a close alignment between the professional development activities and the mathematics curriculum framework and assessment standards. Teachers must be able to see the common goals or objectives shared by all of the following aspects: policy initiatives, pre-service education, professional development activities, assessment strategies, curriculum materials, textbooks, and professional literature. In addition, it is important that teachers gain support through ongoing professional discourse with their peers and other teachers who are engaged in the efforts to reform their assessments in similar ways.
7. *Teacher Outcomes.* According to the teacher learning and professional development literature, there is a need to assess the effects of professional development on the teachers' knowledge and skills. For example, in the intervention study on improving mathematics teachers' assessment literacy as reported in this chapter, the effects of professional development on the teachers' assessment literacy were examined using the quality of the teachers' assessment tasks as one of the teacher outcomes.

CONCLUSION

In conclusion, successful education reform in mathematics is built on and sustained by a coherent alignment between revised mathematics curriculum, new forms of assessment, and teacher professional development. This chapter shows that some good work has started in Singapore classrooms, but there is still a need for more careful and systematic planning and implementation of teacher training courses and professional development programs for mathematics teachers. Fan, a prominent mathematics educator and researcher, has worked with his colleagues in the development of disciplinary tasks to improve the quality of mathematics assessment and pedagogy in Singapore schools. Their initial findings suggest that both teachers and students found the tasks innovative, challenging, and useful in facilitating the teaching and learning of mathematics (Fan & Associates, 2010). Although it is a research project, it is essential for the researchers to consider how mathematics teachers can be actively involved in learning about task design through professional development. This will not only improve the teachers' assessment literacy and pedagogical content knowledge, but also create meaningful opportunities for the teachers' professional growth, which is perceived as embedded in and as a part of their ongoing work.

APPENDIX A

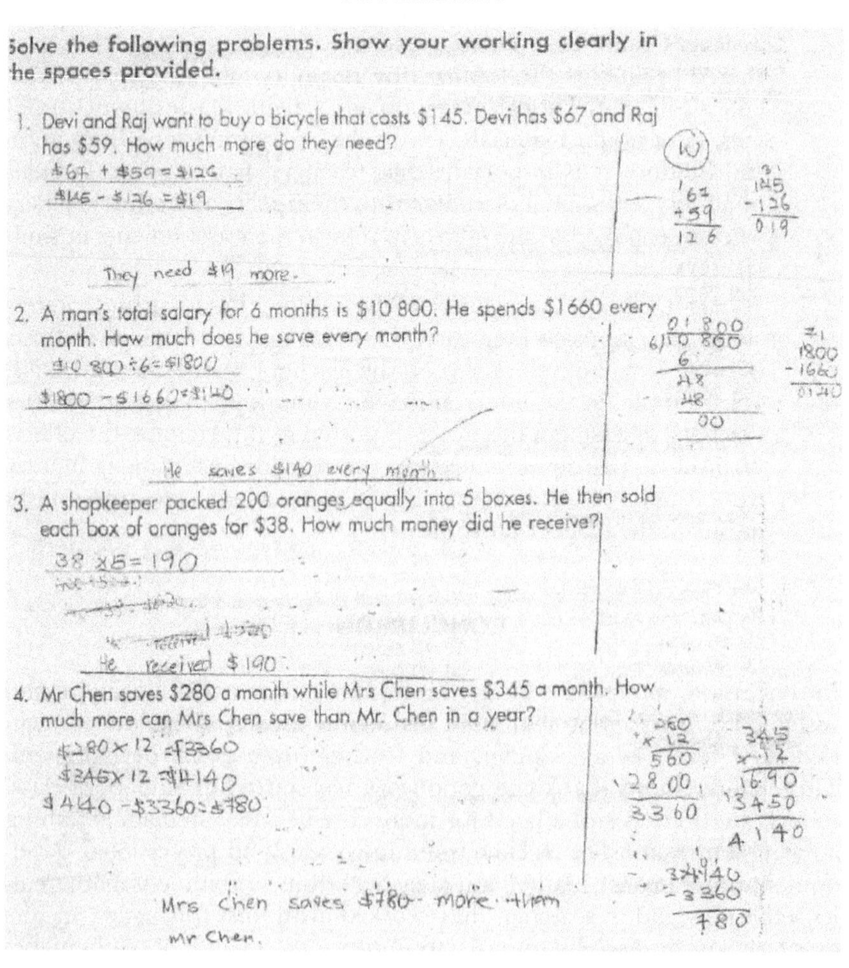

Solve the following problems. Show your working clearly in the spaces provided.

1. Devi and Raj want to buy a bicycle that costs $145. Devi has $67 and Raj has $59. How much more do they need?

   $67 + $59 = $126
   $145 − $126 = $19
   They need $19 more.

2. A man's total salary for 6 months is $10 800. He spends $1660 every month. How much does he save every month?

   $10 800 ÷ 6 = $1800
   $1800 − $1660 = $140
   He saves $140 every month.

3. A shopkeeper packed 200 oranges equally into 5 boxes. He then sold each box of oranges for $38. How much money did he receive?

   38 × 5 = 190
   He received $190.

4. Mr Chen saves $280 a month while Mrs Chen saves $345 a month. How much more can Mrs Chen save than Mr. Chen in a year?

   $280 × 12 = $3360
   $345 × 12 = $4140
   $4140 − $3360 = $780
   Mrs Chen saves $780 more than Mr Chen.

APPENDIX B

S13/P5/MA/-/B/E5/T7/14/3

Primary School
P5 Mathematics – Percentage

P: 2/3
C: 3/3
A: 1/3
(6)/9

Name: ______________ ( )

Class: 5 ______________

Date: 23rd August 2007

Study the advertisement given. The advertisement given to you shows 6 different models of digital camera and their prices.

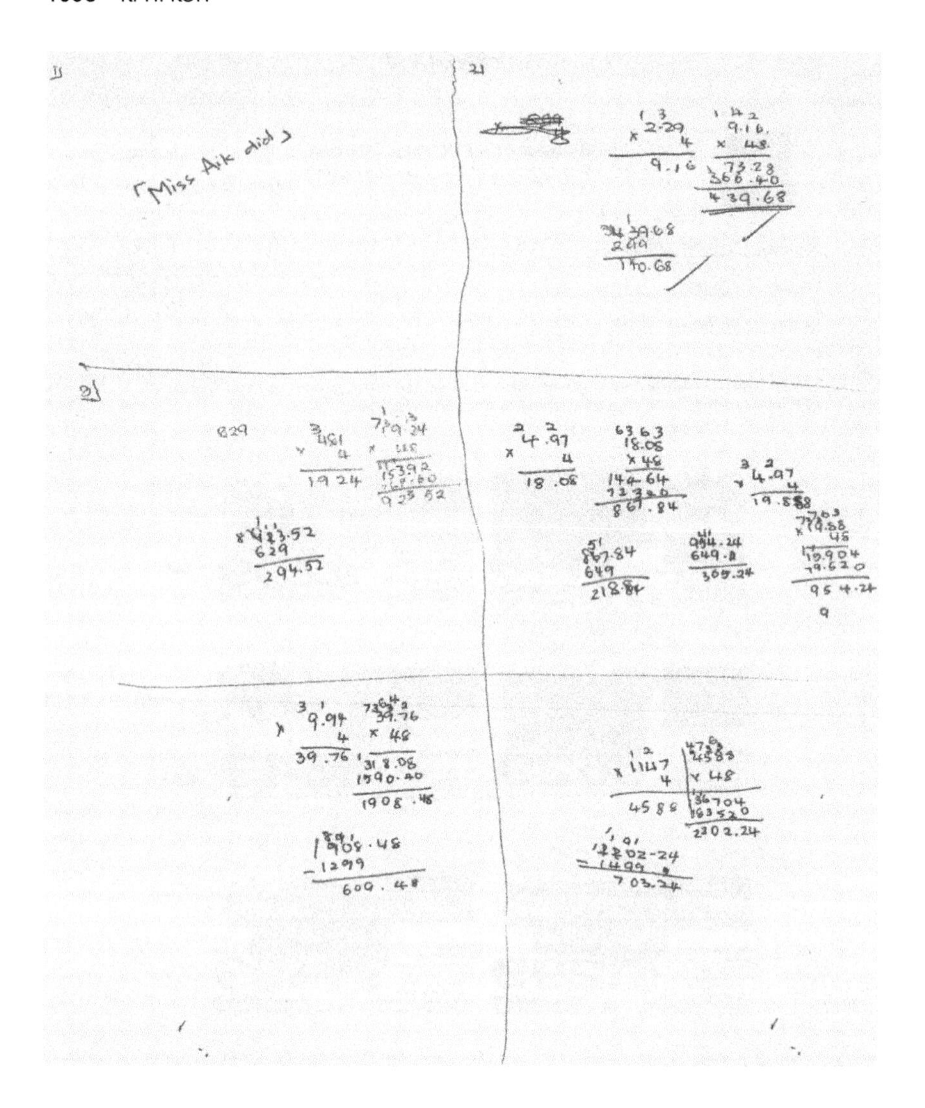

1. Fill in the amount paid for the different models of cameras when
   a) someone pays in full by cash.
   b) someone pays by installments.
2. Find the actual difference in prices between the 2 modes of payment.

Model	Payment by Cash	Payment by Installments	Difference in prices between Payment by Cash & Payment by Installments
Samsung Digimax S630	$ 199 (300)	1.52 × 4 × 48 = 291.84 $	291.84 − 199 = 92.84 92.84 ≈ $93 $ $93
Panasonic DMC LS70	$ 299 (300)	2.29×4×48= = 439.68 $	439.68 − 299 = 140.68 140.68 ≈ 141 $ $141
Nikon Coolpix P5000	$ 629 (630)	4.81 × 4 × 48 = 923.52 $	923.52 − 629 = 294.52 294.52 ≈ 295 $ $295
Samsung i7	$ 649 (650)	4.97 × 4 × 48 = 954.24 $	954.24 − 649 = 305.24 305.24 ≈ 305 $ $305
Olympus E330	$ 1299 (1300)	9.94 × 4 × 48 = 1908.48 $	1908.48 − 1299 = 609.48 609.48 ≈ 609 $ $609
Canon EOS 400D Kit	$ 1499 (1500)	1147 × 4.48 = 2202.24 $	2202.24 − 1499 = 703.24 703.24 ≈ ( ) $ $703

3. Which camera is the best buy if you want to purchase it by installments? Why?

Samsung Digimax S630 because the difference is alot but it is the cheapest than the rest.

REFERENCES

Amir, Y., & Fan, L. (2002). Exploring how to implement journal writing effectively in primary mathematics in Singapore. In D. Edge & B.H. Yeap (Eds.), *Mathe-*

matics education for a knowledge-based era (Proceedings of EARCOME-2) (Vol. 2, pp. 56-62). Singapore: Association of Mathematics Education.

Anderson, L. W., & Krathwohl, D.R. (2001). *A taxonomy for learning, teaching, and assessing: A revision of Bloom's taxonomy of educational objectives*. New York: Longman.

Arter, J. (1999). Teaching about performance assessment. *Educational Measurement: Issues and Practice*. *18*(2), 30-44.

Borko, H., Mayfield, V., Marion, S., Flexer, R., & Cumbo, K. (1997). Teachers' developing ideas and practices about mathematics performance assessment: Success, stumbling blocks, and implications for professional development. *Teaching & Teacher Education, 13*(3), 259-278.

Cockcroft, W. H. (1982). *Mathematics counts.* London: Her Majesty's Stationery Office.

Cumming, J. J., & Maxwell, G. S. (1999). Contextualizing authentic assessment. *Assessment in Education, 6*(2), 177-194.

Darling-Hammond, L., & Adamson, F. (2010). *Beyond basic skills: The role of performance assessment in achieving 21st century standards of learning*. San Francisco, CA: Stanford University, Stanford Center for Opportunity Policy in Education.

Fan, L. (2002). In-service training in alternative assessment with Singapore mathematics teachers. *The Mathematics Educator, 6*(2), 77-94.

Fan, L., & Quek, K. S. (2005, June). Integrating new assessment strategies into mathematics classrooms: What have we learned from A CRPP mathematics assessment project? Paper presented at the International Conference on Education: Redesigning Pedagogy: Research, Policy and Practice, Singapore.

Fan, L., & Yeo, S. M. (2007). Integrating oral presentation into mathematics teaching and learning: An exploratory study with Singapore secondary students. *The Montana Mathematics Enthusiast, Monograph 3*, 81-98.

Fan, L., & Zhu, Y. (2007). From convergence to divergence: The development of mathematical problem solving in research, curriculum, and classroom practice in Singapore. *ZDM International Journal on Mathematics Education, 39*, 491-501.

Fan, L., & Zhu, Y. (2008). Using performance assessment in secondary mathematics: An empirical study in a Singapore classroom. *Journal of Mathematics Education, 1*(1), 132-152.

Fan, L., Zhao, D., Cheang, K. K., Teo, K. M., & Ling, P. Y. (2010). Developing disciplinary tasks to improve mathematics assessment and pedagogy: An exploratory study in Singapore schools. *Procedia – Social and Behavioral Sciences, 2*(2). 2000-2005.

Garet, M. S., Porter, A. C., Desimone, L., Birman, B. F., & Yoon, K. S. (2001). What makes professional development effective? Results from a national sample of teachers. *American Educational Research Journal, 38*(4), 915-945.

Goh, C. T. (1997). Shaping our future: "Thinking Schools" and a "Learning Nation". *Speeches, 21*(3): 12-20. Singapore: Ministry of Information and the Arts.

Hiebert, J., & Carpenter, T. P. (1992). Learning and teaching with understanding. In D. A. Grouws (Ed.), *Handbook of research on mathematics teaching and learning* (pp. 65-97). New York: Macmillan.

Koh, K. (2006). Core 1: Panel 5 Classroom assessment and student performance. *Technical Report*. Singapore: Centre for Research in Pedagogy & Practice, National Institute of Education.

Koh, K. (2011). *Improving teachers' assessment literacy*. Singapore: Pearson Education South Asia.

Koh, K. (2011). Improving teachers' assessment literacy through professional development. *Teaching Education Journal, 22*(3), 255-276.

Koh, K., & Luke, A. (2009). Authentic and conventional assessment in Singapore schools: An empirical study of teacher assignments and student work. *Assessment in Education: Principles, Policy & Practice, 16*(3), 291-318.

Lingard, B., Ladwig, J., Mills, M., Bahr, M., Chant, D., Warry, M., Ailwood, J., Capeness, R., Christie, P., Gore, J., Hayes, D., & Luke, A. (2001). *The Queensland School Reform Longitudinal Study*. Brisbane: Education Queensland.

Marzano, R. J. (1992). *A different kind of learning: Teaching with dimensions of learning*. Alexandria, VA: Association of Supervision and Curriculum Development.

Ministry of Education. (2000). Mathematics syllabus Primary. Singapore: Curriculum Planning and Development Division, Ministry of Education.

Ministry of Education. (2006). *Mathematics Syllabus Primary*. Singapore: Curriculum Planning and Development Division, Ministry of Education.

National Council of Teachers of Mathematics. (1989). *Curriculum and evaluation standards forschool mathematics*. Reston, VA: Author.

National Council of Teachers of Mathematics. (1991). *Professional teaching standards for school mathematics*. Reston, VA: Author.

National Council of Teachers of Mathematics. (1995). *Assessment standards for school mathematics*. Reston, VA: Author.

National Council of Teachers of Mathematics. (2000). *Principles and standards for school achievement*. Reston, VA: Author.

Newmann, F. M., & Associates. (1996). *Authentic achievement: Restructuring schools for intellectual quality*. San Francisco: Jossey Bass.

Ng, L. S. (2004). *Using a mathematics dictionary to improve students' performance in Geometry* (Unpublished master's thesis). National Institute of education, Nanyang Technological University, Singapore.

Organisation for Economic Co-operation and Development. (2005). *Formative assessment: Improving learning in secondary classrooms*. Paris: OECD.

Organisation for Economic Co-operation and Development. (2010). *PISA 2012 Mathematics framework*. Paris: OECD Publications.

Pellegrino, J. W., Chudowsky, N., & Glaser, R. (2001). *Knowing what students know: The science and design of educational assessment*. Washington, DC: National Research Council.

PERI (2009). *Report of the Primary Education Review and Implementation Committee*. Retrieved from http://www.moe.gov.sg/initiatives/peri/files/perireport.pdf

Perkins, D. N. (1998). *What is understanding*? In M.S. Wiske (Ed.), Teaching for understanding: Linking research to practice (pp. 39-57). San Francisco, CA: Jossey-Bass.

Putnam, R. T., Lampert, M., & Peterson, P. L. (1990). Alternative perspectives on knowing mathematics in elementary schools. In C. Cazden (Ed.), *Review of research in education, 16*, 57-152. Washington: American Educational Research Association.

Romberg, T. A. (2001). Understanding the standards-based reform movement in school mathematics. *The Mathematics Educator, 6*(1), 17-21.

Schoenfeld, A. H. (2002). Making mathematics work for all children: Issues of standards, testing, and equity. *Educational Researcher, 31*(1), 13-25.

Seto, C. (2002). Oral presentation as alternative assessment in primary mathematics in Singapore classroom. In D. Edge & B.H. Yeap (Eds.), *Mathematics education for a knowledge-based era* (pp. 33-39). Singapore: Association of Mathematics Educators.

Shanmugaratnam, T. (2005). *MOE Work Plan Seminar ? Speech by Mr Tharman Shanmugaratnam, Minister of Education*, September 22. Singapore: The Ngee Ann Polytechnic Convention Centre.

Shepard, L. (2000). The role of assessment in a learning culture. *Educational Researcher, 29*(7), 4-14.

Shepard, L., Hammerness, K., Darling-Hammond, L., Rust, F., Baratz Snowden, J., Gordon, E., et al. (2005). In L. Darling-Hammond & J. Bransford (Eds.), *Preparing teachers for a changing world* (pp. 275-326). San Francisco: John Wiley.

Silver, E. A., Mesa, V. M., Morris, K. A., Star, J. R., & Benken, B. M. (2009). Teaching mathematics for understanding: An analysis of lessons submitted by teachers seeking NBPTS certification. *American Educational Research Journal, 46*(2), 501-531.

Soh, K. S. (2002). *Real world mathematics — the design of mathematics trails* (Unpublished master's thesis). National Institute of education, Nanyang Technological University, Singapore.

Wiggins, G. (1989). A true test: Toward more authentic and equitable assessment. *Phi Delta Kappan, 70*, 703-713.

Wiliam, D., & Thompson, M. (2008). Integrating assessment with learning: What will it take to make it work? In C.A. Dwyer (Ed.), *The future of assessment*. New York: Lawrence Erlbaum Associates.

Wright, R. E., Palmar, J. C., & Kavanaugh, D. C. (1995). The importance of promoting stakeholder acceptance of educational innovations. *Education, 115*(4), 628-632.

Wyatt-Smith, C., & Cumming, J. (Eds.) (2009). *Educational assessment in the 21st century: Connecting theory and practice*. Dordrecht, The Netherlands: Springer.

CHAPTER 45

A THEORETICAL FRAMEWORK FOR UNDERSTANDING THE DIFFERENT ATTENTION RESOURCE DEMANDS OF LETTER-SYMBOLIC VERSUS MODEL METHOD

Swee Fong Ng
National Institute of Education, Nanyang Technological University, Singapore

INTRODUCTION

In Singapore, the model method (Ng & Lee, 2009) is taught to young children to support them in their problem solving activities. An unexpected consequence of this pedagogical decision is that many students who enter secondary school and are thus expected to use letter-symbolic algebra continue to use the concrete and visual model method to solve related algebra word problems. In letter-symbolic method letters are used to represent the unknowns presented in a problem and the solution is

The First Sourcebook on Asian Research in Mathematics Education: China, Korea, Singapore, Japan, Malaysia, and India, pp. 1011–1043

found by solving a series of algebraic equations. In contrast, in the model method all the relevant information is captured by a series of rectangles in which the relationships of the rectangles are specified and are presented globally. With the aid of the model drawing appropriate sets of step-by-step arithmetic procedures are constructed to solve the problems. Various studies adopting behavioural as well as neuroimaging methodologies have been conducted analysing learners' competencies with the model method and the letter-symbolic method as a problem solving tool. Although both methods activated similar areas of the brain, neuroimaging studies (Lee et al., 2010, 2007) found that the letter-symbolic method was more attentional resource intensive of the two. It could be that more resources were directed towards attention orientation or retrieval of information specific to generating algebraic equations. Why should this be the case?

This chapter reviews related studies investigating how primary pupils used the model method to solve arithmetic and algebraic word problems and secondary students' competencies with the letter-symbolic method to solve algebra word problems. It then offers a theoretical framework to explain why the letter-symbolic method is more resource intensive of the two methods. The final section discusses how current pedagogical practices could be improved to support the transition from the model method to the letter-symbolic method. The term pupils is used to identify primary school participants and the term students, for secondary school participants.

MODEL METHOD AND PRIMARY SCHOOL MATHEMATICS

The Singapore mathematics curriculum puts a premium on students' capacity to solve problems (Curriculum Planning & Development Division (CPDD) 2001, 2006). To cultivate this capacity, students are taught to use various problem-solving heuristics. The policy to introduce problem-solving heuristics at the primary level (CPDD, 1990) has shown positive results. The heuristic 'draw a diagram' known locally as the model method seemed to be particularly effective. With this heuristic, rectangles are used to represent numbers and the resulting schematic representation or model drawing (known as the model), captures the information in a problem. The construction of the model draws on the most important relationship that can be developed about numbers, namely the part-part-whole nature of numbers, (Van de Walle, 2001). For example, 10 can be thought of as a set of 3 and a set of 7 or a set of 4 and a set of 6.The model helps students "visualise" or "see" (Ng, 2003) a problem. The model method can be used to solve arithmetic as well as linear algebraic word problems involving one or two variables. In arithmetic word problems each rectangle represents a given numerical value. Translation of the information captured in the model results in arithmetic equations which may result in the solution of the problem. With algebraic word problems, the rectangles can be used to represent numbers or unknowns. The model drawing becomes a pictorial equation (Ng, 2003). Should the textual information be translated into an algebraic equation, then there is a correspondence between the resulting equation and the model drawing.

Seeing simple arithmetic problems in terms of part-part whole relations is a more powerful way of seeing a problem than interpreting them in terms of arithmetic operations (Neuman, 1997). The part-whole way of seeing works well for any problems in which two or more parts are given and the objective is to find the whole, or when the whole and one of the parts are given and the objective is to determine the remaining part. Thus the model method can be used to solve arithmetic problems with these structures: $x + y = ?$; $x - y = ?$; $? + y = z$; $x + ? = z$; $? - y = z$; and $x - ? = z$.

What Makes the Model Method a Powerful Solving Tool for Solving Simple Arithmetic Problems?

The construction of the model draws on the part-part-whole nature of numbers. That rectangles are used to represent specific numbers or unknowns makes the model method a visual yet powerful representational tool. This is because the construction of the appropriate model representation builds upon pupils' understanding of the part-part-whole relationship of numbers where a whole number can be partitioned into two or more parts. Similarly a whole rectangle can be partitioned into two or more smaller rectangles. The whole rectangle corresponds to the whole number while the smaller rectangles correspond to the parts of the whole number. The value of the whole can be found by summing all the smaller parts. The missing value of any part can be found by finding the difference between the whole and the sum of the remaining parts. Construction of the model representation requires pupils to conceptualise a given problem whereby they have to examine how one part is a function of another. This conceptualisation process requires pupils to think about a problem, consider the appropriate representation and the corresponding solution. The model drawing in Figure 45.1 captures all the various permutations of $x + y = z$

There is no assurance that children who were taught the model method would apply it to solve word problems. Poh (2007) used an instrument comprising three parts to investigate 341 Primary 3 (9+) pupils' competencies with the model method. The first part tested pupils' capacity to

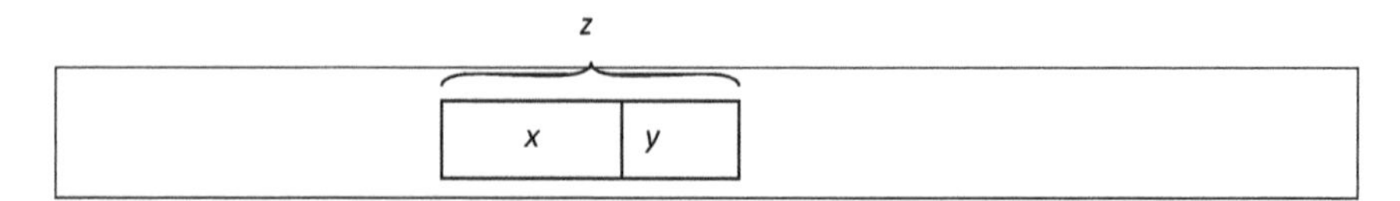

Figure 45.1. Model drawing illustrating the part-part whole nature of numbers and from here the various permutations of the relation: $x + y = ?$; $? + y = z$; and $x + ? = z$.

solve one-step word problems involving any of the four operations and two step word problems requiring the application of addition and subtraction. There were two sets of multiple-choice items. In the first set, given a particular word problem, pupils were required to select from a set of four options the model drawing that best matched the textual information. In the remaining set, model drawings with missing numerical value were provided and again from a set of four options, pupils were to select the arithmetic equation that would result in the resolution of the problem. The study showed that majority of the pupils did not use the model method to solve word problems. Instead they applied their knowledge of real-world experiences of related situations to solve the one- and two-step word problems. Pupils' capacity to solve word problems were affected by the difficulties they had in decoding textual information. In particular relational phrases such as 'more than', 'less than' and 'as many as' were particularly challenging for these pupils. Also pupils were unable to match the given model representation with the correct arithmetic representation that would result in the resolution of the problem. In particular pupils were unable to select the correct model representation when more than one solution step was required.

Lee, Ng, Ng, and Lim (2004) conducted a study with 151 Primary Five (11+) pupils to explore how well working memory, literacy and IQ predicted to children's performance on algebraic word problems. The Working Memory Test Battery for Children (WMTB-C) (Pickering & Gathercole, 2001) which assessed children's working memory in three domains: central executive, and the two slave systems, the phonological loop, and the visuo-spatial sketchpad, was used to assess participants' working memory capacity. Theoretically the central executive controls the flow of information through the working memory, retrieval of information from long term memory and organising multiple cognitive activities. The two slave systems are highly specialised. The phonological loop retains verbal materials but this is subjected to rapid decay. The visuo-spatial sketchpad serves as a store for visual as well as spatial information (For a detailed discussion of working memory and the related components, readers may refer to see works by Baddeley, 1996; Baddeley, Gathercole, & Papagno, 1998). Participants' literacy capacity was measured using the locally normed version of the Wechsler Objective Reading and Language Dimensions instrument (Rust, 2000). Because the intent was to assess participants' capacity to decode information, only the listening component was used in the study. To control for individual differences in intelligence, two subtests from the Wechsler Intelligence Scale for Children – third edition (WISC-III) (Wechsler, 1991), vocabulary and block design, were used. A 10-item mathematics instrument was used to assess pupils' capacity to use the model method to solve these problems. Using a series of path analyses the study found that the overall contribution of working memory as measured by the executive function component was comparable to that of literacy; the effect of executive function was mediated by that of literacy. The study concluded that literacy was the best predictor of success and the one factor that was amenable to educational intervention.

Using the same data set cited in Lee et al. (2004) but applying fine grain analysis to interrogate pupils' written solutions, Ng and Lee (2009) showed that the model method afforded higher ability pupils who typically had no access to letter-symbolic algebra, a means to represent and solve algebraic word problems. High ability pupils were those who were studying their mother tongue at a higher level than the others in the group. Precision in constructing the appropriate model representations and the subsequent arithmetic procedures needed to solve for the unknown values led to the correct solutions. Partly correct solutions suggested that problem representation was not an all-or-nothing process where the model drawing was either completely correct or completely wrong. Instead a wrong solution could be the consequence of a misrepresentation of a single piece of information.

Poh's study (op. cit.) showed that younger children were not proficient users of the model method and they had difficulties with word problems which involved relational terms such as 'more than', 'less than' and 'as many as'. Lee, Ng, Ng, and Lim (2004) and Ng and Lee's (2009) studies using different methodologies differentiated the proficient users from the less proficient users of the model method. Proficient users had good command of language and were precise in their work.

Did pupils' use of the model method as a problem solving tool enable them to perceive the rectangles as representing a specific value or as variables? Khng, Lee, and Ng (2008) hypothesised that if the students perceived the rectangles as variables, then the length of the rectangles drawn to represent unknowns would be independent of the perceived size they represented. A software programme was developed that enabled participants to construct model drawings on the computer. Sixty-eight Secondary 2 students (14 year olds) and 111 Primary 5 pupils (11 year olds) were asked to use this software to solve a set of 12 algebraic word problems, with the same structure as the Marble problem which is shown below, but the total was one of increasing quantum. Half of each set of participants were given questions with the total number of marbles in increasing magnitudes (ones, tens, hundreds and thousands) and with the half, the questions were presented in a randomised manner.

> *Marble problem:* Mary and John have 6 marbles altogether. John has 2 marbles more than Mary. How many marbles does Mary have?

Participants' drawings were recorded by the computer programme. Metric was measured in terms of pixels. The study found that both groups of participants drew longer rectangles when the total of marbles was longer, and shorter rectangles for smaller totals, suggesting that both groups of participants treated the rectangles as variables. This was not the case in a follow-up study by Ng and Lee (2008) who conducted in-depth one-on-one interviews with ten Primary Five pupils (11+) who were proficient with the model method. In addition to solving the same set of tasks as used by Khng, Lee, and Ng (2008), the pupils were also asked to reflect on hypothetical solutions to two questions. They were able to discuss whether they would accept two different but correct model drawings presented for the same problem. The 'mathematical talk' of these primary 5 pupils showed how they oscillated between the different modes of representation to check the accuracy of the solutions. Because lengths of rectangles were used to signify size of numbers, it was possible for pupils to offer descriptions such as "Draw a longer rectangle for a bigger number" or "a shorter rectangle for a smaller number". These pupils were able to justify why the same model drawing could be used to solve different problems that had a similar structure but different numerical quantity. These pupils explained that if the numbers "are put in the right places. Rectangles are drawn correctly. If you understand the question you don't have to change the size of the rectangles" (Ng & Lee, 2008, p. 78). These pupils had abstracted new meanings of how numbers, known or unknown could be represented. This study showed that the logical relation of the rectangles to one another and not the actual length of the rectangles was paramount in the construction of any model representation.

The studies in the preceding section shows how Primary 3 children struggled to use the model method to solve arithmetic word problems but by Primary 5 those proficient in mother tongue were able to use the model method to solve algebraic word problems. The following section discusses how pupils' knowledge of the model method could be an obstacle to their learning to use conventional algebra to solve algebraic word problems once they are in secondary school.

MODEL METHOD AND SECONDARY SCHOOL MATHEMATICS

Their progress from the primary to the secondary mathematics curriculum meant that students would eventually be introduced to formal letter-symbolic algebra. Hence secondary students were expected to use letter-symbolic algebra to solve algebra word problems that would require the construction of no more than two linear simultaneous equations. In Singapore, however, many secondary mathematics teachers considered students' knowledge of the model method affected their willingness to learn to use letter-symbolic method to solve algebra word problems (Ng, Lee, Ang, & Khng, 2006). Although the curriculum expects students to use letter-symbolic algebra, many continued to rely on their prior knowledge of problem solving heuristics, in particular the model method to solve algebra word problems. How accurate were teachers' perceptions that the model method continued to influence students' choice of methods? In 2008, 124 Secondary 2 (14+) students from five schools took part in a study investigating the strategies they used to solve ten algebra word problems. Table 45.1 lists (i) the ten questions used in the study, (ii) the number and the proportion of correct and incorrect responses using letter-symbolic algebra (A) and the model method (M) to solve them. Only 602 of the 1240 solutions applied either of the two methods correctly. Of these, 37% (463) of the correct solutions were found using letter-symbolic method and 22% (139) were solved correctly using the model method. Less than one percent of the students used the guess and check method to solve these problems. Because of the low proportion, their responses are not recorded here.

The study gives credence to teachers' perceptions that students' prior knowledge continued to influence their choice of methods.

The examples in Figure 45.2 demonstrate how letter-symbolic, a semi-algebraic method and the model method were used to solve the Parade Problem. The solution to an algebraic word problem can be found by selecting any of the unknowns. With the Parade Problem the number of children can be found by using any of the unknowns as the generator of the equation. In Figure 45.2, the solution on the extreme left shows a purely algebraic solution which uses the unknown number of children as the generator. The number of children was found directly by solving for the unknown x. The far right panel shows how the model method was used to find the number of children. In this case the unknown number of women was the generator. The solution in the centre panel shows a semi-algebraic method. Here a combination of the letter-symbolic method and the model method was used to solve for the number of children. The solution was accessed by using the unknown number of women as the generator. The letter x into each rectangle suggests that the student knew that

Table 45.1. Word Problems Presented To Secondary 2 Students and the Related Success Rates as a Function of Letter-Symbolic Methods (A) Versus Model Method (MM)

	The Algebra Word Problems	*A (C)*	*A (IC)*	*MM (C)*	*MM (IC)*
1	Parade Problem: There are 900 people at a parade. There are 40 more men than women. There are twice as many children as there are men. How many children are there?	39 52[P]	36 48	11 26	32 74
2	Money Problem: Ann, Ben and Chin have $1000 altogether. Ben has $40 less than Ann. Chin has four times as much as Ben. How much money has Ann?	58 77	17 23	26 59	18 41
3	Games Problem: There were 60 more boys than girls in the basketball team last year. After 30 boys and 24 girls joined the team this year, there were 240 team members altogether. How many girls are there in the team this year?	49 65	26 35	17 43	23 57
4	Spending Problem: Ahmad has four times as much money as Betty. After Ahmad spent $160 and Betty spent $40, they each had equal amount of money. How much money did Ahmad have at first?	59 81	14 19	13 30	30 70
5	Attire Problem: A blouse and 2 shirts cost $24. 5 blouses and 6 shirts cost $92. Find the cost of one blouse.	73 83	15 17	11 38	18 62
6	Florist Problem: A florist sold some roses and 4 times as many lilies. A stalk of rose cost $12 while a stalk of lily cost $8. The florist sold $528 worth of roses and lilies. How many stalks of lilies did the florist sell?	51 67	25 33	8 28	21 72
7	Sharing Problem: A sum of $171 is to be shared between Alex, Bryan and Charles. Bryan will get four times as much money as Alex. Charles will get as much money as Bryan. How much money has Bryan?	36 55	29 45	13 32	27 68
8	Salary Problem: Adrian, Ben and Chan earned a total of $910. Adrian's salary is 3/8 of Ben's salary. Chan's salary is 2/3 of Adrian's salary. What is Ben's salary?	34 62	21 38	19 42	26 58
9	Age Problem: In 4 years' time, Mr Wong will be 3 times as old as his son. 4 years ago he was 5 times as old as his son. How old is Mr Wong now?	22 35	41 65	1 5	21 95
10	Gender Problem: There are as many boys as girls. If there are 2700 children altogether, how many boys are there?	42 75	14 25	20 44	25 56
	Total	463	238	139	241

Note: *C represents correct responses and IC incorrect responses. [P] represents proportions reported as percent who used the method. The total does not sum to 100% because other methods used are not recorded here. Problems are named to facilitate discussion.

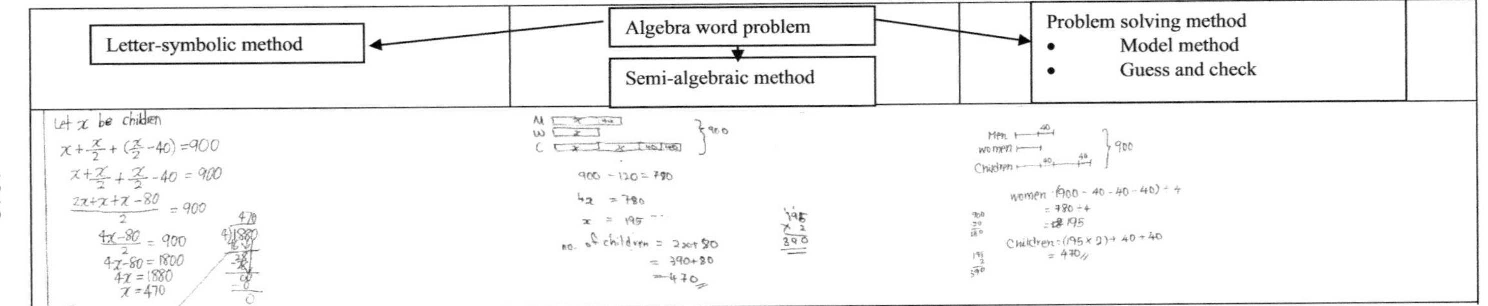

Figure 45.2. Left panel shows how letter-symbolic method was used to solve the Parade Problem, the centre panel, the semi-algebraic method, a combination of the two methods and the far right panel the model method.

the rectangle represented the unknown x. However a mixture of methods comprising doing-undoing and letter-symbolic methods was used to solve for the unknown x. Perhaps the student thought that the solution was algebraic because the letter x was used to announce the answer. This solution suggests that the student may not be clear what made a solution algebraic. It could be that student may be focused on the problem solving process rather than on the choice of the problem solving method.

The solution to the Spending Problem tells a different story. The solution in the left panel of Figure 45.3 shows how the letter-symbolic method was used to solve for the unknown amount of money held by Ahmad. The solution in the right panel shows how two methods were used to help solve the problem. The model drawing correctly depicted the amount held by Ahmad and Betty. To the left of the model drawing the letter x was used to declare the amount to money held by each friend. After this initial use of the letter x, no reference was made of the letter x at all. Instead of constructing a system of equivalent equations to find the value of x, the amount of money held by Ahmad was found by a series of doing-undoing processes.

Either the letter-symbolic method or the guess and check heuristic could be used to solve the Age Problem presented in Figure 45.4. Although it is not wrong for students to use alternative methods other than letter-symbolic algebra to solve these problems, students' continued use of these alternative problem solving strategies may not serve them well in the long term. It is vital for students to acquire a sound knowledge of letter-symbolic algebra and the related procedures and skills should they wish to engage with higher mathematics and other disciplines such as the sciences and economics.

In fact, many secondary students valued the importance of letter-symbolic algebra as a problem solving tool. Ng (2003) asked 145 Secondary 2 students for their perceptions of the model method as compared to letter-symbolic algebra. Comparing and contrasting these two methods, 114 of these students saw the value of the model method as it provided them a powerful means to visualise and conceptualise a problem. Nevertheless construction of the model representation could prove to be tedious as it required care and precision on the part of the students. Any information missing in the model representation meant that the problem could not be solved or that an erroneous solution would be the consequence. These students abstracted the value of letter-symbolic method which they described as simpler of the two methods and it could be "used more extensively when finding out 2 unknowns, and the model method is rather limited" (Ng, 2003, p. 13).

Why did secondary students continue to use the model method to solve algebra word problems, particularly after they were taught letter-symbolic

Figure 45.3. Left panel shows how an example of letter-symbolic method used to solve the Spending Problem, right panel, the model method.

Figure 45.4. Left panel shows application of letter-symbolic method used to solve the Age Problem, guess and check in the right panel.

method and were expected to use the method in all works related to algebra? Khng and Lee (2009) proposed that because they were unable to inhibit the effect of previously learned strategies, these previously learned strategies continued to intrude into students' consciousness during their decision making. Intrusions are likely to occur when there are competing strategies to solve a given problem. According to Siegler's strategy-choice model (Siegler, 1987; Siegler & Shrager, 1984) the strategy of choice to solve a future problem is that which has proven to be effective in the past. Khng and Lee examined whether the choice of problem solving strategies of 157 Secondary 2 students (14+) correlated with their ability to inhibit intrusions from earlier learned strategies. The students were asked to use letter-symbolic algebra to solve nine algebra word problems. After controlling for students' algebraic knowledge as measured by the Chelsea Diagnostic Mathematics Test (Brown, Hart, & Kuchemann, 1984), working memory and intelligence, Khng and Lee found that students' ability to apply letter-symbolic method to solve algebraic word problems was partially related to inhibitory abilities. The study found that knowledge of algebra as measured by the Chelsea Diagnostic Mathematics Test and intellectual ability were the strongest predictors of both arithmetic intrusions and problem solving accuracy. Those who were able to inhibit intrusions from previously learned skills and procedures were more likely to use algebraic methods than problem solving heuristics such as the model method, guess and check or systematic listing. Fifty-two percent (38% arithmetic and 14% mixed method solutions) of the time, students used non-letter symbolic methods to solve the problems. More importantly, letter-symbolic method accounted for a higher success rate (76%) but problem-solving heuristics only 50%.

The occurrence of such intrusions into the consciousness of the students should come as no surprise. By the time primary pupils enrolled into secondary school they would have applied any of the problem solving strategies for at least three years, the better part of their upper primary years. Furthermore all primary pupils in Singapore sit for the Primary School Leaving Examination (PSLE) held at the end of the sixth and final year in primary school. Given that success in this examination is paramount, pupils are primed to perform. Much of the curricular time in Primary 6 is spent solving challenging word problems, often presented as algebraic word problems. Examples of such questions can be found in *PSLE Examination Questions 2004–2008* (Singapore Examinations and Assessment Board, 2008). As secondary students, their experience with letter-symbolic method, however, was much shorter. Their experience with letter-symbolic method would have been at most four weeks of curricular time, 2 weeks in Secondary 1 and 2 weeks in Secondary 2. Thus it is not surprising that although they were specifically asked to use letter-symbolic method, many of the students in the study asked if they could use the model method instead. This was because the problem solving heuristics, in particular the model method were well learned and had been found to be an effective problem solving strategy.

The findings by Khng and Lee (2009) are not unexpected. Research (e.g., Koedinger & Nathan, 2004; Stacey & MacGregor, 200) showed that early learners of algebra continued to use arithmetic methods to solve algebra word problems. Also research (Steinberg, Sleeman, & Ktorza, 1990; Kieran, 1989) showed that poor knowledge of structural properties, transformational and representational skills hinder early learners of algebra. The evidence reported thus far for secondary students showed that perhaps "a focus on problem solving can work against the learning of algebra, because the focus becomes that of 'solving the problem' and not on the method for solving the problem" (Lins, Rojano, Bell, & Sutherland, 2001, p. 10).

RESEARCH UTILISING NEUROIMAGING METHODOLOGIES

Teachers and secondary school students perceived the model method as a childlike method (Ng, 2003; Ng, Lee, Khng, & Ang, 2006). Of the two, letter-symbolic algebra was seen to have a more abstract nature against the more visual and thus more concrete model representation. It is difficult to evaluate such perceptions at the behavioural level. The availability of neuroimaging technologies such as functional Magnetic Resonance Imaging (fMRI) meant new methodologies could be used to investigate the veracity of such perceptions. Two studies using neuroimaging methodologies investigated the nature of the brain responses when participants were asked to respond to different representations. The first study focused on the initial stages of problem solving: that of translating textual information into either model representation or algebraic equation (Lee et al., 2007). Eighteen adults who were proficient and competent users of letter-symbolic and model method were presented with algebraic word problems. They were then asked to re-present the textual information either as algebraic equations or model representations and then validated whether presented solutions matched their representations. The study found that both representations were comparable in terms of area of activation as both activated areas linked to working memory and quantitative processing. In contrast, however, letter symbolic representations imposed greater attentional demands. The authors concluded that although the two representations activated similar processes but letter-symbolic method was more resource intensive. The second study (Lee et al., 2010) focused on the solution phase of the problem solving chain. Here 17 adult participants were asked to find the value of the unknown when the information was presented either as a model representation or as an algebraic equation. The study found that of the two methods, evaluating for the unknown using the algebraic equations required 'greater general cognitive and numeric processing resources" (p. 591). A noteworthy finding from the study suggested that "linguistic processes play a more prominent role when processing symbolic stimuli" (Lee et al., 2010). Taken together findings from the two studies showed that translation from text to representation and evaluating for the unknown activated similar areas of the brain and in both cases letter-symbolic method required more attentional resources.

The fMRI studies confirmed the difference in perceptions teachers and students had for the model representations and letter-symbolic method: the childlike perception could be translated as the model representation as being less cognitively demanding of the two. Policy makers and curriculum designers may wish to take note that only adults who were proficient in the two methods participated in the fMRI studies. These proficient adults found the letter-symbolic method more attentional resource intensive than the model method. Given that children attain the full working memory capacity only in their adolescent years (Gathercole & Alloway, 2008) it is best to leave the teaching of letter-symbolic algebra to secondary years when the working memory capacities of students are sufficiently developed to cope with demands of learning letter-symbolic algebra. More importantly within a particular age group there is a wide variation in working memory capacity between individuals and there could be as much as six year difference between individuals (Gathercole & Alloway, 2008). If this is the case then it is unsurprising that within a class there could be some students who were less challenged by letter-symbolic algebra whilst others found it cognitively demanding.

Three important findings emerged from all the studies summarised in Table 45.2. First, participants' linguistic proficiencies have a contributory role on their word problem solving performance. Although such findings are consistent with the extant literature (Swanson, Cooney, & Brock, 1993; Pape, 2003) the studies reviewed in this chapter did not identify which specific aspect of literacy affects performance. Future studies need to be more focused so as to identify which particular aspect of literacy affects performance so that pedagogical innovations could be more focused. Second, many novice learners and all the adults in the neuroimaging studies found letter-symbolic algebra challenging. Third, many novice learners continued to use problem solving heuristics to solve algebra word problems.

Although behavioural studies and those using neuroimaging methodologies provided empirical evidence to support the conception that letter-

Table 45.2. Summary of Major Findings From the Studies Reviewed in This Chapter

References Age and Number of Participants	*Instrument*	*Findings*
Lee, Ng, Ng and Lim (2004) Primary 5 (11+) 151 from 5 schools.	• 10-item paper and pencil mathematics test • Working memory test battery (WMTB-C) • IQ tests (WIC-III)	• Working memory correlated to success with word problems. • More importantly literacy (decoding information) was the better predictor of success.
Poh (2007) Primary 3(9+) 341 from one school.	• 3- part paper and pencil • (A1) one-step word problems involving any of the four operations • (A2) Two step word problems involving addition and subtraction • (B) MCQ items: Given word problems choose from four options the model drawing that best match the task. • Given model drawing with missing numerical information, choose from four options the arithmetic equation that would resolve the problem.	• Instead of using the model method to solve the problems, primary pupils used knowledge of real-world experiences to solve the problems. • Difficulties: Children had difficulties decoding textual information • Children were unable to make the correct match between model representation and arithmetic equation, in particular with two step word problems.
Ng and Lee (2008) Primary 5 (11+) 151 from 5 schools.	• 10-item -paper and pencil test	• Children who were enrolled at the highest level of mother tongue studies outperformed those who took mother tongue at a lower level. • More than 50 percent of the children were successful with arithmetic word problems. • Less than 50% of the children were successful with algebraic word problems.
Ng and Lee (2008) 10 children from one school	• Production tasks: 24 items divided into 4 categories. Numbers in each category was confined to a particular quantum. Used model to solve these items. • Validation task: Given a pair of models, students were to identify whether these were correct and to justify their choice.	• Children abstracted for themselves new meanings of how numbers, known or unknown could be represented. • Logical relation and not length of rectangles more important criterion for validating accuracy of model.

(Table continues on next page)

Table 45.2. (Continued)

References Age and Number of Participants	*Instrument*	*Findings*
Khng and Lee (2009) Secondary 2 (14+) 157 from 6 schools	• 9 item paper and pencil mathematics task • CDMT test • Inhibition tasks a mix of computerised and non-computerised tasks	• More students used model method than letter-symbolic method to solve 9 algebra word problems. • Those who used letter-symbolic algebra were more successful.
Ng (2003) Secondary 2 (14+) 137 from two schools.	• 3-item paper and pencil mathematics task. • 1 question asking for students' perception	• Students saw model method as a powerful visualising tool. • Algebra was more utilitarian of the two.
Ng (current) Secondary 2 (14+) 124 from 5 schools.	• 10-item paper and pencil mathematics task	• About two-fifths of the correct solutions were found using letter-symbolic algebra, one-fifth with model method.
Ng, Lee, Ang, & Khng, (2006) Teachers 10 primary and 10 secondary mathematics teachers	• 6 item paper and pencil mathematics task to evaluate whether teachers can shift between model method and letter-algebra method.	• Primary rather than secondary teachers were more versatile being able to shift between methods.
Lee et al. (2007) 18 adults	• Neuroimaging study • Presented with questions, participants validated whether the model or letter-symbolic representation was correct	• Both methods activated similar parts of the brain with participants spending more time viewing the model stimuli. • Of the two, letter-symbolic algebra was more attentional resource intensive.
Lee et al. (2010) 17 adults	• Neuroimaging study • Presented with questions, participants evaluated the unknown.	• Similar findings as above.

Note: References related to primary pupils are cited first and in chronological order.

symbolic algebra is more challenging than model method, these studies do not, however, explain this difference. Albeit neuroimaging studies explained that of the two methods, letter-symbolic method was more effortful, they did not explain why that should be the case. Studies such as Ng, Lee, Ang and Khng (2006), Lee et al. (2007) and Lee et al. (2010) offered pedagogical suggestions whereby teachers used students' knowledge of the model method as a scaffold to help them learn letter-symbolic algebra. Given that both representations activated similar areas of the brain, this suggestion may not go amiss. A common suggestion is to relate the role of the rectangle to that of the letter as an unknown. Since the rectangle represents an unknown value, replacing the rectangle with a letter would be a useful scaffold. Other than using letters to replace the role of the rectangles such generic pedagogical suggestion however, is not effective as it does not target specifically at what is algebraic about letter-symbolic algebra and what mathematical knowledge is needed to support the shift from the model method to letter-symbolic algebra. Even Primary 5 children (Ng & Lee, 2008) had abstracted for themselves the role of the rectangle and that other objects such as letters could replace the function of the rectangle.

Hence pedagogical suggestions need to be more focussed. They should highlight what are the congruencies and the divergences between the two methods. Next it is necessary to establish the mathematical knowledge and skills particular to each method. The shift from using the model method to solve arithmetic word problems to algebraic word problems and then to letter-symbolic algebra to solve algebra word problems require the assimilation of new meanings and accommodating the role played by the rectangle in arithmetic word problems. Such shifts require the construction of new knowledge and the production of new meanings (Lins, Rojano, Bell, & Sutherland, 2001). In other words what kinds of abstractions are necessary to make the shift from the model method to letter-symbolic method? Once this abstraction is made it is then possible to shift between the two methods. For these abstractions to be possible, it is hypothesised that new knowledge has to be constructed. Therefore it is necessary to identify the nature of the new knowledge students need to construct that would enable them to discern the commonalities and differences between the two methods and hence identify what makes letter-symbolic method algebraic and model method a problem-solving heuristic. Therefore pedagogical suggestions to help novice learners make the shift from model method to letter-symbolic algebra need to address the following questions.

1. What knowledge do learners need to have to work with each method?

2. What abstractions of letter-symbolic method do students need to make?
3. What new knowledge do students need to construct that would enable them to discern the commonalities and differences between the methods and shift between the two methods?

Hence it is the aim of the remaining section of the chapter to offer a theoretical framework to address the above three questions.

WHY SHOULD LETTER-SYMBOLIC ALGEBRA BE MORE RESOURCE INTENSIVE? A THEORETICAL PERSPECTIVE

This section draws upon the extant literature which analyses the cognitive challenges faced by early learners of letter-symbolic algebra to discuss why, of the two methods, using letter-symbolic algebra to solve algebra word problems may be more resource intensive. Because students' knowledge of the model method was developed in the early primary years where they used the method first to solve arithmetic word problems and then later progressed to algebraic type word problems it is therefore logical to begin the discussion of the theoretical framework with work related to arithmetic word problems progressing to algebraic word problems in upper primary years and finally focusing on work related to algebra word problems in the early secondary years. Figure 45.5 gives the structural aspect of the theoretical framework.

Arithmetic word problems: In primary school the model method is used to solve arithmetic (result unknown) as well as algebraic (start unknown) word problems. In arithmetic word problems, all information required to solve a given problem is provided. The task requires pupils to find the result such as the sum.

> *Compare word problem:* Ali has $250. Peter has $69 more than Ali. How much money do Peter and Ali have altogether?

In this compare word problem, although the amount of money held by Peter is not yet known it can be found by summing the difference between Ali and Peter and the amount held by Ali. The total held by Ali and Peter can be found by summing the amounts held by each individual. Of course this problem can be solved without constructing the appropriate model drawing. This arithmetic equation $\$250 + \$250 + \$69 = \$569$ represents the solution to the problem.

What kind of knowledge would be required to use the model method to solve this problem? Part-whole relationship of numbers and proportional reasoning

Word Problems	Arithmetic word problems (result unknown) Ali has $250. Peter has $69 more than Ali. How much money do Peter and Ali have altogether?	Algebraic word problems (start unknown) There are 900 people at a parade. There are 40 more men than women. There are twice as many children as there are men. How many children are there?	
Method of solution	Model method	Model method	Letter-symbolic algebra
Structure of word problem	Result unknown Given → Given	Result known Start unknown → Start unknown	
Objects of representation	Rectangles for numbers Numerical equations formed with a combination of any of the four operations	Rectangles for unknowns and specific values Numerical equations	Letters and numbers, symbols of the operations Algebraic equations
Methods of solution	Combination of any number of the operations	Doing-undoing	Construct a system of equivalent equations.
Knowledge needed for the construction of the representation	Proportional reasoning: a longer rectangle for a bigger number, a shorter rectangle for a smaller number.	Pictorial equations As long as the model drawing is logical, length of rectangles do not matter.	Letters as unknown numbers Algebraic expressions and algebraic equations as mathematical objects representing information in the text.
New knowledge needed to move from one mode of representation to the next.	To move from arithmetic to algebraic word problems: Abstraction: Rectangles representing specific numbers to rectangles as unknowns as well as numbers.	To shift from model method to letter-symbolic algebra Abstraction: Meaning of letters Accept the lack of closure Conventions related to letters as objects Meanings of structure particularly concepts of equality- equivalence.	

Figure 45.5. This table summarises the objects of representation and the resulting object of representation: model drawing and equations, the nature of knowledge required to move from one mode of representation to another.

are prerequisite knowledge necessary for the construction of models. The whole of Peter's money is a sum of two parts: that of Ali's and the difference. The total amount is the sum of the two parts: Ali and Peter's. A rectangle is used to represent each amount and the rectangles are related to each other. First a rectangle is used to represent the amount of money held by Ali. Next a longer rectangle is used to represent Peter's amount. This rectangle is longer because it comprises two parts: one part representing the original amount compared to (the generator) and also the difference. The total amount is found by summing all the three parts.

What kind of abstraction would be required to construct a model representation? Because the quantum of each individual is given, the length of the individual rectangles should reflect the respective numerical value, a longer rectangle for a bigger number and a shorter rectangle for a smaller number. Nevertheless because the quantum is big it is not possible to represent accurately each amount. In this case it is not possible to represent $250 on a page. Thus the length of rectangle representing the amount held by Peter is just a representation of the actual value. Thus the construction of the model drawing to represent the arithmetic word problem requires a certain amount of abstraction: the length is just an arbitrary representation of the amount.

What new knowledge is constructed when using the model method to solve arithmetic word problems? Once the rectangle known as the generator rectangle (in this case Ali) has been identified, the length of the generator rectangles guides the construction of other related rectangles. Because the difference between Ali and Peter's money is about one-fourth that of Ali's, the difference rectangle should be about one-fourth the length of the generator rectangle. Thus construction of the model drawing requires the construction of new knowledge, in this case that of proportional reasoning. Furthermore the translation of the model representation into the appropriate number sentences requires construction of the knowledge that different operations can be used to relate each of the rectangles and therefore different number sentences can be used to represent the same model drawing. In this example, the total sum of money can be represented by two number sentences: (i) $250 + $250 + $60 or (ii) 2 × $250 + 60. Furthermore the model method could be used to solve structurally more complex word problems whereby there are more than two protagonists and the comparisons between different protagonists could involve relationships such as 'less than' and 'as many as'. Furthermore problems involving fractions could be transformed into problems involving whole numbers. Concepts of fractions which are undergirded by part-whole relationships are needed to construct the model representation but operations involving whole numbers are used to solve for the unknown instead. Thus new knowledge is constructed when pupils are able to apply the model method to solve word problems involving whole numbers as well as fractions. Furthermore when presented with a particular model drawing, pupils should be able to create different examples of word problems exemplified by that model. Pupils could be described as having a good understanding of the role of the rectangles in the model drawing if they are capable of constructing the relevant models for a given problem, "the doing process" and constructing appropriate problems for a given model, "the undoing process".

Algebraic word problems: Algebraic word problems differ from arithmetic word problems in that all relevant information needed to solve the latter are provided. The challenge is to identify the appropriate operations that will lead to the resolution of the problem. With algebraic word problems, however, although the result is provided there are many unknowns in the problem, and the challenge is to solve for a specific unknown. To complicate matters any of the unknowns could be used to start the problem-solving process. Thus the transition from solving arithmetic word problems to algebraic word problems requires a new way of thinking of representation. With arithmetic word problems every rectangle represents a specific number. This is not the case with algebraic word problems. Some rectangles represent specific numbers while others represent the unknown which has to be evaluated. The power of the model method lies in the dual meaning held by a rectangle. The rectangle can represent a known value and it can also be an unknown.

What kind of knowledge would be required to construct a model representation for algebraic word problems? The rectangle, the object of representation can be used to represent a specific number or an unknown value to be evaluated. The capacity to use rectangles for different purposes is indicative whether pupils have abstracted the different roles of rectangles.

What kind of abstraction would be required to construct a model representation for algebraic word problems? Because a rectangle representing the unknown, the value of which has yet to be evaluated, proportional reasoning used to construct rectangles of different lengths in arithmetic word problems has to be suspended. Thus using the model method to represent algebraic word problems requires another level of abstraction: The lengths of the rectangles no longer matter. What matters most is the logical relationship of the different rectangles.

What new knowledge is constructed when using the model method to solve algebraic word problems? Two sets of knowledge can be constructed. The first and more accessible of the two is the realization that there exists relationships between pairs of operations; for example subtraction is the reverse of addition, division is the reverse of multiplication. Solution of the unknown unit can only be found by undoing each specific operation. Such knowledge does not arise when working with arithmetic word prob-

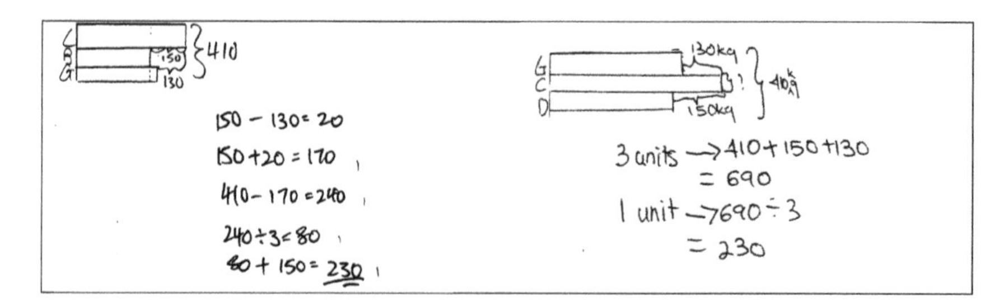

Figure 45.6. Two different model method solutions, each using a different generator to solve for the unknown mass of the cow in the Animal problem: A cow weighs 150 kg more than a dog. A goat weighs 130 kg less than the cow. Altogether the three animals weigh 410 kg. What is the mass of the cow?

lems as their solution requires the direct application of the specified operations. For example solution of the Compare word problem, the solution of the arithmetic equation can be found by directly operating on the numbers as indicated by the operations: $250 + $250 +$69= $569.

The second and more important set of knowledge construction is the realization that any of the unknown can serve as the generator. The solution to the Animal problem in Figure 45.6 can be accessed via the mass of any of three animals. The solution in the right panel shows that choosing the mass of the cow as the generator results in the direct solution of the problem. In contrast, the solution on the left panel shows that the animal problem is resolved by choosing the mass of the dog as the generator. This solution is less direct than the former as the mass the dog has first to be ascertained. Of the two, the direct route is a more strategic choice as it requires more reflective thought by the problem solver about the problem. The problem-solver would have to interrogate the problem by thinking more deeply about the problem. Such metacognitive thoughts and questions may arise.

1. If the mass of the cow is the generator, then the rectangles representing the mass of the other two animals have to be equalized to that of the cow. How could this be done?
2. If the lengths of all the rectangles are equalized to that of the cow, th en what would happen to the current total mass of the three animals?

Such thoughts do not arise naturally but are the consequence of focused questioning during teaching. The awareness that one choice is more strategic than another can be achieved by comparing and contrasting two sets of solutions, one which uses the direct method and the other the indirect method.

Solution of the unknown unit. Ng and Lee (2009) discussed in detail how the value of the unknown unit can be found. A summary of that discussion is presented here. After the information from the text has been decoded into the model drawing, another level of decoding takes place where the information captured in the model drawing is translated into an arithmetic equation. Through the process of undoing the value of the generator is found. If the direct method is used, then the value of the generator is the value of the unknown unit. Further processing, however, is needed to find the unknown unit with the use of the indirect method. In the case of the animal problem, the mass of the cow is found by adding the difference in mass between the cow and the dog to the value of the 'dog' generator.

Moving from primary school to secondary school. At the primary level the model method or other problem-solving heuristics could be the method of choice to solve algebraic word problems. The rectangle is the object used to represent the unknown and the model drawing the representation for a given problem. At the secondary level, however, letters represent unknowns (Usiskin, 1988). They are used to construct mathematical objects such as the algebraic expressions and algebraic equations that best represent the information presented in a problem. With the model method, two sets of representations are required: (i) the model drawing, and (ii) the arithmetic equation that captures the information in the model drawing. The value of the unknown unit is found by the process of undoing.

With the letter-symbolic method, it is no longer necessary to construct the model drawing. Instead information in the text is translated directly into an algebraic equation. Solution of the letter is not found by the process of undoing. Instead the value of the letter representing the unknown is evaluated by construction of a series of equivalent equations. Thus the transition from the model method to letter-symbolic method requires another level of abstraction whereby the role of the rectangle is taken over by the letter. The transition to letters as unknowns requires an expansion of the knowledge related to the use of letters. In particular the research literature has highlighted the following areas of meaning making related to the transition from arithmetic to formal algebra.

- Letters have different meanings in different situations.
- Algebraic expressions are legitimate forms of answers.
- Algebraic representations no longer adhere to the same set of conventions underpinning the use of numbers.
- Knowledge of equality-equivalence of algebraic expressions.

The following section compares and contrasts the difference between the model representation and the letter-symbolic representation and offers reasons why letter-symbolic method is therefore more attentional resource intensive of the two methods.

What kinds of knowledge would be required in order to begin work with letter-symbolic algebra? It is difficult to define school algebra. This quote by Mac Lane and Birkhoff (1967, p. 1) serves as good start.

> Algebra starts as the art of manipulating sums, products, and powers of numbers. The rules for these manipulations hold for all numbers, so the manipulations may be carried out with letters standing for numbers.

Thus formal work with letter-symbolic algebra starts when letters are used to represent unknown numbers. More importantly although letters have different meanings (Küchemann, 1981) it is essential that letters should be interpreted as representing generalized numbers or as variables. Early work with letter-symbolic algebra can begin when letters are treated as representing specific unknowns, namely the letter represents an unknown number to be evaluated. Once it is established that letters can be used to represent unknown numbers, than a collection of numbers, letters, operation symbols and grouping symbols such as brackets letters and the operations could be used to construct an algebraic expression to represent a given phenomenon (Lial & Hornsby, 2000). Basic knowledge of arithmetic and the order of operations or knowledge of surface structures (Kieran, 1989) are necessary to construct algebraic expressions. Furthermore a sound grasp of fractions and integers and how to operate with these numbers are essential knowledge before work with letter-symbolic algebra can begin. Two or more expressions can be used to describe the same situation. Hence these two expressions can be related by an equal sign to construct an equation. An equation is solved when the unknown value represented by the letter, that makes the equation true is found (Lial & Hornsby, 2000).

What kind of abstraction would be required to shift from working with model method to letter-symbolic method? With the model method rectangles are used to represent unknown as well as specific numerical values. The resulting schematic representation functions as a "pictorial equation" (Ng, 2004). If the role of the rectangle as an object representing the unknown is clear, then it does not matter what object is used to replace the rectangle. That any other object or any letter of the alphabet could replace the function of the rectangle (Ng & Lee, 2008) is one form of abstraction. Another level of abstraction is necessary. With model drawings, great care must be taken to construct the logical ordering of the rectangles and that the part-whole and the proportional relationships between the rectangles are accurately specified as illustrated by the model drawing for the Spending Problem (Figure 45.3, right panel). With the use of letters as unknown, the need to construct appropriately sized rectangles can be dispensed. At this level of abstraction, the object chosen to replace the rectangle should result in greater functionality. When letters are used to replace the role of rectangles, the part-whole and proportional relationships between rectangles are to be replaced by mathematical statements instead. For example, the mathematical term $4x$ is equivalent to the four identical rectangles used to represent Ahmad's money.

The shift from the model method to the letter-symbolic method as problem solving tool must necessarily entail the construction of new knowledge. The construction of new knowledge for novice learners is discussed under the following sections.

Letters have different meanings in different situations (Booth, 1984; Kieran, 1989; Küchemann, 1981; Stacey & MacGregor, 2000; Schoenfeld & Arcavi, 1993; Usiskin, 1988). For students to begin work with algebra they need to have the concept that school algebra is generalised arithmetic (Usiskin, 1988). Although letters may have different meanings in different situations (Kieran, 1989; Küchemann, 1981; Usiskin, 1988) such students have to construct for themselves the knowledge that letters stand for specific unknown value which can be evaluated (Usiskin, 1988) and specific procedures are related to specific objects. Solving for the unknown x in any equation may require first simplifying the algebraic expressions on either side of the equal sign and then transforming the simplified expressions into simpler equations. Such knowledge construction is no mean feat as it requires attending knowledge associated with operating on letters.

The acceptance of algebraic expressions as legitimate answers (Collis, 1975; Davis, 1975). When the model method is used to solve algebraic word problems, the information captured by the model is translated into arithmetic expressions which meant that they could be evaluated and a single answer was the result from operating on the arithmetic expression. For example, the number of children participating in the parade was found by first writing down this arithmetic expression: $(195 \times 2) + 40 + 40$ and the resulting sum was 470. When letter-symbolic algebra was used, this was not the case. This algebraic expression $x + x/2 + (x/2 - 40)$ represents the number of children at the parade. Because the operators are still visible; many novice learners of algebra experience a product-process dilemma (Davis, 1975) where they do not treat this algebraic expression as a legitimate answer. Thus students have to accept that the algebraic expression $x + x/2 + (x/2 - 40)$ does represent the number of children at the parade, and this expression is a single entity. Early work in algebra begins with students being able to accept that there is a lack of closure in algebraic forms Collis (1975).

Algebraic representations no longer adhere to the same set of conventions underpinning the use of numbers (Kieran, 1989). In arithmetic, operating on whole numbers can be expressed as a single entity as illustrated by this arithmetic equation: $(195 \times 2) + 40 + 40 = 470$. To complicate matters further the sum of two fractions can be expressed as the concatenation of two numbers. For example the sum of these two numbers $2 + \frac{1}{3}$ can be written side by side: $2\frac{1}{3}$. There is no one-to-one correspondence between how numbers can be expressed and how algebraic expressions can be simplified. For example $x + \frac{x}{2}$ cannot be simplified to $x\frac{x}{2}$. It is wrong to concatenate the two objects. Thus the shift from model method to letter-symbolic algebra as a problem solving tool requires construction of new knowledge, namely that the conventions that underpin the use of numbers do not necessarily transfer to algebraic expressions (Kieran, 1989).

Knowledge of equality-equivalence of algebraic expressions (Kieran, 1989, 1997). Since algebraic equations are necessary for solving algebraic word problems, the shift from the model to letter-symbolic method as a problem solving tool requires construction of knowledge that an algebraic equation is (i) a structure which links two different algebraic expressions and (ii) two or more different algebraic expressions can be constructed to represent the same situation. This then requires sound knowledge of equivalence: reflexive (the same is equal to the same), symmetric (equality of the left and right sides of each equation) and transitive.

In arithmetic, the equal sign is seen as a procedural symbol that announces the answer after a series of operations have been conducted (Kieran, 1981; MacGregor & Stacey, 1999; Stacey & MacGregor, 1997). For example, when the model method was used to solve the amount of money held by Ahmad, the equal sign at the end of the arithmetic expression \$160 – \$40 was used to announce the answer, that is the difference of \$160 and \$40 is \$120. Contrast the meaning of the equal sign in the arithmetic expression with that required when letter-symbolic algebra is used to solve the same problem. Research (Kieran, 1997; Sfard & Linchevski, 1994) shows that there are two main sets of conceptual demands associated with solving equations: simplifying expressions and working with equality-equivalence (Kieran, 1997). The cognitive demands needed to simplify expressions have already been discussed in the preceding section: Hence the acceptance of algebraic expressions as legitimate answers will not be repeated here. Solving algebraic equations require two significantly different conceptualisations of equality-equivalence.

Reflexive-equivalence (equality of the left and right sides of each equation). In a conditional equation, for a specific value of x, the resulting values of the expressions on either side of the equal sign are the same. For example, in the spending problem, the equation $4x - \$160 = x - \$40$, the expression $4x - \$160$ is equal to the expression $x - \$40$ when x has the value of \$40.

Equivalence of successive equations in the system of equations constructed to solve the problem. To solve a given equation, the conventional procedure is to construct the vertical chain of equivalent equations that will result in the resolution of the unknown value. The equivalence is achieved in one of two ways. First equivalence is maintained by replacing an expression with an equivalent expression. In the parade problem, the expression $x + \frac{x}{2} + \left(\frac{x}{2} - 40\right)$ in the equation $x + \frac{x}{2} + \left(\frac{x}{2} - 40\right) = 900$ was transformed into the expression $(4x - 80)/2$ and was replaced by latter to yield the subsequent equation $\frac{4x - 80}{2} = 900$ in the equation solving chain. Second equivalence is achieved by replacing an equation with an equivalent equation (i.e. one having the same solution as the previous equation in the chain) without having to replace the preceding expression with another equivalent expression. The equation $\frac{4x - 80}{2} = 900$ is equivalent to the equation $\left(\frac{4x - 80}{2}\right) \times 2 = 900 \times 2$. In this case equivalence of the previous equation is maintained by multiplying both sides of the equation by the same amount. The expression $\left(\frac{4x - 80}{2}\right)$ was not replaced by any other expression.

Comparing and contrasting the solution of the parade problem using letter-symbolic algebra against the model method suggests that major cognitive adjustments – 'accommodations' rather than assimilation are needed to solve algebraic equations. Such major accommodations may help in some way explain why neuroimaging studies found, of the two methods, the letter-symbolic method tended to activate the procedural part of the brain and why it required more attentional resources. This was because to maintain equality-equivalence of the series of equations in the equation solving chain appropriate rules and procedures specific to operating on letters were operationalized. These rules and procedures were no longer identical to those used to operate on numbers. With the model method, because solving for the unknown values involved working only with numbers there was no conflict in the conventions used and hence may be less procedurally driven. In the model method, solving for the unknown involves the processes of doing and undoing. With the letter-symbolic method the solving for the unknown requires using forward operations and mandates the maintenance of equivalence of each single algebraic equation.

How teachers can students know what is algebraic about the letter-symbolic method. Novice learners of letter-symbolic algebra continue to use problem solving heuristics such as the model method or guess-and-check to solve algebra word problems because the primary objective is to find the solution to a given problem. Teachers, however, have a different objective from their students. They have the responsibility to help students learn letter-symbolic method to solve algebra word problems as the related knowledge and skills are needed for higher mathematics and other disciplines. Fine grained analysis of their written solutions to word problems suggests that to students the difference between the two methods was that letter-symbolic algebra required the use of letters in their solutions. The solution in the right panel of Figure 45.2 is a case in point. Students' solution may begin with the appropriate model drawing with a letter x in the rectangle representing the unknown. Once they have inserted the letter x into the appropriate rectangle, these students continued to apply the doing-undoing process to solve for the unknown unit and the final solution ended by announcing the value of x. Because the value of x was correct, teachers marked the solution as correct. Such practices gave students inappropriate feedback suggesting that a solution was considered algebraic as long as letter x was used to announce the solution.

Strategies Which Teachers Can Use to Sensitize Students to the Differences Between the Two Methods

The theory of variation (Marton & Tsui, 2004) offers a possible way forward to help those students who used the model method or the semi-algebraic methods *successfully* to solve algebra word problems to learn the letter-symbolic method. It may be difficult for these students to discern the differences between the two methods if they are shown only the letter-symbolic method to a given problem. To sensitize students to differences between methods, teachers can offer students different solutions to the same problem. For example teachers can offer students the three different solutions to the Parade Problem. They can then ask students to compare and contrast the three different solutions. How are the solutions alike and how are they different from each other? What object is used to represent unknowns in each of the method and the relationship between these objects? What are the steps used to find the unknown? In the ensuing discussion teachers need to focus students' attention on the nature of the vertical chain of equivalent equations and more importantly to highlight to students that a solution is considered algebraic only when the vertical

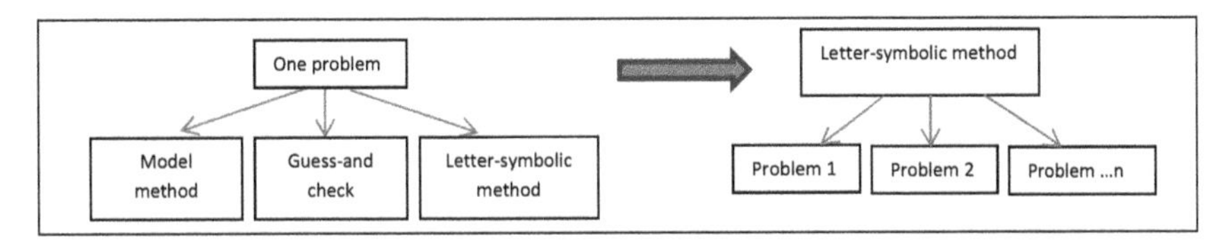

Figure 45.7. Structure of suggested pedagogy to sensitize students to the differences between methods.

chain of equivalent equations that will result in the resolution of the unknown value is constructed.

Secondary students' use of a variety of problem-solving strategies such as the model method and the guess and check method suggests that they may be more concerned about problem solving and getting the right answer rather than the problem solving method itself. Thus the compare and contrast approach of teaching may encourage students to reflect on the differences between the two methods. It is possible that this reflective activity may encourage students to be more discerning about the methods used to solve algebra word problems and thus sensitize them to the difference between problem solving and getting the answer versus that of the problem solving method. After this reflective activity, students should be provided with practice questions where they are now asked to solve algebra word problems using the letter-symbolic method. Figure 45.7 captures the pedagogy to help sensitize students to the difference in methods and to become more discerning in the differences between the methods.

CONCLUSIONS

Difficulties solving algebra word problems are not particular to Singapore. This chapter reviews a suite of studies conducted in Singapore demonstrating how model method could be used to solve arithmetic and algebraic word problems at the primary level. Although they were supposed to solve algebra word problems using letter-symbolic method, behavioural studies showed that many secondary students continued to use the model method instead. Notwithstanding neuroimaging studies found that letter-symbolic method was more resource intensive of the two methods, no reasons however were offered for this difference. A theoretical framework built upon existing literature on the knowledge learners of letter-symbolic algebra should acquire was presented. The aim of this framework is to help teachers focus on the mathematics and the attending representations of each method and how they are related. First, although different modes of representation can be used to represent the same information, the correspondences may not be one-to-one. The translation between different modes of representation means certain information is lost and new information has to be constructed. Such construction of knowledge requires certain degree of abstraction. For example, the concrete and visual nature of the pictorial equation is lost when that information is translated into its corresponding algebraic equation. The new information constructed results in the knowledge that letters represent unknowns and the conventions particular to arithmetic are not generalisable to algebra. Furthermore conceptual and procedural knowledge of equality-equivalences are necessary for the solution of equations. Effective teaching needs to identify the congruencies and divergences between the two methods and their attending representations and the procedures used to solve the specific word problem. It is crucial to sensitize learners to these similarities and differences so that they know why model method is a problem solving tool and what makes letter-symbolic method algebraic. It is hoped by identifying the specifics, students' knowledge of the model method need no longer be a barrier, but instead becomes a gateway to their learning of letter-symbolic algebra.

REFERENCES

Baddeley, A. D. (1996). Exploring the central executive. *Quarterly Journal of Experimental Psychology, 49A*, 5–28.

Baddeley, A. D., Gathercole, S. E., & Papagno, C. (1998). The phonological loop as a language learning device. *Psychological Review, 105,* 158–173.

Booth, S. (1984). Algebra: *Children's strategies and errors.* Windsor, Berkshire: NFER-Nelson.

Brown, M., Hart, K., & Küchemann, D. (1984). *Chelsea Diagnostic Mathematics Tests.* Windsor: NFER-NELSON.

Collis, K. F. (1975). *A study of concrete and formal operations in school mathematics: A Piagetian viewpoint.* Victoria, Melbourne: Australian Council for Educational Research.

Curriculum Planning & Development Division (2001). *2001 Mathematics syllabus: Primary.* Singapore: Ministry of Education.

Curriculum Planning & Development Division (2006). *2006 Mathematics syllabus: Primary. Singapore: Ministry of Education.*

Curriculum Planning Division (1990). *Mathematics syllabus: Primary.* Singapore: Ministry of Education.

Davis, R. B. (1975). Cognitive processes involved in solving simple algebraic equations. *Journal of Mathematical Behavior,* 1(3), 7–35.

Gathercole, S. E., & Alloway, T. P. (2008). *Working memory and learning: A practical guide for teachers.* London, UK: Sage Publications.

Khng, K. H., & Lee, K (2009). Inhibiting interference from prior knowledge: Arithmetic intrusions in algebra word problem solving. *Learning and Individual Differences 19,* 262–268

Khng, K. H., Lee, K., & Ng, S. F. (2008). Children's perception of the mathematical model as representing fixed or variable quantities: 11 year-olds vs. 14 year-olds. Poster presented at the 20th Biennial International Society for the Study of Behavioural Development (ISSBD) meeting in Würzburg, Germany, July 2008.

Kieran, C. (1989). The early learning of algebra: A structural perspective. In S. Wagner & C. Kieran (Eds.), *Research issues in the learning and teaching of algebra* (pp. 33-53). Reston, Virginia: NCTM.

Kieran, C. (1981). Concepts associated with the equality symbol. *Educational Studies in Mathematics, 12,* 317-326.

Kieran, C. (1997). Mathematical concepts at the secondary school level: The learning of algebra and functions. In T. Nunes & P. Bryant (Eds.) *Learning and teaching mathematics: An international perspective* (pp. 133-158). East Sussex, UK: Psychology Press.

Koedinger, K. R., & Nathan, M. J. (2004). The real story behind story problems: Effects of representations on quantitative reasoning. *The Journal of the Learning Sciences, 13*(2), 129–164.

Küchemann, D. (1981). Algebra. In K. Hart (Ed.) *Children's understanding of mathematics: 11–16.* London: John Murray.

Lee, K., Lim, Z. Y., Yeong, S. H. M., Ng, S. F., Venkatraman, V., & Chee, M. W. L. (2007). Strategic differences in algebraic problem solving: Neuroanatomical correlates. *Brain Research, 1155,* 163–171.

Lee, K., Ng, S. F., Ng, E. L., & Lim, Z. Y. (2004).Working memory and literacy as predictors of performance on algebraic word problems. *Journal of Experimental Child Psychology, 89,* 140–158.

Lee, K., Yeong, S. H. M., Ng, S. F., Venkatraman, V., Graham S., & Chee, M. W. L. (2010) Computing Solutions to Algebraic Problems Using a Symbolic Versus a Schematic Strategy. *ZDM - The International Journal on Mathematics Education, 42,* 591–605.

Lial, M. L., & Hornsby, J. (2000). *Beginning and Intermediate Algebra.* Reading, MA: Addison Wesley Longman.

Lins, R., Rojano, T.; Bell, A., & Sutherland, R. (2001). Approaches to algebra. In R. Sutherland, T. Rojano, A, Bell & R. Lins (Eds.) *Perspectives on School Algebra* (pp. 1–11). Dordrecht, The Netherlands: Kluwer Academic Publishers.

Mac Lane, S., & Birkhoff, G. (1967). *Algebra.* New York: Macmillan Co.

MacGregor, M., & Stacey, K. (1999). Learning the algebraic method of solving problems. *Journal of Mathematical Behavior, 18*(2), 149–167.

Marton, F., & Tsui, A. B. M. (2004). *Classroom discourse and the space of learning.* Mahwah. N.J.: Lawrence Erlbaum.

Neuman, D. (1997). Phenomenography: Exploring the roots of numeracy. *Journal for Research in mathematics Education,* Monograph, Vol. 9, Qualitative Research Methods in Mathematics Education, 63–177.

Ng, S. F. (2003). How secondary two express stream students used algebra and the model method to solve problems. *The Mathematics Educator, 7*(1), 1–17.

Ng, S. F. (2004). Developing Algebraic Thinking in Early Grades: Case Study of the Singapore Primary Mathematics Curriculum. Developing Algebraic Thinking in the Earlier Grades from an International Perspective (Special Issue). *The Mathematics Educator, 8*, 39 – 59.

Ng, S. F., & Lee, K. (2008). As long as the drawing is logical, size does not matter. *Korean Journal of Thinking and Problem Solving, 18*(1), 67-80.

Ng, S. F., & Lee, K. (2009). The Model Method: Singapore Children's Tool for Representing and Solving Algebraic Word Problems. *Journal for Research in Mathematics Education, 40*(3), 282-313

Ng, S. F., Lee, K., Ang, S. Y., & Khng, F. (2006). Model method: Obstacle or bridge to learning symbolic algebra. In W. Bokhorst-Heng, M. Osborne, & K. Lee (Eds.), *Redesigning pedagogies: Reflections from theory and praxis* (pp. 227–242). Rotterdam: Sense publishers.

Pape, S. J. (2003). Compare word problems: Consistency hypothesis revisited. *Contemporary Educational Psychology, 28*, 396–421.

Pickering, S., & Gathercole, S. (2001). *Working Memory Test Battery for Children*. London: The Psychological Corporation Ltd.

Poh, B. K. (2007). *Model method: primary three pupils' ability to use models for representing and solving word problems* (Unpublished master's thesis). National Institute of Education, Nanyang Technological University, Singapore.

Rust, J. (2000). *Wechsler objective reading and language dimensions (Singapore)*. London: Psychological Corporation.

Schoenfeld, A., & Arcavi, A. (1993). On the meaning of variable. *Mathematics Teacher, 81*, 420–427.

Sfard, A., & Linchevski, L. (1994). The gains and pitfalls of reification - the case of algebra. *Educational Studies, 26,* 191-228.

Siegler, R. S. (1987). Some general conclusions about children's strategy choice procedures. *International Journal of Psychology, 22,* 729–749.

Siegler, R. S., & Shrager, J. (1984). Strategy choices in addition and subtraction: How do children know what to do? In C. Sophian (Ed.), *Origins of cognitive skills* (pp. 229–293). Hillsdale, NJ: Lawrence Erlbaum Associates.

Singapore Examinations and Assessment Board, (2008). *PSLE examination questions 2004-2008*. Singapore: Educational Publishing House.

Stacey, K. & MacGregor, M. (2000). Learning the algebraic method of solving problems. *Journal of Mathematical Behaviour, 18*(2), 149–167.

Stacey, K., & MacGregor, M. (1997). Ideas about symbolism that students bring to algebra. *The Mathematics Teacher, 90*(2), 110–113.

Steinberg, R. M., Sleeman, D. H., & Ktorza, D. (1990). Algebra students' knowledge of equivalence of equations. *Journal for Research in Mathematics Education, 22*(2), 112–121.

Swanson, H. L., Cooney, J. B., & Brock, S. (1993). The influence of working memory and classification ability on children's word problem solution. *Journal of Experimental Child Psychology, 55,* 374–395.

Usiskin, Z. (1988). Conceptions of school algebra and uses of variables. In A. Coxford (Ed.), *Ideas of algebra: K-12* (pp. 8–19). Reston, VA: NCTM.

Van de Walle, J. A. (2001). Elementary and middle school mathematics: Teaching developmentally (4th edition). New York: Addison Wesley Longman, Inc.

Wechsler, D. (1991). *Wechsler Intelligence Scale for Children—third edition*. San Antonio, TX: Psychological Corporation.

CHAPTER 46

A MULTIDIMENSIONAL APPROACH TO UNDERSTANDING IN MATHEMATICS AMONG GRADE 8 STUDENTS IN SINGAPORE

Boey Kok Leong, Shaljan Areepattamannil, and Berinderjeet Kaur
National Institute of Education, Nanyang Technological University, Singapore

BACKGROUND

Singapore has been participating in the Trends in International Mathematics and Science Study (TIMSS), conducted on a 4-year cycle, ever since its inception in 1995. TIMSS has assessed Grade 4 and 8 students' mathematics and science in 1995, 1999, 2003, and 2007. The performance of Singapore's fourth and eighth grade students on all the four TIMSS assessments was outstanding. At the eighth grade, students from Singapore ranked first on the TIMSS 1995, 1999, and 2003 assessments. The average mathematics scale scores of Grade 8 students in the top five

The First Sourcebook on Asian Research in Mathematics Education: China, Korea, Singapore, Japan, Malaysia, and India, pp. 1045–1058

Table 46.1. Average Mathematics Scale Scores of Grade 8 Students in Top Five Participating Countries

	TIMSS 1995[1]		*TIMSS 1999*[2]		*TIMSS 2003*[3]		*TIMSS 2007*[4]	
Rank	*Country*	*Avg Scale Score*	*Country*	*Avg Scale Score*	*Country*	*Avg Scale Score*	*Country*	*Avg Scale Score*
1	Singapore	643 (4.9)	Singapore	604 (6.3)	Singapore	605 (3.6)	Chinese Taipei	598 (4.5)
2	Korea	607 (2.4)	Korea	587 (2.0)	Korea	589 (2.2)	Korea	597 (2.7)
3	Japan	605 (1.9)	Chinese Taipei	585 (4.0)	Hong Kong SAR	586 (3.3)	Singapore	593 (3.8)
4	Hong Kong SAR	588 (6.5)	Hong Kong SAR	582 (4.3)	Chinese Taipei	585 (4.6)	Hong Kong, SAR	572 (5.8)
5	Belgium (Flemish)	565 (5.7)	Japan	579 (1.7)	Japan	570 (2.1)	Japan	570 (2.4)
	International Average	513	International Average	487 (0.7)	International Average	467 (0.5)	International Average	500

Sources: [1]Table 1.1 in Beaton, Mullis, Martin, Gonzalez, Kelly, & Smith (1996). [2]Exhibit 1.1 in Mullis, Martin, Gonzalez, Gregory, Garden, O'Connor, Chrostowski, & Smith, 2000. [3]Exhibit 1.1 in Mullis, Martin, Gonzalez, & Chrostowski, 2004. [4]Exhibit 1.1 in Mullis, Martin, & Foy, 2008.
Note: ( ) Standard error appears in parentheses.

participating countries on the TIMSS 1995, 1999, 2003, and 2007 assessments are presented in Table 46.1.

In 1995, Singapore was ranked first, and her average mathematics scale score was significantly higher than all other participating countries. Even though Singapore was again ranked first in 1999, its average mathematics scale score was not significantly different from Korea ranked second, Chinese Taipei ranked third, and Hong Kong SAR ranked fourth. Otherwise, Singapore's average mathematics scale score was significantly higher than all other participating countries. In 2003, Singapore was once again ranked first, and its average mathematics scale score was significantly higher than all other participating countries. In 2007, Singapore was ranked third next to Chinese Taipei and Korea. However, there was no significant variation in the mean scores of the top three participating countries.

Given the stellar performance of Singapore's Grade 8 students on the TIMSS assessments, it is imperative to examine the mathematics learning of Grade 8 students in Singapore. Although there are different approaches or perspectives to understand and measure the mathematics learning of students in elementary and secondary schools, one approach that appears particularly promising and pertinent for the study of students' mathematics learning is the SPUR approach (see Thompson, Kaur, & Bleiler, 2010; Thompson & Senk, 2008).

THE SPUR APPROACH TO MATHEMATICS LEARNING

To ensure that both elementary and secondary students experience the multiple aspects of understanding mathematical situations, the University of Chicago School Mathematics Project (UCSMP) propounded the SPUR approach to mathematics learning (Usiskin, 2003, 2007; Viktora et al., 2008). According to the SPUR approach to mathematics learning, there are four important dimensions to understanding in mathematics: skills—having ways to obtain and check solutions; properties—knowing properties that identify and justify reasoning; uses—recognizing situations in which the mathematics is applied; and representations—picturing or otherwise representing the mathematical ideas (see Viktora et al., 2008). Thompson and Senk (2008) described the multiple dimensions of the SPUR approach to mathematics learning as follows:

> Skills represent those procedures that students should master with fluency; they range from applications of standard algorithms to the selection and comparison of algorithms to the discovery or invention of algorithms, including procedures with technology. Properties are the principles underlying the mathematics, ranging from the naming of properties used to justify conclusions to derivations and proofs. Uses are the applications of the concepts to the real world or to other concepts in mathematics and range from routine "word problems" to the development and use of mathematical models. Representations are graphs, pictures, and other visual depictions of the concepts, including standard representations of concepts and relations to the discovery of new ways to represent concepts. At times, there is a fifth dimension, History, which includes cultural understanding of mathematics, such as names and origins of ideas and relationships between cultures and mathematics. (p. 3)

Only a small body of research, however, has examined the relationships between the SPUR objectives and students' achievement in mathematics (e.g., Thompson et al., 2010; Thompson & Senk, 2001, 2008). Thompson et al. (2010), drawing on data from the International Project on Mathematical Attainment (IPMA), examined whether the mathematics achievement of elementary school students in the United States (n = 181) and

Singapore (n = 856) would vary across the multiple dimensions of understanding in mathematics. The authors also explored whether the growth patterns in mathematics achievement would differ according to the dimensions of understanding in mathematics. Results of the study suggested that mathematics achievement of elementary students in the United States and Singapore would vary across the different dimensions of understanding in mathematics. Furthermore, elementary students' mathematics achievement within each dimension of understanding increased progressively in the United States and Singapore.

Thompson and Senk (2001), employing a matched-pair design, explored the effects of a standards-oriented curriculum (i.e., UCSMP) and a traditional curriculum on mathematics achievement of 306 secondary students (150 students in the UCSMP classes and 156 students in the comparison classes) from four high schools in the United States. The students in the UCSMP classes outperformed their peers in the comparison classes on multistep problems and problems involving applications or graphical representations. However, the mean performance of students in the UCSMP classes and comparison classes on the conservative test measuring algebraic skills was more or less equal. More recently, Thompson and Senk (2008) examined the mathematics achievement of 237 Grades 6 and 7 students in terms of the SPUR approach to mathematical learning. The performance of Grades 6 and 7 students was not the same across the four dimensions of understanding in mathematics. Both Grade 6 and 7 students did better on items focusing on properties. Moreover, the majority of Grade 7 students' achievement with representations of concepts was higher than their achievement with skills or uses.

Thus, nurturing students' multidimensional approaches to understanding in mathematics may help them to develop a robust understanding of mathematical concepts and reasoning (Thompson et al., 2010; Thompson & Senk, 2008). Given the dearth of research examining the mathematics achievement of elementary and secondary students in terms of the SPUR approach to mathematical learning, more empirical research is warranted to evaluate the effectiveness of the SPUR approach to mathematics learning. Further, there is sparse research on the impact of student-level and school-level factors on the SPUR approach to mathematics learning. Hence, the present study sought to investigate the following two research questions:

1. To what extent does the mathematics achievement of Grade 8 students in TIMSS 2007 vary across the multiple dimensions of understanding in mathematics, namely skills, properties, uses, and representations?

2. How well do student-level and school-level factors predict Grade 8 students' multidimensional mathematical knowledge—skills, properties, uses, and representations?

METHOD

Data

Data for the present study were drawn from the TIMSS 2007 international database (see Foy & Olson, 2009). A total of 4599 students from 164 schools in Singapore participated in the TIMSS 2007 Grade 8 assessment. To facilitate comparison with Singapore, we selected countries/economies such as Australia, Chinese Taipei, England, Hong Kong SAR, Japan, South Korea, Minnesota, Massachusetts, and the United States to form a comparison group. These countries/economies were selected on the basis of their performance on the TIMSS 2007 mathematics assessment. Countries/economies that formed the comparison group in the current study scored higher than the international average of 500 in mathematics. Moreover, the comparison group included some of the top performing East Asian countries and English speaking countries.

The SPUR Items

A total of 88 released TIMSS 2007 Grade 8 mathematics items were classified using the SPUR approach to mathematics learning (see Table 46.2). The number of items in the groups ranged from 14 to 29. The items comprised both multiple choice items and questions requiring constructed response worth up to 2 points. Student achievement was computed using the partial credit item response theory (IRT) model for multiple responses. IRT scales were created for every dimension of understanding on a scale with a mean of 500 and a standard deviation of 100. For more accurate estimation of results for the subpopulations of students

Table 46.2. Number of Items Coded Across the Dimensions of Understanding at Grade 8

Item Type	*Skills (S)*	*Properties (P)*	*Uses (U)*	*Representations (R)*	*Total*
Multiple choice	17	7	14	12	50
Constructed response	4	7	15	12	38
	21	14	29	24	88

as well as to be consistent with the procedure used in TIMSS, the scaling made use of plausible value technology. Five plausible values each were computed for every dimension of understanding per student record in the database.

Student-Level and School-Level Factors

Student-level factors included student demographic variables—gender and language spoken at home—and students' self-confidence in mathematics, positive affect toward mathematics, valuing of mathematics, and homework (see Appendix A). Prior studies have demonstrated the influences of gender (e.g., Perkins, Quaynor, & Engelhard, 2011), language spoken at home (e.g., Perkins et al., 2011), and students' self-confidence in mathematics (e.g., Areepattamannil, 2011) positive affect toward mathematics (e.g., Areepattamannil, 2011), valuing of mathematics (e.g., Phan, Sentovich, Kromrey, Dedrick, & Ferron, 2010), and homework (e.g., Jeynes, 2011) on students' achievement in mathematics. School-level factors included school resources and school climate (see Appendix A). A large body of research has indicated the critical roles that school resources and school climate may play in the mathematics achievement of students (e.g., Afana & Lietz, 2010; Lubienski, Lubienski, & Crane, 2008).

FINDINGS

Descriptive statistics and correlations are presented in Tables 46.3 and 46.4. Singapore's mean scores for the four SPUR dimensions of understanding in mathematics were 553, 552, 560, and 541, respectively. Grade 8 students from Singapore scored statistically significantly higher than their counterparts in the comparison group countries on the four SPUR dimensions of understanding in mathematics (see Table 46.3). However,

Table 46.3. Average Mathematics Achievement Scores by Dimensions of Understanding at Grade 8

	Mean Score for Dimensions of Understanding											
Country	*Skills (S)*			*Properties (P)*			*Uses (U)*			*Representations (R)*		
Singapore	553	(3.4)	▲	552	(3.0)	▲	560	(3.8)	▲	541	(3.4)	▲
International Mean	500	(1.6)		500	(1.3)		500	(1.6)		500	(1.7)	

Note: Standard error in ( ). ▲ mean significantly higher than international mean.

Table 46.4. Descriptive Statistics and Correlations

			Correlations			
	M	*SD*	*Skills*	*Properties*	*Uses*	*Representations*
Gender	1.50	.50	-0.08**	-0.08**	-0.03**	-0.08**
Home language	1.65	.93	0.18**	0.18**	0.21**	0.18**
Positive affect	1.42	.79	0.26**	0.25**	0.26**	0.23**
Valuing mathematics	1.74	.52	0.08**	0.10**	0.10**	0.10**
Self-confidence	1.17	.79	0.33**	0.31**	0.36**	0.30**
Homework	1.26	.71	0.21**	0.19**	0.23**	0.22**
School resources	1.91	.28	0.02	0.00	0.01	0.01
School climate	1.18	.53	0.26**	0.27**	0.29**	0.27**

Note: $**p < 0.01$.

Singapore's students' mathematics achievement varied across the multiple dimensions of understanding in mathematics. Grade 8 students in Singapore scored higher on items measuring the dimension, uses, and scored lower on items measuring the dimension, representations. Nevertheless, they performed more or less equally well on items measuring the dimensions, skills and properties.

Multilevel analyses were conducted to examine the influences of student-level and school-level factors on the four SPUR dimensions of understanding in mathematics (see Appendix B). The results of the multilevel analyses are shown in Table 46.5. Gender differences favouring female students were found for the SPUR dimensions, skills, properties, and representations dimensions. However, no gender differences were found for the SPUR dimension, uses. The student demographic variable, language spoken at home, was a positive predictor of all the four SPUR dimensions—skills, properties, uses, and representations, suggesting that students who speak the language used on the test at home tend to score higher on the SPUR dimensions of understanding in mathematics than do their peers who do not speak the language of test at home.

Students' positive affect toward mathematics positively predicted all the four SPUR dimensions of understanding in mathematics. In other words, the Grade 8 students in Singapore who reported high positive affect toward mathematics performed significantly better in mathematics—skills, properties, uses, and representations—than did their peers who reported low positive affect toward mathematics. Similarly, students' self-confidence in mathematics was also positively associated with all the four SPUR dimensions of understanding in mathematics, indicating that

Table 46.5. Multilevel Analyses Predicting Dimensions of Understanding in Mathematics

Model	*Parameter*	*Estimate for S*		*Estimate for P*		*Estimate for U*		*Estimate for R*	
Level 1	Intercept	449.28	***	446.61	***	422.18	***	433.29	***
	Gender	-10.40	***	-10.36	***	2.83		-9.47	**
	Home language	5.28	***	5.76	***	6.83	***	5.58	***
	Positive affect	9.72	***	9.63	***	6.91	***	6.45	**
	Valuing mathematics	0.88		3.55		4.30		6.01	*
	Self-confidence	23.13	***	19.89	***	25.73	***	20.11	***
	Homework	14.81	***	12.97	***	17.82	***	16.99	***
	R^2	0.12		0.10		0.13		0.10	
Level 2	School resources	0.33		1.84		-3.56		-2.27	
	School climate	41.04	***	40.11	***	49.07	***	43.29	***
	R^2	0.21		0.27		0.10		0.17	

Note: *$p < 0.05$. **$p < 0.01$. ***$p < 0.001$.

Grade 8 students who perceived themselves to have high confidence in mathematics performed significantly better in mathematics than did their peers who perceived themselves to have low confidence in mathematics. Homework was also a positive predictor of all the four SPUR dimensions of understanding in mathematics. However, students' valuing of mathematics was only predictive of the SPUR dimension, representations. Of the two school-level factors—school resources and school climate—school climate was the sole predictor of all the four SPUR dimensions of understanding in mathematics.

DISCUSSION

The purpose of the present study was to examine whether the mathematics achievement of Grade 8 students in Singapore varied across the four SPUR dimensions of understanding in mathematics, and to explore the influences of student-level and school-level factors on the four SPUR dimensions of understanding in mathematics.

The findings of the study suggested that the mathematics achievement of Grade 8 students in Singapore varied across the four SPUR dimensions of understanding in mathematics—skills, properties, uses, and representations. This finding has far-reaching implications for mathematics teaching and learning in Singapore. The present mathematics curriculum in Singapore does not address the multiple dimensions of understanding in mathematics. Incorporating the SPUR approach to mathematics teaching and learning in the mathematics curriculum may help students in Singapore to develop a robust understanding of mathematical concepts and reasoning, which, in turn, may enhance student achievement in mathematics.

However, revamping the mathematics curriculum to incorporate the SPUR approach to understanding in mathematics alone may not help to enhance students' achievement in mathematics. Concerted efforts should also be directed towards training the pre-service and in-service mathematics teachers employing the SPUR approach to mathematics teaching and learning.

The findings of the current study also suggests the pivotal roles that some of the student-level and school-level factors might play in the mathematics achievement of Grade 8 students in Singapore in terms of the four SPUR dimensions of understanding in mathematics. The finding with regard to students' positive affect toward mathematics suggests that boosting Grade 8 students' positive affect toward mathematics may help to improve their achievement in mathematics. Hence, mathematics teachers in Singapore have a key role to play in improving their students' positive affect toward mathematics. There is mounting evidence that self-concept enhancement intervention programs would help to improve the self-concepts of adolescents in educational settings (e.g., O'Mara, Green, & Marsh, 2006; O'Mara, Marsh, Craven, & Debus, 2006).

Meta-analyses of the effectiveness of self-concept enhancement interventions in schools suggest that self-concept enhancement intervention programs would be more efficient and effective if such interventions incorporated appropriate praise and/or feedback strategies (e.g., attributional feedback, goal feedback, or contingent praise) into the program (O'Mara et al., 2006). Moreover, targeting disadvantaged adolescents with diagnosed problems (e.g., those diagnosed with pre-existing problems, such as low self-esteem, behavioral problems, and learning disabilities) would be more valuable and effective than preventive intervention programs (Haney & Durlak, 1998; O'Mara et al., 2006). Therefore, mathematics teachers in Singapore need to design appropriate self-concept enhancement intervention programs to enhance the mathematics self-concept of their students.

The finding in regard to students' self-confidence in mathematics indicates the crucial role that greater self-perceived competence in mathematics may play in enhancing the mathematics achievement of Grade 8 students in Singapore. Since self-perceived academic competence precedes intrinsic motivation (Losier & Vallenard, 1994), there is growing evidence that greater self-perceived academic competence would enhance students' intrinsic motivation, which, in turn, would promote their academic achievement (Zisimopoulos & Galanaki, 2009). Moreover, unlike students with low perceived academic competence, students who are mastery goal-oriented and who perceive themselves to have high academic competence may employ a more adaptive learning strategy (Cho, Weinstein, & Wicker, 2011). The use of an adaptive learning strategy may help students with high perceived academic competence to perform significantly better than students with low perceived academic competence. Furthermore, students with high perceived academic competence tend to adopt high-performance approach goals, which, in turn, enhance their academic achievement (Cho et al., 2011).

Mathematics homework was also an important determinant of mathematics achievement in all the four SPUR dimensions of understanding in mathematics. Therefore, designing effective mathematics homework in line with the SPUR approach to understanding in mathematics may help Grade 8 students in Singapore to improve their achievement in mathematics. Finally, the findings of the study indicated the need for a positive school climate to promote the mathematics achievement of Grade 8 students in Singapore in terms of the SPUR approach to understanding in mathematics. Hence, school teachers and school administrators in Singapore may need to take up appropriate measures to foster a positive school climate that supports the academic achievement of all students.

In conclusion, the results of the study reiterate the need to incorporate the SPUR multidimensional approach to understanding in mathematics teaching and learning. Further, the positive influences of students' self-confidence in mathematics and positive affect toward mathematics on all the four SPUR dimensions of understanding in mathematics suggest that the SPUR multidimensional approach to understanding in mathematics may not only improve students' engagement with mathematics but also enhance their achievement in mathematics.

APPENDIX A: DESCRIPTION OF LIST OF VARIABLES USED IN THE ANALYSIS

Questionnaire	*Variable*		*Description*
Student	Gender		Coded as 1 = Female and 2 = Male
	Home Language		Recoded as 2 = English and 1 = Others: (English is the language of instruction in schools)
	Positive Affect	*	Items used • I enjoy learning mathematics. • Mathematics is boring. • I like mathematics.
	Valuing Mathematics	*	Items used • I think learning mathematics will help me in my daily life. • I need mathematics to learn other school subjects. • I need to do well in mathematics to get into the university of my choice. • I need to do well in mathematics to get the job I want.
	Self confidence	*	Items used • I usually do well in mathematics. • Mathematics is harder for me than for many of my classmates. • I am just not good at mathematics. • I learn things quickly in mathematics.
	Homework	*	Index created • on the basis of the frequency of mathematics homework students are assigned each week • the amount of time they spend on it.
School	School Resources	*	Index of Availability of School Resources for Mathematics Instruction
	School climate	*	Items used • Teachers' job satisfaction • Teachers' understanding of the school's curricular goals • Teachers' degree of success in implementing the school's curriculum • Teachers' expectations for student achievement • Parental support for student achievement • Parental involvement in school activities • Students' regard for school property • Students' desire to do well in school.

Source: Mullis, Martin & Foy (2008).
Note: *Index created by TIMSS. Each index was recoded as 0 = Low to 2 = High.

APPENDIX B

Multilevel model

For each dimension of understanding, models were constructed to represent level 1 and level 2 of the TIMSS 2007 data. The multilevel model can be represented as follows:

Level 1 Model (student-related variables):

$$Y_{ij} = \beta_{0j} + \beta_{1j}(\text{Gender}) + \beta_{2j}(\text{Home Language}) + \beta_{3j}(\text{Positive Affect}) + \beta_{4j}(\text{Valuing}) + \beta_{5j}(\text{Self Confidence}) + \beta_{6j}(\text{Homework}) + r_{ij}$$

$$\beta_{kj} = \gamma_{k0} + u_{kj}.$$

Level 2 Model (school-related factors):

$$Y_{ij} = \beta_{0j} + \beta_{1j}(\text{Gender}) + \beta_{2j}(\text{Home Language}) + \beta_{3j}(\text{Positive Affect}) + \beta_{4j}(\text{Valuing}) + \beta_{5j}(\text{Self Confidence}) + \beta_{6j}(\text{Homework}) + r_{ij}$$

$$\beta_{kj} = \gamma_{k0} + \gamma_{k1}(\text{School Resources}) + \gamma_{k2}(\text{School Climate}) + u_{kj}.$$

In each model,

Y_{ij} : the score from each dimension of understanding of student i in school j,
β_{0j} : regression intercept of school j,
γ_{00} : the overall average score for every dimension for all schools,
u_{oj} : the random effect of school j,
r_{ij} : the random effect of student i in school j.

Further, for k = 1, 2, 3, 4, 5, 6,

β_{kj} refer to regression slopes of school j,
u_{kj} refer to the random effects,
γ_{k0} to γ_{k3} refer to the level 2 fixed effects.

REFERENCES

Afana, Y., & Lietz, P. (2010). *The relationship between school resources and mathematics achievement at Grade 8: A comparison of Israeli and Palestinian schools in TIMSS*

2007. Retrieved from http://www.iea.nl/user_upload/IRC_2010/Papers/IRC2010_Afana_Lietz.pdf

Areepattamannil, S. (2011). First- and second-generation immigrant adolescents' multidimensional mathematics and science self-concepts and their achievement in mathematics and science. *International Journal of Science and Mathematics Education*. Advance online publication. doi:10.1007/s10763-011-9319-7.

Beaton, A. E., Mullis, I. V. S., Martin, M. O., Gonzalez, E. J., Kelly, D. L., & Smith, T. A. (1996). Mathematics achievement in the middle school years: IEA's third international mathematics and science study (TIMSS). USA: TIMSS International Study Center, Boston College.

Cho, Y., Weinstein, C. E., & Wicker, F. (2011). Perceived competence and autonomy as moderators of the effects of achievement goal orientations. *Educational Psychology, 31*, 393-411.

Foy, P. & Olson, J. F. (Eds.). (2009). *TIMSS 2007 International Database and User Guide*. Chestnut Hill, MA: TIMSS & PIRLS International Study Center, Boston College.

Haney, P., & Durlak, J. A. (1998). Changing self-esteem in children and adolescents: A meta-analytic review. *Journal of Clinical Child Psychology, 27*, 423-433.

Jeynes, W. H. (2011). Parental involvement and academic success. New York, NY: Routledge.

Losier, G. A., & Vallenard, R. J. (1994). The temporal relationship between perceived competence and self-determined motivation. *Journal of Social Psychology, 134*, 793-801.

Lubienski, S. T., Lubienski, C., & Crane, C. C. (2008). Achievement differences among public and private schools: The role of school climate, teacher certification, and instruction. *American Journal of Education, 151*, 97-138.

Mullis, I. V. S., Martin, M. O., Gonzalez, E. J., Gregory, K. D., Garden, R. A., O'Connor, K. M., Chrostowski, S. J. & Smith, T. A., (2000). TIMSS 1999 international mathematics report. USA: International Study Center, Lynch School of Education, Boston College.

Mullis, I. V. S., Martin, M. O., Gonzalez, E. J., & Chrostowski, S. J. (2004). TIMSS 2003 international mathematics report. USA: International Study Center, Lynch School of Education, Boston College.

Mullis, I. V. S., Martin, M. O., & Foy, P. (2008). TIMSS 2007 International Mathematics Report. Boston, MA: TIMSS & PIRLS International Study Centre

O'Mara, A. J., Green, J., & Marsh, H. W. (2006). Administering self-concept interventions in schools: No training necessary? A meta-analysis. *International Education Journal*, 7, 524-533.

O'Mara, A. J., Marsh, H. W., Craven, R. G., & Debus, R. (2006). Do self-concept interventions make a difference? A synergistic blend of construct validation and meta-analysis. *Educational Psychologist, 41*, 181-206.

Perkins, A., Quaynor, L., & Engelhard, G. (2011). *The influences of home language, gender, and social class on mathematics literacy in France, Germany, Hong Kong, and the United States*. Retrieved from http://www.ierinstitute.org/IERI_Monograph_Volume04_Chapter_2.pdf

Phan, H., Sentovich, C., Kromrey, J., Ferron, J., & Dedrick, R. (2010, May). *Correlates of mathematics achievement in developed and developing countries: An analysis of TIMSS 2003*. Paper presented at the annual meeting of the American Educational Research Association, Denver, CO.

Thompson, D. R., & Senk, S. L. (2001). The effects of curriculum on achievement in secondyear algebra: The example of the University of Chicago School Mathematics Project. *Journal for Research in Mathematics Education, 32*, 58-84.

Thompson, D. R., & Senk, S. L. (2008, July). *A Multi-Dimensional Approach to Understanding in Mathematics Textbooks Developed by UCSMP.* Paper presented in Discussion Group 17 of the International Congress on Mathematics Education. Monterrey, Mexico.

Thompson, D. R., Kaur, B., & Bleiler, S. (2010). Using a multi-dimensional approach to understanding to assess primary students' mathematical knowledge. In Shimizu, Y., Sekiguchi, Y., & Hino, K. (Eds.), *Proceedings of the 5th East Asia Regional Conference on Mathematical Education* (Vol. 2, pp. 472-479). Tokyo, Japan: Japan Society of Mathematical Education.

Usiskin, Z. (2003). A personal history of the UCSMP secondary school curriculum: 1960-1999. In Stanic, G. M. A., & Kilpatrick, J. (Eds.), *A history of school mathematics*. (Vol. 1, pp. 673-736). Reston, VA: National Council of Teachers of Mathematics.

Usiskin, Z. (2007). The case of the University of Chicago School Mathematics Project: Secondary component. In C. R. Hirsch (Ed.), *Perspectives on the design and development of school mathematics curricula* (pp. 173-182). Reston, VA: National Council of Teachers of Mathematics.

Viktora, S. S., Cheung, E., Highstone, V., Capuzzi, C. R., Heeres, D., Metcalf, N. A., Sabrio, S., Jakucyn, N., & Usiskin, Z. (2008). *The University of Chicago School Mathematics Project: Transition mathematics*. Chicago, IL: Wright Group/ McGraw Hill.

Zisimopoulos, D., & Galanaki, E. P. (2009). Academic intrinsic motivation and perceived academic competence in Greek elementary students with and without learning disabilities. *Learning Disabilities Research and Practice, 24*, 33-43.

MALAYSIA

SECTION EDITORS
Chap Sam Lim, Liew Kee Kor, and Cheng Meng Chew
Universiti Sains Malaysia

Bharath Sriraman
The University of Montana, USA

CHAPTER 47

INTRODUCTION TO THE MALAYSIAN CHAPTERS

Chap Sam Lim, Liew Kee Kor, and Cheng Meng Chew
Universiti Sains Malaysia

Bharath Sriraman
The University of Montana

Mathematics education research in Malaysia is comparatively new and under explored. However, for the last three decades, there is a growing interest and effort in conducting research studies related to mathematics teaching and learning in schools by both the postgraduate students and higher institution professors and lecturers. However, dissemination of the findings has been confined to a limited group of research community, such as presentation in local or international conferences; publication in academic journals and a majority of these research works were just reported as theses, monograph or project reports. This in a way has curtailed efforts in making improvement in the teaching and learning of mathematics in the classrooms. It is commonly accepted that the basic aim of doing research in mathematics education is to construct theories and to inform practice. Findings from any research studies, if disseminated well, will be able to enlighten and enrich the experience of educa-

The First Sourcebook on Asian Research in Mathematics Education: China, Korea, Singapore, Japan, Malaysia, and India, pp. 1061–1063

tors, teacher practitioners and the policy makers so as to improve the quality of mathematics education.

Therefore, in line with the aim of this source book, that is to "provide the first comprehensive and unified treatment of historical and contemporary research trends in mathematics education in the Asian and South Asian world." We have attempted to compile and document as widely as possible all research studies that have been carried out in Malaysia from the year 1970 till 2011. We hope our attempt of documenting such reviews will provide invaluable information to teachers, parents, policy makers, research students and researchers on the possible happenings and further determine future directions of research in mathematics education.

This Malaysian section consists of nine chapters. The first chapter provides a comprehensive overview on the historical development and contemporary research trends in mathematics education research in Malaysia since 1970. To set the context, Chapter 48 began with a brief description of the curriculum reform in mathematics education in Malaysia as these reforms have directly and indirectly impacted the trend of research studies conducted. In order to provide a more comprehensive scenario of the research trends discussed in Chapter 48, the next eight chapters discussed in detail each of the major research areas. Each chapter was structured in such a way that it will begin with a critical and comprehensive review of related studies, follow with an exemplary study in that particular area, and then end the chapter with implications and suggestions for future research in that area.

For example, Chapter 49 began with a critical and comprehensive review of mathematics learning and understanding studies that have been carried out in Malaysia. Four major themes were identified and discussed, followed by a detailed discussion of an exemplary study on multiplicative thinking in proportion and ratio. The chapter then ended with implications and suggestions for future research.

In a similar manner, Chapter 50 was divided into three subsections. The first section discussed different views on the existing definitions of numeracy and its related components. The second section reviews Primary School Mathematics Curriculum Specifications (in the Year 2002-2007) with regard to the aspects of numeracy in all topics. The third section reports and highlights all studies on numeracy and numeracy related areas in Malaysia.

Chapter 51 focused on research studies related to Malaysian research in geometry, followed by a detailed discussion of a particular study conducted by the author on the teaching and learning of geometry in a Malaysian classroom.

In a slightly different way, Chapter 52 began with a brief discussion of the importance of mathematical thinking in the Malaysian school curriculum, followed by a critical review of the relevant local literature to shed some light on the extent to which these research works have contributed to the development of mathematical thinking in Malaysia.

Although studies of values in mathematics education in Malaysia are still sporadic and in the developing stage, Chapter 53 has taken a brave attempt to explore the development of values education in Malaysian mathematics schools curriculum, discussed problems and challenges in inculcating values in mathematics education and described some values studies that have been conducted by researchers from three major universities in Malaysia. The chapter wraps up with a detailed discussion of a study to explore values in mathematics teaching as espoused by three secondary schools mathematics teachers in Malaysia.

Likewise, Chapter 54 began with a brief historical development of Malaysian examination system, a brief review of school mathematics curriculum and assessment in Malaysia and issues related to mathematics assessment. To elaborate further, a study that focuses on implementing a Mathematical Thinking Assessment (MaTA) Framework to assess students' thinking processes and its limitations were also be discussed. The chapter concludes with some suggestions for future research of the study that will help to assess students learning in a more holistic and reliable fashion.

Chapter 55 mainly focused on the use of graphics calculator in the teaching and learning of mathematics in Malaysia. In particular, it reports chronologically the stages of implementation of graphics calculator in the Malaysian mathematics curriculum. Consequently, it presents the findings of local research studies and gives an account on graphics calculator related events in Malaysia from the year 2002 onwards.

The last chapter focused its discussion on two main strands. First, the situational context of mathematics teacher professional development will be elaborated to provide the background and setting. Second, the research findings on Action Research and Lesson Study in Malaysia will be discussed, primarily to examine their feasibility as an innovative form of teacher professional development.

We acknowledge that these nine chapters might not be exhaustive in reviewing all mathematics education research studies in Malaysia, but we highly appreciate all the authors' effort and time in documenting and writing these chapters. We believe their effort could contribute a lot to make this sourcebook serve as a standard reference for mathematics education researchers, policy makers, practitioners and students both in and outside Malaysia.

CHAPTER 48

MATHEMATICS EDUCATION RESEARCH IN MALAYSIA

An Overview

Chap Sam Lim, Parmjit Singh, Liew Kee Kor, and Cheng Meng Chew
Universiti Sains Malaysia

ABSTRACT

This chapter aims to provide an overview and historical development of mathematics education research in Malaysia since 1970. Research studies in mathematics education in the earlier years were generally limited to post-graduate theses and dissertations. The focus was mainly on cognitive development and teaching approaches in the 70's. The research trend became more diverse in the 80's with various topics ranging from attitudes, problem solving to error analysis. In the 90's, the research areas continued to expand those topics found in the 80's but with a significant focus on scheme of mathematical concepts. However, there were also some studies related to assessment and evaluation. Starting the year 2000, several new research areas such as language and mathematics; beliefs in mathematics; cultural differences and technology were explored. With the setting up of research universities, more and more research grants were awarded to the local

The First Sourcebook on Asian Research in Mathematics Education: China, Korea, Singapore, Japan, Malaysia, and India, pp. 1065–1110

researchers and educators. Consequently many more research projects were undertaken with the majority of the topics focusing on mathematical software and courseware. Generally there was a paradigm shift in research design from scientific to interpretative research approach. Likewise, there was an expansion of sample or respondents focus on students to teachers and preservice mathematics teachers. In order to provide a more comprehensive scenario of the research trend as mentioned above, this chapter also outline and discuss briefly introduction of the various chapters that have been included in this Malaysian section of this book at the end of the chapter.

INTRODUCTION

Mathematics education undertaking the teaching and learning of mathematics is universally important and has gained increasing and undivided concerns from the Ministry of Education, educators and parents globally, likewise in Malaysia. However, mathematics education as a research field in Malaysia is still new and growing. To this date, research studies in mathematics education are mainly comprised of postgraduate student theses and small scale research projects carried out by faculty members of local universities.

This chapter will document and review various studies related to mathematics education that have been carried out in Malaysia so as to provide an overview and historical development of mathematics education research in Malaysia since 1970. So far, there has not been much comprehensive documentation of mathematics education research studies in Malaysia. The earliest report was the *State of the art review of research in mathematics education in Malaysia*, prepared by Lai and Loo (1992) under the commission of the South East Asian Research Review and Advisory Group [SEARRAG]. In their report, Lai and Loo (1992) reviewed and abstracted 37 documents, consisting of 22 master degree theses, three doctoral dissertations, 11 journal articles and one monograph which were published from 1970 to 1990. Later, Lee and Sharifah (2005) as well as Sharifah and Lee (2000) also compiled and reviewed 73 studies (mainly master theses/project report and PhD theses) on mathematics education. However, their compilation were limited to only those conducted in one institution, namely the Faculty of Education, University of Malaya and those have been reported between 1970 to 2000. Comparing these two reports, we notice there is a spur of interest in the number of mathematics education related researches and this is an encouraging development even though the latter focused on postgraduate theses in only one institution. Therefore, we would expect much more research studies on mathematics education that might have been done in the recent 10 years with the growing awareness of the importance of mathematics education in Malaysia. Thus there is an urgent need to expand the documentation and dissemination of these research studies, so as to improve the quality and efficiency of mathematics education, not only in Malaysia but also in other parts of the world.

As Lai and Loo (1992) observed that "Due to the shortage of local experts and researchers, it has been common practice for developing countries to rely heavily on imported ideas and knowledge" (p.1). However, imported ideas and practices might not be practical and effective in local context. Therefore, it is suggested that "To ensure chances of success, the formulation and planning of educational reform should be based on local research studies" (Lai & Loo, 1992, p.1). This suggestion further expedites the importance of systematically documenting and reviewing mathematics education related research done in Malaysia.

Although compilation and documentation are time consuming and tedious jobs, these efforts are worthwhile and important as the outcomes will be useful to inform the future research trend and issues to be addressed. The information obtained will be particularly helpful to future researchers or postgraduate students in identifying the specific areas for their future studies.

To set the context, this chapter will begin with a brief description of the curriculum reform in mathematics education in Malaysia as these reforms have directly and indirectly impacted the trend of research studies conducted.

CURRICULUM REFORMS IN MATHEMATICS EDUCATION IN MALAYSIA

The Malaysian school mathematics curriculum has undergone numerous reforms in the last half of a century which have impacted the research culture on the learning and understanding of mathematics in schools. The earliest reform can be traced back to the Razak Report of 1956 (Federation of Malaya, 1956) following the struggle for independence in Malaysia. The three underlying themes of that report were universalization, democratization and unification of the educational system. The resultant effect in the next decade was the physical expansion in education (Asiah Abu Samah, 1995). During the 60's, the core of the national syllabus of mathematics curriculum was the basic skills of predominant computation. Without any enhancement of teaching methods, the traditional approach of teaching was commonly used, with limited attempt to treat mathematics as an integrated subject (Asiah Abu Samah, 1984). Coincidentally, in the late 60s, numerous educational systems worldwide underwent various reforms. These reforms were sparked by the launch of the space satellite Sputnik on October 4, 1957 by Russia into the earth's orbit, outdoing the most developed nation at that time, the United States, thus scoring a massive propaganda victory. As a result, we see "New Mathematics" which was introduced in the United States being used in Malaysia in the late 1960s and early 1970s (Asiah Abu Samah, 1995). The subsequent result was a second phase of change in mathematics education. In line with the worldwide educational reform, the modern mathematics program (MMP) was introduced in Malaysian schools. A few modern topics such as sets, matrix, vector, transformational geometry and statistics were introduced into the syllabus. Sets, relations and modern geometry were seen as unifying across all topics. Finally, the mathematics curriculum was revised in the national educational reform, the national integrated curriculum for primary schools (KBSR) in the late 80's (which was fully implemented in 1982) and secondary school (KBSM) in 90's. The content of the secondary mathematics syllabus did not differ significantly from the MMP, but its main emphasis was on attaining a good balance between understanding concepts and computational skills (Noor Azlan Ahmad Zanzali, 1995).

The next reform came when Vision 2020 was announced in the 1990s (Mahathir, 1991), where Malaysia aspired to become a technological and industrialized society and to emerge as a fully developed nation by 2020. Taking this into consideration, mathematics educators in particular had to make a quantum leap from 'Meeting Today's Challenges' to 'Fulfilling Tomorrow's Dreams' (Asiah Abu Samah, 1995; Ministry of Education, 1998). Hence, as a means of fulfilling this aim, it requires a strong professional and technical workforce which directly and indirectly places mathematics as the core to basic education. In meeting this challenge, the next wave was the introduction of The Smart School Concept in 1997. Incidentally, this was also one of the flagship applications of the Multimedia Super Corridor. This concept was expected to produce technologically literate, critical thinking citizens who would contribute to the global growth of Malaysia in the 21st century (Smart School Project Team, 1997).

In 2003, another major reform took place with the introduction of PPSMI (the Teaching and Learning of Science and Mathematics in English). This major agenda came about in 2002 when the government announced that for the two subjects, Mathematics and Science, at both the primary and secondary levels, should be taught in English (Ministry of Education, 2002a, 2002b). This move was deemed necessary as it would enable learners to access information using multimedia in order to gather knowledge in the dynamic fields of Mathematics and Science using English, a language in which the two subjects' current and latest knowledge is available. However, after 7 years of its implementation, this policy was announced in the Cabinet in July 9, 2009 that it will revert to requiring the mother-tongue as the language of instruction for Mathematics and Science from year 2012 (Chapman, 2009). The reason given was that "the command of Malaysian students in Science and Mathematics subjects has been on a steady decline forcing the Government to revert to the teaching of these subjects to Bahasa Malaysia and other vernacular languages" (Deputy Education Minister, in The Star, 10 July 2009).

In a nutshell, Malaysia and other countries worldwide have observed certain reforms with new agendas and pursuant actions over a period of time. The resultant effect was a burst of energy channelled into new research studies by the mathematics education community. The Ministry of Education has redefined the desired outcomes of mathematics education to include characteristics such as the ability to think, to reason and to deal confidently with the future; to seek, to process and apply knowledge; innovativeness; a spirit of continual improvement; a lifelong habit of learning and an enterprising spirit in all undertakings (Ministry of Education, 2000).

Having shed some light on the historical perspectives of education reform movements in Malaysia, the next section highlights the changes in the trends of mathematics education research. However, before this, we would like to discuss briefly the method used to access information on research studies conducted mainly in the field of mathematics education.

METHOD AND PROCEDURE OF ACCESSING AND GATHERING INFORMATION

For the purpose of this chapter, we have employed the following procedures to search and gather the information required from four main sources:

a) Defining the Keywords and Criteria for Selection of Research Works

The purpose of this review is to gather information with regard to mathematics education research in Malaysia. Therefore our keywords used for searching were initially "mathematics education" and "mathematics teaching and learning". However, we sensed that we missed many related studies, so we have expanded our keywords to include "mathematics" and "matematik" (in Malay language). Nevertheless, by adding the latter, our search results appeared to include also studies that were not mathematics education per se. Hence, we have to filter our search and setting criteria for selection. Finally, we decided to include in our review only empirical research that fulfill the following criteria:

- Data based report pertaining to mathematics education in the Malaysian education context (that is, data collected in Malaysian schools only);
- Conducted between 1970 to 2011
- Report could be in the form of articles, monograph, Master or Doctoral theses or dissertation, and research project reports.

b) Sources of reference

Based on the above keywords and criteria for selection, we sought to gather and audit research studies data that were done between the period of 1970 to 2011 from the following four sources:

1. *Online database on Malaysian studies:*

    i) **MyAIS**: <http://myais.fsktm.um.edu.my/> which is an open access system for abstracts and indexes of articles published in refereed scholarly Malaysian journals.
    ii) My Thesis online (**MYTO**) <http://www.perpun.net.my/myto/index2.php> which lists all theses compiled from both public and private universities and university colleges.

2. *Official websites of five selected research universities in Malaysia:* As shown in Table 48.1, these five universities were selected based on their status as Research Universities (Ministry of Higher Education, 2011) defined by the Malaysian Ministry of Higher Education and was attributed mainly to their scholarly track achievement. From these websites, we also sought the database (if any) on thesis (masters and PhD level) accomplished by post graduate students in the respective universities.
3. *Archival data sourced from the Web using the Google search engine:* This search was sought for research studies done and published in conference proceedings, journal articles or project reports based on the keywords and filtered by the criteria as mentioned earlier.
4. *Library document search:* Besides the online search, we also made trips to some major libraries, namely libraries at Universiti Sains Malaysia, Universiti Malaya, Universiti Putra Malaysia, Universiti Kebangsaan Malaysia, and the Educational Planning and Research Division [EPRD] of the Ministry of Education. Particularly, the two major reviews done by Lai and Loo (1992) and Lee and Sharifah Norul Akmar (2005) were consulted.

Table 48.1. List of 5 Malaysian Research Universities and its URL links

University	*URL of University*	*URL School of Education/Math Education Department*
University Malaya	www.um.edu.my	http://education.um.edu.my/?modul=Department&pilihan=Mathematics_And_Science_Education
Universiti Sains Malaysia	www.usm.my/	http://web.usm.my/education/
University Kebangsaan Malaysia	http://pkukmweb.ukm.my/v3	http://www.ukm.my/fpendidikan
University Putra Malaysia	http://www.upm.edu.my/	http://www.educ.upm.edu.my/jpst_bi.htm
University Technology Malaysia	http://www.utm.my/	http://www.fp.utm.my/department/jsm/departjsm.html

c) Limitations and challenges faced

In our attempt to collect data, we faced a number of challenges in accessing research studies carried out in Malaysia. First, the online database that was available namely MyAIS (Malaysian access system for abstracts and indexes of articles) and My Thesis online (MYTO) are still at an infant stage. MyAIS, is an open access system containing abstracts and indexes of articles published in scholarly refereed Malaysian journals and the contribution to the database relies on the voluntary contribution from Malaysian academics, academic and professional publishers. Therefore, the database was still not comprehensive enough and we were only able to assess a limited number of studies done by the Malaysian scholars based on articles published in some Malaysian journals. Likewise, My Thesis online (MYTO) was developed at the end of 2005 and was meant to be a central repository of Malaysian theses to be shared electronically with the collection of theses between academic libraries in Malaysia. However, we were not able to assess much data that were related to our area of discussion that is mathematics education research.

Second, the official websites of universities in Malaysia were not updated often enough to track the latest studies conducted by scholars in their respective institutions. Moreover, postgraduate students' thesis database for many of the universities was also incomplete and some were not available online. Thus, it could be surmised to a certain extend that the findings of studies conducted in Malaysia were not shared and utilized for the betterment of mathematics learning and understanding. In short, it is an uphill task and time consuming attempt for those who are interested in exploring the findings of these studies. In line with this, we acknowledge that what we have reviewed here will never be exhaustive. We believe that it is essential that research studies on mathematics education should be continuously compiled and systematically documented.

TRENDS IN MATHEMATICS EDUCATION RESEARCH

Curriculum reform leads to changes in curriculum focus and emphasis; consequently these changes are reflected in the culture and trends in mathematics education research. Much of this cultural change is reflected in the trends of how research is conducted, what methodology is employed, types of research and research questions being pursued. Table 48.2a-e depict the trends and areas of mathematics education research conducted chronologically from 1970 to 2011.

Trend of Research in 1970's

As shown in Table 48.2a, only a limited number of research studies were conducted on mathematics education and the methodology used was mainly experimental design and quantitative assessment test.

Trend of Research in 1980's

Even though the research methods used in the 1980's was still dominated by quantitative research approaches such as assessment test, questionnaires and survey test, qualitative approach such as interview has started to be applied in some of the studies conducted in this decade. The number of studies conducted has also been increased greatly (see Table 48.2b).

Table 48.2a. Areas of Mathematics Education Research Conducted in 1970's

	Area/Theme	*Examples of Study*	*Methodology*
1.	Cognitive development	Khoo (1972)	Experimental study
2.	Teaching approaches	Wong (1975); Ab. Rahim bin Ahmad (1978)	Experimental study
3.	Understanding and conception of mathematical terms and knowledge	Lim (1978)	Quantitative assessment test
4.	Teacher education	Siong (1979)	Quantitative test instrument

Table 48.2b. Areas of Mathematics Education Research Conducted in 1980's

	Area/Theme	*Example of Study*	*Methodology*
1.	Error analysis	Lim (1980)	Quantitative assessment test
2.	Understanding and conception of mathematical terms and knowledge	Lee (1982)	Quantitative assessment test and interview
3.	Mathematics attitude	Swetz (1983)	Quantitative attitude test
4.	Problem solving	Chan (1984); Sufean Hussin (1986)	Interview and quantitative attitude test
5.	Mathematical needs	Tan (1984); Fadzilah Awang (1985); Chia (1989)	Questionnaires; survey test; checklist
6.	Cognitive development	Palanisamy K. Veloo (1986)	Quantitative mathematical test
7.	Understanding and conception of mathematical terms and knowledge	Neo (1989); Seow (1989)	Survey test; observation and interview
8.	Geometry – van Hiele level	Sarojini (1989)	Quantitative mathematical test

Trend of Research in 1990's

Perhaps it was not surprising to see a rapid spur of interest in mathematics education research along with the economic bloom in Malaysia. As shown in Table 48.2c, there were 13 themes or areas of research conducted in the 90's. Particularly, there were more than 10 postgraduate thesis carried out related to the schemes of mathematical concepts and constructivism. Incidentally, majority of the studies in this theme were done in one institution and under the same supervisor. This implies that supervisor's area of interest or expertise could also influence the trend of research in Malaysia.

In terms of methodology, it was also not surprise to see various methods of data collections were used, vary from quantitative to qualitative approaches.

Trend of Research in 2000's

A careful analysis of Table 48.2d shows that the three most researched themes during this period were a) Language of instruction and mathe-

Table 48.2c. Areas of Mathematics Education Research Conducted in 1990's

Area/Theme	*Example of Study*	*Methodology*
1. Schemes of mathematical concepts and constructivism	Noraini Idris (1990); Hasnul Hadi b. Abdullah Sani (1992); Aida Suraya bt Hj Mohd Yunus (1996); Sharifah Norul Akmar bt Syed Zamri (1997); Sutriyono (1997); Fatimah Saleh (1997); Goh (1998); Munisamy Susila (1998); Wan Fatanah (1998); Jayaletchumy a/p R.S Anantham (1999); Sharida bt Hashim (1999); Ng (1999); Othman bin Sodikin (1999); Ding (1999); Wun (1999);	Clinical interview
2. Understanding and conception of mathematical terms and knowledge	Doraisamy Logeswary (1990); Giam (1992)	Quantitative mathematical test
3. Mathematical attitudes and computer literacy	Lim (1991)	Quantitative attitude questionnaire and literacy test
4. Error analysis	Kung (1992); Wong (1994); Hassan Pardi (1998); Lim (1999);	quantitative assessment test
5. Effective Teaching approaches	Mohd. Majid Konting (1997); Tan (1995)	Classroom observation
6. Computer Aided Instruction and ICT related instruction	Azman Abdullah (1993); Tham (1995); Ong (1994); Kor (1995); Rosihan Ali and Ahmad Izani Md. Ismail (1998)	Development of CAI courseware
7. Teacher education	Sinnadurai (1993)	Interview and questionnaires
8. Language and mathematics	Lim (1993)	Quantitative mathematical test and interview
9. Mathematics Achievement	Jamaliah Kamal (1993); Siow (1993); Parmjit Singh (1994);	Quantitative mathematical test
10. Perception about mathematical related factors	Chiu (1994); Lee Molly N. N. et al (1996); Lim (1999);	Survey questionnaires; interview
11. Problem solving	Lim (1993);Sevenesan Raju (1996); Teng (1997)	Test and questionnaire
12. Design and development of mathematical Instruction	Hashim Yusup & Chan(1997);	Instructional model
13. Evaluation and Assessment	Jafri Jaafar (1999);	Test and task

matics; b) ICT related research; and c) Mathematics Achievement and related factors. Linking this trend to our earlier discussion on the historical development of curriculum reform in Malaysia, clearly this research trend is influenced by the curriculum reforms. Incidentally, the much discussed topic in the 2000's was the language policy of teaching mathematics and science in English or better known as PPSMI (short form) launched at 2003 but the policy was reverted back in 2012. In relation to the language policy was the promotion of integrating information and communication technology [ICT] into teaching and learning. Hence, we would not surprise to notice that the above two themes: language of instruction and ICT were at the lime light of the research trend at this period.

Trend of Research in 2010's

Even though there were just two years during the era of 2010's, we already notice the increasing number of research done in mathematics education, particularly related to ICT. There were also a couple of new themes such as Lesson Study and values related research. In terms of methodology, the variety was increasing from experimental study to quantitative survey to qualitative observation and clinical interviews.

An overall examination of the data provided in Table 48.2a to Table 48.2e shows that the trend of research has begun to evolve over the past decade from a preoccupation with quantitative approach to a pervasive use of qualitative approach. In the earlier years, courses in Quantitative Statistical analysis were primarily the focus of data analysis methodology. However, during the last decade with the influence of western research culture, the trend of qualitative data analysis has proliferated within the schools of education in Malaysia. Research in mathematics education has slowly embraced this shift in methodology with some focus on both approaches, though not at the expected rate, with case studies of students and teachers, and ethnographic studies of students in the classroom with a focus on learning practice and classroom discourse. The emphasis has now shifted towards interpreting the meaning of learning and understanding from the perspective of learners. The changes in current trends of research in mathematics education are evident even as the numbers of research titles are increasing in correlation with the increase in the number of post graduates conducting research. Consequently, despite the statements made, we cannot refute the fact that both approaches generate knowledge. This knowledge, of course, takes different forms. As Patton (1990) noted, quantitative research generates knowledge that focuses on outcomes, generalizations, predictions, and causal explanations. On the

Table 48.2d. Areas of Mathematics Education Research Conducted in 2000's

	Area/Theme	*Example of Study*	*Methodology*
1.	Teaching approaches	Gan (2000); Kim (2003)	Experimental study (Gan), Kim?
2.	Schemes of mathematical concepts and constructivism	Haslina Jaafar (2000); Parmjit Singh (2001)	Clinical interview
3.	Understanding and conception of mathematical terms and knowledge	Ong (2000); Parmjit Singh (2006)	Tests; document analysis
4.	Language of instruction and mathematics	Pang (2000); Lim (2001); Pandian and Ramiah (2003); Chan and Tan (2006); Clarkson (2006); Lim, Saleh and Tang (2007); Aziz Nordin (2007); Noraini Idris (2007); Tan (2007); Ong and Tan (2008); Lim & Presmeg (2009)	Survey questionnaires; interview; classroom observation
5.	Error Analysis	Wong M. L (2000); Lim (2008)	Quantitative assessment test
6.	Evaluation and Assessment	Wong O. C. (2000); Munirah Ghazali (2004); Parmjit (2009)	Quantitative assessment test
7.	Mathematics Achievement and related factors	Wong S. E. (2000);Norlia Abd Aziz et al (2006); Hanizah Hamzah (2006); Lelechothy Davrajoo (2007); Effandi Zakaria and Norazah Mohd Nordin (2008); Noor Azlina Ismail (2009)	Survey questionnaires
8.	Cultural difference and mathematics learning	Lim (2002); Zulkifli Mohd Nopiah (2006);	Classroom observation and interview; survey questionnaires
9.	ICT related research	Kor (2005); Teoh and Fong (2005); Su and Hong (2006); Zur'aini Dahlan(2006); Noraini Idris (2006); Rahimi Md Saad et al (2006); Palanisamy (2007); Abdullah Lazim (2007); Nor'ain Mohd. Tajudin et al (2007); Rosnaini Mahmud et al (2009)	Experimental study
10.	Lesson Study	Goh (2007); Chiew (2009)	Classroom observation and interview
11.	Values and mathematics education	Mohd. Uzi Dollah (2007); Sharifah Norul Akmar (2002); Wan Zah et al. (2005)	Classroom observation and interview
12.	Psychological factors and mathematics teachers	Selva and Loh (2008)	Survey questionnaires
13.	Problem solving	Lee (2003); Koay (2006)	Clinical interview; think aloud
14.	Geometry – van Hiele level	Chew (2007)	Interview and observation

Table 48.2e. Areas of Mathematics Education Research Conducted in 2010's

	Area/Theme	*Example of Study*	*Methodology*
1.	Lesson Study	Ong (2010); Lim, Chiew & Chew (2011); Lim & Kor (2010)	Classroom observation and interview
2.	Error analysis	Lim (2010)	Quantitative assessment test
3.	ICT related research	Teoh (2010); Saipunidzam Mahamad (2010); Ashaari & Noraidah Sahari (2011); Malathi and Rohani Ahmad Tarmizi (2011); Norazah Nordin et al. (2010); Rohani Ahmad Tarmizi et al (2010)	Experimental study
4.	Mathematics Achievement and related factors—beliefs	Roslina Radzali, T. Subahan Mohd Meerah & Effandi Zakaria (2010);	Survey questionnaires
5.	Mental computation	Munirah Ghazali (2010)	Clinical interview
6.	Language and mathematics	Parmjit Singh (2010); Lim, Chew, Kor & Tan (2011); Lim & Presmeg (2011); Neo (2011)	Quantitative achievement test; Classroom observation and interview
7.	Evaluation and Assessment	Hwa (2010); Lim, Wun & Idris (2010)	Qualitative assessment test; quantitative test

other hand, qualitative research generates knowledge that emphasises process, extrapolation, understanding and illumination. In either case, the central feature remains that knowledge is generated thus creating a significant bond between the two approaches.

OVERVIEW OF SOME MAJOR AREAS OF RESEARCH

To provide a more detail view of some of the research done so far in Malaysia, the subsequent sections provide an overview of the reviewed studies according to the following focus areas of research: (1) cognitive development; (2) teaching approaches; (3) attitudes and beliefs in mathematics; (4) problem solving abilities; (5) error analysis; (6) schemes of mathematical concepts and constructivism; (7) assessment and evaluation; (8) language and mathematics; and (9) technology in mathematics teaching and learning

Cognitive Development

Examining students' cognitive development form the earliest trend of research studies in mathematics education in Malaysia. There were two studies that examined students' cognitive development: Khoo (1972) and Palanisamy (1986). In the first study, Khoo (1972) explored the relationships between concrete reasoning and the learning of electric circuits and of motion geometry. The sample of his study consisted of 124 primary pupils aged 8-10 years old and they were randomly assigned to three experimental groups and a control group. The experimental groups were given separate training in concrete reasoning using Piagetian tasks in two topics: electric circuits and of motion geometry. He found that concrete reasoning was predictive of the learning of the two topics among the pupils. In addition, canonical and factor analyses indicated that the strongest relationships involved moderately difficult Piagetian tasks and recognition of basic concepts in the two topics that required minimal instruction. Even though all the pupils showed improvement in concrete reasoning on the post-test and transfer test, the pupils who were given training in concrete reasoning showed a consistently marked improvement for all the subgroups. Since it had been demonstrated that the pupils could be successfully trained in these tasks, it was suggested that Malaysian teachers should be guided to take the best advantage of Piagetian materials and procedures.

Likewise, the objectives of Palanisamy's (1986) study were to identify: (1) the stages of cognitive development of a sample of urban secondary school children; (2) the pupils' achievement and their levels of understanding of the concepts of Fraction, Ratio and Proportion; and (3) the relationship between cognitive levels, achievement, and levels of understanding of pupils in the concepts tested. The sample of his study consisted of 295 Form 2 and Form 4 students from two urban secondary schools. The findings of the study showed that: (1) most of the Form 2 and Form 4 Arts students were at the Concrete Operational stage but most of the Form 4 Science students were at the Formal Operational stage of cognitive development. However, there was no significant difference in the stages of cognitive development between male and female students; (2) Students at higher levels of cognitive development performed significantly better than students at lower levels of cognitive development for both Form 2 and Form 4 students; and (3) there was a relationship between students' levels of cognitive development and their levels of understanding of the concepts of Fraction, Ratio and Proportion. Students at higher levels of cognitive development achieved higher levels of understanding of the concepts tested than students at lower levels of cognitive development and the difference in the levels of understanding was statistically significant.

Teaching Approaches

The second common theme that has been much researched in Malaysia was the issue dealing with teaching approaches. A number of studies such as Wong (1975), Ab. Rahim Ahmad (1978), Tan (1995), Gan (2000) and Tay (2003) have examined issues about teaching approaches from various levels – both primary and secondary level. For instance, Wong (1972) investigated the effects of four instructional strategies on mathematics achievement and attitude towards mathematics. The sample of this study consisted of 169 Form One pupils in a secondary school in Malaysia and they were randomly assigned to four groups. Each group was randomly assigned to one of the four instructional strategies: (1) a strictly expository strategy (E); (2) a strictly programmed self-instructional strategy (P); (3) a guided-discovery strategy aided by elements of expository strategy (D); and (4) a hybrid strategy which included all the elements of the strategies described above. The findings of the study showed that: (1) the D and H strategies were more effective than the E and P strategies in improving students' mathematics achievement and attitude towards mathematics. In addition, the H strategy was most effective among low SES, low motivation, unfavourable attitude and 'traditional' students.

Also research on secondary students, Ab. Rahim Ahmad (1978) examined the effects of two teaching methods on achievement in recall and retention. He employed a quasi-experimental design using two intact groups of Form Four students in a secondary school. For one week, the students in the experimental group were taught the concepts of enlargement using a new teaching method while the students in the control group were taught the concepts of enlargement using a traditional teaching method. The results of the two-way between subjects analysis of variance (ANOVA) showed that there was no significant difference in achievement in recall and retention of the concepts of enlargement between the traditional teaching method and the new teaching method. Out of the 24 hypotheses tested only 2 hypotheses showed a significant difference in the Retention Achievement Test: (i) for students whose mathematical abilities were not controlled, the new teaching method is more effective than the traditional teaching method in helping them to answer questions of knowledge type; and (b) for students whose mathematical abilities were high, the traditional teaching method is more effective than the new teaching method in helping them to answer questions of problem solving type.

The aim of Tan's (1995) study was to compare mathematics teaching in Years 1, 2 and 3 in the New Curriculum for Primary Schools (*KBSR*) between a national school (*SK*) and a national type Chinese school (*SJKC*). The qualitative data were collected through classroom observations, interviews and analysis of documents. The findings of the study showed that: (1) the percentage of Teacher Talk category increased with the year of study for *SK* but decreased with the year of study for *SJKC*; (2) the percentage of Silence category decreased with the year of study for *SK* but increased with the year of study for *SJKC*; (3) the problems faced by *SK* teachers were pupils forgot to bring their books and counters as well as pupils in Years 2 and 3 could not recite the multiplication tables but the *SJKC* teachers did not face these problems; (4) on average, the ratio of the total time for marking exercise books in a week for *SK* and *SJKC* teachers was 2:5; and (5) on average, the ratio of the total number of questions given to pupils as an exercise for *SK* and *SJKC* teachers was 1:3.

Gan (2000) examined the effect of programmed instructional material on achievement in Reflection. He employed a quasi-experimental design using two intact groups of Form Two students from a rural secondary school. The students in the experimental group were taught the concepts of reflection using the programmed text, Learn It Yourself Reflection (LIY Reflection) which comprised 4 units whereas the students in the control group followed the normal classroom instruction. The results of the study showed that there was a significant difference in the post-test mean scores for Units 2 and 3 as well as the overall unit between the experimental and control groups but there was no significant difference in the post-test mean scores for Units 1 and 4 between the two groups. The significant difference in the post-test mean score for the overall unit indicates that the LIY Reflection materials did play a major role in improving the students' performance in the post-test.

The objectives of Tay's (2003) study were to examine the effects of a van Hiele-based instruction on geometry achievement and levels of geometric thinking. She employed a quasi-experimental design using two intact groups of Form One students from a public secondary school. The Geometry Achievement Test (GAT) and the Van Hiele Geometry Test (VHGT) (Usiskin, 1982) were used to assess the students' geometry achievement and van Hiele levels of geometric thinking, respectively. The students in the experimental group were taught using the van Hiele-based instruction while the students in the control group were taught using the textbook. The results of the study showed that the students who followed the van Hiele-based instruction obtained a significantly higher geometry achievement than those taught with the traditional approach. Further, the students in the experimental group attained higher levels of geometric thinking than the students in the control group.

Attitudes and Beliefs in Mathematics

Several studies have been conducted on students' attitudes and beliefs in mathematics (Swetz, 1983; Lim, 1991; Arellano and Ong, 2002). Swetz (1983) compared the attitude towards mathematics between Malaysian and Indonesian students. He found that Malaysian students had more favourable attitude towards mathematics than Indonesian students. In addition, urban students had more favourable attitude towards mathematics than rural students and male students had more positive attitude towards mathematics than female students.

The objectives of Lim's (1991) study were to: (1) assess students' computer literacy level (CL) and their attitude towards mathematics (ATM); and (2) ascertain whether there was any relationship between CL and ATM. The sample of her study consisted of 185 boys and 148 girls who had taken computer courses for at least two years at four fully residential schools in Negeri Sembilan. In general, she found that the students had a positive attitude towards computers and supported the use of computers in the education system as well as enjoyed learning about computers. They also had a positive attitude towards mathematics, agreed that mathematics was important in society, enjoyed learning mathematics and were motivated to do so. In addition, students' ATM was positively and significantly correlated to knowledge of computers. For the low and average mathematics achievers, 'Enjoyment in mathematics' and 'Membership in computer clubs' were the best predictors of CL score. But for the high mathematics achievers, 'Motivation in mathematics' and 'Values of mathematics in society' were the best predictors.

Arellano and Ong (2002) sought to assess the science/mathematics locus of control of Malaysian science/mathematics teachers and how this locus of control was influenced by their participation in a seven-week science/mathematics teacher training programme at RECSAM. The sample of the study consisted of 209 science teachers and 123 mathematics teachers from all over Malaysia who were following through RECSAM's in-service teacher training programmes. They found that mathematics teachers had become more internal in their locus of control orientation after their participation in the seven-week teacher training, implying that they had begun to see themselves as responsible for being able to attain or not to attain their teaching goals. The mathematics teachers believed that they could have impacted on the learning of their students.

Tan (2007) examined the policy of teaching science and mathematics in English in the Malaysian educational system by focusing on the attitudes and achievement orientations of secondary school students towards the learning of these two subjects as well as their variations according to four background variables, namely gender, ethnicity, types of feeder school and English achievement grades. He also examined the inter-correlations between the students' attitudes and achievement orientations. The sample of the study consisted of 400 secondary school students selected from four non-premier schools. The results of the study showed that the students' general attitudes and achievement orientations towards learning of science and mathematics in English do not indicate that the policy has achieved its objective. In addition, their attitudes and achievement motivations vary according to the four background variables. There were significant and positive inter-correlations between attitudes and achievement orientations towards learning of science and mathematics which further confirm the causal relationship between these two important dimensions of learning.

Problem Solving Abilities

Sufean Hussin (1986), Lim (1993), Sevanesan (1996), Teng (1997) and Lim, Lourdusamy and Munirah Ghazali (2001) examined students' problem solving abilities. The objectives of Sufean Hussin's (1986) study were: (1) to determine the status of Grade Three and Four pupils' ability to solve multiplication and division open-sentences; (2) to evaluate the pupils' skills in reading mathematical sentences, recalling basic facts of multiplication and division, and understanding the concepts of multiplication and division operations; (3) to investigate possible relationships between pupils' ability to solve multiplication and division open-sentences and their ability to read mathematical sentences, recall basic facts, and understand the concepts of multiplication and division operations; and (4) to investigate possible interactions between some external factors and some mathematical factors; and (5) to identify and categorise pupils' solution methods and strategies in solving elementary problems in multiplication and division, reading mathematical sentences, recalling basic facts and conceptualising multiplication and division operations. The sample of the study consisted of four National Schools and four National Primary Schools. The major findings concerning problem solving abilities of the pupils were as follows: (1) Number-seriation type of open-sentences such as a * b = ( ) and ( ) = a * b were more frequently solved than the reverse-thinking type such as ( ) * b = c, a * ( ) = c, c = ( ) * b and c = a * ( ); (2) Pupils' performance in solving multiplication open sentences was better than their performance in solving division open sentences; (3) For multiplication, an open sentence of the form a x ( ) = c was the most difficult to solve whereas for division, an open sentence of the form ( ) / b = c was the most difficult to solve; (4) Grade, school-type, the symmetric position of the operation and the position of the unknown affected pupils' performance in solving multiplication and division open sentences; (5) There were significant correlations between the ability to solve multiplication and division open-sentences, and the ability to read the open-sentences, recall basic facts of multiplication and division, and understand the concepts of multiplication and division operations; (6) Family income level was related to pupils' ability to solve multiplication and division open-sentences, and to the ability to understand the concept of division operation; and (7) Father's educational level was related to pupils' ability to solve multiplication and division open-sentences, recall basic facts of multiplication and division, and understand the concept of division operation.

Lim (1993) investigated language difficulties of Malaysian Form 4 Arts students in solving mathematics questions. The sample of the study comprised 60 mixed-ability male students in an urban school in Kuala Lumpur. The findings of the study showed that language difficulties were very prominent in the poor performance of the students in solving mathematics questions. In fact, 79% of the causes of initial error leading to failure to solve the questions were related to language, of which 24% were related to reading recognition and 55% were related to comprehension. In each of the four weak skills (recognition of mathematical symbols, comprehension of mathematical terms, comprehension of mathematical symbols, and comprehension of general meaning), both high and medium achievers in mathematics performed significantly better than low achievers. In addition, the high achievers were significantly better than the medium achievers regarding comprehension of the general meaning of questions.

The objective of Sevanesan's (1996) study was to describe the performance of Year Six pupils in problem solving thinking skills in mathematics. The sample comprised 80 pupils from three mixed-ability classes in Ipoh and the pupils were from urban and rural areas of various ethnic backgrounds. The findings of the study indicated that pupils generally performed better in the objective test than in the subjective test in all the thinking skills assessed. In the objective test, pupils generally obtained an achievement rate of more than 70% in all the thinking skills except for formulating sub-problems and selecting strategy to multiple step problems where their achievement rate was 57.5% and 56.3%, respectively. But, in the subjective test, understanding the variables and writing number sentence had the lowest achievement rate of 25% and 30%, respectively.

Teng (1997) examined: (1) the Initial Error Causes of incorrect solutions given by Year Six pupils in solving word problems in relation to pupils with the same total scores in the word problem test administered and the different word problems in the test; and (2) the relationships of mathematical confidence and problem solving behaviour with pupils' performances in solving word problems. The sample comprised 36 Year Six pupils from a national primary school. Data were collected using the Mathematical Word Problem Test, the Mathematical Confidence Scale, and the Problem Solving Behaviour Questionnaire. Additionally, the participants were interviewed using the Newman Interview Procedure. The findings of the study showed that the major Initial Error Causes were transformation (45%), carelessness (30.2%) and comprehension (19.2%). In general, the Initial Error Cause profiles of pupils with the same scores were very different and different word problems had different Initial Error Cause patterns. Further, the distribution of Initial Error Causes by topic differed among the four topics tested.

Lim, Lourdusamy and Munirah Ghazali (2001) aimed to identify some of the factors that affect students' abilities to solve operational and word problems in the topic of negative numbers. In particular, the factors examined were language skills, field dependence-field independence cognitive style, gender, ethnicity and basic mathematical skills. The sample comprised 113 (39 boys and 74 girls) Form Four students from seven secondary schools in the State of Penang. Data were collected using three instruments, namely Test 1 (consisted of 10 operational problems), Test 2 (consisted of three word problems) and the Group Embedded Figures Test (GEFT). They found that basic mathematical skills, field dependence-field independence cognitive style and language skills of the students were significantly related to their abilities in solving operational and word problems in the topic of negative numbers. The regression analysis showed that 43% of the variance of Test 1 scores and 24% of the variance of Test 2 scores were accounted for by the first two factors mentioned above.

Error Analysis

Several studies have been conducted on error analysis (Lim, 1980; Kung, 1992; Wong, 1994; Wong, 2000; Lim, 2010). Lim (1980) identified and analysed the computational errors in the four arithmetic algorithms with the whole numbers and the error patterns were classified in relation to school, gender and achievement. The sample consisted of 237 pupils from six Standard Four classes in three schools. Some of the major findings are as follows:

There were (a) 9.7% systematic, 4.0% random and 14.3% careless errors in the addition algorithm; (b) 15.0% systematic, 10.6% random and 19.4% careless errors in the subtraction algorithm; (c) 20.8% systematic, 17.5% random and 17.4% careless errors in the multiplication algorithm; and (d) 15.6% systematic, 37.5% random and 16.2% careless errors in the division algorithm. Adding the same digit in two columns was the most frequently occurring error pattern in the addition algorithm whereas subtracting minuend from subtrahend was the most frequently occurring error pattern in the subtraction algorithm. While multiplying digits in multiplicand by corresponding digits in multiplier was the most frequently occurring error pattern in the multiplication algorithm, using a remainder larger than the divisor was the most frequently occurring error pattern in the division algorithm.

The error pattern occurring most frequently in the addition algorithm in the national, national primary and Chinese primary schools was adding the same digit in two columns, writing carried number in answer, and adding two digits separately disregarding columns, respectively. The error pattern occurring most frequently in the subtraction algorithm in the national and national primary schools was subtracting minuend from subtrahend, and in the Chinese primary schools was not allowing for having borrowed. The error pattern occurring most frequently in the multiplication algorithm in the national, national primary and Chinese primary schools was not multiplying a digit in the multiplicand, multiplying digits in multiplicand by corresponding digits in multiplier, and errors in multiplication combinations, respectively. The error pattern occurring most frequently in the division algorithm in the national and Chinese primary schools was using a remainder larger than the divisor, and in the national primary schools was using a writing operation.

The error pattern occurring most frequently in the addition algorithm for boys was adding the same digits in two columns whereas for girls was adding the two digits separately disregarding columns. In the subtraction algorithm, the error pattern occurring most frequently for both boys and girls was subtracting minuend from subtrahend. The error pattern occurring most frequently in the multiplication algorithm for boys was multiplying digits in multiplicand by corresponding digits in multiplier whereas for girls was not multiplying a digit. In the division algorithm, the error pattern occurring most frequently for boys was using a remainder larger than the divisor but for girls was using a wrong operation.

For high achievers, the error pattern occurring most frequently in the addition algorithm was using a wrong operation while for low achievers was adding same digit in two columns. The error pattern occurring most frequently in the subtraction algorithm for high achievers was not allowing for having borrowed but for low achievers was subtracting minuend from subtrahend. In the multiplication algorithm, the error pattern occurring most frequently for both high and low achievers was multiplying digits in multiplicand by corresponding digits in multiplier whereas in the division algorithm it was using a remainder larger than the divisor.

Kung (1992) aimed to determine the initial difficulty (Initial Error Cause) which prevents a pupil from obtaining the correct solution to any mathematical tasks. He also investigated the relationships between initial error causes and the variables of gender differences, language achievement and type of primary school attended by pupils. The sample of the study consisted of Form four students. The findings of the study showed that: (1) A large percentage of errors made by Form Four mathematics low achievers on written mathematical tasks were in Newman's categories of 'Comprehension', 'Transformation' and 'Process Skills'; (2) Gender differences correlated with errors due to careless slips and lack of motivation; and (3) Language achievement did affect errors due to weak mathematical language ability and poor mathematical knowledge or skills; and (4) There was a relationship between the type of primary school attended and incidence of errors that occurred due to poor mathematical knowledge or skills.

The main objectives of Wong's (1994) study were: (1) To identify and obtain a list of mathematical skills needed in learning Form Six Economics; and (2) To identify the common errors committed by students in solving quantitative problems in Economics. The sample of the study consisted of 95 Upper Six students from a secondary school in Kuala Lumpur and 11 Economics teachers from six secondary schools. Some of the mathematical skills needed in learning Microeconomics and Macroeconomics are as follows: (1) The most frequent algebraic skills required in Microeconomics and Macroeconomics was 'calculate percentage and percentage change' followed by 'calculate rate of change', 'change subject of equation', 'formulate equations' and 'compute reciprocal'; (2) In Macroeconomics, the most frequent graphical skills required was 'identify linear graphs' followed by 'determine the values from graphs', 'plot graphs' and 'determine the quantity under the curve' while in Microeconomics, all the graphical skills listed in checklist were required; (3) In Microeconomics, the calculus skills required were 'use the symbols Δx, $<$, $>$, ∞' and 'use of slopes of a tangent' whereas in Macroeconomics, the calculus skills required were 'use the symbols Δx, $<$, $>$'; (4) In Microeconomics, the statistical skills required was 'determine the price index' while in Macroeconomics, the statistical skills required were 'determine the price index' and 'draw charts'.

Some of the common errors committed by the students in the Quantitative Economics Test are as follows: (1) Use of wrong deflationary gap and value of multiplier in solving Problem 14; (2) Calculate the wrong total output in Problem 7; (3) Draw the histogram as bar chart in Problem 17; (4) Cannot identify area under the graph which represents consumer surplus in Problem 10c; (5) Cannot identify area under the graph which represents transfer earnings in Problem 6a; (6) Calculate the percentage change in x as percentage change in income in Problem 9a; (7) Use the wrong value for price of x, use of wrong value for ΔP and the use of wrong denominator in Problem 12 which involves the determination of percentage change in price on a budget line; (8) Use the wrong formula of elasticity and use of wrong percentage change of x in Problem 9b which involves the determination of income elasticity of demand; and (9) Calculate the wrong ration for elasticity and substitute the wrong value in the elasticity formula in solving Problem 11 which involves the determination of price elasticity of demand at a point on a demand curve.

Wong, (2000) studied students' computational errors in whole number multiplication. The sample consisted of 40 Year Three pupils from a national primary school in an urban town. The findings showed that: (1) 38.8% were systematic errors, 21.6% were careless errors and 39.6% were random errors; (2) The highest number of systematic errors was in multiplication skill level 5, but the lowest number of systematic errors was in multiplication skill level 1; (3) The major causes of errors were due to the difficulties faced by pupils in basic multiplication facts, intermediate zeros and the carried number; (4) For the multiplication open sentences, 58.3% of the errors were found in the right-multiplication-reverse type whereas only 16.7% of the errors were found in the number-seriation-multiplication type.

Lim, Wun and Idris (2010) investigated the errors made by students in simplifying algebraic expressions. The sample consisted of 265 Form 2 male students and 10 high, medium and low ability students in each group were selected for interviews. The study identified 12 types of errors and these errors might be the result of interference from new learning, difficulty in operating with the negative integers, misconceptions of algebraic expressions and misapplication of rules.

Schemes of Mathematical Concepts and Constructivism

Many studies have been conducted on schemes of mathematical concepts using clinical interviews (Noraini Idris, 1990; Hasnul Hadi, 1992; Aida Suraya, 1996; Fatimah Saleh, 1997; Sharifah Norul Akmar, 1997; Sutriyono, 1997; Ding, 1999; Ng, 1999; Othman Sodikin, 1999; Sharida Hashim, 1999; Wun, 1999; Haslina Jaafar, 2000; Noor Fazilah Abeta, 2000).

Noraini Idris (1990) aimed to identify the whole number addition scheme of primary pupils based on their behaviour and explanations when they solve four whole number addition problems, namely mental picture, open-ended problem, place value concept and 'box problem'. Data were collected through clinical interviews with the pupils. The participants comprised three Standard Two pupils and three Standard Three pupils from a national primary school in Kuala Lumpur. Four types of whole number addition schemes were identified that is count all, count directly, count directly from the largest addend, and algorithmic technique.

The purpose of Hasnul Hadi's (1992) study was to identify the integer addition scheme of secondary students based on their behaviour and explanations when they solve four integer addition problems, namely mental picture, box problem, written problem and word problem. Data were collected through clinical interviews with the students. The participants comprised four Form Two students from a secondary school in Ipoh. In the addition of two small positive integers, two types of schemes were identified, namely scheme involving direct counting from the bigger addend and scheme involving related number concept. In the addition of positive and negative integers, three types of schemes were identified, namely scheme involving number line, scheme involving the concept of debt and repay, and scheme involving subtraction operation by disregarding the sign. In the addition of two negative integers, three types of schemes were identified, namely scheme involving number line, scheme involving the concept of debt and repay, and scheme involving addition operation by disregarding the negative sign.

The study of Aida Suraya (1996) had two purposes: (1) to identify decimal schemes of primary school's Year Five pupils; and (2) to identify how the pupils use their decimal schemes in solving problems on decimals. The participants comprised seven Year Five pupils and data were collected through five successive clinical interviews with the pupils using eight problem tasks. All the sessions were videotaped. Four stages of data analysis were carried out that is transcriptions of the interviews into written form, development of case studies describing each participant's behaviour on certain aspects of decimals, identification of each participant's behaviour patterns and a cross-case analysis to enable the identification of the Year Five pupils' schemes of decimals. She identified five schemes used by the participants in giving meanings to decimals, namely decimals as fractions scheme (F), decimals as pseudo-fractions scheme (PF), decimals as a combination of numbers scheme (CN), decimals based on the number 'number' before and after the decimal point scheme (NBA) and decimals as mixed-numbers scheme (MN).

The participants used F, PF, CN and NBA schemes in giving meanings to decimals less than one whereas they used the F, PF, CN and MN schemes in giving meanings to decimals greater than one. Additionally, six sub-schemes and two sub-schemes were identified for the PF and NBA schemes, respectively.

Fatimah Saleh (1997) aimed to determine the problem solving schemes of Form Two mathematics teachers. The participants comprised eight Form Two mathematics teachers who were teaching mathematics in secondary schools in Kedah. They were selected based on their willingness to participate in the study. Data were collected through clinical interviews with the participants using four tasks, namely mental pictures, mathematics problem solving, belief system and teaching activities. In addition, five mathematics inventories concerning conceptions towards mathematics and problem solving were administered to the participants before the interview sessions. The findings of the study indicated that most of the participants viewed mathematics from two perspectives, that is the perspective of pure mathematics and the perspective of applied mathematics. Most of the participants tended to believe that problem solving is an important component of school mathematics and they agreed that it is difficult to make problem solving as one of the foci of teaching. In addition, the focus of teaching seemed to rely on their conceptions of mathematics and problem solving. Manipulations of numbers and operations, basic concepts, memorisation of basic facts and heuristic which were viewed as similar to algorithmic solutions were among the emphases in their classroom teaching. Further, their main purpose of teaching mathematics was to assist students in passing school examinations although mathematics was said to be important from the utilitarian point of view. Most of the participants relied on their own learning experiences to carry out their teaching activities and they employed the teaching styles that were used by teachers they perceived as model teachers. The participants' conceptions mathematics seemed to be inconsistent with their teaching practices in the classroom.

Sharifah Norul Akmar's study (1997) had two objectives: (1) to identify Form Two students' subtraction of integer schemes; and (2) to identify how they use these schemes in solving subtraction problems involving integers in specific context. Data were collected through clinical interviews with the participants. She identified eight integer subtraction schemes, namely take away, linear movement, difference between two numbers, debt and repay, reverse subtraction, addition, ignoring negative sign and subtraction, and ignoring any sign and addition schemes. While take away scheme only contained one sub-scheme that is taking away inadequate object, addition scheme contained two sub-schemes that is addition and forward movement as well as addition and difference of absolute value schemes. The participants used three types of schemes to solve the problems of $a - b$ where $a > b$: take away, linear movement and difference between two numbers schemes. To solve the problem of $a - b$ where $a < b$, they employed four types of schemes: linear movement, reverse subtraction, taking away inadequate object, and debt and repay schemes. For problems involving -a – b where $|a| > |b|$, and $|a| < |b|$, they used three types of schemes to solve the problems, namely linear movement, debt and repay, and ignoring any sign and addition schemes. For problems involving a – (-b) where $|a| > |-b|$, they used three types of schemes to solve the problems, namely addition, linear movement, and combination of the two schemes. For problems involving -a – (-b) where $|-a| > |-b|$, they used all the types of schemes mentioned above except take away and ignoring any sign and addition schemes to solve the problems.

The objectives of Sutriyono's (1997) study were firstly to identify primary pupils' subtraction schemes and secondly to identify how they use their subtraction schemes to solve subtraction problems involving whole numbers in specific contexts. Data were collected through clinical interviews with the participants. The participants comprised four Grade Two pupils and three Grade Three pupils from an elementary school in Central Java. The findings of the study indicated that there were ten different subtraction schemes: take away, union set, set comparison, enlargement, diminution, right motion with some other motions, right-right motion, separation and reduplication of ten, ordinary subtraction, and ordinary addition schemes. The right motion with some other motions scheme had three sub-schemes, namely right-left motion, right motion and right-right motion, and right-left and right motion schemes. The ordinary subtraction scheme could be divided into two sub-schemes, namely concrete ordinary subtraction and basic fact ordinary subtraction schemes. The ordinary addition scheme could also be divided into two sub-schemes, namely concrete ordinary addition and basic fact ordinary addition schemes. The participants used all the schemes mentioned above except the union set, right-right motion and ordinary subtraction schemes to solve subtraction problem of a – b = ( ). For solving subtraction problem of a – ( ) = c, they used take away, enlargement, diminution and ordinary subtraction schemes. For solving subtraction problem of ( ) – b = c, they used enlargement, right motion, union set, and ordinary addition schemes.

Ding (1999) investigated the understanding of Form Four mathematics teachers on Critical and Creative Thinking skills approach in teaching Statistics. The participants consisted of three Form Four mathematics teachers in two different schools in Selangor. Data were collected through clinical interviews with the participants. The findings of the study showed that the participants viewed critical and creative thinking skills as the capability of the mind to determine the suitable ways and look for new solutions.

Ng (1999) aimed to identify primary school Year Four pupils' understanding of decimal addition and to ascertain how they solve addition problems involving decimals. The participants consisted of Three Year

Four pupils and data were collected through clinical interviews with the participants. The findings of the study showed that the participants used four types of addition involving decimals: (1) the addition of two decimals with the same decimal places (one or two decimal places) and less than one; (2) the addition of two decimals with both one or two decimal places and at least one of them is greater than one; (3) the addition of any two decimals in which one of them has one decimal place and the other has two decimal places; and (4) the addition of decimals (until two decimal places) with a positive integer.

Othman Sodikin (1999) identified the conceptions of primary pupils on the multiplication of whole numbers. The participants consisted of six Year Three pupils and data were collected through clinical interviews with the participants. He identified ten different conceptions in solving multiplication of whole number of which five conceptions were used by the participants to solve problems involving the multiplication of whole numbers, the 'box' situation, problems in written form, problem in the form of story-line and problems in the form of games. These conceptions involved basic facts of multiplication, drawing of squares or circles that contained line markings, counting of visual objects, counting in the normal operational form, and repetitive addition from left to right. The other five conceptions were used by the participants to solve problems in sentence form. These conceptions involved the use of space and its content, equipment or materials and their amount, groups of animals or plants, transportation and the number of wheels, and groups of people with the number of members.

Sharida Hashim (1999) examined the strategies employed by Form Four students in solving problems involving multiplication of two matrices. The sample comprised 30 Form Four students. She identified eight strategies that were used by the students to solve the multiplication of two matrices questions.

Wun (1999) identified secondary teachers' understanding of logical reasoning. The participants consisted of three Form Four mathematics teachers from a secondary school in Hulu Selangor district. Data were collected through clinical interviews with the participants. He identified three types of teachers' understanding of logical reasoning, namely logic, number and operation. The findings of the study also showed that: (1) the teachers' instructions in the classroom had been influenced by their own experience as a student; (2) teaching was conducted in a rote-learning manner as well as drills; and (3) the teaching approaches were based on definition-example-practise and explain-example-practise.

The purpose of Haslina Jaafar's (2000) study was to identify how Form Two students solve linear equations. The participants consisted of three Form two students from a secondary school in Kuala Lumpur. Data were collected through clinical interviews with the participants. She found that students used different methods to solve linear equations and weak students often failed to use the correct method in solving linear equations.

Noor Fazilah Abetah (2000) identified the understanding of Form One students in addition of fractions involving the same denominators, different denominators, mixed numbers as well as the problems faced by students in addition of fractions. The participants consisted of Form One students and data were collected through clinical interviews with the participants. The findings of the study showed that students add fractions with the same denominator better as compared to different denominators and mixed numbers. The difficulty faced by students in adding fractions of different denominator and mixed numbers was in finding the equivalent fraction for a given fraction value.

Language and Mathematics

Malaysia is a multicultural country and the issue of language use in education never ceases to attract the attention of the policy makers, educators and the society at large. In the 1990s, local researchers began to study factors associating secondary schools students' mathematics problem solving abilities and the language of instruction. For example, Lim (1993) found that the under achieving students in mathematics were gravely affected by the language used in presenting the mathematics problems when attempting to answer the questions. She observed that due to poor linguistic ability those less performing students were at a disadvantage when reading mathematics text is concerned. Pang (2000) supported the above finding when her research on the influence of language in mathematics achievement in a Chinese independent high school showed that Chinese students performed poorly in solving mathematics word problems when these problems were presented in Malay language. She attributed the cause to students' poor mastery of the Malay language.

It was observed that the number of studies on language use in mathematics instruction proliferated after year 2003. The main reason for the increase was the implementation of the policy of Teaching Mathematics and Science in English or better known as PPSMI (a Malay abbreviation) in all Malaysian primary schools in 2003. There were studies (Chan, 2006; Pandian & Ramiah, 2003; Tan, 2011) that investigated the practicing as well as trainer teachers' perception and beliefs in the value and achievability of teaching mathematics and science from the existing pupils' own language (POL) to English. Notwithstanding, students' perception on teaching and learning mathematics in English was also studied by Aziz Nordin (2007). He found that lower secondary schools students concur that English Language is important in everyday life and it broadens career opportunity. These students also found that it is easier to learn mathematics than science in English. Clarkson (2006), however, examined both teachers and students flexibility in adapting to the change of language used in teaching mathematics. He found that at this transition period, teachers were adapting but they still need more training incorporated in their professional development. Above all, the positive feedback was students were beginning to use English in their thinking about mathematics. On the subject of the professional preparation of Malaysian teachers in coping with the implementation of teaching and learning of mathematics and science in English, Noraini Idris (2007) reported that secondary school teachers perceived that the training provided is adequate and the immediate need to prime their readiness to teach mathematics and science in English.

The actual happenings in classroom practice of mathematics instruction using English did not go unnoticed. Ong (2008) did an in-depth study on teachers' experiences in teaching mathematics entirely in English found that prior educational background of the teachers, the language culture in school, and the linguistic abilities of students contributed to the language used in teacher talk. Similarly, Lim (2009, 2011a, 2011b) studied the dilemma faced by mathematics teachers in the Malaysian Chinese primary schools found that both expert and novice teachers did not teach mathematics entirely in English to performing and non-performing classes. These teachers practiced code-switching when communicating with students to overcome poor mastery of English language when teaching for understanding.

After six years of PPSMI implementation, researchers were ready to investigate pupils' achievement in mathematics taught in English. Parmjit Singh (2010) examined languages and mathematics achievement among rural and urban primary four pupils. He compared Primary 5 pupils' mathematics achievement in two tests conducted in English and English/ Bahasa Malaysia versions. The results showed that urban pupils' mathematics achievement was influenced by the language used in the test but not the rural pupils. In addition, he found that urban pupils' mathematics achievements in both tests surpassed those of their rural counterparts.

However, in July 8, 2009, the Malaysian government has decided to abolish PPSMI, and gradually revert back to teaching primary mathematics in either Malay language or in pupils' mother tongue by year 2012. In response to the abolition, Faizah Mohammad Nor, Marzilah A. Aziz dan Kamaruzaman Jusoff (2011) have argued about the timeliness to abolish English for teaching mathematics and science in Malaysia based on their research finding.

ICT in Mathematics Teaching and Learning

Like most developed countries, the Malaysian government is promoting and facilitating the integration of ICT in schools with the believe that "ICT supports students' constructive thinking, allows them to transcend their cognitive limitations to engage in cognitive operations that they may not have been capable of otherwise" (Salomon, 1993 cited in Lim, 2007, p. 84). The Malaysian school mathematics curriculum emphasizes the use of technology in the teaching and learning of mathematics to support the nation's aspiration of becoming an industrialized nation. It states that,

> The use of technology especially, Information and Communication Technology (ICT) is much encouraged in the teaching and learning process. Pupils' understanding of concepts can be enhanced as visual stimuli are provided and complex calculations are made easier with the use of calculators. (KBSM, 2006, p. 4)

Research studies on ICT in mathematics in Malaysia include the application of mathematical software and courseware (GSP, CAS, GeoGebra, Autograph, etc.), mathematical analysis tools appliance (computer and graphics calculator), and pedagogical use of ICT in modelling as well as presenting virtual mathematics.

Early research on mathematical courseware focus on the development and evaluation of logo-based geometry package (Kor, 1995) in which a logo-based geometry package was developed and evaluated to investigate the effects of the package on learners' performances in specific geometric skills and concepts. The finding showed that both high and low ability groups showed a significant gain in the subtests on angle estimation, angle drawing, and classification of various quadrilaterals with the usage of logo-based geometry package. Subsequently, the use of multimedia prototype was introduced. Zur'aini Dahlan (2006) promoted in a bilingual model for First Year Mathematics course. The findings showed that language plays an important role in determining the effectiveness of a multimedia presentation. The bilingual model showed that when more complex mathematical notations are presented, students were found to encounter difficulties in explaining their solutions in both English and Malay languages. Furthermore, when more complex challenges were given in mathematics, pseudo-bilingual/multilingual learners were found to select the Malay language as their domain. In 2009, Rosnaini Mahmud et al. developed a CAI courseware 'G-Reflect' and examined the effect of the courseware on students' achievement and motivation in learning mathematics. The result was positive. Similarly, Teoh (2006) investigated the effect of a newly developed interactive courseware in the learning of matrices found that the courseware was most effective to use together with mastery learning strategy. Besides the development of courseware, Ashaari and Noraidah Sahari (2011) also examined the measurement of the usefulness of a mathematics courseware.

Nonetheless, research into the application of mathematical analysis tools such as the use of computer and graphics calculator in mathematics instruction started to emerge in year 2004 when these electronic tools were more affordable in Malaysia. Year 1998 paved the way for the introduction of technology-based mathematics program for the undergraduates (Rosihan M. Ali & Ahmad Izani Md. Ismail, 1998). Problems of implementation were identified and training plans for staff were expounded. Subsequent research includes exploring the impact of the tools such as graphics calculator on students' cognitive development (e.g. Nor'ain Mohd. Tajudin et al., 2008; Teoh & Fong, 2005); mathematics achievement and their affective conducts such as anxiety, motivation, attitudes and other affective domains (e.g. Kor, 2005; Noraini Idris, 2006).

Works on presenting mathematics virtually were also observed from several researchers. For example, Su and Hong (2006) investigated the effectiveness of microworld constructivist learning environment in enhancing the learning of probability as well as in improving students' attitudes toward mathematics. Palanisamy (2007) conducted a comparative study on E-learning for mathematics subjects in two Malaysian secondary smart schools and found that e-learning students achieved a better standard mean score than the conventional learning. Koo (2008) investigated factors affecting the perceived readiness for online collaborative learning (OCL) of mathematics teachers found that time constraint and insufficient access to technology was two impediments to the OCL. There was also a research (Malathi & Rohani Ahmad Tarmizi, 2011) that looked into e-book utilization among students in mathematics to identify the e-book utilization habit of postgraduate and undergraduate students.

Research on the use of mathematical software was prominent especially in year 2007 onwards. Studies conducted on the pedagogical use of geometer's sketchpad (GSP) in teaching and learning mathematics were observed (Lim, 2007; Norazah Nordin, 2010). Other software such as GeoGebra (Kamariah Abu Bakar et al., 2010), Autograph (Rohani Ahmad Tarmizi et al., 2010) and CAS (Abdullah Lazim, 2010) were similarly expounded. Results obtained showed that students exposed to these integrated software performed better than those taught using the conventional method.

Assessment and Evaluation

Research on assessment in mathematics in Malaysia is not as popular as other fields of study as mentioned above. This is probably due to the fact that all public examinations in Malaysia are governed by a centralized examination body, the *Lembaga Peperiksaan*. As such any new findings from local researchers will need time to realize. Munirah Ghazali (2004) conducted an assessment on Malaysian primary pupils' number sense with respect to multiplication and division. This was followed up by Parmjit Singh (2009) who assessed students' achievement in number sense test across levels. The results were students obtained a low percentage of success in achievement in number sense and no significant difference in the results was observed between Secondary 1 and Secondary 2 students. All students in the study faced great difficulty in making sense with numbers. In another study, Lim et al. (2010) developed a superitem test to assess students' algebraic solving ability through interview method. She found that high ability students were more able to seek the recurring linear pattern and identify the linear relationship between variables. However, the low attaining students showed more abilities on counting and drawing method. In addition, there was another study conducted by Hwa (2010) that focuses on implementing a Mathematical Thinking Assessment (MaTA) Framework to assess students' thinking processes and its limitations. Nevertheless, these few studies were observed to make some impact on the area of mathematics assessment.

IMPLICATIONS AND ISSUES FOR FUTURE RESEARCH

The review so far has attempted to identify the changes in the trends and areas of mathematics education research in Malaysia from 1970 to 2011. In general, mathematics education research studies in Malaysia can be grouped under the following categories: (1) cognitive development; (2) teaching approaches; (3) attitudes and beliefs in mathematics; (4) problem solving abilities; (5) error analysis; (6) schemes of mathematical concepts and constructivism; (7) assessment and evaluation; (8) language and mathematics; and (9) technology in mathematics teaching and learning. Moreover, the larger bulk of recent researches were focused on technology in mathematics teaching and learning, language and mathematics, as well as schemes of mathematical concepts. In addition, most of these researches were small scale postgraduate studies or small grant research. Consequently, the findings were limited in its impact and generalisibility. Therefore, future research should attempt to extend and expand from the present categories of studies using different samples, bigger sample size, research designs and methods as well as technologies so as to contribute to the knowledge base of research in mathematics education, particularly in Malaysia.

In line with the latest development in mathematics education such as the introduction of school based assessment and the promotion of using technology in mathematics teaching and learning, more future studies could focus on these two areas. Since the majority of the studies on technology in mathematics teaching and learning involved graphics calculator and courseware, further studies should employ other ICT tools such as GeoGebra, Autograph, CAS, Geometer's Sketchpad, e-learning and e-book.

Apart from that, most of the studies on language and mathematics focused mainly on the implementation of the recent language policy of Teaching Mathematics and Science in English which was reverted back in 2012. Therefore future studies might focus on the impact of latest policy of teaching school mathematics in bilingual. In addition, future studies can also focus on the teaching of mathematics using English as the language of instruction at the tertiary level, which are still very rarely explored in this country.

Another interesting observation from the review of the studies was that the majority of the studies on schemes of mathematical concepts and constructivism focused on primary and secondary students, only a few studies on secondary mathematics teachers. Thus, more studies could be carried out to investigate the mathematical concept schemes of primary mathematics teachers and university students.

Another noteworthy observation is the research methods employed in most of the studies mentioned above are either empirical or designed to test the conceptual/theoretical framework. Further research may engaged grounded theory to develop new perspective in the study of Asian mathematics classroom. In other words, future researchers will need to look beyond the conventional research methodology and dare to explore new and innovative approach to visualize upcoming challenges for the teaching and learning of mathematics in future. Subsequently, new theoretical models may be formulated. In addition, further research works are needed to operationalize the understanding of mathematics teaching and learning process which is still largely theoretically based. Research efforts may attune to developing and establishing more valid and reliable instruments that can accurately measure a construct in practice, such as mathematical thinking. Besides these, research on designing instructional materials that include digital mathematics textbooks, worksheets or modules that meet the needs of our future young technological generation is also recommended.

Doing research is a culture and to instil this culture into schools or higher institutions can be challenging. Currently, research works on mathematics education are still limited to the scholarly work done by professors or lecturers or research theses of postgraduate students in universities and higher institutions. Therefore, it will be aptly to also promote research studies done by both in service and pre service mathematics teachers since these are people who are closest to the students in their profession. We attested that although researchers can learn many things by observing a mathematics classroom, the real happenings are best learnt from those who are at work in the classroom. In brief more qualitative research studies that employ classroom observation in real context and writing of reflective journals are recommended as research tools that will help to strengthen the quality of mathematics education research in Malaysia.

CONCLUSION

It is hoped that this review of studies will provide some guidance for mathematics educators, particularly new researchers and postgraduate students in Malaysia to conduct research and write academic publications in the future. It is also recommended that a similar analysis be repeated every five years in Malaysia. This ultimately will enable mathematics education researchers to review and update the research trends, and propose new research focus directions to work on.

In addition, documentation and dissemination of these findings will provide valuable information to various stake holders such as teachers, parents, policy makers, research students and researchers on the possible happenings as well as determine future directions of research in mathematics education. Moreover, the findings will further assist in identifying factors contributing to effective mathematics teaching and learning. This will serve as the basis for planning and initiating steps to encourage students to learn mathematics. In short, this type of documentation will pave the way and provide the right impetus towards the direction of future research.

REFERENCES

Ab. Rahim Ahmad. (1978), *Perbandingan kesan-kesan dua kaedah mengajar dalam menyampaikan konsep besaran dalam matematik kepada murid-murid Tingkatan IV sekolah menengah kebangsaan [A comparison of the effects of two methods of teaching the mathematical concept of enlargement to Form IV students in a national secondary school]*. Unpublished M.Ed. Thesis, University of Malaya, Kuala Lumpur.

Abdullah, Lazim M. (2007). Procedural knowledge in the presence of a computer algebra system (CAS): Rating the drawbacks using a multi-factorial evaluation approach. *International Journal for Technology in Mathematics Education, 14*(1), 14-20.

Aida Suraya Md Yunus. (1996). *Skim nombor perpuluhan bagi murid Tahun Lima sekolah rendah [Decimal schemes of primary school's Year Five pupils]*. Unpublished Ph.D. Thesis, University of Malaya, Kuala Lumpur.

Arellano, E. L., & Ong, E. T. (2002). Science/Mathematics locus of control of Malaysian Science/Mathematics educators: Implications to teacher-training programmes. *Journal of Science and Mathematics Education in Southeast Asia, 25* (2), 32-47.

Ashaari, Noraidah Sahari. (2011). Mathematics courseware usefulness measurement score based on evaluator's factors.

Asiah Abu Samah (1984) *Perkembangan Kurikulum matematik sekolah di Malaysia sejak zaman penjajah*. Paper presented at the Seminar on mathematics education

Asiah Abu Samah (1995). *Education and human resource development in Malaysia:Yesterday's concerns, today's challenges and tomorrow's dreams*. Keynote address,AARE (Australian Association for Research in Education) Conference, 1995. Seeabstract, retrieved January 19, 2011, from: http://www. aare.edu.au/95pap/abs95.htm

Aziz Nordin. (2007). Students' perception on teaching and learning mathematics in English.

Azman Abdullah. (1993). *The design, development and evaluation of an authoring language for mathematics teachers.* Unpublished M.Ed. Thesis, University of Malaya, Kuala Lumpur.

Begle, E. G., & Gibb, E. G. (1980). Why do research? In R. Shumway (Ed.), *Research in mathematics education*, pp. 3-19. Washington, DC: NCTM

Cai, Z. J. (2004). *MIDAS as an instrument to evaluate multiple intelligences of high school students in china: a validation study.* Unpublished Master Thesis, Universiti Sains Malaysia, Penang.

Chan, L. H. (1984). *Mathematical information gathering ability: A study of a sample of Form Four girls.* Unpublished M.Ed. Thesis, University of Malaya, Kuala Lumpur.

Chan, P. C. (2010). *The use of manipulatives to facilitate the teaching and learning of probabilities in A level course*. Unpublished M.Ed. Thesis, Open University Malaysia, Kuala Lumpur.

Chan, S. H. and Tan, H. (2006). English for mathematics and science: Current Malaysian language-in-education policies and practices. *Language and Education, 20*(4), 306-321.

Chapman, K. (2009, July 9). It is Bahasa again but more emphasis will be placed on learning English, *The Star, N2.* Retrieved August 15, 2011, from http://thestar.com.my/news/story.asp?file=/2009/7/9/nation/4286168&sec=nation

Charngeet Kaur, B. S. (2000). *Understanding of percentage in relation to number sense among Form One students.* Unpublished M.Ed. Project Report, University of Malaya, Kuala Lumpur.

Cheah, U. H. (2001). *The construction of mathematical beliefs by trainee teachers in a teachers college*. Unpublished Ph.D. Thesis, Universiti Sains Malaysia, Penang.

Chew, C. M. (2007). *Form One students' learning of solid geometry in a phase-based instructional environment using the Geometer's Sketchpad*. Unpublished Ph.D. Thesis, University of Malaya, Kuala Lumpur.

Chia, C. F. (1989). *Mathematical needs of Malaysian higher school level physics*. Unpublished M. Ed. Thesis, University of Malaya, Kuala Lumpur.

Chiew, C. M. (2009). *Implementation of Lesson Study as an innovative professional development model among mathematics teachers*. Unpublished Ph.D. Thesis, Universiti Sains Malaysia, Penang.

Chinappan, Mohan. (2009). Malaysian and Australian children's representations and explanations of numeracy problems.

Chiu, K. C. (1994). Persepsi murid Sekolah Menengah Sri Gombak terhadap Matematik KBSM *[Perceptions of Sekolah Menengah Sri Gombak students towards KBSM Mathematics]*. Unpublished M.Ed. Project Report, University of Malaya, Kuala Lumpur.

Clarkson, P. C. (2006). Reverting to english to teach mathematics: How are Malaysian teachers and students changing in response to a new language context for learner?

Ding, H. E. (1999). Kefahaman tiga orang guru matematik Tingkatan Empat terhadap pendekatan kemahiran berfikir secara kritis dan kreatif dalam pengajaran Statistik *[The understanding of three Form Four mathematics teachers on critical and creative thinking skills approach in Statistics]*. Unpublished M.Ed. Project Report, University of Malaya, Kuala Lumpur.

Doraisamy, L.. (1990). *Levels of understanding of probability concepts among Forms Four and Six pupils*. Unpublished M.Ed. Thesis, University of Malaya, Kuala Lumpur.

Dorothy Dewitt. (2000). *Quadratic equations in algebra: Analysis of written responses among Form Four students*. Unpublished M.Ed. Project Report, University of Malaya, Kuala Lumpur.

Dylan, W. (1998) A framework for thinking about research in mathematics and science education. In J. A. Malone, B. Atweh & J. R. Northfield (Eds.), *Research and supervision in mathematics and science education* , pp. 1-18. Mahwah, NJ: Erlbaum.

Effandi Zakaria and Norazah Mohd Nordin. (2008). The effects of mathematics anxiety on matriculation students as related to motivation and achievement. Eurasia Journal of Mathematics, Science & Technology Education, 4(1), 27-30.

Faizah Mohammad Nor, Marzilah A. Aziz and Kamaruzaman Jusoff. (2011). Should English for Teaching Mathematics and Science (ETeMS) in Malaysia be abolished? *World Applied Sciences Journal 12 (Special Issue on Creating a Knowledge Based Society)*, 36-40.

Fatimah Saleh. (1997). *Skim penyelesaian masalah guru Tingkatan Dua [Problem solving schemes of Form Two teachers]*. Unpublished Ph.D. Thesis, University of Malaya, Kuala Lumpur.

Gan, S. P. (2000). *Effectiveness of programmed instructional material in the teaching of Reflection to Form Two pupils in a rural school*. Unpublished M.Ed. Thesis, University of Malaya, Kuala Lumpur.

Gan, W. L. (2008). *A research into year five pupils' pre-algebraic thinking in solving pre-algebraic problems*. Unpublished Ph.D. Thesis, Universiti Sains Malaysia, Penang.

Giam, K. H. (1992). *Understanding of concepts in mechanics among Form Four Science students in the Klang district*. Unpublished M.Ed. Thesis, University of Malaya, Kuala Lumpur.

Goh, B. T. (1997). Kesan Sistem 4MAT terhadap pencapaian matematik pelajar Tingkatan 4 di sebuah sekolah di Pulau Pinang *[The impact of 4MAT System on the mathematics achievement of some Form 4 Students in one of the Penang secondary school]*. Unpublished Master Thesis, Universiti Sains Malaysia, Penang.

Goh, P. C. (1998). Konsepsi Pelajar Tingkatan Dua tentang Nisbah *[Form Two students' conception of ratio]*. Unpublished M.Ed. Thesis, University of Malaya, Kuala Lumpur.

Goh, S. C. (2007). *Enhancing mathematics teachers' content knowledge and their confidence in teaching mathematics using English through Lesson Study process.* Unpublished Master Dissertation, Universiti Sains Malaysia, Penang.

Hamidah Maidinsah. (2003). Kesan kaedah pengajaran metakognisi-inkuiri terhadap prestasi dalam matematik dan penaakulan saintifik di kalangan pelajar Diploma *[The effect of metacognitive-inqury teaching method on mathematics performance and scientific reasoning of Diploma students]*. Unpublished Ph.D. Thesis, Universiti Sains Malaysia, Penang.

Hashim, Yusup., & Chan, C. T. (1997). Use of instructional design with master learning. *Educational Technology, 37*(2), 61-63.

Haslina Jaafar. (2000). Penyelesaian persamaan linear oleh tiga orang pelajar Tingkatan Dua *[Solving linear equations by three Form Two students]*. Unpublished M.Ed. Project Report, University of Malaya, Kuala Lumpur.

Hasnul Hadi Abdullah Sani. (1992). Skim penambahan integer bagi pelajar-pelajar Tingkatan Dua *[Addition integer scheme of Form Two students]*. Unpublished M.Ed. Thesis, University of Malaya, Kuala Lumpur.

Hassan Pardi. (1998). Kajian kes tentang pola kesilapan murid Tahun Tiga yang lemah dalam menyelesaikan masalah bercerita dalam matematik *[Error patterns of weak Year Three students in solving word problems in mathematics: A case study]*. Unpublished M.Ed. Project Report, University of Malaya, Kuala Lumpur.

Hwa, T. Y. (2010). *Development, usability and practicality of a mathematics thinking assessment framework*. Unpublished Ph.D. Thesis, Universiti Sains Malaysia, Penang.

Jafri Jaafar. (1999). Perbezaan markah pencapaian pelajar Tingkatan 4 menggunakan alat penilaian berbentuk ujian berbanding alat penilaian berbentuk tugasan bagi mata pelajaran Matematik *[Differences in Form 4 pupils' achievement in using test assessment as compared to task assessement in Mathematics]*. Unpublished M.Ed. Project Report, University of Malaya, Kuala Lumpur.

Jamaliah Kamal. (1993). Kajian tentang pencapaian matematik Tahun III di Sekolah Rendah Simpang Renggam *[Research on mathematics achievement of Year III pupils in Sekolah Rendah Simpang Renggam]* Unpublished M.Ed. Project Report, University of Malaya, Kuala Lumpur.

Jayaletchumy, R.S.A. (1999). *Secondary School children's understanding of the arithmetic average*. Unpublished M.Ed. Project Report, University of Malaya, Kuala Lumpur.

Kalsom Saidin. (2004), Pola atribusi pelajar terhadap pencapaian matematik *[The students' patterns of attribution towards mathematics achievement]*. Unpublished Master Thesis, Universiti Sains Malaysia, Penang.

Kamariah Abu Bakar, Ahmad Fauzi Mohd Ayub and Rohani Ahmad Tarmizi. (2010). Exploring the effectiveness of using GeoGebra and e-transformation in teaching and learning Mathematics. *Advance Education Technologies*, 19-23.

Khoo, P. S. (1972). *Relationship between cognitive development and the learning of science and mathematics.* Unpublished Ph.D. Thesis, University of Malaya, Kuala Lumpur.

Kilpatrick, J. (1993). Beyond face value: Assessing research in mathematics education. In G. Nissen & M. Blomhoj (Eds.,), *Criteria for scientific quality and relevance in the didactics of mathematics,* pp.15-34. Roskilde, Denmark: University of Roskilde.

Kim, T. S. (2006). *The modality factor in two approaches of abacus-based calculation and its effect on mental arithmetic achievement.* Unpublished Ph.D. Thesis, Universiti Sains Malaysia, Penang.

Kim, T. S. (2003). *Three modes of abacus-based calculation and their effects on mental arithmetic achievement.* Unpublished Master Thesis, Universiti Sains Malaysia, Penang.

Koay, C. Y. (2006). *Mathematical problem solving solution path solution path of trainees in a teacher training college*. Unpublished Ph.D. Thesis, Universiti Sains Malaysia, Penang.

Koo, A. C. (2008). Factors affecting teachers' perceived readiness for online collaborative learning: A case study in Malaysia. *Educational Technology & Society, 11*(1), 266-278.

Kor, A. L. (1995). The development and evaluation of logo-based geometry package. Unpublished M.Ed. Thesis, University of Malaya, Kuala Lumpur.

Kor, L. K. (2005), *The Impact of the use of Graphic Calculator on the culture of statistics learning.* Unpublished Ph.D. Thesis, Universiti Sains Malaysia, Penang.

Kung, W. C. (1992). *Errors in written mathematical tasks of Form Four pupils in a Malaysian secondary school.* Unpublished M.Ed. Thesis, University of Malaya, Kuala Lumpur.

Lai, K. H., & Loo, S. P. (1992). *State-of-the Art review of research in mathematics education in Malaysia*. South East Asian Research Review and Advisory Group [SEARRAG], Penang, Malaysia.

Lee, Molly N.N. et al (1996). *Students' orientation towards science and mathematics: Why are enrolments falling?* Monograph Series No.1/ 1996. School of Educational Studies, Universiti Sains Malaysia, Penang, Malaysia.

Lee, S. E. & Sharifah Norul Akmar bt Syed Zamri. (2005). Studies of mathematics education at the Faculty of Education, University of Malaya: Focus and direction. *Journal of Education* 2005, 7-28.

Lee, S. E. (1982). *Understanding of basic concepts in transformation geometry - a study of a sample of pupils in Forms IV and VI*. Unpublished M.Ed. Thesis, University of Malaya, Kuala Lumpur.

Lee, S. E. (2003). *Cognitive and metacognitive process in solving geometric problems.* Unpublished Ph.D Thesis, Universiti Sains Malaysia, Penang.

Lim, B. L. (1999). *Division of whole number: Analysis of computational errors of Form One students.* Unpublished M.Ed. Project Report, University of Malaya, Kuala Lumpur.

Lim, C. B. (1993). *Language difficulties of Form Four Arts students in solving mathematics questions.* Unpublished M.Ed. Thesis, University of Malaya, Kuala Lumpur.

Lim, C. P. (2007). Effective integration of ICT in Singapore schools: pedagogical and policy implications. *Education Tech Research Dev, 55,* 83–116. DOI 10.1007/s11423-006-9025-2

Lim, C. S. (1991). *Relationship between computer literacy and mathematics attitude.* Unpublished M.Ed. Thesis, University of Malaya, Kuala Lumpur.

Lim, C. S., Lourdusamy, A., & Munirah Ghazali (2001). Factors affecting students' abilities to solve operational and word problems in mathematics. *Journal of Science and Mathematics Education in South East Asia, 24*(1), 84-95

Lim, C. S. (2003). Cultural Differences and Mathematics Learning in Malaysia. *The Mathematics Educator,* 7(1), 110-122.

Lim, C. S. & Kor L. K. (2010). *Innovative use of GSP through Lesson Study collaboration.* Penang: Basic Educational Research Unit (UPPA), School of Educational Studies, Universiti Sains Malaysia.

Lim, C. S., & Presmeg, N. (2011). Teaching Mathematics in two languages: A teaching dilemma of Malaysian Chinese Primary Schools. *International Journal of Science and Mathematics Education, 9*(1), 137-161.

Lim, C. S., Chew, C. M., Kor, L. K., & Tan, K. E. (2011). *Communication and Language use in Primary mathematics Classroom Discourse.* Penang: Basic Educational Research Unit (UPPA), School of Educational Studies, Universiti Sains Malaysia.

Lim, C. S., Chiew, C. M., &Chew, C. M. (2011). *Promoting Mathematical Thinking and Communication through Lesson Study Collaboration.* Penang: Basic Educational Research Unit (UPPA), School of Educational Studies, Universiti Sains Malaysia.

Lim, C. S., Lourdusamy, A., & Munirah Ghazali (2001). Factors affecting students' abilities to solve operational and word problems in mathematics. *Journal of Science and Mathematics Education in Southeast Asia, XXIV* (1), 84-94.

Lim, H. L. (2007). Penggunaan model SOLO dalam penilaian kebolehan penyelesaian persamaan linear pelajar Tingkatan 4 *[The use of SOLO model to evaluate the ability of Form 4 students in solving linear equations].* Unpublished Ph.D. Thesis, University of Malaya, Kuala Lumpur.

Lim, H. L., Wun, T. Y., & Idris, N. (2010). Superitem Test: An alternative assessment tool to assess students' algebraic solving ability. *International Journal for Mathematics Teaching and Learning*. October 2010. EJ904892

Lim, K. S. (2008). *An analysis of errors made by Form 2 students in simplifying algebraic expressions*. Unpublished Ed.D. Dissertation, Universiti Sains Malaysia, Penang.

Lim, K. S. (2010). An error analysis of Form 2 (Grade 7) students in simplifying algebraic expressions: A descriptive study. *Electronic Journal of Research in Educational Psychology, 8* (1), 139-162.

Lim, S. K. (1978). *A study of the relationship between of perspective mathematical vocabulary and mathematics achievement in selected remove classes*. Unpublished M.Ed. Thesis, University of Malaya, Kuala Lumpur

Lim, T. S. (1980). *Analysis of computational errors of Standard Four pupils in three selected primary schools.* Unpublished M.Ed. Thesis, University of Malaya, Kuala Lumpur.

Logesh, K. S. (2000). *Understanding of the limit of functions among college students*. Unpublished M.Ed. Project Report, University of Malaya, Kuala Lumpur.

Mahathir Mohamad. (1991). *The way forward. Paper presented to the Malaysian Business Council. Office of the P.M.*, 28 February. Retrieved January 11, 2011, from http://vlib.unitarklj1.edu.my/htm/w2020.htmMinistry of Education, 2000

Malathi, L. and Rohani Ahmad Tarmizi. (2011). E-book utilization among mathematics students of Universiti Putra Malaysia. *Library Hi Tech, 29*(1), 109-121.

Ministry of Education Malaysia (1998). *Sekolah ke kerjaya*. Kuala Lumpur: Pusat Perkembangan Kurikulum.(unpublished)

Ministry of Education, Malaysia. (2002a). *Integrated curriculum for secondary schools: Curriculum specifications Mathematics Form 1.* Kuala Lumpur: Pusat Perkembangan Kurikulum.

Ministry of Education, Malaysia. (2002b). *Integrated curriculum for primary schools: Curriculum specifications Mathematics Year 1.* Kuala Lumpur: Pusat Perkembangan Kurikulum.

Ministry of Education. (2000). *Huraian sukatan pelajaran sains, Tingkatan 4.* Kuala Lumpur. Pusat Perkembangan Kurikulum

Ministry of Higher Education, (2011). Official Portal. Retrieved February 12, 2011, fromhttp://www.portal.mohe.gov.my/portal/page/portal/ExtPortal/IPT/ResearchU

Mohd Uzi b. Dollah. (2007). Penerapan nilai dalam pengajaran guru matematik menengah: Satu kajian kes [*Inculcating values in mathematics teaching among secondary school mathematics teachers: A case study*]. Unpublished Ph.D. Thesis, Universiti Sains Malaysia, Penang.

Mohd. Majid Konting. (1997). In search of good practice: A case study of Malaysian effective mathematics teachers classroom practice. *Journal of Science and Mathematics Education in South East Asia, vol. XX,* no. 2, 8-20.

Ghazali, M., Idros, S. N., & McIntosh, A. (2004). From doing to understanding: an assessment of Malaysian primary pupils' number sense with respect to multiplication and division. *International Journal of Science and Mathematics education in Southeast Asia, 27*(2), 92-111.

Ghazali, Munirah, Rohana Alias, Noor Asrul Anuar Ariffin, & Ayminsyafora Ayub (2010). Identification of students' intuitive mental computational strategies for 1,2 and 3 digits addition and subtraction: Pedagogical and curricular implications. *International Journal of Science and Mathematics education in Southeast Asia, 33*(1), 17-38.

Neo, K. S. (1989). Pemahaman istilah matematik umum di peringkat menengah atas *[Understanding of general mathematical terms at the upper secondary level].* Unpublished M. Ed. Project Report. University of Malaya, Kuala Lumpur

Neo, K. S. (2011). *Code switching in primary mathematics classrooms.* Unpublished Ed.D. Dissertation, Universiti Sains Malaysia, Penang.

Ng, S. G. (1999). Kefahaman penambahan nombor perpuluhan bagi murid Tahun Empat *[Understanding of the addition of decimals by Primary School Year 4 pupils]*. Unpublished M.Ed. Project Report, University of Malaya, Kuala Lumpur.

Nik Azis Nik Pa. (1987). *Children's fractional schemes*. Unpublished Ed.D. Dissertation University of Georgia

Nik Aziz Nik Pa & Ng, S. N. (1992). *Research on mathematical learning and teaching in Malaysia.* Paper presented at the International Seminar: State of the Art of Research in Science and Mathematics Education in Southeast Asia and the Pacific, SEAMEO-RECSAM, Penang, Malaysia.

Noor Azina Ismail. (2009). Understanding the gap in Mathematics achievement of Malaysian students. *The Journal of Educational Research, 102*(5), 389-394.

Noor Azlan Ahmad Zanzali (1995) *Histroy of Mathematics Education In Malaysia: Coping With The Demands Of Curriculum Change*. Paper Presented At The International Seminar On Education And Change, University Of South Africa, Pretoria, September 1995.

Noor Fazilah Abetah. (2000). Penambahan pecahan oleh pelajar Tingkatan Satu di sebuah sekolah di Selangor *[Addition in fraction by Form One students of a school in Selangor]*. Unpublished M.Ed. Project Report, University of Malaya, Kuala Lumpur.

Nor'ain M. Tajudin, Rohani A. Tarmizi , Wan Z.W. Ali, Mohd.M.Konting (2007). Effects of using graphic calculators in teachong and leanning of mathematics. *Malaysian Journal of Mathematics Sciences*, 1(1): 45-61

Noraini Idris. (1990). *Skim penambahan nombor bulat bagi murid-murid Darjah Dua dan Tiga [Whole number addition scheme of Standard Two and Three pupils].* Unpublished M.Ed. Research Report, University of Malaya, Kuala Lumpur.

Noraini Idris. (2006). Exploring the effects of T1-84 plus on achievement and anxiety in mathematics. Eurasia Journal of Mathematics, Science and Technology Education, 2(3), 66-78.

Noraini Idris, Loh Sau Cheong, Norjoharuddeen Mohd. Nor, Ahmad Zabidi Abdul Razak and Rahimi Md. Saad (2007). The professional preparation of Malaysian teachers in the implementation of teaching and learning of mathematics and science in English. *Eurasia Journal of Mathematics, Science & Technology Education, 3*(2), 101-110.

Norazah Nordin, Effandi Zakaria Nik Rahimah Nik Mohamed Nik Rahimah Nik Mohamed, Mohamed Amin Embi. (2010). Pedagogical usability of the geometer's sketchpad (GSP) digital module in the Mathematics teaching. TOJET: The Turkish Online Journal of Educational Technology, 9(4), 113-117.

Norlia Abd. Aziz, T. Subahan M. Meerah, Lilia Halim and Kamisah Osman. (2006). Hubungan antara motivasi, gaya pembelajaran dengan pencapaian matematik tambahan pelajar Tingkatan 4. *Jurnal Pendidikan, 31*, 123-141.

Ong, C. K. (1994). *Utilization of television in teaching and learning of lower secondary mathematics*. Unpublished M.Ed. Thesis, University of Malaya, Kuala Lumpur.

Ong, E. G. (2010). *Changes in mathematics teachers' questioning techniques through lesson study process*. Unpublished Ph.D. Thesis, Universiti Sains Malaysia, Penang.

Ong, S. H. (2000). *Understanding of algebraic notation and its relationship with cognitive development among Form Four students*. Unpublished M.Ed. Research Report, University of Malaya, Kuala Lumpur.

Ong, S. L. and Tan, M. (2008). Mathematics and Science in English: Teachers Experience inside the classroom. *Jurnal Pendidik dan Pendidikan, 23*, 141-150.

Othman b. Sodikin. (1999). Konsepsi pendaraban nombor bulat bagi pelajar Tahun Tiga *[Year Three pupils' conceptions of multiplication of whole numbers]*. Unpublished M.Ed. Thesis, University of Malaya, Kuala Lumpur.

Palanisamy, K. V. (1986). *Cognitive development and acquisition of the mathematical concepts of fraction, ratio and proportion: A study of a sample of Malaysian urban secondary school pupils*. Unpublished M.Ed. Thesis, University of Malaya, Kuala Lumpur.

Palanisamy, S. (2007). A comparative study on e-learning for mathematics subjects in two Malaysian smart schools.

Pandian, A. and Ramiah, R.. (2003). Mathematics and Science in English: Teacher Voices. Paper presented at ELTC ETeMS Conference 2003: Managing Curricular Change, 2-4 Dec. 2003

Pang, S. T. (2000). Pengaruh bahasa dalam pencapaian matematik di sebuah sekolah persendirian Cina [*The influence of language in mathematics achievement in a Chinese independent school]*. Unpublished M.Ed. Research Report, University of Malaya, Kuala Lumpur.

Parmjit, S. (1994). Kesan Kekerapan pengujian terhadap pencapaian dan sikap dalam matematik bagi pelajar Tingkatan 4 *[Effects of assessment on achievement and attitudes towards mathematics of Form 4 students]*. Unpublished M.Ed. Project Report, University of Malaya, Kuala Lumpur.

Parmjit, S. (2001). Understanding the concepts proportion and ratio constructed by two grade six students. Educational Studies in Mathematics, 43(3), 271 – 292.

Parmjit, S. (2006). An analysis of word problems in school mathematics texts: operation of addition and subtraction. *Journal of Science and Mathematics Education in South East Asia, 29*(1), 41-61.

Parmjit, S. (2009). An assessment of number sense among secondary school students. International Journal for Mathematics Teaching and Learning. Oct 2009, 1 – 29 (ISSN 1473 – 0111). - http://www.cimt.plymouth.ac.uk/journal/singh.pdf

Parmjit, S. (2010). Languages and Mathematics achievement among rural and urban primary four pupils: A Malaysian Experience. *Journal of Science and Mathematics Education in Southeast Asia, 33*(1), p. 65-85.

Patton, M. Q. (1990). *Qualitative evaluation and research methods*. Thousand Oaks, CA: Sage.

Pusparani, S. (1998). Keperluan beberapa orang guru dalam pengajaran matematik KBSR di sebuah sekolah rendah Tamil di Sg Besi *[The needs of mathematics teachers in teaching mathematics in Tamil primary school]*. Unpublished M.Ed. Project Report, University of Malaya, Kuala Lumpur.

Rahimi Md Saad, Noraini Idris, Loh, S. C., Ahmad Zabidi Abdul Razak, Norjoharuddeen Mohd Nor. (2006). Penilaian guru terhadap koswer Matematik dan Sans dalam Bahasa Inggeris tingkatan satu [Teachers' evaluation of Form

One Mathematics and Science coursewares in English]. *Jurnal Pendidikan 2006, Universiti Malaya*, 93-106.

Rohani Ahmad Tarmizi, Ahmad Fauzi Mohd Ayub, Kamariah Abu Bakar and Aida Suraya Yunus. (2010). Effects of technology enchance teaching on performance and cognitive load in calculus. *International Journal of Education and Information Technologies, 4*(2), 109-120.

Rosihan M. Ali and Ahmad Izani Md. Ismail. (1998). An undergraduate technology-based applied mathematics program at Universiti Sains Malaysia. Proceedings of the Third Asian Technology Conference in Mathematics (ATCM ' 98), August 24-28, Tsukuba, Japan, 75-81.

Roslina Radzali, T. Subahan Mohd Meerah and Effandi Zakaria. (2010). Hubungan antara kepercayaan Matematik, metakognisi dan perwakilan masalah dengan kejayaan penyelesaian masalah Matematik [The relationship between mathematical beliefs, metacognition and problem representation with mathematical probel solving performance]. *Jurnal Pendidikan Malaysia, 35*(2), 1-7.

Rosnaini Mahmud, Mohd Arif Hj. Ismail and Lim, A. K. (2009). Development and Evaluation of a CAI courseware 'G-Reflect' on Students' Achivement and Motivation in Learning Mathematics. *European Journal of Social Sciences, 8*(4), 557-568.

Saipunidzam Mahamad, Mohd Noor Ibrahim and Shakirah Mohd Taib. (2010). M-Learning: A new paradigm of learning mathematics in Malaysia. *International Journal of Computer Science & Information Technology (UCSIT), 2*(4), 76-81.

Sarojini, D. A. (1989). *van Hiele levels of geometric thought in relation to similarity and congruence in triangles.* Unpublished M.Ed. Thesis, University of Malaya, Kuala Lumpur.

Selva, R. S. and Loh, S. C. (2008). Emotional intelligence of science and mathematics teachers: A Malaysian experience. *Journal of Science and Mathematics Education in Southeast Asia, 31*(2), 132-163.

Seow, S. H. (1989). *Conceptions of Mathematics teaching: Case studies of four trainee teachers*. Unpublished M.Ed. Thesis, University of Malaya, Kuala Lumpur.

Seow, S. H. (2001). Kajian tentang penggunaan pendekatan strategi konflik untuk mengatasi miskonsepsi dalam sistem nombor perpuluhan di sekolah rendah Malaysia *[A study of the use of conflict strategy approaches in overcoming misconception in decimal number system in Malaysian primary schools]*. Unpublished Ph.D. Thesis, Universiti Sains Malaysia, Penang.

Sevanesan, R. (1996) *Performance of Primary Six pupils in mathematical problem solving thinking skills.* Unpublished M.Ed. Research Report, University of Malaya, Kuala Lumpur.

Sharida Hashim. (1999). Kefahaman pelajar Tingkatan Empat tentang pendaraban dua matriks *[The understanding of a class of Form Four on the multiplication of two matrices]*. Unpublished M.Ed. Project Report, University of Malaya, Kuala Lumpur.

Sharifah Norul Akmar Syed Zamri, & Lee, S. E. (2000). *State-of-the-art review of research in mathematics education in the Faculty of Education, University of Malaya reported up to 2000.* Kuala Lumpur: Faculty of Education, University of Malaya.

Sharifah Norul Akmar Syed Zamri. (1997). *Skim penolakan integer pelajar Tingkatan Dua [Integer subtraction schemes of Form Two students]*. Unpublished Ph.D. Thesis, University of Malaya, Kuala Lumpur.

Sharifah Norul Akmar, S. Z. (2002). Incorporating values in the teaching of mathematics in Malaysian primary school. Paper presented at the Invitational Conference on Values in Mathematics and Science Education, Monash University, Australia, 2 – 5 October.

Sinnadurai, W. (1993). *Conceptions, perceptions and practices of year three mathematics teachers with respect to mathematically gifted pupils*. Unpublished M.Ed. Project Report, University of Malaya, Kuala Lumpur.

Siong, S. T. (1979). *Selected behavior-characteristic of perspective mathematics teachers and their relationship to teaching practice performance: A study of Malaysian Teachers' College Penang*. Unpublished M. Ed. Thesis, University of Malaya, Kuala Lumpur.

Siow, C. F. (1993). *Achievement of Form Five students in selected aspects of logical reasoning in mathematics.* Unpublished M.Ed. Project Report, , University of Malaya, Kuala Lumpur.

Siti Aishah bt. Sheikh Abdullah. (2006). Skim fungsi gubahan di kalangan pelajar Tingkatan Empat *[Composite function schemes of Form Four Students]*. Unpublished Ph.D. Thesis, Universiti Sains Malaysia, Penang.

Smart School Project Team. (1997). *Smart School flagship application: The Malaysian Smart School – A conceptual blueprint*. Kuala Lumpur: Ministry of Education.

Su, D. K. and Hong, K. S. (2006). WPE: A mathematical microworld for learning probability. Journal of Science and Mathematics Education in Southeast Asia, 29(2), 43-68.

Sufean Hussin. (1986). *Solving open-sentences in multiplication and division: Factors related to performance of Primary III and IV pupils.* Unpublished M.Ed. Thesis, University of Malaya, Kuala Lumpur.

Supiah Saad. (1995). Proses membentuk dan melaksanakan rancangan pengajaran individu dalam pengajaran pemulihan matematik: Satu kajian kes *[The process of planning and implementing individual lesson plan in teaching remedial mathematics: A case study]*. Unpublished M.Ed. Project Report, University of Malaya, Kuala Lumpur.

Sutriyono. (1997). *Skim penolakan nombor bulat Darjah Dua dan Tiga [Whole number subtraction schemes of Standard Two and Three pupils]*. Unpublished Ph.D. Thesis, University of Malaya, Kuala Lumpur.

Swetz, F. J. (1983). Attitudes toward mathematics and school learning in Malaysia and Indonesia: Urban-rural and male-female dichotomies. *Comparative Education Review,* 27(3), 394-402.

Tan, B. C. (2004). *An exploratory study to develop a mathematical creativity rating scale for secondary school students*. Unpublished Master Thesis, Universiti Sains Malaysia, Penang.

Tan, C. T. (1995). Pengajaran Matematik KBSR: Perbandingan dua buah sekolah rendah di Daerah Petaling, Selangor *[KBSR Mathematics teaching: A comparison of two primary schools in Petaling District of Selangor]* Unpublished M.Ed. Thesis, University of Malaya, Kuala Lumpur

Tan, F. B. (2010). Jenis kesilapan murid Tahun 5 SJKC semasa menyelesaikan masalah matematik berayat. [*Types of errors made by SJKC Year 5 pupils when solving mathematical word problems*]. Unpublished M.Ed. Thesis, Open University Malaysia, Kuala Lumpur.

Tan, H. K. (1986). *The use of hints to solve problems in mathematics*. Unpublished M. Ed. Thesis, University of Malaya, Kuala Lumpur.

Tan, H. L. (1984). *The mathematical needs of Sijil Pelajaran Malaysia level Physics.* Unpublished M.Ed. Thesis, University of Malaya, Kuala Lumpur.

Tan, M. and Ong, S. L. (2011). Teaching Mathematics and Science in English in Malaysian Classrooms: The impact of teacher beliefs on classroom practices and student learning. *Journal of English for Academic Purposes, 10*(1), 5-18.

Tan, Y. S. (2007). Attitudes and achievement orientations of students towards learning of science and mathematics in English. *Kajian Malaysia, XXV* (1), 15-39.

Tay, B. L. (2003). *A van Hiele-based instruction and its impact on the geometry achievement of Form One students*. Unpublished Master's Dissertation, University of Malaya, Kuala Lumpur.

Tay, P. K. (1994). *Reading for learning in mathematics: The classification of Lower Secondary KBSM text and the used of directed activities related to text.* Unpublished M.Ed. Thesis, University of Malaya, Kuala Lumpur.

Teng, P. K. (1997). *Analysis of errors of Year Six pupils in solving arithmetic word problems.* Unpublished M.Ed. Research Report, University of Malaya, Kuala Lumpur.

Teng, S. L. (2002). Konsepsi alternatif dalam persamaan linear di kalangan pelajar Tingkatan 4 *[Alternative conception of linear equation among Form 4 students].* Unpublished M.Ed. Thesis, Universiti Sains Malaysia, Penang.

Tengku Zawawi Tengku Zainal. (2009). Pengetahuan pedagogi isi kandungan guru matematik bagi tajuk pecahan: kajian kes di sekolah rendah.

Teoh, B. T. and Fong, S. F. (2005). The effects of Geometer's sketch pad and graphic calculator in the Malaysian mathematics classroom. *Malaysian Online Journal of Instructional Technology*, 2(2), 82-96.

Teoh, S. H. (2006). *The effectiveness of three learning strategies in the teaching of matrices among Malaysian secondary students*. Unpublished Ph.D. Thesis, Universiti Sains Malaysia, Penang.

Teoh, S. H. (2010). Effect of an interactive courseware in the learning of matrices.

Teoh, S. H. (2010). Extracting factors for students' motivation in studying mathamtics.

Tham, Y. M. (1995). *Development and evaluation of a computer-assisted instruction (CAI) package on Introductory Statistics*. Unpublished M.Ed. Project Report, University of Malaya, Kuala Lumpur.

Thayalarani, S. (1998). Kefahaman tiga orang pelajar Tingkatan Tiga tentang Ungkapan Algebra *[The understanding of three Form III students on algebraic expressions].* Unpublished M.Ed. Project Report, University of Malaya, Kuala Lumpur.

Thiruchelvam, K.. (2002). Kesan strategi penyelesaian masalah algo-heuristik terhadap prestasi matematik sekolah menengah rendah *[The impact of algo-heu-*

ristic strategy of problem solving on lower secondary school students performance]. Unpublished M.Ed. Thesis, Universiti Sains Malaysia, Penang.

Velayudhan, D. (1985). *Understanding of number operations among Malaysian Standard Six children.* Unpublished M. Ed. Project Report, University of Malaya, Kuala Lumpur.

Wan Fatanah. (1998). Konsepsi Pelajar Tahun Satu mengenai beberapa stilah matematik *[Year One pupil's conception of selected mathematical terms].* Unpublished M.Ed. Research Report, University of Malaya, Kuala Lumpur.

Wan Zah, W. A., Sharifah, K.S.H., Habsah, I., Ramlah, H., Rofa, I., Majid, K., & Rohani, A.T. (2005). Teacher's understanding of values in mathematic. *Jurnal Teknologi Malaysia, 43*(E), 45-62.

Wong, A. B. (1994). Keperluan Matematik dalam Ekonomi Tingkatan Enam dan kesilapan pelajar dalam menyelesaikan masalah kuantitatif *[Mathematical needs of Form Six Economics and students' errors in solving quantitative problems].* Unpublished M.Ed. Research Report, University of Malaya, Kuala Lumpur.

Wong, M. L. (2000). *Analysis of computational in whole number multiplication of Year Three pupils.* Unpublished M.Ed. Project Report, University of Malaya, Kuala Lumpur.

Wong, O. C. (2000). *Peer assessment of a mathematics group project of a pre-university programme.* Unpublished M.Ed. Project Report, University of Malaya, Kuala Lumpur.

Wong, S. E. (2000). *Relationship between achievement in fraction and mathematics anxiety of Form Two students.* Unpublished M.Ed. Project Report, University of Malaya, Kuala Lumpur.

Wong, S. M. (1975). *Comparing four strategies in teaching selected topics in modern mathematics.* Unpublished M.Ed. Thesis, University of Malaya, Kuala Lumpur.

Wun, T. Y. (1999). Kefahaman tiga orang guru tentang penakulan mantik *[The understanding of logical reasoning among three teachers].* Unpublished M.Ed. Project Report, University of Malaya, Kuala Lumpur.

Yee, C. T. (1994). Kesan Pembelajaran Koperatif terhadap pencapaian matematik dari segi akademik dan kemahiran penyelesaian masalah di kalangan pelajar Tingkat Empat *[Impact of cooperative learning in mathematics achievement of Form IV students with respect to both academic and problem solving skills].* Unpublished M.Ed Thesis, Universiti Sains Malaysia, Penang.

Yudariah Mohammad Yusof. (1999). Changing attitudes to university mathematics through problem solving.

Zaini, Z.H (2010). A study on students' motivation in learning mathematics using multimedia courseware.

Zulkifli Zur'aini Dahlan, Noraini Shafie and Rozeha A. Rashid. (2006). Multimedia prototype of bilingual model within technology based learning environment: An implementation of a Mathematics Learning Framework. In A. Iglesias and N. Takayama (Eds). *ICMS 2006, LNCS 4151*, pp. 320-330, Springer-Verlag Berlin Heidelberg

CHAPTER 49

RESEARCH STUDIES IN THE LEARNING AND UNDERSTANDING OF MATHEMATICS

A Malaysian Context

Parmjit Singh and Sian Hoon Teoh

ABSTRACT

This chapter provides a critical and comprehensive review of studies on mathematics learning and understanding that have been carried out in Malaysia. Four main themes were identified and discussed, namely: schemes of learning in mathematical concepts; understanding and conception of mathematical terms and knowledge; development of concepts in mathematics instruction; and performance and achievement in mathematics. This is followed by a detailed discussion of a study on multiplicative thinking in proportion and ratio. Finally, the implications and suggestions for future research conclude this chapter.

The First Sourcebook on Asian Research in Mathematics Education: China, Korea, Singapore, Japan, Malaysia, and India, pp. 1111–1138

INTRODUCTION

The corpus of edification in Mathematics, specifically research in mathematics education, lies at the intellectual "crossroads of many well-established domains such as mathematics, psychology, sociology, epistemology, cognitive science, and it may be concerned with problems imported from those fields" (Sierpinska et al., 1993:276). However, within the Malaysian context, two disciplines have had seminal influence on research in mathematics education. The first is the discipline of mathematics that focuses on the content of mathematics which deals primarily with teaching and learning. The second is the discipline of psychology that centres mainly on how content is taught and learned (Nik Azis, 1989; Parmjit, 1999; Sharifah, 2003). Research in mathematics education, then refers to a regimented systematic inquiry into the teaching and learning of mathematics. This is based on our firm conviction that the ultimate goal of research in mathematics education is to improve teaching and learning of mathematics, especially in schools. Without research on mathematics topics, there would be little possibility that any substantive and systematic improvement could ever occur. Any success then would be the result of chance alone and even then, might fail to be recognized as such. Therefore, the fundamental question that one would ask of any research conducted would surely be whether it adds value to our understanding of mathematics teaching and learning.

FOCUSED AREA AND METHODOLOGY EMPLOYED IN THE LEARNING AND UNDERSTANDING OF MATHEMATICS

The contemporary analysis on research development in Malaysia shows that the focused area and methodology employed in research into the learning and understanding of mathematics is in line with the direction of the Malaysian mathematics curriculum that gears towards mathematical thinking. Researchers (e.g. Parmjit, 1999; Sharifah, 2003) emphasized that there were two main focused areas of research in mathematics. The first domain concerns with the discipline of mathematics itself which focuses on the content of mathematics in learning and understanding, such as mathematical thinking. The second focused area comprises the discipline of psychology which centres on how the contents are taught and learnt, such as teacher knowledge and practice on certain concepts of mathematics. Presently, we see a growing body of research intellectual capacity about the nature of teaching towards students' understanding in learning mathematics. For example, Nik Azis Nik Pa (1989) in his seminal work stressed a need for more studies to focus on the learning and under-

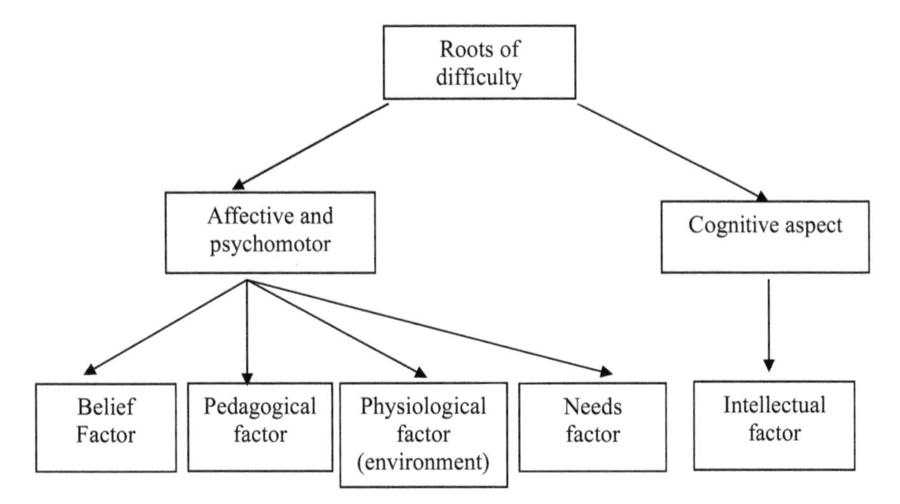

Figure 49.1. Difficulties in mathematics learning and understanding.

standing of mathematics. In further analyzing scopes of research, he distinctly divided the basic reasons causing difficulty in mathematics learning and understanding into two tenable categories of the Cognitive Aspects as well as the Affective & Psychomotor Aspects (Figure 49.1).

In further elaborating the temperament of these major fields of research, the affective component basically emphasizes the emotional aspects focused on assessing students' attitudes and beliefs regarding mathematics. On the other hand, the cognitive aspect deals with the mental processes required to acquire knowledge and are related to conscious intellectual activities such as thinking, reasoning, judgment and intuition. Despite the fact that the outcomes of the research done in these areas have been disseminated among educators, there has been little concerted effort made to rectify or remedy those aspects in the learning and understanding of mathematics that are lacking with regards to the promotion of students' mathematical thinking.

In the past quarter of a century, the school mathematics curriculum in Malaysia has undergone several reformations, in tandem with the global movement that has significantly impacted the research climate related to the learning and understanding of mathematics. Much of this culture is reflected in trends of how research is conducted, the methodology employed, types of research and questions being pursued. This trend of research has begun to evolve over the past decade from a preoccupation with the quantitative approach to a pervasive use of the qualitative approach. Courses in statistical analysis were previously the primary focus on data analysis methodology. However, during the last decade with the influence of western research culture, it has become the contemporary trend to use qualitative data analysis in researches conducted within the schools of education in Malaysia. More intriguingly is the growing popularity of a certain creed in the field of research in mathematics education that has slowly embraced this shift in methodology with some eclectic focus on both approaches, though not at the expected rate, with case studies of students and teachers, and ethnographic studies of students in the classroom with a focus on learning practice and classroom discourse. Similarly potent is the trend of research in which the emphasis has now shifted towards interpreting the meaning of learning and understanding from the perspective of learners. The changes in current trends of research in mathematics education are evident even as the numbers of research titles are increasing in correlation with the increase in the number of postgraduates conducting research. Consequently, despite the statements made, we cannot refute the fact that both approaches generate knowledge although the specific strand of its form, nature and usability may vary. Quantitative research on one hand generates knowledge that focuses on outcomes, generalizations, predictions, and causal explanations (Patton 1990). On the other hand, qualitative research generates knowledge that emphasizes process, extrapolation, understanding and illumination. In either case, the dominant feature remains: ultimately, knowledge is generated thus creating a significant bond between the two approaches.

Documentation of mathematics education studies in Malaysia had been carried out in the past by researchers such as Lai and Loo (1992) as well as Lee and Sharifah (2005) who have contributed to the local trend and development of types of studies. Alas, these studies are not free from certain setbacks. The study by Lai and Loo (1992) only compiled a limited number of studies on mathematics education in the country while Lee and Sharifah (2005) compiled and reviewed research studies on mathematics education that focused only on one education institution. Herein lies two difficult problems. First, the findings of these studies are rich but they are also scattered (Nik Azis & Ng, 1992) not utilizing a certain expectation of validity. The second setback is that these findings are—to a certain extent – confined to the research community and have not been channeled and made accessible to the community at large. Deductive reasoning beckons that this improper distribution and disclosure of information to the community, in a way, has curtailed efforts in the improvement of mathematics learning and teaching in the classrooms since it is well-established that informing practice and the construction of theory are the basic aims of mathematics education research. Subsequently, findings from these studies can enlighten and enrich educators' experiences to fulfill their educational objectives (Begle & Gibb, 1980; Dylan, 1998; Kilpatrick, 1993).

The implication then is that the attempt to document such reviews will provide invaluable information to teachers, parents, policy makers, research students, and researchers on determining potential areas and future directions of research in mathematics education. The outcomes of these studies can thus assist in identifying factors contributing to effective mathematics instruction that can serve as the basis for planning and initiating steps to encourage students' learning and understanding of mathematics. Therefore, documentation of studies done is important in terms of providing a real depiction of the overall pattern of research which will eventually provide direction for future studies especially in Malaysia.

Based on the discussion of the roots of difficulties in learning mathematics, the focused studies areas carried out in Malaysia, and to provide a more comprehensive overview, the following section reviews the major studies in learning and understanding of mathematics undertaken in Malaysia in the past decades. The discussion of these studies is based on major research done in Malaysia and is limited to the learning and understanding of mathematics, focusing on mathematical concepts. The findings of these reviewed studies will generally paint a picture of the trends and accomplishments of the Malaysian mathematics education research community in mathematics learning and understanding.

Schemes of Learning in Mathematical Concepts

Studies on the schemes of mathematics concepts have been thoroughly explored since its inception in the 1990s (with the most number being conducted between 1995 to 1998. Table 49.1 illustrates the topics of studies). Among the topics studied were whole numbers (Noraini Idris, 1990; Hasnul Hadi b. Abdullah Sani, 1992; Sharifah Norul Akmar bt Syed Zamri, 1997; Sutriyono, 1997), decimals (Aida Suraaya, 1996); ratios and proportions (Palanisamy, 1988; Parmjit, 1998), problem solving and mathematical thinking (Fatimah Salleh, 1997) and integer subtraction (Sharifah Norul Akmar, 1997).

The importance of learning and mastering the fundamental and crucial concepts of the topics outlined above is evident from the trend of research done. Researchers claim that knowledge of fractions, decimals, ratios/proportions and algebra form the very fabric of fundamentals in learning problem solving or number sense (Norton, 2005). More amusing is the fact that, this claim seems to correlate with the growing number of studies which were completed concurrently with problem solving and mathematical thinking. Presently, more topics, such as algebra, are being explored by researchers (Gan & Munirah, 2007) to further investigate conceptual understanding and schemes in the learning of mathematics. Simultaneously and rightly so, cognition and thinking skills have also been emphasized (Parmjit, 2005). Of late, the studies on these vital topics by some Malaysian scholars (e.g. Aida Suraya, 1996; Fatimah Salleh, 1997; Sharifah Norul Akmar, 1997; Sutriyono, 1997; Parmjit, 1998) were done as part of their doctoral studies. However, very few studies seem to focus on schemes of learning mathematical concepts, which we believe is the cornerstone for improving instruction in the learning of mathematics. These types of studies are of paramount importance because students' achievement and understanding are significantly improved when teachers: i) become aware of how students construct knowledge; ii) familiar with the intuitive solution methods that students use when they solve problems; and iii) utilize this knowledge when planning and conducting instruction in mathematics. The trend seems to point to the imminent rethinking and reevaluation of the role as the teachers as the *Omniscient Spoon-feeder* to that of a *Constantly Learning Facilitator.* The ranges of mathematical domains covered by these studies are quite vast and thus it is important to catalogue them and comprehensively notice them. Presented below is such tabulated documentation – with reference to all the studies cited, and the samples comprised primary school and secondary school students – on the studies that have been engaged in elucidating educational endeavor in studying mathematics learning and understanding of 'Schemes of mathematical concepts':

Understanding and Conception of Mathematical Terms and Knowledge

Mental comprehension and conception is not a simplistic area that could be defined in linear reasoning. It can however, be perceived in certain areas of focus. These areas of focus can be referred to Sierpinska's (1994) inference in her book on understanding mathematics which is: "how to teach so that students understand? What exactly don't they understand? What do they understand and how?"

Unlike the earlier days on mathematical researches, studies on understanding mathematical knowledge have always been emphasized by Malaysian researchers. Many researchers believe that the abundant inputs and discoveries of mathematical knowledge are needed to promote the understanding of mathematical knowledge (Ball & Bass, 2000). Along with the attention to develop a desirable framework for learning and understanding of mathematical knowledge, a few researchers have studied the mathematical knowledge needed in the development of numeracy

Table 49.1. Review of Studies in Mathematics Learning and Understanding of 'Schemes of Mathematical Concepts'

Schemes of Mathematical Concepts in:		*Author(s)*	*Year*	*Samples*
Learning of Whole Numbers	Addition of whole number schemes of Primary Two and Three pupils	Noraini bt ldris	1990	Lower Primary pupils
Whole number schemes	Addition of integer schemes of Form Two students	Hasnul Hadi b. Abdullah Sani	1992	Lower Secondary pupils
Integer	Integer subtraction schemes of Form Two students	Sharifah Norul Akmar bt Syed Zamri	1997	Lower Secondary pupils
Whole number	Whole number subtraction schemes of Standard Two and Three students	Sutriyono	1997	Lower Primary pupils
Learning of Fractions and Decimals	Decimal schemes of primary school year five students	Aida Suraya bt Hj Mohd Yunus	1996	Upper Primary pupils
Fraction	An Investigation Of Fraction Sense Among Form One Students: A Qualitative Study	Tee Sean Sean & Md Nor Bakar	2005	Form One students
Learning of Ratio and Proportion	Mathematical concepts of Fraction, Ratio and Proportion	Palanisamy, Veloo.	1988	Secondary school students
Learning of Ratio and Proportion	Understanding the concepts of ratio and proportion among primary six students	Parmjit Singh	1998	Upper Primary pupils
Learning of Ratio and Proportion	Understanding the concepts of proportion and ratio constructed by two grade six students	Parmjit Singh	2001	Upper primary pupils
Learning of Algebra	Solution Strategies, Modes of Representation and Justifications of Primary Five Pupils in Solving Pre Algebra Problems: An Experience of Using Task-Based Interview and Verbal Protocol Analysis	Gan We Ling & Munirah Ghazali	2007	Primary Five pupils
Problem solving	Problem solving schemes of Form Two Mathematics teachers	Fatimah Salleh	1997	Secondary school teachers
Multiplicative thinking	Multiplicative thinking in children's learning at early	Parmjit Singh	2005	Primary school pupils

(Teoh, 1991), basic number concepts (Yoong, Santhiran, Fatimah, Lim, Munirah, 1997), fractions (Amy Chin, & Pumadevi, 2010) and other aspects.

Teoh's (1991) study for example, identified errors and the corresponding cognitive activities undertaken by the Years 3 and 6 primary school students, among the different cultural groups, in solving word problems for the topic of numeracy. The design of the study was quantitative in nature using Newman Error Analysis method in determining the types of errors made by the students. The findings showed a large portion of the errors were due to lack of understanding, followed by transformation skills process skills, encoding and reading. This study identified the types of errors students made due to the nature of mathematical knowledge conception but stops short in proposing the types of activities needed to overcome these errors. Meanwhile Yoong, Santhiran, Fatimah, Lim, Munirah (1997) in their quantitative exploratory study on the other hand recognized cross-cultural differences in learning and understanding of number concepts. Their findings indicate that differences in the nature of mathematics knowledge in cross cultural contexts and language seem to play a major role in students' acquisition of number concepts. Can we say then that cultural differences are a significant factor in determining mathematics achievement? If so, what are the factors within the cultural contexts that give rise to these differences? Perhaps as Yong and colleagues (1997) posit "this is due in part to the reason that cultural influences are difficult to study as much of it appear tacit, implicit and hidden from awareness" (p. 6). However, we believe that there are a limited number of exploratory studies that are yet to fill this gap.

Consequently, the problem of understanding is closely linked to how the nature of mathematical knowledge is conceived. The lack of content knowledge has been stressed by a number of researchers (Aida Suraya, 2006; Munirah Ghazali & Noor Azlan Ahmad Zanzali, 1999; Parmjit, 2003c; Parmjit , Singh & Allan, 2006; Parmjit, Rosmawati & Rusyah, 2006). Findings by Parmjit, et. al. (2006) revealed that although students obtained "A" grades in national examinations, most of them still had very poor conceptual knowledge and understanding of concepts such as ratios and proportions. This clearly indicates that the majority of students who attained good mathematics results were only well versed in computational skills. They become confident and competent in procedural knowledge of computation through 'practice' (Munirah Ghazali & Noor Azlan Ahmad Zanzali, 1999; Parmjit, 2003c; Parmjit , Singh & Allan, 2006). In further lending support to this point, Effandi Zakaria & Norliza Zaini's research (2009) also highlighted the same outcome. They inferred that trainee teachers were too dependent on algorithms and memorization of formulas, tips and rules but providing explanations or justifications on how to obtain a particular answer posed a problem for them. The question we need to ponder would be "what type of understanding are we focusing on: *instrumental knowledge or relational knowledge as proposed by Skemp (1978)*? It has been found that successful mathematics students do indeed construct a fairly large number and variety of algorithms in order to continue to achieve good results in mathematics examinations. However, what is the quality of this mathematical knowledge? The research studies cited seem to focus purely on instrumental knowledge as a means of understanding which can be generally measured by quantitative methods. It is our belief that we have to move forward and use qualitative methods in measuring students' understanding of relational knowledge in the learning and understanding of mathematics. All the studies cited here only involved primary school pupils as the samples of study and the research design was mainly quantitative in nature.

Mathematics Achievement

There has been an increase of late, in the number of research studies conducted on mathematics achievement on students learning (Table 49.2). Research on students' achievement is important as it is seen as the first step in identifying students' difficulties in learning. Data from these studies reveal that most of the mathematics class time is spent practicing routine procedures, whilst the remainder of the time is generally spent applying procedures in new situations. The dominant pattern seems to be teach "how to" but never "why is it like that?".

The issue of lack of content knowledge has been stressed by various researchers (Aida Suraya, 2006; Noor Azlan Ahmad Zanzali & Lui, 2002; Parmjit, 2003c; Parmjit , Singh & Allan, 2006; Parmjit, Rosmawati & Rusyah, 2006). For example, Parmjit, Rosmawati & Rusyah (2006) in their number sense study discovered that although students obtained "A" grades in national examinations, most of them had very poor conceptual knowledge and understanding of concepts such as ratios and proportions. This clearly shows that students who obtained good mathematics results were only "good" in computational skills. They became confident and competent in procedural knowledge of computation through 'practice' (Munirah Ghazali & Noor Azlan Ahmad Zanzali, 1999; Parmjit, 2003c; Parmjit, 2006; Parmjit, 2009; Parmjit , Singh & Allan, 2006). Apart from this, mathematics achievement studies were also carried out for the purpose of making comparisons between schools with different cultural backgrounds (Lim & Chan, 1993).

Recently, the trend seems to indicate that there are also plenty of studies examining the external existential effects of teaching-aids, and meth-

Table 49.2. Review of Studies in Mathematics Learning and Understanding of 'Mathematics Achievement'

Mathematics Achievement	*Author(s)*	*Year*	*Samples*	*Research Design*
• A case study comparing the learning of mathematics among Malay pupils in primary national schools and primary national type [Chinese] schools.	Lim, S. K., & Chan, T. B.	1993	Primary school students	Quantitative
• Relationship between achievement in fraction and mathematics anxiety of Form Two students	Wong Soo Eet	2000	Lower Secondary Students	Quantitative
• Evaluating the levels of problem solving skills of secondary school students.	Noor Azlan Ahmad Zanzali & Lui Lai Nam	2002		Quantitative
• Procedural orientation of school mathematics in Malaysia.	Parmjit, S.	2003	Lower Secondary Students	Quantitative
• Number sense and mental computation among secondary students in Selangor.	Parmjit, S., Rosmawati, A. H., and Rusyah, A. G.	2006	Secondary school students	Quantitative
• Ability in logical reasoning and problem solving characteristics of form two and form four students. 2000 – 2001	Aida Suraya Md. Yunus.	2006	Secondary school students	Quantitative
• Usage of graphing calculator Ti-83 Plus: Motivation and achievement	Noraini Idris	2006	Secondary Form Four students	Quantitative
• Effects of use of graphic calculators on performance in teaching and learning mathematics	Nor'ain M. Tajudin, Rohani A. Tarmizi , Wan Z.W. Ali, Mohd. M. Konting	2007	Secondary school students	Quantitative
• The effects of mathematics anxiety on matriculation students as related to motivation and achievement	Effandi Zakaria and Norazah Mohd Nordin	2008	Matriculation students	Quantitative
• Motivation in the learning of mathematics	Aida Suraya Md.Yunus Wan Zah Wan Ali	2009	University students	Quantitative
• Variation in first year college students' understanding on their conceptions of and approaches to solving mathematical problems	Parmjit, S.	2009	University students	Quantitative
• Unpacking first year university students' mathematical content knowledge through problem solving.	Parmjit, S., and Allan White	2006	University students	Quantitative
• Development and evaluation of a CAI courseware 'G-Reflect' on students' achievement and motivation in learning mathematics	Rosnaini Mahmud, Nohd Arif hj Ismail, Lim Ai Kiaw	2009	Lower secondary school students	Quantitative
• The effects of cooperative learning on students' mathematics achievement and attitude towards mathematics	Effandi Zakaria, Lu Chung Chin and Md. Yusoff Daud	2010	Lower Secondary school students	Quantitative

ods or learning factors on students' achievements in mathematics (Noraini, 2006; Nor'ain, Rohani, Wan Zah & Konting, 2007; Effandi & Norazah, 2008; Aida Suraya & Wan Zah, 2009; Rosnaini, Mohd Arif, & Lim, 2009; Effandi, Lu & Md. Yusoff, 2010). Another popular area of research that has drawn considerable interest relates to the impact of technology such as the usage of graphic calculators on student learning. Numerous studies (Noraini, 2006; Nor'ain et. al., 2007) have quite consistently shown that thoughtful use of calculators in mathematics classes improves students' mathematics achievement and attitudes towards mathematics. In particular, they found improvement in students' understanding of arithmetical concepts and problem-solving skills. It is noted that data analysis and interpretations of these studies were examined by means of quantitative instruments.

The following section presents a detailed discussion on multiplicative thinking schemes in proportion and ratio, a study related to the topic of discussion of this chapter.

The Study: The Construction of Multiplicative Thinking Schemes in Ratio and Proportion

Mathematics educators worldwide have often quipped that mathematics should be taught as a thinking activity. In order to make this paradigm shift, teachers need to adopt a conceptual approach as opposed to using traditional and routine computation. The latter involves teaching of rules and procedures rather than the actual learning of mathematics. The conceptual paradigm requires that teachers obtain information concerning students' thinking activities, their efforts at understanding and their conceptual difficulties. Understanding of mathematics involves far more than accurate computation and rote memorization (Munirah et. al., 2010; Yoong et al., 1997; Liew & Wan, 1991; Mack, 1990). A number of researchers (Palanisamy, 1988; Peck and Jencks, 1981) have reported that children were able to follow rules for operating on fundamental concepts such as fractions, additive and multiplicative thinking but were unable to understand why the rules worked. They recommended shifting emphasis in instruction from learning mere rules for operations to understanding concepts. It is believed that the development of conceptual understanding can be attained by getting children to actually think – in an empirical and logical manner – about mathematics and represent topics in ways other than procedures. Successful mathematics students are seen to construct a fairly large number and variety of algorithms according to the immediate needs and problems in order to continue to achieve good results in mathematics exams. Yet, we should ask: what is the quality of this knowledge?

There has been considerable research done in Malaysia identifying students' mathematical understanding and approaches to learning school mathematics (Munirah et. al., 2010; Yoong et al.,1997; Noraini & Norjoharuddeen, 2008; Parmjit, 2009; Nik Azis, 1987). Results from research conducted over the years reveal that there is a cause for alarm with regards to the quality of students' mathematical learning and understanding in schools. These studies have shown that students who scored well on standardized tests are often, unable to successfully use memorized facts and formulae in real-life application outside the classroom (Parmjit, 2000; Parmjit, 2009; Yager, 1991). For example, a study by Parmjit (2000) found that only a small percentage of students who did well in the Lower Secondary Examination (Penilaian Menengah Rendah (PMR) were able to solve complex proportional problems and the grades obtained in this exam were not indicative of their knowledge of ratios and proportions. He noted that:

> the more we focus on raising test scores, the more instruction is distorted and the less credible are the scores themselves. Rather than serving as accurate indicators of students' knowledge and performance, the tests become indicators of the amount of instructional time and attention paid to the narrow range of skills assessed. (p.107)

The key milestone here is that the quality of students' mathematics knowledge is vital. The crucial factor determining the quality of knowing is the quality of the students' experiences in constructing their knowledge. Very often, we consider only aspects of knowledge that focus attention upon the concepts identified, the generalization recalled, the problem solved, the theorem proven or the procedure extended. The emphasis is upon the mathematical result, rather than the process of construction of the idea. We believe that construction of mathematics ideas by a child is the most effective way to empower children mathematically. It helps to reduce a child's memory load, broaden understanding of and appreciation for mathematical concepts, provide connection across the discipline and reveal the beauty of mathematics. The contrast only being in the efforts to "learn" in which many students are forced to take refuge in memorization rather than understanding. Wittrock (1974) emphasized that learning with understanding is a generative process and Piaget (1973) asserted that to understand is to be able to invent.

One of the most fundamental mathematics structures that children develop early in life is that of addition with multiplication as discussed in the earlier section of this chapter. These two important concepts, additivity and multiplicativity, develops progressively throughout children's mathematics education years after the counting process. The former relates to addition and subtraction activities while the latter to multiplicative and division activities. Children in Malaysia are introduced to the learning of multiplication tables of two's, three's, four's and five's as early as in grade 2 (Kementerian Pendidikan Malaysia, 2003). Steffe in his study that investigated children using number sequences as the initial point argues that: "For a situation to be established as multiplicative, it is necessary to at least coordinate two composite units in such a way that one of the composite units is distributed over the elements of the other composite units" (1994:19). His key to children's meaningful dealings with multiplication is the ability to iterate abstract composite units. This involves taking a set as a countable unit while maintaining the unit nature of its element. Using Steffe's benchmark, Parmjit (2003a; 2003b) introduced the construct of iteration of number sequences as an early introduction in multiplicative learning. His qualitative analysis study suggests that multiplicative reasoning expressed by using iterative composite units can help children construct meaning for multiplication thinking in various multiplicative settings.

We will be using proportional reasoning problems as the settings to evaluate children's understanding because in the field of mathematics education, the concept of proportion is fundamentally important in schools and is widely used in real-world situations. Lesh, Post and Behr (1988) recognize proportional reasoning as the conceptual watershed which plays at the borderline that separates elementary from more advanced concepts. In fact, proportional reasoning is a capstone of elementary arithmetic, number and measurement concepts. Once understood, it becomes the culminating point wherein students view concepts as interwoven. Furthermore, to a large extent, proportional reasoning involves algebraic understanding, having to do with equivalence, variables and transformations. Beyond algebra, performing proportional reasoning tasks is a pre-requisite to almost all mathematical skills and goes beyond mechanical computation (Post, 1986). Accordingly, in order to understand how children think about these concepts, it is necessary to understand the schemes children use in attempting to solve a proportionality problem which builds on their mathematical understanding and thinking from additive, multiplicative to proportionality. These analyses of schemes will provide a framework for understanding the strategies that they use to solve proportion problems. By analysing their schemes in learning, we can detail the evolution of children's strategies for solving problems in the domains of additive and multiplicative reasoning, as the focus of this section. The write up reported here is not drawing conclusions on how to integrate multiplicative schemes into children's learning. In fact, it is intended to be conjectural, that is, to outline an approach to instructions based on a broad definition of mathematical knowledge that seems to be worth trying.

For the purpose of this chapter, the selected data and analysis is based on a longitudinal study. The discussions presented here is the voice of a single child aged 9 whom we will symbolize as S. The episodes below give a descriptive analysis of the meanings that S gave to the tasks and the conceptual advances she made as she dealt with proportion and ratio tasks over a one and a half year period using clinical interview protocols. All the clinical interviews were conducted at her home and the data below describes her development in multiplicative thinking schemes in giving meaning to the ratio and proportion tasks.

R(Researcher)
S (Interviewee)

Episode 1 (Middle of Primary 2)

R: If 2 students share 3 pizzas equally, how many pizzas should 10 students share?
(After a few minutes working on her working sheet)

S: 15

R: Can you explain?

Pointing to her working paper, she explained the following steps:

$$\begin{array}{ccc} 2 & = & 3 \\ 2 & = & 3 \\ 2 & = & 3 \\ 2 & = & 3 \\ \underline{2} & = & \underline{3} \\ 10 & & 15 \end{array}$$

S: You see, two (students) will share three (pizza), another two (students) will share three, another two 2 (students) will get three and it goes on until I get for ten (students). That gives 15 pizzas.

From the ratio unit (simplest form of the ratio) of 2 to 3, she iterated it by counting (and adding) as a singleton unit rather than coordinate its distribution over the elements of the other composite units of 4 to 6, 6 to 9, 8 to 12 and 10 to 15.

Episode 2 (End of Primary 2)

R: How many two's are there in twelve?
S: 1, 2; 3, 4; 5, 6; 7, 8; 9, 10; 11, 12 ... there are six

Here, S was coordinating units of units as she keeps track of the paired counting acts with her fingers. She counts by two's to twelve (2, 4, 6, 8, 10, 12), keeping track of how many times she counted. This keeping track involves coordination of unit items at two levels called unit of unit, as shown below.

	Unit	Unit
1		1, 2
2	3, 4	
3		5, 6
4	7, 8	
5	9, 10	
6	11, 12	

Here, her fingers help her in the coordination iteration process but it seemed she iterated it additively rather than multiplicatively. This is because there was a pause in each finger for her to do the counting. At this stage of competency, her thinking was based on additive reasoning rather than multiplicative.

Episode 3 (Early Primary 3)

R: To make coffee, Jenny needs exactly 3 cups of water to make 4 small cups of coffee. How many cups of coffee can she make with 12 cups of water?
S: Can I use the paper?

After working on her worksheet:

S: 16 cups
R: Can you please explain?
S: (Showing her worksheet) 3 cups you make 4, 6 you make 8, nine make 12 and 12 makes 16.

3 water	–	4 coffee
6	–	8
9	–	12
12	–	16

She coordinated the two number sequences of 3, 6, 9, 12 with 4, 8, 12, 16. This shows her ability to coordinate two number sequences. (In the following episode, S has been introduced to the concept of multiplication in school)

The following question posed is similar to the one posed in Episode 3.

Episode 4 (Mid Primary 3)

R: Mariam needs exactly 3 cups of water to make 4 small cups of coffee. How many cups of coffee can she make with 12 cups of water?

After a while working on her worksheet, she responded:

S: 16 cups
R: How did you get that?

Showing me her worksheet, as appended below:

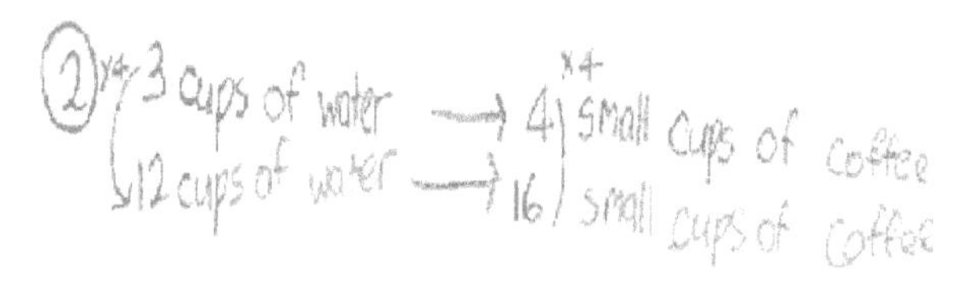

R: What does the arrow mean?
S: It means three cups of water make four cups of coffee.
R: I see
R: Why did you times four here (pointing at her worksheet)?
S: Because in 3, 6, 9, 12 there are four … so times four!
R: Good

S did not use the iteration process as in the previous episode and instead utilized the relationship as a scalar function. We believe that she is beginning to conceptualize the iteration action of the composite unit to make sense of multiplicative facts in a problem.

Episode 4 (Mid Primary 3)

R: Jenny needs exactly 4 eggs to bake 6 medium sized chocolate flavoured cakes. How many medium sized cakes can she bake with 10 eggs?
S: Can I use the paper?
R: Yes

After a while working on it quite intensely she responded:

S: It is difficult, ... I cannot get ... because four, eight, twelve ... but I want ten!

After a while working on her work sheet again, she responded:

S: 15 cakes

(the actual scanned work could not be reproduced here because of difficulties in downloading the scanned file).

In her work sheet, she wrote:

4 eggs --- 6 cakes	4 eggs --- 6 cakes		2 eggs --- 3 cakes	
10 eggs ---	8 eggs --- 12 cakes	x 5		x 5
	10 ---		10 eggs --- 15 cakes	
	12 cakes --- 18 cakes			

R: How did you get 2 eggs to 3 cakes (pointing at her worksheet)?
S: I cannot go from four to ten...(meaning she was unable to iterate from four to ten)
R: I see, but how did you get 2 eggs to 3 cakes?
S: You see in two, four there are two, in three six also there are also two ... so, I get two eggs for three cakes.

This reasoning that in two, four there are two times indicates that she divided the composite unit of 4 to 6 with 2 to get the ratio unit of 2 to 3. It shows she was able to unitize the composite unit from 4 to 6 to 2 to 3, and then iterate it as she had done previously.

This shows S's abstraction in decomposing the composite units into ratio units with her iteration scheme.

R: What do you multiply with five?
S: I times (multiply) five to get the answer.

R: Why?
S: Because to get for 10 eggs, you times 5 … then 3 times 5 gives 15 cakes of my favourite chocolate flavoured cake!

In these problems, S curtailed the iteration process by using known multiplication facts to aid in determining the total number of iterations. She then correctly multiplied the relevant composite units by that total. This curtailment required S to sufficiently abstract the iteration action so that she could reflect on it and anticipate that the result of several iterations could be captured by a known multiplication fact. It is quite obvious that she was beginning to represent the problem symbolically. This reflects S's ability to move from the iteration schemes to a more abstract level of understanding in multiplicative thinking. This level of competency from an early 3rd grade child represents Vergnaud's (1988) representations of isomorphism of measures to illustrate the structure of multiplicative problems. The data from the episodes above suggest that multiplicative reasoning expressed by using iterative composite units (coordination of number sequences) helped S construct meaning for multiplication thinking in proportionality settings. Her earliest ideas of multiplication grow out of the counting processes (episode 1 and episode 2), and the action of grouping and sharing. However, iterative process scheme (number sequences) seems to be the pre-knowledge schemes that led her to the counting, grouping and sharing processes in multiplicative settings.

As S moved from additive to multiplicative reasoning with whole numbers, there are two significant related changes. There are changes in what the numbers are and changes in what the numbers are about. In relation to this, Steffe (1988) traces children's construction of numbers from the construction of single entities with singleton units to coordination with composite units that signals the onset of multiplication. It is not a trivial shift, because it represents a change in what counts as a number. The ability to use operations with composites (units of units) from the data presented seemed to involve three essential components (Figure 1). First, children needed to explicitly conceptualize the iteration action of the composite unit to make sense of multiplicative problems (Level 1). Second, one needs to have sufficient understanding of the meaning of multiplication and division so that children can see their relevance in the iteration process (Level 2). Third, and finally, children need to have sufficiently abstracted the iteration process so that they can reflect on it. Then re-conceptualize it in terms of their knowledge of the multiplication and division operations (Level 3).

Example: If 2 students share 3 pizzas equally, how many pizzas should 8 students share? (see Figure 49.2).

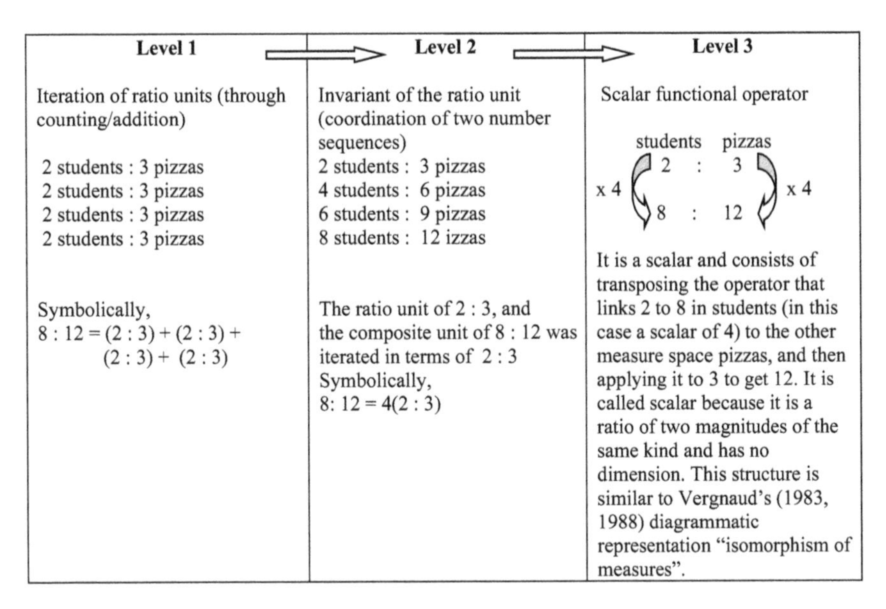

Level 1	Level 2	Level 3
Iteration of ratio units (through counting/addition) 2 students : 3 pizzas 2 students : 3 pizzas 2 students : 3 pizzas 2 students : 3 pizzas Symbolically, 8 : 12 = (2 : 3) + (2 : 3) + (2 : 3) + (2 : 3)	Invariant of the ratio unit (coordination of two number sequences) 2 students : 3 pizzas 4 students : 6 pizzas 6 students : 9 pizzas 8 students : 12 izzas The ratio unit of 2 : 3, and the composite unit of 8 : 12 was iterated in terms of 2 : 3 Symbolically, 8: 12 = 4(2 : 3)	Scalar functional operator students pizzas 2 : 3 x 4 x 4 8 : 12 It is a scalar and consists of transposing the operator that links 2 to 8 in students (in this case a scalar of 4) to the other measure space pizzas, and then applying it to 3 to get 12. It is called scalar because it is a ratio of two magnitudes of the same kind and has no dimension. This structure is similar to Vergnaud's (1983, 1988) diagrammatic representation "isomorphism of measures".

Figure 49.2.

Findings from earlier studies (Parmjit, 2000; Behr, et al., 1992; Lamon, 1994) indicates that children are reluctant (or unable) to use multiplication facts to solve proportional problems because they are unable to comprehend the multiplicative structure embedded in these problems. Compounding children's inability to utilize multiplicative structures is their failure to recognize the isomorphism between proportion problems (Tournaire, 1986). This is because the objective and sound understanding of a proportion is to keep the value of the ratio unit invariant under the iteration. It is conjectured that introducing children to derive the iterative strategies that emphasizes thinking patterns which will provide them with meaningful models for generating all of the multiplicative structures facts is of utmost importance. The main idea for generating the iterative units is thus to "think of the multiplication facts" embedded in proportional reasoning situations. In this context, it is significant to highlight that the data above indicates that S's conceptual competencies (understanding about the dimensions of multiplicative reasoning) were far greater than the symbolic competence of students in grades Six (Parmjit, 2001). Rich stores of informal knowledge as shown above and sense making tools invite the development of insightful methods for solving multiplicative problems. This is because at this level, prior to instruction, this child brings a great deal of knowledge into the classroom, and instruction needs to take this child's voice into account for meaningful learning in multiplicative thinking.

The question that comes to mind is, is elementary mathematics so simple that teaching it requires knowing only the "math facts" and a handful of algorithms? The premise of this discussion based on the data above is that, quite to the contrary, this early content is rich in important ideas. It is during their elementary years that young children begin to lay down these habits of reasoning upon which later achievement in mathematics will crucially depend. We must focus on how children learn and understand mathematics and base instructional decisions on this knowledge. It is crucial for the educational research community to identify bridging methods that will help children move from additive perspectives and methods to multiplicative perspectives and methods in ways that permit children to integrate all their knowledge in these two areas. Understanding effective implementations of such approaches will allow multiplicative domains to become widely accessible to all children in their learning.

SUMMARY OF FINDINGS

There has been a remarkable increase in research studies focused on mathematics learning and understanding over the past twenty years in the Malaysian context. These studies continue to span many different disciplines and differ in focus, theory, methodology, and supporting literature. However, when reviewed together as a single entity, they vividly present the difficulties students have in learning mathematics and the need to revise traditional methods of teaching. Numerous research studies as discussed above have shown that traditional methods of teaching mathematics are not only ineffective but also seriously inhibit the growth of students' mathematical learning and understanding (Noraini, 2002; Munirah Ghazali, Sharifah Norhaidah Idros & McIntosh, 2004; Parmjit, Rosmawati & Rusyah, 2006; Lim & Noraini Idris, 2006). For example, Munirah Ghazali, Sharifah Norhaidah Idros & McIntosh (2004) claimed that the learning of mathematics should be focused more on thinking and reasoning about mathematics rather than merely memorizing the algorithms mechanically. Data from their study reveal that children have difficulty understanding basic number concepts. This was reflected in the low scores in tests of number concepts, particularly those related to the effects of operations and multiple representations. The study further disclosed that children perform better on items that are written in computation format compared to the same item written in number sense format. Similarly, findings by Parmjit, Rosmawati & Rusyah, (2006) found that although students obtained a grade 'A' in national examinations, most of them still exhibited very poor conceptual knowledge and understanding of concepts such as ratio and proportion. Further findings by Aida Suraya (2006) highlighted that students also lacked skills in decision making in learning mathematics, particularly in implementing and verifying solutions. These skills involve the ability to carry out solution plans optimally, monitor its effects, troubleshoot if the solution is not effective, and self-reinforce if the outcome is satisfactory.

On the whole, these research findings indicate that to a certain extent in the Malaysian school climate, children's natural thinking seems to be gradually replaced by attempts at rote learning, resulting in disaster. Why is this so? This could be due to the following reasons: Firstly, the focus on computation is so myopic that few students develop any understanding at all of why the computations work or when they should be applied. Second, the grades obtained in the national examination for mathematics do not indicate their mathematical knowledge in conceptual understanding (Parmjit, 2000; Omar, Aziz. 2002; Zakaria, Mohd Johan. 2002; Parmjit et al., 2005) as the examination is largely void of contextual and conceptual elements. Third, for many children, school mathematics seems to be a sequence of memorizing and forgetting facts and procedures that make little sense to them (Noraini, 2003; Munirah Ghazali and Noor Azlan Ahmad Zanzali, 1999; Noraini, 2005). Though the same topics are taught and re-taught year after year, the children do not internalize them and seem to make the same mistakes year in and year out. What is the root of the difficulties that hinders or inhibits children's development in mathematics learning and understanding? The question remains to be explored further.

IMPLICATION AND SUGGESTION FOR FUTURE RESEARCH

This review of studies (in mathematics learning and understanding) indicates that the influx of different forms of research by the mathematics education community in Malaysia since the 1990's has had positive consequences on mathematics education, especially in the area of learning and understanding. The findings of the studies are also consistent with the literature. In brief, we found that most of the studies were conducted either with primary students, secondary students or pre-service teachers, but rarely with university students, kindergarten students, or teachers. Furthermore, most studies cited used the survey methodology which was quantitative in nature. Most published articles collected data using tests, followed by questionnaire, while only a few (comparatively) used interviews and observations. Theoretical material and instrument development, and review articles were, however, rarely cited in our local reviews.

The cognitive and affective domains were the most frequently studied. The following section details the implications of the findings of the research studies surveyed in this chapter.

First, we notice that studies of areas of "concepts in mathematics instruction" were rarely investigated in Malaysia and thus needs more attention. This is based on our firm belief that research activities must focus on how children learn and understand mathematics so as to inform instructional decisions. An abundance of research indicates that mathematics instruction that focuses on student cognition, problem solving, and personal sense-making - especially that which is guided by research on students' (or children's) construction of meaning for particular topics-produces powerful mathematical thinkers who not only will compute, but also possess strong mathematical conceptualizations and will eventually become skilled problem solvers. It is essential to examine and gain insight into how children learn mathematics, particularly in analyzing student representations, procedures, and pre-instructional knowledge in relation to fundamental concepts in mathematics learning.

Second, the review disclosed that most of the researchers of the studies cited were categorized as university instructors. Very often from the western perspective, research is a collaborative activity where practitioners work together to help one another design and carry out investigations in improving mathematics learning and understanding. However, from our reviews we found that the majority of studies reviewed in this chapter were conducted based on individual / group researchers from the same organization and most of the authors of the studies cited were categorized as university instructors with very little (if any) collaboration with fellow researchers from different organizations. In other words, these researches are largely remote. Furthermore, no studies were cited where school teachers participated as co-researchers. We believe that teachers can play a bigger role in improving their teaching and learning processes if they are roped in as co-researchers with fellow researchers from organizations that conduct research. This is because teachers are learning and relearning together with their students in the classroom and they are the ones who actually mould their students in the learning process. In view of this, it is of paramount importance to collaborate with teachers to help university-based researchers in order to fully comprehend what happens in a classroom. Furthermore, as these teachers (as researchers) grow to better understand their own practices, they will be better able to effect change in instruction to meet the needs of their students.

Third, research on mathematics education is subject to many changes; views on mathematics education are changing, and at the same time, research paradigms are shifting. In the community of mathematics education, the view of mathematics as a system of definition, rules, principles, and procedures that must be taught is being exchanged for the concept of mathematics as a process in which students must understand why numbers work in a certain system under certain conditions. The majority of the studies reported in this chapter seemed to be inclined towards quantitative methodology as opposed to the qualitative approach. No doubt, many thoughtful people are critical of the quality of research in mathematics education in Malaysia. They look at tables of statistical data and they say "so what!" They feel that vital questions go unanswered while means, standard deviation, and t-tests pile up. There is too great a reliance on statistics, while an in-depth look at process is avoided. Statistics are valuable in their place. They can suggest hypotheses in preliminary studies and help to test them in well-designed experimental studies. But if we want to understand what goes on in students' heads when they solve problems, we have to watch them solving problems (Schoenfeld, 1981, 1985). Thus, we strongly believe that a particular fruitful approach for research that focuses on learning and understanding is to concentrate on the students themselves, and the ways in which they individually construct knowledge. Mathematics education research should concentrate on intensive studies of individual students and the ways in which they construct viable mathematical knowledge. The research community in Malaysia is slowly but surely beginning to tilt towards this qualitative approach.

On a final note, the research community must assume the responsibility to improve the ways in which it communicates what has been learned, its context, and its significance, to the many audiences concerned with mathematics teaching and learning. No person is better situated than the researcher to explain the multiple facets of significance held by a piece of research.

REFERENCES

Aida Suraya Hj Md.Yunus & Wan Zah Wan Ali (2009). Motivation in the Learning of Mathematics. *European Journal of Social Sciences*, 7(4), 93-101.

Aida Suraya Hj Mohd Yunus (1996). *Skim nombor perpuluhan bagi murid Tahun Lima sekolah rendah*. Unpublished doctoral dissertation, Universiti Malaya.

Aida Suraya Md. Yunus.(2006). *Ability in Logical Reasoning and Problem Solving Characteristics of Form Two and Form Four Students*. A completed research at Universiti Putra Malaysia.

Amy Chin, Y.S. & Pumadevi, S.(2010). *Examples and conceptual understanding of proper fractions among primary school students.* Paper presented at Third International Conference on Science and Mathematics Education, Penang, Malaysia, 10-12 Nov 2009.

Ball, D.L., & Bass, H. (2000). Interweaving content and pedagogy in teaching and learning to teach: Knowing and using mathematics. In J. Boaler (Ed). *Multiple perspectives on the teaching and learning of mathematic,* pp. 83-104. Westport, CT: Ablex.

Begle, E. G., & Gibb, E. G. (1980). Why do research? In R. Shumway (Ed.), *Research in mathematics education* , pp. 3-19. Washington, DC: NCTM

Behr, M., Harel, G., Post, T., & Lesh, R. (1992). Rational number, ratio and proportion. In D. Grouws (Ed.), *Handbook on research of teaching and learning* , pp. 296-333. New York: McMillan.

Dylan, W. (1998) A framework for thinking about research in mathematics and science

education. In J. A. Malone, B. Atweh & J. R. Northfield (Eds.), *Research and supervision*

Effandi Zakaria & Norazah Mohd Nordin (2008). The effects of mathematics anxiety on Matriculation students as related to motivation and achievement. Eurasia *Journal of Mathematics, Science & Technology Education*, 2008, 4(1), 27-30.

Effandi Zakaria & Norliza Zaini. (2009). Conceptual and Procedural Knowledge of Rational Numbers in Trainee Teachers. *European Journal of Social Sciences,* 9 (2): 202-217

Effandi Zakaria & Normah Yusoff (2009). Attitudes and problem-solving skills in Algebra among Malaysian Matriculation college students. *European Journal of Social Sciences*, 8(2), 232-246

Effandi Zakaria, Lu Chung Chin & Md. Yusoff Daud (2010). The effects of cooperative learning on students' mathematics achievement and attitude towards mathematics. *Journal of Social Sciences* 6 (2): 272-275

Fatimah Salleh (1997). *Skim penyelesaian masalah guru matematik Tingkatan Dua. (Problem solving schemes of Form Two Mathematics teaches).* Unpublished doctoral dissertation, Universiti Malaya.

Gan We Ling & Munirah Ghazali (2007). Solution Strategies, Modes of Representation and Justifications of Primary Five Pupils in Solving Pre Algebra Problems: An Experience of Using Task-Based Interview and Verbal Protocol Analysis. *Journal of Science and Mathematics Education in Southeast Asia*, 30(1), 45-66.

Hasnu Hadi b. Abdullah Sani (1992). *Skim penambahan integer bagi pelajar-pelajar Tingkatan Dua.* Unpublished master dissertation, Universiti Malaya

Kementerian Pendidikan Malaysia (2003). *Huraian sukatan pelajaran matematik tahun 2.* Kurikulum Bersepadu Sekolah Rendah. Kuala Lumpur: Pusat Perkembangan Kurikulum.

Kilpatrick, J. (1993). Beyond face value: Assessing research in mathematics education. In G. Nissen & M. Blomhoj (Eds.,), *Criteria for scientific quality and relevance in the didactics of mathematics,* pp.15-34. Roskilde, Denmark: University of Roskilde.

Lai, K. H., & Loo, S. P. (1992). *State-of-the-art review of research in mathematics education in Malaysia.* South East Asian Research Review and Advisory Group. SEMEO, RECSAM. Penang, Malaysia.

Lamon, S. (1994). Ratio and proportion: Cognitive foundations in unitizing and norming. In H. Guershon & J. Confrey (Eds.), *The development of multiplicative reasoning in the learning of mathematics* (pp. 89-120). Albany, NY: State University of New York Press.

Lee Siew Eng & Sharifah Norul Akmar (2005). Studies of mathematics education at the faculty of education, University of Malaya: Focus and Direction. *Journal of Education* 2005, 7-28.

Lesh, R., Post, T., & Behr, M. (1988). Proportional reasoning. In J. Hiebert & M.Behr (Eds.), *Numbers concepts and operations in the middle grades*, pp. 93- 118. Reston, VA. : National Council of Teachers of Mathematics.

Liew Su Tim & Wan Muhamad Saridan (1991). Ke Arah Memahami Dan Mengurangkan Kesukaran Dalam Pembelajaran Matematik (Towards Understanding and Minimising Students' Diffciulties in Mathematics Learning). *Berita Matematik*. 38, 22 - 29.

Lim, S. K., & Chan, T. B. (1993). A case study comparing the learning of mathematics among Malay pupils in primary national school and primary national type (Chinese) schools. *Journal of Science and Mathematics Education in South East Asia, 16*(2), 49-53.

Mack, N. K. (1990). Learning fractions with understanding: Building on informal knowledge. *Journal for Research in Mathematics Education*, 21 (1), 16-32.

Munirah Ghazali & Noor Azlan Ahmad Zanzali (1999). Assessment of school children's number sense. *Proceedings of The International Conference on Mathematics Education into the 21st Century: Societal Challenges: Issues and Approaches*. Cairo, Egypt.

Munirah Ghazali, Rohana Alias, Noor Asrul Anuar Arifin & Ayminsyadora (2010). Identification of students intuitive mental computational strategies for 1, 2 and 3 digits addition and subtraction. Pedagogical and curricular implication. *Journal of Science and Mathematics Education in Southeast Asia*, 33(1), 17 – 38.

Munirah Ghazali, Sharifah Norhaidah Idros & McIntosh, A. (2004). From doing to understanding: an assessment of Malaysian Primary Pupils' number sense with respect to multiplication and division *Journal of Science and Mathematics Education in Southeast Asia*, 27(2), 92-111.

Nik Azis Nik Pa (1989). Satu persepsi tentang diagnosis dan pemulihan dalam Pendidikan matematik dan sains. *Berita Matematik* 13 : 91-105.

Nik Azis Nik Pa. (1987). *Children's fractional schemes*. Unpublished PhD dissertation, University of Georgia.

Nik Pa, N. A., & Ng, S. N. (1992). *Research on mathematical learning and teaching in Malaysia*. Paper presented at the International Seminar: State of the Art of Research in Science and Mathematics Education in Southeast Asia and the Pacific, SEAMEORECSAM, Penang, Malaysia.

Noor Azlan Ahmad Zanzali & Lui Lai Nam (2002). *Evaluating The Levels of Problem Solving Skills Of Secondary School Students*. Retrieved February 12, 2011, from http://math.unipa.it/~grim/Jzanzalinam.PDF

Nor'ain M. Tajudin, Rohani A. Tarmizi , Wan Z.W. Ali, Mohd.M.Konting (2007). Effects of use of graphic calculators on performance in teaching and learning mathematics.*Educationist*, 1(Jan 2007), 1-9.

Noraini Idris (1990). *Skim penambahan nombor bulat bagi murid-murid Darjah Dua dan Tiga*. Unpublished master dissertation, Universiti Malaya.

Noraini Idris (2006). Usage of graphing calculator Ti-83 Plus: Motivation and achievement. *Jurnal Pendidian*, 31, 143-156

Noraini Idris (2009). Enhancing students' understanding calculus through writing. *International Electronic Journal of Mathematics Education,* 4(1), 36-55.

Noraini Idris, & Norjoharuddeen Mohd. Nor. (2008). *Assessing mathematical thinking in primary schools in Malaysia*. Kuala Lumpur: McGraw-Hill

Noraini Idris, (2003). *To be literate, know Maths.* The STAR Online Sunday October 26, 2003.

Noraini Idris, (2005). *Teaching and learning of mathematics: making sense and developing cognitive ability.* Kuala Lumpur: Utusan Publication.

Norton,S.J. (2005). The construction of proportional reasoning. In Chick,H.L. & Vincent,J.L. (Eds). *Proceedings of the 29th Conference of the International Group for the Psychology of Mathematics Education*, Vol 4. pp. 17-24. Melbourne:PME.

Omar, Aziz. (2002). *Transformation of word problem solving and fraction word problem among Year Five pupils.* Master of Education, Research Project. Bangi: Universiti Kebangsaan Malaysia.

Palanisamy K. V. (1988). Cognitive development and acquisition of the mathematical concepts of fraction, ratio, proportion: A study a sample of Malaysian urban secondary school pupils.

Palanisamy, K. V. (1988). *Cognitive Development and Acquisition of the Mathematical Concepts of Fraction, Ratio and Proportion: A Study of a Sample of Malaysian Urban Secondary School Pupils.* University of Malaya.

Parmjit (2005). *Multiplicative thinking in children's learning at early.* Paper presented at ICMI Regional Conference ----The Third East Asia Regional Conference on Mathematics Education (ICMI-EARCOME3).

Parmjit, Singh and Allan, White (2006) Unpacking first year university students' mathematical content knowledge through problem solving. *Asean Journal University of Education*, 2 (1). pp. 33-56. ISSN 1823-7797

Parmjit S. (2001). Understanding the concepts of proportion and ratio constructed by two grade six students. *Educational Studies in Mathematics, 43*, 271 – 292.

Parmjit S. (2003a). Schemes of children's learning in additive and multiplicative structures. *Proceedings of the International Seminar on Best Practices and Innovations in the Teaching and Learning of Science and Mathematics at the Primary School Level,* pp. 52-66, Asia Pacific Economic Cooperation & Ministry of Education, Malaysia.

Parmjit S. (2003b). An iterative process schemes in children's multiplicative thinking and learning. In H.S Dhinsa, S. B Lim, Achleitner, P. & Clements K. (Eds.), *Studies in Science, Mathematics and Technical Education*, pp. 239-248. Brunei: University Brunei Darussalam.

Parmjit, S. (2009). Variation in First Year College Students' Understanding on their Conceptions of and Approaches to Solving Mathematical Problems. *Asian Journal for University Learning and Teaching*. 5(1), 95-118.

Parmjit, S. (2000). Understanding the concepts of proportion and ratio among grade nine students in Malaysia. *International Journal of Mathematical Education in Science and Technology*, 31(4), 579-99.

Parmjit, S. (2003c). Procedural orientation of school mathematics in Malaysia. *Educators and Education Journal*, 18, 58-64. Penang: University Science Malaysia.

Parmjit, S. (2005). *Multiplicative thinking in children's learning at early* . Paper presented at ICMI Regional Conference ----The Third East Asia Regional Conference on Mathematics Education (ICMI-EARCOME3).

Parmjit, S. (2007). Roots of difficulties in proportional reasoning. *Journal of the Primary Association for Mathematics, Square One,* 17(4), 15 – 20, ISSN 0725-1092, AUSTRALIA

Parmjit, S., Rosmawati, A. H., and Rusyah, A. G. (2006). *Number sense and mental computation among secondary students' in Selangor.* Paper resentation in Conference on Scientific & Social Research (CSSR 2006). 7-9 July 2006, Pulau Langkawi.

Parmjit, S (1999). Research in Mathematics Education and its Paradigm. *Science, Mathematics & Technical Education Journal* (SMTE), 15, pp. 37-40, University Brunei Darussalam.

Parmjit, S. (1998). Tabla and Mathematics. In Presmeg, N. C. (1998). *Teachers Own Cultural Mathematics* (Ed) (part of chapter six), Reston, Virginia, NCTM.

Parmjit, S. (1998). Tabla and Mathematics. In Presmeg, N. C. (1998). *Teachers Own Cultural Mathematics* (Ed) (part of chapter six), Reston, Virginia, NCTM.

Parmjit, S. (2003). Procedural orientation of school mathematics in Malaysia. *Educators and Education Journal*, 18, 58-64, University Science Malaysia.

Parmjit, S. (2006). An analysis of word problems in school mathematics texts: Operation of addition and subtraction. *Journal of Science and Mathematics Education in South East Asia*, *29*(1), 41-61.

Patton, M. Q. (1990). *Qualitative evaluation and Research methods*. Thousand oaks, CA: Sage.

Peck, D., & Jencks, S. (1981). Conceptual issues in the teaching and learning of fractions.. *Journal for Research in Mathematics Education, 12*, 339-348.

Piaget, Jean. (1973). *To Understand is to Invent. New York: Grossman.* Retrieved Jan 21, 2011, from http://unesdoc.unesco.org/images/0000/000061/006133eo.pdf

Post, T. R. (1986). The learning and assessment of proportional reasoning abilities. In G. Lappan, & R. Even (Eds.), *Proceedings of the Eight Annual Meeting of the North American Chapter of the International Group for the Psychology of Mathematics Education,* pp. 342-351. Noordwijkerhout, The Netherlands.

Rosnaini Mahmud, Nohd Arif hj Ismail, Lim Ai Kiaw (2009). Development and evaluation of a CAI courseware 'G-Reflect' on students' achievement and motivation in learning mathematics. *European Journal of Social Sciences*, 8(4), 557-568

Schoenfeld, A.H. (1981). *Episodes and executive decisions in mathematical problem solving.* Paper presented at the annual meeting of the American Educational Research Association, April, 1981, Los Angeles.

Schoenfeld, A.H. (1985). *Mathematical problem solving.* Orlando, FL: Academic Press.

Sharifah Norul Akmar bt Syed Zamri (1997). *Skim penolakan integer pelajar Tingkatan Dua*. Unpublished doctoral dissertation, Universiti Malaya.

Sharifah, M. S. Z. (2003). *Implementing reform in science and mathematics education: The teaching and learning of science and mathematics in English.* International Conference on Science and Mathematics Educations: Which Way Now? University of Malaya.

Sierpinska, A. (1994). *Understanding in mathematics*. London: The Falmer Press.

Sierpinska, A., Kilpatrick, J., Balacheff, N., Howson, A. G., & Steinbring, H. (1993). What is research in mathematics education and what are its results? *JournalforResearch in Mathematics Education,* 24, 274-278.

Skemp, P. R. (1978). *Relational understanding in mathematics*. London: The Falmer Press.

Steffee, L. P. (1988). Children's construction of number sequences and multiplying schemes. In M Behr and J. Hiebert (eds.), *Number Concepts and Operations in the Middle Grades*, pp. 41-52. National Council of Teachers and Mathematics, Reston, VA.

Sutriyono (1997). *Skim penolakan nombor bulat murid Darjah Dua dan Tiga*. Unpublished doctoral dissertation, Universiti Malaya.

Tee Sean Sean & Md Nor Bakar (2005). *An Investigation Of Fraction Sense Among Form One Students: A Qualitative Study.* The 3rd International Qualitative Research Convention 2005 di Sofitel Palm Resort, Senai, Johor Bahru, Johor pada 21 – 23 Ogos 2005.

Teoh Sooi Kim (1991). *An investigation into three aspects of numeracy among pupils studying in Year three and Year six in two primary schools in Malaysia*. Penang: SEAMEO-RECSAM.

Tournaire, F. (1986). Proportion in elementary school. *Educational Studies in Mathematics, 17,* pp. 410 – 412.

Vergnaud, G. (1988). Multiplicative structures. In J. Hiebert & M. Behr (Eds.), *Numbers concepts and operations in the middle grades* (pp.141-161). Reston, VA: National Council of Teachers of mathematics.

Wittrock, M. C. (1974). Learning as a generative process. *Educational Psychologist, I I*, 87-95.

Wong, S. E. (2000). *Relationship between achievement in fraction and mathematics anxiety of Form Two students*. Unpublished master dissertation, Universiti Malaya

Yager, R. (1991). The constructivist learning model, towards real reform in science education. *The Science Teacher,* 58 (6).

Yoong S., Santhiran R., Fatimah Salleh, Lim C. S., Munirah Ghazali (1997). *Basic number concepts acquisition in mathematics learning: An exploratory cross- cultural study.* Programme Innovation, Excellence and Research (PIER) grant. (Research report)

Zakaria, Mohd Johan. (2002). *Relationship between learning approach and problem solving on the topic of fraction*. Unpublished doctoral dissertation, Universiti Kebangsaan Malaysia.

CHAPTER 50

NUMERACY STUDIES IN MALAYSIA

Munirah Ghazali and Abdul Razak Othman

ABSTRACT

This chapter will report on numeracy and numeracy related research from Malaysia. The chapter comprises four sections; (a) definition of numeracy; (b) numeracy in Malaysian Mathematics' curriculum and (c) research on students' numeracy and (d) teachers' numeracy.

One of the critical issues is the definition of numeracy. This section discusses different views on the existing definitions of numeracy and its related components. Since there is yet to be an official definition of numeracy in official mathematics curriculum in Malaysia mathematics education, the discussion focuses on elements that lead to the suggested definition of numeracy that fit the practices in mathematics education. Some of the elements are on knowledge that is required to be numerate including numbers, basic mathematics operation, data representation as well as shapes and space. The discussion on the meaning of numeracy will look at both the cognitive ability as well as the affective ability such as confidence and perseverance in situations involving numbers.

The second section looks into primary school mathematics curriculum specification (in the Year 2002-2007) that highlights the related aspects of numeracy. Based on the discussion on the definition of numeracy in the first

The First Sourcebook on Asian Research in Mathematics Education: China, Korea, Singapore, Japan, Malaysia, and India, pp. 1139–1161

section, we report the review on the examination of the elements of numeracy in all topics with regard to learning outcomes and suggested teaching activities from the Malaysian curriculum specification. We then concluded that even though there is no specific mention of numeracy skills, these skills are embedded in the topics in the primary school mathematics curriculum used in the Year One to Year Six Malaysian Integrated Curriculum for Primary Schools Mathematics.

The third section reviewed research on aspects of numeracy focus on students' understanding of number operations, multiplicative thinking, factors that affect secondary school students' abilities to solve operational and word problems. Whether a student has conceptual understanding or merely carrying out procedural algorithms of mathematics problem is key component of numeracy. The third section too has a special review on research on students number sense as well as the assessment of students' number sense. Mental computation strategies also reflect students' number sense and the third section discuss research on students mental computation strategies. Finally, the fourth section reviews a research on teachers' teaching and instructional practices that support and develop students' number sense in Malaysia.

DEFINITION OF NUMERACY

Skills in numeracy are key determinants of success in schooling, post-secondary education and, increasingly, in the workforce (Leitch, 2006). The early years provide the foundation for numeracy skills. Many of the students who do not reach an acceptable standard in numeracy during their school years might face greater challenges in the future. This failure would results in unhappiness and low self-esteem and finally it leads to unemployable and denied opportunities throughout their lives. Our analysis of definitions of numeracy could be categorized as focusing on what knowledge and skills are needed to be numerate in life. In terms of knowledge, Evans (2000) suggest that numeracy is not about the acquisition of even a large number of decontextualised mathematical facts and procedures and discuss that numeracy is about practical knowledge. The term 'practical knowledge' here should not be confused with low level, 'hands on' or procedural knowledge. Practical knowledge refers to knowledge which has its origins and/or importance in the physical or social world rather than in the conceptual field of mathematics itself. In other words, numeracy is underpinned by some 'basic mathematics' that is needed for everyday life. Numeracy refers to the ability to apply mathematics and the social contexts. Numeracy too is seen as the effective use of mathematics to meet the general demands of life at home, in paid work, and for participation in community and civic life. (Australian Association of Mathematics Teachers, 1997).

Further, while knowledge of mathematics is necessary for numeracy, having that knowledge is not in itself sufficient to ensure that learners become numerate. A numerate person has the capacity to bridge the gap between mathematics and the real world through their ability to choose and use the mathematical skills they have in the service of other things. The application of mathematics also rely on higher order thinking. Students need to learn mathematics in ways that enable them to recognize when mathematics might help to interpret information or solve practical problems, apply their knowledge appropriately in contexts. In other words, numerate students will have to use skills including mathematical reasoning processes, choosing mathematics that makes sense in the circumstances, making assumptions, resolving ambiguity and judging what is reasonable (Westwood, 2000).

Number sense is also a related concept to numeracy. However, number sense, according to Reys, Reys, McIntosh, Emanuelsson, Johansson, and Der (1999) focuses on the "general understanding of number and operations, along with the ability and inclination to use this understanding in flexible ways to make mathematical judgments and to develop useful and efficient strategies for managing numerical situations (p.61)" Numeracy on the other hand does not imply mastering of numbers and basic operations alone (Fleer & Raban, 2007), rather it encompasses the broad areas of measurement, spatial knowledge and the many different aspects of number. Furthermore, numeracy is seen as the ability to process, interpret and communicate numerical, quantitative, spatial, statistical, even mathematical information, in ways that are appropriate for a variety of contexts, and that will enable a typical member of the culture or subculture to participate effectively in activities that they value (Evans, 2000; Askew, Brown, Rhodes, John & William, 1997). Numeracy does not specifically describe cognitive ability alone, it covers confidence and competence with numbers and measures (DfEE, 1998). Numeracy also involves intuition and experience whenever mathematical knowledge and its understanding is needed in everyday life (Department of Education and the Arts Tasmania, 1995).

Similarly, attitude is also considered as one of the key aspects of numeracy (Kelleher, Nicol, Martin, & Anderson, 2003). Key aspects to assessing numeracy according to Kelleher et al (2003) are:

- **Number Skills**—the basic tools of numerate thinking, including counting, reading and writing numerals, and recognizing visual-spatial quantities without counting.

- **Number Concepts**—the "understanding" part of numerate thinking, through sorting, comparing, ordering, patterning, estimating, quantifying, seeing part/whole relationships, joining, separating, grouping, sharing and representing.
- **Visual-Spatial Thinking—**the ability to make sense of visual spatial information.
- **Attitudes**—the dispositions that support numerate activity: the inclination to make sense, to try hard, to persevere, to see mathematics as fun and relevant

Based on our discussion on the various definition of numeracy, we propose that a numerate person has basic mathematical knowledge which encompasses the broad areas of numbers, measurement, spatial knowledge and is able to apply these knowledge in everyday life. A numerate person too has confidence and competence and an attitude to make sense and try hard in experiences involving numerical situations.

NUMERACY IN PRIMARY SCHOOL MATHEMATICS SPECIFICATION

Our discussion in the earlier part takes into account different ways of defining numeracy across the world. This section will describe some elements of numeracy as stated in the Malaysian primary school mathematics curriculum since 2002 up to the current syllabus.

Since 2003, the Malaysian Integrated Curriculum for Primary School (Mathematics) for Year One till Year Six was published in English version in line with the national policy, The Teaching of Science and Mathematics in English (PPSMI). The latest curriculum (Kurikulum Standard Sekolah Rendah or KSSR) then switches the medium of instruction into Bahasa Melayu, the national language of Malaysia. The implementation of KSSR starts in the year 2010 for Year One and will complete in the year 2015.

First, we review the Malaysian Integrated Curriculum for Primary School (Mathematics) and the KSSR's curriculum with regard to aspects of numeracy in these documents. It is interesting to note that none of the six Malaysian Integrated Curriculum for Primary School (Mathematics) documents mentioned the term "numeracy" in the introductory section of the curriculum specification. However, in describing the rationale of primary mathematics education, the latest KSSR for Year One highlights the numeracy as one of the important skills as follows,

> Mathematics becomes the best way to develop an individual intellectual proficiency in performing logical reasoning, visualizing space, analyzing and thinking abstractly. Pupils will develop **numeracy skills**, reasoning, ways of

thinking, and solving problem through learning and application of mathematics. (Kementerian Pelajaran Malaysia, 2010, p. ix)

Even though numeracy is mentioned as one of the important skills, the document does not provide a clear definition of numeracy. The first explicit definition of numeracy is introduced in a new programme known as LINUS (Literacy and Numeracy Screening). The LINUS program which was launched in 2010 by the Ministry of Education involves a numeracy screening programme. The screening programme will help to identify low-attaining students two months into the school year. LINUS is a remedial programme designed to ensure pupils acquire basic literacy and numeracy skills at the end of three years of primary education. In the programme numeracy is defined as "the ability to do basic arithmetic operations, understanding the simple mathematical ideas, and applying the knowledge and skills in mathematics in daily life (Kementerian Pelajaran Malaysia, 2010). The definition of numeracy is found only in Year One to Year Three curriculum. The programme considers three main areas in the basic numeracy skills - numbers, basic operations and application. In learning numbers, the pupils are required to master the skills in reading and writing numbers, and count the numbers in order involving numbers up to 1 000. In mastering basic operations, the Year One pupils are required to master addition and subtraction while all four basic operations are exposed to Year Two and Year Three pupils. Other important topics include money up to RM 100, time, and basic measurement. Table 50.1 shows the basic numeracy skills with respect to years of schooling.

As mentioned earlier, the Malaysian Integrated Curriculum for Primary School (Mathematics) did not use the term numeracy in discussing the acquisition of important skills in primary school mathematics. We assume that the terminology numeracy is still new to local mathematics educators even though the essence of numeracy has long been embedded in the curriculum as discussed above.

While research which specifically examines numeracy is scarce in the literature, there has always been research on mathematics education in Malaysia that addresses the issue of students' understanding of mathematics in general and number concepts in particular. Among research that specifically address issues related to numeracy are research on number sense which began in early 2000 (Munirah 2001, Parmjit, 2005; Munirah, Ayminsyadora, Abdul Razak & Rohana, 2009; Mohini & Jacinta; 2010).

Since there was no definition of numeracy, we prefer to study the numeracy skills in Malaysian Integrated Curriculum for Primary School (Mathematics) as defined in the LINUS programme. First, we focus on the same three areas – numbers, basic operation and application. Table

Table 50.1. Basics Numeracy Skills in the LINUS Programme

Numeracy	*Year One*	*Year Two*	*Year Three*
Whole number	• Number up to 50	• Number up to 100	• Number up to 1 000
Basic operations	• Basic facts for Addition • And subtraction	• Basic facts for Addition, • Subtraction and Multiplication • Division (2,10,5,4)	• Basic facts for Addition • Subtraction and Multiplication • Division (3,6,7,8,9)
Application	• Malaysian currency up to RM 10 • Say parts of a day • Recognize 3-D and 2-D shapes	• Malaysian currency up to RM50 • Say time in hour • Classify 3-D and 2-D shapes	• Malaysian currency up to RM 100 • Say time in hour and minutes • Name 3-D and 2-D shapes

50.2 shows basic numeracy skills in the Year One to Year Six Malaysian Integrated Curriculum for Primary School (Mathematics)

In learning numbers, the Malaysian Integrated Curriculum for Primary School (Mathematics) has listed a sequence of skills from Year One to Year Six. There are five major skills. As shown in Table 50.2, it comprises of use, read and write numbers up to 1 000 000; comparing and arranging numbers; recognize the place value of numbers; comparing numbers and rounding off the numbers. It is interesting to note that the term "number sense" was use in the year 5 and year six syllabus to describe some skills related to seven digit numbers. Here, to "develop number sense" as mentioned in the documents refers to the following skills: name and write seven digit numbers, determine the place value of the digit given, expressing whole number and fraction of a million, comparing number value and rounding off numbers.

Mastering the four basic mathematics operations (addition, subtraction, multiplication and division) was emphasized in the Malaysian Integrated Curriculum for Primary School (Mathematics). Table 50.2 shows the development of skills in addition and subtraction from Year One to Year Six. For addition, pupils learn to use vocabulary involved in addition, understand addition as combining groups of objects, recall basic facts of addition, adding numbers with and without regrouping up to seven digit numbers and applying knowledge of addition in real life. For subtraction, the skills involved are, using vocabulary involved in addition, understanding subtraction as take away or difference, recall basic facts of subtraction, subtracting numbers with and without regrouping less than 100 000 and applying knowledge of subtraction in real life.

The next operation is multiplication. In the Malaysian Integrated Curriculum for Primary School (Mathematics), multiplication was first introduced in the Year Two curriculum. In Table 50.2, there are five major skills in multiplication – understand addition as repeated addition, known by heart the multiplication tables, multiplying up to six digit numbers with one one-digit number, a two-digit numbers and 10, 100 and 1 000, and solving problem of multiplication in real life.

Similar to multiplication, division was first introduced in the Year Two curriculum. Five major skills in division are given in Table 50.2 – understand division as sharing equally or grouping, known by heart the division facts, dividing up to seven digit numbers with one one-digit number, a two-digit numbers and 10, 100 and 1 000, and solving problem of division in real life up to seven digit numbers.

The definition of numeracy also considers application of mathematics in Malaysian currency, time and shapes and space. In the Malaysian Integrated Curriculum for Primary School (Mathematics), skills related to Malaysian currency were first introduced in the Year One curriculum.

Table 50.2 shows five major skills involving money – recognizing coins (up to RM 5) and notes (up to RM 50), represent money (in RM and Sen), exchange coins (up to RM 5) and notes (up to RM 50), adding and subtracting money up to RM 100 000, multiplying money to the highest RM 10 000, dividing money with dividend not more than RM 100 000 , rounding of money to the nearest Ringgit. In Year Six curriculum, the document stated "use and apply number sense in real context involving money" which includes the skills - perform mixed operation up RM 10 million and solving problem of money in real life up to RM 100 000.

In learning time (see Table 50.2), Malaysian primary school pupils engage with understand and use vocabulary related to time (days, sequence of the day, weeks, months), understand read, and write related to time (hour, five minutes, hour and minutes, day and hour, the half and quarter hour, 12-hours and 24-hours systems) construct a simple schedule, reading a calendar, and understanding relationship between units of time. The curriculum also includes skills in adding, subtracting, multiplying and dividing units of time, and applying knowledge of time to find the duration (in minute, hour, day) and as well as solving problems involving time duration infractions and/or decimals of hours, minutes and seconds.

The numeracy involving Shapes and Space starts from Year One. As shown in Table 50.2. It requires the pupil to understand and use the vocabulary related to 3-D and 2-D shapes, describe and classify common 3-D and 2-D shapes, building 3-D shapes and recognise and sketch lines of symmetry. To acquire the higher skills the pupils also have to understand the perimeter and area of a two-dimensional shape, understands

Table 50.2. Basic Numeracy Skills in the Year One to Year Six Malaysian Integrated Curriculum for Primary School (Mathematics)

Area	*Numeracy Skills*
Numbers	Use, read and write numbers up to 1 000 000; comparing and arranging numbers; recognize the place value of numbers; comparing numbers and rounding off the numbers
Basic operations; Addition	Addition as combining groups of objects; recall basic facts of addition, adding numbers with and without regrouping up to seven digit numbers and applying knowledge of addition in real life
Subtraction	Using vocabulary involved in addition; understanding subtraction as take away or difference; recall basic facts of subtraction; subtracting numbers with and without regrouping less than 100 000 and applying knowledge of subtraction in real life.
Multiplication	Understand addition as repeated addition; known by heart the multiplication tables; multiplying up to six digit numbers with one one-digit number, a two-digit numbers and 10, 100 and 1 000, and solving problem of multiplication in real life.
Division	Understand division as sharing equally or grouping; known by heart the division facts, dividing up to seven digit numbers with one one-digit number; a two-digit numbers and 10, 100 and 1 000, and solving problem of division in real life up to seven digit numbers.
Currency	Recognizing coins (up to RM 5) and notes (up to RM 50); Represent money (in RM and Sen), Exchange coins (up to RM 5) and notes (up to RM 50); Adding and subtracting money up to RM 100 000; Multiplying money to the highest RM 10 000; Dividing money with dividend not more than RM 100 000; Rounding of money to the nearest Ringgit;Use and apply number sense in real context involving money
Time	Understand and use vocabulary related to time (days, sequence of the day, weeks, months); Understand read, and write related to time (hour, five minutes, hour and minutes, day and hour, the half and quarter hour, 12-hours and 24-hours systems); Construct a simple schedule, reading a calendar, and understanding relationship between units of time;Adding, subtracting, multiplying and dividing units of time, and Applying knowledge of time to find the duration (in minute, hour, day) and Solving problems involving time duration infractions and/or decimals of hours, minutes and seconds
Shapes and space	Understand and use the vocabulary related to 3-D and 2-D shapes; describe and classify common 3-D and 2-D shapes; building 3-D shapes and recognise and sketch lines of symmetry; Understand the perimeter and area of a two-dimensional shape, understands and find the volume for cubes and cuboids; Find the perimeter and the area of composite 2-D shapes;Find the perimeter and area of composite two-dimensional shapes, Find the surface area and volume of composite three-dimensional shapes; Solving problems related to perimeter, area, and volume of the shapes

and find the volume for cubes and cuboids, find the perimeter and the area of composite 2-D shapes, find the perimeter and area of composite two-dimensional shapes, find the surface area and volume of composite three-dimensional shapes. Solving problems related to perimeter, area, and volume of the shapes are also taught at this level.

In summary, numeracy skills are embedded in these topics in the primary school mathematics curriculum even though the term numeracy was not specifically used in the Year One to Year Six Malaysian Integrated Curriculum for Primary School Mathematics.

RESEARCH ON STUDENTS' NUMERACY IN MALAYSIA

Based on our discussion for the definition of numeracy, we reviewed research on numeracy in Malaysia. Since the concept of numeracy appears to be embedded in the mathematics skills, we could not find the exact studies on numeracy. Therefore we reviewed research on aspects of numeracy specifically on students' understanding of number operation including word problems. We also reviewed research focusing on students' number sense, assessment of number sense and mental computation.

One of the aspects of numeracy is understanding operations and ability to solve word problems. Here, we describe studies with regard to students' understanding of number operations (Aida Suraya, 2001; Parmjit, 2005), and factors that affect students' abilities to solve word problems (Lim, Lourdusamy & Munirah, 2001; Chinappan and Ambigapathy, (2009). Aida Suraya (2001) studied the decimal multiplication schemes specifically in multiplying decimal number with whole number (e.g, 3 x 4.7, 3.6 x 4, 0.5 x 28, 0.62 x 100) and multiplying decimal numbers with decimal number (e.g. 0.5 x 0.23, 0.4 x 0.3). She identified seven algorithms that were used by Form One students. The algorithms used were to determine the decimal point by looking at the decimal point of the coefficient, to multiply whole numbers first without considering decimal point, to execute repeated addition, to convert one of the coefficients into fraction, to shift the position of decimal point to the right based on the decimal point of the coefficient, to arranging the numbers so that the decimal points are in line and then multiply the digits at the correspond decimal point, and lastly to shift the decimal point of the product to the left based on the decimal point of the coeficient.

In his study of multiplicative thinking, Parmjit (2005) found that in multiplying numbers, children started the progression from counting by ones or counting each individual item and then to the counting of groups. He identified children's "iterative additive multiplication" that bridge from addition to multiplication through transitional thinking method in conceptualizing the underlying problem situation and carrying out the numerical solution process. Thus, he suggested that multiplicative reasoning expressed by using iterative composite units (coordination of number sequences) can help children construct meaning for multiplication thinking in various multiplicative settings. Parmjit's study focus on primary school children's abilities to solve multiplication problems presented as both operational and word problems.

Lim, Lourdusamy and Munirah (2001) study was designed to identify factors that affect secondary school students' abilities to solve operational and word problems in mathematics. The factors examined were language skills, field dependence-field independence cognitive style, gender, ethnicity and basic mathematical skills. Factors such as basic mathematical skills, language skills, field dependence-field independence cognitive style and ethnicity have significant correlations with the students' abilities in solving operational and word problems. However, basic mathematical skills seem to be the best predictor of the students' abilities, followed by the field dependence-field independence cognitive style. The authors continued to suggest that measures should be taken to upgrade students' basic mathematical skills as mathematics is a hierarchical subject. Lim, Lourdusamy and Munirah (2001) further deliberated that higher order concepts are dependent on lower order concepts while the acquisition of basic mathematical skills is always a pre-requisite of problem solving skills.

An interesting comparative study on Malaysian and Australian students' numeracy was investigated by Chinappan and Ambigapathy (2009). The study aimed to make learner's cognition more visible by immersing a cohort of Malaysian and Australian students in arithmetic word problems. Students were required to explain and justify their solutions. The students' readings, explanations, representations and reflections about a given set of problems were analyzed through the frameworks of *dialogue,* representation and schema (Chinappan & Ambigapathy, 2009). Results showed that both the groups experienced difficulty in representing far-transfer numeracy problems. However, Australian children tended to develop longer and more varied explanations in comparison to their Malaysian counterparts. However, the researchers in this study did caution that the study reported here was based on problem-solving attempts by a small sample of Malaysian and Australian students. The kind of representations and explanations that could be generated by these two cohorts of students may lead to overgeneralization. The discussions above looked into different aspects of numeracy. Since number sense is also an important aspect of numeracy, the next section will review research on number sense.

RESEARCH ON NUMBER SENSE

Number sense is seen as a subset of numeracy (Australian Association of Mathematics Teachers (1997). Number sense research has received considerable interest in the Malaysian context for it provide a framework to examine students' understanding or lack of it as they solve mathematics problem. For example, when children are asked why they say 9 is more than 5, they respond 'it just it' or 'don't know'. Many students know the answer to 6 x 6 but need to recite the 6 multiplication all over again in order to answer 7 x 6. In another example, Kamii and Livingston (1993) discuss children's inability to answer 7 x 6 when they know the correct answer for 6 x 6. The reason was that the children were unable to relate the product of 6 x 6 with the product of 7 x 6 or how the product of 7 x 6 is actually 6 more than 36 based on the multiplication order. Students are good rule followers but unfortunately do not always understand the procedures they learned (Hiebert, 1986). National Council of Teachers of Mathematics (1989) had initiated among the early efforts to define number sense with greater objectivity and had listed five factors as indicators of having number sense; understand the meaning of numbers; able to recognise the connections between numbers; recognize the relative magnitude of a number; to know the effects of operations on numbers, and ; construct a reference to measure a normal object or situation in everyday life.

About a decade ago, Munirah Ghazali (2001) investigated primary school Year Five students' number sense by using McIntosh, Reys, Reys, Bana, and Farrell's, (1997) number sense framework. A computation test and similar test in number sense format was administered to 406 students and six students were chosen for further in depth interview. Finding from the study indicated that the students in this survey did not exhibit satisfactory number sense with students having better score in computing and calculation questions and struggling in effect of operations, multiple representations and number concepts questions. For example, Munirah's (2001) study showed the percentage of students who were able to answer problems on fractions decreased to 16.5% when the same questions were asked in the form of number sense. In the study, Munirah (2001) asked students to circle the correct answer to the problem $\frac{5}{6} + \frac{8}{9}$ without doing any calculation. The choices given were A. 1, B. 2, C. 19 and D. 21. 16.5% of the students chose the correct answer B. 2. The students too were requested to compute $\frac{5}{6} + \frac{8}{9}$ and 70.9% of the students were able to give the correct answer. Findings from students in depth interview revealed that there were qualitative differences to their response on fraction problems. For example students who had scored high marks in the number sense test represented fractions using shaded regions while students who performed poorly on the number sense test represented fraction as numbers.

In another study on number sense, Mohini and Jacinta (2010) investigated the relationship between student performance in number sense and mathematics achievement and explore the components of number sense that students are weak in. Findings from their study indicated that despite having a high level of competency in performing algorithms in the classroom, these students were generally weak in understanding the meaning of number and operations, relative number sizes, composing numbers and recognising effect of operations on numbers. Despite obtaining high scores in school examinations, the data obtained from the sample shows that 47% of the respondents scored within the range of 6 to 9. Students who were able to perform excellently in the classroom were facing difficulty in giving meaning to number sense questions that was administered to them. This comparison once again raises the issue of students demonstrating little understanding of numerical situations in solving number problems (Munirah & Noor Azlan, 1999). In summary, the focus of the previous research on number sense more on the students' nature of number sense and their understanding of numerical situations.

ASSESSMENT OF NUMBER SENSE

Mathematics educators have for some time been concerned that many students demonstrate little understanding of numerical situations in which they have to solve number problems. The acquisition of computational skill remains at the centre of the mathematics curriculum in many primary schools. The traditional goals of primary school mathematics have been, and often still are, the development of knowledge of basic addition and multiplication facts and the skills of pencil-and-paper addition, subtraction, multiplication and division. Therefore, assessment of number sense involves not only an indication whether the answer is 'correct' or 'incorrect' but also knowledge of the strategies used and their relative sophistication, as well as indications of hesitancy and the exact nature of errors and misunderstandings revealed by the student. There has been considerable interest to assess Malaysian students' number sense as reported below.

Parmjit (2009) investigated the number sense proficiency of secondary school students ages 13 to 16. A total of 1756 students, from thirteen schools in a state in Malaysia participated in this study. A majority (74.9%) of these students obtained an A grade for their respective year-end school examinations. Data on student's sense of numbers was collected using Number Sense Test adapted from McIntosh et al (1997). The results from this study indicated that students obtained a low percentage of success rate ranging from 37.3% to 47.7% across the levels. There was no significant difference in the results between Form One students and Form Two students and also between Form Three students and Form Four students. In terms of gender comparison, although the male students obtained a higher score than their female counterparts, this difference was only significant among the Form One student's. The result reported in this study reveals a cause for concern. Firstly, it indicates students from ages 13 (Form One) to 16 (Form Four) faced great difficulty in making sense of numbers. Given that these items in the Number Sense Test needed very little computation, with more of making sense of numbers, it is troubling that the percentage of correct responses ranged only from 37.3% to 47.7% and the mean score from 18.65 to 23.38 (with a maximum of 50) across the levels. Secondly, students performance on this number sense test did not increase dramatically as one would expect from Form One to Form Four and the level of performance leaves much to be desired. There was no significant difference in the mean scores between students in Form One and Form Two and also between Form Three and Form Four. One would expect that as students move up to a higher level (especially Form Four students) they should become facile in working with fractions, decimals, and percents meaningfully.

Parmjit (2009) echoes Munirah's (2001) concern that an over reliance on paper and pencil computation at the expense of intuitive understanding of numbers is taking place among these students. The result of Parmjit (2009) study seems to indicate an existence of a gap between the ability to do paper-and-pencil calculations and intuitive understanding. Majority of the students (74.9%) obtained an A grade for their year-end school examination but there seems to be a vast disparity between the grade scores and the Number Sense test, as the low score indicates.

Findings from three research studies by Munirah (2001), Parmjit, (2009) and Mohini and Jacinta (2010) respectively in Malaysia that used McIntosh et al (1997) number sense framework are in agreement that the effect of operation strand proves the most challenging strand for the students whether they are in primary or secondary schools. When highlighting number sense, students should focus on their solution strategies rather than on a "right answer," on thinking rather than on the mechanical application of rules, and on student-generated solutions rather than on teacher-supplied answers.

Earlier on Munirah, Shafia, Sharifah Norhaidah, Zurida and Fatimah, (2003) reported on a research project that has developed four separate modules aimed at assessing students number sense in the four components: counting, addition and subtraction, multiplication and division and place value. The framework developed in the study by Munirah et al (2003) was developed based on existing framework from other researches and modified to suit local context. The items in each separate component were constructed to fulfill three main numerical representations that is contextual, pictorial/objects and symbolic. Details of the project have been reported in the Journal of Science and Mathematics Education in South East Asia (Munirah , Sharifah Norhaidah & McIntosh, 2004).

Munirah, Sharifah Norhaidah, Fatimah and Shafia (2006) reported that the four modules produced by this project have several positive impacts on classroom assessment, particularly as a diagnostic tool for teachers to assess students' number sense. The modules can be utilized in the following manners:

- Clusters of items defining specific understandings are used to identify difficulties in particular areas thereby making prescriptions possible.
- To separate out children lacking understanding from those who just need practice.
- Enabling on-the-spot assessment by providing prepared possible alternatives.
- As professional development for teachers in becoming more informed and sensitive towards varying strategies as well as learning difficulties in children.

As discussed above, research on assessment of number sense in Malaysia has focussed on number sense proficiency, strategies used by students and developing modules to assess number sense.

MENTAL COMPUTATION

Numeracy involves making sense of numerical situation. Therefore, mental compuatation ability is seen as indicator of both number sense and numeracy. Mental computation is an important aspect of number sense (McIntosh, 1994; Sowder, 1992; Reys, 2006). Research on mental computation and number sense has proposed connections among mental computation and number sense, particularly number facts, computational estimation, numeration, and properties of number and operation; social and affective issues including attributions, self efficacy, and social context (e.g. classroom and home); and metacognitive processes (Heirdsfield, 2000). It has been posited that when children are encouraged to formulate their own mental computation strategies, they learn how numbers work, gain a richer experience in dealing with numbers, develop number sense, and develop confidence in their ability to make sense of number operations (Kamii and Dominick, 1998; Reys and Barger, 1994; Sowder, 1990).

There is an emerging trend to focus on mental computation in the mathematics education curriculum of many countries (Harnett, 2007). Curricular reform documents in the US, Australia, UK, New Zealand and the Netherlands has included mental computation as an important aspect of primary mathematics (National Council of Teachers of Mathematics, 2000, DfES,2007; Treffers & DeMoor, 1990; Australian Education Council, 1991).

In Malaysia, mental computation is documented in the primary school Mathematics curriculum as evidenced in the curriculum specification documents (Kementerian Pendidikan Malaysia, 2003). For example the Year One curriculum specification actually specify 'Emphasize mental calculation' under 'points to note' for the curriculum specification on the topic whole numbers under the learning area addition with the highest total of 10, 18, and subtraction within the range of 18. The curriculum specification documents recommend and encourage teachers under 'points to note' to emphasize on mental calculation for all the discussion on the operations addition, subtraction, multiplication and division from years one to six. However, very few of the mental computation strategies are mentioned in the document.

Munirah, Rohana, Asrul and Amysyahdora (2010) reported a study that examined mental computation strategies used by Year One, Year Two, and Year Three students to solve addition and subtraction problems. The participants in this study were twenty five 7 to 9 year-old students identified as excellent, good and satisfactory in their mathematics performance from a school in Penang, Malaysia. Findings from this study showed a range of strategies employed by the students even though addition and subtraction mental strategies are not formally taught to Year One, Year Two and Year Three students. The different types of addition and subtraction strategies used were highlighted. The strategies for 1-digit, 2-digit and 3-digit mental computation of both addition and subtraction problems were compared. Further, the students employed intuitive mental computation strategies and there are qualitative differences in these strategies which may prove invaluable to help teachers in strengthening students' understanding of the number concepts. Overall, a variety of strategies were used to solve the addition problems. These strategies were, counting all, counting on from first number, counting on from larger, bridging through ten, separation (left to right and right to left), wholistic-compensation, mental recall, and mental image of the pen and paper algorithm. It should be noted here that there were three responses (7%) from the students who requested to use the Standard written method where the student uses pen and paper to get their answers. The strategy for one of the responses was categorised as Unidentified even though the answer was correct. Examples of the strategies used identified from the sample of this study are as follows:

Wholistic-compensation 37+19: 10+30=40; 9+1=10; 40+10=50; 7-1=6; 50+6=56

Separation-left to right 37+19: 3+1=4; 9+7=16; 4+1=5; 56

Separation-right to left 106+228: 6+8=14; 2+1=3; 334

This research revealed that, first, students invent or use their own intuitive strategies when asked to solve problems using mental computation even when mental computation may or may not have been formally taught to them. Secondly, while some students did invent their own intuitive strategies, there were other students who did not display their ability for mental computation. While this study did not connect actual teachers' teaching strategies with students' strategies, the findings from the study raised questions whether students' do invent their own strategies or whether their use of intuitive strategies were indirectly encouraged by modeling teachers' own mental computation strategies. Moreover, for students who did not display the ability for mental computation, the question raised was whether they could be encouraged to invent their own strategies if the teachers encouraged them to do so. These questions give some suggestions for further research into teachers' pedagogical practice when teaching mathematics. Another finding from this research is that there was a tendency for Year Two and Year Three students to resort to algorithmic computation rather than mental computation especially when the number magnitude for the questions increases. While this research argued that students' ability to carry out mental computation reflect their understanding of numbers, the question that arises is whether the trend for older children to resort to algorithmic computation be seen as a concern, and in what ways could the curriculum address such situation. Ideally, children should be encouraged to invent their own strategies when solving problems as a reflection of them acquiring number sense.

The use of representation also reflects the students' number sense. Representation is the ways in which mathematical ideas are represented is fundamental to how people can understand and use those ideas (National Council of Teachers of Mathematics, 2000). A representation cannot describe fully a mathematical construct and that each representation has different advantages. Using multiple representations for the same mathematical situation is at the core of mathematical understanding (Duval, 2002). Behr, Lesh, Post, and Silver (1983), Lesh (1981) and Lesh, Landau, and Hamilton (1983) have identified five distinct types of representation systems that occur in mathematics learning and problem solving. These are written symbols, static pictures, manipulative models, real scripts and spoken language.

A research to investigate Malaysian primary students' number sense in solving problem presented as pictorial representation was carried out by Rohana, Munirah and Muhamad Faiz (2009). The research aimed to investigate students number sense when presented with problems presented as pictures or pictorial representation. Specifically, this study investigates students' strategies when solving addition and subtraction problems presented in pictorial forms. The addition and subtraction problems were given in three different form of problems - combine unknown outcome, part-part whole unknown part and combine unknown change (Carpenter, 1999).

Regardless whether a student were able to give the correct answer or not, the strategies used to answer counting questions were counting by one's, two's, three's, four's, five's and tens. There were students in this study who were able to subitize the correct number of objects presented. While the counting strategies refer to different levels of counting expertise, the ability to subitize is seen as an indicator of number sense. The strategies used by students to solve addition and subtraction problems presented as pictures were mental computation, counting on, modelling using actual objects (marble), and drawing. It is also interesting to note that these strategies were employed by both students were able to give the correct answer or not. Data from this study showed that when solving addition and subtraction problems presented in pictorial forms, the students actually translate the numbers presented in pictures to numerical symbols first before solving the problem. Once the students see the numerical representation of the numbers, their strategies differ. There were students who use mental computation strategies but many resort to solving using standard addition and subtraction algorithm.

Findings from the interview too indicated that for the sets of pictured objects majority of the students used counting strategies and a few of the Year Two and Year Three students was able to obtain the answers by subitizing. Although many of Year One students used the counting strategies to solve the addition and subtraction problems, Year Two and Year Three students translated the pictorial representation into symbolic representation before solving algorithmically. Data also indicated that Year One students prefer to use counting strategies and Year Three students prefer algorithmic procedures. Moreover, the excellent students in this study appeared to use other counting strategies such as counting by two's and five's while the average students normally use counting by one's.

In summary, the discussions in this section provide the explanation of mental computation which is one part of number sense. The focus was on students' mental computation strategies and their use of representation when solving mathematical problems.

TEACHERS' NUMBER SENSE

The importance of teacher's knowledge in the teaching is well documented (Graeber, 1999; Munby, Russell, & Martin, 2001). On the same note, teachers' knowledge and instructional practices into teaching numeracy and number sense is also an important component to ensure that the students themselves develop numeracy and number sense. The Malaysian National Education Blueprint 2006-2010 identified that improving the teaching profession is one of the important components of education excellence in Malaysia. Research have shown that teachers knowledge place a great influence on student's achievement (Hill, Rowan, and Ball, 2005). Since the last two decades, the efforts to provide accurate explanation on teacher's knowledge were given equal attention ranging from "what kind of knowledge need for teaching", assessing knowledge base for teaching, or professional knowledge (Carlson, 1990) to practical knowledge (Clandinin, 1985). Research on teachers teaching for number sense is being emphasized in many countries such as United Kingdom, Australia and United State (Askew et al, 1997; Reys, 2006).

In Malaysia, Munirah, Abdul Razak, Rohana and Fatimah (2010) conducted a research study to develop standards for teachers' teaching and instructional practices that support and develop students' number sense. Data was collected through interview of primary school teachers, classroom observation of teachers' teaching and examination of artifacts such as teachers' lesson plan, teaching aids, relevant resources used in teaching. All the classroom observations were videotaped for analysis. Findings from the study indicated the emergence of certain aspects of teaching such as good teacher characteristics and effective student involvement. However, other aspects of teaching such as teacher's pedagogical content knowledge and connection of the mathematics to the content were observed as posing a challenge to the teachers. In their analysis, they found that selected teachers developed certain criteria and instructional practices in developing students' number sense. In most of the lessons mathematics teachers designed the activities to include organizational and management strategies, teaching strategies for numbers, teachers' questioning techniques, and classroom interactions. Each aspect has an important link to number sense. During the organizational and management strategies, teacher used a simple daily problem with a solution through modeling. Questioning plays an important role that guides the students to reach at the answer meaningfully. The students could "see" both the problem and mathematical idea that is embedded in the problem. The students' understanding was enhanced through active participation. In group work as well as individual task, teacher provided ample opportunities to enhance number sense through discussion on the problems. Teachers also supported them in their own group in more detail. The study found that the teachers emphasized on the effort to make the problems easily understood and reorganizing students' understanding. The teacher used effective means of representing the ideas through examples and modeling. The study also found that the teacher did not present the number sentence as a mathematical problem in isolation of its' context. On the other hand, the teacher used a series of developmental steps starting from modelling of the concrete objects, representing the problems via pictorial representation and finally to developing the mathematical sentence.

In another research, Effandi and Norliza (2009) examined trainee teachers' conceptual and procedural knowledge regarding rational numbers. The findings revealed that the trainees' level of conceptual and procedural knowledge is high-average. The trainees displayed competence in representing a fraction as part of a set, a region and a ratio. They also demonstrated conceptual knowledge in sketching one whole when given a fraction and in solving word problems that involved fractions. However, they were too dependent on algorithms and on the memorization of formulas, tips and rules, and they were unable to provide explanation or justification on how they obtained a particular answer. Research revealed that most trainee teachers display average levels of conceptual knowledge regarding rational numbers, while others merely depend on knowledge of and skills in algorithms. Their weak conceptual knowledge will cause their teaching to be ineffective and thus later affect the students whom they teach.

CONCLUSION

In summary, a numerate person has basic mathematical knowledge which encompasses the broad areas of numbers, measurement, spatial knowledge and is able to apply these knowledge in everyday life allowing them to participate in community and civic life. Being numerate too involves confidence and competence and an attitude to make sense and try hard in experiences involving numerical situations. Review of the Malaysian Integrated Curriculum for Primary School Mathematics curriculum year 2002 - 2007 documents revealed that even though numeracy is not presented explicitly, elements and essence of numeracy abound in the curriculum. However, the latest mathematics curriculum document that is the KSSR, 2010 or Standard Curriculum for Primary School highlights that numeracy is one of the important skills in Malaysia primary mathematics. Research reviewed also address students, pre service teachers and teachers' understanding of number concepts and elements of numeracy even when the specific terminology is not mentioned. Since numeracy is becoming an important aspect in Malaysian mathematics education, we would suggest that a proper definition of numeracy at different levels of schooling need to be developed. If this definition is well accepted, it might lead to more research on numeracy in the future.

REFERENCES

Aida Suraya (2001). Algoritma pendaraban nombor perpuluhan dari perspektif pelajar tingkatan satu. *Pertanika Journal Social Sciences and Humanities, 9*(1), 21-33.

Askew, M., Brown, M., Rhodes, V., Johnson, D., & William, D. (1997). *Effective teachers of numeracy.* London: School of Education, King's College.

Australian Association of Mathematics Teachers (1997). *Numeracy = everyone's business. report of the numeracy education strategy development conference.* Adelaide: Australian Association of Mathematics Teachers.

Australian Education Council (1991). *A national statement on mathematics for Australian schools.* Melbourne: Curriculum Corporation.

Behr, M. J., Lesh, R., Post, T. R., & Silver, E. A. (1983). Rational number concepts. In R. Lesh & M. Landau (Eds.), *Acquisition of mathematical concepts and processes* (pp. 91-126). New York: Academic Press.

Carlson, R. E. (1990). Assessing teachers' pedagogical content knowledge: Item development issues. *Journal of Personnel Evaluation in Education, 4*, 157-173.

Carpenter, T. P. (1999). *Children's mathematics : cognitively guided instruction.* Portsmouth, NH: Heinemann.

Chinnapan, M., & Pandian, A. (2009). Malaysian and Australian children's representations and explanations of numeracy problem. *Educational Research for Policy and Practice 8*(3), 197-209.

Clandinin, D. J. (1985). Personal practical knowledge: A study of teachers' classroom images. *Curriculum Inquiry 15*(4), 362-385.

Department of Education Arts (1995). *Numerate student-numerate adults.* Hobart: Department of Education Arts, Tasmania.

DfEE (1998). *The implmentation of national numeracy strategy : The final report of numeracy task force* Sudbury, Suffolk: Department for Education and Employment.

DfES (2007). *Primary national strategy.* London.

Duval, R. (2002). The cognitive analysis of problems of comprehension in the learning of mathematics. *Mediterranean Journal for Research in Mathematics Education,, 1*(2), 1-16.

Effandi, Z., & Norliza, Z. (2009). Conceptual and procedural knowledge of rational numbers in trainee teachers. *European Journal of Social Sciences, 9*(2), 202-217.

Fleer, M., & Raban, B. (2007). *Early childhood literacy and numeracy: Building good practice*. Australia: Commonwealth of Australia.

Graeber, A. O. (1999). Form of knowing mathematics: What preservice teacher should learn. *Educational Studies in Mathematics, 38*, 189-208.

Hartnett, J. E. (2007). Categorisation of mental computation strategies to support teaching and to encourage classroom dialogue. In J. Watson & K. Beswick (Eds.), *Mathematics: Essential research, essential practice (Proceedings of the 30th annual conference of the Mathematics Education Research Group of Australasia, Hobart)* (pp. 345-352). Sydney: MERGA.

Heirdsfield, A. M. (2000). Mental computation: Is it more than mental architecture? *Australian Association for Research in Education*.

Hiebert, J., & Wearne, D. (1986). Procedures over concepts: The acquisition of decimal number knowledge. In J. Hiebert (Ed.), *Conceptual and procedural knowledge: The case of mathematics* (pp. 199-223). Hillsdale, NJ: Lawrence Erlbaum Associates.

Hill, H. C., Rowan, B., & Ball, D. L. (2005). Effects of Teachers' Mathematical Knowledge for Teaching on Student Achievement. *American Educational Research Journal, 42*(2), 371-406.

Kamii, C., & Dominick, A. (1998). The harmful effects of algorithms in grades 1-4. In L. J. Morrow & M. J. Kenney (Eds.), *The teaching and learning of algorithms in school mathematics, 1998 yearbook* (pp. 130-140). Reston, VA: NCTM.

Kamii, C., Lewis, B. A., & Livingston, S. J. (1993). Primary arithmetic: Children inventing their own procedures. *Arithmetic Teacher, 41*(4), 200-203.

Kelleher, H., Nicol, C., Martin, L., & Anderson, A. (2003). *Assessing early numeracy : BC Early Numeracy Project (K-1)*. British Columbia: Ministry of Education, Province of British Columbia.

Kementerian Pelajaran Malaysia (2010). *Kurikulum standard sekolah rendah; Matematik tahun satu*. Kuala Lumpur: Kementerian Pelajaran Malaysia.

Kementerian Pendidikan Malaysia (2003). *Kurikulum bersepadu sekolah rendah: Huraian sukatan pelajaran matematik tahun 3*. Kuala lumpur: Pusat Perkembangan Kurikulum.

Leitch, S. (2006). *Prosperity for all in the global economy: World class skills. (The Leitch Review of Skills)*. London: HM Treasury.

Lesh, R. (1981). Applied mathematical problem solving. *Education Studies in Mathematics, 12* 235-264.

Lesh, R., Landau, M., & Hamilton, E. (1983). Conceptual models in applied mathematical problem solving. In R. Lesh (Ed.), *The acquisition of mathematical concepts and processes*. New York Academic Press.

Lim, C. S., Lourdusamy, A., & Munirah, G. (2001). Factors affecting students' abilities to solve operational and word problems in mathematics. *Journal of Science and Mathematics Education in Southeast Asia, 24*(1), 84-94.

McIntosh, A., Reys, B., Reys, R., Bana, J., & Farrell, B. (1997). *Number sense in school mathematics: Student performance in four countries*. Perth: MASTEC, Edith Cowen University.

McIntosh, A. J., De Nardi, E., & Swan, P. (1994). *Think mathematically*. Melbourne: Longman.

Mohini, M., & Jacinta, J. (2010). Investigating Number Sense Among Students. *Procedia Social and Behavioral Sciences, 8*, 317–324.

Munby, H., Russell, T., & Martin, A. K. (2001). Teachers' knowledge and how it develops. In V. Richardson (Ed.), *Handbook of research on teaching* (pp. 433-436). New York: Macmillan.

Munirah, G. (2001). *Kajian kepekaan nombor murid tahun lima*. Unpublished Doctoral Thesis, Universiti Teknologi Malaysia, Johor Bahru.

Munirah, G., Abdul Razak, O., Rohana, A., & Fatimah, S. (2010). Development of teaching models for effective teaching of number sense in the Malaysian primary schools. *Procedia Social and Behavioral Sciences, 8*, 344-350.

Munirah, G., Amynsyadora, A., Abdul Razak, O., & Rohana, A. (2009). *From research to classroom: Development of modules that support primary mathematics learning focus on at risk students.* Paper presented at the *Proceedings of the 5th Asian Mathematical Conference, Malaysia 2009*, Kuala Lumpur.

Munirah, G., & Noor Azlan, A. Z. (1999). Assessment of school children's number sense *Proceedings of The International Conference on Mathematics Education into the 21 Century: Societal Challenges: Issues and approaches*. Cairo, Egypt.

Munirah, G., Rohana, A., Asrul, & Ayminsyahdora (2010). Identification of students' intuitive mental computational strategis for 1, 2 and 3 digits addition and subtraction: pedagogical and curricular implications. *Journal of Science and Mathematics Education in Southeast Asia 33*.(1), 17 - 38.

Munirah, G., Shafia, A. R., Sharifah Norhaidah, I., Zurida, I., & Fatimah, S. (2003). *Development of a framework to assess primary students' number sense in Malaysia: Counting*. Paper presented at the SEMT '03: International Seminar For Elementary Mathematics, Charles University, Prague, Czech Republic.

Munirah, G., Sharifah Norhaidah, Fatimah, S., & Shafia, A. R. (2006). The development of a framework to assess primary school children's number sense. In T. Abdul Aziz & Lim Koon Ong (Eds.), *Fundamental research at USM 2002-2003* (Vol. 3 (Arts), pp. 140-160). Penang: Penerbit Universiti Sains Malaysia.

Munirah, G., Sharifah Norhaidah, I., & MacIntosh, A. (2004). From doing to understanding: An assessment of Malaysian primary pupils' number sense with respect to multiplication and division. *Journal of Science and Mathematics Education in Southeast Asia, 27*(2), 92-111.

National Council of Teachers of Mathematics (1989). *Curriculum and evaluation standards for school mathematics*. Reston, VA: Author.

National Council of Teachers of Mathematics (2000). *Principles and standards for school mathematics*. Reston, VA: NCTM.

Parmjit, S. (2005). An exploration into children mathematics thinking: Implication for teaching. In S. Parmjit & C. S. Lim (Eds.), *Improving teaching and learning: From research to pratice* (pp. 81-108). Shah Alam: Pusat Penerbitan Universiti.

Parmjit, S. (2005, Aug. 7 - 12). *Multiplicative thinking in children's learning at early grades.* Paper presented at the ICMI East Asia regional Conference on Mathematics Education, China.

Parmjit, S. (2009). An assessment of number sense among secondary school students. *International Journal for Mathematics Teaching and Learning, Oct 2009*, 1 - 29.

Reys, B. J. (2006). *The intended mathematics curriculum as represented in state-level curriculum standards: Consensus or confusion?* Charlotte, NC: Information Age Publishing.

Reys, B. J., & Barger, R. H. (1994). Mental computation: Issues from the United States. In R. E. R. N. Nohda (Ed.), *Computational alternatives for the twenty-first Century: Cross cultural perspectives from Japan and the United States* (pp. 31-47). Reston, VA: The National Council of Teachers of Mathematics.

Reys, R., Reys, B., McIntosh, A., G, E., Johansson, B., & Der, C. Y. (1999). Assessing number sense of students in Australia, Sweden, Taiwan, and the United States. *School Science and Mathematics, 99*(2), 61-70.

Rohana, A., Munirah, G., & Muhamad Faiz (2009, 10-12 November). *Students' number sense when solving problem presented in pictorial representations.* Paper presented at the 3rd International Conference on Science and Mathematics Education (CoSMEd 2009), Penang Malaysia.

Sowder, J. (1990). Mental computation and number sense. *Arithmetic Teacher, 37(7),* 18-20.

Sowder, J. (1992). Estimation and number sense. In D. Grouws (Ed.), *Handbook of research on mathematics teaching and learning*. New York: MacMillan.

Treffers, A., & de Moor, E. (1990). *Towards a national mathematics curriculum for the elementary school. Part 2. Basic Skills and Writing Computation*. Zwijssen, Tilburg, The Netherlands.

Wood, K., & Frid, S. (2005). Early childhood numeracy in a multiage setting. *Mathematics Education Research Journal, 6*(3), 80-89.

CHAPTER 51

MALAYSIAN RESEARCH IN GEOMETRY

Cheng Meng Chew
Universiti Sains Malaysia

ABSTRACT

This chapter begins with a discussion of the importance of geometry and its implications for the primary and secondary mathematics curricula in Malaysia. Next, it provides a critical and comprehensive literature review of mathematics education studies in geometry that have been carried out in the country. These studies can be broadly classified into seven categories according to their main research purpose, namely teaching and learning of geometry without using computer, teaching and learning of geometry using computer, development and evaluation of courseware, assessment of van Hiele levels of geometric thinking, assessment of cognitive levels of geometric understanding, assessment of problem solving behaviours, and assessment of subject matter knowledge. Lastly, the conclusions and implications of the findings of these studies for future research in geometry in the country are discussed.

The First Sourcebook on Asian Research in Mathematics Education: China, Korea, Singapore, Japan, Malaysia, and India, pp. 1163–1191

INTRODUCTION

This chapter has three main objectives. The first objective is to discuss the importance of geometry in general and its implications for the Malaysian primary and secondary mathematics curricula in particular. The second objective is to provide a critical and comprehensive literature review of mathematics education studies in geometry that have been carried out in the country. The last objective is to provide conclusions and implications of the findings of these studies for future research in geometry in the country.

IMPORTANCE OF GEOMETRY

Geometry is recognized as a basic skill in mathematics (Hoffer & Hoffer, 1992; National Council of Supervisors of Mathematics [NCSM], 1977; National Council of Teachers of Mathematics [NCTM], 1989, 2000) for several important reasons. First, geometry is an important aid for communication. We use geometric terminology in describing shapes of objects and giving directions. A basic geometric vocabulary allows us to communicate to others observations that we have made about the natural and synthetic world around us (Sherard, 1981).

Second, geometry has important applications to real-life problems (Sherard, 1981). Many practical experiences involve problem-solving situations that require an understanding of geometric concepts and procedures, such as making frames, planning a garden, arranging a living room, determining the amount of wallpaper, carpet, paint, grass or fertilizer to buy, and other work situations (Hatfield et al., 2000; Hoffer, 1979; van de Walle, 2001). In addition, geometric concepts, such as points, line segments, curves and grids, are necessary for basic map reading skills (Sherard, 1981).

Third, geometry has important applications to topics in basic mathematics (Sherard, 1981). Geometric examples or models are frequently used to help students understand mathematical concepts. For example, the number line is helpful to illustrate various number concepts and operations (Sherard, 1981) and distances (Hoffer & Hoffer, 1992). Geometric regions and shapes are useful for teaching fractional numbers, equivalent fractions, ordering of fractions, computing with fractions (Sherard, 1981), as well as decimals and percents (Hatfield et al., 2000). Rectangular arrays are helpful in teaching properties of natural numbers (e.g., odd versus even or prime versus composite) or in teaching multiplication of natural numbers (Sherard, 1981). Linear, area, and volume measurements are directly related with geometric concepts (Sherard, 1981; van de

Walle, 2001). Coordinate systems provide us with the means for relating algebra and geometry (Sherard, 1981). Bar, line or circle graphs enable us to interpret and better understand statistical concepts (Sherard, 1981; Usiskin, 1980). Thus, "[g]eometry is a unifying theme to the entire mathematics curriculum and as such is a rich source of visualization for arithmetical, algebraic and statistical concepts" (Sherard, 1981, p. 20).

Fourth, geometry gives valuable preparation for courses in higher mathematics and the sciences and for a variety of careers requiring mathematical skills (Sherard, 1981). Geometry is essential for learning functions and calculus (Sherard, 1981; Usiskin, 1980), and thus all other fields of study that have calculus as a prerequisite (Sherard, 1981). For example, the derivative of a function can be visualized as the slope of the tangent line to the graph of the function or the definite integral as the area under a curve (Sherard, 1981; Usiskin, 1980). The study of groups, linear algebra, and graph theory received much motivation from symmetry groups, geometric pictures of the algebra, and networks respectively (Usiskin, 1980). In all these fields of study, "without the pictures we would be lost, and reasoning from the pictures using geometry provides help in understanding" (Usiskin, 1980, p. 419). Further, geometry is a foundation for study in such fields as architecture (Kennedy & Tipps, 2000; Sherard, 1981; van de Walle, 2001), engineering (Hoffer, 1979; Lam, 1994; Sherard, 1981), physics (Sherard, 1981; Tan, 1985), chemistry and biology (Serra, 1997; Sherard, 1981), geology (Sherard, 1981) and astronomy (Hoffer, 1979; Sherard, 1981). Geometric skills are also essential in art, design, graphics, animation, and dozens of other vocational and recreational settings (Kennedy & Tipps, 2000) as well as in various aspects of construction work (Sherard, 1981).

Fifth, geometry provides opportunities for developing spatial perception and visualization which is becoming recognized as an extremely important skill for success in mathematics and the sciences. We all need the ability to visualize objects in space and the relationships among objects in space as well as the ability to read two-dimensional representations of three-dimensional objects (Sherard, 1981).

Sixth, geometry can serve as a vehicle for stimulating and exercising general thinking skills and problem-solving abilities. Geometry provides students with opportunities to observe, compare, measure, guess, generalize and abstract, which can help them to learn how to discover properties of and interrelationships among classes of shapes for themselves and to become better problem solvers (Sherard, 1981). Furthermore, geometric explorations can develop problem-solving skills, and problem solving is one of the major reasons for studying mathematics (Malaysian Ministry of Education, 2003a; NCSM, 1977; NCTM, 1989, 2000; van de Walle, 2001).

Finally, there are cultural and aesthetic values to be derived from the study of geometry (Hoffer, 1979; O'Daffer & Clemens, 1976; Sherard, 1981; van de Walle, 2001). Geometry helps students to better understand and appreciate the world they live in. Geometric form and structure can readily be found in the natural universe, such as in the structure of the solar system, in geological formations, in rocks and crystals, in plants and flowers, and even in animals. It also pervades our synthetic universe: Art, design, graphics, animation, architecture, cars, machines, and virtually everything that humans create have elements of geometric form and structure (Sherard, 1981; van de Walle, 2001). Thus, a solid understanding and a complete appreciation of both the natural and synthetic universe can contribute to a full and well-rounded life of students (Sherard, 1981).

As such, it is important for every student to acquire an understanding of geometric concepts and skills as well as to develop an ability to think geometrically and apply geometric properties and relationships effectively in solving problems, both in mathematics and other disciplines, and in everyday life so as to function successfully in a rapidly changing technological society (Malaysian Ministry of Education, 2003a; NCTM, 1989, 2000). Most importantly, learning geometry will enable students to gain in what the NCTM refers to as mathematical power: problem solving, communication, reasoning, representing and connections (Cathcart et al., 2003).

Recognition of geometry as a basic skill in mathematics and therefore is an integral part of the mathematics curriculum has also resulted in an increased emphasis on geometry in the revised mathematics curriculum by the Malaysian Ministry of Education. In the past, geometry was neglected at the primary school level in favour of teaching arithmetic to allow children more time to develop and perfect computational skills. In fact, learning geometry was only formally introduced in primary education in Year 4 in the previous mathematics curriculum in which primary school pupils at the age of 10 learned two- and three-dimensional shapes (Malaysian Ministry of Education, 1998a). In contrast, the geometric concepts of two- and three-dimensional shapes have been formally introduced in Year 1 (at the age of 7) in the revised Primary School Mathematics Curriculum (Malaysian Ministry of Education, 2002).

LITERATURE REVIEW

In view of the importance of geometry in general and its implications for the Malaysian primary and secondary mathematics curricula in particular, many mathematics education studies in geometry have been conducted in the country. These studies can be broadly classified into seven categories according to their main research purpose: (1) teaching and learning of geometry without using computer (Ab. Rahim Ahmad, 1978; Gan, 2000; Tay, 2003); (2) teaching and learning of geometry using computer (Chew, 2007; Chew & Lim, 2010; Chew & Noraini Idris, 2006; Chong & Noraini Idris, 2002; Koo, Ahmad Rafi Mohamed Eshaq, Khairul Anuar Samsudin & Balachandher Krishnan, 2009; Kor & Lim 2009; Kor, Tan, & Lim, 2009; Lim, Chew, Chiew & Goh 2008; Noraini Idris, 2007; Nurul Hidayah Lucy, 2005); (3) development and evaluation of courseware (Kamariah Abu Bakar, Ahmad Fauzi Mohd Ayub & Rohani Ahmad Tarmizi, 2010; Kor, 1995; Lam, 2004; Rosanini Mahmud, Mohd Arif Hj Ismail & Lim, 2009); (4) assessment of van Hiele levels of geometric thinking (Chew & Lim, 2009; Chew, Lim & Noraini Idris, 2009; Chew & Noraini Idris, 2009; Noraini Idris, 1999; Sarojini, 1989); (5) assessment of cognitive levels of geometric understanding (Lee, 1982); (6) assessment of problem solving behaviours (Lee, 2002); and (7) assessment of subject matter knowledge (Wun, 2010). Each of these categories of studies is discussed next.

Teaching and Learning of Geometry Without Using Computers

Ab. Rahim Ahmad (1978), Gan (2000) and Tay (2003) investigated the teaching and learning of geometry without using computer. The objectives of Ab. Rahim Ahmad's (1978) study were to investigate whether there was a significant difference in achievement in recall and retention of the concepts of enlargement between a traditional teaching method and a new teaching method, as well as between male and female students, and whether there was a significant interaction effect between teaching method and gender. He employed a quasi-experimental design using two intact groups of Form Four (the fourth year of secondary school) students from a public national secondary school. For one week, the students in the experimental group were taught the concepts of enlargement using a new teaching method whereas the students in the control group were taught the concepts of enlargement using a traditional teaching method. The achievement in recall of the concepts of enlargement by the two groups of students was measured by administering the Recall Achievement Test immediately after the treatment. The achievement in retention of the concepts of enlargement by the two groups of students was measured by administering the Retention Achievement Test five weeks after the treatment.

In general, the results of the two-way between subjects analysis of variance (ANOVA) showed that was no significant difference in achievement in recall and retention of the concepts of enlargement between the traditional teaching method and the new teaching method. Out of the 24 hypotheses that were evaluated only 2 hypotheses showed a significant difference in achievement in retention of the concepts of enlargement: (a) For students whose mathematical abilities were not controlled, the new teaching method is more effective than the traditional teaching method in helping them to answer questions of knowledge type in the Retention Achievement Test; and (b) For students whose mathematical abilities were high, the traditional teaching method is more effective than the new teaching method in helping them to answer questions of problem solving type in the Retention Achievement Test.

Gan (2000) aimed to determine the effectiveness of programmed instructional material as complementary aids in the teaching of Reflection to Form Two (the second year of secondary school) students. He employed a quasi-experimental design using two intact groups of Form Two students from a rural secondary school. The students in the experimental group were taught the concepts of reflection using the programmed text, Learn It Yourself Reflection (LIY Reflection) which comprised 4 units while the students in the control group followed the normal classroom instruction. There was no significant difference in the students' pre-requisite test performance between the experimental and control groups. In addition, for both the experimental and control groups the mean score for each unit in the post-test was higher than the mean score for each unit in the pre-test. Further, the post-test mean score for each unit in the experimental group was higher than the post-test mean score for each unit in the control group. The results of the independent-samples t test showed that there was a significant difference in the post-test mean scores for Unit 2, Unit 3 as well as the overall unit between the experimental and control groups but there was no significant difference in the post-test mean scores for Unit 1 and Unit 4 between the two groups. The significant difference in the post-test mean score for the overall unit indicates that the LIY Reflection materials did play a major role in improving the students' performance in the post-test. Moreover, the students responded positively towards the use of the LIY Reflection materials with 100% of them stating that it was interesting to use the LIY Reflection materials in learning reflection and 85% of them recommending the use of programmed instructional material as complementary aids in the teaching of other topics in mathematics and in other subjects as well as for other students.

Tay (2003) examined the effects of a van Hiele-based instruction on the geometry achievement of Form One (the first year of secondary school) students and their van Hiele levels of geometric thinking. She employed a quasi-experimental design using two intact groups of students from a public secondary school in Temerloh, Pahang. The Geometry Achievement Test (GAT) and the Van Hiele Geometry Test (VHGT) (Usiskin, 1982) were used to assess the students' geometry achievement and van Hiele levels of geometric thinking respectively. A pilot test for the GAT registered a .88 test-retest correlation while the VHGT showed a test-retest reliability of .60, .63, .28 and .23 for the first four van Hiele levels. Tay designed 37 van Hiele-based lessons and instructional materials with a reorganized sequence of geometry concepts to be used over ten weeks of treatment in the experimental group. The lessons and materials emphasized hands-on activities and making conjectures in order to develop geometric thinking. The content and scope of these lessons and materials closely followed the Form One Geometry Syllabus and three Form One teachers with between 16 to 20 years of experience validated the lessons and materials. Fifteen students who were not in the actual study piloted ten of the instructional materials. The students in the experimental group were taught using the van Hiele-based instruction which was more student-centered whereas the students in the control group were taught using the textbook which was more teacher-centered. The results of the study showed that the students who followed the van Hiele-based instruction obtained a significantly greater geometry achievement than those taught with the traditional approach at $p < .01$. In addition, the experimental group showed higher levels of geometric thinking than the control group. Eleven students in the experimental group advanced from Level 1 to Level 2, while only four students in the control group showed a similar improvement.

Teaching and Learning of Geometry Using Computer

There are relatively more studies on the teaching and learning of geometry using computer than those without using computer. While the majority of these studies used The Geometer's Sketchpad (Chew, 2007; Chew & Lim, 2010; Chew & Noraini Idris, 2006; Chong & Noraini Idris, 2002; Kor & Lim 2009; Kor, Tan, & Lim, 2009; Lim, Chew, Chiew & Goh 2008; Noraini Idris, 2007; Nurul Hidayah Lucy, 2005), one study employed online collaborative learning activity for the teaching and learning of geometry (Koo, Ahmad Rafi Mohamed Eshaq, Khairul Anuar Samsudin & Balachandher Krishnan, 2009). Each of these categories of studies is discussed next.

There are several studies on the teaching and learning of geometry using The Geometer's Sketchpad (GSP). Some studies employed a quasi-experimental design (Chong & Noraini Idris, 2002; Noraini Idris, 2007; Nurul Hidayah Lucy, 2005) while others employed a case study research design (Chew, 2007; Chew & Lim, 2010; Chew & Noraini Idris, 2006; Kor & Lim 2009; Kor, Tan, & Lim, 2009; Lim, Chew, Chiew & Goh 2008).

Chong and Noraini Idris (2002) investigated the effect of GSP using graphic calculator (TI-92 Plus) on secondary students' van Hiele levels of geometric thinking. The study adopted a quasi-experimental design using two intact groups of students from a public secondary school in Kuala Lumpur. The experimental group (32 students) was taught circles using TI-92 Plus whereas the control group (33 students) was taught using the traditional method without using TI-92 Plus. The van Hiele Geometry Test (Usiskin, 1982) and a questionnaire were used to assess the students' van Hiele levels of geometric thinking and perceptions of using TI-92 Plus to learn circles, respectively. The results showed that TI-92 Plus had a significant effect on students' van Hiele levels of geometric thinking. In general, the number of students in the experimental group who progressed from lower levels to higher levels was higher than those in the control group. In addition, the students in the experimental group showed positive perceptions of using TI-92 Plus to learn circles as they could understand better the concepts of radius, diameter and circle. In fact, more than 70% of the students agreed that the use of TI-92 Plus could improve their van Hiele levels of geometric thinking.

Similar results were obtained by Noraini Idris (2007) who investigated the effects of GSP on secondary students' geometry achievement and van Hiele levels of geometric thinking. This quasi-experimental study was carried out in one of the secondary schools in Kuala Lumpur. A total of 65 Form Two students from the school were chosen for this research. The treatment group (N = 32) was taught geometry using GSP for ten weeks. At the same time the control group (N = 33) was taught by using the traditional approach without using GSP. The students' geometry achievement and van Hiele levels of geometric thinking were assessed using the Geometry Test and the van Hiele Geometry Test (Usiskin, 1982), respectively. In addition, a questionnaire was used to assess the students' perceptions of using GSP to learn geometry. The results showed that GSP had a significant effect on students' geometry achievement and van Hiele levels of geometric thinking. In addition, most of the students in the experimental group showed positive perceptions of using GSP to learn geometry.

Nurul Hidayah Lucy (2005) investigated the effects of GSP on secondary students' geometry achievement and van Hiele levels of geometric thinking. She found that 40 secondary school students who had undergone use of the GSP instructional program gained higher mathematics achievement scores and achieved higher geometric thinking levels as compared to their counterparts in the control group.

Unlike the three quasi-experimental studies, Chew and Noraini Idris (2006), Chew (2007), as well as Chew and Lim (2010) employed a case study research design to investigate primary pupils' or secondary students' learning of geometry in a phase-based instructional environment using GSP based on the van Hiele theory of geometric thinking. For example, Chew and Noraini Idris (2006) employed a case study research design to assess Form One students' learning of solid geometry in a phase-based instructional environment using manipulatives and GSP based on the van Hiele theory of geometric thinking. The study comprised three sessions. Session 1 consisted of individual pre-interviews and a pretest prior to instruction. Session 2 consisted of phase-based instruction using manipulatives and GSP. The phase-based instruction comprised five phases of learning, namely information, guided orientation, explicitation, free orientation and integration. Additionally, participant observations, audio- and video-tapings as well as student worksheets were used to assess students' learning of solid geometry during the five phases of learning. Session 3 consisted of individual post-interviews, a posttest and a survey after the instruction. Pre- and post-interviews were individually conducted with the students using the researcher-devised interview instrument which was based on the structure of Mayberry's (1981) interview instrument to determine their van Hiele levels of geometric thinking for each concept of cubes and cuboids prior to and after the instruction. The interview instrument comprised 12 items and required about 15 to 20 minutes to complete. Level 1 (recognition)-items assessed students' ability to recognise and name cubes and cuboids as well as to discriminate cubes and cuboids from rhomboids and parallelepipeds. Level 2 (analysis)-items assessed students' ability to identify properties of cubes and cuboids. Level 3 (informal deduction)-items assessed students' ability to understand definitions, class inclusion, similarity, congruence and implications regarding cubes and cuboids. Pre- and post-test were also administered to all the students using the Solid Geometry Achievement Test prior to and after the instruction to determine their achievement in solid geometry. The test comprised 25 items, that is, 8 factual items (5 multiple-choice and 3 short-answer), 12 application items (10 multiple-choice and 2 short-answer) and 5 short-answer problem-solving items. The Student Survey Form was used to elicit students' perceptions of using manipulatives and GSP to learn solid geometry.

The findings of the study showed that phase-based instruction using manipulatives and GSP could enhance students' van Hiele levels of geometric thinking of cubes and cuboids from Level 1 to Level 3. That is, the students progressed from recognizing cubes and cuboids as whole shapes as well as discriminating cubes and cuboids from rhomboids and parallelepipeds to an understanding of definitions, class inclusion, similarity, congruence and implications regarding cubes and cuboids. The phase-based instruction using manipulatives and GSP also enhanced students' achievement in content knowledge of solid geometry. Additionally, the results also revealed that students had positive perceptions of using manipulatives and GSP to learn solid geometry because they could understand better the properties of and relationships between cubes and cuboids.

Likewise, Chew (2007) employed a case study research design to investigate Form One students' learning of solid geometry in a phase-based instructional environment using GSP based on the van Hiele theory of geometric thinking. Specifically, it sought to examine the students' initial van Hiele levels of geometric thinking regarding cubes and cuboids, how their geometric reasoning about cubes and cuboids changed during Learning Periods 1 and 2, and how their van Hiele levels regarding cubes and cuboids changed after Learning Periods 1 and 2. The researcher employed purposeful sampling to select six case study participants from a class of mixed-ability Form One students. Individual interviews were conducted with the participants to determine their van Hiele levels regarding cubes and cuboids before and after Learning Period 1, and after Learning Period 2. All the interviews were audio- and videotaped. Documents (student worksheets accompanying the phase-based GSP instructional activities) and observations (participant-observations, audio- and videotapings of the three pairs of participants and researcher's field notes) were used to gather data on the participants' geometric reasoning while learning cubes and cuboids during Learning Periods 1 and 2. The duration of study was approximately four weeks. Data were analysed using within-case and cross-case analyses.

The researcher found that the participants' initial van Hiele levels regarding cubes and cuboids ranged from Level 0 to Level 2. During Learning Period 1, their geometric reasoning about cubes progressed from Level 1 to Level 2 or predominantly Level 2 whereas their geometric reasoning about cuboids progressed from Level 0 to predominantly Level 2, or from Level 1 to Level 2 or predominantly Level 2. After Learning Period 1, their van Hiele levels regarding cubes and cuboids progressed from Level 0 to Level 2, Level 1 to Level 2, or remained at Level 2. During Learning Period 2, their geometric reasoning about cubes progressed from Level 2 to predominantly Level 3, predominantly Level 2-Level 3 and predominantly Level 3, or predominantly Level 2. But, their geometric reasoning about cuboids progressed from Level 2 to predominantly Level 3, predominantly Level 2-Level 3 and predominantly Level 3, predominantly Level 2 and predominantly Level 2-Level 3, or predominantly Level 2. After Learning Period 2, their van Hiele levels regarding cubes and cuboids advanced from Level 2 to Level 3, or remained at Level 2.

While Chew and Noraini Idris (2006) and Chew (2007) investigated secondary students' learning of solid geometry, Chew and Lim (2010) investigated the progression of primary pupils' van Hiele levels of geometric thinking in a phase-based instructional environment using GSP. More specifically, the study sought to examine the levels and patterns of the pupils' van Hiele levels of geometric thinking of selected regular polygons before and after phase-based instruction using GSP. The researchers employed a case study research design and purposeful sampling to select a class of 26 mixed-ability Year Four pupils (the fourth year of primary school) from a rural primary school in Malaysia. A researcher-devised instrument based on Mayberry's (1981) instrument and scoring criteria was administered to the participants before and after the intervention to assess their van Hiele levels of geometric thinking of equilateral triangles, squares, regular pentagons, and regular hexagons. The results of the pretest showed that the participants' initial van Hiele levels were predominantly at Level 0 (Pre-recognition) for regular pentagons and regular hexagons but at Level 1 (Recognition) for equilateral triangles and squares. However, the results of the post-test revealed that the participants' van Hiele levels after the intervention were predominantly at Level 2 (Analysis) for all the selected regular polygons indicating that they had progressed from Level 0 to Level 2 or from Level 1 to Level 2. In addition, the participants with different levels of achievement in both Mathematics and English Language exhibited different patterns of progression after the intervention ranging from Level 0 to Level 1, Level 0 to Level 2, Level 1 to Level 2, or remained at Level 2 for the selected regular polygons.

Apart from these van Hiele theory-based studies, Lim, Chew, Chiew and Goh (2008), Kor and Lim (2009), as well as Kor, Tan and Lim (2009) employed a case study research design to promote secondary school mathematics teachers' innovative use of GSP through Lesson Study (LS) collaboration. Lim, Chew, Chiew and Goh (2008), for instance, reported how the LS process had enhanced the innovative use of GSP in one LS group. The LS group which consisted of 3 females and one male mathematics teachers completed two LS cycles during June 2007 through May 2008. Qualitative data were collected through written lesson plans, videotaped teaching and participants' individual interview. The findings of the study indicated positive changes in the mathematics teachers' basic knowledge and skills in using GSP to teach "lines and planes in three dimensions" to Form Four students in the first LS cycle and "loci in two dimensions" to Form Two students in the second LS cycle. These changes were evidenced in their mathematics lesson plans, GSP sketches, worksheets and videotaped teaching observations. Analysis of their interview transcripts also reveals positive acceptance and encouraging feedback about LS process that promotes peer support and collaboration. Consequently this has enhanced their confidence in using GSP to teach geometry at the secondary school level.

In a related study to the above, Kor and Lim (2009) reported how the LS process had enhanced the innovative use of GSP in two LS groups. Group A consisted of 4 mathematics teachers and 24 students (6 males and 18 females) whereas Group B consisted of 4 mathematics teachers and 22 students (10 males and 12 females). The students in both groups were around 16 years old. Six of the eight teachers were beginners in terms of their GSP knowledge and skills, and the students did not have any GSP knowledge or skills prior to the study. Qualitative data were collected through written lesson plans and interviews with individual participants. Additionally, the students in both groups answered a questionnaire consisting of 36 items which were adapted from PISA 2003 Student Questionnaire (PISA, 2003) after the first LS cycle. The purpose of the questionnaire was to examine the students' perception of the topic taught using GSP. The three subscales of the questionnaire were (a) general perception and interest about the topic (10 items, reliability coefficient α = .8977), (b) cognition or knowledge about the topic (14 items, reliability coefficient α = .8559), and (c) perceived classroom atmosphere or classroom setting during the lesson (12 items, reliability coefficient α = .8810). The scoring was in Likert scale ranging from 1 (*strongly disagree*) to 4 (*strongly agree*). Findings of the teacher interviews revealed that LS enabled teachers to learn new GSP skills effectively despite time conundrum. These teachers indicated that collegial support was the motivating factor that sustained the LS process. Feedbacks from questionnaires and student interviews also revealed that students perceived positively towards the GSP learning environment. In addition, all the means of the three subscales (*Mgeneral* = 3.60, *SD* = .31; *Mcontent* = 3.46, *SD* = .40; *Mclassroom* = 3.51, *SD* = .33) were above 3.00, indicating that the students had a positive perception of the overall GSP learning environment. Further, prior to the first lesson, a pre-test on the topic of "plans and elevations" was administered to the students in Group B. Next, they were given 50 minutes to explore the plans and elevations of the given solids using GSP. A similar post-test was administered to the students immediately after the intervention. The results showed that there was an improvement in the students' achievement. The mean score for the pre-test was 13.6 and the post-test recorded 28.4 out of the possible score of 48 (*M*pretest= 13.6, *SD* =6.95; *M*posttest = 28.4, *SD* = 9.14). The total average gain was 14.8 or a 31% increase from the pre-test score.

Similarly, in another related study to the above, Kor, Tan and Lim (2009) reported that a case study that aimed to motivate secondary mathematics teachers to innovatively integrate GSP in their mathematics classroom teaching through LS collaboration. A total of five teachers and a group of 24 students participated in this study. Qualitative data were collected from the teachers through open-ended questionnaires and individual interviews. Quantitative data were collected from the students using a Likert-scale questionnaire to elicit their perceptions of the new teaching approach. The LS group was mentored by a senior mathematics teacher who was more knowledgeable in GSP and therefore acted as the leader of the group. The LS group members who were initially novices in using GSP received constant training and support from the leader. The group selected the topic "plans and elevations" as the focus of their study. The results of the study indicated that the participating teachers managed to master the adoption, adaptation and appropriation of GSP integration in teaching the topic within a period of less than four months despite their hectic teaching periods. The teachers collaboratively developed a content template for the topic using GSP and several other templates with modified diagrams based on the past year examination questions were also created using GSP for reinforcement and revision purposes. The findings of the individual interviews with teachers revealed that they were motivated and excited to participate in the next cycle of LS collaboration. Furthermore, the students perceived positively towards the use of GSP in learning the topic.

Unlike the above GSP-based studies, Koo, Ahmad Rafi Mohamed Eshaq, Khairul Anuar Samsudin and Balachandher Krishnan (2009) aimed to: (a) to explore and investigate the nature of students' interaction and participation in an online collaborative learning (OCL) activity for teaching and learning of geometry called Diary of Discovering Geometry or Diary; and (b) investigate whether Diary was able to promote positive effects in cognitive and affective learning aspects as perceived by the students. They designed Diary based on the three principles of Engagement Theory (Kearsley & Shneiderman, 1999), namely relate, create and donate. The scope of geometry content in Diary comprised the basic concepts of geometry such as points, lines, planes and spaces in real life, and their attributes. In addition, other concepts and applications like shapes, polygons, symmetry, tessellations, perimeter, area, angle and volume were also suggested as topics for exploration.

The researchers employed a case study research design and the participants comprised 32 secondary school students. The majority of the students (that is 24 students or 75%) were lower secondary students, namely Form 1 (aged 13) and Form 2 (aged 14) students. A few Form 4 (aged 16) students (that is 8 students or 25%) were also invited to join Diary so that they could share their knowledge and guide their juniors in learning and exploring geometrical concepts and applications using Diary. There were 23 male and 9 female students in the sample and they had successfully subscribed to the Diary electronic group system or mailing list and continued to participate in Diary until the end of the activity. These participants were selected from four schools using purposive sampling and the selection criteria were: (1) able to access the Internet either at home or at schools; (2) possess a functional email address; (3) secure parental consent; and (4) volunteer to participate in this activity. The research instruments consisted of: (1) a project entry form to collect students' particulars before they start joining Diary; (2) survey questionnaires for the beginning and post-Diary stage that contained questions related to students' background information and various aspects on online collaboration; (3) interview questions to collect students' feedback after participating in Diary; (4) e-mails and electronic messages posted to the electronic group; and (5) research journals to record the activities and the process of implementing Diary from the beginning until the finishing stage.

The study was conducted for six and a half months in three stages, namely (1) the Beginning Stage, (2) the Icebreaking Stage and finally (3) the Collaborative Stage. The progress and development of Diary were observed and recorded in research journals and the electronic mails sent to the group and teacher coordinator were recorded. When Diary was introduced in the schools, project entry forms were given to each student and collected on the next day. After three weeks of exploration in the Beginning Stage, the participants were asked to complete the survey questionnaires mainly relating to the problems encountered in using Diary and their suggestions to overcome the problems. After the Collaborative Stage, the participants were interviewed via telephone at their preferred timeslot. Additionally, they were asked to complete the Post-Diary questionnaires by providing their demographics information and feedback on the perceived knowledge acquisition, interest gained in geometry and learning outcomes. The quantitative data collected were analyzed using descriptive statistics such as frequencies and percentages, and were represented in chats. Qualitative data were also used to complement the quantitative data.

The findings of the study indicated that the participants tended to follow rather than being pro-active in Diary online interaction. The participants' overall geometry knowledge gained in through using Diary was minimal. That is they were unable to: (1) demonstrate through online discussion, online project submission and interviews that they learnt geometry; (2) define the meaning of geometry; and (3) observe and report critically what kind of geometry concepts (shapes, areas, symmetry, etc) were used in their surrounding. This lack of geometry knowledge among the participants could be due to their overall low or passive participation and interaction in Diary. Only several participants were active or moderately active and relatively most of the exchanged messages were categorized as administrative purposes and socialization matter. In fact, all of these did not indicate that they had acquired the geometry knowledge discussed. Nevertheless, the majority of the participants perceived Diary positively such as "useful, interesting, promote knowledge sharing and increased their interest in geometry and computer." However, their motivation and interest in Diary decreased over time. In addition, individual participant motivation and personal quality did affect the process or outcomes of Diary. Only those participants who were responsible, committed and had high interpersonal skills reacted positively in Diary. These weaknesses could be remedied by introducing some interesting activities in order to maintain their interest throughout the activities in Diary.

Development and Evaluation of Courseware

There are several studies that developed and evaluated geometry courseware. These studies include Kor (1995), Lam, 2004; Rosanini Mahmud, Mohd Arif Hj Ismail and Lim (2009) and Kamariah Abu Bakar, Ahmad Fauzi Mohd Ayub and Rohani Ahmad Tarmizi (2010).

Kor (1995) developed and evaluated a Logo-based geometry package. She also examined the effects of the Logo-based geometry package on students' performances in specific geometrical skills and concepts. The students were divided into two groups based on their first semester mathematics monthly average results, namely high ability and low ability groups. Then 8 students were chosen from each group to represent the high ability and low ability groups, respectively. All the 16 students were not told which group of mathematical ability they were in. The students were given six days of instruction for a total of six sessions using the Logo-based geometry package. The total amount of time required for the completion of all the lessons was 8½ hours. The students were given a pre-test prior to and a post-test after the treatment to determine whether they had acquired the geometrical skills and concepts of angle, estimation, angle drawing, definition of quadrilaterals, and classification of quadrilaterals. The results of the dependent-samples *t* test on the pre-test and post-test scores showed that there was a gain in the scores of the subtests on angle, estimation, angle drawing, and classification of quadrilaterals for both the high ability and low ability groups. However, only the pre-test and post-test scores of the high ability group were significantly different.

Lam (2004) examined the effectiveness of structured cooperative computer-based instruction (CBI) in enhancing geometry achievement and metacognition. In addition, he also investigated the verbal interactions that characterized the CBI in comparison with a non-structured group CBI. Lam developed and evaluated two versions of a CBI courseware on the topic of Polygon in Form One Mathematics used by the experimental and control groups. He developed two versions of the CBI courseware using the authoring software Macromedia Authorware 5 in the Windows platform. One version of the CBI courseware was designed for cooperative learning of dyads with structured interaction for the experimental group while the other was designed without structured interaction for the control group. The structured group version was provided with generative or metacognitive prompts to promote verbal interactions whereas the unstructured group version was not provided with the prompts. The researcher employed a quasi-experimental pretest-posttest research design. The sample of the study consisted of four intact classes of Form One students with mixed-gender and mixed-ability composition in a fully residential school located in a sub-urban area of Kuantan, Pahang. Two classes each were randomly assigned as experimental and control groups. The results of the study showed that: (1) there were significant differences in geometry achievement and metacognition between the experimental and control groups; (2) there were no significant interaction effects between treatment and mathematics achievement of the students for both the geometry achievement and metacognition; (3) the experimental dyads exhibited some patterns of interaction that suggested better quality of generative verbal interactions than the four control dyads, particularly those interactions related to questioning, explaining, reviewing, giving examples and predicting; and (4) other factors like prior knowledge, terminology, clarity of questions and the nature of the task might influence the quality of these verbal interactions besides the factor of structuring the interaction between members of the experimental dyads.

The focus of Rosanini Mahmud, Mohd Arif Hj Ismail and Lim's (2009) study was different from Kor's (1995) study. Rosanini et al. aimed to develop and evaluate a CAI courseware called 'G-Reflect' using GSP. They employed a quasi-experimental design to examine the effects of 'G-Reflect' on students' achievement and motivation in learning reflections. The 'G-Reflect' was developed based on the ADDIE instructional system design model. The three instruments used were Courseware Evaluation Checklist, Instructional Materials Motivation Survey (IMMS) questionnaire and Diagnostic Test which were validated by a panel of subject matter and instructional design experts. The reliability for 'G-Reflect' evaluation checklist and the IMMS was 0.79 and 0.72, respectively. The sample consisted of 34 students of Form 2DR (treatment group) and 34 students of Form 2GD (control group). The quantitative data were analyzed using descriptive and inferential statistics and inferentially. Based on the results of the Courseware Evaluation Checklist, the 'G-Reflect' courseware was assessed as a good courseware to teach the topic of reflections. The results of the Diagnostic Test showed that there was a significant difference in the mean scores obtained, t (67) = 10.162, $p < .05$, indicating that the treatment group performed better in the test as compared to the control group. In terms of motivation, the results of the IMMS questionnaire showed that the students from the treatment group were highly motivated in learning topic of reflections.

Kamariah Abu Bakar, Ahmad Fauzi Mohd Ayub and Rohani Ahmad Tarmizi (2010) compared the effects of three courseware (that is, GeoGebra, E-transformation and V-transformation) on Form Two students' achievement in three topics of transformations, namely reflections, translations and rotations. GeoGebra is an open source software that combines geometry, algebra and calculus into a single easy-to-use package for teaching and learning mathematics from primary to tertiary level and it is available for free at www.geogebra.org under General Public License. The second courseware, E-transformation was developed by the researchers using a software called Lecture-Maker. E-transformation consisted of a video showing a teacher explaining the topics in transformations followed by animations to help students understand the topics. The third courseware, V-transformation was developed by a group of researchers using Macromedia Flash and Lecture-Maker to help students visualize the concepts of reflections, translations and rotations by using animations.

The researchers employed a true experimental design and the sample comprised 101 Form Two students studying in a national secondary school in Malaysia who were randomly assigned into three separate groups. One group underwent instruction on reflections, translations and rotations utilizing GeoGebra while the other two groups underwent instruction on reflections, translations and rotations using E-transformation and V-transformation, respectively. However, the researchers did not use a control group due to the fact that all the three groups underwent computer-based instruction. All the three groups of students underwent four phases of the study, namely: (1) testing phase for the pre-test, (2) introduction to the courseware; (3) integrated teaching and learning of reflection, translation and rotation using the respective courseware and a Learning Activity Module; and (4) testing phase for the post-test.

The results of the study showed that there were significant differences between the pre- and post-test achievement scores of each group that underwent instruction on reflections, translations and rotations utilizing GeoGebra, E-transformation and V-transformation, respectively. The results indicated that students who learned reflections, translations and rotations using GeoGebra, E-transformation or V-transformation showed an increase in their achievement scores after using the courseware. But, there was no significant difference in the post-test achievement scores of the three groups, indicating that students who used GeoGebra, E-transformation or V-transformation performed just as well on the post-test regardless of which courseware was utilized in learning reflections, translations and rotations. In addition, there was no significant difference in the post-test achievement scores of each of the three topics of transformations, suggesting that for each topic included in the post-test, students who used GeoGebra, E-transformation or V-transformation had the same skills when answering questions related to the topics in transformations. In general, the results of this study had shown that computer technology was effective in teaching transformations at Malaysian secondary school level. But, the results did not indicate which software was more effective in the teaching and learning of reflections, translations and rotations.

Assessment of Van Hiele Levels of Geometric Thinking

There are several studies that assessed students' geometric thinking. These studies include Chew and Lim (2009), Chew, Lim and Noraini Idris (2009), Chew and Noraini Idris (2009), Noraini Idris (1999) and Sarojini (1989). Each of these studies is discussed next.

Chew and Lim (2009) assessed pre-service secondary mathematics teachers' geometric thinking using the van Hiele Geometry Test. The 25-item, multiple-choice, paper-and-pencil test was developed by the Cognitive Development and Achievement in Secondary School Geometry Project (Usiskin, 1982) based on the van Hiele theory of geometric thinking. The participants comprised 147 pre-service secondary mathematics teachers who attended a mathematics teaching methods course in the academic year 2008/9 in a local public university. The data were analysed based on the '4 of 5 criterion' to minimise the chance of a student being at a level by guessing (Mason, 1997). The results showed that 16 (10.9%) of the participants were at Level 0, 52 (35.4%) were at Level 1, 62 (42.2%) were at Level 2, 9 (6.1%) were at Level 3, 1 (0.7%) were at Level 4, none (0.0%) was at Level 5, and 7 (4.8%) could not be assigned a van Hiele level because their responses did not fit the '4 of 5 criterion.' The majority (88.5%) of the participants were at or below van Hiele Level 2. This result is worrying and signals the urgency to enhance these pre-service mathematics teachers' levels of geometric thinking so that they can promote higher levels of geometric thinking among their students.

In a related study to the above, Chew, Lim and Noraini Idris (2009) examined the differences in the modal response patterns among the pre-service secondary mathematics teachers with different van Hiele levels of geometric thinking. The response patterns of the participants in the five subtests of the van Hiele Geometry Test were represented by the letters, *c* (correct response) and *i* (incorrect response). Thus, there were five letters for each subtest which represented the correct or incorrect responses to the five items in each subtest. For example, for the Level-0 thinkers: the response patterns for the five Level-1 items, *cccii* (25.0%) showed that 25.0% of the Level-0 thinkers correctly answered Items 1, 2 and 3 but incorrectly answered Items 4 and 5; the response patterns for the five Level-2 items, *cicci* (18.8%) showed that 18.8% of the Level-0 thinkers correctly answered Items 6, 8 and 9 but incorrectly answered Items 7 and 10; and so on. Further, they also examined whether there was a significant correlation between the pre-service secondary mathematics teachers' van Hiele level and their mathematics teaching methods course grade.

The results of this study indicated that for Level-1 items, the modal response patterns for the Level-0 thinkers were *cccii* (25.0%) and *cicci* (25.0%). The former modal response pattern showed that 25.0% of the Level-0 thinkers incorrectly answered Items 4 and 5 while the latter modal response pattern indicated that 25.0% of the Level-0 thinkers incorrectly answered Items 2 and 5. This implied that 25.0% of the Level-0 thinkers were not able to recognize triangles (Item 2) and squares (Item 4), and 50.0% of the Level-0 thinkers were not able to recognize parallelograms. For Level-2 items, the modal response patterns for the Level-0 thinkers were *cicci* (18.8%) and *ciicc* (18.8%). The former modal response pattern showed that 18.8% of the Level-0 thinkers incorrectly answered Items 7 and 10 while the latter modal response pattern indicated that 18.8% of the Level-0 thinkers incorrectly answered Items 7 and 8. This indicated that 37.6% of the Level-0 thinkers were not able to identify the correct properties of rectangles (Item 7), 18.8% of the Level-0 thinkers were not able to identify the correct properties of rhombuses (Item 8) and kites (Item 10). But, for Level-2 items, the modal response patterns for the Level-1 thinkers were *ccici* (13.5%) and *ciici* (13.5%). The former modal response pattern showed that 13.5% of the Level-1 thinkers incorrectly answered Items 8 and 10 while the latter modal response pattern indicated that 13.5% of the Level-1 thinkers incorrectly answered Items 7, 8 and 10. This implied that 13.5% of the Level-1 thinkers were not able to identify the correct properties of rectangles (Item 7), 27.0% of the Level-1 thinkers were not able to identify the correct properties of rhombuses (Item 8) and kites (Item 10). The results of this study also showed that there was a weak positive correlation between the pre-service secondary mathematics teachers' van Hiele level and their mathematics teaching methods course grade but it was not significant. This was probably due to the fact that no geometry lesson was given to them before and during the mathematics teaching methods course or none of them had taken a formal course in geometry prior to taking the mathematics teaching methods course.

In a another related study to the above, Chew and Noraini Idris (2009) examined whether there was a significant difference in the pre-service secondary mathematics teachers' van Hiele levels of geometric thinking in terms of gender and the highest level of school mathematics studied before entering university. The results of this study indicated that there was no statistically significant difference in the pre-service secondary mathematics teachers' van Hiele levels of geometric thinking in terms of gender. The study also found that there was no statistically significant difference in van Hiele levels of geometric thinking between the pre-service secondary mathematics teachers who studied Form Six Mathematics and those who studied Matriculation Mathematics before entering university. This is not surprising as the grades for Form Six Mathematics and Matriculation Mathematics are awarded by the Malaysian Ministry of Education and are recognized by the Malaysian government as the equivalent entry requirement to Malaysian public universities.

Noraini Idris (1999) attempted to identify secondary school students' van Hiele levels of geometric thinking of triangles and quadrilaterals using Burger and Shaughnessy's (1986) drawing, identifying and defining, as well as sorting tasks. The sample of this study consisted of two Form Two students, selected by their class teachers based on their performance in the First Term Examination. Videotaped clinical interviews were used to administer the experimental tasks in the school's library. The two participants, P and Q, were asked to complete the three tasks using think aloud technique. On the drawing tasks, P who seemed to focus on the components of the shapes and who realized that the components could be varied was assigned Level 2. But, Q who drew shapes that showed the sides and angles and could inter-relate shapes was assigned Level 3. On the identifying tasks, P who contrasted shapes and identified them explicitly by means of their properties was assigned to Level 2. In contrast, Q who gave minimal characterization of the shapes by using other types of shapes was assigned Level 3. On the third tasks, P sorted triangles and quadrilaterals by using their properties were considered indicative of Level 2 thinking. In addition, P seemed to have difficulty with the use of terminology. However, Q who referred explicitly to a variety of types of triangles and quadrilaterals in the sorting tasks was considered indicative of Level 3 thinking. Further, Q remembered terminology such as rectangle, parallelogram, and square, and used them appropriately. Moreover, Q became more fluent in talking geometry as she moved through the tasks. Q also displayed the ability to reason well both inductively and deductively.

Sarojini (1989) investigated the validity of the van Hiele theory of geometric thinking using two geometric concepts, namely similar triangles and congruent triangles. The sample comprised students from two Form Four Arts classes and two Form Four Science classes. Two paper-and-pencil achievement tests (Test 1 for similar triangles and Test 2 for congruent triangles) and one test on van Hiele levels of geometric thinking (Test 3 for both concepts) were administered to the participants. The highest level of geometric thinking tested was Level 2. The results of the Guttman's scalogram analysis indicated that the levels of geometric thinking as exemplified by the Test 3 items did form a hierarchy for both similar triangles and congruent triangles. Correlation analyses and Chi square tests of independence between Test 3 and the two achievement tests showed a moderately strong linear relationship for the Science students. This means that if a Science student had a high level of geometric thinking in either similar triangles or congruent triangles, he or she would have a high level of achievement in either similar triangles or congruent triangles as well. However, there was no such relationship between Test 3 and the two achievement tests for the Arts students. Generally, the Arts students appeared to perform better in the achievement tests than in Test 3. In addition, the results of the study showed that if a student performed well in one concept, there was a moderate chance of the student performing equally well in the other concept.

Assessment of Cognitive Levels of Geometric Understanding

Lee's (1982) study is the only study that assessed students' cognitive levels of geometric understanding. She explored the relationship between a priori logical development of six basic concepts in transformation, and the cognitive levels of understanding of Malaysian secondary school pupils. The six basic concepts in transformation are Reflection, Rotation, Translation, Enlargement, Stretching and Shearing. She systematically developed a series of six tests. The aim of each test was to assess students' levels of understanding of each concept in transformation. A panel of judges who were experienced in mathematics teaching was requested to provide opinions on specific aspects of the six tests. The tests were then administered to a sample of 256 Form Four and Lower Six students. In addition, interviews were conducted on 36 Form Four pupils.

The findings of the study showed that the judges had different opinions regarding the a priori sequence of cognitive levels of understanding. The analysis of facility indices and hierarchies of levels indicated that the observed sequences of levels of understanding differed from those based on the a priori logical classification but the observed sequences were consistent among the two groups of students. The students' incorrect responses varied across the six basic concepts with the concept of Shearing appeared to be the most difficult for the students. The students generally followed a step-by-step procedure to obtain the resulting figure of an object after a transformation but they had difficulties with transformations involving slanting invariant lines as well as invariant points not on the object. The findings of the study also revealed that there was a mismatch between a priori logical development of the six basic concepts in transformation and the cognitive levels of understanding of the students in the sample. Thus the findings of the study are limited in generalizability. Lee recommended that further in-depth studies that take into account the differences due to rural and urban location, gender, streams as well as other differences in Malaysian students should be conducted. Furthermore, the observed difficulties encountered by the students in the study should be examined further.

Assessment of Problem Solving Behaviours

Lee (2002) examined secondary school students' cognitive and metacognitive processes in solving geometric problems. The sample consisted of 30 Form Four students from a secondary school in the Klang Valley. The students performed the Geometry Construction Activities individually whereby they were required to construct geometric figures with specific conditions and they were encouraged to use the "Think-aloud" approach while solving the problems. Next, retrospection and interview were conducted with the participants. The problem solving and the interview sessions were video-recorded, transcribed and parsed into episodes of problem solving activities: reading, analysis, exploring, planning, implementing and verifying. The data were analyzed using protocol analysis and time-line graphs to facilitate interpretation of the participants' problem solving behaviours. The findings of the study showed that in solving a routine problem, the participants' problem solving process had become habitual without much need for recalling of geometric knowledge and skills. For an unfamiliar problem, the successful problem solvers could easily recall, retrieve and relate the concepts to the task and the process was further enhanced by self-questioning, paraphrasing the problem and making sketches to visualise the situation. While they spent a considerable amount of time in planning and exploring, implementing the solution was much shorter. Although the unsuccessful problem solvers showed their knowledge of geometry concepts, they felt short of establishing the inter-relationships of these concepts to meet the requirement of the problem resulting in their failure to accomplish the task. Their problem solving activities were mostly single-tracked and non-directional in nature. They were also indecisive in giving up despite having selected non-fruitful alternatives. However, she could not establish a single step-by-step procedure for solving problems of the successful problem solvers and the common activities for the unsuccessful problem solvers. Nevertheless, Lee proposed the inclusion of two additional elements of "solver's mathematical knowledge" and the ability to "retrieve" them in a problem solving model.

Assessment of Subject Matter Knowledge

Wun (2010) investigated pre-service secondary mathematics teachers' five basic types of subject matter knowledge (SMK) of perimeter and area, namely conceptual knowledge, procedural knowledge, linguistic knowledge, strategic knowledge and ethical knowledge. He also aimed to investigate the participants' levels of SMK of perimeter and area. Data were collected using clinical interviews which were audio- and videotaped. The sample consisted of 8 pre-service secondary mathematics teachers who enrolled in the 4-year Bachelor of Science with Education program and attending a Mathematics Teaching Methods course at a public university in Peninsular Malaysia. They were chosen based on their majors and minors in mathematics, biology, chemistry or physics education. The main findings of this study showed that: (a) Most of the participants did not know that there is no relationship between perimeter and area, and none of them were able to develop the formula for the area of a rectangle; (b) The majority of the participants had adequate procedural knowledge of calculating perimeter and area of composite figures; (c) Most of the participants used appropriate mathematical symbols to write the formula for the area of a rectangle, parallelogram, triangle and trapezium. All of them understand the general measurement convention that perimeter is measured in linear units while area is measured in square units. But, they had limited knowledge about the conventions pertaining to writing and reading of Standard International area measurement units; (d) The participants used the cut and paste strategy, partition strategy and algebraic method to develop the formula for the area of a parallelogram, triangle and trapezium, respectively; and (e) All the participants had taken the effort to justify the selection of shapes that have a perimeter and an area. However, the majority of them did not check the correctness of their answers for the perimeters and areas. In addition, only one participant had a high level of subject matter knowledge, six had a medium level and one had a low level. Wun concluded that most of them lacked SMK of perimeter and area that they are expected to teach when they graduate.

CONCLUSIONS AND IMPLICATIONS

Geometry, without doubt, is a basic skill in mathematics. For this reason, it is essential that geometry receives greater attention in both instruction and research. At present, the Malaysian mathematics education studies in geometry can be classified into seven broad categories based on their main research purpose: (1) teaching and learning of geometry without using computer; (2) teaching and learning of geometry using computer; (3) development and evaluation of courseware; (4) assessment of van Hiele levels of geometric thinking; (5) assessment of cognitive levels of geometric understanding; (6) assessment of problem solving behaviours; and (7) assessment of subject matter knowledge. Thus, future research should attempt to extend from the present categories of studies using different samples, topics, research designs and methods as well as technologies in order to contribute to the knowledge base of research in mathematics education in general and geometry education in particular. In addition, future research should attempt to explore other areas such as the development of geometric understanding from other theoretical perspectives like Piaget and cognitive science besides the van Hieles, spatial abilities, spatial visualisation, geometric language, representations of geometric ideas and technological pedagogical content knowledge.

Presently, there are relatively more studies on the teaching and learning of geometry using computer than those without using computer. In addition, the studies on the teaching and learning of geometry without using computer were conducted at the secondary school level (e.g. Ab. Rahim Ahmad, 1978; Gan, 2000; Tay, 2003) using quasi-experimental design. Thus, further studies on the teaching and learning of geometry without using computer should be conducted using other research designs such as case study especially at the primary school level and involving concrete or physical manipulatives like three-dimensional models of solid geometry, pattern blocks, colour tiles, tangrams, geoboards, snap cubes, and so on.

Since the majority of the studies on the teaching and learning of geometry using computer utilised GSP as an instructional tool, further studies should employ other dynamic geometry software such as Geogebra. GeoGebra is an open-source software that combines geometry, algebra and calculus into a single easy-to-use package for teaching and learning mathematics from primary to tertiary level and it is available for free at www.geogebra.org under General Public License. In Geogebra, constructions can be made with points, vectors, segments, lines, polygons, conic sections, and functions and all of them can be changed dynamically. Further, more studies on the teaching and learning of geometry using computer should be carried out at the primary school level as the majority of the previous studies were conducted at the secondary school level (e.g. Chew, 2007; Chew & Noraini Idris, 2006; Chong & Noraini Idris, 2002; Koo, Ahmad Rafi Mohamed Eshaq, Khairul Anuar Samsudin & Balachandher Krishnan, 2009; Kor & Lim 2009; Kor, Tan, & Lim, 2009; Lim, Chew, Chiew & Goh 2008; Noraini Idris, 2007; Nurul Hidayah Lucy, 2005).

Likewise, the studies on the development and evaluation of courseware were mainly conducted at the secondary school level (e.g. (Kamariah Abu Bakar, Ahmad Fauzi Mohd Ayub & Rohani Ahmad Tarmizi, 2010; Kor, 1995; Lam, 2004; Rosanini Mahmud, Mohd Arif Hj Ismail & Lim, 2009). Therefore, further studies on the development and evaluation of courseware should be conducted at the primary school level. These studies should also employ other instructional theories and geometry software to further enhance the effectiveness of the courseware in the teaching and learning of geometry.

In assessing the van Hiele levels of geometric thinking, only a relatively narrow spectrum of geometric concepts has been studied, that is mainly plane geometry especially polygons. Thus, there are vast areas left open for further investigation such as measurement, transformations, loci in two dimensions, geometric construction, lines and planes in three dimensional geometry, plan and elevation, and the earth as a sphere. In terms of educational levels, more research should be conducted on primary pupils' and pre-service primary school teachers' van Hiele levels of geometric thinking as the samples of the previous studies were secondary students or pre-service secondary school teachers (e.g. Chew & Lim, 2009; Chew, Lim & Noraini Idris, 2009; Chew & Noraini Idris, 2009; Noraini Idris, 1999; Sarojini, 1989). Besides, more studies should be conducted to examine the cognitive levels of geometric understanding in other geometry topics besides transformations (e.g. Lee, 1982) at both primary and secondary school levels.

As Lee's (2002) study only focused on geometrical constructions and with the continuing emphasis in problem solving in the current school mathematics curriculum, more research on problem solving in geometry should be conducted especially in other topics such as perimeter and area, solid geometry, loci in two dimensions, transformations, lines and planes in three dimensions, plan and elevation, and the earth as a sphere. A noted behavior in her study was the spontaneous realization of the solution or insight set by a few students without any apparent reason which enabled then to succeed in solving the geometric problems. Thus, future research could consider this aspect to gather more information of this phenomenon in geometric problem solving. In addition, mathematicians should be selected as a sample and the 'think-aloud' approach should be employed to investigate their problem solving behaviours because they are good problem solvers and role models for the students.

Wun's (2010) study only focused on pre-service secondary school mathematics teachers' subject matter knowledge of perimeter and area. Hence, further research should examine subject matter knowledge of other measurement concepts such as time, length, mass, surface area, volume and other geometry concepts as well. Further research should examine pre-service secondary school mathematics teachers' subject matter knowledge of perimeter and area through the observation of a series of microteachings of the entire topic of perimeter and area as well as analysis of documents such as lesson plans and accompanying worksheets. Further, pre- and post-instructional clinical interviews should be conducted to examine the changes in the participants' subject matter knowledge of perimeter and area. In addition, further research should examine pre-service secondary school mathematics teachers' belief about perimeter and area and teaching and learning of the topics as well.

An additional area of further research is Mathematics textbooks analysis. Kor (1995) analysed two Malaysian secondary Mathematics textbooks as part of her study. She found that the textbooks did not help students learn geometry effectively because they did not have the opportunity to explore the properties of quadrilaterals or to develop spatial thinking and geometry problem solving skills. Thus, further research should examine the presentation of geometry contents and teaching approaches in primary and secondary Mathematics textbooks particularly in terms of the development of students' geometric reasoning in order to improve the teaching and learning of geometry.

REFERENCES

Ab. Rahim Ahmad (1978). *Perbandingan kesan-kesan dua kaedah mengajar dalam menyampaikan konsep besaran dalam matematik kepada murid-murid Tingkatan IV sekolah menengah kebangsaan [A comparison of the effects of two methods of teaching the mathematical concept of enlargement to Form IV students in a national secondary school]*. Unpublished M.Ed. thesis, University of Malaya, Malaysia.

Cathcart, W.G., Pothier, Y.M., Vance, J.H., & Bezak, N.S. (2003). *Learning mathematics in elementary and middle schools* (3rd ed.). Upper Saddle River, NJ: Merrill Prentice Hall.

Chew, C. M. (2007). *Form one students' learning of solid geometry in a phase-based instructional environment using The Geometer's Sketchpad.* Unpublished PhD thesis, University of Malaya, Malaysia.

Chew, C. M., & Lim C. S. (2009) Assessing pre-service secondary mathematics teachers' geometric thinking. In Mohd Tahir Ismail, & Adli Mustafa (Eds.), *Proceedings of the 5th Asian Mathematical Conference* (Vol. III, pp. 685 – 692). Penang: Universiti Sains Malaysia.

Chew, C. M. & Lim, C. S. (2010). Developing primary pupils' geometric thinking through phase-based instruction using The Geometer's Sketchpad. In Y. Shimuzu, Y. Sekiguchi & K. Hino (Eds.), *Proceedings of the Fifth East Asia Regional Conference on Mathematics Education (EARCOME 5)* (Vol. 2, pp. 496-503). Tokyo, Japan: Japan Society of Mathematical Education.

Chew, C. M., Lim, H. L., & Noraini Idris (2009). Pre-service secondary mathematics teachers' geometric thinking and course grade. In U. H. Cheah, Wahyudi, D. R. Peter, K. T. Ng, P. Warabhorn, & A. Julito (Eds.), *Proceedings of the Third International Conference on Science and Mathematics Education (CoSMEd 2009)* (pp. 255-262). Penang: SEAMEO, RECSAM.

Chew, C. M., & Noraini Idris (2006). Assessing Form One students' learning of solid geometry in a phase-based instructional environment using manipulatives and The Geometer's Sketchpad. *Proceedings of the Third International Conference on Measurement and Evaluation in Education (ICMEE 2006)* (pp. 533-543). Penang: Universiti Sains Malaysia.

Chew, C. M., & Noraini Idris (2009). Preservice secondary mathematics teachers' van Hiele levels of geometric thinking. *Proceedings of the First International Conference on Educational Research and Practice (ICERP 2009)* (pp. 1191-1198). Selangor: Universiti Putra Malaysia.

Chong, L. H. & Noraini Idris (2002). Pembelajaran geometri menggunakan perisian Geometer's Sketchpad (TI-92 Plus) dan kaitannya dengan tahap pemikiran van Hiele dalam geometri. *Prosiding Persidangan Matematik Teknologi Asia (ACTM 2002): Sesi Khas Bahasa Melayu* (pp. 73-83). Melaka: Universiti Multimedia Melaka.

Gan, S. P. (2000). *Effectiveness of programmed instructional material in the teaching of Reflection to Form Two pupils in a rural school.* Unpublished master's dissertation, University of Malaya, Malaysia.

Hatfield, M. M., Edwards, N. T., Bitter, G.G., & Morrow, J. (2000). *Mathematics methods for elementary and middle school teachers* (4th ed.). New York: John Wiley & Sons, Inc.

Hoffer, A. R. (1979). *Geometry: A model of the universe.* Menlo Park, California: Addison-Wesley Publishing Company.

Hoffer, A. R., & Hoffer, S. A. K. (1992). Geometry and visual thinking. In T. R. Post (Ed.), *Teaching mathematics in grades K-8: Research-based mathematics* (2nd. ed., pp. 249-227). Boston: Allyn and Bacon.

Kamariah Abu Bakar, Ahmad Fauzi Mohd Ayub and Rohani Ahmad Tarmizi (2010). Utilization of computer technology in learning transformation. *International Journal of Education and Information Technologies, 4*(2), 91-99.

Kearsley, G. and Shneiderman, B. (1998). Engagement theory: A framework for technology-based teaching and learning. *Educational Technology, 38*(5), 20-23.

Kennedy, L. M., & Tipps, S. (2000). *Guiding children's learning of mathematics* (9th ed.). Belmont, CA: Wasworth/Thompson Learning.

Koo, Ahmad Rafi Mohamed Eshaq, Khairul Anuar Samsudin and Balachandher Krishnan (2009). An evaluation of a constructivist online collaborative learning activity: A case study of geometry. *The Turkish Online Journal of Educational Technology, 8*(1), 15-25.

Kor, A. L. (1995). *The development and evaluation of a logo-based geometry package.* Unpublished M.Ed. thesis, University of Malaya, Malaysia.

Kor Liew Kee & Lim Chap Sam (2009). Lesson Study: A Potential Driving Force behind the Innovative Use of Geometer's Sketchpad, Journal of Mathematics Education, 2 (1), pp.69-82

Kor, L. K., Tan, K. A., & Lim, C. S. (2009). Use of Geometer's Sketchpad (GSP) in teaching "Plan and Elevation". In U. H. Cheah, Wahyudi, R. P. Devadason, K. T. Ng, W. Preechaporn, J. C. Aligaen (Eds.), *Proceedings of the 3rd International Conference on Science and Mathematics Education (CoSMED 2009)*, pp. 336-342, 10-12 November 2009 at SEAMEO-RECSAM, Penang, Malaysia.

Lam, K. K. (2004). *Effects of structured cooperative computer-based instruction on geometry achievement and metacognition.* Unpublished Phd thesis, University of Malaya, Malaysia.

Lam, S. Y. (1994). *Spatial ability, formal reasoning ability, and field dependence-independence as predictors of Form IV students' achievements in geometry and engineering drawing.* Unpublished Masters Thesis, University Malaya, Malaysia.

Lee, S. E. (1982). *Understanding of basic concepts in transformation geometry: A study of a sample of pupils in Forms IV and VI.* Unpublished M.Ed. thesis, University of Malaya, Malaysia.

Lee, S. E. (2002). *Cognitive and metacognitive processes in solving geometric problems.* Unpublished Phd thesis, Universiti Sains Malaysia, Penang, Malaysia.

Lim, C. S., Chew, C. M., Chiew, C. M., & Goh, S. I. (2008). *Promoting Innovative Use of Geometer's Sketchpad (GSP) through Lesson Study Collaboration: A case study.* Paper presented at Seminar Kebangsaan Pendidikan Sains dan Matematik 2008, anjuran Persatuan Pendidikan Sain dan Matematik Johor dengan kerjasama Jabatan Pelajaran Negeri Johor dan Fakulti Pendidikan UTM pada 11-12 Oktober 2008 di Universiti Teknologi Malaysia.

Malaysian Ministry of Education. (1998a). *Huraian sukatan pelajaran matematik KBSR Tahun 4* [Year 4 KBSR Mathematics Syllabus]. Kuala Lumpur: Curriculum Development Centre.

Malaysian Ministry of Education. (2002). *Integrated curriculum for primary schools: Curriculum specifications, Mathematics Year 1.* Kuala Lumpur: Curriculum Development Centre.

Malaysian Ministry of Education. (2003a). *Integrated curriculum for secondary schools: Curriculum specifications, Mathematics Form 1.* Kuala Lumpur: Curriculum Development Centre.

National Council of Supervisors of Mathematics. (1977). National Council of Supervisors of Mathematics Position paper on basic skills. *Arithmetic Teacher, 25*(1), 19-22.

National Council of Teachers of Mathematics. (1989). *Curriculum and evaluation standards for school mathematics.* Reston, VA: Author.

National Council of Teachers of Mathematics (NCTM). (2000). *Principles and standards for school mathematics.* Reston, VA: Author.

Noraini Idris (1999). Students' intellectual growth in geometry: A case study of Form Two students. *Jurnal Pendidikan, 20,* 71-82.

Noraini Idris. (2007). The effect of Geometer's Sketchpad on the performance in geometry of Malaysian students' achievement and van Hiele geometric thinking. *Malaysian Journal of Mathematical Sciences, 1*(2), 169-180.

Nurul Hidayah Lucy Bt Abdullah (2005). *The effectiveness of using dynamic geometry software on students' achievement in geometry.* Unpublished master's thesis, University Malaya, Kuala Lumpur, Malaysia.

O'Daffer, P. G., & Clemens, S. R. (1976). *Geometry: An investigative approach.* Menlo Park, California: Addison-Wesley Publishing Company, Inc.

Rosanini Mahmud, Mohd Arif Hj Ismail & Lim, (2009). Development and evaluation of a CAI courseware 'G-Reflect' on students' achievement and motivation in learning mathematics. *European Journal of Social Sciences, 8*(4), 557-568.

Sarojini, D. A. (1989). *Van Hiele levels of geometric thought in relation to similarity and congruence in triangles.* Unpublished M.Ed. thesis, University of Malaya, Malaysia.

Serra, M. (1997). *Discovering geometry: An inductive approach* (2nd ed.). California: Key Curriculum Press.

Sherard, W. H. (1981). Why is geometry a basic skill? *The Mathematics Teacher, 74*(1), 19-21.

Tan, H. L. (1985). *The mathematical needs of SPM-level Physics.* Unpublished Masters Thesis, University of Malaya, Malaysia.

Tay, B. L. (2003). *A van Hiele-based instruction and its impact on the geometry achievement of Form One students.* Unpublished master's dissertation, University of Malaya, Malaysia.

Usiskin, Z. (1980). What should *not* be in algebra and geometry curricula of average college-bound students? *Mathematics Teacher, 73*(6), 413-424.

Usiskin, Z. (1982). *Van Hiele levels of achievement in secondary school geometry.* (Final report

of the Cognitive Development and Achievement in Secondary School Geometry Project). Chicago, University of Chicago. (ERIC Document Reproduction Service No. ED 220 288).

van de Walle, J. A. (2001). *Elementary and middle school mathematics: Teaching developmentally* (4th ed.). New York: Addison Wesley Longman, Inc.

Wun, T. Y. (2010). *Preservice secondary school mathematics teachers' subject matter knowledge of perimeter and area.* Unpublished PhD thesis, University of Malaya, Malaysia.

CHAPTER 52

RESEARCH IN MATHEMATICAL THINKING IN MALAYSIA

Some Issues and Suggestions

Shafia Abdul Rahman

ABSTRACT

The objective of this chapter is to provide an overview of research in mathematical thinking carried out in Malaysia. It begins with a brief introduction of the place of mathematical thinking in Malaysian curriculum, followed by a critical review of the relevant literature by local researchers to shed some light on the extent to which these pieces of research have contributed to the development of mathematical thinking in Malaysia. The chapter concludes with some implications and suggestions for future research.

INTRODUCTION

The Malaysian school mathematics curriculum stipulates that mathematical thinking be developed in every student with the aim of producing children who can think mathematically. The new Standard Curriculum for

The First Sourcebook on Asian Research in Mathematics Education: China, Korea, Singapore, Japan, Malaysia, and India, pp. 1193–1208

Primary School aims to develop children's understanding of number concepts, basic skills in calculation, understanding simple mathematical ideas and apply the knowledge and skills in daily life. The curriculum emphasizes on *fikrah* (thinking), which focuses on the development of mathematical thinking through problem solving, communicating, reasoning, making connections and representations and application of technology (Curriculum Development Centre, Ministry of Education Malaysia, 2011). The learning area includes number and operation, measurement and geometry and statistics and probability.

Similarly, the secondary school curriculum also stresses that mathematical thinking is infused throughout the teaching and learning process. The curriculum states:

> The mathematics curriculum for secondary school aims to develop individuals who are able to *think mathematically*, and can apply mathematical knowledge effectively and responsibly in solving problems and making decisions. (Curriculum Development Centre, Ministry of Education Malaysia, 2006, p. x)

At the secondary level, students learn topics such as Numbers, Shape and Space and Relationships. The secondary curriculum also emphasizes on problem solving, communicating, reasoning, making connections and representations and application of technology in producing children who can think mathematically through the acquisition of mathematical concepts and skills. However, the reality in Malaysian mathematics classrooms seems to show that much emphasis is placed on algorithmic manipulation. Parmjit (2009) claims that in the Malaysian school climate, "children's natural thinking becomes gradually replaced by attempts at rote learning", with disastrous consequences as "grades obtained in the national examination for mathematics do not indicate children's mathematical knowledge" (p. 4). The Trends in International Mathematics and Science Study (TIMSS) indicates a gradual decrease of Malaysian 8th graders performance in Mathematics, with average scores of 519, 508 and 474 in 1999, 2003 and 2007, respectively. The trend shows a widening gap within Malaysian students' outcomes in recent years, with about 20% failing to meet the minimum TIMSS benchmarks for Mathematics and Science in 2007, as opposed to 5 – 7% in 2003.This situation calls for a scrutiny into the implementation of the Mathematics curriculum and the implications of research in the field of mathematical thinking in Malaysia.

Mathematical Thinking

What is mathematical thinking? Polya (1962) regards mathematical thinking as problem solving. For Polya, "to have a problem means to search consciously for some action appropriate to attain a clearly conceived, but not immediately attainable, aim. To solve the problem means to find such action" (p. 117). Mathematician Paul Halmos (1980) says problem solving is "the heart of mathematics". According to Burton (1984), mathematical thinking is not thinking about the subject matter of mathematics but a style of thinking that is a function of particular operations, processes, and dynamics recognizably mathematical. She argues that because mathematical thinking becomes confused with thinking about mathematics, there has been little success in separating process from content in the classroom presentation of the subject.

Mason, Burton & Stacey (1982) contend that to think mathematically, one needs to engage in the act of thinking for oneself. They suggest that "three kinds of involvement are required [to think mathematically]: physical, emotional and intellectual" (Mason, Burton & Stacey, 1982, pp. ix - x). Children may go back and forth within mathematical representation before they acquire formal understanding. Transitions between or moving back and forth within these forms of representation were recast as movements up and down a spiral of *Manipulating–Getting-a-sense-of–Articulating* (Mason, Burton & Stacey, 1982):

> Picture mathematical thinking on a helix which loops round and round. Each loop represents an opportunity to extend understanding by encountering an idea, an object, a diagram, or symbol with enough surprise or curiosity to impel exploration of it by manipulating. ... Tension provoked by the gap which opens between what is expected and what actually happens provides a force to keep the process going and some sense of pattern or connectedness releases the tension into achievement, wonder, pleasure, further surprise or curiosity which drives the process on. While the sense of what is happening remains vague, more specialisation is required until the force of the sense is expressed in the articulation of a generalisation.... And achieved articulation immediately becomes available for new manipulating, and the wrap-around of the helix. (Mason et al., 1982, p. 155)

Manipulating affords learners opportunities for seeing new ideas and experiencing previously met ideas, and themes, and for mastering techniques. Recognising patterns, multiple relationships and attributes offers learners opportunity to 'get a sense of' the concept. Articulating these attributes and relationships leads to formalisation through discerning properties independent of the objects being manipulated and, hence, to appreciation of generality.

Mason et al. (1982) went on to announce that everyone *can* think mathematically, and that mathematical thinking is *provoked* by contradiction, attention and surprise, *supported* by an atmosphere of questioning, challenging and reflecting, and can be improved by *practice with reflection*, and that mathematical thinking helps in understanding *oneself* and the world.

> Thinking mathematically is not an end in itself. It is a process by which we increase our understanding of the world and extend our choices..... However, sustaining mathematical thinking requires more than just getting answers to questions, no matter how elegant the solution or how difficult the question. (ibid., p. 154)

RESEARCH IN MATHEMATICAL THINKING IN MALAYSIA

Research in the field of mathematical thinking in Malaysia is still at infant stage. Many pieces of research focus on different aspects of mathematical thinking through getting learners to solve (non-routine) problems and *labeling* them as thinking mathematically or otherwise, without exploring their potential power to engage in mathematical activities meaningfully. In what follows, I present a collection of research done in Malaysia in mathematical thinking based on categories which best describe their work.

Problem Solving and Mathematical Thinking

Noraini (2002) in her article "*Fostering Mathematical Thinking in Higher Institution: Use of Problem-Based Learning*" advocated the use of problem-based learning to enhance thinking in mathematics. According to Noraini, the ability to solve problems in mathematics depends on the thinking level of the student. In the process of solving problems in a problem-based learning (PBL) environment, the student needs to be able to think to understand the problem, devise a solution plan, carry out the plan and relate the problem to previous and current experience. PBL is capable of bringing out the relationship between facts, concepts, theories in mathematics, algorithm and real life issues. She lists five thinking skills that might help the instructor (teacher) to plan ways to enhance students' thinking: *focusing skills* (the ability to focus their thinking on specific issues and temporarily ignore other things), *information-gathering skills* (the ability to gather information beyond that which the instructor provides), *organizing skills* (the ability to organize that information into a form that will enable them to interpret the information and put it to best use), *analyzing and integrating skills* (the ability to identify key elements and relationship between the various pieces of information they have gathered in order to analyze and integrate information into existing knowledge structures) and *evaluating skills* (the ability to know that non-trivial problems rarely have a single answer or a single method and the merit or logic of ideas and apply these criteria objectively as well as emotionally). She raised some of the challenges to create PBL environment in the university setting, mainly issues related to manpower and understanding on the part of management personnel to implement PBL.

Lau and Hwa (2003) studied Form Four students' (aged 15 – 16) problem-solving skills and their thinking processes while solving mathematics problems of different levels of difficulty. Stratified sampling method was used to select 412 Form Four students from a total of 2962 students from the 16 secondary schools in Sarawak. Four sets of problems, one set for each level of difficulty were designed and administered. Three sets of questionnaire were administered to gather information on the students' personal background and the mathematics skills and problem-solving skills employed while answering the problems of the various levels of difficulty. Another set of questions equivalent to the third or fourth level of difficulty of mathematics problems were also designed and were posed to 18 selected students to identify their thinking processes while solving these problems. An interview was conducted with theses students to gather more information on these identified issues. The researchers found that the overall performance of students declined drastically as the level of difficulty of problems increased. There was a statistically significant gender-related difference in mathematics ability in favor of females, based on the higher overall mean score. Racial-ethnic differences in mathematics achievement were very pervasive; large differences remain between the mathematics achievements of Chinese, Malay and Iban students. There was also a consistent disparity in mathematics achievement that is related to the socioeconomic status of these students.

Chan and Mousley (2005) studied the influence of Malaysian Chinese students' schooling in a tradition of abstract, technical mathematics and rote learning on ways that they responded to mathematical word problems. The research aimed to find out whether the students who apparently preferred "surface learning" of mathematics were able to appreciate deeper concepts and contexts in mathematical word problems. The students were encouraged to engage in discussion, peer-group activities and reflection to bring about "deep" learning and not usually adopted in traditional Malaysian education environments. A total of 290 secondary school leavers from a Chinese school background aged 17 – 18 years participated in the study. Theses students were enrolled in the first semester of a computing and information technology diploma course. Seven 14–week cycles of action research were carried out over two and a half years using seven cohorts of students. Data on students' achievements, interest levels, beliefs and attitudes, and mathematical performance were collected using questionnaires, interview schedules, journal notes of conversations and observations, as well as mathematics work and exams. Data from these sources were sorted under headings including the use of word problems, collaborative and reflective learning, learning mathematics in a second language, and the incorporation of values into mathematical concepts and practices. The authors concluded that most of the students in the project felt a need to practice sufficient examples before they develop adequate confidence and curiosity for more independent and diverse ways of solving problems. It seemed that what could be termed *surface approaches* can be used to build a foundation for the use of deeper learning approaches. Technical problems did not appear to be inferior to word problems in terms of their ability to lead to deeper learning. However, what appeared to matter most was how the students approached the problems, and how their mathematical thinking developed as a result of having tackled a number of problems. They suggested looking beneath the surface and having a deeper understanding of how repetitive practice and deeper learning intertwine was essential.

These pieces of research focus on the process of solving problems as indications of children's mathematical thinking. Although mathematical thinking can be inferred through problem solving activity, a problem in itself is *static* and does not allow much opportunity for children to make use their natural powers to encounter important mathematical themes in the problems.

Communication and Mathematical Thinking

Language plays an important role in the learning of mathematics and is therefore, a prerequisite to developing mathematical thinking. As generalizations play a central role in mathematics, language acts as a vehicle to express it. In this aspect, Cheah (2008) carried out a case study on a lesson conducted as part of a project to introduce the Lesson Study approach to a group of teachers. The lesson which was planned collaboratively by three teachers to focus on mathematical communication and thinking was then carried out by one of the teachers. The lesson was videotaped and the communication that took place was transcribed and interpretively analyzed. The findings of the study revealed that the lesson tasks designed by the teachers were generally able to stimulate active pupil participation in the lesson. However, the communication in the lesson was mostly focused on the teacher's attempt to lead the pupils to arrive at the teacher's answers. While the study raised some issues on the way mathematical communication is carried out in the Malaysian primary classrooms, he contends that the Lesson Study method was a suitable approach for teachers to improve and further develop mathematical communication in the classroom.

Lau, Parmjit and Hwa (2009) investigated the nature of teacher-student interaction as they worked on different mathematics activities in a single classroom over a 10-month period. Socio-cultural theories and Vygotsky's Zone of Proximal Development (ZPD) provided the main framework for examining the teaching and learning processes and explaining the incorporation of a four-phase lesson plan (whole-class discussion, group work, reporting back and summing up) as increasing participation of the teacher and students in the teaching and learning process. Analyses of discourse yielded five different types of interactions that emphasized mathematical sense-making and justification of ideas and arguments – offering clarification, inviting students' participation, maintaining students' focus on the activity, reinforcing key features of the activity and evaluating students' understandings.

Cheah's study shows that children are not given much opportunity to express themselves mathematically and teacher often dominates classroom discussion. Lau et. al.'s research focused on characterizing the nature of teacher-student interaction in a mathematics classroom. A key aspect in analyzing children's (and teacher's) interaction and communicative competencies in mathematics is expression of generality. Drury (2007) studied the complexity of the process of expressing generality in mathematics classroom and found that teachers' use of language can affect students' appreciation of generality in mathematics.

Reasoning and Mathematical Thinking

Gan (2008) investigated Year 5 pupils' (aged 10 – 11) pre-algebraic thinking using numerical and geometrical patterns, and solving for unknown quantities in word problems. Thirteen Year Five pupils from a rural primary school in Kota Samarahan, Sarawak were given ten pre-algebraic problems consisting of numerical patterns, geometrical patterns and word problems involving unknown quantities. Participants' written solutions, think-aloud verbal protocols, retrospection through task-based interview and videotaping of their solution processes comprised the data for this study. Findings suggested that recursive strategy and 'based on shape of figure' strategy were most frequently used in solving problems involving number patterns and geometric patterns, respectively. 'Unwinding' and arithmetic strategies were most frequently used for word problems. Participants' pre-algebraic thinking was inferred through their strategies in identifying and extending patterns and identifying relationships between quantities. Arithmetic-symbolic representation was predominant in the participants' solutions for generalization involving number patterns and the word problems. Pictorial representation was predominant for generalization involving geometric patterns. In justifying their solutions for generalization problems involving number and geometric patterns, participants demonstrated inductive reasoning in identifying the characteristics of the patterns given. In justifying their solutions for word problems, they were inclined to explain why a particular method worked based on reasons for operations used and verification of answers. Findings also indicated early signs of algebraic thinking among some participants.

Yackel (1997) argues that in order to develop the foundations for algebraic reasoning, teachers should consider activities that encourage children to move beyond numerical reasoning to more general reasoning about relationships, quantity, and ways of notating and symbolizing. Although Gan's study focused on numerical reasoning, the study could be further extended to include other aspects of algebraic reasoning as suggested by Yackel (1997).

Technology and Mathematical Thinking

Kamariah, Rohani, Ahmad Fauzi & Aida Suraya (2009) explored the effect of utilizing Geometer's Sketchpad (GSP) on performance and mathematical thinking of secondary mathematics learners. They attempted to explore the effects of integrating the GSP on the mathematical performance in secondary mathematics and students' attitudes towards the respective approaches used to teach the groups were investigated. The mean overall mathematical performance for the group using the GSP was 11.78 while the mean overall performance for traditional teaching strategy group was 13.03. Independent samples t-test results showed that there was no significant difference in mean mathematical performance between the GSP group and the traditional teaching strategy group. Findings also indicated that the use of GSP induced higher mathematical thinking process amongst the GSP group. These findings showed that the use of GSP had an impact on both mathematical thinking process and performance. However, these findings provided evidences of limited and deficient use of the technology, specifically in the teaching of mathematics at the Malaysian secondary level.

Chew (2009) investigated Form One students' learning of solid geometry in a phase-based instructional environment using Geometer's Sketchpad (GSP) based on the van Hiele theory. He examined the students' initial van Hiele levels of geometric thinking about cubes and cuboids, and how their van Hiele levels changed after phase-based instruction with GSP. This case study design used six participants from a class of mixed-ability Form One students. Findings revealed that the participants' initial van Hiele levels ranged from Level 0 to Level 2. After phase-based instruction with GSP, their van Hiele levels either increased or remained the same.

Both Kamariah et al. and Chew's studies investigated the impact of using technology on students' performance in mathematics at the end of a technology-based instruction. While these pieces of research showed positive outcome, future research could focus on the development of mathematical thinking *whilst* using technology such as Geometer's Sketchpad, SMART Board and Graphic Calculators so that the full potential of these technologies can be realized.

Assessment and Mathematical Thinking

In an attempt to construct a framework to assess mathematical thinking, Hwa & Lim (2008) believe there are three components which contribute to the construction of the framework: mathematical content knowledge, attitudes or disposition and mental operations of problem solving and decision making. They proposed The Mathematical Thinking Assessment (MaTA) framework to assess mathematical thinking in a school-based assessment setting which consists of four components: (a) performance assessment, (b) Metacognition Rating Scale, (c) Mathematical Dispositions Rating Scale, and (d) Mathematical Thinking Scoring Rubric. The performance assessment is used to elicit students' thinking process while solving the mathematical problem; the Metacognition Rating Scale is used to look for students' awareness during the problem solving process; the Mathematical Dispositions Rating Scale is used to indicate students' predisposition toward learning of mathematics; and the Mathematical Thinking Scoring Rubric is used to score and grade students' mathematical thinking according to the domains defined.

Kargar, Rohani & Bayat (2010) investigated the relationships between mathematics anxiety, attitudes toward mathematics and mathematical thinking among university students. A 60-item questionnaire on mathematical thinking rating scale, mathematics anxiety rating scale and attitudes toward mathematics rating scale were completed by 203 university students from the Faculty of Science, Engineering, Food Science, and Human Ecology in one of the public universities in Malaysia. The students' mathematical thinking was assessed using a questionnaire called Mathematical Thinking Rating Scale (MTRS), which was adopted from the Diagnostic Instrument of Mathematics Learning (DIML), a self-report instrument consisting of 32 items with seven subscales: skill, concept, ability to apply knowledge, understanding mathematical relation, use of mathematical language, judgment and evaluation and cognitive process of learning. Students' mathematical anxiety was measured using a 16-item questionnaire known as the Mathematics Anxiety Rating Scale (MARS-R) developed by Richardson and Suinn (1972) and revised by Plake and Parker (1982). It had two subscales – Learning Math Anxiety (LME) and - Math Evaluation Anxiety (MEA). A correlation analysis was used to establish relationship between the three constructs and an independent t-test was used to investigate differences between the two gender groups and two race groups on their mathematical anxiety, attitudes toward mathematics and mathematical thinking. The results indicated a significantly high positive correlation between mathematical thinking and attitudes toward mathematics. Also, there was a negative moderate correlation between mathematical thinking and mathematics anxiety. There was also a negative correlation between mathematics anxiety and attitudes toward mathematics. The findings suggest that level of mathematics anxiety is related to mathematical thinking and attitudes toward mathematics.

Nurul Hidayah (2011) investigated Year Six students' levels of mathematical thinking using performance tasks. The research attempted to discover how students use mathematics in six constructed response items. It assessed Year Six students' mathematical thinking processes in four domains: conceptual understanding, mathematical representation, computational skills and mathematical explanation. A total of 157 Year Six students enrolled in four schools participated in the study. Students' responses in each task were rated independently using six scoring rubrics with four score levels. Data consisted of students' scripts and interviews. Quantitative analysis showed strong correlation between the mathematical thinking domains. Findings also indicated that students had diverse and unrelated mathematical facts and terminologies.

Can understanding be assessed or evaluated? Understanding is difficult to capture in words. All that is available for the tester is students' answers, which may not be a true indication of their ability since learning is a process of maturation that happens over time. The SOLO taxonomy (Biggs & Collis, 1982) can be considered for gauging possible understanding from responses to test items.

Getting learners to make up their own questions (and answers) at the end of each lesson can often reveal what dominates their attention and what they are aware of. In this regard, Shafia (2006; 2008) investigated learners' understanding of integration. Understanding is a complex process, extending beyond learners' ability to reproduce mathematical techniques in familiar, slightly modified situations. Understanding is not only the ability to use knowledge to solve routine problems correctly but as the ability to act creatively in unfamiliar situations. In her study, she prompted learners to construct several mathematical examples meeting specified constraints, what they choose to change reveals dimensions, depth and quality of their awareness. Data consisted of semi-structured interviews and construction tasks. Learners ranging from A-level students to graduates and, a spectrum of first year undergraduates in between, were interviewed and invited to construct relevant mathematical objects. Responses in the interview were compared with responses in the construction tasks to say something about the nature of their awareness and thus, understanding. Findings showed that the construction tasks revealed a greater deal about learners' awareness and understanding than their responses in the interview. The study provided a description of different dynamics and depths of awareness which contribute to the different forms of understanding displayed by learners with different backgrounds.

Lesson Study and Mathematical Thinking

Lim (2007) conducted an exploratory study that aimed to develop mathematical thinking in a primary mathematics lesson. Mathematical thinking was set as one of the goals of an existing Lesson Study group in a Chinese primary school. Two lesson study cycles were carried out with a result of two mathematics lessons planned and observed. A total of five mathematics teachers took part in the study. Analysis of data showed that the teachers agreed that it was easier to learn new teaching ideas such as developing mathematical thinking through Lesson Study collaboration. Although they did not understand what mathematical thinking was and how to help students develop it, after two lesson study cycles, teachers in this study did gain a deeper understanding and developed confidence in promoting mathematical thinking in their classrooms. However, time constraints and heavy workloads remained the main challenges to integrating any new teaching ideas and strategies.

Cheah (2010) worked on a theoretical framework for assessment in the primary mathematics classroom, principally to be used by teachers and researchers conducting Lesson Study as an approach to improve the teaching and learning of mathematics. In the Malaysian context, assessment is viewed as an integral part together with other teaching and learning activities in the classroom. It begins with planning rich mathematical tasks that would enable students to actively construct mathematical ideas. In the classroom the teacher assesses students' learning in order to further facilitate the construction of mathematical ideas. The actual implementation of this model of classroom didactics however would entail the further development the teachers' facilitating skills which are essential to assist students learn mathematics meaningfully. This teacher development program could perhaps be best realized through a collaborative school-based teacher development program such as the Lesson Study.

CONCLUDING REMARKS

Producing children who can think mathematically must be the main goal of schooling and is important as a way of learning and teaching mathematics (Stacey, 2007). Through schooling, students at best acquire the ability to think mathematically and to use mathematical thinking to solve problems. Students who have an understanding of the components of mathematical thinking will be able to use these abilities independently to make sense of mathematics that they are learning. To solve problems correctly and confidently requires learners having a repertoire of skills, techniques and familiarity that comes with practice. Although an ancient Chinese idiom says "practice makes perfect", it has to come with understanding, lest children will be engaging in instrumental understanding ("rules without reasons") rather than relational understanding ("knowing both what to do and why") (Skemp, 1976). The way in which children approach a mathematics problem has significant implications on their ability to engage in the tasks effectively and meaningfully and can indicate their ability to think mathematically.

Gattegno's (1987) asserts that 'knowing' means stressing awareness of something and that this awareness is what is educable. Behaviour is trained through practice but training alone renders the individual inflexible. Flexibility arises from awareness which informs and directs behaviour. Therefore, behaviour which is to be flexible and responsive to subtle changes *must* be guided by active awareness. Learning (and therefore, thinking) involves educating awareness which in turn directs appropriate behaviour. Awareness of mathematical themes such as *doing and undoing*, *invariance in the midst of change* and *freedom and constraint* (Mason & Johnston-Wilder, 2006) can illuminate learners in their journey through mathematical thinking.

The richness of learners' mathematical experience depends on the opportunities afforded for the learners to use and develop their own powers and opportunities to make significant mathematical choices (Mason, 2002). Mathematical thinking proceeds through exercising mathematical powers such as imagining and expressing, specializing and generalizing, conjecturing and convincing, stressing and ignoring, ordering and characterizing, seeing sameness and seeing difference, assenting and asserting, among others. Learners' ability to engage in such activities effectively depends on what sense they make of mathematical topics and which aspects of the topic they emphasize or dominate their attention. If children were to be encouraged to think mathematically, they need to be given the opportunity to come into contact with their own mathematical powers and realize their choices. A task in and of itself does not elicit mathematical thinking. Opportunities for various possibilities that make use of children's natural powers can be generated through careful selection of tasks and by the way they are developed in the classroom. Learners need to engage in mathematical tasks in such a way that their awareness of such powers is present. They need to be given the opportunity to articulate what they are thinking. A conjecturing atmosphere (Mason, 2002) needs to be created so as to bring out these mathematical powers.

I contend that teachers can encourage mathematical thinking by asking the right questions – questions that are mathematically developmental in nature. Teachers need to realize the importance of mathematical thinking not only for the learning of mathematics but also for teaching mathematics – to analyze subject matter, to plan lessons and to understand student thinking, questions and solutions – as rightly suggested by Stacey (2007) and supported by Gattegno (1988).

> [Teachers need to] make themselves vulnerable to the awareness of awareness, and to mathematization, rather than to the historical content of mathematics. They need to give themselves an opportunity to experience their own creativity, and when they are in contact with it, to turn to their students to give them the opportunity as well. (Gattegno, 1988, p. 167)

Implications and Suggestions for Future Research

Research in the field of mathematical thinking in Malaysia has a long and challenging way ahead. If research in this field were to move in the right direction, it has to reevaluate policy makers' and stakeholders' conceptions of mathematical thinking in the first instance. Mathematical thinking is not about solving problems correctly alone; it is about empowering children with the necessary attunements to *experience generality* through a spiral process of manipulation and sense-making. The way forward for research in mathematical thinking in Malaysia is to move away from *behavior-driven* research to more *awareness-driven* research. By this I mean rather than focusing on what learners *can do*, research in mathematical thinking should, at best, focus on what learners are *capable of doing* and how to bring that out in learners.

One of the hindrances to conducting such research is the inability of the researcher/teaching practitioners to *see* seeds of mathematical thinking in learners and to *allow* (let alone promote) such thinking to take place in this highly *assessment-oriented* society. The very sense of assessment in our society measures not the student's mathematical thinking ability but how well a student can *reproduce* teacher's method of working out a problem. Therefore, research should focus on evaluating teachers' conception/understanding of mathematical thinking and their readiness/

obstacles to incorporate it in their teaching and learning. Malarvili and Shafia (in press) found that although teachers perceived highly of the importance of critical and creative thinking skills in teaching and learning, their knowledge of it was moderate and the implementation of it in the classroom was low. Besides that, research should also concentrate on developing tasks that promote mathematical thinking and ways of incorporating such tasks in the teaching and learning.

REFERENCES

Biggs, J. & Collis, K. (1982). *Evaluating the quality of learning: The SOLO taxonomy.* New York: Academic Press.

Burton, L. (1984). Mathematical thinking: The struggle for meaning. *Journal for Research in Mathematics Education, 15* (1), 35 – 49.

Chan, K. Y. & Mousley, J. (2005). Using word problems in Malaysian mathematics education: Looking beneath the surface. In Chick, H. L. & Vincent, J. L. (Eds.). *Proceedings of the 29th Conference of the International Group for the Psychology of Mathematics Education, 2,* 217 – 224.

Cheah, U. H. (2008). Refining communication to improve mathematics didactics: A case study. Paper presented at the APEC–KHON KAEN International Symposium 2008.

Cheah, U. H. (2010). Assessment in primary mathematics classrooms in Malaysia. Paper presented at the Fourth APEC-Tsukuba International Conference: Innovation of Mathematics Teaching and Learning through Lesson Study, 17 – 21 February 2010.

Chew, C. M. (2009). Enhancing students' geometric thinking through phase-based instruction using Geometer's sketchpad: A case study. *Jurnal Pendidik dan Pendidikan, 24,* 89 – 107.

Chiew, C. M. (2009). *Implementation of Lesson Study as an innovative professional development model among mathematics teachers.* Unpublished PhD thesis, Universiti Sains Malaysia, Penang, Malaysia.

Gan, W. L. (2008). A research into Year 5 pupils' pre-algebraic thinking in solving pre-algebraic problems. Unpublished PhD thesis, Universiti Sains Malaysia, July 2008.

Gattegno, C. (1988). *The science of education, Part 2B: The awareness of mathematization.* New York: Educational Solutions.

Kamariah Abu Bakar, Rohani Ahmad Tarmizi, Ahmad Fauzi Mohd Ayub & Aida Suraya Md.Yunus (2009). Effect of utilizing Geometer's Sketchpad on performance and mathematical thinking of secondary mathematics learners: An initial exploration. *International Journal of Education and Information Technologies, 1*(3), 20 – 27.

Lau, P. N. K. & Hwa, T. Y. (2003). The thinking processes of mathematics problem solving of Form Four secondary school students. *Social and Management Research Journal, 1,* August 2003.

Lim, C. S. (2007). Developing mathematical thinking in a primary mathematics classroom through Lesson Study: An exploratory study. *Proceedings of the APEC-Khon Kaen International symposium on Innovative Teaching Mathematics through Lesson Study (II): Focusing on mathematical thinking*, August 16 – 20, 2007, Khon Kaen, Thailand, 141 – 153.

Malarvili Sellaparma and Shafia Abdul Rahman (in press). Penerapan unsur-unsur kemahiran berfikir secara kritis dan kreatif dalam pengajaran dan pembelajaran matematik sekolah rendah (*The incorporation of elements of critical and creative thinking skills in the teaching and learning of primary mathematics). Malaysian Journal of Education.*

Maryam Kargar, Rohani Ahmad Tarmizia & Sahar Bayat (2010). Relationship between mathematical thinking, mathematics anxiety and mathematics attitudes among university students. *Procedia International Conference on Mathematics Education Research (ICMER 2010).*

Mason, J. & Johnston-Wilder, S. (2006). *Designing and using mathematical tasks 2nd edition*. St. Albans: Tarquin.

Mason, J. (2002). *Researching your own practice: The discipline of noticing*. London: RoutledgeFalmer.

Mason, J. (2008). Closing speech at the MA-ATM-NANAMIC-AMET Combined Conference, Keele University, 2 – 5 April, 2008.

Mason, J., Burton, L. & Stacey, K. (1982). *Thinking mathematically*. London: Addison Wesley.

NCTM. webref. http://standards.nctm.org/document/appendix/process.htm

Noraini Idris (2002). Fostering mathematical thinking in higher institution: Use of problem-based learning. *Issues in Education, 25*, 117 – 128.

Nurul Hidayah Lucy Abdullah (2011). *Assessing Mathematical Thinking Levels of Year Six Students Using Performance Tasks*. Unpublished PhD thesis, Universiti Sains Malaysia, Penang, Malaysia.

Parmjit, S. (2009). An Assessment of Number Sense Among Secondary School Students. *International Journal for Mathematics Teaching and Learning*, Oct 2009, 1 – 29.

Plake, B. S. & Parker, C. S. (1982). The development and validation of a revised version of the Mathematics Anxiety Rating Scale. *Educational and Psychological Measurement, 42*, 551 – 557.

Richardson, F. C. & Suinn, R. M. (1972). The mathematics anxiety rating scale: Psychometric data. *Journal of Counselling Psychology, 19*, 551 – 554.

Shafia Abdul Rahman (2006). Probing learners' awareness and understanding through example construction: The case of integration. *British Society for Research in Learning Mathematics (BSRLM)*, 17 June 2006, Bristol, England.

Shafia Abdul Rahman (2008). *Exploring learners' understanding of integration using structured interviews and construction tasks*. Unpublished PhD thesis, The Open University, Milton Keynes, United Kingdom.

Stacey, K. (2007). What is mathematical thinking and why is it important? *Proceedings of APEC-Tsukuba International Conference 2007: Innovative Teaching Mathematics through Lesson Study (II) – Focusing on Mathematical Thinking,* 2 – 7 Dec, 2006. Tokyo and Sapporo, Japan.

Yackel, E. (1997). A foundation for algebraic reasoning in the early grades. *Teaching Children Mathematics*, *3*, 276 – 280.

CHAPTER 53

STUDIES ABOUT VALUES IN MATHEMATICS TEACHING AND LEARNING IN MALAYSIA

Sharifah Norul Akmar Syed Zamri and Mohd Uzi Dollah

ABSTRACT

This chapter describes some of the studies about values in mathematics teaching and learning in Malaysia. Studies of values in mathematics education in Malaysia can be considered quite sporadic and still in the expansion stage. The discussions in this chapter include development of values education in Malaysian schools, values in the Malaysian mathematic curriculum, problems and challenges in inculcating values in mathematics education and some studies that has been conducted regarding values in Malaysia. Studies conducted by researchers from three different universities were also discussed. The chapter wraps up with a detailed discussion of a study to explore values in mathematics teaching as espoused by three secondary schools mathematics teachers in Malaysia.

The First Sourcebook on Asian Research in Mathematics Education: China, Korea, Singapore, Japan, Malaysia, and India, pp. 1209–1226

INTRODUCTION

Value is one of the affect components in mathematics education research besides beliefs, attitudes and emotions. Review of literature indicated that most of the research on values in mathematics education were conducted mostly at the end of 1990s and early 2000 (see example, Lim & Earnest, 1997; Bishop, FitzSimons, Seah, & Clarkson, 1999; Bishop, 2007; Chin & Lin, 2001; Seah et.al, 2001; Keitel, 2003). It was also reported that value was the least used in research and thus can be considered in its developmental stage. The issue of values has been a longstanding concern of mathematics education in Malaysia. Since its adoption in the education system more than two decades ago, specifically since the nation-wide implementation of the Integrated Primary School Curriculum (KBSR) in 1983 and subsequently in the Integrated Secondary School Curriculum (KBSM) in 1988, it is only timely that we take a closer look at the studies in this area. In Malaysia, the growing interest of research on values in mathematics education since early 2000 (see example, Nik Azis et al., 2007; Wan Zah et al., 2005; Sharifah Norul Akmar; 2002). This chapter describes some of the studies about values in mathematics teaching and learning in Malaysia. Research studies on values in mathematics education in Malaysia are still sporadic and in the exploring phase. The discussions in this chapter will include definition of values adopted by mathematics education researchers in their studies, development of values education in Malaysian schools, and some studies that has been conducted regarding values in mathematics education in Malaysia.

LITERATURE REVIEW

Definition of Values

There is a wide range of definition and categorization of values adopted in mathematics education research in Malaysia. The diverse choice of the meaning and categorization of values was partly influenced by the epistemological perspectives of the researchers themselves. For example, Lim and Ernest (1997) suggested three categories of values after consulting the literature. These categories are epistemological values, social and cultural values and personal values. Another researcher, Sharifah Norul Akmar (2002) in her study investigated two categories of values namely the sixteen noble values shared by all religions and agreed upon by all cultural backgrounds in Malaysia and intrinsic values in mathematics such as logical thinking, estimation and approximation, accurate com-

Table 53.1. Sixteen Noble or Shared Values

Compassion	*Self-Reliance*	*Humility*	*Respect*
Justice	Freedom	Courage	Physical and mental cleanliness
Honesty	Diligence	Cooperation	Moderation
Rationality	Love	Public spiritedness	Gratitude

putation, making clear and accurate statements (see Table 53.1 for the 16 noble values).

Meanwhile, Nik Azis Nik Pa (2007a) integrated some spiritual aspects to the meaning of values in his study. He argued that the current view of values is devoid of spiritual aspect and that limit the discussion of values within empirical experience and rational thinking only. His Universal Integrated Approach view places values in a hierarchical multi-dimensional concept and not in a single dimension thus affirming the fact that human beings are created from imperceptible spiritual element and tangible physical element, both of which are different entities and real. To facilitate the discussion of values, however he adapted the categorization made by Bishop, FitzSimons, Seah, and Clarkson, (1999), and classified values in mathematics education into three categories: general educational values, mathematics educational values and mathematical values. General education values is further sub divided into four namely basic, core, main and expanded values (see Table 53.2). Mathematics educational values meanwhile refer to the values in the teaching and learning of mathematics such as values related to the learning approaches, types of knowledge, levels of questions, application of knowledge, application of technology and learning process (see Table 53.3). Mathematical values are values related to mathematical knowledge and this includes values related to philosophy, reality, sentiments, and society.

EMPHASIS ON VALUES IN THE MALAYSIAN PRIMARY AND SECONDARY SCHOOL MATHEMATICS CURRICULUM

Value education is integral in Malaysian schools as society sees school as perpetuating and influencing the future development of society, particularly in achieving the country's vision of attaining the status of a fully developed nation in terms of economic development, social justice, and spiritual, moral and ethical strength, towards creating a society that is united, democratic, liberal and dynamic (Education Act 1996, p. 9). It is stipulated the National Curriculum as:

Table 53.2. General Educational Values in Four Hierarchical Categories

Categories	*Values (+ For positive, - for negative)*	
Basic	+ Faithfulness	- Infidel
Core	+ Good Conduct	- Bad behavior
	+ Courage	- Cowardice
	+ Wisdom	- Suspicion, skeptical
	+ Fairness/Justice	- Unfairness, cruelty
Main	+ Concentration	- Laziness, procrastination
	+ Obedience	- Disobedience, rebelliousness
	+ Gratefulness	- Ungratefulness
	+ Moderation	- Wastefulness, exaggeration
	+ Conscious	- Unconscious, recklessness
	+ Patience	- Impatience
	+ Generous	- Stinginess, greed
	+ Forgiveness	- Unforgiving
	+ Cooperation	- Egocentric, isolation
	+ Truth	- Pretence, fakeness, False
	+ Honesty	- Dishonesty
	+ Dedication	- Heedlessness
	+ Trustfulness	- Lie, pretence
	+ Cleanliness	- Untidiness, disorganization
	+ Knowledgeable	- Ignorance, inexperience
	+ Trustworthiness	- Corruption, untrustworthiness
Expanded	+ Enjoy working	- Work as a burden
	+ Ready to sacrifice	- Selfish to one own self
	+ Develop talent	- Not confident
	+ Appreciate time	- Waste time
	+ Be creative	- Conservative and not creative
	+ Interaction	- Reclusive, isolated
	... many more ...	

> An educational program that includes curriculum and co-curricular activities that encompasses all the knowledge, skills, norms, values, cultural elements and beliefs to help develop a pupil fully with respect to the physical, spiritual, mental and emotional aspects as well as to inculcate and develop desirable moral values and to transmit knowledge. (Ministry of Education, Education Act 1996, p. 120)

Historically, in identifying what values are desirable to inculcate and develop in Malaysian school system, the Ministry of Education had set up several committees and conducted a nationwide survey on the values as identified by various religions, traditions and beliefs of various communities. A technical committee on Moral education then finalized the list of 16 values that were considered to be most desirable and acceptable by

Table 53.3. Contexts of Values (Ten of the Contexts Belong to Mathematics Educational Values)

Contexts	*Explanation*	*Focus*
Personal value	Emphasizes self development or activities as students or individuals	The development of oneself or personal life
Interaction value	Emphasizes social interaction and discussion with others	Interaction or discussion
Epistemology value	Emphasizes the building-up, expansion and evaluation of mathematical knowledge	Epistemology or learning
Teaching value	Emphasizes the teaching of mathematics	Pedagogy or teaching
Reasoning value	Emphasizes reasoning and proving	Reasoning, proving
Problem solving value	Emphasizes solving mathematical problems	Mathematical problem solving
Representation value	Emphasizes mathematical representation	Mathematical representation
Communication value	Emphasizes communication in mathematics	Communication in mathematics
Connection value	Emphasizes making mathematical relationship	Mathematical connection
Enculturation value	Emphasizes mathematics knowledge as part of culture	Mathematical enculturation
History value	Emphasizes the historical background of topics and subtopics in mathematics	The history of mathematics
Technology value	Emphasizes the use of technology in mathematics education	The use of technology

Malaysians (Ministry of Education, 1999). Altogether 16 values known as noble values of Malaysian society were identified. The values include compassion, self-reliance, humility, respect, love, justice, freedom, courage, physical and mental cleanliness, honesty, diligence, co-operation, moderation, gratitude, rationality and public spiritedness. These 16 noble values are implemented in the national school curriculum at the secondary school level.

The values were implemented in the school systems in the following ways: (a) as a formal school subject, namely Moral Education for non Muslim pupils and as a component in Islamic Education for Muslim Pupils; (b) as an inclusion of values across subjects; and (c) as part of school co-curricular activities. These values emphasized the spiritual, humanitarian, and social development of the individual and considered essential to ensure healthy interaction between the individual and his or her family, peers, society, and the institutions of which he or she was a member (Ministry of Edu-

Table 53.4. Values According to the Seven Areas

Related Area	*Related Values*
Self development	Belief in God, trustworthiness, self esteem, responsibility, humility, tolerance, self reliance, diligence, love, justice, rationality and moderation
Family	Love for family, respect and loyalty for family members, preservation of family traditions, and responsibility towards the family
Environment	Love and care for the environment, harmony between man and environment, sustainability or environment, and sensitivity towards environmental issues
Patriotism	Love for the nation, loyalty to the King and nation, willingness to die or sacrifice for the nation
Human Rights	Protection of child's right, respect for women's right, protection of labor's rights, respects right of disabled persons, and protection of consumer right's
Democracy	Respect rules and regulations, freedom of speech and expression, freedom of religion, participation in nation building, and open mindedness
Peace and harmony	Living together in harmony, mutual help and cooperation, and mutual respect among nations

cation, Moral education syllabus for secondary schools, 1988).To facilitate the incorporation of values in school subjects including mathematics, the Ministry of Education has also undertaken several steps such as producing general guidelines on inclusion of specific values in all subjects as well as samples lesson plans with tapes on how values are incorporated in each of the subject, inclusion of values in textbooks and the teaching of values across the curriculum in both pre-service and in-service teacher training. In 2000, a nation wide review of the curriculum after a 10 year cycle of curriculum implementation more values were added to the 16 noble values making thirty six values which were organized in terms of seven learning areas were introduced at the secondary school level. The values can be categorized into seven areas namely self-development, family, environment, patriotism, human rights, democracy and peace and harmony. Table 53.4 shows the values according to the seven areas.

STUDIES OF VALUES IN MATHEMATICS EDUCATION IN MALAYSIA

Problems and Challenges Faced by Teachers

Although inculcating values is one of the important goals in the Malaysian mathematics curriculum, many researchers question the effectiveness in implementation through teaching and learning. Habib (1997) highlighted a few challenges in inculcating values in classroom teaching. First, there were multiple assumptions about teaching values in the classroom because of different backgrounds, such as religion, culture, education and family background among teachers. Second, the duty to inculcate values in class is always misunderstood by teachers, which is claimed to be done by certain teachers such as Religion and Moral teachers only. Third, the focus on academic excellence in Malaysian schools might deny other aspect like inculcating values in classroom.

Another research to study the integration of noble values in the secondary school teaching and learning by the Malaysian Ministry of Education (1999) also found similar findings. The nation-wide research comprises 327 school principals, 1145 teachers from 130 schools and 1563 students from 65 schools. It was reported that teachers generally had no difficulty in incorporating the noble values but they had, among others, faced difficulty relating them with the subject they are teaching, have clear understanding on the definitions of the noble values, specifically justice, diligence, physical and mental cleanliness and rationality, finding suitable techniques to incorporate values, time factor, sustaining students' interest and incompatibility between values taught and social values in the community. The teachers and principles agreed that incorporating values in teaching and learning is very important, beneficial and needed to increase their knowledge on the noble values and techniques in teaching values.

Sharifah Norul Akmar (2003) conducted a study which is situated within a larger research project funded by the Ministry of Science, Technology and Environment which investigates the teaching of values by understanding what and how primary school teachers plan, implement and reflect on the incorporation of values within and across six subjects (Chang et.al., 2003). Specifically the objectives of the study were to determine what and how mathematics teachers incorporate the 16 noble values and intrinsic values in teaching mathematics at the primary school level and to examine problems and challenges encountered by mathematics teachers in incorporating the 16 Malaysian noble values and intrinsic values in teaching mathematics. Seventy three upper primary level mathematics teachers from 73 selected schools in the northern, central, south and east coast of Peninsular Malaysia with 10 urban and 10 rural schools from each region were involved in the study. The survey results showed that the noble values most often incorporated by teachers in primary mathematics teaching is honesty (95.9%), followed by cooperation (95%), diligence (94.6 %) and self reliance (93.2) The teachers also reported that not all topics in the primary mathematics syllabus are suitable for direct inclusion of noble values. Although the teachers generally had no great difficulty in incorporating the 16 values and the intrinsic values, the teachers identified various problems they encountered and offered a wide spectrum of suggestions to how to deal with the problem. The teachers faced difficulty in terms of relating values with certain topics they are teaching and understanding the definitions of certain values such as freedom, moderation and rational. The teachers also indicated that they should be given an opportunity to attend specific courses on understanding the values as the explanation of the values in the text were limited.

The teachers recommended guide lines and resources, more in- house training, courses, workshops on the process of incorporating values, establish relationship with parents and society, include examples of teaching values in the text book, begin teaching value early on in life at the kindergarten and create awareness among pupil, teachers, parents and society.

Values From the Teachers' Perspectives

Lim and Ernest (1997) investigate the planned values as explicitly and implicitly documented in the Malaysian mathematics curriculum and to compare them with mathematics teachers' perception of what values are appropriate to be taught through mathematics. They came out with three categories of values after consulting the literature namely epistemological values, social and cultural values and personal values. The observation was that most of the explicitly stated values by teachers were epistemological followed by personal values. A comparison of explicit and implicit values revealed that not all intended values mentioned by the mathematics teachers were explicitly and implicitly expressed in the curriculum. Another study by Lim and Mohd Uzi(2007) is discussed at length in the following section.

Wan Zah and her colleagues (2005) used a different approach in her study. They explored values in mathematics in terms of different school of thought: Logicism, formalism, intuitism and Kuhnism. Teachers were interviewed to see which school of thought they belonged to. Formalism and logicism are the most popular views among the teachers. However, teachers are still unable to perceive the inter-connection among these values and view mathematics as value-free subjects. In general, the researchers suggested that more exposure of values education is necessary to ensure that teachers understand more about values in mathematics.

Values in Mathematics Textbooks

Nor Afizah and Sharifah Norul Akmar (2009) compared values in Malaysian and Singaporean textbooks. The aim of this research was to explore the explicit and implicit values in the Form One Mathematics textbook in Malaysia and Secondary One Mathematics Textbooks in Singapore. Only four chapters namely chapters on the topic *percentage, algebra, angles and perimeter* were chosen. Three categories of values, namely general education values, mathematics education values and mathematical values, were the focus of this study. The analysis included the identification of all value signals in textbook pages followed by the partitioning of these values into individual excerpts. Three cycles for the three categories of values were conducted and 1429 signals were identified for the eight chapters in both textbooks. At the end of the analysis, comparisons were made and all frequencies presented in percentages, tables and bar charts. Results have shown that both textbooks portrayed almost similar trend in values employment and distribution with little differences to highlight. Most of the general educational values were implicitly embedded while most of the mathematics education values were explicit in nature. Problem solving values were paramount in both textbooks; enculturation values, however, were not given enough attention in both textbooks.

The following section gave a detailed discussion of a study by Lim and Uzi (2007).

A DETAILED DISCUSSION OF A STUDY TO INCULCATE VALUES IN MALAYSIA

This research is aimed at exploring values in mathematics teaching as espoused by three secondary schools mathematics teachers in Malaysia. This study seeks to answer three research questions: what are the values inculcated in mathematics teaching?; are these values planned before mathematics teaching?; and are there conflicts of values inculcated in mathematics teaching? Figure 53.1 illustrates the conceptual framework of this research. It shows the relationship between values of the teachers and the values inculcated in teaching. Three categories of values are involved, beside the influence factors of teaching namely students, curriculum, school management and parents. There is also conflict between values as intended by teachers and the values implemented in the teaching.

The study was conducted using the qualitative approach. Three Form Four mathematics teachers from two secondary schools in Malaysia participated in the study. Data were collected mainly through classroom observations, individual interviews and field notes and documents collected during observations. All teaching sessions were video-taped while the interviews were audio-taped. Each participant was observed three times. Interviews were carried out before and after each observation, and finally at the end of the data collection for each participant. Video-tapes and

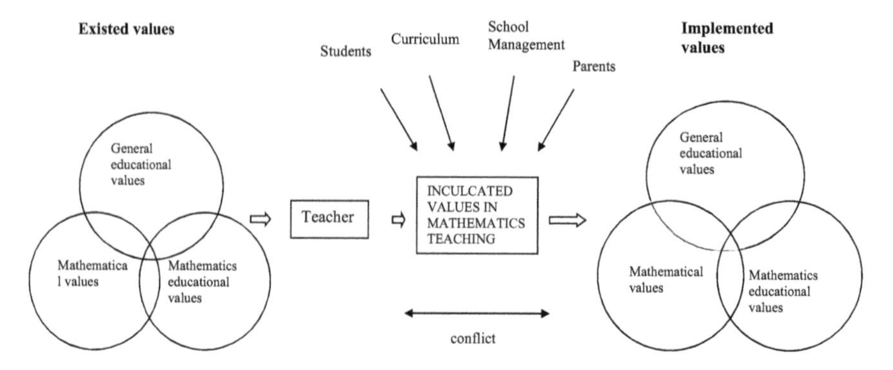

Figure 53.1. Conceptual framework of inculcated values in mathematics teaching.

audio-tapes were transcribed for analysis. Analysis for individual participant was done before analysis across all the three participants was made.

RESULT

Generally, there are three categories of values inculcated by teachers in mathematics teaching. They are general educational values, mathematical values, and mathematics educational values (refer to Table 53.5).

Plan to Inculcate Values in Teaching

All participants did not explicitly plan to inculcate values in their teaching. Most of them believed that their principals and authorities in education were not concerned with the importance of writing the values explicitly in lesson plans. One category of values that had been planned and focused by all participants was general education value. Based on pre-teaching interview on all participants, twelve general education values were planned to be inculcated in teaching. Eleven of them were related to self development values namely: careful, concern, diligence, self-reliance, responsibility, courage, confidence, proactive, perseverance, truthfulness, and gratitude. Only one general educational value (co-operation value) related to peace and harmony values as suggested by the Ministry of Education (2003) has been planned by participants (refer to Table 53.5).

There were participants, such as Aza dan Ciko, who were not consistent in planning to inculcate values in teaching. For example, gratitude value

Table 53.5. Values Inculcated by Aza, Bob and Ciko

		Aza			*Bob*			*Ciko*		
Values Categories	*Types of Values*	*P1*	*P2*	*P3*	*P1*	*P2*	*P3*	*P1*	*P2*	*P3*
General educational values	Careful	X^{1}								
	Concern	X^{1}								
	Diligence^{s}		X^{1s}							
	Self-reliance		X^{1}							
	Responsibility		X^{1}	X^{1}	X^{2}					
	Co-operation^{s}		X^{2s}	X^{1s}						X^{3}
	Courage					X^{1}	X^{3}			
	Confidence					X^{1}				
	Creative^{s}				X^{2s}					
	Honesty				X^{2}					
	Proactive						X^{3}			
	Never give up							X^{1}	X^{1}	X^{1}
	Truthfulness								Y^{1}	
	Gratitude							Y^{2}	Y^{2}	Y^{1}
	Respect									Y^{2}
	Systematic	X^{2}						X^{4}		
Mathematical values	Objectism		X^{4}		X^{4}		X^{4}			X^{4}
	Openness	X^{4}							X^{4}	
Mathematics educational values	Formalistic	X^{4}	X^{4}	X^{4}	X^{4}	X^{4}	X^{4}	X^{4}	X^{4}	X^{4}
	Instrumental				X^{4}	X^{4}				X^{4}
	Relevance				X^{1}					
	Theoretical	X^{4}		X^{4}		X^{4}	X^{4}			
	Accessibility				X^{4}	X^{4}				X^{4}
	Evaluation	X^{4}		X^{4}	X^{4}			X^{4}	X^{4}	

Key: P–Teaching (for example, P1 meaning Teaching 1). X^{1}–inculcating implicitly values which was planned before teaching. X^{2}–inculcating implicitly values which teachers were aware of although not the planned values. X^{3}–values which were planned but were not implemented in teaching. X^{4}–inculcating values implicitly, but teachers were not aware of them. Y^{1}–values which were planned and explicitly implemented in teaching. Y^{2}–inculcating values explicitly and teachers were aware of them although not the values which were planned before teaching. ^{s}–inculcating diligence, co-operation and creativity values were "superficial".

has been planned in one teaching lesson, but not in another teaching lesson. From the interview, one participant mentioned about failing to plan, due to the value which is commonly used in daily life. The readiness of participants to plan mathematics educational values before teaching was very low. No participant planned to inculcate mathematical values before teaching (see Table 53.6).

Table 53.6. Values Planned Before Teaching by Aza, Bob and Ciko

Categories of Values	*Field of Values*	*Types of Values*		
		Aza	*Bob*	*Ciko*
General educational values	Self development values	Careful, concern, diligence, self-reliance, responsibility,	Courage, confidence, proactive	Perseverance, truthfulness, and gratitude
	Peace and harmony values	Co-operation		Co-operation
Mathematics educational values			Relevance	

Inculcating Values in Teaching

Table 53.5 lists all values inculcated and not inculcated in teaching. Three categories of values, whether planned or not planned by the participants, were general educational values, mathematical values and mathematics educational values.

Formalistic value has been inculcated by all participants, in all teaching lessons. There were values, such as objectism and evaluation values, inculcated by all participants but not in all teaching lessons. Two categories of values which were planned and implemented in teaching were general educational values and mathematics educational values. From twelve values which were planned, only proactive values failed to be implemented by any participants. Nine general educational values, related to self development, were planned and implemented. They were careful, concern, self-reliance, responsibility, courage, confidence, perseverance, truthfulness, and gratitude. Relevance was the only mathematics educational value implemented in teaching.

There were also values implemented in teaching with awareness by participants, although the values were not planned before teaching. The implementation of the not planned values was described by participants only after being asked by the researcher post-interview. Generally, those values were general educational values, such as responsibility, honesty, gratitude, respect and systematic. Awareness of the participants to inculcate mathematical values and mathematics educational values is questionable. Only one of mathematics educational values was implemented in awareness by participants. None of the participants were aware of the implementation of mathematical values in teaching. Three values were implemented in teaching superficially, namely diligence, co-operation and creativity.

Generally, values can be implemented in teaching either explicitly or implicitly (refer to Table 53.5). Most of the implementations were implicit. Only one out of two participants inculcated values in teaching explicitly. Inculcating values explicitly in teaching happened in three situations. First, gratitude value which was implemented through utterance by teacher after any particular student finished answering a teacher's questions. The second situation involved the inculcating of "truthfulness" value by uttering the word with opposite meaning when the participant warned the students not to "bluff" in class. The third situation was to implement the value of "respect" at the beginning of the teaching lesson when the teacher asked students to be silent while their friends recite a prayer.

Inculcating general education values implicitly happened in teaching through three situations. The first situation was when participants inculcated values through explanations of concept and problem solving in class teaching. For example, the value of carefulness was inculcated by participants through teacher's carefulness to identify mistakes made by the students while teaching. The second situation was related to the inculcation of two values by participants, namely confidence and courage, by randomly asking students to show answers on the blackboard. The third situation was related to the implementation of inculcating values, such as confidence, honesty, independence and responsibility, through asking students to answer mathematical problems individually.

Two mathematical values inculcated in teaching by participants were objectism and openness. Objectism value was inculcated by participants through symbols and diagrams to represent abstract mathematical problems or situation. All participants agreed that the use of symbols and diagrams would help students to answer mathematical problems easily.

Inculcating values of openness, done by two participants, was through asking students to answer mathematical problems openly, with freedom to choose suitable strategy. For example, Aza gave freedom to students to use alternative strategy besides the strategy given in class.

Six mathematics educational values were implemented in class teaching, namely formalistic, instrumental, relevance, theoretical, accessibility and evaluation. One implementation obviously done by all participants was related to formalistic values through concept explanations, followed by examples and problem solving practices.

Inculcating instrumental value was through two situations. The first situation was through explanations about how to use formulas without concern for conceptual understanding. Another situation was through procedural problem solving without understanding the related mathematical concepts. For example, Bob used algebra procedural mechanically to get the quadratic equation general form ($ax^2 + bx + c = 0$) from ($5x^2 = 3x + 2$).

Two participants, Aza and Bob, were inculcating theoretical value through explanation and mathematical problem solving without contexts. Inculcating relevance values by participants was through use of contexts in presenting mathematics problems in class. Three situations were identified to inculcate accessibility values in teaching. The first situation was through helping students to solve mathematical problem. Second situation was through explanation of difficult mathematical concepts at the beginning of teaching. Third was through asking questions verbally, beginning with open question to all students, followed by specific students. Inculcating evaluation value was through observing and correcting works and problem solving answers done by students in class. By doing this, opportunity would be given to students to evaluate themselves in understanding teaching.

Values That Failed to be Implemented

Three values failed to be implemented, even though they were planned before teaching. Those values were courage, proactive and cooperation values. Two major causes accounted for failure of participants to implement the planned values. Firstly, the lesson teaching plan was not properly prepared by participants. Because of this, participants wasted a lot of time writing notes on the board rather than explaining or discussing mathematical concepts. Another major cause was change of focus by participants during teaching. The change of focus during teaching by participants was from group to individual. This change was done because participants thought the lesson was too easy to discuss in groups.

SUMMARY OF THE STUDY

Mathematics teaching is a cultural activity that is value laden (Lim & Mohd Uzi, 2007). The nature of mathematics teaching is very much influenced by the cultural background of the teachers and students involved. The cultural values, personal philosophies and attitudes of the mathematics teachers will affect the nature of mathematics teaching, learning and content chosen. Each mathematics teacher brings with him or her a set of values which will be used to decide what to teach, what not to teach and how to teach. It is through decision making and negotiation between teachers and students that values are being learned or inculcated constantly and continuously. This process occurs both at interpersonal and intrapersonal levels. Through interpersonal communication between teachers and students as well as students and students, certain values are preferred and certain values are being confirmed and reinforced. Likewise, during intrapersonal discourse such as when one is thinking or deciding, certain values are preferred, certain values are being re-examined and certain values are being rejected.

Planning to inculcate values in teaching is not a major concern by all participants of this study. Findings of the study showed that all three participants did not plan to explicitly inculcate values in mathematics teaching. Most of them believed that their principles and authorities in education were not concerned with the importance of writing values explicitly in lesson plans. This is in line with the earlier research by Wan Zah et al. (2005) which found that the term 'value' was unordinary to most of the teacher participants. In many cases, participants were unprepared and shocked upon being asked by the interviewer about inculcating values in mathematics teaching. We believe, without such a proper planning, inculcating values in mathematics teaching will become less effective.

Three values were planned and implemented in teaching by participants, namely general educational values, mathematical values and mathematics educational values. Most values implemented in awareness by participants are general educational values especially those related to self-development values, such as gratitude, respect and confidence. Although there were values inculcated explicitly, most of the values were inculcated implicitly by participants. Many researches in mathematics teaching have made similar findings, whereby teaching noble values was implicitly highlighted by most of their participants (Sharifah Norul Akmar, 2003; Wan Zah at al., 2005).

Inculcating mathematical values and mathematics educational values was a very challenging task to the mathematics educationists. In all cases, the participants had no knowledge and were unaware about values related to mathematics and mathematics education. That was why, in most of the cases, the inculcation of the mathematical values and mathematics educational values was only identified by the observer or researcher, although not aware by the participants.

Generally, two mathematical values, namely objectism and openness were inculcated implicitly in mathematics teaching by the teacher participants. All participants were keen to help students as soon as the students faced difficulties or failed to solve mathematical problems. One of the problems most highlighted by the participants in their teaching was giving explanation on meaning of the problems through new representation, such as drawing a suitable diagram or showing a mathematical relationship. Most of them believed that using representation would help students to understand and answer the problems better.

The challenge in inculcating value of openness was to encourage freedom in using strategy to solve mathematical problems among students. This finding supports previous research done by others in the field.In case of inculcating mathematics educational values, five values keen to be implemented by participants were formalistics, instrumental, accessibility and teoretical, besides very little to implement relevance value. The implementation of those values in teaching were more likely to bring students to passive learning, whereby students were becoming mere receivers of instruction. Opportunities for the students to critically and creatively learn mathematics were limited. Given this situation, it will probably be hard to consider teaching mathematical values and mathematics educational values in the classroom as a successful task.

CONCLUSION, IMPLICATIONS AND SUGGESTIONS FOR FUTURE RESEARCH

A new trend in research in mathematics education in Malaysia during the recent years has been the focus on values integration in the teaching and learning of mathematics. Research related to values in mathematics education in Malaysia is rather limited, disconnected and sporadic. There are three implications of research in this area. First implication of the research is to highlight values as an integral factor to be considered in mathematics teaching. Hence, teachers will be aware that teaching is more holistic and focus on both affective and cognitive matters. Second implication of the research is to raise teachers' awareness about three categories of values, namely general education values, mathematical values and mathematics educational values, rather than only focus on general education values or noble values. Third implication of the research is to uncover two possible ways to inculcate values in teaching: explicit and implicit. But, the study shows that the most common way to inculcate values in mathematics teaching is implicit rather than explicit.

Future studies with regard to values in mathematics education should include various dimensions of research to gain more insight and understanding of values in mathematics teaching and learning. One dimension to be studied is about values in mathematics learning, preferably through a study from the students' perspective. Most of the studies reviewed explored values from the secondary school teacher's perspective thus future research should consider primary school teachers' perspective. The studies from different dimensions of participants will help the teacher, educator, researcher or related parties to broadly understand the inculcation of values in mathematics teaching and learning, from primary to secondary schools. There is also a need to enhance collaborative research and sharing of best practices between researchers in mathematics and values education globally.

REFERENCES

Australian Government (2005). *National Framework for Values Education in Australian Schools.* Retrieved from http://www.valueseducation.edu.au/verve/_resources/Framework_PDF_version_for_the_web.pdf

Bishop, A. J. (1988). Mathematics education in its cultural context. *Educational Studies in Mathematics, 19,* 179-191.

Bishop, A. J. (2007). Values in mathematics and science education: Exploring students' values. In C. S. Lim et al. (Eds.), *Proceedings of the 4th East Asia Regional Conference on Mathematics Education [EARCOME4]:* Meeting Challenges of Developing Quality Mathematics Education (pp. 368-374). Penang: Universiti Sains Malaysia.

Bishop, A., FitzSimons, G., Seah, W. T., & Clarkson, P. (1999). *Values in mathematics education: Making values teaching explicit in the mathematics classroom.* Australia: Values and Mathematics Project (VAMP). Retrieved from http://www.aare.edu.au/99pap/bis99188.html

Chang et al. (2003). *Teaching of values in Malaysian Primary Education Project (Phase One),* IRPA (Intensification of Research in Priority Areas) report.

Chin, C. & Lin, F. L., (2001). Value-loaded activities in mathematics classroom. In M. v. d. Heuvel-Panhuisen (Ed.), *Proceedings of the 25th conference of the International Group for the Psychology of Mathematics Education,* Vol. 2. Utrecht. The Netherlands: Freudenthal Institute. (pp. 249-256). Retrieved from http://www.education.monash.edu.au/projects/vamp/publications.html

Habib, M. S. (1997). Nilai murni merentas kurikulum: Satu sorotan terhadap pelaksanaannya. *Masalah Pendidikan, 20,* 1–11.

International law book services (1999). *Education Act 1996 (Act 550)* & selected regulations. Kuala Lumpur: International Law Book Services

Keitel, C. (2003). *Values in mathematics classroom practice: The students' perspectives.* Paper presented as part of the conference of Learner's perspective study international research team, University of Melbourne, December 1-3.

Ministry of Education (Malaysia) (2003). *Huraian Sukatan Pelajaran Pendidikan Moral Tingkatan 5.* Kuala Lumpur: Dewan Bahasa dan Pustaka.

Lim, C. S., & Ernest, P. (1997). Value in mathematics education: What is planned and what is espoused? In British Society for Research into Learning Mathematics (BSRLM), *Proceedings of the Day Conference held at University of Nottingham,* 1st March 1997 (pp. 37–44).

Lim, C. S., & Fatimah, S.(2002). *Culture differences and values in mathematics education.* Paper presented at Invitational Conference on Values in Mathematics and Science Education 2002, Monash University, Australia. October 2-3.

Lim, C. S., & Mohd Uzi, D. (2007). Inculcating values in Mathematics teaching: Constraints and conflicts. In *Proceeding of International Seminar on Development of Values in Mathematics and Science Education*, University of Malaya, Kuala Lumpur, Malaysia, 3-4 August, 2007.

Nik Azis Nik Pa et.al. (2007a). *Values development in secondary school mathematics and science education in Malaysia*. Fundamental Research Grant Scheme (FRGS).

Nik Azis, N.P. (2007). Research about values in mathematics and science education : where and which way? In *Proceeding of International Seminar on Development of Values in Mathematics and Science Education*, University of Malaya, Kuala Lumpur, Malaysia, 3-4 August, 2007

Nor Afizah, & Sharifah Norul Akmar, S. Z. (2009). *Values in Form One mathematics textbooks in Malaysia and Singapore.* Paper presented at the National Seminar on the Development of Values in Mathematics and Science Education, Faculty of Education, University of Malaya, August 2009.

Seah, W. T., Bishop, A. J., FitzSimons, G. E., & Clarkson, P. C. (2001). *Exploring issues of control over values teaching in the mathematics classroom*. In Values And Mathematics Project (VAMP). Retrieved from http://www.education.monash.edu.au/projects/vamp/publications.html

Sharifah Norul Akmar, S. Z. (2002). *Incorporating values in the teaching of mathematics in Malaysian primary school*. Paper presented at the Invitational Conference on Values in Mathematics and Science Education, Monash University, Australia, 2 – 5 October.

Singapore Ministry of Education (2007). *Mission and vision statement* [Online]. Retrieved from http://www.moe.gov.sg/

Wan Zah, W. A., Sharifah, K.S.H., Habsah, I., Ramlah, H., Rofa, I., Majid, K., & Rohani, A.T. (2005). Teacher's understanding of values in mathematic. *Jurnal Teknologi Malaysia, 43*(E), 45-62.

CHAPTER 54

TRANSFORMATION OF SCHOOL MATHEMATICS ASSESSMENT

Tee Yong Hwa, Chap Sam Lim, and Ngee Kiong Lau

ABSTRACT

This chapter will begin with a brief historical development of Malaysian examination system. This is followed by an analysis of the school mathematics curriculum and assessment in Malaysia and issues related to mathematics assessment. Next, a review of local literatures on mathematics assessment will be presented. To elaborate further, a study that focuses on implementing a Mathematical Thinking Assessment (MaTA) framework to assess students' thinking processes and its limitations will be discussed. This chapter concludes with some suggestions for future research of the study that will help to assess students' learning in a more holistic and reliable fashion.

INTRODUCTION

The school Mathematics Curriculum in Malaysia is a continuum stretching from primary school to secondary school. It has been reviewed and revised in order to fulfill the National Education Policy and 2020 Vision.

The First Sourcebook on Asian Research in Mathematics Education: China, Korea, Singapore, Japan, Malaysia, and India, pp. 1227–1248

The ultimate aim is to produce students with the ability to think critically and creatively, acquire the mathematical knowledge and skills, and become skilled workers that are able to fulfill the country's vision in becoming a developed nation (Ministry of Education Malaysia, 2005).

HISTORY OF MALAYSIAN EXAMINATION SYSTEM

The examination system in Malaysia has gone through different stages of development. According to the Malaysian Examination Syndicate (MES) official website, a national policy on education was yet to be formalized during pre-independence stage. Hence, the examinations were carried out according to the needs of the school or by external examinations such as the *Overseas Senior Cambridge* (OSC). Upon independence, the National Education Policy was drafted through Razak Report in the year 1956. In this education policy, it proposed that all the public examinations were to be conducted by the MES. As a result, OSC examination was abolished.

The first public examinations conducted by the Malaysian Examination Syndicate in 1957 were Malayan Secondary School Entrance Examination (MSSEE), Lower Certificate of Education (LCE), School Certificate (SC) and Federation of Malaya Certificate (FMC). However, the examination system in Malaysia has been revised and changed several times over the years. To date, there are three major public examinations conducted by the MES: UPSR (*Primary School Achievement Test*) for 11 – 12 years old students, PMR (*Lower Secondary Examination*) for 14 – 15 years old students, and SPM (*Malaysian Certificate Examination*) for 16 – 17 years old students.

MATHEMATICS CURRICULUM AND ASSESSMENT IN MALAYSIA

In view of the importance of mathematics in the Malaysian school education curriculum, Mathematics is a compulsory subject and taught to all Malaysian children from pre-school to upper secondary level. According to the Malaysian Mathematics Curriculum, one of the major aims is to cultivate mathematical thinking of students:

> The Mathematics curriculum for secondary school aims to develop individuals who are able to think mathematically and who can apply mathematical knowledge effectively and responsibly in solving problems and making decisions. This will enable individuals to face challenges in everyday life that arise due to the advancement of science and technology. (Ministry of Education Malaysia, 2005, p. 2)

Mathematical thinking is not only required during mathematics problem-solving, but is also important for tackling workplace problems (Ministry of Education Malaysia, 2005, 2007). This is in concurrence with the following statement in *Principles and Standards for School Mathematics* (NCTM, 2000):

The need to understand and able to use mathematics in everyday life and in the workplace has never been greater and will continue to increase... Just as the level of mathematics needed for intelligent citizenship has increased dramatically, so too has the level of mathematical thinking and problem solving needed in the workplace, in professional areas ranging from health care to graphic design. (p. 4).

In order to promote mathematical thinking among the students, the evaluation principle stated in the Malaysian Mathematics Curriculum highlights that one of the focuses of mathematics assessment is formative assessment:

> Evaluation or assessment is part of the teaching and learning process to ascertain the strengths and weaknesses of students. It has to be planned and carried out as part of the classroom activities. Different methods of assessment can be conducted. These maybe in the forms f assignments, oral questioning and answering, observations and interviews. Based on the response, teachers can rectify students' misconceptions and weaknesses and also improve their own teaching skills. Teachers can then take subsequent effective measures in conducting remedial and enrichment activities in upgrading students' performance. (Ministry of Education Malaysia, 2005, p. 6)

However, school teachers tend to perceive formative assessment as of importance in the school assessment (Cheah, 2010). To make the situation worse, the current examination system does not seem to measure what we value as proposed by the Malaysian Mathematics Curriculum (Noor Azlan & Lui, 2000). As a matter of fact, the examination system is more to assess students' mathematics achievement by using paper-and-pencil test. For all the public examinations, Mathematics is tested by two different question papers with different examination formats. Paper 1 focuses on multiple-choice questions whereas short-answer questions make up Paper 2. The students are required to perform well in both Mathematics papers if they are to obtain good grades in these government examinations.

Even though mathematics project work is introduced to Additional Mathematics curriculum for Science students at the upper secondary level, the focus of additional mathematics teachers and students is still on paper-and-pencil test. This is because the performance of the project work does not affect students' mathematics result at the government examination. Difficulty in scoring, reliability of their work, as well as extra workload might have hindered the Malaysian teachers from using this alternative assessment to improve the teaching and learning of mathematics in the classroom (Hwa, 2010; Suzieleez Syrene, Venville & Chapman, 2009).

ISSUES OF MATHEMATICS ASSESSMENT IN MALAYSIA

Generally, Malaysian school mathematics teachers prepare students for these government examinations by using "typical lessons" (Lau, Hwa, Lau & Liew, 2006). A typical lesson consists of teacher presenting students with step-by-step instructions, such as introducing or reviewing a new mathematical concept through examples. This is followed by other activities such as giving of notes, assigning of working problems from the textbook and monitoring students while they are working on those problems. After some time, students are given a monthly test or a semester examination, which is a standardized test in nature, to evaluate their progress on the concepts taught. They are said to have learned those concepts if they can answer the problems in the test successfully.

Given the fact that the Malaysian government examinations are high-stake in nature (Ong, 2010), school teachers tend to practise "teach-to-test" culture by focusing on producing students with good grades (Lim, 2010). Students' learning is stressed more on drilling and practising rather than their understanding of the concepts taught. This practice has resulted in students' inability to communicate mathematical ideas and skills during problem solving process and in the real world context (Herman, 1992).

Rosnani and Suhailah (2003), in their book *The Teaching of Thinking in Malaysia*, claimed that "many studies have begun to reveal symptoms of decline in students' ability to think well, especially when schools begin to focus on the mastery of subject content [more] than the process of deriving the products" (p. 1). Asmah (1994) commented that students practising the forecast questions and memorizing the answers being one of the reasons they are unable to apply thinking skills in solving problem. The corpus of observations and assertions stated above indicate that it is vital to revise and refine our assessment system where more emphasis should be placed on the thinking process, rather than the academic achievement or performance.

Despite numerous efforts made by the Malaysian Ministry of Education to promote thinking curriculum, such as introducing Critical and Creative Thinking Skills (*KBKK*) since 1993, the results to date do not seem satisfactory. This is evidenced when the Malaysian grade-eight students' mathematics results in the Trends in International Mathematics and Science

Study (TIMSS) recorded a gradual decline over the years 1999, 2003 and 2007 with average scores of 519, 508, and 474 respectively. The result in the year 2007 was below the TIMSS average score (TIMSS scale average = 500) and was classified in the "Intermediate" category. Falling in this category implies that most of the Malaysian grade-eight students could only apply basic mathematical knowledge in straightforward situations. This achievement was below par as compared to students from countries that fell in the "Advanced" category, where these students were able to organize and draw conclusions from information, make generalizations, and solve non-routine problems (Gonzales, Williams, Jocelyn, Roey, Kastberg, & Brenwald, 2008).

Mathematical thinking is important particularly in the process of acquiring mathematical concepts and skills. However, teachers in schools are not aware of the importance of thinking in mathematics and hence they do not emphasize it in the development of students' intellectual growth (Ministry of Education Malaysia, 1993). Thus, many students fail to engage thinking skills in solving complex real life problems. In the words of Von Glaserfeld (1995):

> [Educators] have noticed that many students were quite able to learn the necessary formula and apply them to the limited range of textbook and test situation, but when faced with novel problem, they fell short and showed that they were far from having understood the relevant concepts and conceptual relations. (p. 20)

One of the causes for this situation is the assessment format. The current traditional assessment formats incline towards giving emphasis to recalling content knowledge, and hence provide little indication about students' level of understanding or quality of thinking (Nickerson, 1989). For this reason, students practise very little act of cognition during the assessment since they only memorize what is imparted to them by their teachers. On top of this, students are "bombarded with exercises, which function only to give them training on the rules or procedures that they have just learnt. They give students no training in calling to mind possible strategies for a solution and discriminating between them" (Lau, Hwa, Lau & Shem, 2003, p. 3). Consequently, the examinations results, particularly Mathematics results, do not reflect students' mathematical thinking process and their ability to communicate mathematical ideas and skills during the problem solving process.

Another drawback as to why students fail to engage thinking in solving complex real life problems is the examination-oriented type of learning in most of the Malaysian schools. Since the results of the government examinations determine students' future in higher learning, school education tends to focus on producing students with good grades (Ong,

2010). How students engage in the process of acquiring knowledge and how students apply this knowledge in the real world, however, are not being stressed in the current Malaysian schools practices. This observation was noted by the former director of the Malaysian Examination Syndicate (MES), Adi Badiozaman Tuah (2006). He commented that assessments in Malaysian schools emphasized more on students' achievement. As a result, school teachers tend to place greater attention on preparing students for the government examinations rather than the learning processes.

Mismatch between Malaysian school Mathematics Curriculum and assessments is also one of the reasons why the students are unable to demonstrate thinking skills and problem solving skills at a satisfactory level. A closer analysis of the intended aim of primary and secondary school mathematics curriculum shows that there are three components which contributed to the construction of a mathematical thinking model developed by Hwa (2010). These are content knowledge (i.e., mathematical knowledge), disposition (i.e., to learn effectively and responsibly) and mental operations during problem-solving and decision making. A close analysis done by the researcher indicates that these three components are well integrated into both the Malaysian primary and the secondary school mathematics curriculum documents as displayed in Table 54.1.

Even though these three important components are well documented in the Malaysian school Mathematics Curriculum, they are not being emphasized and assessed accordingly. This is evidenced by the initial analysis done by Hwa (2010) on the questions in the 2004 and 2005 Malaysian Certificate of Education examination Mathematics papers. He found that only approximately 10% of the questions tested students' abilities above application in Bloom's Taxonomy Cognitive Domain (Bloom, 1984). This implies that the current education assessments focus more on students' understanding of content knowledge and conversely, very little on students' thinking processes (Rosnani and Suhailah, 2003). Furthermore, the component of disposition of the mathematics curriculum is totally overlooked in current assessment approaches.

To date, there is no efficient assessment framework to assess mathematical thinking. Beyer (1984) claimed that most of the tests on thinking skills suffer from two flaws: conceptual inadequacy and inadequate definition of the components of the skills that are being tested. He commented that most tests "measure discrete skills in isolation, ignoring, by large, students' ability to engage in a sequences of cognitive operations" and in many circumstances, "items on tests of thinking skills bear no relation to the skills these tests suppose to evaluate" (p. 490). Hence, it is not surprising that despite much effort to promote a thinking curriculum through

Table 54.1. Comparison of Curriculum Objectives between Primary and Secondary School Mathematics Curriculum

Component	*Primary School Mathematics Curriculum (MOE[a], 2003)*	*Secondary School*	
		Mathematics Curriculum (MOE, 2005)	*Additional Mathematics Curriculum (MOE, 2004)*
Mathematical Content Knowledge	**Objective 1:** know and understand the concepts, definition, rules and principles related to numbers, operations, space, measures and data representation	**Objective 1:** understand definition, concepts, laws, principles and theorem related to Number. Shape and Space, and Relationships	**Objective 1:** widen their ability in the field of numbers, shapes and relationships as well as to gain knowledge in calculus, vector and linear programming
	Objective 2: master the basic operations of mathematics: addition; subtraction; multiplication; division	**Objective 2:** widen application of basic fundamental skills such as addition, subtraction, multiplication and division related to Number. Shape and Space, and Relationships	
	Objective 3: master the skills of combined operations		
	Objective 4: master basic mathematical skills, namely: making estimates and approximates; measuring; handling data; representing information in the form of graphs and charts	**Objective 3:** acquire basic mathematical skills such as: making estimation and rounding; measuring and constructing; collecting and handling data; representing and interpreting data; recognizing and representing relationship mathematically; using algorithm and relationship; solving problem; and making decision.	
Mental Operations	**Objective 6:** use the language of mathematics correctly	**Objective 4:** communicate mathematically	**Objective 7:** debate solutions in accurate language of mathematics

(Table continues on next page)

Table 54.1. (Continued)

Component	*Primary School Mathematics Curriculum (MOE[a], 2003)*	*Secondary School*	
		Mathematics Curriculum (MOE, 2005)	*Additional Mathematics Curriculum (MOE, 2004)*
Mental Operations (continued)	**Objective 8:** apply the knowledge of mathematics systematically, heuristically, accurately and carefully	**Objective 5:** apply knowledge and the skills of mathematics in solving problems and making decisions	**Objective 2:** enhance problem solving skills
			Objective 4: make inference and reasonable generalization from given information
			Objective 3: develop the ability to think critically, creatively and to reason out logically
			Objective 6: use the knowledge and skills of mathematics to interpret and solve real-life problems
		Objective 6: relate mathematics with other areas of knowledge	**Objective 5:** relate the learning of mathematics to daily activities and careers
			Objective 8: Relate mathematical ideas to the needs and activities of human beings

Mathematical Dis-position	**Objective 5:** use mathematical skills and knowledge to solve problems in everyday life effectively and responsibly.	**Objective 8:** cultivate mathematical knowledge and skills effectively and responsibly	**Objective 10:** Practice intrinsic mathematical values
	Objective 9: Participate in activities related to mathe-matics	**Objective 9:** Inculcate positive attitudes towards math-ematics	
	Objective 10: appreciate the importance and beauty of mathematics	**Objective 10:** appreciate the importance and beauty of mathematics	

Note: One objective related to the use of ICT in mathematics is excluded from comparison. (These are Objective 7 for both Primary and Secondary School Mathematics Curriculum and Objective 9 for Secondary School Additional Mathematics Curriculum). [a]MOE – Ministry of Education Malaysia.

Critical and Creative Thinking Skills (Ministry of Education Malaysia, 1993), this has not come to fruition in the Malaysian classroom.

To sum up, the current and past assessment approaches may have failed to evaluate students' thinking during mathematics problem-solving in Malaysian schools. Hence, there is a need to reform education assessments in Malaysia so that teachers and students do not over-emphasize examination results. In addition, school examinations should allow students to demonstrate their knowledge and skills learnt in real world contexts if the Malaysian Mathematics Curriculum intends to produce a skilled workforce as a requirement in realizing Vision 2020.

The Malaysian Government is aware of the current problems in our examination system. The question now is: In which direction should our education assessments be heading? The Ministry of Education has given us their answer by introducing school-based assessment, as stated in the National Educational Blueprint (Ministry of Education Malaysia, 2007). With this school-based assessment, all grading will be holistic in nature and criterion-based. It is the Malaysian Ministry of Education's hope that by revising the grading and assessment system, students' schooling will become less examination oriented. At the same time, the human capital that is produced by our education systems will be able to "think critically and creatively, master problem solving skills, have the ability to create new opportunities, and possess the endurance and capability to face the ever changing environment in the global world" (Ministry of Education Malaysia, 2007, p. 52). According to the blueprint, school-based assessment is supposed to be implemented throughout the nation by the year 2010. However, up to this date, the Malaysian Examination Syndicate has yet to announce the definite time frame for this assessment to be materialized.

PAST RELATED STUDIES ON MATHEMATICS ASSESSMENT IN MALAYSIA

Even though Malaysian Mathematics Curriculum is emphasizing on the cultivation of students' thinking during teaching and learning, not many researches on assessment of students' thinking have been done by the Malaysian mathematics educators. For example, an extensive search on the research studies on "thinking" through the University of Malaya Theses and Dissertations website reveals that out of twenty nine "thinking" research studies, only two are related to mathematics over the years 1966 to 2007. Worse still, none of these two were studied on assessing "thinking" in mathematics curriculum.

However, the latest development shows that more and more Malaysian mathematics educators see the importance of "thinking" assessment in mathematics curriculum. There were evidences shown by the research study done on algebraic thinking which focuses on linear pattern (pictorial), direct variations, concepts of function and arithmetic sequence (Lim & Noraini Idris, 2006); identified different levels of algebraic thinking of pre-service teachers, namely prestructure, unistructural, multistructural, lower relational, relational, upper relational and extended abstract (Lim, Chew, Wun & Noraini Idris, 2009); and examine pre-algebraic thinking of primary students based on the solution processes encompassed the solution strategy, mode of representation and mathematical justification (Gan, 2009).

As for geometry thinking, the van Hiele model of learning is used as the assessment framework. For example, Chew and Noraini Idris (2007) used van Hiele five phases of learning, namely information, guided orientation, explicitation, free orientation and integration, to assess geometric thinking of Form One students. Noraini Idris (2009) and Chew (2009) used van Hiele five levels of learning: recognition, analysis, ordering, deduction and rigor to investigate the impact of using geometer's sketchpad on students' geometry achievement.

Hwa (2010), NurulHidayah and Ong (2009) and Kamariah Abu Bakar, Rohani Ahmad Tarmizi, Ahmad Fauzi Mohd. Ayub, Aida Suraya Md. Yunus (2009) used different assessment framework or method to capture students' mathematical thinking. Hwa (2010) defined and assessed students' mathematical thinking based on six domains: conceptual knowledge, procedural knowledge, thinking strategies, thinking skills, metacognition and mathematical disposition; NurulHidayah and Ong (2009) used performance assessment to assess students' mathematical thinking based on their conceptual understanding, mathematical representation, computational skills and mathematical explanation; whereas Kamariah Abu Bakar, et al. (2009) utilizing geometer's sketchpad to investigate students' mathematical thinking in the aspects of mathematics achievement and attitude towards integrating the geometer's sketchpad and the traditional teaching strategy.

Other than assessing "thinking", another important area of research study which draws increasing interest of Malaysian mathematics educators is problem solving (Lee, 2001). For example, Noor Azlan and Lui (2000) assessed students' problem solving abilities in using basic knowledge, standard procedures and problem solving skills based on students' written responses. Lau, Hwa, Lau and Shem (2003), assessed students' problem solving abilities based on four levels of difficulties: one rule under your nose, application, choice of a combination and approaching research level proposal by Polya. Lee (2005) assessed Malaysian secondary students' problem solving abilities based on project work in four aspects: the presentation of the report, assessing the problem/task, applying the problem solving heuristics, and making decisions.

Due to increasing usage of computer in teaching and learning of mathematics in Malaysian schools, multiple choice questions in computer-based assessment has also caught the attention of Malaysian mathematics educators. Ramesh, Manjit Sidhu and Watugala (2005) explored the potential of multiple choice questions in computer-based assessment to assess student learning. Mean while, Noraini, Loh, Norjoharudden, Rasidah, Wan Rohaya, Hamidah and Mohd Yusuf Saad (2008) developed a computer-based assessment system, Classroom Assessment System for Teaching and Learning (CASTLe), to assess students' mathematics attitude and achievement. Another example for computer-based assessment is a Web-Based Computer-Adaptive Mathematics Multiple Choice Assessment (CAAS) developed by Lau, Hong, Lau, and Hasbee (2008). The CASS scores students' mathematics performance by using NRET scoring method, which is a hybrid of Conventional Number Right (NR) method and Elimination Testing (ET) method. This computer-based assessment was proposed to be able to provide feedback on the student's knowledge state as an alternative to feedback on degree of realism.

TRANSFORMATION OF MATHEMATICS ASSESSMENT

In order to answer the call of the Ministry of Education, Hwa (2010) carried out a study on developing a Mathematical Thinking Assessment (MaTA) framework, and its usability and practicality in the Malaysian school context. The Mathematical Thinking Assessment (MaTA) framework consists of four components: (a) a Performance assessment, (b) a Metacognition Rating Scale, (c) a Mathematical Disposition Rating Scale, and (d) a Mathematical Thinking Scoring Rubric. The MaTA framework is intended to be implemented by teachers with the aim of assessing students' mathematical thinking.

The Performance Assessment component is administered by the classroom teacher to assess students' mathematical knowledge and skills (e.g. conceptual, procedural, strategies and skills) while solving particular mathematical problems in one or more content areas that has been the focus of classroom instruction. The Metacognition Rating Scale is used, also by the teacher, to elicit students' cognition awareness, such as monitoring and regulation during the problem solving process. The Mathematical Disposition Rating Scale is used by the teacher to indicate students' disposition toward learning of mathematics. Finally, the Mathematical Thinking Scoring Rubric is used to score and report students' mathematical thinking performance according to the domains defined in this study. This section presents only the feedback from students and teachers after the MaTA framework had been implemented in the school context.

Participants and Instrumentation

The study involved secondary school mathematics teachers and secondary students in the State of Sarawak. Six mathematics teachers from four different schools and a total of 203 Form Four students from five Science classes and two Arts classes participated in the study. At the end of the study, the Mathematical Thinking Assessment (MaTA) framework Questionnaire, which was adapted from the questionnaire used by Parke and Lane (2007) in Maryland School Performance Assessment Program (MSPAP), was administered to the participating students. All the six teachers involved were interviewed to elicit their views towards the implementation of the MaTA framework.

Procedures

The following summarizes how teachers could be expected to implement the MaTA framework in their home school context.

Step 1:Designing Performance Assessment

Based on the procedures or guidelines provided in the MaTA framework, the teachers have to design the performance tasks (i.e. test items or questions) and then administer these to their students. During the assessment, teachers need to ensure that the students use appropriate approaches to perform the tasks, such as explaining and justifying the answers obtained in their solutions, as required by the MaTA framework. Usually, this is achieved by including specific prompts in questions, such as asking students to explain their thinking or to justify their solutions.

Step 2: Scoring Students' Performance

1. By referring to the scoring criteria and scoring guide for each of the domains in Mathematical Thinking Scoring Rubric, namely conceptual knowledge, procedural knowledge, thinking strategies and thinking skills, teachers are able score their students' levels of performance respectively based on their written solutions.
2. After scoring students' written solutions, the teachers then use the Metacognition Rating Scale to rate students' levels of metacognition based on teachers' classroom observations.

3. Similarly, the levels of performance for students' mathematical disposition can be determined by the teachers through a Mathematical Deposition Rating Scale.

Step 3: Reporting Students' Mathematics Performance

After scoring students' written solutions and rating their metacognition and mathematical disposition, students' levels of performance for each domain are summarized into a standard report, entitled Teacher's Report on Student's Mathematical Thinking Performance. This report contains band scores and comments from the teacher for each domain of mathematical thinking. This report can then be given to students as feedback on each of the three areas indicating the quality of their performance, based on their written solutions and on their teacher's classroom observations.

Findings and Discussions

Students' Feedback on the MaTA Framework

Nearly three quarters of the students perceived that the MaTA framework had a positive impact on their learning of mathematics. This finding is supported by the data in Table 54.2 where 72.3 percent of the students somehow or very much agreed with the claim. Even though a high percentage of students acknowledged the positive impact of MaTA framework, only 64.4 percent of the students admitted that the MaTA framework had "Somewhat" helped them or helped them a lot ("Very") in improving their mathematics performance.

An open-ended question asked students to elaborate the positive impact of the MaTA framework on their learning of mathematics. The results of this question are summarized in Table 54.3. With respect to the responses of "Not at all" and "A little", the most common comment was "lack of understanding" of MaTA framework (34.6 percent). Students also described MaTA framework as a type of troublesome (30.8%) assessment to them, as one student commented, "*I think it just stress us and make us headache*" (Implementation/Student 13). A small percentage of students responded that the MaTA framework had no or little positive impact on their learning because performance tasks in the MaTA framework were like a normal test (11.5 percent) in the classroom. These groups of students ("Not at all" and "A little") said that they could not see much difference between the tasks given in the MaTA framework and the tasks given in current classroom assessment.

Students who gave their responses as "Somewhat" or "Very" perceived that MaTA framework had a positive effect especially on their problem

Table 54.2. Frequency and Percentage Showing the Impact of MaTA Framework on Students' Learning

	No. of Responses	
Item	*"Not at all" and "A little"*	*"Somewhat" and "Very"*
How much positive impact has MaTA framework had on your learning of mathematics?	56 (27.7%)	146 (72.3%)
How much do you think the MaTA framework can help you to improve mathematics performance?	72 (35.6%)	130 (64.4%)

Table 54.3. Frequency and Percentage of Students' Written Feedback on Positive Impact on Learning

	No. of Responses		
Group	*Top Three Reasons*	*Frequency*	*Percentage*
"Not at all" and "A little" (n = 26)	Lack of understanding	9	34.6%
	Troublesome	8	30.8%
	Like normal tests	3	11.5%
"Somewhat" and "Very" (n = 75)	Improve problem solving skills	23	30.7%
	Promote thinking	15	20.0%
	Understanding mathematics better	15	20.0%

solving skills (30.7 percent, see Table 54.3). The students' commented, "Yes, because this assessment makes me more careful when answer a question. Example we must check the answer after we get the answer" (Implementation/Student 42), and "MaTA [framework] helps me to use various types of methods to solve mathematics problems" (Implementation/Student 43)

Besides improving their problem solving skills, this group of students also perceived that promoting thinking (20.0 percent) was another important positive impact of MaTA framework. One student said, "It helps me to think further and logically about the mathematical problems and not just learning the concepts and formula as usual in the class" (Implementation/Student 30), another student commented "MaTA [framework] teaches me to think, more exquisitely, independently when working on question" (Implementation/Student 64).

The third positive effect on students' learning is that the MaTA framework allowed students to understand mathematics better (20.0 percent). In the MaTA framework, students are required to justify how they derived the solution to the problem given. Students need to think about the steps used and have to explain their thinking in words. By explaining, it is hoped that students would know and understand better the mathematical concepts learned. One student said, "because we need to explain the answer and this makes me understand how and why we get the answer" (Implementation/Student 133). Other favorable comments were, "Understand mathematics better" (Implementation/Student 69), "I can understand more about the concepts" (Implementation/Student 71), "I can understand clearly of the questions" (Implementation/Student 81) and "It helps me to know mathematics in a different way" (Implementation/Student 85).

Teacher's Feedback on the MaTA Framework

From the structured interview, the teachers revealed that one of the major setbacks of the MaTA framework was due to the nature of performance tasks where justifications or explanations to the solutions were required. According to the teachers, explaining the solution strategies during mathematical problem-solving was not a common practice in the classroom activities. Their students were used to the types of questions where they could obtain the answers straight away without providing any reasoning. Hence, explaining or justifying the solutions in words during problem-solving created a certain degree of nervousness among students. This was supported by the following comments given by the teachers:

> Because the main problem for them is they are not used to it, suddenly they are asked to do like this, plan, for example and then justify their answer, explain, they are not used to it. (Pilot/Interview/Teacher 2)

> Most of the students do not like the mathematical task, like the one in the MaTA because they said that the tasks need more working. They have to write essay [explain], but most of the students like simple activity, simple question. (Pilot/InterviewTeacher 3)

Another problem raised during teachers' interviews was English proficiency. Even though Mathematics was taught in English, many students, especially those from Arts classes, were unable to understand the tasks given to them. Hence, the teachers had to "*translate the meaning of the tasks to Bahasa Melayu (Malay Language)*", remarked Teacher 6 (Interview/Teacher 6). These students were able to understand and solve the task if the task was in Malay Language. Their limited English proficiency caused them to perform below their actual performance under the MaTA way of assessment as compared to normal classroom assessment. According to one teacher, this was because "most of the students can calculate their answers correctly but when goes to explanation, it is really a problem. They cannot write even a clear English sentence!" (Interview/Teacher 5).

If the students gave no or improper justification or explanation to the solutions, they were unable to score well under the scoring system provided by the MaTA framework.

Poor basic understanding of mathematics was another reason why students did not like the tasks given in the MaTA framework, commented one teacher. This was probably due to "their level of Mathematics [understanding] hasn't reached that standard, so they cannot understand the questions" (Interview/Teacher 1). He further commented that "if the very basic [mathematical concepts] they cannot do, how can they do the one that is related to real-life and needs explanation". Furthermore, the students were used to multiple-choice type of questions. Even if they did not understand the multiple-choice question, they could still "choose the answer" (Interview/Teacher 5). Hence, performance tasks in the MaTA framework which were open-ended and context-based in nature, appeared to be very difficult to those students who were weak in basic mathematical concepts.

While many students did not like the performance tasks, there were also some positive feedbacks from the teachers. The performance tasks were able to: (a) promote deep understanding of problem ("I think for this one, mathematical tasks given in the MaTA [framework] can train them to understand what the question wants, relate [the question] to real-life and make them to explain what they are doing" (Interview/Teacher 5)), (b) cultivate thinking skills ("it helps them to develop their thinking skills because they have to think more" (Interview/Teacher 3)), and, (c) help teachers to know their students better ("We can know how much the students have mastered" (Interview/Teacher 6)).

Summary of the Study

The findings of students' beliefs about the type of assessment tasks used and whether these beliefs affected their mathematical thinking performance could be summarized as follows: (a) students who said that performance tasks were more interesting, who liked performance tasks better and who agreed that performance tasks let them show what they knew, clearly performed better than those who preferred traditional multiple-choice tests; and (b) students who said that performance tasks in the MaTA framework made them think harder and were easier to know what to do as compared to traditional standardized tests performed neither badly nor better in the assessment. In addition, more than two thirds of students perceived that the MaTA framework had a positive impact, helping them to improve their learning of mathematics. According to the students' own comments, the MaTA framework generally enabled them to sharpen their problem solving skills, promoted their thinking and let them understand mathematics better. Based on these findings, the MaTA framework appears to be an alternative way of assessing students' thinking processes that fosters more effective and holistic mathematical thinking.

Several aspects of the MaTA framework seem to be the key to fostering these gains. They are the design of the assessment tasks themselves with the explicit inclusion of components that require students to carry out a chain of mathematical thinking and a requirement to explain and/or justify their conclusions or the processes they have used. Equally important is the written feedback that students receive from teachers (Ong, 2010) about the quality of their mathematical thinking, how it might be improved, and the fact that teachers are asked explicitly to include in their feedback to students' comments relating to students' metacognition and disposition. These features of students' performance are rarely communicated using standard assessment tasks. The MaTA framework shines a spotlight on these important components of students' mathematical thinking performance, bringing them to the attention of both teachers and students. That appears to be one of its great promises. Further research is clearly needed on how well these potential benefits are realized when the MaTA framework is used by teachers and students over a sustained period of time.

Limitations of the Study

They were few limitations found in this study. However, this section highlights only those related to the students and teachers. In the MaTA framework, students were asked to show not only the solutions to the problems, but also the explanations on how the solutions were derived. However, many students were unable to express their thinking processes through written explanation due to their limited English proficiency, even though they were told that grammatical errors would be ignored in this assessment. Hence, this group of students was unable to obtain grades corresponding to their thinking abilities during mathematical problem solving.

The MaTA framework required teachers to design performance tasks, mark and score students' written solutions, observe students' learning behavior and prepare a descriptive report for each of the students in the class. Hence, a lot of time was needed to implement this performance assessment as compared to the current traditional assessment. Furthermore, the classes in Malaysian schools normally exceed 35 students. This big number of students not only made the teachers feel exhausted in implementing the MaTA framework, but also created problems pertaining to the subjectivity of assessment based on classroom observations. The teachers faced difficulties in observing and recording accurately each of the students' learning behavior, particularly the students' metacognition and mathematical disposition.

Suggestions for Future Research

In this study, the development of the MaTA framework was mainly done based on the comments and suggestions of the school mathematics teachers. However, the principals' and students' views on the development of this framework were not taken into consideration due to the constraints of time and resources. As acknowledged by Parke and Lane (2007), other than key school personnel, students are the essential group of stake-holders in the assessment process. Hence, it is proposed that in future studies, both principals' and students' views are to be collected to form a more comprehensive picture on how classroom practices reflect the standards and assessment of the curriculum.

The validity of the MaTA scale in the study was established mainly based on the review of existing assessment frameworks and expert validation. Hence, in any future study, it is proposed to use psychometric properties to examine the validity of the MaTA scale. The psychometric properties are able to provide evidence of discriminant validity of the MaTA scoring system by showing the relations of the inter-correlated of these six domains in the MaTa framework, and to ensure that classifications of the skills or performance levels are continuous and appropriate.

Another aspect which can enhance the validity of this type of assessment framework is through think aloud protocol analysis. As it is proposed by Silver and Lane (1993), think aloud is able to provide "rich information from a relatively small number of students regarding the degree to which the tasks evoke the content knowledge and complex processes..., and allows for additional probing regarding the processes underlying student performance" (pp. 28-29). Hence, think aloud not only can be used to capture students' thinking processes during their attempt on a performance assessment, but also to triangulate and validate the constructs of the assessment model developed in future study.

The application of technology to enhance assessment in mathematics education is increasingly gaining its importance. Therefore, it is proposed to research on turning this assessment guideline into a computerized assessment tool in future study. By using this computerized assessment tool, teachers' workload can be lightened particularly in the process of scoring and reporting students' mathematical thinking performance.

REFERENCES

Adi Badiozaman Tuah (2006). Improving the quality of primary education in Malaysia through curriculum innovation: Some current issues on assessment of students performance and achievement. *Proceedings 3rd International Conference on Measurement and Evaluation in Education* (pp. 16–26). Penang: Universiti Sains Malaysia.

Asmah Othman. (1994). *Critical thinking skills across the curriculum: A survey of teachers' knowledge, skills and attitudes in secondary schools in Kuching, Sarawak.* Unpublished master's thesis, University of Houston, US.

Beyer, B. K. (1984). Improving thinking skills: Defining the problem. *Phi Delta Kappan, 65*(7), 486-490.

Bloom, B. S. (1984). *Taxonomy of Educational Objectives*. Boston: Allyn and Bacon.

Cheah, U. H. (2010, February). *Assessment in Primary Mathematics Classrooms in Malaysia.* Paper presented at the meeting of the Fourth APEC - Tsukuba International Conference: Innovation of Mathematics Teaching and Learning through Lesson Study - Connection between Assessment and Subject Matter, Tsukuba, Japan. Retrieved from http://www.criced.tsukuba.ac.jp/math/apec/apec2009/doc/pdf_20-21/CheahUiHock-paper.pdf

Chew, C. M. & Noraini Idris (2007). Assessing Form One students' learning of solid geometry in a phased-based instructional environment using The Geometer's Sketchpad. In Noraini Idris (Ed.), *Classroom assessment in mathematics education* (pp. 103-122). Shah Alam: McGraw-Hill (Malaysia).

Chew, C. M. (2009). Enhancing students' geometric thinking through phase-based instruction using geometer's sketchpad: A case study. *Jurnal Pendidik Dan Pendidikan, 24,* 89–107.

Gan, W. L. (2007). *A research into Year Five pupils' pre-algebraic thinking in solving pre-algebraic thinking*. Unpublished doctoral thesis, Univesiti Sains Malaysia, Penang.

Gonzales, P., Williams, T., Jocelyn, L., Roey, S., Kastberg, D., & Brenwald, S. (2008). *Highlights from TIMSS 2007: Mathematics and science achievement of u.s. fourth- and eighth-grade students in an international context (NCES 2009–001 Revised).* Washington, DC: National Center for Education Statistics, Institute of Education Sciences, United States Department of Education.

Herman, J. (1992). *What's happening with educational assessment?* (Report No. TM-019-067). Los Angeles: National Center on Evaluation, Standard, and Students Testing. (Eric Document Reproduction Service No. ED 349342)

Hwa, T. Y. (2010). *Development, Usability and Practicality of a Mathematical Thinking Assessment Framework.* Unpublished doctoral thesis, Univesiti Sains Malaysia, Penang.

Kamariah Abu Bakar, Rohani Ahmad Tarmizi, Ahmad Fauzi Mohd. Ayub, Aida Suraya Md. Yunus (2009) Effect of utilizing Geometer's Sketchpad on performance and mathematical thinking of secondary mathematics learners: An initial exploration. *International Journal of Education and Information Technologies , 1* (3). 20-27.

Lau, P. N. K., Hwa, T. Y., Lau, S. E., & Liew, C. Y. (2006). *Successful Mathematics Problem Solving Using Heuristics* (Research Report). Malaysia: Universiti Teknologi MARA, Institute of Research Development and Commercialization

Lau, P. N. K., Hwa, T. Y., Lau, S. E., & Shem, L. (2003). *The thinking processes of mathematics problem solving of Form Four secondary school students* (Research Report). Malaysia: Universiti Teknologi MARA, Institute of Research Development and Commercialization

Lau, S. H., Hong, K. S., Lau, P. N, K., & Hasbee, U (2008). Web based Computer-Adaptive Mathematics Multiple Choice Assessment (CAAS) Using the New NRET Scoring Method. *Electronic Proceedings of the Thirteenth Asian Technology Conference in Mathematics*, Bangkok, Thailand. Retrieved from http://atcm.mathandtech.org/EP2008/ papers_full/2412008_15339.pdf

Lee, S. E. (2001). Studies on problem solving in school mathematics in Malaysia. In E. Pehkonen (Ed.), *Problem solving around the world*. Faculty of Education Report Series C:14. University of Turku.

Lee, S. E. (2005, August). *Assessing mathematical problem solving abilities of Malaysian secondary students*. Paper presented at the 3rd East Asia Regional Conference on Mathematics Education (EARCOME 3), Shanghai, China. Retrieved from http://math.ecnu.edu.cn/earcome3/TSG6/EARCOME-TSG6-LSE-Assessing%20PS.doc

Lim, C. S. (2010, February). *Assessment in Malaysian School Mathematics: Issues and Concerns*. Paper presented at the meeting of the Fourth APEC - Tsukuba International Conference: Innovation of Mathematics Teaching and Learning through Lesson Study - Connection between Assessment and Subject Matter, Tsukuba, Japan. Retrieved from http://www.criced.tsukuba.ac.jp/math/apec/apec2009/doc/pdf_20-21/LimChapSam-paper.pdf

Lim, H. L., Chew, C. M., Wun, T. Y., & Noraini Idris (2009, November). Assessing a hierarchy of pre-service teachers' algebraic thinking of equation. Paper presented at the *the 3rd International Conference on Science and Mathematics Education*, Reasam, Penang.

Lim, H.L., & Noraini Idris (2006). Assessing algebraic solving ability of form four students. *International Electronic Journal of Mathematics Education, 1*(1), 55 -76.

Ministry of Education Malaysia (1993). *Thinking skills: Concept, model and teaching-learning strategies*. Curriculum Development Centre, Malaysia.

Ministry of Education Malaysia (2005). *Mathematics syllabus for integrated curriculum for secondary school*. Curriculum Development Centre, Malaysia.

Ministry of Education Malaysia (2007). *Blueprint of Education Development*. Curriculum Development Centre, Malaysia.

National Council of Teachers of Mathematics (2000), *Principles and Standards for School Mathematics*, Reston, VA: NCTM.

Nickerson, R.S. (1989). New directions in educational assessment. *Educational Researcher, 18*(9), 3-7.

Noor Azlan A. Z., & Lui, L. N. (2000). Evaluating the levels of problem solving abilities in mathematics. In R, Alan (ed.). *Proceedings of the International Conference on Mathematics Education into the 21st Century: Mathematics for living*. Amman, Jordan. Retrieved from http://math.unipa.it/~grim/Jzanzalinam.

Noraini Idris (2009). The Impact of Using Geometers' Sketchpad on Malaysian Students' Achievement and Van Hiele Geometric Thinking. *Journal of Mathematics Education 2*(2), 94-107

Noraini Idris, Loh, S. C., Norjoharudden, M. N., Rasidah, H., Wan Rohaya, W. Y., Hamidah, S., & Mohd Yusuf Saad. (2008). Classroom Assessment System for Teaching and Learning (CASTLe). In Naraini Idris, Loh, S. C., & Palaniappan, A. K. (Eds.), *Proceeding of the International Conference on Educational Innovation* (pp. 17 – 28), Kuala Lumpur.

NurulHidayah, L. A., & Ong, S. L. (2009). Use of Performance Task in Assessing Year Six Student's Levels of Mathematical Thinking. In Cheah, U. H., Wahyudi, Devadason, R. P., Ng, K. T., Preechaporn, W., & Aligaen, J. C. (Eds.), *Proceedings of the 3rd International Conference on Science and Mathematics Education* (pp. 386 – 394), Reasam, Penang.

Ong, S. C. (2010). Assessment profile in Malaysia: High-stakes external examination dominant. *Assessment in Education: Principles, Policy & Practice, 17*(1), 91-103.

Parke, C. S., & Lane, S. (2007). Students' perceptions of a Maryland state performance assessment. *The Elementary School Journal*, *107*(3), 306–324.

Ramesh, S., and Sidhu, S. Manjit, and Watugala, G.K., (2005) Exploring the Potential of Multiple Choice Questions in Computer-Based Assessment of Student Learning. *Malaysian Online Journal of Instructional Technology (MOJIT), 2* (1). Retrieved from http://pppjj.usm.my/mojit/articles/pdf/April05/04-Ramesh&Manjit.pdf

Rosnani Hashim & Suhailah Hussein. (2003). *The Teaching of Thinking in Malaysia*. Kuala Lumpur: Research Centre, IIUM.

Silver, E. A., & Lane, S. (1993). *Balancing Considerations of Equity, Content Quality, and Technical Excellence in Designing, Validating and Implementing Performance Assessment in the Content of Mathematical Instructional Reform: The Experience of the QUASAR Project*. (Contract. No. 890-0572). NY: Ford Foundation.

Suzieleez Syrene, A. R., Venville, G, & Chapman, A. (2009, December). *Classroom Assessment: Juxtaposing Teachers' Beliefs with Classroom Practices*. Paper presented at the 2009 International Education Research Conference, Australian Association for Research in Education. Retrieved from http://www.aare.edu.au/09pap/abd091051.pdf

Von Glasersfeld, E. (1995). *Radical Constructivism: A Way of Knowing and Learning*. London: Falmer Press.

CHAPTER 55

MATHEMATICS INCORPORATING GRAPHICS CALCULATOR TECHNOLOGY IN MALAYSIA

Liew Kee Kor
Universiti Teknologi MARA Malaysia

ABSTRACT

Graphics calculator technology encompasses more than just plotting of graphs. Programming capabilities of graphic calculator feature a variety of mathematical computational skills from manipulating symbolic expressions to manoeuvring analytic mathematical problems. This chapter begins with an overview on the use of graphics calculator in the teaching and learning of mathematics in Malaysia. In particular, it reports chronologically the stages of implementation of graphics calculator in the Malaysian mathematics curriculum. Consequently, it presents the findings of local research studies and gives an account on graphics calculator related events in Malaysia from year 2002 onwards.

INTRODUCTION

The invention of new technologies and advancement of the existing ones in the modern society are deemed perpetual and continuous. When com-

The First Sourcebook on Asian Research in Mathematics Education: China, Korea, Singapore, Japan, Malaysia, and India, pp. 1249–1264

puters became more affordable to schools in the 1990s, there was much optimism about the potential of the use of computers in mathematics teaching to acquire higher-level mathematics thinking skills. It is envisaged that the use of ICT as a teaching and learning tool, as part of a subject, or as a subject by itself will in some way revolutionize learning as well as enhances pedagogies that consequently, enable the development of mathematical knowledge. Eventually, the ICT-enabled education is targeted to develop a pool of knowledge workers and to produce more technocrat workforce within the economy.

As a result of readily available resources, more teachers are anticipated to embark on the integration of technology to enhance teaching and learning strategy. However, in 2007 the Malaysia Ministry of Education reported that there was only a minimal and inappropriate use of technological applications in mathematics teaching and learning at all levels (Ministry of Education, 2007). This finding is not peculiar as such phenomenon was observed much earlier in the United Kingdom that only a minority of both primary and secondary teachers of mathematics were using computers to support their teaching and learning (Ball, Higgo, Oldknow, Straker & Wood, 1987, cited in Thomas, Tyrrell & Bullock, 1996).

Despite various benefits technology can bring to enhance the process of mathematics learning, whether mathematics courses that employed graphing technology in the Malaysian mathematics classroom attained its intended purpose remain as a question. This chapter aims to consolidate research findings and reports on the roles of Malaysian mathematics teachers implementing graphics calculator in the mathematics courses. The purpose is to ascertain if there exists a gap between "rhetoric and reality" (Thomas, et al., 1996) in technology implementation in the Malaysian scene.

THE USE OF TECHNOLOGY IN MALAYSIAN MATHEMATICS CURRICULUM

Mathematics for the secondary schools in Malaysia consists of the core-subject Mathematics (KBSM, 2004) for both upper and lower secondary schools, and the Additional Mathematics (KBSM, 2006) which is an elective subject for the upper secondary school pupils. The curriculum for both of these mathematics subjects aims to develop mathematical knowledge, competency and inculcate positive attitudes towards mathematics and to enable pupils to cope with daily mathematics challenges.

The mathematics curriculum emphasizes the use of technology in the teaching and learning of mathematics to support the nation's aspiration of becoming an industrialized nation. Beginning year 2003, English was used as the medium of instruction for Science and Mathematics subjects. The curriculum stresses that the teaching of mathematics in English together with implementation of ICT provide superior opportunities for their mathematical development (KBSM, 2006). It states that,

> "The use of technology especially, Information and Communication Technology (ICT) is much encouraged in the teaching and learning process. Pupils' understanding of concepts can be enhanced as visual stimuli are provided and complex calculations are made easier with the use of calculators."(KBSM, 2006, p. 4)

Technology in education is believed to support the mastery and achievement of the desired learning outcomes. Students are expected to be able to "use hardware and software to explore mathematics with the caution that "... calculators, are to be regarded as tools to enhance the teaching and learning process and not to replace teachers" (KBSM, 2004, p. 4).

GRAPHICS CALCULATOR MOVEMENT IN MALAYSIA

The types of technology employed in the Malaysia mathematics classroom can be classified into three broad categories: the spreadsheet, mathematics software and mathematics courseware. These technologies are executed via the use of mathematical analysis tools (Pierce & Stacey, 2009) such as calculators in the classroom or computers in the computer lab. Graphics calculator, a type of handheld calculator was first introduced to the international educational market in year 1985 and emerged in the Malaysian institutions of higher learning in year 2001. The calculator is equipped with computer algebra system (CAS) and is capable of plotting graphs, solving simultaneous equations, as well as performing mathematical tasks displaying symbolic and numerical output. Its unique programming feature allows students to create customized programs for particular mathematics and scientific applications. Many studies abroad dated late 1980s and onwards reported that judicious use of graphics calculator technology improved university or school mathematics achievement (e.g. Brolin, 1987; Barkatsas, et al. 2010, Chiappini & Bottino, 1999; Graham & Thomas, 2000; Hembree & Dessart, 1986; Pierce, Stacey & Barkatsas, 2007; Quesada & Maxwell, 1994).

Despite positive acclamation of the use of graphing technology in mathematics by many mathematics educators worldwide, the inclusion of graphics calculator technology in the mathematics curriculum in Malaysia is slow in pace. Year 2001 signified the beginning of the graphics calculator movement in Malaysia. Markedly there were two separate interest groups that led the movement. In year 2001 a project named the "Kalkulator Grafik untuk Matematik"(translated as Graphics Calculator for Mathematics) or popularly known as KaGUM was initiated by a local university to investigate the effectiveness of using TI-83 Plus in the teaching and learning of secondary school mathematics (Noraini Idris et al., 2003). Concurrently the School of Mathematical Sciences at another local university set off a special course that integrated handheld technology in mathematics in the 2001/2002 academic year (Suraiya Kassim, Rosihan M. Ali, Seth, Zarita Zainuddin and Mokthar Ismail, 2002). The repercussion of both initiatives sees the graphics calculator gradually gained its popularity and impact in the secondary schools. Starting from year 2003 until mid 2011, the Malaysia Ministry of Education (MOE) sponsored TI graphics calculators and probing tools to 473 secondary schools across Malaysia (Kanniapan, 2011). A few graphics calculator companies also offered to sponsor and organize workshops to train school mathematics teachers to use the tool.

Increasing interest in graphics calculator has led to the materialization of the National Conference on Graphing Calculator (NCGC) on the national level. The Faculty of Education of Universiti Malaya had taken the lead to host the First NCGC in year 2003. The School of Mathematical Sciences of Universiti Sains Malaysia hosted the Second NCGC in year 2004. After a lapse of four years the third NCGC was held in year 2008. These conferences had opened up avenues for the calculator interest group to interact locally and internationally. Above all, the conferences provided a platform for the participants to present their research findings as well as to participate in the seminar debating major issues concerning graphics calculator. Continuous supports from the calculator interest group have made possible the fourth NCGC scheduled in 2011. The following sections describe the themes and focus as well as a summary report from the proceedings of each major graphics calculator related conferences held in Malaysia from the year 2002 till present.

(a) The 7th Asian Technology Conference in Mathematics held in Malacca, Malaysia on 18-21 Dec 2002
   - Area of focus: Graphics calculator, CAS and mathematics
   - Summary of reports from the conference proceedings:
     - Students showed positive attitude and interest towards graphics calculator. Their performance was also improved after graphics calculator instruction (Suraya Kassim et al., 2002).

- Graphics calculator helps students who previously could not cope with algebra to learn algebra. Students are motivated to work actively and cooperatively with their peers in the presence of graphics calculator (Khairiree, 2002).
- With the help of graphics calculator, students can interpret data more efficiently without worrying of making calculation mistake. Prior to attending any graphics calculator lab course that uses graphics calculator excessively in exploring data, students should be equipped with some conceptual basics for a more meaningful learning (Haili & Sulaiman, 2002).

Source: Yang, W-C., Chu, S-C., de Alwis, T., & Bhatti. F. M. (Eds.) (2002). *Proceedings of the Seventh Asian Technology Conference in Mathematics*. Melaka, Malaysia: ATCM Inc.

(b) The 1st National Conference on Graphing Calculators held in Kuala Lumpur, Malaysia on 11-12 July 2003

- o Area of Focus: Incorporating ICT in mathematics classroom teaching, curriculum and technology, teacher training and development, and pedagogical issues.
- o Summary of reports from the conference proceedings:
  - In graphics calculator instruction, students acquire better understanding of the concepts related to Straight Lines. Achievements of weaker ability students also showed improvement (Noraini et al., 2003).
  - Students' attitudes towards the use of graphics calculator were expressed in metaphors such as graphics calculator as a useful tool, graphics calculator as a status symbol and graphics calculator as a partner (Kor & Lim, 2003).
  - ICT tool such as graphics calculator provides opportunities for students to learn mathematics through learning from feedback, observing patterns, seeing connections, working with dynamic images, exploring data and programming (Rohini, 2003).
  - Graphics calculator can be engaged in laboratory explorations in teaching Calculus as well as integrated in a graphics calculator Lab Course in Mathematics (Rosihan Ali, et al., 2003).
  - Graphics calculator helps to equip students with in-depth understanding of mathematics concepts thus in examination that does not allow the use of graphics calculator, they can perform equally well (Khairiree, 2003).

Source: *Proceedings of First National Conference on Graphing Calculators*. Petaling Jaya: University Malaya.

(c) The 2nd National Conference on Graphing Calculators (NCGC 2) held in Penang, Malaysia on 4 – 6 Oct 2004

- o Area of focus: Pedagogical strategies, assessment, curriculum, and professional development for teachers.
- o Summary of reports from the conference proceedings:
  - Appropriate use of graphing technology enhances presentation of mathematical concepts and support mathematics understanding (Rosihan Ali & Kor, 2004).
  - Despite the advantages, the acceptance of the use of graphing calculators in the mathematics classrooms is greatly impeded by false beliefs. The theory of distributed cognition is used to dispel these beliefs and highlight the advantages of using graphing calculators in the learning of mathematics (Pumadevi, 2004).
  - An example was given to show how graphics calculator can be used to help students develop meaningful understanding of the concept of solving linear inequalities (Norjoharuddeen Mohd Nor, 2004)
  - The TI-83 Graphic Calculator as an essential tool enables students to focus on exploration, decision making, reflection, reasoning, and problem solving (Noraini Idris, 2004).

Source: Rosihan M. Ali, Anton Abdulbasah Kamil, Adam Baharum, Adli Mustafa, Ahmad Izani Md. Ismail & V. Ravichandran (Eds.) (2004). *Proceedings of the 2nd National Conference on Graphing Calculators*. USM Penang, Malaysia: School of Mathematical Sciences.

(d) The 3rd National Conference on Graphing Calculators (NCGC 3) held in Kula Lumpur, Malaysia on 16 -18 April 2008

- o Area of focus: Attitudes and performance, assessment, and technology use in teaching particular mathematics topic.
- o Summary of reports from the conference proceedings:

Undergraduates' attitudes towards statistics improved after graphics calculator intervention. Also, attitudes towards statistics correlated positively with attitudes towards the use of graphics calculator (Kor, 2008).

  - Graphics calculator enhances school students' performance and induced better levels of their metacognitive awareness.

The average mathematics ability group benefited most from the graphics calculator instruction (Nor'ain et al., 2008).

- Designing examination questions incorporating graphics calculator allows teacher to access the higher order thinking skills (Sundram, 2008).
- Graphics calculator is an option to assess higher-order thinking in school students' project paper (Muthiah, 2008).
- Best practices in using graphics calculator to teach functions and graphs, straight lines and probability.

Source: Adam Baharum, Adli Mustafa, & Zarita Zainuddin (Eds.) (2008). Proceedings of *The 3rd National Conference on Graphing Calculators, NCGC 2008*. Malaysia: School of Mathematical Sciences.

(e) The 15th Asian Technology Conference in Mathematics held in Kula Lumpur, Malaysia on 17 -21 Dec 2010

- o Theme: Linking applications with mathematics and technology
- o Summary of reports from the conference proceedings:
  - A case study of the use of the Socratic Method to develop understanding of the relationship between the standard form and the general algebraic form of the equation of a circle was mediated with the use of the graphing calculator as the primary feedback tool. Students from a Malaysian teacher training institute participated in this study. The study also revealed the process of thinking of the students, the realization of the importance of a systematic record of data to aid the construction of new knowledge from prior knowledge and the need of an efficient feedback tool to explore students' conjectures which lend to progress in the learning process (Pumadevi Sivasubramaniam, 2010).
  - Teachers and pre-service teachers in a local university at a master's level course on mathematics and technology participated in a 5- week lab exploration during the 2009/2010 Academic Session. Topics were mostly from linear algebra and calculus and activities were done based on materials found in the Texas Instrument (TI)'s websites and mathematics books. At the end of the 5-week exploration, students were asked to reflect their experiences with the hand-held tool by answering survey questions on perception towards the graphic calculator technology before and after the exploration (Hajar Sulaiman, 2010).

Source: Majewski, M., Yang, W-C. , de Alwis, T., & Hew, W. P. (Eds.) (2010). *Proceedings of the 15th Annual Conference of the Asian Technology Conference on Mathematics*. Kuala Lumpur, Malaysia: ATCM Inc.

(f) The 4th National Conference on Graphing Calculators (NCGC 4) held at Penang, Malaysia on 21 – 23 June 2011
- o Theme: Realizing innovative teaching-learning-assessment through graphing technology
- o Summary of reports from the conference proceedings:
  - Situated in a traditional classroom setting where there is more teaching and less hands-on, undergraduate students of an ODE course enriched with only an hour a week laboratory practice with MAPLE Computer Algebra System (CAS) were given worksheets with printed MAPLE commands alongside each question to practice. Results shows that it is possible to incorporate technology in learning mathematics using the improvised 2-in-1 approach for under-prepared students in technology. In other words students in an exam-oriented environment or in institutions where there is a shortage of computer lab, time constraint, lack resources or infrastructure to receive proper tool training can still benefit richly from the use of computer or graphics calculator (Kor, 2011).
  - Teaching and learning with the use of TI-84 Plus graphics calculator as well as Geometer's Sketchpad was discussed respectively using the Technological Pedagogical and Content Knowledge (TPACK) and the theory of cognition (Pumadevi Sivasubramaniam, 2011)
  - Topical teachings using the graphics calculator were also discussed. Raja Lailatul Zuraida and Shafia (2011) found that graphics calculator helps students to visualize conceptual image of functional graphs in Calculus. Munintaran Sundram (2011) found exploring real data with graphics calculator in teching statistical concepts beneficial to the students' understanding of sampling distribution.

Source: Hailiza Kamarulhaili, Hajar Sulaiman, Mohd. Tahir Ismail & Suraiya Kassim (Eds.) (2011). *Proceedings of the 4th National Conference on Graphing Calculators*. USM Penang, Malaysia: School of Mathematical Sciences.

RESEARCH FINDINGS ON THE USE OF GRAPHICS CALCULATOR TECHNOLOGY IN MALAYSIA

Research on the use of graphics calculator in mathematics in Malaysia generally covers two aspects. The first aspect studied on students' mathematics achievement level, attitudes and behavior towards using the tool to learn mathematics. The second aspect explored and investigated the mathematical functional features afforded by the technological tool pedagogically.

According to the former head of Mathematics Division of the South East Asian Ministry of Education Organization and the Regional Centre for Science and Mathematics (SEAMEO RECSAM) in Malaysia, since year 1998 until year 2002 the centre had conducted numerous researches on graphics calculator as well as regularly organizing workshops and training courses for mathematics teachers in the Southeast Asia (Khairiree, 2002). Particularly in year 2000, RECSAM offered a six-week course on exploring secondary mathematics with graphics calculator to ten SEAMEO member countries (Brunei Darussalam, Cambodia, Indonesia, Lao PDR, Malaysia, Myanmar, Philippines, Singapore, Thailand and Vietnam). The survey on the perceptions of Malaysian mathematics teachers towards the use of graphics calculator (TI-83 Plus) found that handheld technology creates interest in learning mathematics, allows students' active participations in the classroom discourse as well as improves students' understanding of concepts and procedures through conventional and discovery method of learning,

Apart from the research interest and efforts shown by the SEAMEO RECSAM, mathematics educators from the local universities were keen on experimenting new mathematics courses incorporating graphics calculator technology. Suraiya Kassim et al. (2002) made an attempt to address issues related to effective integration of graphics calculator technology into the mathematics and science curriculum. They designed a special laboratory course requiring the use of graphics calculator and conducted the course on 41 mathematics graduate students at the university. The result obtained revealed that graphics calculator enhanced students' mathematical understanding and improved their performance. The team advised that it is necessary to equip the students with the theoretical and conceptual understanding of the topics first before engaging graphics calculator in the lesson. The extension of research in conceptual and procedural knowledge with graphics calculator was conducted by Nor'ain Mohd. Tajudin et al. (2008) on a group of 16 year-old Malaysian secondary school students to investigate the effects of using graphics calculator to teach Straight Line. Similar to Suraiya Kassim et al. (2002), they found that students in the graphics calculator strategy group had gained better conceptual knowledge performance as compared to conventional instruction strategy group. In addition, they claimed that students using graphics calculator to learn Straight Lines did not underperform in their procedural knowledge performance test.

In a topical teaching of linear algebra with TI-92 Plus, Hailiza Kamarul Haili and Hajar Sulaiman (2002) concluded that graphics calculator technology could enhance students' understanding of the concepts of linear algebra. They too agreed that students should be taught concepts first in the course before the implementation of graphics calculator in a mathematics lesson. In another study, Kor and Lim (2003, 2004) used graphics calculator to teach "Introductory to Statistics" to 69 diploma students. Their findings showed that in a graphics calculator-enhanced classroom, students were more active and participated more in group work. They reported further that the use of graphics calculator had stimulated students' enjoyment in statistics as well as helping them to improve their mathematical skills and conceptual understanding.

Besides conducting research on university students, KaGUM project in 2001 was initiated with the aim to investigate the effectiveness of using TI-83 Plus to teach secondary school mathematics (Noraini Idris et al., 2003). This project sampled a total of 712 sixteen-year-old students from eleven secondary schools over seven states in Malaysia to participate in a quasi-experimental study. The study was carried out over ten weeks with the support from the Ministry of Education and the respective State Education Departments. The research findings showed that graphics calculator had helped the students to understand the mathematical concepts better, improved their attitudes towards mathematics learning, cultivated their interest for the subject, motivated them to learn, and improved significantly their logical thinking abilities as well as their abilities to dissemble information from a given data set. Along with the project mentioned, Ding, Anis Sabrina and Suraini Mohamad (2003) conducted an action research on teaching "Straight Line" to the Form Four students with graphics calculator. They reported that graphics calculator had helped these students to acquire a better conceptual understanding and improved the weak students' achievements. In another study, Chong and Noraini Idris (2002) explored the teaching of "Transformation" using Geometer's Sketchpad available in TI-92 Plus on 32 Form Two (equivalent to grade eight) students. They found that there was an improvement in the level of thinking among the sample in the experimental group who used TI-92 Plus compared to the control group who did not use the calculator. They reported further that a majority of their sample in the experimental group was positive that the use of technology had helped them to learn geometry better.

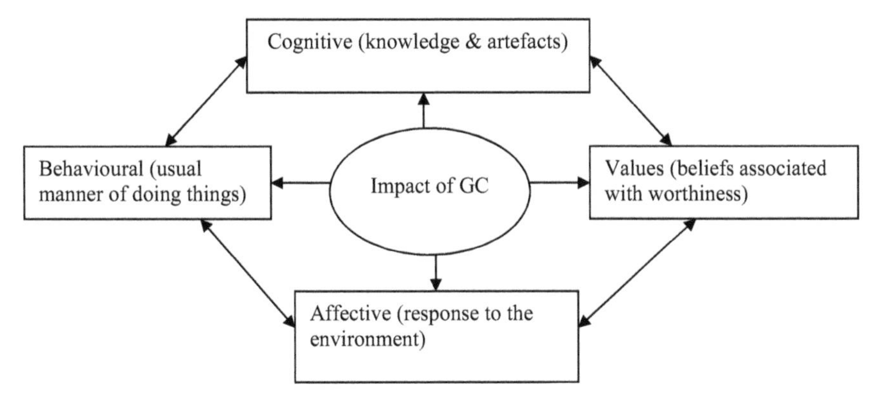

Figure 55.1. Four-factor model contributing to the culture of learning statistics learning.

Kor (2005) investigated the impact of graphics calculator on the culture of learning statistics found four major factors contributing to the development of the learning culture in the presence of technological tool. These identified factors are cognitive, affective, behavioural and values. The relationship between the four factors is depicted in Figure 55.1.

A total of 72 diploma students participated in the above study. Quantitative data were collected using a survey questionnaire comprising of (a) Survey of Attitudes towards Statistics (SATS, adapted from Gal, Ginsburg, & Schau, 1997) and (b) Students' attitudes towards using graphics calculator to learn Statistics (STech, adapted from Forgaty, Cretchley, Harman, Ellerton, & Konki, 2001).

The four subscales in SATS are the "Cognitive Competence", "Affect", "Values" and "Easiness". Both SATS and STech reported a reliability of $\alpha = .7762$ and $\alpha = .7536$ respectively. Mean scores for SATS and STech showed significant improvement at $p < .05$ after the graphics calculator engagement. SATS and its subscales were correlated positively and moderately with STech. "Easiness" scored lower after the graphics calculator intervention than before. Comparison of mean scores between the pre and post-test tests showed that the post-test mean score had improved significantly after the graphics calculator engagement at $p < .05$. Results of qualitative analysis showed that there is a change in the pattern of learning and doing statistics among the respondents. Learning through multiple representations, working in partnership with graphics calculator, peer teaching, collaboration and group discussions, were evident in the graphics calculator -enhanced classroom. Respondents also claimed that they could understand statistical data better with the use of graphics calculator.

In 2008, Cheah presented a conceptual paper that outlined the use of technology in facilitating the new reform of mathematics didactics that focuses on mathematical processes. He construes that technological tool which offers quick and accurate computations as well as dynamic visuals such as those found in geometry and graphs allows students and teachers more time to concentrate on the mathematical processes in the classroom. A study comparing the perceived usefulness of three different types of dynamic mathematical software was conducted by Mohd Ayub, Ahmad Tarmizi, Abu Bakar, and S. Mohd Yunus (2008). They investigated students' perception of the ease of use and usefulness of using Autograph and GSP and graphics calculator in learning Quadratic Functions on a total of 124 secondary school students in Malaysia. Findings showed that students' mean score of perceived ease of use of the graphics calculator was higher compared to Autograph and GSP. No significant difference in the mean scores of perceived ease of use and perceived usefulness was recorded between the software.

Nor'ain Mohd. Tajudin et al. (2009) acted on the premise that graphics calculators can help to assist students in acquisition of knowledge in-depth conducted a three-phases of quasi-experimental study with non-equivalent control group post-test only design to investigate the effects of using graphics calculator on Form Four Malaysian secondary school students' mathematics performance. Their findings showed that students in the graphics calculator instruction group invested less mental effort during the learning and test phases. The graphics calculator instruction group of students with low and average mathematics ability also showed an increase in 3-dimensional instructional efficiency index in learning of Straight Lines. They claimed that the graphics calculator instruction is more superior to the conventional instruction, and benefited especially the average mathematics ability students. The effort to explore the use of calculator technology in Malaysia is a continual endeavor. In year 2011, a survey on the use of TI graphics calculator was again carried out on a group of 30 trained science teachers in selected Malaysia secondary schools (Kanniapan, 2011). Questionnaires were used to survey teachers' attitude, knowledge and skills in the teaching and learning of elective science with graphics calculator.

Last but not least, researchers continued to explore into the feasibility of introducing graphics calculator-based performance assessment in school mathematics. Noraini Idris and Chew (2011) found that Form Four students' (15-year old) mathematics achievement improved significantly in the two topics of Straight Line and Statistics in graphics calculator-based performance assessment. They attributed the success factor to students' use of graphics calculator that had directly increased their use of graphic, numeric and symbolic solution strategies and hence improved their understanding of the mathematical topics studied. They concluded that performance assessment that incorporated the use of graphics calculator supports students' visualization and effectively increase their problems solving opportunities.

IMPLICATIONS AND CONCLUSION

The analysis of research on graphics calculator usage in education shows that most of these research focus on the quantitative data using questionnaires and survey method. To date, research that examines the association between learning theories and the pedagogy of technology integration in a mathematics lesson are relatively rare in Malaysia. An excellent example to model will be Goo's (2010) theoretical frameworks designed "to understand the teacher's role in technology-integrated learning environments and to discover the relationships between factors influencing teachers' use of digital technologies" (p. 174). Result of a similar study like this in Malaysia may help to explain why integration of digital technologies into mathematics teaching and learning is lackadaisical and has progressed slower than expected. It is hoped that more researchers of technology in education in the future will continue to explore the innovative use of graphics calculator, extract factors associated to learning theories and develop new theoretical framework that can best describe a novel and yet effective way of learning mathematics with technology.

Another research area that worthy to note is mathematics assessment in the presence of hand-held technology. Although there is no universal acceptance of allowing the use of a particular type of graphics calculator in the examination, mathematics educators need to take extra effort to discuss, modify and revise existing test items as well as to develop new test items which measure precisely students' analytical, graphical and numerical understanding of mathematical concepts with graphing technology. More research is definitely needed to pave way for assessing technology enabled mathematics subjects.

Nevertheless, an important point that needs to be highlighted is that the literature search on the studies of graphics calculator conducted by the Malaysian researchers revealed that almost all of these papers cited heavily on works conducted by their Western counterparts in their literature review. The reasons for not citing the findings carried out by their fellow Malaysian researchers could due to lack of distinguished local expertise in the calculator-related teaching activities. More importantly, it could be the lack of an effective mechanism that coordinates the accessibility to all researches on graphics calculator conducted in Malaysia. This chapter has taken the initiative to report chronologically activities and achievement contributed by graphics calculator in the Malaysian mathematics classroom. Constantly updating the progress of the implementation of the graphics calculator to the calculator interest group is vital in order to advocate dynamically for a full acceptance of the graphics calculator into the mathematics curriculum.

REFERENCES

Barkatsas, A. (2010). Secondary students attitudes toward using Computer Algebra Systems (CAS) calculators in mathematics: Gender, achievement, year level and years of CAS use differences. Paper to be presented at the Mathematics Education Research Group of Australasia (MERGA) conference to be held in July 2010, in Fremantle, Australia.

Brolin, H. (1987).Mathematics in Swedish schools. In I. Wirszup and R. Streit (eds). *Developments in School Mathematics Around the World*. National Council of Teachers of Mathematics. VA: Reston, 130-155.

Cheah, U. H. (2008). A Practical Framework for Technology Integration in Mathematics Education. Paper presented at the *SEAMEO 2008 ICT Conference held in conjunction with the 43rd SEAMEO Council Conference and the 3rd ASEAN Education Minister Meeting* on 10-12 March 2008 in Kuala Lumpur, Malaysia.

Chiappini, G. & Bottino, R. M. (1999). *Visualisation in teaching-learning mathematics: the role of the computer.* Retrieved August 10, 2001 from http://www.siggraph.org/education/conferences/GVE99/papers/GVE99.G.Chiappini.pdf

Chong, L. H. & Noraini Idris (2002). Pembelajaran geometri menggunakan perisian Geometer's Sketchpad (TI-92 Plus) dan kaitannya dengan tahap pemikiran van Hiele dalam Geometri. *Prosiding Persidangan Matematik Teknologi Asia Ke*-7. ATCM 2002. Melaka, Malaysia: ATCM, Inc., 73-83.

Ding, H. E., Anis Sabrina, & Suraini Mohamad. (2003). Action research on teaching and learning using graphing calculators. *Proceedings of First National Conference on Graphing Calculators*. Petaling Jaya: University Malaya, 27-32.

Forgaty, G., Cretchley, P., Harman, C., Ellerton, N., & Konki, N. (2001). Validation of a questioonaire to measure mathematics confidence, computer confidence, and attitudes towards the use of technology for learning mathematics. *Mathematics Education Research Journal. 13*(2), 154-160.

Gal, I., Ginsburg, L., & Schau, C. (1997). Monitoring attitudes and beliefs in statistics education. In I. Gal & J. B. Garfield (Eds.). *The Assessment Challenge in Statistics Education*. Amsterdam: IOS Press, 37-51.

Goos, M. (2010). A sociocultural framework for understanding technology integration in secondary school mathematics. PNA, 5(1), 173-182.

Graham, A. T., & Thomas, M. O. J. (2000). Building a versatile understanding of algebraic variables with a graphic calculator. *Educational Studies in Mathematics. 41*(3), 265-282.

Hailiza Kamarul Haili, & Hajar Sulaiman. (2002). Learning linear algebra using TI-92. In W-C. Yang, S-C. Chu, T. de Alwis & F. M. Bhatti (Eds). *Proceedings of the Seventh Asian Technology Conference in Mathematics*. Melaka, Malaysia: ATCM, Inc., 324-331.

Hembree, R., & Dessart, D. J. (1986). Effects of hand-held calculators in precollege mathematics education: A meta-analysis. *Journal for Research in Mathematics Education. 17,* 83-99.

Kanniapan, A. (2011). A survey on the usage of Texas Instruments (TI) graphing calculators in the learning and teaching of elective science subjects. In Hailiza Kamarulhaili, Hajar Sulaiman, Mohd. Tahir Ismail & Suraiya Kassim (Eds). *Proceedings of the 4th National Conference on Graphing Calculators* (p.51). USM Penang, Malaysia: School of Mathematical Sciences.

KBSM (2004). Malaysian Secondary School Integrated Curriculum. PTK.

KBSM (2006). Malaysian Secondary School Integrated Curriculum. PTK.

Khairiree, K. (2001). Mathematics teachers' perception on the use of graphing calculator (TI-83): Southeast Asian Perspective. In Wei-Chi Yang (Ed). *The Sixth Asian Technology Conference in Mathematics Proceedings.* (pp. 263-272). USA: ATCM Inc.

Khairiree, K. (2002). Enhancing students' understanding in secondary mathematics through the use of graphing calculators. . In W-C. Yang, S-C. Chu, T. de Alwis & F. M. Bhatti (Eds). *Proceedings of the Seventh Asian Technology Conference in Mathematics* (pp. 304-313). Melaka, Malaysia: ATCM Inc.

Kor, L. K. (2011). A socio-cultural research of mathematics learning in a partially technology incorporated environment. In Hailiza Kamarulhaili, Hajar Sulaiman, Mohd. Tahir Ismail & Suraiya Kassim (Eds). *Proceedings of the 4th National Conference on Graphing Calculators* (pp.11-20). USM Penang, Malaysia: School of Mathematical Sciences.

Kor, L. K., & Lim, C. S. (2003). Learning statistics with graphics calculator: A case study. *Proceedings of First National Conference on Graphing Calculators* (pp. 18-26). Petaling Jaya: University Malaya.

Kor, L. K., & Lim, C. S. (2004). Learning Statistics with graphics calculators: Students' viewpoints. In Yahya Abu Hassan, Adam Baharum, Ahmad Izani M. Ismail, H. C. Koh & H. C. Chin (Eds). *Integrating Technology in the Mathematical Sciences* (pp. 69-78). USM, Pulau Pinang: Universiti Sains Malaysia.

Ministry of Education (2007). Integrating ICT-based content in teaching and learning: mathematics. Retrieved April 2, 2011 from http:www.mscmalaysia.my/codenavia/portals/

Mohd Ayub, Ahmad Tarmizi, Abu Bakar, & Mohd Yunus (2008). A Comparison Of Malaysian Secondary Students Perceived Ease Of Use And Usefulness Of Dynamic Mathematical Software. *International Journal of Education and Information Technologies, 2*(3).

Nor'ain Mohd. Tajudin, Rohani Ahmad Tarmizi, Wan Zah Wan Ali, & Mohd. Majid Konting (2009).Graphing calculator strategy in teaching and learning of mathematics: effects on conceptual and procedural knowledge performance and instructional efficiency. Retrieved April 22, 2011 from http://www.atcm.mathandtech.org/EP2009/papers_full/2812009_17077.pdf

Nor'ain Mohd. Tajudin, Rohani Ahmad Tarmizi, Wan Zah Wan Ali, & Mohd. Majid Konting (2009). The effects of using graphing calculators in teaching and learning of mathematics on students' performance. *Jurnal Sains dan Matematik* , *1* (2),1-28.

Noraini Idris & Chew, C. M. (2011). Effect of graphic calculator-based performance assessment on mathematics achievement. *Academic Research International, 1*(1), 5 – 14.

Noraini Idris & Norjoharuddeen Mohd Nor (2011). Graphing calculator usage in mathematics assessment in STPM and matriculation level in Malaysia. In Hailiza Kamarulhaili, Hajar Sulaiman, Mohd. Tahir Ismail & Suraiya Kassim (Eds). *Proceedings of the 4th National Conference on Graphing Calculators* (pp. 21-38). USM Penang, Malaysia: School of Mathematical Sciences.

Pierce, R., & Stacey, K. (2009). Researching principles of lesson design to realize the pedagogical opportunities of mathematics analysis software. *Teaching Mathematics and Its Applications, 28*, 228-233. doi:10.1093/teamat/hrp023

Pierce, R., Stacey, K. & Barkatsas, A. N. (2007). A scale for monitoring students' attitudes to learning mathematics with technology. *Computer and Education, 48*(2), 285-300.

Quesada, A. R., & Maxwell, M. E. (1994). The effects of using graphing calculators to enhance college students' performance in precalculus. *Educational Studies in Mathematics.* 27(2), 205-215. Retrieved June 8, 2002 from http://www.t3ww.org/pdf/therole.pdf

Sivasubramanim, P. (2011) Innovative teachers and technological pedagogical and content knowledge. In Hailiza Kamarulhaili, Hajar Sulaiman, Mohd. Tahir Ismail & Suraiya Kassim (Eds). *Proceedings of the 4th National Conference on Graphing Calculators* (pp. 39-48). USM Penang, Malaysia: School of Mathematical Sciences.

Suraiya Kassim, Rosihan, M. Ali, Seth, D., Zarita Zainuddin and Mokthar Ismail (2002). Addressing the issues underlying hand-held technology use in the classroom. In W-C. Yang, S-C. Chu, T. de Alwis & F. M. Bhatti (eds). *Proceedings of the Seventh Asian Technology Conference in Mathematics.* (pp. 274-283).Melaka, Malaysia: ATCM Inc.

Thomas, M, Tyrrell, J., & Bullock, J. (1996). Using computers in the mathematics: The role of the teacher. *Mathematics Education Research Journal, 8* (1), 38-57.

CHAPTER 56

MATHEMATICS TEACHER PROFESSIONAL DEVELOPMENT IN MALAYSIA

Chin Mon Chiew, Chap Sam Lim, and Ui Hock Cheah

ABSTRACT

This chapter aims to provide an overview and the scenario of mathematics teacher professional development in Malaysia. Literature review on teacher professional development [TPD] in Malaysia suggests that the nature of TPD can be divided into two main strands: a) top-down teacher professional development programme initiated and conducted by the Ministry of Education through in-service courses or workshops with the objective to introduce innovations in teacher's teachings or prepare teachers for curriculum change; and b) research based TPD such as Action Research and Lesson Study initiated by researchers and teachers with an intention of local interest to improve teaching practices. Hence, the discussion in this chapter consists of two main parts. First, the situational context of mathematics teacher professional development is elaborated to provide the background and setting. Second, some research findings on Action Research and Lesson Study in Malaysia are discussed; primarily to examine their feasibility as an option and alternative form for teacher professional development. Finally, some

The First Sourcebook on Asian Research in Mathematics Education: China, Korea, Singapore, Japan, Malaysia, and India, pp. 1265–1284

implications and future issues related to mathematics teacher professional development are also addressed.

INTRODUCTION

In Malaysia, the pre-service teacher education programmes for both primary and secondary schools is provided by the 27 Teacher Education Institutes and the 11 public universities. Mathematics teachers serving in both primary and secondary schools can be regarded as *specialist* as they were trained to teach the subject during their pre-service teacher education programme. Generally, the public universities provide the bulk of teacher education for secondary school teachers while the Teacher Education Institutes train primary school teachers at diploma level. However, since 2007, the teacher education programme for primary school teachers was upgraded to degree level; extending the course programme to four years. The Malaysia government has targeted that by 2015, it will be able to achieve at least 50% of primary and 100% of secondary school teachers with a minimum qualification of a bachelor degree. The objective is to provide a higher standard and a better quality of teacher education that aims to raise the standard of education ultimately.

In order to sustain and continue improving the quality of teachers' teaching, the Malaysian Ministry of Education [MOE] acknowledged that continuous professional development [CPD] of teachers is relevant, important and significant (Wong, 1998; Lee, 2000). In this chapter, the term of *professional development* is also referred and reserved for *continuous professional development* but more broadly, it is the body of systematic activities to prepare teachers for their job that includes pre-service training, induction course, in-service training, and continuous professional development within school settings (OECD, 2010). More precisely, Guskey (2000) defined professional development as "those processes and activities designed to enhance the professional knowledge, skills, and attitudes of educators so that they might, in turn, improve the learning of students" (p. 16). To date, many programmes have been initiated and promoted by the MOE with much effort but the CPD among the school teachers in Malaysia remains lackadaisical.

A study by the International Association for Evaluation of Educational Achievement [IEA] on "Teacher Education and Development Study: Learning to Teach Mathematics" [TEDS-M] in 2008 has revealed that the Malaysian primary and secondary pre-service mathematics teachers were rated slightly below average in terms of mathematics content and pedagogical content knowledge (Kaiser & Bloemeke, 2010). In addition, the lack of pedagogical content knowledge among mathematics teachers was also exposed by Tengku Zawawi (2005) and Noorshah (2006) in their post-graduate research studies. As subject matter and pedagogical content knowledge are fundamental and essential elements in teacher's teaching as described by Shulman (1987), there is urgency for the MOE to take appropriate and effective actions.

One main viable strategy is to adopt and promote the continuous professional development among school teachers as "teachers are key figures in changing the ways in which mathematics is taught and learned in schools" (NCTM, 1991, p.2). Hence, the MOE has little choice, but to put much more effort and concerted initiative to curb any weaknesses in the teacher professional development programmes that aim to improve teachers' teaching in their classroom.

TEACHER PROFESSIONAL DEVELOPMENT IN MALAYSIA

Teacher In-Service Programmes

As in other countries, the MOE acknowledges and realises the importance and significance of teacher professional development in teachers' teaching. As a result, since 2008, all practising school teachers are mandated by the MOE to attend at least seven days of professional development in a year. This directive was regarded as a major boost for professional development of teachers. However, the feedback and responses of the implementation of this initiative thus far have suggested that it lacks focus and direction as the school administrators seem to hold different priorities and views pertaining to the directive. Due to lack of impact and focus, the seven days of professional development programmes are carried out primarily to abide and fulfill the requirement, rather than to achieve the underlying objectives of the MOE.

In Malaysia, the Teacher Education Division [TED] under the MOE is responsible and entrusted to improve the teachers' quality in teaching through various in-service courses and teacher professional development programmes. In the document, entitled *Education in Malaysia: A Journey to Excellence* (Ministry of Education, 2001), six types of programmes are enlisted: (i) The Special Degree Programme, (ii) One-year Specialist Certificate Course, (iii) Professional Development Programme, (iv) The Malaysian Trainer Development Programme, (v) Smart School Courses, and (iv) The Computer Maintenance Course. Besides, there are also many other short courses and workshops conducted jointly by the TED or the MOE with other educational authorities such as the Curriculum Development Division, State Education Departments and District Education Offices from time to time in view of the situational needs and demands. In the context of professional development programmes for mathematics teachers, this includes abacus, graphic calculators, geometer's sketchpad as well as pedagogical aspects such as constructivism, problem solving, cooperative learning, creativity and problem based learning.

Top-Down Approach of TPD

The approach of professional development can be broadly divided into two forms: teacher self initiative and top-down TPD programmes through courses and workshops. However, the main bulk of the in-service programmes conducted were mainly to inform and update teachers' knowledge pertaining to the changes or innovations introduced in the curriculum or policies of the education system (Lourdusamy & Tan, 1992; Lee, 2002). In general, whenever there is any revision of the curriculum, teachers are requested to be exposed to new information and changes by attending short courses which are also regarded as professional development. The TALIS (Teaching And Learning International Survey) prepared by the OECD [Organisation for Economic Cooperation and Development] in 2010 has provided two informative findings for Malaysia: (i) the number of days of professional development for teachers is 11.0 days, well below the TALIS average of 15.3 days, and (ii) 83% of teachers in Malaysia demanded more professional development than they received, against the TALIS average of 55%. Hence, based on this international survey which involved 23 participating countries, the MOE should garner more effort in providing professional development activities and opportunities to the teachers. In other words, the practising teachers felt inadequacy of their teaching knowledge and information to carry out their tasks in school.

Indeed, the MOE has been working diligently to improve teachers' teaching in the classroom. In fact, there is no lack of concerted efforts or initiatives by the MOE to conduct in-service training for teachers' professional development, especially by the TED. However, its implementation often fell short in achieving the targeted or desirable results. For instance, massive in-service courses were conducted nationwide for teachers to cater for the needs and changes when the *New Primary School Curriculum* (Kurikulum Baru Sekolah Rendah [KBSR]) was introduced in 1983 and likewise for the *Integrated Secondary School Curriculum* (Kurikulum Bersepadu Sekolah Menengah [KBSM]) in 1989. Subsequently, a quantitative survey (see Hussein, 1990) was administered to the 400 participating primary school teachers to evaluate the effectiveness of these in-service courses. One of the weaknesses reported was that the courses conducted were superficial and lack of content-depth.

Similarly in 1997, all teachers involved in the Smart School project were required to attend the *14-week In-service Smart School Teacher Training* before its implementation. Based on observations conducted on 87 pilot schools, the MOE reported several weaknesses (KPM, 2001). One of them was teachers are still lack of pedagogical knowledge despite of the intensity of the trainings and this includes the use of innovations and Information Communication Technology [ICT] in the teaching. Another massive nationwide in-service programme was the English for Teaching Mathematics and Science [ETeMS] course in 2003 which aimed to adopt the policy of Teaching Mathematics and Science in English. Various in-service programmes and initiatives had been conducted to support and assist teachers to be competent in English language. However, despite the intensity of the ETeMS course, teachers were found to be lacking the competency in English language. Consequently, this has become one of the factors that the language policy of teaching Mathematics and Science in English was reverted. The above chronological episodes evidently reflected the weaknesses of the in-service programmes conducted for teachers in Malaysia. In 2010, the *Standard Primary School Curriculum* (Kurikulum Standard Sekolah Rendah [KSSR]) was introduced to replace the KBSR and one significant element is to cultivate creativity among the primary school children in mathematics learning. As anticipated, the approach to disseminate information among the school teachers remains unchanged to date.

Issues in In-Service Teacher Programmes

When analysed, the inefficiency in implementation of these in-service courses and programmes might be one of the root causes that many of the aspirations and intentions in the curriculum were not transmitted in the actual practices. For instance, Noor Azlan (1987) disclosed the discrepancy of the Modern Mathematics Programme: the activity-based, student-centred and guided discovery approach advocated in the school mathematics curriculum was replaced with mainly teacher-centred and *chalk and talk* approach in the actual teachings. Some years later, Poon (2004) exposed similar findings in schools that there are significant difference between the intended curriculum and the actual mathematics teaching in classroom. Based on a qualitative study involving four Form 4 mathematics teachers, she deduced that the constraint was due to teachers' attitude and the weaknesses of the curriculum developer in conveying information through the in-service courses or workshops. Hence, it could be deduced that the in-service programmes and courses conducted barely made any impacts as there was little change in teachers' teaching practices over the years. Probably, this prolonged problem has affected the teachers to perceive in-service training or professional development programmes as ineffective because it made little impact to their teaching practices (Guskey, 2000).

In fact, based on the context in Malaysia, one particular issue that need to be seriously addressed is the approach and implementation of the in-service programmes widely adopted by the MOE. A network of master trainers or key personnel is created at national and state levels to coordinate the in-service programmes. These trainers are usually school teachers or lecturers identified and specially trained at national level by the educational authorities. In turn, they conduct the courses and workshops at state, district and school levels. In actual practice, a teacher from each school will attend the course and on his/her return to the school, the teacher will conduct the same course content to his/her peers. This approach was known as *cascade strategy* or *multiplier effect* and has been widely adopted due to logistic, time and financial constraint. One major criticism of the approach is the *information dilution*. For example, a two or three-day workshop or course is reduced merely to two or three-hours by the teacher at school level. Certainly, the impacts received at the lower end are greatly diluted and reduced.

Another issue is the approach used to conduct the teacher in-service programmes and courses is usually *one-shot* and *top-down*. According to Dawson (1999), such approach for teacher professional development was deemed ineffective as it takes little consideration of teachers' perceptions of teaching and classroom realities in context. Moreover, the information conveyed through the in-service courses may not be relevant and supported by the school authorities (Darling-Hammond & McLaughlin, 1995). Subsequently, teachers may then perceive professional development activities as irrelevant when it is unconnected to the realities in the actual classroom (Lieberman, 1995; Ball & Cohen, 1999). In fact, there were much criticism of the approach and literatures in the 1990s have then outlined several principles of effective professional development. Literature pertaining to effective teacher professional development (such as Darling-Hammond & McLaughlin, 1995; Loucks-Horsley et al., 1996; Guskey, 1997; Hawley & Valli, 1999) have proposed that teachers be active learners in their school context. In short, teachers need to be self initiate and be responsible for their own professional development. There is a need for teachers themselves to play an active role and be part of the professional learning community in school. Indeed, these principles proposed a paradigm shift in the context of teacher professional development.

Another strand in teacher professional development is the reflective practice. It was based on the work of John Dewey but was promoted by

Schon in the 1980s which later established itself as a powerful metaphor in the context of professional development (Ball, 1996; Calderhead & Gates, 1993). Calderhead and Shorrock (1997) emphasised the need for teachers to analyse, discuss and evaluate their teaching to improve their own teaching practices. According to Lee (2003), the reflective practitioner concept was introduced into the Malaysian teacher education curriculum in 1989 when clinical supervision was implemented as part of the student teaching component. Later in 1996, the TED conceptualised a model of teacher development emphasising knowledge construction based on student teachers' reflection. However, using a qualitative study on six trainee teachers, Lee (2003) reported that the reflective practice in the teaching practicum was not implemented by the supervising lecturers, mainly due to their own lack of understanding about reflective practice. Hence, the vision on professional development framework in reflective practice as proposed by the TED seemed not materialised.

The above has highlighted several issues pertaining to TPD in Malaysia. Indeed, there is critically lack of research based-evidence as some documents and reports related to TPD were classified as confidential by the MOE. Nonetheless, teacher professional development remains a priority by the MOE to uplift the standard of education and of late, there are plans to promote school-based professional learning community as the thrust for TPD.

RESEARCH-BASED TEACHER PROFESSIONAL DEVELOPMENT IN MALAYSIA

Action Research

Brief Historical Development of Action Research in Malaysia

Our literature search shows that prior to 1993, there were already attempts to introduce action research in Malaysia on a small scale. For example, the *Thinking in Science and Mathematics* (TISM) *Project* was conducted by the Southeast Asia Ministers of Education Organisation Regional Centre for Education in Science and Mathematics (SEAMEO RECSAM) primarily to introduce action research on problem solving in the classroom. The project, first conceived in 1987, began as a pilot study in only one secondary school and focuses on improving teaching in science and mathematics classrooms (SEAMEO RECSAM, 1990). The TISM project was collaborative, school-based project which involved researchers from SEAMEO RECSAM as well as teachers from the pilot school. One main feature of the project was the sharing of ideas and collaborative reflection among the researchers and teachers in order to bring about teacher change and professional growth. The project also received support from visiting consultants from Australia and New Zealand. The visiting consultants provided input in the form of workshops that were conducted. As it was funded by the Australian government, some elements and procedures of the project were similar to the *Project for Enhancing Effective Learning* (PEEL) which was founded by a group of researchers from Monash University (Baird & Mitchell, 1986). The classrooms procedures adopted were aimed at building understanding of subject content and skills, restructuring students' existing views, extending school knowledge to new situations, improving communication, improving problem-solving and assessment. However, as Kim (1997) noted, many of these small-scale action research projects were short lived and did not go beyond two to three years.

Thereafter, action research became more prominent through the launch of the *Programme for Innovations, Excellence and Research* (PIER) in 1993. The main purpose of PIER was to improve four main educational areas: Innovations in science and mathematics, small and isolated schools, distance education and educational research. Funded by the World Bank, the PIER program was completed in 1996 (Educational Planning and Research Division, 1997). In order to fulfill the objective of improving educational research, one of the main aims of PIER was to inculcate the culture of research among school educators and school administrators through action research. Through the services of Australian as well as local consultants of the PIER project, the idea of developing teachers as researchers through action research was disseminated to teachers throughout the country. By 1996, a total of 741 action research projects had been completed (Educational Planning and Research Division, 1997) of which 632 were projects pertaining to science and mathematics (Madzniyah Md. Jaafar, 1998).

In 1995, the action research programme was reorganised and restructured by the Ministry of Education so that the effective impact of the programme was expanded to include all primary and secondary schools in Malaysia. The Malaysian Council of Educational Research (Majlis Penyelidikan Pendidikan MAPPEMA) was formed to activate and to coordinate research and assessment activities in education in Malaysia (Kementerian Pendidikan Malaysia, 1997). This was followed by the formation of the Educational Research Forum in all states in 1996. One of the strategies of MAPPEMA was to sustain action research activities in schools particularly teachers who were involved in the PIER project.

At the teacher training level, action research was initially started when the six lecturers from various teacher training colleges were sent to the United Kingdom for a short four weeks course relating to reflective teaching in 1994. In 1995, the Sultan Idris Teacher's College initiated a module called "Reflecting Teaching in Teacher Education" (Hanipah Hussin, 2004). Inputs from prominent action research experts such as Stephen Kemmis from Deakin University, Australia and John Elliot from University of East Anglia, United Kingdom were also sought. Thereafter action research was introduced at the Ministry level. Subsequently, the concept of "teachers as researchers" was introduced into the pre-service teacher education curriculum in all teacher training colleges in Malaysia. The rationale for introducing action research in the teacher training curiculum was to develop: (i) critical attitudes; (ii) as reflective practitioners and research into teaching; (iii) accountability; (iv) self-evaluate and (v) upgrade teaching professionalism (cited in Chee, 2005).

As seen from the engagement of Australian consultants at various levels in the various projects, the action research model adopted in Malaysia also seems to carry some semblance of influence of the Australian models, in particular, Kemmis and Taggart's model of action research cycle which has been adopted and practiced by schools and teaching colleges. The main research cycle adopted was based on the Plan-Act-Observe-Reflect cycle (Kemmis & McTaggart, 1988).

As with other educational innovations, the action research programme in Malaysia was faced with its own set of challenges. Subahan, Abd. Rashid and Jamil (2002) conducted a survey to look into the constraints faced by teachers as they attempted to inculcate a research culture in the schools. The survey sample covered 679 teachers in Terengganu, a state in Malaysia that had actively participated in the action research project. They found that there were many similarities between teachers who had attended action research courses and those who had not. Knowledge in action research, skills in action research, use of research findings and disseminating research findings were moderate whilst in both groups the teachers reported low skills in conducting research. The study also found that teachers who had attended courses in action research reported a higher level of critical attitude. Qualitative data from the study also revealed several emerging issues among the teachers. The interviews with the teachers revealed that over-emphasising knowledge and skills during the courses did not change the teachers' behavior to do action research.

Kim (1997) in a state-of-the-practice review of action research noted several challenges faced by action research teams in the conduct of their projects. Among the main challenges were the lack of knowledge of research skills and methodologies and the constraints of time and motivation. These same themes were also highlighted in the interviews of teachers in the study conducted by Subahan, Abd. Rashid and Jamil (2002). While teachers perform best in teaching and learning activities in the classroom, research skills and knowledge are often not in the repertoire of their classroom expertise. Moreover, action research is carried out over and above the teachers' normal school duties.

Mindful of the challenges faced by teachers, the MOE presented several recommendations, particularly to increase support for the continuation of action research in schools (NIER, 2006). There were two areas of support. First, financial and management support for the teachers, and second a monitoring mechanism and standardizing training modules for more effective training.

Thus, the action research program continues to receive much support from the MOE. At the school level, every teacher is encouraged to carry out at least one action research in the classroom each year (Kementerian Pelajaran Malaysia, 2008). As an incentive, teachers are encouraged and sponsored to present their action research write up reports in action research seminars and conferences organised by the Ministry of Education whether at the school, state, national or even at the international levels. The best action research reports were also published in the EPRD website which could be accessed by the public.

Action research has also found its way into teacher preparation programmes. In 2010, all trainee teachers undergoing the Bachelor of Education Degree Programme in all the 27 teacher training institutes in Malaysia were required to take the Action Research Course which is one of the core components in their major of the Programme (Chee, 2010). This course was conducted for a full one year where trainee teachers were not only exposed to the theoretical aspects of action research in the teacher training institutes for the first semester, but also the practical aspect of action research in the classroom during their 12 weeks final year practicum in schools in the second semester (in the final-year). They were also required to fullfil the action research examinations at the end of the first semester and present their action research findings in a seminar organised by each cohort after practicum. After making the necessary amendments prior to the comments given by their lecturers and colleagues, they were required to submit their mini action research dissertation to the institute before they could graduate.

Currently, the main concern of the action research programme is whether it can be sustainably carried out by teachers at the school level. While there has been evidence of concerted efforts at various level both in pre-service and in-service training to provide teachers with skills and knowledge of doing action research, there remains many challenges to overcome at the schools level. Motivating teachers remains a key element besides the teachers' workload of administrative tasks that has yet to be resolved. Extrinsic motivation may be provided through various incentives by the MOE through various forms of teacher support but it is only through intrinsic motivation via a bottoms-up approach and teacher empowerment that the inculcation of the research culture can be seen.

Even as action research seems to have reached its peak with regards to teacher awareness of the teacher-as-researcher movement, another teacher development model popularised in Japan began to make its way into the Malaysian scene.

Lesson Study

The Development of Lesson Study in Malaysia

Similar to Action Research, the growth of Lesson Study in Malaysia began in 2003 with small scale research projects mainly carried out by Lim and her colleagues (see e.g. Chiew & Lim, 2003; Lim, White & Chiew, 2005; Lim, 2006; Lim, 2007; White & Lim, 2008; Kor & Lim, 2009; Lim, Chew, Chiew & Goh, 2009; Cheah & Lim, 2010; Ong, Lim & Munirah Ghazali, 2010; Lim & Kor, 2010; Lim, Chiew & Chew, 2011) and post-graduate studies (Goh, 2007; Chiew, 2009; Ong, 2010). Originated from Japan, Lesson Study as an innovative model of teacher professional development began to gain popular in the United States since 2000, and thereafter spread to other parts of the world.

Adopting the Japanese model of Lesson Study, Chiew and Lim (2003) first piloted the Lesson Study with a group of five trainee mathematics teachers who were undergoing teaching practicum in a secondary school. Positive feedbacks were given by the trainee teachers who claimed that they have gained much more confidence and their pedagogical content knowledge was enhanced through the Lesson Study process. Encouraged by this positive outcome, as reported in Lim, White & Chiew (2005), they intiated a Lesson Study research project in two secondary schools in a district at Northern Malaysia. The aim of the project was to explore the influence of Lesson Study as well as the feasibility of implementing Lesson Study as a professional development programme for mathematics teachers in the Malaysian school context. Their findings indicated both positive and negative responses. Among the positive responses were through group discussions and observing other teachers teach, the participating teachers claimed that they have gained and enhanced both their mathematical content knowledge and pedagogical knowledge. They also expressed that their reflective practice was enhanced when engaging in the Lesson Study process. In addition, the participating mathematics teachers also reported that Lesson Study has promoted a collaborative culture that enhances professional collegial bonds within their mathematics colleagues. However, several constraints faced by the participating teachers in implementing Lesson Study were (i) time factor, (ii) heavy school workload, (iii) shyness to be observed by colleagues and (iv) teachers' attitude and commitment. Subsequently, Lim (2006) also introduced Lesson Study to her 86 pre-service mathematics teachers with the aim to promote peer collaboration. The finding was similar with time factor as the major constraint in Lesson Study implementation.

In 2006, under the Asia Pacific Economic Cooperation [APEC] – Lesson Study project, 11 APEC countries were initially invited to pilot Lesson Study in their respective economies and Malaysia was one of the participating economies. This project was further expanded to include 19 economies in 2007 and subsequently 21 economies in 2008. In the first year, the theme was "Innovation and Good Practice for Teaching and Learning Mathematics through Lesson Study"; the theme for the second year was "Innovative Teaching mathematics through Lesson Study: Focusing on Mathematical Thinking" while the third year's theme was "Focus on Mathematical Communication". In line with this international project, a number of small scale research studies were carried out in a number of primary and secondary schools in Northern Malaysia and the findings were shared through international symposium organised by the Khon Kaen University of Thailand.

Lesson Study Research Findings

As mentioned earlier, Lesson Study started with small scale research projects initiated by researchers as well as post-graduate students for their dissertations. For examples, Lim (2007) explored how to develop mathematical thinking through Lesson Study collaboration in a Chinese Primary school. After experiencing Lesson Study in their respective schools, Goh, Tan and Lim (2007) share their insights about engaging in Lesson Study in a regional conference. Goh taught mathematics in a Chinese primary school while Tan taught mathematics in a secondary residential school. Both Goh and Tan could be regarded as pioneer teachers that involved in Lesson Study projects in Malaysia. Their reflection about Lesson Study was: "It is really not an easy matter to run and sustain Lesson Study but it is worthwhile when we are seeing the result and benefits gained" (p. 578).

In 2008, with collaboration between Universiti Sains Malaysia (USM) and SEAMEO-RECSAM, another Lesson Study project had taken off which involved 10 primary schools in Penang. As reported in Cheah and Lim (2010), a series of workshops were given to familiarize the participants about the concept of Lesson Study, mathematical thinking and communication before the actual implementation of Lesson Study in the respective schools. Even though at the end of the project, four of the ten primary schools withdrew half-way, all the remaining six schools completed one to two cycles of Lesson Study. Their responses spelled out positively about Lesson Study despite of the time constraint and other challenges that remain a hurdle for implementation of Lesson Study.

Another small scale research project that adopted Lesson Study as a tool for promoting the innovative use of Geometer's Sketchpad [GSP] in mathematics teaching and learning was carried out by Lim and her team during 2007-2009. Despite several GSP workshops conducted by the educational authorities, Kasmawati (2006) claimed that only 2% of mathematics teachers use GSP in their classroom teaching. For this project, three secondary schools were involved and the challenges and constraints encountered in the implementation were within expectation. Similar to the earlier studies, the findings of this study reflected positive outcomes in several areas: (i) increase teachers' knowledge and skills, (ii) encourage sharing among teachers, (iii) making mathematics lessons more interesting, (iv) motivate teachers to explore new teaching ideas, (v) fostering collegial relationship, and (vi) opportunities for mentoring among the teachers. Compiling the teachers' experiences and the products of GSP templates, a book edited by Lim and Kor (2010) was published together with a sample video of how to carry out Lesson Study.

Besides these research projects, two doctoral dissertations (see Chiew, 2009; Ong, 2010) and one Master degree research (Goh, 2007) on Lesson Study were completed. Based on his doctoral study on two groups of eight secondary school mathematics teachers each in two different schools, Chiew (2009) revealed the positive influences of Lesson Study process on teachers' content and pedagogical content knowledge as well as reflective practice. Likewise, Ong's (2010) doctoral study involved ten mathematics teachers in two different schools and data was collected over a period of fifteen months. She found that teachers' questioning techniques were enhanced through the Lesson Study process. The study by Goh (2007) also reported an increase of subject matter knowledge and confidence in teaching mathematics using English as the medium among the eight participating teachers through Lesson Study process.

Nonetheless, the degree of impacts or influences were somehow varied and limited due to the constraints faced in the Lesson Study implementation. It was noted that participating teachers who were committed and ready to embrace in Lesson Study process were found to gain far more benefits compared to other participants in the group (Chiew, 2009; Ong, 2010). Thus far, the impacts of Lesson Study on teachers' teaching in the classroom could not be easily ascertained in short term. Due to limited time in his study, Chiew (2009) acknowledged that changes on teachers' teaching should be limited to the lessons discussed in the Lesson Study process. The change or impacts on teachers' teaching practices in general, would require much more time and engage in a few Lesson Study cycles over a period of two or three years before any changes or enhancement could become apparent.

Takahasi (2010) identified two aspects of Lesson study, namely the improvement of teacher practice and the promotion of collaboration among teachers as crucial in making it a successful teacher development program and this has been "credited to the dramatic success in improving practices in the Japanese elementary school" (Takahasi, 2010, p.169). In a similar ways, Lesson study would provide a potential and alternative model for professional teacher development in Malaysia.

Challenges and Constraints

Although the findings thus far have been impressive and encouraging, the impacts of Lesson Study actually been cushioned by constraints in the implementation. Based on the studies conducted on Lesson Study, there were much consistency found with regards to the constraints and challenges. Chiew and Lim (2010) pointed out three main challenges: (i) time constraint, (ii) teacher's heavy workload, and (iii) teacher's perception of teaching observation. In fact, these factors were somehow expected due to the misconceptions and cultural factors in one's education system (Chokshi & Fernandez, 2004). Chiew (2009) argued that much of the constraints were actually due to the lack of awareness about the importance of professional development among the teachers. It was indeed paradoxical to observe that these teachers expressed to have gained much benefit through the Lesson Study process, yet at the same time, they were not willing to work beyond the school working hours. As with any model of teacher professional development, the success of Lesson Study lies heavily on the teachers' commitment and attitude change towards teacher professional development. In many cases, the Lesson Study practice is not able to sustain among the participating teachers after the researchers completed their projects

Nonetheless, despite these challenges, there is optimism for Lesson Study to thrive in the Malaysian schools. One of the supporting factors for smooth implementation of Lesson Study is the strong support from the school administrative (Chiew, 2009). Lesson Study as a model for TPD made great strides in 2011 when the TED officially reckoned and introduced the innovation to schools. The TED has directed all 20 schools with *high performing status* to adopt and initiate Lesson Study as one of the teacher's development programme in their schools. Beginning May 2011, 289 schools in the *low performing school* category were also required to carry out Lesson Study under close monitoring of the TED, state education department and district education office. While it is at infancy stage to gauge its change or impact, it is apparent that the innovation of Lesson

Study had managed to gain attention from the MOE through various Lesson Study research conducted.

Reviewed on the local efforts in promoting Lesson Study has resulted in both optimism as well as scepticism. On one hand, there was significant research evidence that suggested that mathematics teachers' classroom pedagogy has changed and improved through peer support as result of Lesson Study collaboration. The teachers' collegiality was also enhanced through professional learning. However, on the other hand, the lack of awareness of professional development among teachers has not been properly addressed, both at teacher and school administrative level, and this has affect the success of Lesson Study implementation to a certain extent.

Nevertheless, the latest recognition of Lesson Study by the MOE such as the adoption of Lesson Study as a strategy of developing professional learning community is an encouraging sign. With support from the top administration, it will certainly facilitate the implementation of Lesson Study in schools, particularly in terms of flexibility of time tabling and arrangement for group discussion.

IMPLICATIONS AND FUTURE ISSUES

Our review thus far indicates that in general, the Malaysian Ministry of Education is serious and acknowledged the importance and significance of teacher professional development as a move to improve teacher teaching quality and standard. This was evidenced in the numerous in-service training courses and workshops conducted to the teachers from time to time. Yet, the success of implementation remains much to be desired in terms of scope and impact on teacher teaching quality. Several weaknesses identified as with typical or traditional model of teacher professional development were *one-shot, short term* and *top-down*. As pointed out by Guskey (1997) and Dawson (1999) that effective model of teacher professional development needs to consider teachers' conceptions of effective teaching and the actual classroom context. More importantly, teachers need to have continuous support after attending the courses.

Perhaps, as highlighted by Chiew (2009) that another main issue was the examination oriented education system as perceived by school teachers. While the mathematics curriculum promotes innovative teaching strategies through problem solving, student-based activities and the use of ICT to transpire the learning process, teachers are held accountable and pressured to show results in public examinations. Consequently, *drill and practice* of past-year examination questions dominated the mathematics teaching scenario, making many teachers to perceive professional development activities as unrealistic and not practical.

From our observations, the top-down in-service courses and workshops, organised and conducted by the MOE may have directly or indirectly impacted on teachers to a certain degree. Due to situational context, it is difficult to evaluate the effectiveness of any professional development programme as the nature of TPD is rather long-term and developmental. Action Research and Lesson Study posed to be alternative model of TPD in Malaysia in long-term but these required a paradigm shift in teacher professional development which is yet to be observed in the Malaysian context. Currently, excellent teachers are encouraged to carry out Action Research as these carry some weight for promotion in future. The aspects of being voluntary, life-long learning and self-initiated by the teachers to improve and learn innovative teaching strategies remain as challenges that are yet to be realised.

Indeed, there are still much more areas to be explored through Lesson Study. Due to the examination-oriented education system, some aspects that need to be emphasized in the mathematics curriculum such as problem solving, reasoning, mathematics thinking and communication are largely ignored and neglected by most teachers. To resolve this long-due issue, Lesson Study might provide an opportunity and platform for mathematics teachers to collaborate and work together in integrating such aspects in their teaching practices. In short, it is an attempt to merge the theoretical and practical aspect in mathematics teaching. However, more research efforts are needed to explore and to ensure how Lesson Study can be incorporated into mathematics teachers' professional development activity to improve and enhance teachers' teaching qualities in those aspects which ultimately lead to quality mathematics education.

REFERENCES

Baird, J., & Mitchell, I. (Eds.) (1986). *Improving the quality of teaching and learning: An Australian case study – the PEEL project.* Melbourne: Monash University Press.

Cheah U. H., & Lim, C. S. (January, 2010). *Disseminating and popularising lesson study in Malaysia and Southeast Asia.* Paper presented at the APEID Hiroshima Seminar "Current Status and Issues on Lesson Study in Asia and the Pacific Regions", Hiroshima University, Japan from 18th to 21st January 2010.

Chee, K. M. (2005). How does action research enhance in-service primary mathematics teachers' teaching practice? In *Proceedings International Conference on Science and Mathematics Education* (pp 170-175). Penang, Malaysia: SEAMEO RECSAM.

Chee, K. M. (2010). *Kajian tindakan: Dari Proses ke produk.* Pulau Pinang: UPPA, Universiti Sains Malaysia.

Chiew, C. M., & Lim, C. S. (October, 2003). *Impact of lesson study on mathematics trainee teachers*. Paper presented at the International Conference for Mathematics and Science Education, 14-16 October 2003. University of Malaya, Kuala Lumpur.

Chiew, C. M., & Lim, C. S. (2010). Challenges and Insights Gained in Conducting Lesson Study: A case in Malaysia. In Yoshinori Shimizu, Yasuhiro Sekiguchi, & Keiko Hino (Eds.), *Proceedings of the 5th East Asia Regional Conference on Mathematics Education (EARCOME5),* Theme: In Search of Excellence in Mathematics Education, vol 2, 684-690. Tokyo: Japan Society of Mathematics Education.

Chiew, C.M. (2009). *Implementation of Lesson Study as an innovative professional development model among mathematics teachers.* Unpublished Ph.D thesis, Universiti Sains Malaysia, Penang.

Chiew, C.M. & Lim, C. S. (2007). Changing teachers' teaching through Lesson Study. In Lim, C. S. et al., (Eds.), *Proceedings of the 4th East Asia Regional Conference on Mathematics Education [EARCOME4], pp.540-546*, 18-22 June 2007, Universiti Sains Malaysia, Penang, Malaysia.

Goh S. C., Tan, K. A. & Lim, C. S. (2007). Engaging in Lesson Study: Our experience. In Lim, C. S. et al., (Eds.), *Proceedings of the 4th East Asia Regional Conference on Mathematics Education [EARCOME4], pp.574-579*, 18-22 June 2007, Universiti Sains Malaysia, Penang, Malaysia.

Goh, S. C. (2007). *Enhancing Mathematics Teachers' Content Knowledge And Their Confidence In Teaching Mathematics Using English Through Lesson Study Process.* Unpublished M.Ed thesis, Universiti Sains Malaysia, Penang.

Guskey, T.R. (2000). *Evaluating Professional Development*. Thousand Oaks: Corwin Press.

Hanipah Hussin (1999). *Learning to be reflective : The content and nature of primary student teachers' thinking during field experience in Malaysia.* Unpublised PhD Thesis. Universiti of Sydney, Australia.

Hanipah Hussin. (2004). *Learning to be reflective: From theory to practices Malaysia experiences.* Tanjung Malim: Penerbit Universiti Pendidikan Sultan Idris.

Hussein Ahmad (1990). Keberkesanan Penyediaan Guru dan Aktiviti Meningkatkan Profesionalisme Mereka. *Seminar Kebangsaan Penilaian Pelaksanaan KBSR.* Pusat Perkembangan Kurikulum, Kementerian Pendidikan Malaysia, 35-56.

Institut Pendidikan Guru Malaysia. (2007). *Pro Forma Kursus Penyelidikan Tindakan: Program Ijazah Sarjana Muda Perguruan Dengan Kepujian (Matematik Pendidikan Rendah) Institut Pendidikan Guru.* Putrajaya: Kementerian Pelajaran Malaysia

Kaiser, G. & Bloemeke, S. (2010). What the Eastern and the Western debate can learn from each other using the case of mathematics teacher education. In Y. Shimizu, Y. Sekiguchi & K. Hino (Eds.), *Proceedings of The Fifth East Asia Regional Conference on Mathematics Education (Vol. 1).* Japan: Japan Society of Mathematical Education, Tokyo, 1-13.

Kasmawati Che Osman (2006). *Meninjau penggunaan Geometer's Sketchpad di kalangan guru matematik sekolah menengah Pulau Pinang.* Unpublished M.Ed thesis, Universiti Sains Malaysia, Penang.

Kementerian Pelajaran Malaysia. (2007). *Pelan Induk Pembangunan Pendidikan (PIPP) 2006-2010*. Putrajaya: Kementerian Pelajaran Malaysia.

Kementerian Pelajaran Malaysia. (2008). *Buku manual kajian tindakan (Edisi Ketiga)*. Putrajaya: Bahagian Perancangan Dan Penyelidikan Dasar Pendidikan.

Kementerian Pelajaran Malaysia. (2010). *Garis panduan praktikum latihan perguruan praperkhidmatan*. Putrajaya: Bahagian Pendidikan Guru.

Kementerian Pendidikan Malaysia (2001). *Laporan pemantauan kolaboratif pelaksanaan projek rintis sekolah bestari*. Kuala Lumpur: Bahagian Sekolah, Kementerian Pendidikan Malaysia.

Kementerian Pendidikan Malaysia. (1995). *Kertas Dasur Untuk Mesyuarat Pertama Majlis Penyelidikan Pendidikan Mulaysia (MAPPEMA)*. Kuala Lumpur: Bahagian Perancangan dan Penyelidikan Dasar Pendidikan.

Kementerian Pendidikan Malaysia. (1996). *PIER: Programme for Innovation, Excellence and Research* (brosur). Kuala Lumpur: Bahagian Perancangan dan Penyelidikan Dasar Pendidikan.

Kementerian Pendidikan Malaysia. (1997). *Majlis Penyelidikan Pendidikan Malaysia (MAPPEMA)* (brosur). Kuala Lumpur: Bahagian Perancangan dan Penyelidikan Dasar Pendidikan.

Kemmis, S. & McTaggart, R. (1988). *The action research planner.* Geelong, Vic: Deakin University.

Kim, P. L. (1994). *Developing field inquiry models for investigating student teacher professional learning: A multidimensional approach*. Unpublished PhD thesis, University of Sydney.

Kim, P. L. (2007). The environments of action research in Malaysia. In S. Hollingsworth (Ed.), *Action research: A casebook for educational reform* (pp. 238-243). London: Falmer Press.

Kim, P. L. (Ed). (1995). *Modul kajian tindakan*. Kuala Lumpur: Bahagian Perancangan dan Penyelidikan Dasar Pendidikan, Kementerian Pendidikan Malaysia.

Kor, L. K & Lim, C. S. (2009). Lesson study: A potential driving force behind the innovative use of Geometer's Sketchpad. Journal *of Mathematics Education, 2* (1), 69-82.

Lee, M.N.N. (2000). The development of teacher education in Malaysia: Problems and challenges. *Asia-Pacific Journal of Teacher Education & Development*, 3(2), 1-16.

Lee, M.N.N. (2002). Teacher education in Malaysia: Current issues and future prospects. In E. Thomas (Ed.), *Teacher Education: Dilemmas and Prospects* (pp. 57 – 68). London: Routledge.

Lee, S. G. (1997). Guru sebagai penyelidik: Faktor-faktor yang mempengaruhi kekerapan aktiviti kajian tindakan di sekolah [Teacher as a researcher: factors affecting frequency of action research activities in school]. Unpublished Master of Science (Management) degree project paper, Universiti Utara Malaysia.

Lee, W.H. (2003). Reflective practice in Malaysian Teacher Education: Assumptions, Practices and Challenges. *Unpublished Ph.D thesis,* Universiti Sains Malaysia.

Lim, C. S. & Kor, L. K. (2010). *Innovative use of GSP through Lesson Study collaboration*. Penang: Penerbit UPPA, USM.

Lim, C. S. (2006). Promoting Peer Collaboration among Pre-service Mathematics teachers through Lesson Study Process. In S. Yoong et al. (Eds.), *Proceedings of XII IOSTE symposium: Science and technology in the service of mankind* (pp. 590-593). Penang, Malaysia: School of Educational Studies, Universiti Sains Malaysia.

Lim, C. S. (2007). Developing Mathematical Thinking in a Primary Mathematics Classroom through Lesson Study: An Exploratory Study. *Proceedings of the APEC-Khon Kaen International symposium on Innovative Teaching Mathematics through Lesson Study (II): Focusing on mathematical thinking*, August 16-20, 2007 at Kosa Hotel, Khon Kaen, Thailand, pp.141-153.

Lim, C. S., White, A., & Chiew, C. M. (2005). Promoting Mathematics Teacher Collaboration through Lesson Study: What Can We Learn from Two Countries' Experience. In A. Rogerson (Ed.), *Proceedings of the 8th International Conference of The Mathematics Education into the 21st Century Project:"Reform, Revolution and Paradigm Shifts in Mathematics Education"*, pp. 135-139, Nov 25-Dec 1, 2005 at Hotel Eden Garden, Johor Bharu, Malaysia.

Lim, C. S., Chew, C. M, Chiew, C.M. & Goh, S. I. (2009). Mathematics Teachers' and Students' Perspectives on the Innovative Use of the Geometer's Sketchpad through Lesson Study Collaboration. Digest Pendidik, *9*(1), 55-67

Lim, C. S., Chiew, C.M. & Chew, C. M. (2008).Promoting mathematical thinking and communication in a Bilingual classroom. *Proceedings of the APEC-Khon Kaen International symposium on Innovative Teaching Mathematics through Lesson Study (III): Focusing on mathematical communication,* pp.92-108, August 25-29, 2008 at Khon Kaen University, Thailand.

Lim, C. S., Chiew, C.M. & Chew, C. M. (2011). *Promoting Mathematical Thinking and Communication through Lesson Study Collaboration.* Penang: Basic Educational Research Unit (UPPA), School of Educational Studies, Universiti Sains Malaysia.

Lim, C. S., Chiew, C.M., & Chew, C. M.. (2010). Assessing for Improvement of Teaching and Learning through Lesson Study Collaboration. Paper presented and published in *Proceedings of APEC Chiang Mai International Conference IV: Innovation of Mathematics Teaching and Learning through Lesson Study – Connection between Assessment and Subject Matter*, Faculty of Education, Chiang Mai University, 2-6 November 2010, pp. 73-85

Lourdusamy, A., & Tan, S.K. (1992). Malaysia. In H.B. Leavitt (Ed.), *Issues and Problems in Teacher Education: An International Handbook*. New York: Greenwood Press, 179-191.

Ministry of Education (2001). Teacher Education. *Education in Malaysia: A Journey to Excellence*. Ministry of Education, 87-105.

National Institute for Educational Policy Research (NIER). (2006). *Best practices in professional learning of science and mathematics teachers: Final report of the seminar.* Tokyo, Japan: National Institute for Educational Policy Research. Retrieved from http://www.apecknowledgebank.org/resources/downloads/BestPracticeMSTeacher_NIER2005Nov.pdf

Noorshah bin Saad (2006). Pengetahuan Pedagogi Kandungan Dan Amalannya Di Kalangan Guru Matematik Sekolah Rendah. *Unpublished Ph.D thesis,* Universiti Pendidikan Sultan Idris.

Ong E. G., Lim, C. S., & Munirah Ghazali (2007). Enhancing Communication skills in mathematics teachers through the Lesson Study Collaboration: A pilot study. *Proceedings of the Second International Conference on Science and Mathematics Education [CoSMEd2007]*, pp. 282-289, 13-16 November, 2007, SEAMEO-RECSAM.

Ong, E. G. (2010). *Changes in Mathematics Teachers' Questioning Techniques Through Lesson Study Process.* Unpublished Ph.D thesis, Universiti Sains Malaysia, Penang.

Ong, E. G., Lim, C. S., & Munirah Ghazali (2010). Examining the changes in novice and experienced mathematics teachers' questioning techniques through lesson study process. *Journal of Science and Mathematics Education in South East Asia, 33(1)*, 86-109.

Organisation for Economic Cooperation and Development (2010). *Teachers' Professional Development: Europe in international comparison.* Luxembourg: European Union. Retrieved from http://ec.europa.eu/education/school-education/doc1962_en.htm

SEAMEO RECSAM. (1990). *Workshop on thinking in science and mathematics (TISM) project.* Penang, Malaysia: SEAMEO regional Centre for Education in Science and Mathematics.

SEAMEO RECSAM. (1991). *Thinking in science and mathematics (TISM) project: Interim report.* Penang, Malaysia: SEAMEO regional Centre for Education in Science and Mathematics.

Shulman, L.S. (1987). Knowledge and teaching: Foundations of the new reform. *Harvard Educational Review*, 57(1), 1-22.

Subahan, M. M., Abd. Rashid, J., & Jamil, A. (2002). What motivates teachers to conduct research? *Journal of Science and Mathematics Education in Southeast Asia, 25*(1), 1-24.

Takahashi, A. (2010). Lesson Study: An Introduction. In Y. Shimizu, Y. Sekiguchi, & K. Hino (Eds.), *Proceedings of the 5th East Asia Regional Conference on Mathematics Education (EARCOME5) Vol. 1* (pp. 169-181). Tokyo: Japan Society of Mathematics Education.

Tengku Zawawi bin Tengku Zainal (2005). Pengetahuan Pedagogi Isi Kandungan Bagi Tajuk Pecahan Di Kalangan Guru Matematik Sekolah Rendah. *Unpublished Ph.D thesis,* Universiti Kebangsaan Malaysia.

White, A. L. & Lim, C. S. (2007). Lesson Study in a global world. In Lim, C. S. et al., (Eds.), *Proceedings of the 4th East Asia Regional Conference on Mathematics Education [EARCOME4], pp.567-573,* 18-22 June 2007, Universiti Sains Malaysia, Penang, Malaysia.

White, A. L. & Lim, C. S. (2008). Lesson study in Asia Pacific classrooms: local responses to a global movement. *ZDM: International Journal for Mathematics Education, 40*(6), 915-939

White, A. L. & Lim, C. S. (2007). Lesson Study: Local, Global or a Glocal Strategy for Teacher Professional Learning? *Proceedings of the Second International Conference on Science and Mathematics Education [CoSMEd2007*], pp. 239-245, , 13-16 November, 2007, SEAMEO-RECSAM

Wong, L.T. (1998). The continuous professional development of teachers. *Jurnal Maktab Perguruan Rajang Bintangor, 9*, 76-82.

JAPAN

SECTION EDITOR
Yoshinori Shimizu
University of Tsukuba

CHAPTER 57

MATHEMATICS EDUCATION RESEARCH IN JAPAN

An Introduction

Yoshinori Shimizu
University of Tsukuba

ABSTRACT

This chapter provides an overview of those areas of mathematics education research in Japan reviewed and discussed in this section. While a brief introduction of each chapter is provided, it is emphasized that we need to take into account the following contexts to understand mathematics education research in Japan. First, in order to examine the trends and issues in the areas of mathematics educataion research in Japan, we cannot neglect their connections with the goals and emphases described in the national curriculum standards. Second, the mathematics education community in Japan has a long tradition of Lesson Study as practical research methodologies in a particular form of action research. Then, to grasp the ongoing research agendas and questions, we need to pay careful attentions to their accumulated findings related to the curriculum and teaching materials, students' thinking, and mode of teaching in each research area. Third, developments of mathematics education research in Japan have been influenced by west-

The First Sourcebook on Asian Research in Mathematics Education: China, Korea, Singapore, Japan, Malaysia, and India, pp. 1287–1295

ern educational theories in various areas of inquiry, while educational activities themselves are rooted in East Asian cultural tradition.

Keywords: Research and practice, lesson study, culture, Japan Society of Mathematical Education

THE JAPANESE CONTEXT

Japan Society of Mathematical Education (2010) has recently published the first handbook of research in mathematics education that includes comprehensive reviews of various research areas in mathematics education in Japan for several decades. The volume, which is composed of forty-eight chapters, illustrates that it is not an easy task to identify the key areas in research in mathematics education in Japan. In fact, the organization of chapters found in the research handbook differs from other similar research handbooks such as those edited by Lester (2007) and by Clements, Bishop, Keitel, Kilpatrick, & Leung (2012).

Most notably, a certain amount of pages are devoted to analyses of teaching materials (kyozai-kenkyu) as well as models and modes of teaching and learning in classrooms. This is a reflection of how Japanese community of mathematics education has been formed and research has been conducted in the past.

In this chapter the author intends to provide an overview of those areas of mathematics education research in Japan reviewed and discussed in this section. We need to take into account the following contexts to understand the trends and issues in mathematics education research in Japan.

First, in Japan we have national curriculum standards (The Course of Studies), which have been revised roughly every 10 years. In order to examine the trends and issues in most areas of mathematics research in Japan, we cannot neglect their connections of research questions with the goals and content described in the national curriculum standards. In the current national curriculum standards (Ministry of Education, Culture, Sports, Science and Technology, 2008), for example, such classroom activity like communication, discussion, explaining, and writing are strongly valued and emphasized as 'activities with languages' in all the subject areas. Then, ongoing research focuses on introducing peer dialogues, small-group discussions, writing, and so on, in addition to the study of certain subject matters. Also, the revision of national curriculum standards that introduces some changes in the scopes and sequences of mathematical content eventually influences the choice of the research questions to be pursued. The introduction of an earlier conceptualization of a common fraction, for example, is taught in the second grade in the new curriculum, while in the former national curriculum standards the concept of a common fraction was introduced in fourth grade. Teaching common fractions throughout the elementary school curriculum can then be a 'hot topic' to be examined. Learning to cope with these new visions in mathematics curriculum, teaching methods, and general emphasis in education is one of catalysts for researchers and teachers to do research together.

Second, the mathematics education community in Japan has a long tradition of Lesson Study (*see Chapter 67 in this volume*). Despite the long history of lesson study in their own country, Japanese mathematics educators, and researchers in other areas have not been much interested in studying lesson study itself until recently. After the publication of *The Teaching Gap* (Stigler & Hiebert, 1999), followed by a Japanese translation (Minato, 2002), Japanese educators, often deeply involved in lesson study, 'found' the importance of this particular cultural activity. Lesson Study, which is a literal translation of 'jugyo kenkyu', means a Japanese approach to develop and maintain quality classroom instruction through a particular form of activity by a group of teachers (Fernandez & Yoshida, 2004). Lesson Study serves as an approach to professional development whereby a group of teachers collaboratively develop and conduct lessons to be observed: examining the subject matter to be taught, how their students think and learn the particular topic in the classroom, how to incorporate new methods for improving classroom instruction, and so on. The activity of lesson study includes planning and implementing the 'research lesson' as a core of the whole activity, followed by a post-lesson discussion and reflection by participants, with revisions of conducted lessons. Researchers and teachers work closely within the community with local theories of students' learning in their perspective. Thus, to grasp the ongoing research agendas, we also need to pay careful attentions to their accumulated findings on curriculum and teaching materials, students' thinking, and mode of teaching in each research area.

Third, developments of mathematics education research in Japan have been influenced by Western educational theories in various areas of inquiry, while educational activities themselves are rooted in East Asian cultural tradition. From the very beginning of the modern era in Japan, the Western educational system has been introduced with confusion. Then, as Sekiguchi points out (*See Chapter 59 in this volume*), influence of educational research from outside Japan became very distinct in many areas. Research on problem solving in Japan, for example, has constantly been influenced by research in the U.S. and European countries, on the one hand, while is has been affected by its tradition which values mathematical nature of problem solving and mathematical richness of thinking, on the other (*See Chapter 60 in this volume*). Giving attention to those mathematical nature and richness in their research on problem solving, Japanese researchers elaborated the theoretical components of problem solving processes to some extent, so that those components can reflect mathematical thinking more explicitly and work as facilitators of students' thinking in mathematics classes. In sum, when we review the development of mathematics education in Japan, we need to look into the influences from both the East and the West.

WORKING WITH AND LEARNING FROM TEACHERS

For more than a decade, educators and researchers in the field of mathematics education have been interested in Lesson Study as a promising source of ideas for improving education. For a Japanese mathematics educator who has been deeply involved in lesson study for more than two decades, this 'movement' has provided an opportunity for reflecting on how Lesson Study as a cultural activity works as a system embedded in the entire society as well as local community of teachers with shared values and beliefs. Given the tradition of lesson study, it is very important for Japanese mathematics educators to work with and to learn from teachers. The relationship between research and practice may be seen differently in other countries.

Among five crucial relationships in research in mathematics education that he identified as important, Bishop (1992) lists the relationship between the teacher and the researcher as a particularly significant one. He characterizes three theoretical traditions, pedagogue, empirical scientist, and scholastic philosopher, and each tradition has the goal of enquiry, role of evidence, and role of theory in different ways. If the goal of study is direct improvement of teaching, and role of theory is accumulated and sharable wisdom of expert teachers, the study is in the pedagogue tradition. The evidence presented is usually highly selective and exemplary here. He noted that in both empirical scientist and scholastic philosopher traditions, the roles of teacher and researcher are incompatible. He wrote:

> The teacher is the practitioner whose practice, it is felt, needs to be informed by the research of the researcher. So, we have a clear hierarchy involved, with the researcher informing the teacher, but not necessary vice versa. (p. 717)

Bishop (1992) noted that the analysis and study of mathematics teaching from both these perspectives can make the teacher an object – not a subject – in the research.

An essential characteristic of the field of mathematics education is that its questions and concerns are deeply tied to matters related to the teaching and learning of mathematics (Silver & Herbst, 2007). Research in mathematics education has not only scientific goal of building theories but also the goal of improving of teaching and learning of mathematics. However, there has been a concern in mathematics education community, at least in Japan, with the relevance and usefulness of the results of research. Also, it is often argued that there is a long-standing issue of the separation between research and practice in education community, in general, and mathematics education community, in particular. As the classroom practices are socially and culturally situated, and shared values and beliefs by teachers are key for continuous development and of the quality teaching, research in mathematics education is socially and culturally situated. Continuous working with, and learning from, teachers raises the issues and shapes the research questions originated in the efforts of improvement of teaching and learning mathematics in the classroom. The problem tackled by teachers has rooted in the reality of the school and the classroom. Research questions can be posed in responding to problems derived from teachers' works. In working with teachers and learning from teachers, mathematics educators can have an opportunity for identifying implicit wisdom and accommodating craft knowledge to scholarly knowledge.

ORGANIZATION OF THE SECTION

Historical Perspectives on the Development of Mathematics Education Research in Japan

The Japanese section includes ten chapters except for this introduction. First, Makinae (Chapter 58) provides a historical perspective on mathematics education in Japan. After the introduction of a new school system in the modern era, school mathematics as a school subject was established. Referencing to western mathematics education, curriculum, textbooks, teaching methods and teacher training system were developed. Key players for the establishing school mathematics were mathematicians, professors in teacher training colleges and teachers in attached schools. Then, colleges for teacher training were reorganized as new universities in postwar educational reform. Becoming one of the divisions in university and professional researchers of mathematics education held positions in university with academic standing point in research field. Following the review of historical development of mathematics education in Japan,

Sekiguchi (Chapter 59) discusses the development of mathematics education research as a scientific inquiry. These two chapters are mutually related and they together provide the warp and weft of the research in mathematics education in Japan. Sekiguchi points out that research in mathematics education in Japan have been shaped by socioeconomic situation, educational policy, research in other countries and other sciences, and so forth. Discussing on how mathematics education has developed as a research field in Japan, he identifies three major sources of Japanese mathematics education research are explored; (1) lesson study tradition, (2) influence from Western countries, and (3) the existence of national syllabus.

Trends and Issues in Key Research Areas

In the second part of the section, several key research areas are identified as have been studied actively in recent years with historical backgrounds of efforts by teachers to tackle with longstanding issues in Japanese mathematics education. These include research on proportional reasoning (Chapter 60), students' understanding of algebra and use of literal symbols (Chapter 61), proof and proving as an explorative activity (Chapter 62), teaching and learning problem solving (Chapter 63), use of ICT tools in mathematics classrooms (Chapter 64), the role of metacognition in learning mathematics (Chapter 65). Then, two chapters follow with a focus on teaching and learning as cultural activities; cross-cultural studies on classroom practices (Chapter 66), and professional development of teachers through lesson study (Chapter 67).

In her review of research on proportional reasoning in Japan, Hino (Chapter 60) notes that teaching of ratios, rates, proportional relationships and related concepts and skills has been a large interest in the community of mathematics educators. She synthesizes a body of research on proportional reasoning with a focus on summarizing trends of research, together with research that has influenced ongoing research agendas.

Studying students' difficulties in understanding of algebra and use of literal symbols is one the major topic in Japanese mathematics education research for a long time. Fujii (Chapter 61) reports on studies of students' understanding of school algebra in Japan. He first reviews accumulated research findings in order to probe Japanese students' understanding of school algebra focusing on the literal symbols, based mainly on research papers published in the journals of Japan Society of Mathematical Education. He then proposes an alternative way of teaching of school algebra by focusing on how the curriculum of the elementary and also secondary school can offer better opportunities for young people to think algebraically. Utilizing the potentially algebraic nature of arithmetic is one way of building a stronger bridge between early arithmetical experiences and the concept of a variable. He uses the terms generalizable numerical expressions or quasi-variable expressions to make a case for a needed reform to the curriculum of the elementary and secondary schools.

In Chapter 62, Miyazaki and Fujita examine proof and proving as an explorative activity. With brief general background information about the current teaching and learning of proof in Japan, they summarize the findings recent research into the teaching and learning of proof in Japan. Then they classify research studies presented and published in the past decade into 'global studies' and 'local studies', with a critical review of the studies in each category. Finally, they discuss needed directions for further studies in order to improve the current situation of the teaching and learning of proof.

In Chapter 63, Nunokawa overviews trends and issues in research on problem solving in Japan and illustrates that it has constantly been affected by its tradition which values mathematical nature of problem solving and mathematical richness of thinking, even after Japanese researchers learned much about problem solving research from the U.S. and European researchers. He then discusses that, giving the attention to those mathematical nature and richness in their research on problem solving, Japanese researchers elaborated the theoretical components of problem solving processes to some extent, so that those components can reflect mathematical thinking more explicitly and work as facilitators of students' thinking in mathematics classes.

Iijima in Chapter 64, illustrates aspects of research and practice of teaching and learning mathematics with Information and Communication Technology (ICT) in Japan by focusing on 'Geometric Constructor', one of the popular dynamic geometry software in Japan. After his reflection on the historical development of computer use in mathematics classrooms in Japan, the potentials and pitfalls are examined of the use of various devices in mathematics classrooms as well as the Internet for providing a platform for teachers' discussion of problems and students' difficulties. The current status of teaching and learning mathematics with ICT is illustrated with examples with a particular style of classroom practices. Implication for classroom practices and needed studies are examined.

In Chapter 65, Shigematsu overviews research on metacognition in Japanese mathematics education that are related to the improvement of the teaching and learning mathematic in classrooms. A special attention is given to the original conceptualization of metacognition as "Inner Teacher" which assumes that a part of students' metacognition related learning mathematics derived from the interaction with their teacher in the classroom. Based on the review of the current status of research on metacognition as well as the development in the methodologies, issues for the further research are discussed. R

Two chapters focus on cultural aspects of teaching and learning of mathematics and professional development of mathematics teachers. Shimizu in Chapter 66 reviews the selected findings of large-scale international studies of classroom practices in mathematics which include Japan as a participating country and discusses the uniqueness of how Japanese teachers structure and deliver their lessons and what Japanese teachers value in their instruction. Particular attention is given to the ways lessons are structured with an emphasis on presenting and discussing students' thinking on alternative solutions to the problem. While characteristics of Japanese mathematics classrooms are explored through the review of research findings, recent development in technological advances and advanced methodological techniques were examined which enable researchers to conduct more detailed analyses of learning taken place in mathematics classrooms.

In Chapter 67, Takahashi discusses how the Japanese systematic support for teacher professional learning has taken place and what Japanese teachers have learned though Lesson Study. He emphasizes that Lesson study does not follow a uniform system or approach and that it is a part of the Japanese teachers' cultural activity taken place in many different forms based on the purpose of the professional development. He then reviews a study on the differences in knowledge and expertise that exist among prospective teachers, novice teachers, and experienced teachers to discuss how the school system supports teachers to learn continuously through their careers and why it is important even after completing formal teacher training at universities. The issues in the study of the implementation of Japanese professional development structure in different cultures are discussed as a possible future research.

CONCLUDING REMARKS

For better understanding of accumulated research findings in mathematics education in Japan, we need to see how those issues are identified in the community of teachers and how research are conducted based on the efforts by teachers. For example, research on proportional reasoning in the Japanese context has a historical development of efforts through Lesson Study in the community of teachers to explore how and in what extent a teacher can provide a learning opportunities for the students to reason proportionally by using particular teaching materials and setting particular tasks. Research in mathematics education in Japan is based on 'research' findings accumulated and tested against the classroom practices of many teachers for many years in Lesson Study which can be considered not only as the way professional developments are conducted, as discussed in Chapter 65 but also as a particular approach to research on teaching and learning.

REFERENCES

Bishop, A. (1992). International perspectives on research in mathematics education. In D. A. Grouws (ed.) *Handbook of research on mathematics teaching and learning*. Reston, VA: National Council of Teachers of Mathematics.

Clements, M. A. (Ken), Bishop, A., Keitel, C., Kilpatrick, J., & Leung, F. (Eds.) (2012). *Third international handbook of mathematics education*, New York, NY: Springer.

Fernandez, C. & Yoshida, M (2004). *Lesson study: A Japanese approach to improving mathematics teaching and learning*. Mahwah, NJ: Lawrence Erlbaum Associates.

Japan Society of Mathematical Education (2010). *The handbook of research in mathematics education*. Tokyo: Toyokan (in Japanese)

Lester, F. K., Jr. (Ed.) (2007). *Second handbook of research on mathematics teaching and learning*. Charlotte, NC: Information Age Publishing.

Makinae, N. (2010). *The origin of lesson study in Japan*. Y. Shimizu, Y. Sekiguchi & K. Hino (eds.) *The proceedings of the 5th east Asia regional conference on mathematics education: In Search of Excellence in Mathematics Education*, Tokyo: Japan Society of Mathematical Education.

Minato, S. (2002). *Learn from Japanese mathematics education: Jugyou-kenkyuu as becoming the focus in the United States. A translation with annotations of The teaching gap (Stigler & Hiebert, 1999)*. Tokyo: Kyoiku Shuppan.

Ministry of Education, Culture, Sports, Science and Technology (2008). *National curriculum standards for kindergarten, elementary school, lower and upper secondary school.* Tokyo: The Ministry.

Sliver, E.A. & Herbst, P. G. (2007). Theory in mathematics education scholarship. In F. K. Lester (ed.) *Second handbook of research on mathematics teaching and learning*. Reston, VA: National Council of Teachers of Mathematics & Information Age Publishing.

Stigler, J. W., & Hiebert, J. (1999). *The teaching gap: Best ideas from the world's teachers for improving education in the classroom*. New York, NY: Free Press.

CHAPTER 58

A HISTORICAL PERSPECTIVE ON MATHEMATICS EDUCATION RESEARCH IN JAPAN

Naomichi Makinae
University of Tsukuba

ABSTRACT

The current chapter provides a historical perspective on mathematics education research in Japan. After the introduction of a new school system in the modern era in the 1860s, mathematics was established as a school subject in Japan. With references to Western mathematics education, curriculum, textbooks, teaching methods and teacher training systems were developed. During this process the key players for establishing school mathematics were mathematicians, professors in teacher training colleges and teachers in the attached schools. Colleges for teacher-training were then reorganized as new universities during postwar educational reform. After mathematics education became one of the divisions in the universities, professional researchers of mathematics education held positions in universities with academic standing.

Keywords: Historical perspective, research field, school mathematics

The First Sourcebook on Asian Research in Mathematics Education: China, Korea, Singapore, Japan, Malaysia, and India, pp. 1297–1313

INTRODUCTION

In this chapter, I will illustrate the history of mathematics education research as a field of academic inquiry in Japan. The Japanese modern school system started in the early Meiji era (1868-1912). Prior to this era, only "Terakoya [primarys school for the general public]" and "Hankou [regional government schools for feudal clans]" were available, both of which were not managed by the national government. Thus the modern school system in Japan began in this period. In 1872, a new school system named "Gaku-sei [School System]" was started by the government. In "Gaku-sei," modern mathematics education started in Japan. This period was the starting point for this history. However, it is difficult to point out exactly when mathematics education research was established. We can see how mathematics education research became the research field it is today through a careful examination of its historical development.

First, mathematics education was incorporated into the new modern school system. With reference to Western mathematics education, we developed curriculum, textbooks, teaching methods, and a teacher-training system. Among these developments, the selection of teaching contents was the most important issue of mathematics education research at that time. Mathematicians in universities, professors in teacher training colleges and teachers in attached schools worked for these developments.

Second, the teacher-training colleges were reorganized as universities during postwar educational reform. After the subject became one of the divisions in the universities, mathematics education was required. Theories of learning and teaching mathematics were objects of research. Professional researchers of mathematics education now had positions in universities.

ESTABLISHMENT OF SCHOOL MATHEMATICS

In the early Meiji era, there was an urgent need to introduce Western civilization and technology, and to build a modern country similar to Western countries. The schools undertook the responsibility to provide the country with appropriately trained human resources. For mathematics education, the Meiji government directed schools to introduce Western mathematics instead of continuing to teach Japanese mathematics, because they needed to import Western technology. The subject Sanyo, or "Arithmetic" was established for elementary mathematics, as well as

Daisu, or "Algebra," and Kika, or "Geometry," for secondary mathematics.

For this mathematics education, the Meiji government directed schools to use translated Western mathematics book and invited American teachers, with experience in teacher training, to teach in the normal school [teacher-training college]. The Meiji government established a normal school in Tokyo, to ensure the pre-service teachers would learn the new teaching methods. Under the guidance of American teachers, new textbooks and teaching manuals were written and published in the normal school. Since the translated books were not suitable for Japanese schools, new textbooks were created with reference to Western textbooks. For Sanyo, *Sho-gaku Sanjutsu-syo* [*Elementary Arithmetic Textbook*] (Monbusyo, 1873) was published. The textbook was based on the American object lesson, with reference to Pestalozzian theory. Through the object lesson, the textbook aimed to teach the concepts of numbers and operations.

The Meiji government not only invited foreigners into Japan but also dispatched Japanese students to Western countries. Dairoku Kikuchi studied in England and became a mathematics professor at Tokyo University. He wrote many geometry textbooks used in secondary schools such as *Shoto Kika-gaku Kyokasyo* [*Elementary Geometry Textbook*] (Kikuchi, 1888). His textbooks were strictly logical and followed Euclid geometry. His teaching principle for these textbooks was written in *Kika-gaku Kogi* [*Lecture on Geometry*] (Kikuchi, 1897).

> Geometry and Algebra are different subjects. In Geometry, we have geometrical solutions. So, we should not use algebraic solutions in Geometry. (Kikuchi, 1897; p. 20, translated)

He insisted on separate teaching for the different subjects in mathematics. This separation affected mathematics education in Japan for a long time.

Additionally, his pupil Rikitaro Fujisawa also studied in England and Germany, and became a professor at Tokyo University. He wrote many algebra and arithmetic textbooks such as *Shoto Dasisu-gaku Kyokasyo* [*Elementary Algebra Textbook*] (Fujisawa, 1898a) and *Sanjutsu Sho Kyokasyo* [*Arithmetic Textbook*] (Fujisawa, 1898b). His teaching principle for these textbooks was written in his book *Sanjutsu Kyoju Yomoku oyobi Kyoju-ho* [*Syllabus and Teaching Method for Arithmetic*] (Fujisawa, 1895). He supported Kikuchi's insistence on the strict separation of subjects and made a distinction between Arithmetic and Algebra:

> The specialty of Arithmetic is dealing with the calculation of fixed numbers. The specialty of Algebra is dealing with the characters that can represent any numbers. (Fujisawa, 1895; p. 8, translated)

He also insisted that using figures and graphs was an incorrect way to study Arithmetic and Algebra. They were not proper solutions for subject. His insistence on the strict separation was based on the formal character-building approach at that time. He thought that there were two aims for mathematics education: substantial training and formal character-building. The students who needed mathematical calculations for their future jobs were few. Most of the students did not need to learn the contents that were taught in school mathematics. Thus, substantial training in mathematics was not an aim for everyone. Formal building was the reason why all students had to study mathematics. For formal character-building, students had to use original solutions for the subject and experiment with alternatives, such as solutions from other subjects. Figures or graphs were shortcuts and a way to avoid proper thinking.

The philosophies of Kikuchi and Fujisawa influenced the national curriculum and textbooks. In 1902, the national syllabus for secondary education was announced by Monbusyo. In the syllabus, teaching contents were strictly separated by subjects such as Sanjutsu or "Arithmetic," Daisu, Kika and Sankaku-ho or "Trigonometry." In 1905, the first national textbook in Arithmetic *Jinjo-Sho-gaku Sanjutsu-syo* [*Arithmetic Textbook for Elementary School*] (Monbusyo, 1905) was published. In this era, there was another philosophy in mathematics education, *Riron Sanjutsu* [*Theoretical Arithmetic*] by Hisashi Terao who studied in France. After discussion, the philosophies of Kikuchi and Fujisawa were accepted as the national standards, and other philosophies were rejected, such as "Riron Sanjutsu" and *Sugaku Sanzen Dai* [*Skill Development by Drill Practice*] by Sekyu Ozeki, which was based on creationism. It seemed that Kikuchi had the authority and power to make this decision because he became a minister of education in 1901.

In this era, the school system was established and regulated under the national government, and Western mathematics was imported. One of the most urgent tasks of mathematics education was providing syllabi and textbooks that specified what contents should be taught in school. During these developments textbooks, a group of mathematicians played a leading role in mathematics education. The works of Kikuchi and Fujisawa played important roles in the establishment of contents and standards for school mathematics. School mathematics in Japan was introduced by Fujisawa during the International Congress of Mathematics (ICM) in Cambridge, England in 1912. Fujisawa had been at the ICM in Paris and reported on Japanese Mathematics in 1900 when Japanese culture received attention from Western artists at the international exhibition hold in Paris.

CURRICULUM DEVELOPMENT AND IMPROVEMENT OF THE TEACHING METHOD

As previously stated, Japanese school mathematics was established in the 1900s under the leadership of Kikuchi and Fujisawa. However, at the same time in Europe and the United States, the mathematics education that served as model for Japan was facing criticism and targeted for improvement. From the end of the Meiji era to World War II (1900s-1945), under the influence of the movement to improve mathematics education in the early twentieth century in Europe and the United States, curriculum development and the improvement of teaching methods were studied from a practical viewpoint. Overseas studies had been translated and introduced to Japan.

The first step in curriculum development was to consider separating the different subjects in mathematics. Minoru Kuroda, who was a teacher in an attached school for the Tokyo Koto Shihan Gakko [Tokyo Higher Normal School] emphasized integrated mathematics. He planned to connect the contents of different subjects in mathematics such as Arithmetic and Algebra, Arithmetic and Geometry, or Algebra and Geometry. By making such connections, students would achieve profound understandings and find better solutions. He studied in Germany under Felix Kline from 1910 onwards, and after returning to Japan he wrote textbooks and addressed mathematics education philosophy abroad. In his textbook, there is a geometric explanation for the algebraic formula "$(a+b)^2=a^2+2ab+b^2$" (Kuroda, 1916; p. 209). This explanation was inappropriate according to Fujisawa's philosophy. Additionally, he emphasized the use of functions across different subjects. His syllabus for the attached school mathematics had an influence on national curriculum development. Additonally, Gaisaburo Mori, who had studied in Germany, also translated the German mathematics textbook *Lehrbuch der Mathematik nach modernen Grrundastzen* under the supervision of Kline. This was published as *Sin-Syugi Sugaku* [*Mathematics under New Principle*] by Monbusyo in 1915 and 1916 (Behrendson & Gotting 1908, 1912/1915, 1916). On the other hand, there was theoretical criticism for separation. For example, Kinnosuke Ogura, who was a professor at Osaka Medical University, also raised objections (Ogura, 1924). He emphasized "Kagaku-teki Seisin [Scientific Mind]," which meant the quest for understanding relationships among national phenomena. To explain the relationship, we need mathematical views and approaches, especially in terms of functions. Other

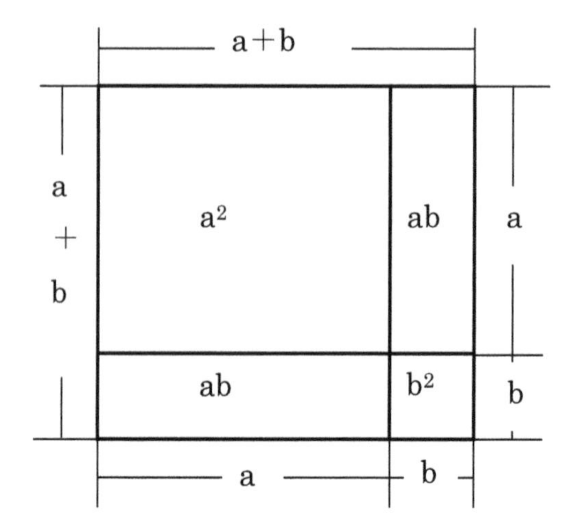

Figure 58.1.

overseas studies were translated and introduced as well. Turuichi Hayashi, who was a professor at Tohoku Imperial University translated Kline's lecture (Hayashi et al., 1921). Nobutaro Nabeshima, who was a professor at Tokyo Higher Normal School translated the papers of John Perry in England, and Eliakim Hastings Moore and John Wedley Young in the United States (Nabeshima, 1926).

Second, the contents of geometry were addressed. Teaching logical construction according to Euclid's Elements was not suitable for children's learning in geometry. All proof statements had to be based on a definition, axiom or theorem that had been previsouly proved, and the students needed to write them. This was needed to avoid a circular argument, but it required students to remember all the definitions, axioms and theorems associated with their numbers. Thus, studying geometry became memorizing all the proof statements. Instead of this type of geometry teaching, intuitive geometry became the focus. Before studying proofs, students learned concepts about figures from concrete objects. The notion of space is acquired from students' intuition. Tokuro Kunimoto who was a teacher at the Gakusyuin secondary school for girls, emphasized intuitive geometry:

> What are the targets to study in geometry? They are the shape, size, and positional relationship of the concrete objects. Through such consideration, students can acquire a deep understanding, enrich their concepts of figures, and build up experiences in thinking. Intuition lead to the concepts of figures. In teaching geometry, therefore, we should start from students' intuition about concrete objects. (Kunimoto, 1925, p. 21 translated)

The focus on intuitive geometry influenced the elementary school curriculum. In elementary school, the only subject taught in mathematics was Sanjustu, or "arithmetic." The contents of geometry could be expanded in elementary school, because elementary students can study figures using their intuition before the secondary geometry proof. In the fourth national textbook *Jinjo-shogaku-Sanjutsu-syo* (Monbusyo, 1927), geometry was added to the teaching contents.

Third, to improve the teaching method in the classroom, of child-centered education was introduced in this era. Teaching approaches that were based on proper subjects in mathematics were not suitable for children's learning. Instead, "Seikatsu-Sanjutu [Life Arithmetic]" was proposed. In this new teaching method, teaching from the prespective of children's life experiences was emphasized. Mathematics was introduced into the child's life or activities. The teachers in the attached schools and private schools wrote papers and books about their teaching. In this era, educational magazines for teachers started to be published. Meetings about lesson study using the new teaching methods were held in the attached schools and private schools. The following are some examples of these lesson studies. Takeshi Sato, who was a teacher at an attached elementary school of the Tokyo Higher Normal School, proposed developmental education, which means learning that is based on children's natural thinking instead of training by the teacher (Sato, 1924). Jingo Shimizu, who was a teacher at an attached elementary school of the Nara Higher Normal School for Girls, proposed arithmetic teaching whereby students generated the problems (Shimizu, 1924). Yoshie Iwashita, who was a teacher at an attached elementary school of the Tokyo Higher Normal School for Girls, proposed arithmetic education through work (Iwashita, 1930). Yoshinori Katori, who was a teacher at the Seikei Elementary School, proposed arithmetic education that is based on students' everyday experiences (Katori, 1933). Yoshio Ikematsu, who was a teacher at an attached elementary school of the Tokyo Higher Normal School, proposed arithmetic education that is based on the local culture in which students live (Ikematsu, 1935). These approaches were proposed from a practical point of view, and they usually included lesson plans or actual lesson reviews.

As a result of this curriculum development and improvement in teaching methods the Ministry of Education revised the syllabus for secondary schools in 1931. In the syllabus, the integrated treatment of mathematics was announced clearly. In 1935, the fifth national textbook in arithmetic, *Jinjo-Sho-gaku Sanjutsu* [*Arithmetic Textbook for Elementary School*] (Monbusyo, 1935), was published. In compiling the textbook, there were many contributions from Naomichi Shiono, who was a compiler of national textbook, in the Ministry of Education. The textbook was completely revised from the previous textbook. To make it child-friendly, the pages were printed in multicolor, and the problems and situations and the added contents of geometry, were introduced with intuitional teaching. At the ICM in Oslo, Norway in 1936, the textbooks were introduced by Motoji Kunieda, who was a professor at Tokyo Liberal Arts University. He reported the trends in the development of mathematics teaching in Japan.

BEGINNINGS OF ACADEMIC RESEARCH IN MATHEMATICS EDUCATION

During the improvement of school mathematics under the influence of the movements in Europe and the United States, we can see the early signs of academic research in mathematics education. The first teachers' society of mathematics education, called "Nihon Tyuto Kyoiku Sugaku-kai [The Mathematical Association of Japan for Secondary Education (MAJSE)]," was established in 1919. The president was Hayashi. The society aimed to mainly improve mathematics and its teaching methods in secondary schools. An annual meeting was held, and the society's journal *Nihon Tyuto Kyoiku Sugaku-kai-shi* [*Journal of the Mathematical Association of Japan for Secondary Education*] was published five times a year. Mathematicians and teachers of mathematics joined the society. They made proposals for the national syllabus and new teaching methods were discussed. However, in the early years, there were many articles about pure mathematics in the journal.

An unavoidable issue in this era is the controversy concerning formal character-building. This was originally addressed by Arata Osada at Hiroshima University in 1922, and a report was published in the journal (Hirosima Koshi Futyu Sugaku Kenkyu-kai, 1923). He described the ability to transfer thinking skills, which is acquired thorough exercise or training, and advocated that mathematics should be taught according to this theory. After his address, the famous pedagogist, psychologist, and mathematics educator discussed the possibilities and limitations of formal building in mathematics. Ogura agreed with with Osada's denying formal character-building (Ogura, 1924). But the president Hayashi and Kunieda who was vice-president of the MAJSE, opposed Ogura and Osada (Hayashi, 1924, Kunieda, 1924). Both sides of the controversy, including Western psychologists' theories were cited, but the discussion was not settled. Formal character-building has been the main reason to study mathematics. The aim of solving the mathematical problems and elaborating was to theories improve one's logical thinking. The controversy over formal character-building made the foundation of the main reason to teach mathematics unstable for mathematicians and teachers. This situation triggered a rethinking of the aims, contents, and teaching methods in mathematics education.

During this era, we can find arguments about mathematics education from academic perspectives, namely *Sugaku Kyoiku Shi* [*History of Mathematics Education*] (Ogura, 1932). In this book, the author considered the cultural history of mathematics education. He described mathematics education in terms of the social backgrounds, educational systems, and the theories of famous philosophers and education reformers. In the postwar period, when mathematics education became one of the divisions at universities, his book provided evidence that there had been academic research on mathematics education.

The "Sai-Kosei Undo [Reconstruction Movement]" played an important role in national mathematics education. The movement dealt with the issue of how to support students who graduated from elementary school, and responded to the new textbook *Jinjo-Sho-gaku Sanjutsu*. The national syllabus was revised in 1931, however it was still insufficient to improve mathematics education. The MAJSE needed to develop a sufficient curriculum to follow *Jinjo-Sho-gaku Sanjutsu*. During the 22nd annual meeting of the MAJSE in 1940, research subcommittees were established, one of which focused on the reconstruction of secondary mathematics education. At the end of the meeting, the head of the subcommittee reported that this research should be continued and decided to launch a study group divided into three branches nationwide (MAJSE, 1940). Their findings were reported at the following annual meeting, and summarized in the journal (Sugimura et al., 1941). The findings were reflected in the revised syllabus in 1942 and the textbook compilation in 1943. Under the concept of integrated mathematics, the teaching contents consisted of not only mathematical contents but also critical thinking skills and functions of mathematics (Tyuto-gakko Kyokasyo Kabushiki Kaisya, 1943). The textbooks introduced students to mathematics with real situations and problem solving. The ideas and philosophies of education were explored and reflected in the curriculum and textbook.

CHANGES AND DEVELOPMENTS IN THE OF RESEARCH FIELD

After World War II, the Japanese educational system changed rapidly and dramatically. Normal schools were reorganized to universities of education or as faculties of education within universities in 1949 and mathematics education research was developed in the universities. A bachelor's degree was required as a qualification basis to obtain a first class teaching credential. Teacher training was held in faculties of education in universities. Professors of specialized areas such as mathematics and mathematics education gave lectures on teaching credentials. The universities created divisions for the professional research of mathematics education. Moreover, in 1953, graduate schools of education (master's degree and doctoral degree) were established at the Tokyo University of Education and Hirosima University. The doctoral courses especially focused on professional development for researchers. At the time, the graduate schools of education (master's degree) were established in education universities and faculties of education throughout Japan.

In 1956, Yasuo Akitsuki, who was the president of Tokyo Liberal Arts University, attended the ICM in Bombay, India to report on the new postwar mathematics education in Japan. This was the third time that Japan reported on its national mathematics. The Ministry of Education had established committees to compile the national report. At that time, Yoshinobu Wada, who was a professor of Tokyo University of Education, took on the role of compiling the national report for Akitsuki. The leaders of mathematics education had changed mathematicians to researchers of mathematics education in universities.

The MAJSE was renamed twice in 1943 and 1970. After the first renaming to the "Nihon Sugaku Kyoiku Kai [Japan Society of Mathematical Education (JSME)]," the society broadened the scope of its study to the entire range of school mathematics. Starting in 1952, journals published by the JSME were divided into two titles: *Sugaku Kyoiku* [*Mathematics Education*] and *Sansu Kyoiku* [*Elementary Mathematics Education*]. In 1961, a new journal, *Sugaku Kyoiku Gaku Ronkyu* [*Full Discussion of Mathematics Education Research*], started to publish. This journal aimed to publish articles on scientific research in mathematics education and to develop academic research. In addition to the annual meeting, a new academic conference "Sugaku Kyoiku Ronbun Happyo Kai [Conference for the Presentation of Academic Papers on Mathematics Education]," was held in 1966. The conference was for researchers in universities, and the papers were required to be of the same quality as those published in journals such as *Sugaku Kyoiku Gaku Ronkyu*. In 1970, the JSME was renamed as "Nihon Sugaku Kyoiku Gakkai" [Japan Society of Mathematical Education (JSME)]. In Japanese, changing "Kai" to "Gakkai" represented a clarification of the characteristics of the academic society; however, the name in English did not change. The JSME was an academic society of mathematics education that kept a focus on practical teaching.

DEVELOPMENT OF TEACHING AND LEARNING THEORY

Following World War II, American progressive education was introduced to Japan under the occupation. Mathematics education was introduced as "Tangen Gakusyu [Life Unit Learning]," which started with problem solving in the everyday lives of children. However, this approach received strong criticism because of the fragmentation of the mathematical contents in students' learning. This emerging criticism triggered the shift to more integrated learning and an emphasis on problem solving not necessarily originating from children's everyday life experiences. Problem solving became a key word in mathematics education. Problem solving was seen as a process of interaction between the teacher and learner and among the learners in which the teacher attempts to provide learners with access to mathematical thinking in to solve the given problems. Learning theory was required to support this practical shift in mathematics education.

First, traditional historical research and the development of theories related to teaching materials took place. The historical study consisted of a review of past mathematics education. Through this study the background and fundamentals of current educational approaches were considered, and a policy on mathematics education was established. Typically, the theories related to teaching materials were written following the historical study. Shiono (1947, 1961) and Ogura and Nabeshima (1957) are studies that were frequently cited in later work. Some studies presented and summarized historical material such as old textbooks and statements in syllabi (Matsubara, 1982, 1983, 1985, 1987; JSME, 1987, 1988, 1995). Sasaki (1986) summarized the history of mathematics education in a chronological table. On the significances of historical study in mathematics education, Yoshinobu Wada stated the following:

> It is possible that the statements in previous studies do not exactly represent the truth. This is an unavoidable possibility in scientific research. However, if this is the case, it does not mean that the statement is meaningless. Our current studies build on the inquiry and intentions of our predecessors, who are the foundations of our research.... Academic research intends to contribute to our understanding of human beings. For this aim we need to love all humans.... By following such intentions we can foster goodwill and sincerity as we build on the original resources. This is the significance of historical study. (Wada, 1961; pp. 4-5, translated)

Historical study played a role in the consideration of the objectives, curriculum and teaching methods for mathematics education.

Theories about teaching material were the main contents in the commentary books about mathematics education. These books were written to help teachers understand mathematical contents, the curriculum and teaching standards. A series of commentary books was published approximately every ten years starting in the 1940s to support the revision of the course of study, which is the national curriculum in Japan. In particular, the "Sugaku Kyoiku Gendai-ka [New Math Movement]" had important implications. The New Math Movement in Western countries from the 1950s to the 1960s reflected the development from modern mathematics to mathematics education for the advancement of science and technology. This movement was influential in Japan. New contents such as set, topology, transformation and statistics were introduced in school mathematics. JSME also edited a series of 31 commentary books for New Math (JSME, 1966-1971). The theory of teaching materials, focused on not only mathematics contents but also students' understanding of the meaning of mathematic contents. Those studies emphasized the importance of understanding why something is true rather than what is true, or why the solution is correct rather than how to find a solution. Theories about the meaning of multiplication of rational numbers, rational thinking (Nakajima, 1981), and the connotation of geometrical figures (Maeda, 1979) were developed in this context and many articles about them have been published in the journal.

In addition to the teaching perspective, various approaches to teaching mathematics were developed. Around 1970 in Western countries, the discovery method was studied, and this was also true in Japan. It was emphasized in the modernization of mathematics teaching, which was influenced by the New Math Movement in Western countries. The discovery method was an improvement of Young's Heuristic Learning which aimed to help learners understand and master what they discovered or constructed through learning-by-doing and heuristic strategies (Ito, 1972). The axiomatic method is a systematic way of presenting knowledge about a domain of discipline by forming a logical system from a set of axioms. It conveys the knowledge effectively by presenting an axiomatic system. It summarizes two principles of teaching. First, it makes students understand why something is true after solving a problem. Second, it asks students if something is true and what can be found next. Through such activities, students confirm the essential construction of mathematical fact and their relationships (Sugiyama, 1986). Around 1980, the open-ended approach was developed. In problem solving activities, students were given problem that were non-routine, of represented insufficient conditions or life situations. Students had to decide on the approach and method to solve the problem. Through discussions of their various solutions to the problem, students would discover better ways of thinking. This approach originated from research on higher thinking among children at the National Institute of Education. The open-ended approach was introduced to Western countries as Japanese style of original mathematics teaching (Shimada, 1977; Nohda, 1991; Becker and Shimada, 1997).

Third, cognitive studies of students' learning were conducted by adopting the framework and methodologies of cognitive science. The focus of this research was on students' understanding and thinking. In the 1960s and 1970s the research aimed to clarify students' understandings of the contents of the New Math and to corroborate Piaget's experiment. Many of these findings did not mention practical implications for mathematics education, or they remained only concerned with general theory. Starting in the 1980s and later, we can see the perspectives from constructivism. Hirabayashi (1987) reviewed the historical origin of activism in Pestalozzi's idea of "insight" and sought the idea's psychological foundation in the work of John Dewey and its modern interpretation in Piaget's epistemology. He summarized these ideas into a modern didactical philosophy of mathematics education which is the background of constructivism. Following these constructivist perspectives, research on mathematics education has been diverse. To understand the process of children's construction of knowledge, various points have been examined, such as their inner thoughts or the reality of the individual (Takahashi, 1994), and the representational system that is shared between the teacher and students (Nakahara, 1995). Research on problem solving in mathematics has contributed to progress in cognitive studies. The processes of problem solving are considered as processes of construction or changes in the students' representational system (Nunokawa, 1990; Yamada, 1996). There was also research from the perspective of meta-cognition (Y. Shimizu, 1995; N. Shimizu 1995). On the other hand, research from sociocultural perspective has supported cognitive studies. According to Vygotsky's theory, children's cognition can be separated from the environment and culture around them (Otani, 1995; Sekiguchi, 1997).

Fourth, the development of ICT (information and communications technology) and their use in mathematics education led to a new research field. Beginning with CAI (Computer Assisted Instruction) in the 1970s, programming was introduced to mathematics education, and in the 1990s computers and calculators were introduced as an educational tool in school. Technological tools such as computers, graphic calculators, and geometrical software were useful not only for teachers' explanation of mathematical contents but also for facilitating students' thinking (Iijima, 1990; Shimizu, 1991). Teaching theories and teaching materials have developed in the field.

INTERNATIONAL PERSPECTIVES ON MATHEMATICS EDUCATION

Since 1980, Japanese mathematics education research has been carried out in international contexts. International conferences were held in Japan. In 1983, the ICMI-JSME Regional Conference on Mathematical Education was held in Tokyo. In 1993, PME17 (Psychology of Mathematics Education) was held in Tsukuba. In 2000, ICME9 (International Congress of Mathematical Education) was held in Makuhari and PME24 was held in Hiroshima. In 2010, EARCOME5 (East Asia Regional Conference of Mathematical Education) was held in Tokyo.

In 1986, Japan-U.S. seminar was held. In the seminar, the processes of children's problem solving were compared, and the characteristics of each country were clarified. A continued extension of the project led to comparative research of Japan-US mathematics and science education from 2002 to 2005. The foci of the project were on new teaching methods such as the use of electronic blackboards and digital resource, social education in science museums, and education programs for gifted students.

From other perspectives, Japanese lessons and lesson study have been popular in the world. From other countries' viewpoints, the high performance of Japanese students on international achievement tests such as TIMSS (Trends in International Mathematics and Science Study) suggests that the quality of Japanese school education is very high. The comparative research focuses on not only students' achievement but also the lessons themselves. Japanese researchers have joined international research projects such as the TIMSS video study and the Learner Perspective Study. Moreover the maintaining the quality of Japanese lessons depends on the professional development of approaches to lesson study. Originally, lesson study was not considered a research subject because it had been considered only as a practical issue in schools. As a result of the new attention given by researchers in other countries, however, lesson study became one of the key research subjects in mathematics education.

REFERENCES

Becker, J. P. & Shimada, S. (1997). *The open-ended approach: A new proposal for teaching mathematics*, Reston, VA: National Council of Teachers of Mathematics.

Behrendson, D. & Gotting, E., Mori, G. (Trans.) (1908,1912/1915,1916). *Sin-Syugi Sugaku* [Mathematics under New Principle], Monbusyo, Tokyo, (in Japanese), (Lehrbuch der Mathematik nach modernen Grrundastzen, Leipzig: B.G. Teubner.)

Fujisawa, R. (1895). *Sanjutsu Kyoju Yomoku oyobi Kyoju-ho [Syllabus and Teaching Method for Arithmetic]*, Maruzen, Tokyo, (in Japanese).

Fujisawa, R. (1898a). *Shoto Daisu-gaku Kyokasyo [Elementary Algebra Textbook]*, Dai-Nihon-Tosyo, Tokyo,(in Japanese).

Fujisawa, R. (1898b). *Sanjutsu Sho Kyokasyo [Arithmetic Textbook]*, Dai-Nihon-Tosyo, Tokyo, (in Japanese).

Hayashi, T. et al. (1921). *Doitsu ni okeru Sugaku Kyoiku* [Mathematics Education in Germany], Dai-Nihon-Tosyo, Tokyo, (in Japanese).

Hayashi, T. (1924). Kaikai no Ji [Opening Speech], *Journal of the Mathematical Association of Japan for Secondary Education, 6*(4,5), pp. 177-181, (in Japanese).

Hirabayashi, I. (1987). *Sugaku Kyoiku no Katsudo-shugi-teki Tenkai* [Mathematics Education from the Viewpoint of Activism], Toyo-kan-Syuppan, Tokyo, (in Japanese).

Hirosima Koshi Futyu Sugaku Kenkyu-kai (1923). Keishiki Toya ni kansuru Saikin no Ronso [Recent Controversy about the Formal Character-Building], *Journal of the Mathematical Association of Japan for Secondary Education, 5*(2), pp. 60-74, (in Japanese).

Iijima, Y. (1990). Computer ni okeru Zukei no Do-tekina Atsukai nit suite [On Dynamic Handling of Geometry by Computer], *Tsukuba Journal of Educational Study in Mathematics, 9*(A), pp.105-117, (in Japanese).

Ikematsu, Y. (1935). *Kyodo-syugi to Sanjutsu Kyoiku* [Arithmetic Education with Local Culture], Shihan-Daigaku Koza, Tokyo, (in Japanese).

Ito, T. (1972). *Sansu Hakken Gakusyu no Riron to Jissai* [Theory and Practice of Discovery Method in Elementary Mathematics], Meiji-Tosyo, Tokyo, (in Japanese).

Iwashita, Y. (1930). *Sagyo-syugi Sanjutus Kyoiku no Genri to Jissai* [Principle and Practice of Arithmetic Education through Work], Jinbun-Syobo, Tokyo, (in Japanese).

Japan Society of Mathematical Education (1966-1971). *Gendai-ka no tameno Sido Sirizu [Teaching for New Math]*, Meiji-Tosyo, Tokyo, (in Japanese).

Japan Society of Mathematical Education. (1987, 1988). *Tyu-Gakko Sugaku Kyoiku Shi [History of Mathematics Education in Junior High School]*, Sinsu-Sya, Tokyo, (in Japanese).

Japan Society of Mathematical Education. (1995). Sengo 50 nen no Sansu Sugaku Kyoiku [Mathematics Education of Japan for the Last Fifty Years after the War], *Journal of Japan Society of Mathematical Education, 77(6,7)*, (in Japanese).

Katori, Y. (1933). *Seikatsu Sanjutsu no Shin Kenkyu: Soshiki-teki Keito-teki [New Study of Life Arithmetic: Organizational and Systematic]*, Monasu, Tokyo, (in Japanese).

Kikuchi, D. (1888). *Shotou Kika-gaku Kyokasyo [Elementary Geometry Textbook]*, Dai-Nihon-Tosyo, Tokyo, (in Japanese).

Kikuchi, D. (1897). *Kika-gaku Kogi [Lecture on Geometry]*, Dai-Nihon-Tosyo, Tokyo, (in Japanese).

Kunieda, M. (1924). Sugaku Kyoiku Zakkan [Miscellanea about Mathematics Education], *Journal of the Mathematical Association of Japan for Secondary Education, 6(4,5)*, pp.131-150, (in Japanese).

Kunimoto, T. (1925). *Chokan-Kiki no Riron to Jisai [Theory and Practice in Intuitive Geometry Teaching]*, Baifu-kan, Tokyo, (in Japanses).

Kuroda, M. (1916). *Kika-gaku Kyokasyo Heimen [Geometry Textbook: Plane Geometry]*, Baifu-kan, Tokyo, (in Japanese).

Maeda, R. (1979). *Sansu Kyoiku Ron [Theory of Elementary Mathematic Education]*, Kaneko-Syobo, Tokyo, (in Japanese).

Mathematical Association of Japan for Secondary Education (1940). Dai 22 kai Sokai Kiji [Assembly Articles in 22nd Annual meeting], Journal of the Mathematical Association of Japan for Secondary Education, 22(5), pp.198-241.

Matsubara, G. (1982). Nihon Sugaku Kyoiku Shi Sansu(1) [History of Mathematics Education in Japan: Elementary(1)], Kazama-Syobo, Tokyo, (in Japanese).

Matsubara, G. (1983). Nihon Sugaku Kyoiku Shi Sansu(2) [History of Mathematics Education in Japan: Elementary(2)], Kazama-Syobo, Tokyo, (in Japanese).

Matsubara, G. (1985). Nihon Sugaku Kyoiku Shi Sugaku1) [History of Mathematics Education in Japan: Secondary(1)], Kazama-Syobo, Tokyo, (in Japanese).

Matsubara, G. (1987). Nihon Sugaku Kyoiku Shi Sugaku(2) [History of Mathematics Education in Japan: Secondary(2)], Kazama-Syobo, Tokyo, (in Japanese).

Monbusyo (1873). Sho-gaku Sanjutsu-syo [Elementary Arithmetic Textbook], Shihan-Gakko, Tokyo, (in Japanese).

Monbusyo (1905). *Jinjo-Sho-gaku Sanjutsu-syo* [Arithmetic Textbook for Elementary School], Dai-Nihon-Tosyo, Tokyo, (in Japanese).

Monbusyo (1927). *Jinjo-Sho-gaku Sanjutsu-syo* [Arithmetic Textbook for Elementary School], Dai-Nihon-Tosyo, Tokyo, (in Japanese).

Monbusyo (1935). *Jinjo-Sho-gaku Sanjutsu* [Arithmetic Textbook for Elementary School], Dai-Nihon-Tosyo, Tokyo, (in Japanese).

Nabeshima, N. (1926). *Sugaku Kyoiku no Kakushin* [Innovation of Mathematics Education], Meguro-Shoten, Tokyo, (in Japanese).

Nakahara, T. (1995). *Sansu Sugaku Kyoiku ni okeru Kosei-teki Apurochi no Kenkyu* [A Study on Constructivism in Mathematics Education], Seibun-Sya, Tokyo, (in Japanese).

Nakajima, K. (1981). *Sansu Sugaku Kyoiku to Sugaku-tekina Kangaekata* [Mathematics Education and Mathematical Thinking], Kaneko-Syobo, Tokyo, (in Japanese).

Nohda, N. (1991). *Sansu Sugaku-ka no Opun Apurochi niyoru Shido no Kenkyu* [A Study on Mathematic Teaching by Open Approach], Toyo-kan-Syuppan, Tokyo, (in Japanese).

Nunokawa, K. (1990). *Kaiketsu Katei ni oite Mondai Bamen no Kozo no Kosei wo Sasaeru Mono* [Support of Construction of the Problem Situation in Problem Solving Process], Dai 23 kai Sugaku Kyoiku Ronbun-Hapyo-kai Ronbun-Syu, pp. 291-296, (in Japanese).

Ogura, K. (1924). *Sugaku Kyouku no Konpon Mondai* [Fundamental Problem in Mathematics Education], Idea-Syoin, Tokyo, (in Japanese).

Ogura, K. (1932). *Sugaku Kyoiku Shi* [History of Mathematics Education], Iwanami-Shoten, Tokyo, (in Japanese).

Ogura, K., & Nabeshima, N. (1957). *Gendai Sugaku Kyoiku Shi* [History of Mathematics Education in Modern], Dai-Nihon-Tosyo", Tokyo, (in Japanese).

Otani, M. (1995). "Hattatsu no Sai-kisetus Ryoiki" Ron niyoru Sansu-ka Jugyo Katei no Bunseki [Analysis of Process of Elementary Mathematics Class with "Closest Area of Development" Theory], *Dai 28 kai Sugaku Kyoiku Ronbun-Hapyo-kai Ronbun-Syu,* pp. 225-230, (in Japanese).

Sasaki, G. (1986). *Gendai Sugaku Kyoiku Shi Nenpyo* [Chronological Table of Mathematics Education], Seibun-Sha, Tokyo, (in Japanese).

Sato, T. (1924). *Hassei-teki Sanjutsu Shin Shido-ho* [New Teaching Method by Children's Development], Bunkyo-Shoin, Tokyo, (in Japanese).

Sekiguchi, Y. (1997). Ninchi to Bunka [Cognition and Culture], *Journal of Japan Society of Mathematical Education, 79*(5), pp.14-23, (in Japanese).

Shiono, N. (1947). *Sugaku Kyoiku Ron* [Theory of Mathematics Education], Kawade-Syobo, Tokyo, (in Japanese).

Shiono, N. (1961). *Sansu Sugaku Kyoiku Ron* [Theory of School Mathematics], Keirin-Kan, Osaka, (in Japanese).

Shimada, S. (1977). *Sansu Sugaku-ka no Opun Endo Apuroch* [Open-Ended Approach in Mathematics Education], Mizu-umi-Shobo, Tokyo, (in Japanese).

Shimizu, K. (1991). Conputa no Riyo niyote Gutai-ka sareru Atarashi Kika Gakusyu [New Geometry Learning realized by Using Computer], *Nihon Kagaku Kyoiku Gakai Dai 15 kai Nenkai Ronbun-syu*, pp.17-20, (in Japanese).

Shimizu, J. (1924). *Sanjutsu no Jihatsu Gakusyu Shido-ho: Jikken Jissoku Sakumon Tyushin* [Method of Spontaneous Learning in Arithmetic: Focus of Experience, Actual Measurement and Making Problem], Meguro-Shobo, Tokyo, (in Japanese).

Shimizu, N. (1995). Sugaku-teki Mondai Kiketsu ni okeru Horyaku-teki Norhoku ni kansuru Kenkyu [A Study on Strategically Solving Ability in Problem Solving in Mathematics], *Research in Mathematics Education, 1*, 101-108 (in Japanese).

Shimizu, Y. (1995). Bunsu no Joho ni kansuru Jido Seito no Ninshiki [Students' Cognition in Division of Fractions], *Sugaku Kyoiku-gaku Ronkyu, 64, 65*, pp. 3-25, (in Japanese).

Sugimura, K. et al. (1941). Sugaku Kyoiku Sai-Kosei Kenkyu-kai Hokoku [Reports of Study Groups for Reconstruction of Mathematics Education], *Journal of the Mathematical Association of Japan for Secondary Education, 23*(6), pp.233-296, (in Japanese).

Sugiyama, Y. (1986). *Kori-teki Hoho ni motoduku Sansu Sugaku no Gakusyu Shido* [Mathematics Teaching based on Axiomatic Method], Toyo-kan-Syuppan, Tokyo, (in Japanese).

Takahashi, H. (1994). Suryou ni kansuru Kodomo no Informal Knowledge wo Sasaeru Mono no Sonzai [The Existence of Support for Children's Informal Knowledge], *Dai 27 kai Sugaku Kyoiku Ronbun-Hapyo-kai Ronbun-Syu*, pp. 149-154, (in Japanese).

Tyuto-gakko Kyokasyo Kabushiki Kaisya (1943). *Sugaku* [Mathematics], Monbusyo, Tokyo, (in Japanese)

Wada, Y. (1961). Sugaku Kyoiku Shi Kenkyu no Hitsuyo [Significance of Historical Research in Mathematics Education], *Sansu Sugaku Kyoiku Kenkyu, 2*(2), pp.4-5, (in Japanese)

Yamada, A. (1996). Sugaku-teki Mondai Kaiketsu ni okeru Ninchi Purocesu ni kansuru Kenkyu [A Study on Cognitive Process in Problem Solving in Mathematics]. *Research in Mathematics Education, 2,* pp.79-89, (in Japanese).

CHAPTER 59

THE DEVELOPMENT OF MATHEMATICS EDUCATION AS A RESEARCH FIELD IN JAPAN

Yasuhiro Sekiguchi
Yamaguchi University

ABSTRACT

For research in mathematics education in Japan, various issues have been pursued from practical to theoretical standpoints (Japan Society of Mathematical Education, 2010). They have been shaped by socioeconomic situation, educational policy, research in other countries and other sciences, and so on. This chapter discusses how mathematics education has developed as a research field in Japan. In particular, three major sources of Japanese mathematics education research are explored; (1) lesson study tradition, (2) influence from Western countries, and (3) existence of national syllabus.

Keywords: Research field, lesson study, national syllabus, Japan Society of Mathematical Education

The First Sourcebook on Asian Research in Mathematics Education: China, Korea, Singapore, Japan, Malaysia, and India, pp. 1315–1321

LESSON STUDY TRADITION: THE ORIGIN OF JAPANESE MATHEMATICS EDUCATIONAL RESEARCH

At the beginning of the modern era, Japanese government established normal schools to train teachers. One of the main activities in normal schools was lesson study, where teachers (including pre-service teachers) set goals of lesson, prepared experimental lessons, conducted those lessons in actual classrooms while other teachers were observing them, and discussed them after the lessons (Fernandez & Yoshida, 2004). Lesson study has been conducted in all subjects, and has come to be held in many schools.

In lesson study, its purpose is twofold: (1) School-based professional development, and (2) Classroom-based action research. Teachers in public schools were required to make efforts to improve their teaching constantly throughout their professional lives. Lesson study practice has been institutionalized, and has been considered to be an important place for teachers to reflect their understanding of teaching materials and improve their teaching skills.

The unique feature of lesson study as research practice is that teachers play both practitioners and researchers. At the beginning of the modern era, normal school teachers explored possibilities of teaching materials or methods in actual classrooms with other teachers. This style of professional life gradually permeated throughout the nation's school teachers.

While lesson study activities became spreading over Japan, school teachers formed societies of educational research. In mathematics education, the Japan Society of Mathematics in Secondary Education (JSMSE) was established in 1919 based on the decision made by the Japan Association of Mathematics Teachers. The aim of the society was to investigate and improve mathematics teaching for secondary schools. This society later became the Japan Society of Mathematical Education (JSME), which aims to research and improve mathematics education in Japan, and include teachers from kindergartens to universities, administrators and researchers:

> The main activity of the JSME is the annual meeting of mathematics teachers which is held every summer, the biggest meeting of mathematics education in Japan. It includes several workshops and lectures focusing on up-to-date teaching and learning issues. The annual meeting involves teachers at all levels, two-thirds of whom are non-members. The JSME also organizes a research-oriented annual conference which is held every autumn. (JSME)

From the tradition of lesson study, Japanese mathematics education research has been closely connected to the classroom practices. As a result, it has held relatively high relevance to classroom practice. Though this is a great advantage, this may have hindered methodological argument, and theoretical development of research in Japan (Hirabayashi, 1989). This does not necessarily mean that lesson study research is weak in validity or generalizability. Recently, lesson study in Japanese mathematics education has received international recognition, and its potential to produce valid knowledge among teachers is pointed out (Stigler & Hiebert, 1999):

> The kind of knowledge produced by lesson-study groups is quite different from that traditionally produced by education researchers. .... Because lesson study is carried out in classrooms, the problem of applying the findings to classrooms disappears. The application is direct and obvious. But we cannot immediately know how generalizable lesson-study results are across different teachers, schools, and children.
>
> The fact that lessons linked with curricula are sharable, however, gives us a means of assessing the generality of findings. A first test concerns the ease with which lessons can be shared when they are passed through the filter of language. If teachers can describe lessons to other teachers in sufficient detail so that the other teachers can actually use the lessons, we can be fairly certain that both groups understand the essential characteristics of the lesson. Lessons that work across diverse groups of teachers can be said to generalize simply because they can be replicated. (p. 165)

Since a single lesson study is conducted in only a few classrooms, its results have very limited validity at first. But, when those results are shared among other teachers, and replicated in many other classrooms, they could increasingly obtain higher validity (generalizability). Therefore, a larger community of teachers, like JSME, has important responsibilities to share, replicate and accumulate valid lesson studies.

INFLUENCE FROM WESTERN COUNTRIES: SEARCHING FOR SCIENTIFIC METHODS

At the beginning of the modern era in Japan, the introduction to the modern Western educational system caused a lot of confusion because it was totally different from the system of the Edo era: Goals, contents, textbooks, teaching methods, teacher training, school management, and so on, all of them had to be newly constructed. The normal schools made intensive efforts in order to understand the Western system, to study how to adapt it to Japanese tradition, and to implement it all over the country. Many books of education were imported from Western countries, translated into Japanese, and published.

Also, Western educational research reports were discussed at teachers' meetings or journals. They sometime shook Japanese foundations of mathematics education. Soon after JSMSE was established, the society faced a big controversy on "formal discipline." At the beginning of the 20th century the doctrine of formal discipline was harshly criticized in the United States by educators and psychologists. Those criticisms were discussed in meetings of Japanese mathematics educators in 1920s. Since in those days the doctrine of formal discipline was taken for granted in teaching of mathematics in Japan, this caused a serious debate among mathematics educators and teachers (cf. Shiomi, 1967, p. 7). Not only their theoretical base but also their research methodology was seriously questioned:

> Without doubt, the methods of experiment in psychology and education are imperfect. Also, I am confident that the theory of statistical correlation is imperfect. ... However, before such methods were discovered, on what base did we make discussions? Except philosophical thoughts, common sense, and imagination, what, in the world, did we have? We should not laugh at educational experiment. No, I believe that we mathematics teachers should make efforts to improve its methods together with psychologists. (Ogura, 1924/1973, p. 94)

And, the experimental psychology came to play an important role in the development of mathematics education research (Nabeshima, 1933, pp. 64-65).

After World War II, influence of educational research of the United States became very distinct in many areas. Since in the United States the behaviorist psychology and statistic analysis played important roles in educational research (Campbell & Stanley, 1966; Howson, Keitel, & Kilpatrick, 1982), Japanese mathematics educators also came to use such research methodology (Takagi, 1964).

However, in Japan there appeared also many other kinds of study articles which belong to historical studies, philosophical studies, curriculum material studies, Piagetian clinical studies, classroom action research, and the like in mathematics education research (Sawada, 1963). Sawada warned there would be limitations in applying so-called "scientific research method" to educational phenomena: "What we should keep in mind here, however, is that we must not expect establishing of clear-cut causal laws like in natural sciences" (Sawada, 1963, p. 7). Hirabayashi (1989) emphasized the need of developing research methodologies designed for mathematics education:

> I think that the most scientific methods among the research methods of mathematics education in Japan are probably those from the experimental

> psychology. I do not mean that they are bad ones. Rather, young scholars should learn more from psychologists, and master their methods.
>
> Nevertheless, when we consider that the research results of so-called experimental psychology so far have not brought very big impacts on practice and research of arithmetic and mathematics education, we should not expect so great results from imitation of psychologists' methodology in the areas proper to mathematics education. (p. 434)

In 1980s research based on the behaviorist psychology and statistical analysis came to be harshly criticized in mathematics education in the United States, and constructivist research and qualitative research methodology came to rapidly receive wide recognition in mathematics education in the world. In Japan also since 1990s they have come to be valued in mathematics education (Sekiguchi, 1993; Ito, 1995).

Currently, in Japanese mathematics education research community, various issues have been discussed, and various methodologies have been employed (Okazaki, 2007; cf. Kelly & Lesh, 2000). For any research methodology, how to enhance the validity and relevance of research is a very important issue. For survey research and experimental research, there have been the standard methods to establish their validity under the positivist paradigm. On the other hand, newly proposed methodologies under a new research paradigm have limited ways to convince their validity. It is urgent to discuss the research methodology of mathematics education to enhance the quality of research.

NATIONAL SYLLABUS AND RESEARCH PRACTICE

Japanese education system has been controlled by the central government since the beginning of the modern era. The national syllabus for elementary education has been enforced since 1881, and for secondary education since 1902. They determine what to teach in each subject for each grade. During the period from 1903 to 1945 even textbooks (so-called, "Black covers") for elementary schools were authored by the government. Though after World War II school textbooks have been published by private companies, they have been inspected by the government whether they follow the national syllabus.

The national syllabus has overshadowed lesson study. Lesson study by public school teachers has played an important role to implement the national syllabus effectively. After World War II the national syllabus has been revised nearly once in a decade. Every time the national syllabus reform was addressed from the government, lesson study activities on reform ideas flourished all over the country because teachers were interested in how to implement a coming reform (Kato, 1974). As a result, lesson study often functioned as a tool of the government to enforce the national syllabus.

Because of the national syllabus, school curriculum is almost the same throughout the country. School teachers, therefore, are relatively easy to communicate and share their ideas in lesson study. When the national syllabus is reformed, huge number of lesson study projects on the same topic are conducted over the country, reported in local meetings of teachers, journals, books, websites, and so on, and shared among teachers. This then would contribute to improvement of not only teaching but also lesson study practice.

Mathematics education researchers at universities also have been affected by the national syllabus. They are often challenged by school teachers whether research conducted by researchers are useful to implement a mathematics reform curriculum in actual classrooms. Also, university researchers try to make those research proposals that support the implementation of the national syllabus because the government is considered to be interested in those studies and fund them.

On the other hand, the national syllabus is considered partly a product of lesson studies and professional research, too. For example, at the beginning of the modern era, the national syllabus was based on the curricula of schools attached to normal schools (Ogura & Nabeshima, 1957, p. 188). Those curricula were results of research and lesson study of normal schools. Furthermore, JSMSE, and later, JSME have worked closely with the Ministry of Education for reforming the national syllabus of mathematics.

CONCLUDING REMARKS

This chapter argued that mathematics education as research field in Japan has been formed from the lesson study tradition, with influence of mathematics education in the Western countries, under the guidance of the national syllabus. As indicated above, those three sources have been importantly related to each other, and directly or indirectly have affected Japanese research activities in mathematics education. Currently, Japanese mathematics educators have increasingly participated in international scenes of research. They should be aware of what are behind their research, and need to communicate it to researchers in other countries in order to facilitate mutual understanding.

REFERENCES

Campbell, D. T., & Stanley, J. C. (1966). *Experimental and quasi-experimental designs for research*. Chicago: Rand McNally.

Fernandez, C., & Yoshida, M. (2004). *Lesson study: A Japanese approach to improving mathematics teaching and learning*. Mahwah, NJ: Lawrence Erlbaum Associates.

Hirabayashi, I. (1989). Situations of mathematics education research: On TME program. *Japan Society of Mathematical Education: 22nd Presentation of Theses on Mathematics Education*, 431-436. (in Japanese)

Howson, G., Keitel, C., & Kilpatrick, J. (1982). *Curriculum development in mathematics*. Cambridge: Cambridge University Press.

Ito, K. (1995). On qualitative research method in mathematics education: Its assumptions and procedures. *Journal of Japan Society of Mathematical Education*, 77(9), 2-12. (in Japanese)

Japan Society of Mathematical Education (JSME) (2010). *Handbook of research in mathematics education*. Tokyo: Toyokan.

Japan Society of Mathematical Education. (n.d.). *JSME is active in*. Retrieved July 20, 2011, from http://www.sme.or.jp/e_index.html

Kato, K. (1974). What the practice and research for modernizing arithmetic education should be. *Journal of Japan Society of Mathematical Education, 56*(4), 46-49. (in Japanese)

Kelly, A. E., & Lesh, R. A. (Eds.) (2000). *Handbook of research design in mathematics and science education*. Mahwah, NJ: Lawrence Erlbaum Associates.

Nabeshima, S. (1933). On scientific research method of mathematics education. *Journal of Secondary Education Research Society, 2*(3), 64-71. (in Japanese)

Ogura, K. (1924/1973). Fundamental problems in mathematics education. *Ogura Kinnosuke Collected Works, Vol. 4* (pp. 3-160). Tokyo: Keiso Shobo. (in Japanese)

Ogura, K. & Nabeshima, S. (1957). *Modern history of mathematics education*. Tokyo: Dai Nippon Tosho.

Okazaki, M. (2007). The place and the tasks of a design experiment methodology in mathematics education: From a viewpoint of a balance between scientific status and practical usefulness. *Journal of Japan Academic Society of Mathematics Education: Research in Mathematics Education, 13*, 1-13. (in Japanese)

Sawada, N. (1963). What is scientific thinking? (Notes of the third Lecture on Promoting Basic and Scientific Research in Mathematics Education). *Journal of Japan Society of Mathematical Education, 45*(3), 29-34. (in Japanese)

Sekiguchi, Y. (1993). For ethnographic research in mathematics education, *Tsukuba Journal of Mathematics Education, 12A*, 1-9. (in Japanese)

Shiomi, K. (1967). A study on the nature and field of mathematics education research. *Journal of Japan Society of Mathematical Education: Reports of Mathematical Education, 14*, 1-9. (in Japanese)

Stigler, J. W., & Hiebert, J. (1999). *The teaching gap*. New York: The Free Press.

Takagi, K. (1964). Behavioral sciences and mathematics. *Journal of Japan Society of Mathematical Education, 46*(3), 2-9. (in Japanese)

CHAPTER 60

RESEARCH ON PROPORTIONAL REASONING IN THE JAPANESE CONTEXT

Keiko Hino
Utsunomiya University

ABSTRACT

Research on proportional reasoning has been conducted worldwide for years. In Japan, the teaching of ratios, rates, and proportional relationships as well as related concepts and skills has also been a subject of great interest in the community of mathematics educators. Therefore, this chapter synthesizes the body of research on proportional reasoning in Japan. More specifically, it summarizes the trends and results that have influenced ongoing research agendas, examines the change in curricular emphasis and its relationship to research on proportional reasoning, and focuses on the connection between research and classroom practices on ratios and proportional relationships. Then it summarizes and identifies five current research agendas. This chapter observed that, in Japan, research on proportional reasoning is based on the results of previous research that, being influenced by curricular emphasis, have been accumulated and tested against the classroom practices of numerous teachers for many years.

The First Sourcebook on Asian Research in Mathematics Education: China, Korea, Singapore, Japan, Malaysia, and India, pp. 1323–1352

Keywords: Proportional reasoning, ratios and proportions, Japanese National Course of Study, textbook, number lines

INTRODUCTION

According to Lesh, Post, & Behr (1988), "Proportional reasoning is a form of mathematical reasoning that involves a sense of co-variation and of multiple comparisons, and the ability to mentally store and process several pieces of information"(p. 93). They also state that proportional reasoning is highly concerned with inference and prediction and it involves both qualitative and quantitative methods of thought. Lamon (2005) refers to proportional reasoning as "the ability to scale up and down in appropriate situations and to supply justifications for assertions made about relationships in situations involving simple direct and inverse proportions" (p. 3).

Proportional reasoning is related to various mathematical concepts, skills or ways of thinking. Lesh et al. (1988) view proportional reasoning as a watershed concept that is both the capstone of elementary arithmetic and the cornerstone of higher level areas of mathematics. Consequently, it includes close connections with different areas of mathematics, such as fractions, long division, percentages, measurements, ratios, and rates, as well as algebraic functions and other higher level areas of mathematical learning. Lamon (2005) also recognizes the critical components of powerful proportional reasoning that are thought to develop in a web-like fashion (i.e., rational number interpretations, measurements, quantities and co-variation, relative thinking, unitizing, sharing and comparing, and reasoning up and down).

In this chapter, following the line of thought in previous studies, proportional reasoning refers to reasoning (or making assertions) based on the proportional relationships between two quantities that simultaneously vary. With respect to the related mathematical concepts or skills, this chapter concentrates on the areas of ratios and proportional relationships at the elementary school level.

When describing research in the Japanese context, several factors need to be considered. First, Japan includes a national curriculum standard (referred to as the "Course of Study"), which has been revised approximately after every 10 years since the end of World War II. To survey research agendas and classroom practices on proportional reasoning, their connections with the goals and emphases in the Course of Study cannot be neglected. Second, Japan has had a long history of focusing on the professional development of its school teachers through lesson studies. To understand the ongoing research agendas, it is important to focus on their accumulated findings in terms of teaching materials and methods regarding ratios and proportional relationships. Considering these factors, this chapter summarizes the research findings on proportional reasoning in Japan from three perspectives: 1) change in curricular emphasis and its relationship to research on proportional reasoning; 2) connection between research and classroom practices on ratios and proportional relationships; and 3) current research agendas.

Terminology concerning ratios, rates or proportions is complex and there have been continuous discussions on this particular subject (e.g., Lamon, 2007). In the Japanese context, researchers have been struggling with the appropriate translations of Japanese terms related to ratios, rates or proportions into English (e.g., Watanabe, 2010). Although the present author does not delve into the complexities of this process, it is important to clarify the meaning of several words to avoid confusion when reading this chapter.

In this chapter, "ratio" means the quantification of a multiplicative relationship between two numbers, magnitudes or quantities. In the ratio of A to B, A and B refer to both the quantities of the same kind (e.g., A (km):B (km)) and the quantities of different kinds (e.g., A (km):B (min)). The latter type of ratio is known as "rate." Since representations of ratios can vary, this chapter particularly focuses on mathematical expressions such as A:B, A/B (i.e., the value of ratio[1]), percentage, and per-unit. "Proportion" is an equation in which two ratios (a/b, c/d) are equal. In this sense, proportion is beyond the scope of this paper since it is a topic covered in lower secondary school. However, the adjective "proportional" is used to state the problem when a multiplicative relationship is involved, and the formation of two equal ratios occurs. For such a case, this paper employs the term "proportional relationship." In addition, the term "idea of proportionality" is used to mean the basis of creating methods for solving proportional problems.

CHANGE IN CURRICULAR EMPHASIS AND ITS RELATIONSHIP TO RESEARCH ON PROPORTIONAL REASONING

Goals and standards of the Course of Study have reflected the interests and needs of the Japanese society as well as its educational requirements. In the process of revising and shaping the Course of Study for more than 60 years, different emphases in mathematics education can be found, which is based on Japanese philosophy and on worldwide reforms and movements. In recent revisions, the results of both national and international surveys on students' performance and attitudes toward mathematics have had a strong influence on decision making.

Ratios and Proportional Relationships in the Progress of the Course of Study

Ratios have been an important curricular content in elementary schools since the Meiji era. However, the goals of teaching ratios have differed between pre- and post-World War II. For example, goals changed from children becoming well-versed in calculations to children developing ways of thinking that are important for understanding the functions of numbers, the meaning of multiplication/division, and quantitative relationships. This section briefly describes the change of curricular emphasis after 1958 and the treatment of ratios and proportional relationships in the elementary school curriculum (Shimizu, 2003; Nao, 1990, 1991). After 1958, the course of study fostering children's mathematical thinking has been emphasized as the goal of mathematics education (Nagasaki, 1990). Since then, principles and standards in teaching have been developed in conjunction with the curricular emphasis on mathematical thinking.

It was the 1958 Course of Study in which the strand of "quantitative relationship" was developed together with the three additional strands of "number and calculation," "quantity and measurement," and "geometrical figure." Furthermore, the quantitative relationship strand consisted of three sub-strands, one of which was *wariai*[2] in Japanese: a colloquial expression for the relative value of a compared quantity with the size of a base quantity which is concerned with the multiplicative relationships of quantities.[3] Ratio is a typical mathematical term that corresponds to *wariai*. In the teaching of mathematics, the word *wariai* has been used without a clear definition since the Meiji era. Thus, *wariai* was adopted as the name of the strand since emphasis was placed on fostering children's thinking in terms of ratios through the learning of different contents, such as numbers and calculations, measurements, and geometrical figures. *Wariai* was considered to play a principle role since it was thought of as a basic way of examining the relationship between quantities (Nao, 1990). The sub-strand *wariai* was prescribed from grades four to six. Table 60.1 presents the content included in the sub-strand of *wariai* in each grade level.

As shown in Table 60.1, the *wariai* sub-strand includes a variety of topics: meaning of numbers, meaning of multiplication/division, functional viewing and thinking, *buai*[4] and percentages, and ratios regarding quantities of different kinds. In the lower grade levels, matters that formed the basis of *wariai* were also prescribed in the strand of number and calculation, as well as quantity and measurement.

In the 1968 Course of Study, due to the modernization of mathematics education, mathematical thinking and the ideas of set, structure, and

Table 60.1. Mathematical Content in the *Wariai* Sub-Strand

Grade Level	*Content*
4	• To deepen the understanding of *wariai* between two quantities. • To understand the ways of calculating *wariai* in simple cases.
5	• To represent *wariai* between two quantities of the same kind, for understanding the use of natural numbers, decimal numbers or fractions, and to understand basic calculations concerning *wariai* (three usages of ratio[5]). • To understand the meaning of percentage and *buai*. • To represent *wariai* between two quantities of different kinds, to understand the use of the amount of one of the two quantities by fixing the amount of the other quantity, to use the idea of "per-unit quantity," and to foster the ability to use such ideas for examining the relationship between quantities.
6	• To deepen the understanding of the three usages of ratio and to develop the ability of utilizing them effectively. • To understand the three usages of ratio when the *wariai* is represented by percentage and *buai*. • To understand the meaning and the ways of representing ratios (A:B) and to foster the ability to use it for the purpose of representing the relationship between quantities. • To understand the idea of proportionality and to foster the ability of using it in simple cases.

Source: Monbusho (1958).

function were emphasized. Many topics included in the *wariai* sub-strand were moved to different strands, and as a result, the name *wariai* was changed to "function" by focusing on "functional viewing and thinking." In the fifth grade, the sub-strand function included making a distinction between the quantity that changed and the quantity that remained constant as well as examining the ways of change for the relationship expressed by mathematical representations such as $A + B = C$ or $A \times B = C$. In the sixth grade, the sub-strand function included the effective treatment of problems by focusing on the proportional relationship. The teaching of ratios on the quantities of different kinds was moved to the strand of quantity and measurement since the idea of creating a new quantity (intensive quantity), such as speed or population density, was stressed. In this case, the roles of functional thinking and proportional relationship for deducing the concept of speed, for example, were also made more explicit. The meaning of multiplication based on *wariai* ($B = A \times p$) had been emphasized since the 1958 Course of Study and at that time, figural representations, such as number lines, were incorporated for the purpose of making the idea of *wariai* more concrete and manipulative for children (see Figure 60.1). Here $A \times p$ (= B) was clarified as the repre-

A ribbon costs 30 yen for 1 meter.

(1) How much does the ribbon cost for 2 meters and 4 meters?

(2) How much does the ribbon cost for 1/5 meter and 3/5 meters?

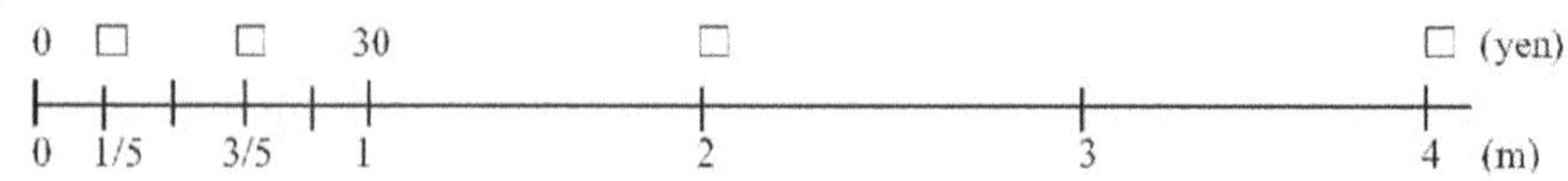

Source: Nakajima (1981, p. 77).

Figure 60.1. Meaning of multiplication on a number line.

sentation of the amount that corresponds to p when amount A is considered as 1. In other words, B is the corresponding quantity that includes the specified relative size p with respect to A.

In the 1977 Course of Study, in the effort to reduce the overall content toward basics, the teaching of ratios concentrated on grades five and six. At this time, percentages became the primary content and *wariai* was taught together with the formula (*wariai*) = (compared quantity) ÷ (base quantity). Moreover, the majority of the *buai* content was eliminated.

In the 1989 Course of Study, combined with the emphasis of cultivating children's attitudes toward utilizing mathematics in daily life situations, advantages of mathematical approaches to viewing and handling objects were concretely presented (Shimizu, 1989). It can be stated that mathematical ideas and thinking were emphasized in terms of their advantages, such as conciseness, clearness, accuracy, efficiency, consistency, or unification. In the teaching of ratios on quantities of different kinds, the advantages of using ratios were stated in terms of their conciseness or accuracy as indicators of population density or speed. Since proportional relationships can be easily applied in everyday life, such utility or applicability was stressed through problems that included measuring the weight of a thin piece of paper by examining the proportional relationship between the number of pieces of paper and their weight.

In the 1998 Course of Study, the term "mathematical activity" was included in the goal statement of mathematics education at the elementary school level. One reason for its inclusion was to direct classroom teaching to focus more explicitly on thinking and acting. Mathematical activity continues to be the focal point of curriculum and instruction in the latest Course of Study released in 2008. In this Course of Study, mathematical activity expands its area to include aspects of communication (e.g., explaining the reason logically by using mathematical signs and symbols) and applications of learned knowledge both in everyday life and in the further learning of mathematics (Monbukagakusho, 2008). It also became a more solid principle for organizing and teaching mathematical content at each grade level. For example, in the teaching of multiplication of fractions, children are encouraged to explain how to calculate the multiplication (based on their previous knowledge) by using number lines or mathematical expressions. In addition, finding quantities that have proportional relationships and solving problems by positively using such relationships are also considered as important mathematical activities in the classroom.

Approaches to Developing Proportional Reasoning

As previously described, the Japanese mathematics education community had been discussing the development of children's proportional reasoning based on the *wariai* concept. To focus on *wariai,* it is necessary to assume the proportional relationship between the two quantities in question. Children who have the ability to reason proportionally are those who can handle the idea of *wariai.* This means that they are able to determine the relationship between the two amounts by being aware of the distinction between relative size and absolute size. It also means that they understand ratios (A:B) and the value of ratios (A/B) as two sides of the same coin for representing *wariai*. Based on this understanding, they can use them flexibly and effectively according to the needs of the problem (Shimada & Nakajima, 1960, p. 119). Shimada and Nakajima also state that by using the idea of *wariai,* children can gain a deeper understanding of the proportional relationship between the two quantities and examine and manage the quantitative relationship in real-world situations.

Another related viewpoint for the development of children's proportional reasoning is functional thinking. Although the teaching and learning of functions is part of the lower secondary school curriculum, at the elementary school level, the idea of functions and functional thinking are considered as an important part of the curriculum. The main ideas of functions include: catching some quantity that is closely related to the quantity under examination, clarifying the relationship between the two quantities, and utilizing the relationship for examination (Monbusho, 1977; Nakajima, 1981). Thus, an inquiry into co-variation and correspondence between the two quantities is considered fundamental. Searching for appropriate variables by conjecturing the influence of the change of one quantity on the change of another is also an important activity for children. Proportional relationship is thought of as the most primitive and basic function, especially the aspect of co-variation (i.e., if one quantity increases two or three times as much as the original quantity, then the other quantity increases in the same manner). In this case, before introducing the term "proportional relationship" to children, the idea of proportionality (based on the aspect of co-variation) is incorporated into the teaching of different mathematical content such as multiplication/division of whole numbers, measurements, or formulas of areas. In other words, it is intended to familiarize children with the idea through the learning of different mathematical concepts and skills.

Still another viewpoint is the focus of proportional reasoning as a tool for investigating real-world problems. This is a rather new area of study, considering that the history of research on mathematical thinking continues to focus on the process of creation and extension within the world of mathematics. The results of Japanese students' low values toward mathematics, based on international assessments of mathematics (e.g., Mullis et al., 2000, 2004), have been spurring efforts in the area of solving real-world problems. Indeed, proportional relationship is a powerful mathematical model for the purpose of investigating such problems. Shimizu (1989) points out the advantage of mathematizing real-world problems through idealization and abstraction. In his example, he uses the problem of calculating the time for walking 10 kilometers by assuming that the constant rate of speed was "walking 4 km per hour" (p. 47). In lower secondary schools, the ability to make assumptions regarding the proportional relationship between two quantities and utilizing the results derived from such assumptions to tackle real-world problems has become the subject of focus by researchers, such as Nagata (2004) and Seino (2004).

RELATIONSHIP BETWEEN RESEARCH AND CLASSROOM PRACTICE

In Japan, there is the tradition of teachers conducting practice-based research, which contributes to the realization of a close relationship between research and classroom practice. In this case, teachers often form groups (that can include university researchers and mathematics educators) and conduct lesson study based on their research interests. They can also present the results of their investigations in both regional and national meetings as well as publish their findings in mathematics education journals and books. Periodicals have also been available for teachers in elementary and lower secondary schools. In addition to reports of classroom practices, such periodicals include special issues on curricular emphases and specific articles written by university researchers or superintendents on how to implement them in the classroom.

Based on these circumstances, the teaching of ratios and proportional relationships in the classroom has been proposed and discussed from diverse viewpoints. One topic that repeatedly appears is how to introduce the concepts of ratios and proportional relationships in the classroom. In general, the introductory phase of a specific concept is considered especially important since it includes a strong influence on both the understanding and motivation toward the subsequent learning content. Relative value or ratio is the idea that is created when two quantities are compared. However, there are two major methods of comparison: to compare the quantities additively and to compare them multiplicatively. In the teaching of relative value or ratio, varieties of problem situations that inspire children to compare different cases multiplicatively (not additively) have been proposed and studied. One such example included comparing basketball shots in which the number of shots were used as data (e.g., data of successful shots among the total number of shots) and the different ways of presenting the problem to the children (e.g., asking them to first work on simpler cases, such as the case in which the total number of shots are equal, and then asking them to focus on harder cases such as the case where both the total number of shots and the number of successful shots are different).

In the introductory teaching phase, open-ended problems (Becker & Shimada, 1997) have also been developed and implemented in the classroom. For example, in the introduction to proportional relationship in the sixth grade, different situations in which two quantities vary simulta-

Please find two quantities where if one quantity changes, then the other quantity also changes in everyday life:

- **When the elapsed time of filling water in a bathtub becomes longer, the depth of the water in the bathtub becomes…**
- **When the number of pages in a book that you are reading becomes greater, the number of remaining pages becomes...**
- **When the diameter of a circle becomes longer, the circumference of the circle becomes…**
- **When the length of daytime becomes longer, the length of nighttime becomes…**
- **When the number of the same type of nails becomes greater, the weight of all of the nails becomes…**

- **Let the birthdays of two siblings be the same. When the age of the younger brother increases, the age of the older sister becomes…**
- **In a rectangle whose area is 18 cm^2, when the length of the vertical side becomes longer, the length of the horizontal sides becomes…**

Which are the cases where if one quantity increases, then the other quantity decreases?
Which are the cases where if one quantity increases, then the other quantity also increases

Source: Sugiyama, Iidaka, and Ito (2005).

Figure 60.2. Different situations in the introduction of proportional relationships in the sixth grade.

Problem: We are going on a school trip. We need to decide on the number of students who will share a hotel room. Out of the three rooms (A, B, and C), which room is the most crowded?

Room	A	B	C
Area (m^2)	10	12	12
Number of Students	7	7	9

Students first compare B and C and realize that C is more crowded than B since C includes more students than B in the area of 12 m^2. Students must decide on which room is more crowded between A and C in which neither the areas nor the numbers of students are equal. Four solutions are anticipated:

Solution a: Making the areas of A and C equal (60) and comparing the numbers of students that correspond to 60 m^2: 42 students in A and 45 students in C. Therefore, C is most crowded.

Solution b: Making the areas of A and C equal (1) and comparing numbers of students that correspond to 1 m^2: 0.7 students in A and 0.75 students in C. Therefore, C is most crowded.

Solution c: Making the number of students in A and C equal (63) and comparing the areas that correspond to 63 students: 90 m^2 in A and 84 m^2 in C. Therefore, C is most crowded.

Solution d: Making the number of students in A and C equal (1) and comparing the areas that correspond to 1 student: 1.43 m^2 in A and 1.33 m^2 in C. Therefore, C is most crowded.

After the students present these solutions, they discuss which method can be used in a more general and convenient manner. In class, they examine the four solutions by solving the problem shown below. Finally, they realize that it is easier to deal with different problems by making areas (or the number of students) a "per-unit quantity (1)."

Room	A	B	C
Area (m^2)	10	15	25
Number of Students	15	20	30

Source: Koto et al. (1998, p. 155).

Figure 60.3. A planned lesson on the introduction to population density in 5 grade.

neously are often provided to children. The children are also asked to compare/contrast similarities and differences among the situations as well as classify the situations according to the increase/decrease of the quantities. Based on this classification, the children are further led to study features of proportional relationships in subsequent lessons. Figure 60.2 presents typical situations and questions applied in an introduction to proportional relationships.

Methods of nurturing students' strategies toward learning a specific lesson concept have also been a major subject of focus in research. For example, in the introduction to ratios on different quantities (population density), children's different ways of comparing crowdedness in the two rooms are anticipated, as shown in Solutions a to d in Figure 60.3 (Koto and elementary school mathematics study group in Niigata, 1998). They also planned the lesson and studied how to nurture children's various approaches toward the concept of population density (Solutions b and d).

Recently, there has been considerable focus on the curricular emphasis of communication skills, the ability to utilize mathematics in everyday life, and the further learning of mathematics. In addition, studies on how to prepare verbal and written communication activities for children have been conducted. With respect to ratios and the multiplication/division of numbers, the number line representation has attracted the attention of teachers as an effective tool of communication (e.g., Society of Elementary Mathematics Education, 2011). In the teaching of proportional relationships in the classroom, examples of utilizing the idea of such relationships for societal and environmental issues, such as waste treatment problems or the rise in the sea level have been developed and studied.

The determination of the research focus by a group of schoolteachers often reflects strongly on the curricular emphasis of the time. However, as the present emphasis of curriculum on mathematical activity indicates, their informed efforts and trials also influence the direction of curriculum and research in Japan. Furthermore, these informed efforts and trials as well as their accumulated results have also been published in textbooks, which are used in elementary schools.

CURRENT RESEARCH AGENDAS

As outlined in the previous two sections, research on proportional reasoning in Japan has been actively conducted from the perspective of mathematical thinking (e.g., its fundamental role as mathematical thinking and ways of fostering such reasoning through classroom teaching). Currently, researchers have extended their focus to mathematical activities and the ability to use mathematics in everyday life. In this section, five research agendas are described, together with the perspectives and findings of related studies.

Studying the Curricular Sequence of Ratios and Proportional Relationships

Studying the sequence of mathematical content in the curriculum with respect to ratios and proportional relationships is one of the latest

Table 60.2. Contents on Ratio and Proportion and Grade Levels Taught in Elementary School

	1989	*1998*	*2008*
Wariai and percentage	5	5	5
Ratio (A:B)	6	6	6
Per-unit quantity	5	6	5 and 6
Proportional relationship	6	6	5 and 6
(Multiplication/division of fractions)	(6)	(6)	(6)
(Enlarged and reduced drawings)	(6)	(8)	(6)
(Introduction to literal symbols)	(5)	(7)	(6)

Note: Topics with parentheses are those that have close relationship with ratios and proportional relationship.

research agendas. Table 60.2 presents the content sequence related to ratios and proportional relationships in the 1989, 1998, and 2008 Courses of Study.

Since the revision of the Course of Study in 1958, the curricular sequence on ratios and proportional relationships has been "*wariai* and percentage" in grade 5, "ratios (A:B)" in grade 6, and "proportional relationships" in grade 6. However, there are variations concerning the time of teaching per-unit quantity (ratio of quantities of different kinds). Moreover, since ratios and proportional relationships are related to different mathematical content, the change of such content influences their teaching method. In this case, the teaching of "enlarged and reduced drawings" is one example, whereas the introduction to literal symbols in elementary school is another example in which an influence is found in the treatment of mathematical expressions, while teaching proportional relationships.

Since the curricular sequence is determined by the Course of Study, it is often regarded as the norm by the majority of teachers. Conversely, some researchers and teachers have proposed a more suitable sequence of teaching these topics. The order of teaching the ratio of quantities of the same kind and different kinds was discussed by Tabata (2002a). He described the expected differences among children's strategies when the ratio on the quantities of the same kind was taught first and when the ratio on the quantities of different kinds was taught first. He further insisted that the ratio on the quantities of different kinds should be taught first. Similarly, within the same grade level, the order of the teaching content raises certain issues, especially in the sixth grade where the mathematical content is congested with ratio-related content such as multiplication/division with fractions, ratios (A:B and A/B), enlarged/reduced figures, and proportional relationships. Thus, more research is necessary regarding the sequence of these topics from the perspective of developing proportional reasoning.

Connection Between Ratio-Related Mathematical Content in Elementary and Lower Secondary Education

In relation to the first issue, the connection between the elementary school curriculum and the lower secondary school curriculum is an important issue (e.g., Morozumi, 2002; Sasaki, 2003; Nagasaki, 2005). In Japan, the teaching of ratios is included in the elementary school curriculum and as a result, it is not included in the lower secondary school curriculum. This has raised problems due to students' insufficient understanding of the concept of ratio. According to the Trends in International Mathematics and Science Study (TIMSS), Japanese eighth graders received low scores on the items that required them to use percentages. Based on the results of large-scale international studies (such as the TIMSS and the Program for International Student Assessment (PISA)), Nagasaki (2005) proposes several recommendations. Among them, he strongly recommends the need for including the teaching of ratios as well as the multiplication/division of decimals and fractions not only in the elementary school curriculum but also in the lower secondary school curriculum.

Conversely, the subject of proportional relationships has been taught in a spiral method both in elementary and lower secondary education. In the latest Course of Study, the topic of proportional relationships is taught from the fifth through the seventh grades. In the fifth grade, children are introduced to proportional relationships in simple cases, e.g., "the relationship between the width and the area of a rectangle when the length is fixed at six centimeters." In the sixth grade, together with the introduction to literal symbols, equations of proportional relationships are formulated (e.g., as $x \times$ (fixed number) $= y$) and its graphical representations are introduced. Inverse proportional relationships are also introduced for the purpose of deepening the understanding of proportional relationships by comparing the similarities and differences between the two. In the seventh grade, proportional relationships are defined in the form of equations such as $y = ax$. Moreover, the extension of domains from positive to negative and the introduction to linear graphs on the coordinate plane are also presented.

In this development of content, one of the matters in question is how to facilitate children's shift of focus from within ratios to between ratios (Nunokawa, 2010, Hino, 2011). In this case, ratios are classified as "within" or "internal" if their constituent magnitudes share the same measured space, and as "between" or "external" if they are composed of magnitudes from different measured spaces (e.g., Freudenthal, 1983). In elementary school the definition of proportional relationships is based on the within-ratio concept, whereas in lower secondary school, such a definition is based on the between ratio concept. Accordingly, a shift in the dominant pattern of proportional reasoning, which is required for students to learn about proportional relationships, is enabled. The nature of this shift should be from the one that children predominantly focus on the within-ratio type to a more hybrid one in which children can distinguish the within ratios from the between ratios (Nunokawa, 2010).

Another matter in question is how to foster children's abilities to symbolize proportional relationships. Children are introduced to proportional relationships, based on tables, in the fifth grade. Subsequently, equations with algebraic letters become their major symbolic means for exploring proportional relationships. Moreover, in the seventh grade, students deal with the connections among equations, graphs, and tables. The impact of these symbolic means on their development of proportional reasoning needs to be studied to derive more information on the teaching of ratios and proportional relationships.

In the studies by Ohtani and Nakamura (2002, 2004), and Ohtani, Kanno, & Nakamura (2002), three members (a university researcher, an elementary school teacher, and a lower secondary school teacher) were asked to realize the smooth transition from elementary to lower secondary school by using the framework of Realistic Mathematics Education (RME) theory. More specifically, they applied four levels for progressing in the level of thinking: a situation that includes a sense of reality (Level 1); a model underlying students' informal procedures (Level 2); the model itself as the target and tool for inference (Level 3); and formal mathematical knowledge (Level 4). Ohtani & Nakamura (2002, 2004) also present the results of their teaching experiment in elementary schools (see also Ohtani, 2005) in which they formulated four attainment levels of symbolizing proportional relationships by tables, equations, and graphs, and developed a teaching unit for sixth graders that consisted of five sub-units (A to E) and a total of 12 class-hour lessons (① - ⑫) (Figure 60.4).

As a result of implementing the lessons, they point out that it was effective to pose problems and organize discourses in which two adjacent stages intermingle. In the lessons, although the teacher posed concrete problems, she asked questions regarding the mechanism, rather than simply asking questions for the answer. To lead children from fragmented information found in the table toward a more generalized property, the teacher also utilized signs, such as arrows, and the children's created words. Concerning the equations of proportional relationships, they were intended to lead children from simply viewing them as formulas for calcu-

Subunit	Class Hours	Topics Covered
A	①②	Motivation
B	③④⑤⑥	Symbolizing table
C	⑦⑧⑨	Symbolizing graph
D	⑩⑪	Symbolizing formula
E	⑫	Summarize

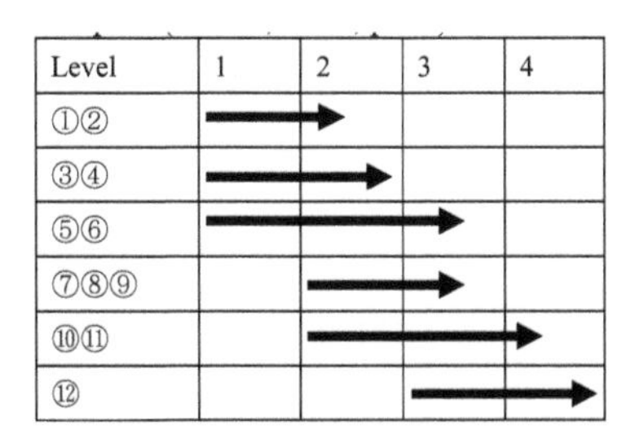

Source: Ohtani (2005, p. 37).

Figure 60.4. Teaching unit plan and attainment levels in the unit plan.

lations to seeing them as a form that commonly expresses a variety of proportional situations. They also argue that these points should be implemented in the teaching of proportional relationships at the elementary school level for the purpose of creating a smooth transition into the lower secondary school level.

Development of Teaching Materials on Multiplication/Division and Ratios

As described in the previous sections, understanding ratios and proportional reasoning were almost identical. Nevertheless, some researchers have focused explicit attention on the connection between the two. Tabata (2002b) proposes students' implicit proportional reasoning in the teaching of multiplication/division and ratios. In problems that can be solved by using multiplication/division, the proportional relationship between the two quantities in the situation is assumed. His proposal includes teaching children how to surface the proportional relationship to solve the problem.

In doing so, Tabata (2003a) employs a number line to visualize proportional reasoning for determining the answers for multiplication/division problems (see also Nakamura, 1996). In Japanese textbooks, linear visual representations, such as tape diagrams, number lines, and double number lines, play a central role in supporting children's mathematical thinking (Watanabe, Takahashi, & Yoshida, 2010). Tabata's proposal stresses these roles in teaching and learning with focused and coherent curricula. As the three problems and solutions in Figure 60.5 show, multiplication/division of whole numbers can be visualized in a unified fashion by using a number line. Thus, a number line can be used across the content of multiplications/divisions with whole numbers, decimal numbers, and fractions. Tabata (1989) also points out that in the statement of the problem,

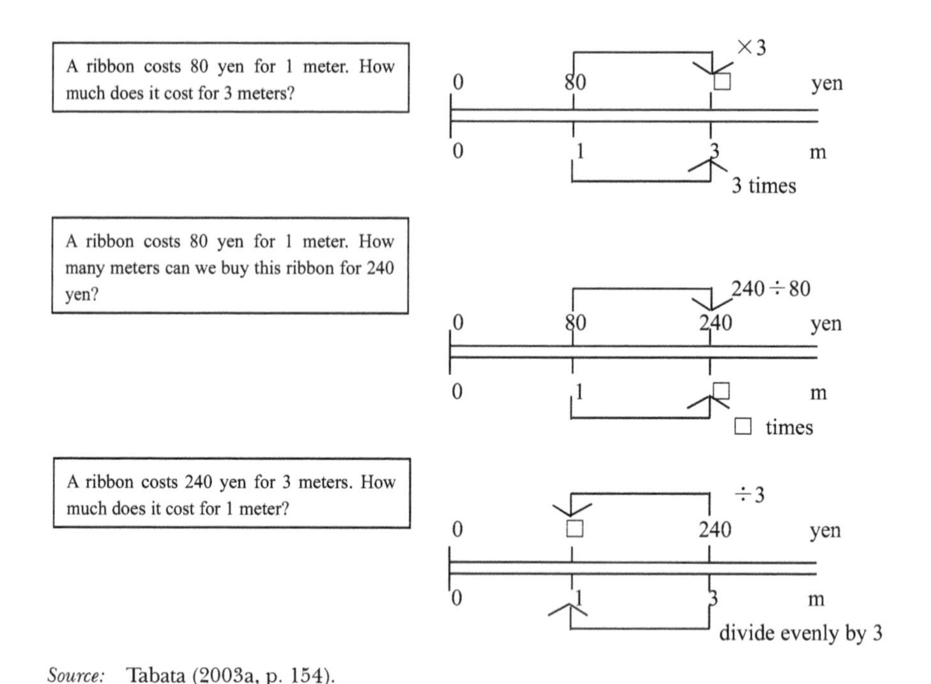

Source: Tabata (2003a, p. 154).

Figure 60.5. Number line representation of three types of multiplication/division of whole numbers.

not unit rates (e.g., 100 yen for 1 meter) but compound units composed of two non-unit quantities (e.g., 400 yen for 3 meters) facilitated children's proportional reasoning. The reason being that if "1" is stated in the problem, then they simply calculate the answer by using multiplication without paying attention to the proportional relationship (see also Nakamura, 2011).

Multiplication/division of fractions is one of the areas in which children face difficulty in determining the most suitable approach to find the answer. Therefore, this is one of the opportunities to show children the effectiveness of number line representation together with the use of proportional reasoning (see Figure 60.6). Moreover, the answer can be obtained in different ways by using previously learned knowledge regarding the division of decimal numbers and multiplication of fractions. For example, by comparing Solutions A and B, it is possible to obtain $3/4 \div 2/5 = 3/4 \times 5/2$.

As shown in Figures 60.5 and 60.6, the common procedure between multiplication/division is to read scales on the two number lines that are

A wire weighs 3/4 gram in 2/5 meter. How much does it weigh in 1 meter?

Solution A:

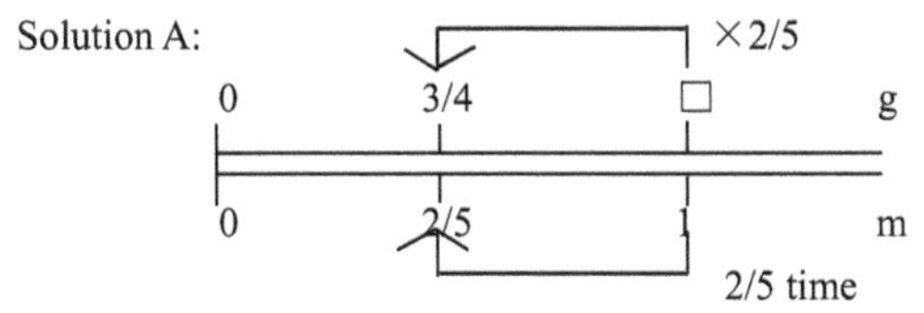

Since $1 \times 2/5 = 2/5$, $\square \times 2/5 = 3/4$.
Therefore, from the relationship between multiplication and division, $\square = 3/4 \div 2/5$.
$\square = 0.75 \div 0.4 = 75 \div 40 = 75/40 = 15/8$ (g).

Solution B:

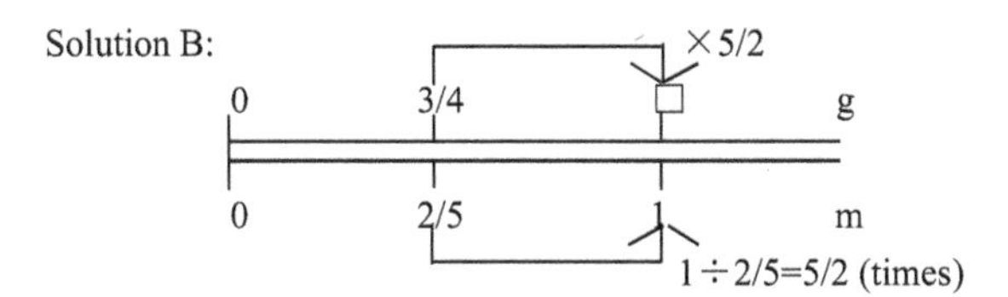

$1 \div 2/5 = 1 \div 0.5 = 2.5 = 5/2$. So 1 meter is 5/2 times of 2/5
By using proportional relationship between the weight and the length, $\square = 3/4 \times 5/2 = 15/8$.

Source: Tabata (2003a, p. 158).

Figure 60.6. Number line representations regarding the division of fraction (3/4÷2/5).

proportional to one another. What children only need is to read values on the number lines by taking into account the correspondence between the two values. By using it across different multiplication/division situations, children can identify where to apply proportional reasoning, determine the number of times, and make the necessary decisions (multiplication/division) to find the answers by considering the direction of the arrows. This repeated experience will help develop proportional reasoning in children. Tabata (2003a) points out that, although the number line has been recognized as an effective way of grasping quantitative relationships and determining the reasons for appropriate arithmetical operations, the number line model also provides children with the opportunity to focus their attention on the mathematical structure of the situations.

Tsuchiya (2002) and Tabata (2003b) also propose utilizing children's proportional reasoning in the teaching of ratios regarding quantities of the same kind. For example, they propose that children make equivalent pairs of numbers that represent a certain ratio. In the lesson by Tsuchiya, the fifth graders compared the ratio of the weight of paper garbage to the entire weight of garbage in three schools. To make the comparison easier, they were allowed to make the entire weight of the garbage from the three schools the same. In this case, the children scaled up and down the weights of the paper garbage and the entire garbage, and developed a table. After the activity, they discussed what number that represented the whole was the easiest one to deal with based on a variety of cases between the "part" (paper garbage) and the "whole" (entire garbage). The children favored 10, 5, and 1, and concluded that 1 would be the easiest.

Yoshida and Kawano (2003) and Yamaguchi (2007) investigate the effect of teaching ratios regarding quantities of the same kind based on children's informal idea of ratios and developed "a model of *wariai*" (see Figure 60.7) using concrete materials. One advantage of this model was that it represented *wariai* (or percent) even if it was greater than 1 (100%). In the model, the "part" could be extended outside the "whole." They also found that this generality contributed to children's understanding of the relationship between the part (compared quantity) and the whole (base quantity).

Based on this model, Yamaguchi (2007) developed "a *wariai* meter" (see Figure 60.8) in which the shaded part could move freely, as the arrow

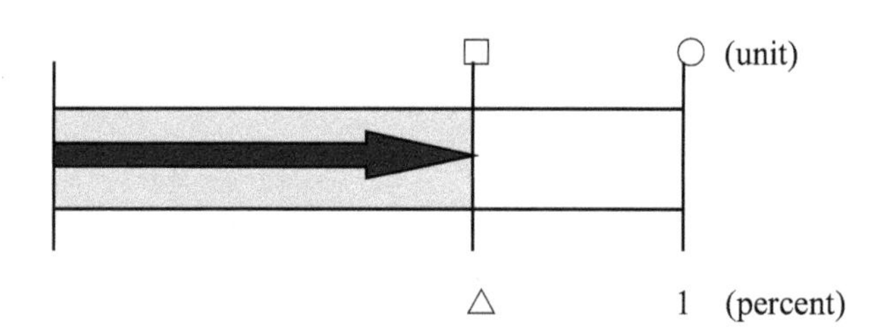

Source: Yoshida and Kawano (2003).

Figure 60.7. A model of *wariai*

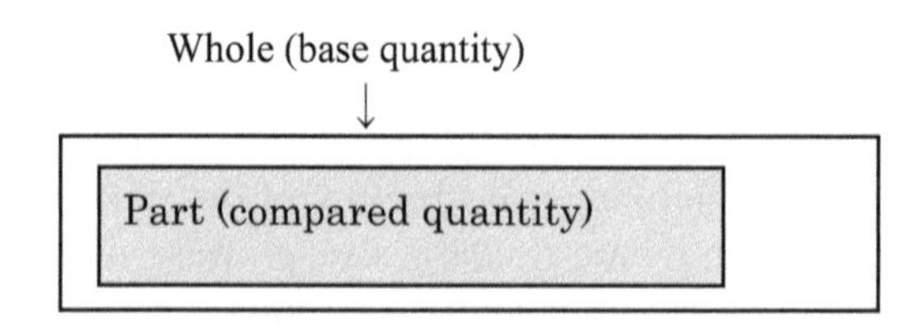

Source: Yamaguchi (2007).

Figure 60.8. A *wariai* meter.

shows in Figure 60.7. In this way, it can dynamically represent the way that the part gradually extends to fill up the whole. In his experiment, he showed the children a cylinder and then asked them to pour water up to the 50% level. Next, he asked them to compare the scores (e.g., 80 versus 40 or 100 versus 50) by expressing the two in the *wariai* meter. Finally, he continued using the *wariai* meter in the teaching of percentages and the three usages of ratio.

These aforementioned studies can also be thought of as attempts to utilize children's proportional reasoning in the teaching of ratios. By using the visual representations of *wariai,* the researchers employed children's senses and imagery of part–whole ratio relationship and proportional relationship together with the related everyday expression. However, additional studies are necessary in order to develop effective teaching materials that can promote children's understanding of ratios through their proportional reasoning abilities.

Proportional Reasoning From the Perspective of Mathematical Literacy

The ability to proportionally reason from the perspective of mathematical modeling and mathematical literacy is an important issue for future research. Such abilities include distinguishing between the situations in which assuming proportional relationships can be considered as meaningful or not. They also include the ability to interpret the results obtained by assuming proportional relationships appropriately and critically. Children are also encouraged to use tools such as diagrams or number lines in their reasoning processes. Studies regarding number lines in the teaching/learning of ratios and multiplication/division of numbers, especially on the utility of number lines as tools for thinking and communication, are expected to extend their focus into mathematical literacy.

As pointed out in the previous sections, the importance of "assuming" or "hypothesizing" the proportional relationship between two quantities in complicated real-world situations has been noted. However, in elementary schools, studies and practices that aim at proportional reasoning (as a powerful tool of reading information and applying it to make action plans) are limited. As one exception, Ikeda (2007) suggests the importance of clarifying the assumptions in solving the problems and gradually fostering this ability even in elementary school. He proposes three ways of incorporating the activity: (i) Make an assumption to simplify the problem and utilize this assumption when interpreting the result; (ii) Make a decision by clarifying the boundary within which the assumption is granted; and (iii) Make and test the assumptions. His examples contain problems that require multiplication/division and the knowledge of ratios. It is expected that the certain aspects of mathematical modeling should be incorporated into the teaching of ratios and proportional relationships at the elementary school level.

Oda (2011) proposes a lesson on per-unit quantity that fosters children's mathematical literacy. She considers mathematical literacy in elementary school as the ability to actively connect mathematics children learn in school to their daily and social life situations. While doing so, children encounter values and they realize that the ways of connecting and using mathematics in these situations can differ according to certain values and priorities. She also expresses the importance of their individual opinions and the ability to share such opinions. In her lesson, information regarding three types of potato chips was offered, which included price, weight, calories, and weights of the different ingredients. Children were required to read this information, use their knowledge of per-unit quantity, compare the three types of potato chips, and make a decision on which one they would most likely purchase. As a result, some children compared the prices per gram (since they wanted to buy the cheapest one), whereas the others compared either the calories per 100 grams or the fat per 100 grams. By listening to the solutions of their peers, the children realized that different priorities had an influence on the final decisions.

Clinical Approach to Children's Learning Processes in the Classroom

The fifth agenda is to investigate children's processes of learning in the classroom in which the teacher introduces concepts and skills related to ratios and proportional relationships. By focusing on the quality of children's thinking during the lessons, this approach attempts to improve classroom teaching from the perspective of the learner. In Japan, teaching materials and methods have been developed through the *Kyozai-Kenkyu* activity by teachers.[6] In this case, it is essential to examine how children actually experience teaching materials during classroom activities and what they are learning from such activities. However, interpreting the information of children's mathematical activities and modifying the lessons can be difficult since it is not always apparent what children are actually learning during the lesson. Thus, in the past two decades, together with the introduction to the qualitative research methodology in the community of Japanese mathematics educators, some researchers have investigated individual children's actual processes of learning during the lessons.

Nunokawa (2005) proposes a clinical approach to learning processes in the classroom based on the studies in which the learning processes and progressions of a small number of children are examined. It illustrates what is taking place in the midst of their learning, how the classroom event is interpreted and built on by the children, in what way the children are grasping the mathematical content in different learning moments, and how their understanding is changing as the lessons progress (p. 29). Nunokawa and graduate students at the Joetsu University of Education have been actively researching children's activity and learning in the lessons on multiplication/division and ratios. They selected several children and collected detailed data by using a camera for each child throughout the lessons. In addition, their results were based on the interpretations of the children's written worksheets and verbal communications with their friends as well as the teacher. The results show that children's proportional reasoning is actually used to acquire the knowledge and skill in multiplication/division and ratios.

First, the findings of these studies repeatedly show that children's activities on unitizing and norming the quantities vary and that they account for the progressions that the children make. For example, Takahashi (2000) analyzes children's creative activity based on their acquired knowledge of multiplication in the lessons regarding the multiplication of decimal numbers. Before introducing the multiplication of decimal numbers, the fifth graders were engaged in the activity of connecting small pieces of tape to form a longer one. Subsequently, the children were asked to solve the problem by applying their knowledge of the multiplication of whole numbers, repeated addition, and proportionality as well as making a comparison between the two solutions. Takahashi compares the learning processes of two children during the series of lessons. Based on the analysis, he argues that the children became aware of the proportional relationship between the two quantities through the tape diagrams (in which they formed compound units) and reviewed the units from the perspective of part–whole relationship to effectively manage the problem situation.

At the beginning of the lesson, the two children simply connected the tape and repeatedly added the price. However, one child, who later changed his reasoning from repeated addition to multiplication, regarded six meters as one unit and became aware of the part–whole multiplicative relationship. In the problem of multiplication by decimal numbers, this child also regarded 0.1 meter as one unit and utilized a multiplicative relationship. The child's solutions and reasoning are presented in Figure 60.9.

Forming and using units have been pointed out as an important mechanism for the development of increasingly sophisticated multiplicative solutions (e.g., Steffe, von Glasersfeld, Richards, & Cobb, 1983). Those who are adept at building and using compound units when the context suggests its efficiency is also considered as a salient feature of proportional reasoning (e.g., Lamon, 1993, 2005, 2007). However, it has been

Problem 1: A particular tape whose price is 240 yen for 6 meters. How much is the price for 30 meters?

A child's behavior:

Her first reaction to this problem is 240×30=7200. The teacher distributed pieces of the tape to each child and led her to line the tape. She lined the pieces of the tape together by saying, "6m, 12m, 18m, ..." Then, she counted twice by mumbling, "1, 2, 3, 4, 5." After a little while, she developed 36÷6 = 5, and 5×240=1200.

Her diagram in the worksheet:

Her reasoning in the interview:

"If the entire length is 30 meters, and since the length is 5 times, if I do 5 times 240 yen, then it becomes the whole."

Problem 2: A ribbon costs 180yen for 1 meter. *I bought the ribbon for 2.8 meters. How much is it?"*

The child's diagram and explanation in the worksheet:

(English translation of Japanese text: First calculate 180×2 and find the answer. Then, add it to the answer of the calculation 18×8.)

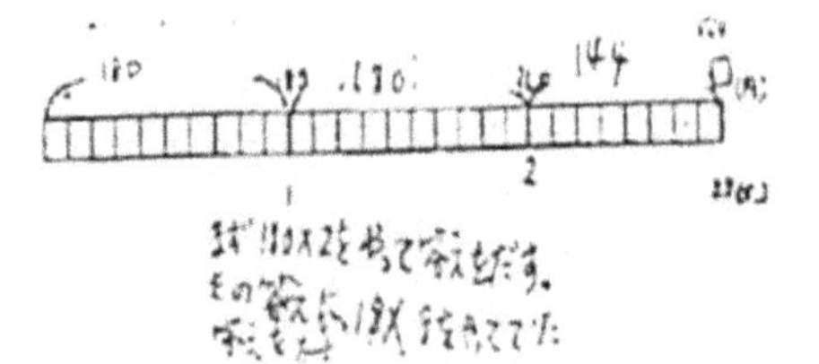

Her reasoning in the interview:

"The reason of 18×8 is that 0.8 is difficult to calculate, so I did 18×8." "The reason of multiplication is since is 0.1 meter, I regard this 0.1 meter as 1 meter, and there are 8 of them, so I thought I should use multiplication."

Source: Takahashi (2000).

Figure 60.9. One child's solutions and reasoning for two problems in the lessons.

observed that Japanese children have the tendency to directly rely on multiplication/ division by using the numbers given in the problem rather than intentionally constructing units (Hino, 1996). Furthermore, Hino (2008) investigates fifth graders' proportional reasoning over a one-year period during which they were taught the multiplication/division of decimal numbers and ratios. Based on the results, she argues that Japanese children only applied the knowledge and procedures that they were taught in class, rather than developing a flexible use of proportional reasoning strategies. Takahashi's study above suggests that children can use and improve their proportional reasoning in multiplication/division lessons if the activity provides children with the opportunity to elicit their ability to build and use compound units.

Nunokawa (2006, 2007) examines the unitizing and norming processes of children in grades 3 and 4 and shows that children can develop the ability even in lower grade levels. To foster their awareness of proportional relationships, he claims that it is important to construct sub-units from the original unit, to interpret the sub-units they constructed, and to use the sub-units as information to solve the problem. He also states that there are two directions of mathematical activity toward fostering their awareness of proportional relationships in the lessons. One is to construct various sub-units from the original unit by making use of halving, whereas the other is to express the process of construction (by using pictures or diagrams) and to clarify the relationship between the units being constructed.

Second, the results of studies based on the clinical approach point out the importance of the interaction between children and their inscriptions. Nunokawa (2007) states that children's diagrams need to be varied beyond formal number-line representations and it is essential that children interact with the diagrams without losing their own interpretation of the problem situation. Hino (2002a) observed the behaviors of five 4thgraders in a series of lessons on area and points out that their use of multiplication changed as the lesson evolved. Such change was characterized conceptually as "using multiplication as a label," "using it positively to approach problems which have not been solved before," and "using it effectively to achieve the goal of measuring areas." These three phases show how the children's understanding of multiplication in the context of measuring areas changes from a secondary accompaniment into a powerful tool of thinking.

These three phases were also studied in a series of lessons on per-unit quantity (density and speed) (Hino, 2002b, 2003). Based on the results of case studies of five 5thgraders, she states that children also change the meaning that they attach to mathematical notations. She also indicates that children differ in the process of change. For example, some children go from "label use" to "positive use," whereas some children go from "positive use" to "effective use." Moreover, children differ in their attractions to certain notations. In other words, there seems to be certain notations for each child that especially contribute to their reasoning processes. Such notations are not always the ones taught in the lessons. For example, one child, Kawaguchi, was attracted by the number-line notation taught by the teacher, which resulted in her changing the use of the notation. Conversely, another child, Himukai, developed the notation "🯅 = 1 m^2" in the early part of the lessons and extended its use in different problems that he faced in the subsequent lessons. Similar to Kawaguchi, he positively interacted with his "○ = Δ" notation in an attempt to determine the relationship between the two quantities. While doing so, he repeatedly engaged in the process of constructing compound units by considering the suitableness of the units to approach the problems

A clinical approach to children's learning processes in classroom in general, and a clinical approach to children's development of proportional reasoning in the teaching of ratio-related content in particular, are expected to provide detailed information on the ways that children's proportional reasoning actually functions in their acquisition of knowledge and skills. As Nunokawa (2005) highlights, such perspective will offer rich information to support children's understanding of ratios and proportional relationships since its focus is to understanding the "change" in the child. Furthermore, it offers detailed information regarding, e.g., "how the understanding and meaning of ratio is changing," "what problems children approached have the effect of change," and "what interactions and feedbacks trigger the change." However, additional studies are necessary in this field and while doing so; methods of data collection and analyses need to be developed to enable productive collaboration between school teachers and children.

IMPLICATIONS AND SUGGESTIONS FOR FUTURE RESEARCH

Based on the research trends described earlier, it is possible to make several suggestions for future research. First, it is important to make more connections among different research findings and utilize the network of knowledge to propose informed classroom practices. As the preceding sections show, current research agendas are spread over different aspects, including teaching materials, ways of teaching, and perspectives of learners. Thus, a healthy merger of these accumulated findings in various studies should be one of the major research tasks in the future.

Second, the ability to reason proportionally needs to be investigated in more diverse populations. Since ratios and proportional relationships is mainly taught in grades 5–7 in Japan, the research only concentrates on children in these grade levels. Research that examines early childhood or lower grades in elementary school as well as proportional reasoning in adults is limited. Thus, since it has been determined that proportional reasoning is one of the key operations on the formal level of thought, it is important to focus on the essence of proportional reasoning and how to assist such reasoning.

Third, the roles of teachers should be more focused on the teaching of ratio-related content in the classroom, especially since the teachers are the ones who actually provide learning opportunities for children by using certain teaching materials. Hence, it is important to know more about teachers' understandings and conceptions on proportional reasoning and how to assist them in realizing classrooms in which children actively develop their proportional reasoning abilities.

Finally, it should be noted that, viewed from the international research on proportional reasoning, there are areas that have received less attention from the community of mathematics educators in Japan. One is the problem of learners' overuse of linearity (e.g., Van Dooren, et al. 2009) or the metacognitive aspect of proportional reasoning (Modestou & Gagatsis, 2010). Another is the studies from the aspects of language (e.g., Moschkovich, 1998; Davis, 2007). Along with the development of proportional reasoning, how does the language (and thinking) of learners become more detailed? Furthermore, in the classroom, in what way do teachers and children manage both formal and informal expressions of ratios and proportionality?

In the research on ratios and proportional relationships in Japan, proportional reasoning has been considered as a powerful "idea" of mathematics. However, in future research, more attention must be paid to the aspect of "reasoning," which enables us to predict, decide, and communicate in a variety of problem situations.

NOTES

1. The value of ratio A:B is $A \div B = A/B$.
2. The Chinese character of *wariai* is "割合."
3. *Wariai* is an umbrella term that covers a variety of multiplicative relationships. For instance, it refers to both the ratio regarding quantities of the same kind and quantities of different kinds, both the relationship itself and to a number (quotient), and both a comparison between two specific values (the relationship in which there is more local property of the referent entities) or the equivalence of ratios (the relationship in which the quantities as a whole and as a referent are considered as homogeneous).

   Currently, in the textbooks for elementary school mathematics, *wariai* is defined as a number p that expresses the number of times when quantity A is compared with quantity B, where A and B are the same kind of quantities. Namely, in the relationship $A \div B = p$, B is called the "base quantity" and A is referred to as the "compared quantity." This definition of *wariai* has been shared since the 1977 Course of Study (Nao, 1991).
4. To represent *wariai,* numbers such as decimals and fractions are used. In addition, percentages are often used to represent the relative value of a compared quantity. For example, A uses whole numbers as much as possible by setting the size of base quantity B at 100. *Buai* [the ratio based on 10

percent] is another way of representing *wariai* in which the size of the base quantity B is considered as 10.

5. If B is the base quantity, then A is the compared quantity, and p is the *wariai*. As a result, the relationship (called "three usages of ratio") is as follows:

$$p = A \div B \qquad A = B \times p \qquad B = A \div p.$$

6. *Kyozai-Kenkyu* (教材研究) is an investigative activity in which teachers deepen their knowledge and increase their teaching skills. It is also considered as the "preparation" stage in lesson studies. According to Baba (2007), it is the process of transforming a planned curriculum into a curriculum that can be implemented in the classroom. He also states, "This process begins with finding and selecting materials relevant to the purpose of the class, and is then followed by refining the class design, based on the actual needs of the students and tying all of this information together into a lesson plan" (p. 2).

REFERENCES

Baba, T. (2007). How is lesson study implemented? In M. Isoda, M. Stephens, Y. Ohara, & T. Miyakawa (Eds.), *Japanese lesson study in mathematics: Its impact, diversity and potential for educational improvement* (pp. 2-7). Hackensack, NJ: World Scientific.

Becker, J. P., & Shimada, S. (Eds.). (1997). *The open-ended approach: A new proposal for teaching mathematics.* Reston, Virginia: National Council of Teachers of Mathematics. (Original work published 1977)

Davis, J. (2007). Real-world contexts, multiple representations, student- invented terminology, and Y-intercept. *Mathematical Thinking and Learning,* 9(4), 387418.

Freudenthal, H. (1983). *Didactical phenomenology of mathematical structures.* Dordrecht, The Netherlands: D. Reidel.

Hino, K. (1996). Ratio and proportion: A case study of construction of unit in U.S. and Japanese students. *Journal of Science Education in Japan, 20*(3), 159-173

Hino, K. (2002a). Acquiring new use of multiplication through classroom teaching: An exploratory study. *Journal of Mathematical Behavior,* 20(4), 477-502.

Hino, K. (2002b). Jyugyo ni okeru ninchiteki henyo to sugakuteki hyoki no yakuwari [Cognitive change of individual pupils through classroom teaching and roles of mathematical notations]. *Journal of Japan Society of Mathematical Education*, 79, 3-23.

Hino, K. (2003). *Kyoshitsu ni okeru kodomo no gakushu process wo sizatosuru hireitekisuiron no sido unit no kaihatsu* [Development of teaching unit on proportional reasoning from the viewpoint of learning processes of children in the classroom] (JSPS Grants-in-Aid for Scientific Research (C) No. 12680178)). Nara University of Education, Japan.

Hino, K. (2008). Shido wo kaisiteno hireitekisuiron no hattatu ni kansuru ichikosatsu [A study on development of proportional reasoning through classroom teaching] *Tsukuba journal of Educational Study in Mathematics*, 27, 1-10.

Hino, K. (2011). Students' uses of tables in learning equations of proportion: A case study of a seventh grade class. In B. Ubuz (Ed.), *Proceedings of the 35th Conference of the International Group for the Psychology of Mathematics Education* (vol. 3, pp. 25-32). Ankara, Turkey: PME.

Ikeda, T. (2007). Sugakuteki modelling to sansu kyoiku [Mathematical modelling and mathematics education in elementary school]. *Journal of Japan Society of Mathematical Education,* 89(4), 2-10

Koto, S., & Elementary school mathematics study group in Niigata (Eds.). (1998). Communication de tsukuru atarashii sansu gakushu [New ways of learning elementary school mathematics created with the idea of communication]. Tokyo: Toyokan.

Lamon, S. J. (1993). Ratio and proportion: Children's cognitive and metacognitive processes. In T. P. Carpenter, E. Fennema, & T. A. Romberg (Eds.), *Rational numbers: An integration of research* (pp. 131-156). Hillsdale, NJ: Erlbaum.

Lamon, S. J. (2005). *Teaching fractions and ratios for understanding (Second edition).* Mahwah, New Jersey: Lawrence Erlbaum Associates.

Lamon, S. J. (2007). Rational numbers and proportional reasoning: Toward a theoretical framework for research. In F. K. Lester, Jr. (Ed.), *Second handbook of research on mathematics teaching and learning* (pp. 629-667). Charlotte, NC: Information Age Publishing Inc.

Lesh, R., Post, T., & Behr, M. (1988). Proportional reasoning. In J. Hiebert, & M. Behr. (Eds.), *Number concepts and operations in the middle grades: Research agenda for mathematics education* (pp. 93-18). Reston, Virginia: National Council of Teachers of Mathematics.

Modestou, M. & Gagatsis, A. (2010). Cognitive and Metacognitive Aspects of Proportional Reasoning. *Mathematical Thinking and Learning*, 12, 36-53.

Monbukagakusho. (2008). *Shogakko gakushu shido yoryo sansu hen* [Instruction manual of the Course of Study, Elementary mathematics]. Tokyo: The Ministry of Education, Culture, Sports, Science and Technology.

Monbusho. (1958). *Shogakko Gakushu Shido Yoryo* [Course of study in elementary school]. Tokyo: The Ministry of Education.

Monbusho. (1977). *Shogakko Gakushu Shido Yoryo* [Course of study in elementary school]. Tokyo: The Ministry of Education.

Morozumi, T. (2002). Chugakko sugaku ni okeru kansu ryoiki no kenkyu doko to hokosei nit suite [Research trends and agendas on the area of function in lower secondary school mathematics]. In Japan Society of Mathematical Education (Ed.), *Proceedings of the 35th conference of the Japan Society of Mathematical Education* (Summary volume of topic groups, pp. 138-144). Tottori, Japan: Authors.

Moschkovich, J. (1998). Resources for refining mathematical conceptions: Case studies in learning about linear functions. *The Journal of the Learning Sciences*,7(2), 209-237.

Mullis, I., Martin, M., Gonzalez, E., & Chrostowski, S. (Eds.). (2004). *TIMSS 2003 International Mathematics Report.* Chestnut Hill, MA: TIMSS & PIRLS International Study Center.

Mullis, I., Martin, M., Gonzalez, E., Gregory, K., Garden, R., O'Connor, K., Chrostowski, S., & Smith, T. (2000). *TIMSS 1999 International Mathematics Report.* Chestnut Hill, MA: International Study Center.

Nagasaki, E. (1990). Problem solving. In Sin Sansu Kyoiku Kenkyukai (Ed.), *Sansu kyoiku no kiso riron* (Basic theory of elementary mathematics education) (pp. 134-146).Tokyo: Toyokan.

Nagasaki, E. (Ed.). (2005). *Sansu/Sugaku ni kansuru hyoka/bunseki report* [A report of evaluation/analysis on the results of mathematics of PISA 2003/TIMSS 2003] Tokyo, Japan: National Institute for Educational Policy Research.

Nagata, J. (2004) "Hireisuru to minasu" koto no yosa ni tsuite no kosatsu [On the merit of "assuming proportionality"]. *Journal of Japan Society of Mathematical Education*, 86(3), 13-20.

Nakajima, K. (1981). *Sansu sugaku kyoiku to sugakuteki na kangaekata* [Mathematics education and mathematical thinking].Tokyo: Kaneko Shobo.

Nakamura, T. (1996). Shosu no joho no wariai ni yoru imizuke [Exploitation of ratios for explaining multiplication of decimal numbers]. *Journal of Japan Society of Mathematical Education*, 78(10), 7-13.

Nakamura, K. (2011). Seisu no joho, joho no mondaibamen deno 4 nensei no kodomo no hireiteki suiron no jittai [Students' proportional reasoning on multiplication and division problem situations]. *Journal of Japan Society of Mathematical Education,* 93(6), 2-10.

Nao, Y. (1990). Shogakko ni okeru "wariai" shido no hensen (I) [Progress of the teaching of "wariai" in elementary schools (I)]. *Journal of Japan Society of Mathematical Education*, 72(12), 22-27.

Nao, Y. (1991). Shogakko ni okeru "wariai" shido no hensen (II) [Progress of the teating of "wariai" in elementary schools (II)]. *Journal of Japan Society of Mathematical Education,* 73(2), 2-9.

Nunokawa, K. (2005): Mondai keiketsu katei no kenkyu to gakushu katei no tankyu [Research on problem solving processes and exploration of learning processes]. *Journal of Japan Society of Mathematical Education*, 54(2), 22-34.

Nunokawa, K. (2006). Hireitekisuiron no jyugyo niokeru shogakko 4 nensei no gakjshu no yoso [Learning processes of fourth grade children for lessons on proportional reasoning]. *Joetsu Journal of Mathematics Education*, 21, 1-12.

Nunokawa, K. (2007). Shogakko 3 nensei ni yoru hireiteki suiron no kadai no kaiketsu [Solving proportional reasoning tasks by the 3rd grade children in elementary school]. *Joetsu Journal of Mathematics Education*, 22, 1-10.

Nunokawa, K. (2010). Suryo kankei no gakushu to haigo no gensho ya kyohensei no ishikika [Learning of quantitative relationships and the awareness of phenomenon behind and co-variation]. *Joetsu Journal of Mathematics Education*, 25, 1-10.

Oda, Y. (2011). Sugakuteki literacy wo ikusei suru jyugyo [Mathematics lessons that foster children's mathematical literacy]. In E. Nagasaki. (Ed.), *Sugaku kyoiku ni okeru literacy ni tsuiteno systemic approach niyoru sogoteki kenkyu* [A comprehensive research on literacy in mathematics education from systemic approach] (JSPS Grants-in-Aid for Scientific Research (B) No. 20300262)). Shizuoka University, Japan. Ohtani, M. (2005). Designing unit for teaching proportion based on cultural-historical activity theory: Process of symbolizing through collective discourse. In Woo, J. H., Lew, H. C., Park, K. S. & Seo, D. Y. (Eds.), *Proceedings of the 31th Conference of the International Group for the Psychology of Mathematics Education,* Vol. 4, pp. 33-40. Seoul: PME.

Ohtani, M., Kanno, Y., & Nakamura, M. (2002). Zokusei kara tokusei eno iko wo mezasu hirei no shido [Teaching of proportion that foster the transition from property to characteristic]. *Proceedings of 35th Conference of Japan Society of Mathematical Education,* 325-330.

Ohtani, M. & Nakamura, M. (2002). Chugakko tono setsuzokusei wo hairyo shita hirei no gakushu shido [Teaching proportion for the articulation with junior secondary level]. *Journal of Japan Society of Mathematical Education,* 84(6), 1122.

Ohtani, M. & Nakamura, M. (2004). Hirei no shido ni okeru suhyo, graph, siki no symbol ka katei [Symbolizing process of number tables, graphs, and expressions in teachings of proportion]. *Journal of Japan Society of Mathematical Education, 86*(4), 3-13.

Sasaki, T. (2003). Chugakko sugaku ni okeru kansu ryoiki no kenkyu seika no matome to kadai [Sumary and future tasks of the research results on the area of function in lower secondary school mathematics]. In Japan Society of Mathematical Education (Ed.), *Proceedings of the 36th conference of the Japan Society of Mathematical Education* (Summary volume of topic groups, pp. 160165). Hokkaido, Japan: Authors.

Seino, T. (2004) "Katei no ishikika" wo jyushi shita sugakuteki model ka no jyugyo [Learning and teaching of mathematical modeling with emphasis on "the awareness of assumptions"]. *Journal of Japan Society of Mathematical Education,* 86(1), 11-21.

Shimada, S., & Nakajima, K. (Eds.). (1960). *Suryo kankei no shido* [Instruction of numerical relationships]. Tokyo: Kaneko Shobo.

Shimizu, S. (Ed.). (1989). *Sansuka no kaisetsu to tenkai* [Description and development of elementary school mathematics]. Tokyo: Kyoiku Kaihatsu Kenkyujo.

Shimizu, S. (Ed.). (2003). *Sengo gakko sugaku no hensen* [Progress of school mathematics after the World War II]. Tsukuba, Japan: Department of Mathematics Education, University of Tsukuba.

Society of Elementary Mathematics Education (Ed.). (2011). *Tokushu: Shikoryoku/ Hyogenryoku no hyoka no arikata* [Special issue: The proper way of assessment on thinking and representation]. Elementary Mathematics Teaching Today, 482.

Steffe, L. P., von Glasersfeld, E., Richards, E., & Cobb, P. (1983). *Children's counting types: Philosophy, theory, and application.* New York: Praeger.

Sugiyama, Y., Iidaka, S., & Ito, S. (2005). *Atarashi Sansu* [New mathematics for elementary school]. Tokyo: Tokyo Shoseki.

Tabata, T. (1989). Joho no imishido no ichikosatsu [A study of teaching the meaning of multiplication: Teaching the meaning of multiplication as the situation of fostering the ability of proportional reasoning]. In Japan Society of Mathematical Education (Ed.), Proceedings of the 22th conference of the Japan

Society of Mathematical Education (Summary volume of topic groups, pp. 297-300). Ishikawa, Japan: Authors.

Tabata, T. (2002a). Doshu no ryo to ishu no ryo no wariai no shidojyunjo ni kansuru kosatsu [A study on the teaching order of ratios: Which to teach first? (a) ratios of similar types of quantities or (b) ratios of different types of quantities] *Journal of Japan Society of Mathematical Education,* 84(8), 22-29.

Tabata, T. (2002b). Sansukyoiku ni okeru hireitekisuiron no yakuwari nit suite [On the role of proportional reasoning in mathematics education of elementary schools]. In Japan Society of Mathematical Education (Ed.), *Proceedings of the 35th conference of the Japan Society of Mathematical Education* (Summary volume of topic groups, pp. 130-137). Tottori, Japan: Authors.

Tabata, T. (2003a). Suchokusen wo katsuyo shita kessonchi mondai no kyoju gakushu katei [Teaching and learning process of missing value problems by utilizing number lines]. In Japan Society of Mathematical Education (Ed.), *Proceedings of the 36th conference of the Japan Society of Mathematical Education* (Summary volume of topic groups, pp. 152-159). Hokkaido, Japan: Authors.

Tabata, T. (2003b). Doshu no ryo no wariai no donyu ni kansuru ichikosatsu [A study on introduction to the concept of ratio]. *Journal of Japan Society of Mathematical Education,* 85(12), 3-13.

Takahashi, H. (2000). Shosu no joho no jyugyo kosei ni kansuru kosatsu [A study on lesson organization in the teaching of multiplication of decimal numbers]. *Joetsu Journal of Mathematics Education,* 15, 85-94.

Tsuchiya, T. (2002). Hirei no mikata wo mochiita "wariai" no shidojissen [A case study of instruction on ratio that incorporates the concept of proportion. *Journal of Japan Society of Mathematical Education,* 84(8), 30-37.

Van Dooren, W., De Bock, D., Evers, M., & Verschaffel, L. (2009). Students' overuse of proportionality on missing-value problems: How numbers may change solutions. *Journal for Research in Mathematics Education,* 40(2), 187-211.

Watanabe, T. (2010). Ratio, rate, proportion no chigai [Differences among ratio, rate and proportion]. *Elementary Mathematics Teaching Today,* 473, 34-35.

Watanabe, T., Takahashi, A., & Yoshida, M. (2010). Supporting focused and cohesive curricula through visual representations: An example from Japanese textbooks. In B. Reys, R. Reys & R. Rubenstein (Eds.), *2010 Yearbook: Contemporary issues in mathematics curriculum* (pp. 131-143). Reston, VA: National Council of Teachers of Mathematics.

Yamaguchi, J. (2007). Wariai ni okeru jido no gakushukatei ni kansuru kenkyu [A study of children's learning processes on the concept of ratio]. *Joetsu Journal of Mathematics Education*, 22, 101-112

Yoshida, H. & Kawano, Y. (2003). Informal na chisiki wo moto ni shita kyojyu kainyu [Instructional intervention based on children's informal knowledge: The case of ratio concepts]. *Journal of Japan Society for Science Education,* 27(2), 111-119.

CHAPTER 61

JAPANESE STUDENTS' UNDERSTANDING OF SCHOOL ALGEBRA

Algebra, Literal Symbols, and Quasi-variables

Toshiakira Fujii
Tokyo Gakugei University

ABSTRACT

This chapter reports on studies of students' understanding of school algebra in Japan from two aspects. The first presents accumulated research results in order to probe Japanese students' understanding of school algebra focusing on the literal symbols, based mainly on research papers published in the journals of Japan Society of Mathematical Education. In order to identify the difficulties of learning of school algebra, the author uses the scheme of use of symbolic or mathematical expressions which consist the three processes: to express a situation using symbolic expressions, to transform the symbolic expressions and to read or interpret the result of transforming of the symbolic expressions. The resulting analysis shows that many Japanese

The First Sourcebook on Asian Research in Mathematics Education: China, Korea, Singapore, Japan, Malaysia, and India, pp. 1353–1374

students in junior high schools appear to have good skills of manipulating equations or calculating algebraic expressions, but as a contrast, they seem to have a very poor grasp of what literal symbols denote and how they are to be treated in mathematical expressions. In the second part, an alternative way of teaching of school algebra is proposed. An attempt is made to show how the curriculum of the elementary and also secondary school can offer better opportunities for young people to think algebraically. Utilizing the potentially algebraic nature of arithmetic is one way of building a stronger bridge between early arithmetical experiences and the concept of a variable. In this chapter the author uses the terms generalizable numerical expressions or quasi-variable expressions to make a case for a needed reform to the curriculum of the elementary and secondary schools.

Keywords: School algebra, literal symbols, early algebra, quasi-variable expressions

INTRODUCTION

Understanding of algebra in school mathematics is one of the most important goals for secondary mathematics education. On the other hand, algebra in curriculum may differ among countries. In fact, some may be surprised to see that there isn't an algebra strand in the Japanese Course of Study of elementary and even of lower secondary school mathematics. This fact is contrasting to the American curriculum recently released as the Common Core State Standards[1] which includes "Operations and Algebraic Thinking" even from Kindergarten to Grade 5.

The Course of Study in Japan is an official document for curriculum that is published and revised in about every ten years by the Ministry of Education, Culture, Sports, Science, and Technology (hereafter referred to as the Ministry). Therefore if there is no strand concerning with algebra in the Course of Study, it is likely that there isn't a unit titled as "algebra" in the elementary and lower secondary mathematics textbook which need to be officially approved by the Ministry although the textbooks are commercially published.

Instead of being an independent domain in the curriculum, algebra is systematically included in various parts of the mathematics curriculum, particularly at elementary level. Watanabe has made a list by analyzing Japanese textbooks from a foreigner's eyes and identified examples to show much of what is considered algebra exists in Japanese elementary mathematics textbooks (Watanabe, 2007). The main source of "algebra" is the Quantitative Relations strand in the Course of Study. That strand consists of ideas of functions, writing and interpreting mathematical expressions, and statistical manipulations. The contents of the Quantitative

Relations strand have strong relations with the other three strands, Numbers and Calculations, Quantities and Measurement, and Geometric Figures (Takahashi, Watanabe, and Yoshida, 2004). In other words, these strands also include ingredients of algebra and algebraic thinking. At lower secondary level, it is much easier to identify what is considered algebra.

As background information of school algebra in Japan, let me describe the contents of or related contents of algebra in the Corse of Study focusing on the algebraic expressions and equations and functions.

In the Japanese curriculum, the frame word box(□) used for unknown number is introduced at 3rd grade, and in the 4th grade, an additional frame word such as circle(○) is introduced. At the 4th grade, these frame words can be used for variables in a mathematical expression such as □+ ○= 10. In the 6th grade, literal symbols such as "x" and "a" are introduced to stand for numbers previously represented by the frame words. However at the 6th grade there is no place to learn manipulations of literal symbols. That is left to lower secondary school.

In lower secondary school, in the 7th grade, translations of concrete situations expressed in ordinary language to mathematical expressions using literal symbols and vice versa are emphasized. Calculations of algebraic expressions with one variable required for solving linear equations are studied. In the 7th grade students learn to solve linear equations using attributes of equality, and here the literal symbol x can be used for an unknown number. Proportions as a special case of linear functions are introduced and the definition of variable such as "Letters that take various values are called variables" (Kodaira, 1992, p.97) is given. In the 8th grade students are required to further develop their ability to find quantitative relationships and to express such relationships in a formula by using literal symbols. Computations of a simple formula using literal symbols and the four fundamental operations are emphasized. Simultaneous linear equations in two variables are studied. The linear function is also studied. In the 9th grade, multiplication of linear expressions and factorizations of simple quadratic trinomials are studied, as well as solving quadratic equations using factorization and formula of solution. The simple quadratic functions such as $y=ax^2$ and their graphical representations are studied.

Although there isn't an exact unit titled as "algebra" in the elementary and lower secondary mathematics textbook, understanding of algebra in school mathematics is an important goal for mathematics education. However, many reports identify specific difficulties of learning of algebra. From international perspective, difficulties are labelled such as the cognitive obstacles (Herscovics, 1989), lack of closure (Collis, 1975), name-process dilemma (Davis, 1975), letter as objects (Kuchemann, 1981), misapplication of the concatenation notation (Chalouh & Herscovics, 1988), misinterpretation of order system in number (Dunkels, 1989) and so on. Matz (1979) also has identified inappropriate but plausible use of literal symbols in the process of transforming algebraic expressions.

In Japan, we are facing with the same problem that many students in lower secondary school are still confusing unknown numbers and variables. However we need to be careful of diagnosing of their nature of understanding, simply because students seem to be good at solving conventional school type problems. Although ratios of correct answers in mathematics achievement tests such as IEA results and PISA results are high, Japanese mathematics educators suspect that limited understanding may coexist with this apparent success story.

This chapter will report on accumulated research results in order to probe Japanese students' understanding of school algebra focusing on the literal symbols, based mainly on research papers published in the journals of Japan Society of Mathematical Education. Then, three detailed discussions follow: The first focuses on the conventions or rules in the context of expressing and interpreting of mathematical expressions. The second focuses on research results could probe the understanding lying behind Japanese students' apparent procedural efficiency. The third focuses on a recent and alternative way of teaching of school algebra. That is made to show how the curriculum of the elementary and also secondary school can offer better opportunities for young people to think algebraically. Utilizing the potentially algebraic nature of arithmetic is one way of building a stronger bridge between early arithmetical experiences and the concept of a variable. The terms generalizable numerical expressions or quasi variable expressions are to use in this chapter to make a case for a needed reform to the curriculum of the elementary and secondary school.

LITERATURE REVIEW

In order to classify the accumulated research held in Japan concerning with learning and teaching of school algebra, this chapter uses the scheme of use of symbolic expressions proposed by Tatsuro Miwa (2001), as "algebra" in school mathematics can be described as learning how to use symbolic expressions.

The Scheme of Use of Symbolic Expressions

Miwa (2001) has illustrated the process as the scheme of use of symbolic expressions as shown in Figure 61.1.

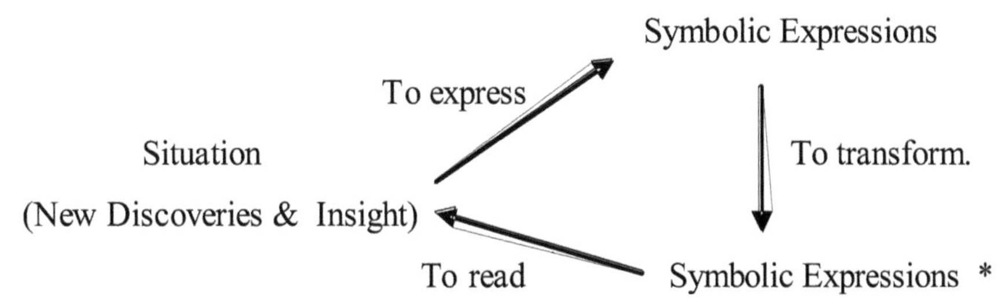

Figure 61.1. Scheme of use of symbolic expressions.

According to Miwa (2001), the symbolic expressions are composed of numerals and mathematical signs together with alphabetical letters. The scheme of use of symbolic expressions is consistent with a mathematical modelling process. That is, starting from a situation, we express the situation in terms of mathematical expressions, and then transform them to get a mathematical conclusion ("Symbolic Expression*" in Fig.1). Finally we need to read or interpret the mathematical conclusion into the original situation to get insight or new interpretation or discoveries.

In order to identify the difficulties of learning of school algebra, this paper uses the scheme of use of symbolic or mathematical expressions which consists the three processes: to express a situation using symbolic expressions, to transform the symbolic expressions and to read or interpret the result of transforming of the symbolic expressions. These three processes are closely related each other, so that research papers would show a combination of these three aspects. This chapter will try to identify papers in terms of which process may mainly focus on.

Students' Understanding of Expressing of Literal Symbols

According to the Japanese Course of Study, the frame words such as □, ○ are used as a foundation of literal symbols. Maruyama(1978) argued an appropriateness of using the frame words at 3rd and 4th grade before introducing the literal symbols. Based on his own survey Maruyama (1978) finally agreed to introduce the frame words as a preparation of literal symbols. Naito(1989) also supported to use the frame words, and Hiroma (1990) and Ishida (1994), based on their survey, suggested to use the frame words even at 2nd grade that is one year earlier than the Course of Study.

On the other hand, Fujii (1999, 2002, 2006), Saito (1989), Hoshino (2002) and Tsunamoto (2008) have suggested to use numerals instead of the frame words. They consider numerals are the quasi-variables. The notion of the quasi-variable will discuss in the later part of this chapter.

Many of the research held in Japan concerning with students' understanding of expressing of literal symbols are focusing on the transition from the words problems to mathematical expressions using literal symbols. The research results are basically consistent with the international research outcomes: focusing on the words problem, Kondo (2007) is trying to identify the secondary school students' difficulty by using the words problems which are hard for elementary school children. Takashita (2009), Koyama (2008) and Kyogoku (1999) are analysing the detail process of expressing situation lying behind words problems into literal symbols focusing on one or a few children. Suzuki (2005) and Morita (2005,2006) are focusing on the linear equation, and Naomi (1987) is on the simultaneous equations, and also Shimada (1991) is on the quadratic equations, they all trying to identify the critical process of expressing a situation mathematically by using literal symbols.

There are kinds of follow-up surveys: Relating to C. Kieran's study (1981) concerning with the equality symbol, Mizoguchi (1998), Takizawa(2003) and Soeda (1992) are identified basically similar tendency of children's usage of equality symbol. Relating to the study of Harper (1987), Fujii (2000) identified a parallelism between Japanese students and English students. Both students seem to be able to express easily an unknown number using the literal symbols, but uneasy to express a known number into literal symbols. Relating to the Letter as Object misconception (Kuchemann, 1981; Booth,1984), Kume and others(1990) and Yokota(1994) identified basically same tendency using the famous problem such as " Students and Professor Problem"(Clement,1982; Clement, Lochhead & Monk,1981). The results show that the reverse error appeared particularly in the context of proportion, although the percentage of the reverse error among Japanese students is lower than international survey results.

Students' Understanding of Interpretation of Literal Symbols

Mathematical expression, such as "a+b" implies both a process and result. In other words, mathematical expression may have two aspects: procedural and structural aspects. The research papers include the term such as "duality" or "process-product" is focusing on this nature of literal symbols. In Japan, Yokota (1994), Makino(1996), Itagaki (1997), Tanaka (2003), Koiwa (2004) and Furukawa (2005) conducted a follow-up survey and concluded the basically same results appeared in the international research papers. Shimizu (1997) identified the difficulty for students to see mathematical expression from structural point of view, giving the task such as "if a+3b+5c=25, then a+3b+5c–10= ?". This task is basically same to the task in CSMS (Concepts in Secondary Mathematics and Science, reported in Hart, 1980, 1981). Fujii (1992a, 1992b, 1994) focused on the conventions or rules in mathematical expression using literal symbols. These studies will be discussed later in this chapter.

On the other hand, a theoretical consideration was given by Sugiyama (1990). He clarified a role of literal symbols in terms of interpreting mathematical expressions. His results gave a deeper thought and rich examples of the role of literal symbols descried in the teaching guide (Ministry of Education, Culture, Sports, Science, and Technology, 1999). The Teaching Guide edited by the Ministry of Education and published by a private publisher in 1999. The Teaching Guide showed that interpreting mathematical expressions is thought of as involving the following:

(a) Interpreting the concrete situation from the expression.
(b) Generalizing the phenomena or the relations that the expression represents.
(c) Widening the range of numbers in expressions, that is, to interpret expansively.
(d) Understanding the thinking process that is the background of the mathematical expression.
(e) Interpreting mathematical expressions in correspondence with a model such as a number line.

Students' Understanding of Transforming of Literal Symbols

The process of transforming of literal symbols relates closely to the other two processes: expressing and interpreting process. Morozumi(1993) particularly focused on the expression of 5ab – 3a = 2ab, and tried to elicit students' understanding of transforming process of literal symbols. Fujimoto (2004) studied on the difficulty in transforming of fractional expressions. T. Tsukada (1992) focused on upper secondary students' difficulty concerning with transforming of literal symbols.

Focusing on the linear equation and linear inequalities, Fujii (1989) showed the nature of Japanese students' efficiency and difficulty of manipulating of literal symbols. Students could solve the conventional manipulative tasks, but their understanding revealed shallow one: they know how to do but they can't see why it works. The detail consideration will be shown in the later part of this paper.

Concerning with as simultaneous equations, Hirano (1986) implemented her lesson plan to let students discover how to solve simultaneous equation by themselves. Concerning with the quadratic equation, Tobioka (1972) reflected his 25 year-teaching career and gave a profound consideration of changes in teaching of quadratic equations in secondary school in Japan, and Sakamoto (2001) investigated students' performance when they use a factorisation method. Concerning with upper secondary level, Anada and Hirono(2005) surveyed upper secondary school students' logical thinking lying behind of classifying the polynomials of degree two. Also, Yokoyama (1957), Suzuki (1989), and Tanaka (1978) were considering upper secondary school students' difficulties dealing with inequalities, and Nihira studied on the extension of $x+y=xy$.

Iijima(2010) analysed in detail on the Chinese Nine Chapter focusing on the procedural strategies used. He tried to find a relationship between the ancient Chinese methods and the contemporary transforming process of linear equation and simultaneous equation.

Mathematicians also contributed to Japan Society of Mathematical Education community. In fact, Sato(1983) revised and rewrote his theoretical paper concerning with the functional equation which appeared in the scientific journal at 1927. Yoshizawa (2001) suggests to use computer to solve the algebraic equations, since equations which have rather special solutions are tend to be introduced into school mathematics.

DISCUSSION

Students' Understanding of Conventions in the Mathematical Expression

The process of expressing and interpreting of literal symbols are closely related to students' understanding of conventions or rules of mathematical expressions. Namely, in order to express or to interpret a mathematical expression with literal symbols, students may need to understand conventions in the expression. Concerning with the conventions or rules in the expression, one of the well-documented misconceptions internationally is that different letters must represent different values. This misconception is illustrated by students' responses of "never" to the following question:

When is the following true – always, never or sometime?

$$L+M+N=L+P+N$$

Kuchemann (1981), from England, reported in the CSMS project that 51% of students answered "never" and Booth (1984) reported in SESM project that 14 out of 35 students (ages 13 to 15 years), namely 40%, gave this response on interview. Olivier (1988), from South Africa, reported that 74% of 13 year olds also answered "never". He suggested that the underlying mechanism for not allowing different literal symbols to take equal values stems from a combination with other valid knowledge, that is, the correct proposition that the same literal symbols in the same expression take the same value. In other words, some students who are aware of the proposition that the same letter stands for the same number, they tend to think that the converse of this proposition is also correct.

Fujii (1993, 2001) claims that the convention, the same letter stands for the same value, is not grasped well by students based on a survey conducted with Japanese and also American students. Studies revealed that in some situations, students conceive that the same letter does not necessarily stand for the same number.

The reason behind of this phenomenon could interpret in terms of students' conception of variable. The concept of variable has been discussed for a long time in mathematics education community. The definition of variable given in the SMSG (School Mathematics Study Group) Student's Text was "the variable is a numeral which represents a definite through unspecified number from a given set of admissible number" (School Mathematics Study Group, 1960, p.37). Although the ideas definite and unspecified appear to be in tension, the concept of variable needs to include these different aspects (Van Engen, 1961a, 1961b).

Data from surveys conducted by Fujii(1993, 2001) showed that students appear to lack one or both aspects. The "definite" aspect of the concept of variable is most clearly embodied in the convention that the same letter stands for the same number. Students' misconceptions described as "x can be any number" emphasizes only the "unspecified" aspect of a variable. On the other hand, the misconception, different letters stand for the different numbers, could be characterized as an unduly strict interpretation of the "definite" aspect of variable by students who persistently reject substituting the same number for different literal symbols. Although the range of variable does not depend on the literal symbol itself, the survey revealed that students tend to focus on the surface character of literal symbols, such as differences in letter, within the range of variables.

Students' Understanding Underlying Procedural Efficiency

In Japan, a country where students face high-stakes exams to enter upper secondary schools or universities, students have no choice about mastering skills to solve problems within a certain fixed time. As an outcome, Japanese students seem to be good at solving mathematical problems presented in school algebra. But is this really any indication that students have a deep understanding of the subject matter, or is it only superficial understanding? Skemp (1976) called this Instrumental Understanding. The Instrumental understanding means "knowing what to do but without knowing why". On the other hand, the Relational Understanding means "knowing what to do and why". Although the instrumental understanding is shallow, it can still work effectively in almost all conventional school mathematical problems.

Fujii(1993) has been developing set of cognitive conflict problems, where cognitive conflict is regarded as a tool to probe and assess the depth and quality of students' understanding. He developed problems on linear equations and inequalities. In solving linear equalities and inequalities in which the solution set contain all numbers, clearly the 'disappearance' of x was expected to provoke cognitive conflict in students. By analysing how students went about resolving this conflict, it was possible to identify the nature of their understanding behind procedural efficiency. The example used at his survey and interview is shown in Figure 61.2.

Students' responses were further classified into five categories. Category A (13%) consisted of responses where the conflict was able to resolve by giving the correct answer. Among lower secondary second graders (n =

Mr. A solved the inequality $1 - 2x < 2(6 - x)$ as follows:

$$1 - 2x < 2(6 - x)$$

$$1 - 2x < 12 - 2x$$

$$-2x + 2x < 12 - 1$$

$$0 < 11$$

Here Mr. A got into difficulty.

1 Write down your opinion about Mr. A's solution.

2 Write down your way of solving this inequality $1 - 2x < 2(6 - 2x)$ and your reasons.

Figure 61.2. The cognitive conflict problem: the inequality.

123), very few were included in this category. Other students' rationales reflected two ways of resolving the cognitive conflict produced by the disappearance of x. The first was exhibited in the students' persistence of coming up with an answer that contained x. This group comprised Category B (34%). Category B was further sub-divided into two groups B1 (26%) and B2(8%). Students in B1group, persisted in having x in the final answer by using irrelevant procedures, while students in B2 who expected to get an answer containing x but couldn't retain an x finally give up by concluding that "there is no solution" 解なし in Japanese, see Figure 61.4). Category C (18%) consisted of students who reached a final answer not containing x. Category D (3%) gave no answer or solution (Fujii, 1989).

This analysis appeared to be effective in revealing huge gaps in students' understanding. These gaps may be masked in much of their classroom work by procedural efficiency in solving standard problems. These are not results we should be happy about.

On the other hand, students who can think of x as a variable can come up with the correct answer by interpreting x to take a definite but unspecified value. Student I wrote the expression: 1–2x < 12–2x, replacing –2x with □, then re-expressing the original expression as 1 + □<12 + □. Student I explained as follows: "The sign of the inequality remains the same even if we add the same number to, or subtract it from both sides of the expression. Any number will do for □; hence the same applies for x." Note that this student focuses on the calculation of adding –2x to both

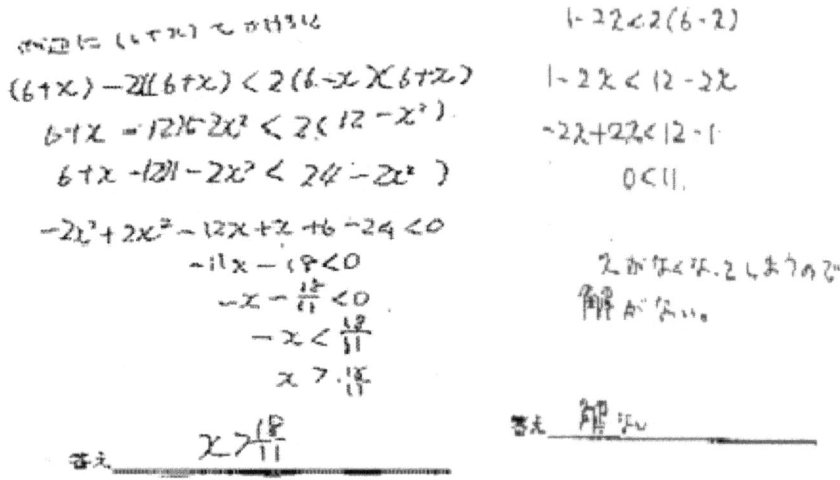

Figure 61.3. Typical example in Category B1

Figure 61.4. Typical example in Category B2

sides without seeing any need to find a concrete number for –2x or x. By re-expressing the original expression, this student seemed able to pay more attention to the operation itself and to the structure of the expression than to the objects of calculation such as –2x, 1–2x and 12–2x. This approach is clear evidence of understanding of x as a variable.

Using Number Sentences to Introduce the Idea of Variable

There is a reluctance to introduce children to algebraic thinking in the early years of elementary school where the focus for almost all teaching of early number is on developing a strong foundation in counting and numeration. Yet Carpenter and Levi (1999) draw attention to "the artificial separation of arithmetic and algebra" which, they argue, "deprives children of powerful schemes for thinking about mathematics in the early grades and makes it more difficult for them to learn algebra in the later grades" (p. 3).In their study, they introduced first and second-grade students to the concept of true and false number sentences. One of the number sentences that they used was 78 – 49 + 49 = 78.When asked whether they thought this was a true sentence, all but one child answered that it was. One child said, "I do because you took away the 49 and it's just like getting it back".

It was never the intention of Carpenter and Levi to introduce first and second-grade children to the formal algebraic expression, a – b + b = a. These children will certainly meet it and other formal algebraic expressions in their later years of school. What Carpenter and Levi wanted children to understand is that the sentence 78 – 49 + 49 = 78 belongs to a type of number sentence which is true whatever number is taken away and then added back. This type of number sentence is also true whatever the first number is, provided the same number is taken away and then added back. Fujii (2000) and Fujii & Stephens (2001) refer to this use of numbers as quasi-variables. By this expression, we mean a number sentence or group of number sentences that indicate an underlying mathematical relationship which remains true whatever the numbers used are. Used in this way, our contention is that generalizable numerical expressions can assist children to identify and discuss algebraic generalisations long before they learn formal algebraic notation.

The idea behind the term "quasi-variable" is not a new one in the teaching of algebra. In his history of mathematics, Nakamura (1971) introduces the expression "quasi-general method" to capture the same meaning. At the practical level, Ohta (1990, 1992) conducted series of lessons concerning with literal symbols as he considered the teaching of literal symbols should be integrated into each domain of school mathematics. He set a game called "Guessing a Number", then discovered that the 7th grade students' could express the nature of generality by using a frame word such as □ and also using a number. In fact, in the lesson students used the number 14 that was a representative of all possible numbers. Here the number 14 is a quasi-variable. Ohta (1990,1992) also suggested that students could see and interpret the literal symbols in mathematical expression from the Gestalt psychology point of view. That is, which part of the mathematical expression is focused or emphasised is a critical evidence of understanding of variables. The insight given by Ohta (1990,1992) is consistent with a idea of Matsubara (1977) who insisted an importance of the Gestalt psychology in mathematics education. Although, there is few research referring to the Gestalt psychology in the context of analysing students' understanding of literal symbols, Matsubara's idea would suggest an interesting future research field.

The concept of a quasi-variable provides an essential counterbalance to that treatment of algebra in the elementary and junior high where the concept of an unknown often dominates students' and teachers' thinking. As Radford (1996) points out, "While the unknown is a number which does not vary, the variable designates a quantity whose value can change" (p. 47). The same point is made by Schoenfeld and Arcavi (1988) that a variable varies (p. 421). The use of generalizable numerical sentences to represent quasi-variables can provide a gateway to the concept of a variable in the early years of school.

Fujii & Stephens (2001) conducted an interview with children in Year 2 and 3 in Australia and Japan using an interview-dialogue based on a method actually used by a student called Peter in subtracting 5. The purpose of the interview is to see how readily young children are able to focus on structural features of Peter's Method. In other words, can they engage in quasi-variable thinking and how do they express that thinking?

The interview-dialogue starts with Peter subtracting 5 from some numbers.

> $37 - 5 = 32$, $59 - 5 = 54$, $86 - 5 = 81$
>
> He says that these are quite easy to do. Do you agree?
>
> But some others are not so easy, like:
>
> $32 - 5$, $53 - 5$, $84 - 5$
>
> Peter says, "I do these by first adding 5 and then subtracting 10, like
>
> $32 - 5 = 32 + 5 - 10$. Working it out this way is easier."
>
> Does Peter's method give the right answer? Look at the other two questions Peter has. Can you use Peter's method? Rewrite each question first using Peter's method, and then work out the answer.

Some children have difficulty re-writing the questions in a form that matches Peter's Method. They go straight to the answer. When asked how to explain why Peter's method works, they say it works because it gives the right answer. The interview does not point children in one direction or the other. But if children follow this kind of thinking, where their focus is on following a correct procedure for subtraction, the interview does not continue any further.

On the other hand, Alan (Australian, 8 years and 10 months, at end of Year 2) gives a quite different explanation when he says:

> "Instead of taking away 5, he (Peter) adds 5 and then takes away 10. If you add 5 you need to take away 10 to equal it out."

This explanation appears to attend more closely to the structural elements of Peter's Method, and suggests that Peter's Method is generalizable. Those children who give an explanation which attends to the structural features of Peter's Method are asked to create some examples of their own for subtracting 5 using Peter's Method, and are then asked to consider how Peter might use his method to subtract 6. The interviewer asks:

> What number would Peter put in the box to give a correct answer?
>
> $73 - 6 = 73 + \square - 10$

If students answer this question successfully, they are asked to create some other examples showing how Peter's Method could be used to subtract 6. Finally, students are told: "Peter says that his method works for subtracting 7, and 8 and 9." They are then asked to show how Peter's Method could be used to re-express subtractions, such as, 83 –7, 123 – 8 and 235 – 9.

The final part of the interview invites students to explain how Peter's Method works in all these different cases. Alan quoted earlier, said:

> "For any number you take away, you have to add the other number, which is between 1 and 10 that equals 10; like 7 and 3, or 4 and 6. You take away 10 and that gives you the answer."

Alan's thinking seems very clearly to embody quasi-variable thinking. He sees that Peter's Method does not depend in any way on the initial number (83, 123, or 235). Alan's explanation also shows that Peters' Method can be generalised for numbers between 1 and 10.

Japanese student, Kou, (age 9 years and 6 month at the start of Year 3) says:

> "It does not matter what number is taken way, when (the) adding number makes a ten the answer is always the same whatever the subtracting number is increasing or decreasing."

These students are able to 'ignore' for the purposes of their explanation the value of the 'starting number'. They recognize that it is not important for their explanation. In this sense, they show that they are comfortable with "a lack of closure".

Their explanations focus on describing in their own language the equivalence between the expressions that experts would represent as a – b and a + (10 – b) - 10 where b is a whole number between 1 and 10. These children show algebraic thinking in so far as they are able to explain how Peter's Method always works "for any number you are taking away" (Alan), "whether the subtracting number is increasing or decreasing" (Kou).

On the other hand, other students needed to close the sentence, by first deciding to calculate the results of 83 –7, 123 – 8, and 235 – 9, and then tried to calculate the number to place in the + □ on the right hand side. Eventually, some came up with a correct number, but interestingly, none could answer the question which asked them to explain how this method always works. Those who first calculated the left side of the equal sign seemed unable to ignore the 'starting number' and unable to leave the expression in unexecuted form. There were clear differences between these students and those who were comfortable with "a lack of closure". The present elementary school curriculum does little to shift students who are inclined to "close" away from this thinking.

Implications and Suggestions for Future Research

Three processes—expressing, transforming, and to reading—are all important elements of mathematical activity, and need to be related each other in how mathematics is described in curriculum documents and in how it is taught and learned. Particularly, the process of transformation needs to connect with the expressing and reading process. The research data have illustrated students' tendency to transform literal symbols without reading them carefully. This appears also to be true for numerical expressions. When students are dealing with generalizable numerical expressions or quasi-variable expressions, teachers have to assist students not to read these expressions as commands to calculate. Identifying the critical numbers and the relational elements embodied in these expressions requires students to focus especially on expressing and transforming the underlying structure. This has important implications for teaching and learning.

Many reports have confirmed that school algebra is difficult for students to understand. The problem should not be construed simply in terms of the cognitive demands that pertain to algebraic thinking as opposed to arithmetical thinking. Important as those cognitive elements are, there is also a serious problem in the way that algebraic thinking and arithmetical thinking have been separated in the school curriculum, especially in the elementary school. In a mathematics curriculum for elementary and secondary schools of the 21st century, we need to study and develop teaching approaches to connect these three processes of mathematical activity. Starting in the elementary years, this can be achieved by exploring the potentially algebraic nature of arithmetic.

Aspects or levels of understanding of literal symbols revealed by surveys held in Japan may serve to help teachers see clearly the diverse conceptual demands of teaching school algebra from its beginnings. It is important for teachers to use teaching approaches that help to integrate the "definite" and "unspecified" aspects of variable.

Any reform of the curriculum of the elementary and secondary school must consider the role of algebra as a tool for mathematical thinking about numerical expressions long before children are introduced to formal symbolic notation. The latter particularly can provide a stronger bridge to algebra in the later years of school, and can also strengthen children's understanding of basic arithmetic.

NOTE

1. http://www.corestandards.org/the-standards/mathematics

REFERENCES

Booth,L.(1984). *Algebra: children's strategies and errors: A report of the Strategies and Errors in Secondary Mathematics Project.* NFER-NELSON

Carpenter, T. P. & Levi, L. (1999). Developing conceptions of algebraic reasoning in the primary grades. *Paper presented at the Annual Meeting of the American Educational Research Association,* Montreal, Canada.

Chalouh, L.,& Herscovics, N.(1988). Teaching algebraic expressions in a meaningful way. In XX (ed.) *The Ideas of Algebra, K-12.* NCTM yearbook (pp.33-42).

Clement, J., Lochhead, J. & Monk, G.S. (1981). Translation difficulties in learning mathematics. *American Mathematical Monthly* 88, 286-290.

Collis, K.F. (1974). Cognitive development and mathematical learning. *Paper presented at the Psychology of Mathematics Workshop, Center for Science Education,* Chelsea College, London. Quoted from Herscovics, N.(1989). Cognitive obstacles encountered in the learning of algebra. *Research Issues in the Learning and Teaching of Algebra*, NCTM, pp.60-86

Davis, R. (1975) Cognitive Processes Involved in Solving Simple Algebraic Equations. *Journal of Children's Behaviour.* 1(3), Summer. 7-35.

Dunkels, A. (1989). What's the next number after G? *Journal of Mathematical Behaviour.8(1)*, 15-20.

Ministry of Education, Culture, Sports, Science, and Technology (1999). *Elementary school teaching guide: Arithmetic*. Tokyo: The Ministry

Fujii, T. (1993). Japanese students' understanding of school mathematics focusing on elementary algebra. In G. Bell (Ed.) *Asian Perspectives on Mathematics Education.* The Northern Rivers Mathematical Association, Lismore, NSW Australia. 70-89.

Fujii, T. (1993). Japanese students' understanding of school mathematics focusing on elementary algebra. In Garry Bell (Ed.) *Asian Perspectives on Mathematics Education.* The Northern Rivers Mathematical Association, Lismore, NSW Australia. 70-89.

Fujii, T. (2001). American students' understanding of algebraic expressions: focusing on the convention of interpreting literal symbols. *Journal of Science Education in Japan.* Vol.25 No.3 167-179

Fujii, T. and Stephens, M. (2001). Fostering an understanding of algebraic generalization through numerical expressions: The role of quasi-variables. In H. Chick, K. Stacey, J. Vincent, & J. Vincent (Eds.), *Proceedings of the 12th ICMI Study Conference. The future of the teaching and learning of algebra* (pp.258-264). Melbourne: University of Melbourne.

Fujii, T. (2003). Probing students' understanding of variables through cognitive conflict problems: Is the concept of a variable so difficult for students to understand? *Proceedings of the 2003 Joint Meeting of PME and PMENA*, 1, 49-65

Herscovics, N. (1989). Cognitive obstacles encountered in the learning of algebra. *Research issues in the learning and teaching of algebra* (pp. 60-86), Reston, VA: National Council of Teachers of Mathematics.

Kieran, C. (1981). Concepts associated with the equality symbol. *Educational Studies in Mathematics*, 12, 317-326.

Kieran, C. (1992). The learning and teaching of school algebra. In D. Grouws (Ed.) *Handbook of research on mathematics teaching and learning* (pp. 390-419). New York: Macmillan.

Kodaira, K. (1992). *Japanese Grade 9 Mathematics.* UCSMP Textbook Translations. Chicago: University of Chicago School Mathematics Project. p.97

Kuchemann,D.(1981). Algebra. *Children's Understanding of Mathematics:11-16*, General editor K.M. Hart & John Murray, pp.102-119.

Takahashi, A, T. Watanabe and M. Yoshida(2004). *Elementary School Teaching Guide for the Japanese Course of Study: Arithmetic (Grade 1-6).* Madison, N.J. Global Education Recourses.

Matz, M. (1982). Toward a computational theory of algebraic competence. *Journal of Mathematical Behavior,* 3(1), 93-166.

Miwa, T. (2001). Crucial issues in teaching of symbolic expressions. *Tsukuba Journal of Educational Study in Mathematics*, 20, 1-22.

Olivier, A. (1988). The construction of algebraic concept through conflict. *Proceedings of the 12th Conference of the International Group for the Psychology of Mathematics Education.* 511-519

School Mathematics Study Group. (1960). *First Course in Algebra Part I.*

Radford, L. (1996). The role of geometry and arithmetic in the development of algebra: Historical remarks from a didactic perspective. In N. Bednarz, C. Kieran & L. Lee (Eds.) *Approaches to algebra,* (pp. 39-53). Dordrecht: Kluwer Academic Publishers.

Schoenfeld, A & Arcavi, A. (1988). On the Meaning of Variable. *Mathematics Teacher,* 81, 420-427.

Skemp, R.R. (1976). Relational understanding and instrumental understanding. *Mathematics Teaching* No.77 December, 20-26.

Takahashi, A, T. Watanabe and M. Yoshida (2004). *Elementary School Teaching Guide for the Japanese Course of Study: Arithmetic (Grade 1-6).* Madison, N.J. Global Education

Watanabe, T. (2007). Algebra in elementary school: A Japanese perspective. *Algebra and algebraic thinking in school mathematics, Seventieth Yearbook.* Reston, VA: National Council of Teachers of Mathematics.

Van Engen, H. (1961a). A Note on "Variable". *The Mathematics Teacher* March, 172-173.

Van Engen, H. (1961b). On "Variable"?a rebuttal. *The Mathematics Teacher* March, 175-177.

REFERENCES IN JAPANESE

Anada, Y and Hirono, N (2005). A survey on students' logical thinking: Focusing on their way of classifying polynomial expressions of second degree. *Proceedings of the 38th annual conference of Japan Society of Mathematical Education,* 61-66.

Fujii, T. (1989). An analysis and assessment of student's mathematical understanding through cognitive conflict problems: Understanding of the linear equation and inequality. *Journal of Japan Society of Mathematical Education, Reports of Mathematical Education,* 53, 3-31.

Fujii, T. (1992a). The Clinical and Tutoring Interview on Understanding of Literal Symbols I n School Mathematics. *Journal of Japan Society of Mathematical Education, Reports of Mathematical Education,* 74, 3-27.

Fujii,T(1992b). Cognitive conflict and the compartmentalization in mathematical understanding (?). *Proceedings of the 25th annual conference of Japan Society of Mathematical Education,* 83-88.

Fujii, T.(1994). American students' of understanding of literal symbols. *Proceedings of the27th annual conference of Japan Society of Mathematical Education,* 113-118.

Fujii, T. (1999). From numeral to literal symbols: Teaching and learning of quasi-variables. In *Toward new mathematics education,* Tokyo: Tokyokan.

Fujii, T. (2001). Study on Understanding of Literal Symbols in School Mathematics : An Analysis of Student's Understanding through Cognitive Conflict

Problems. Journal of Japan Society of Mathematical Education, Reports of Mathematical Education, 76, 19-24

Fujii, T. & Stephens, M. (2002)?The role of quasi-variable in the Number and Calculation Strand? *Proceedings of the35th annual conference of Japan Society of Mathematical Education*?163-168

Fujii, T. & Stephens, M. (2006)?Fostering children's algebraic thinking at elementary school? *Proceedings of the39th annual conference of Japan Society of Mathematical Education*?307-312

Fujimoto, K (2004). A teaching method based on students' thinking types: A case study on teaching the 7th graders of literal symbols using cognitive conflict problems? *Proceedings of the37th annual conference of Japan Society of Mathematical Education*?271-276

Furukawa, S.(2005). A fundamental elements behind students' understanding of literal symbols? *Proceedings of the38th annual conference of Japan Society of Mathematical Education*?253-258

Hirano, S. (1986). Teaching students to find the way to solve simultaneous equation: Applying scoops of super balls. *Journal of Japan Society of Mathematical Education*, 68(1), 14-25

Hiroma, Y. (1990). Investigating efficiency of the formulas that use ? as strategy : - What can be learned from teaching 2nd graders-. *Journal of Japan Society of Mathematical Education*, 72(8), 198-202

Hoshino, M (2002). Educational values and meanings of literal symbols: Introduction stage at 7th grade. *Proceedings of the35th annual conference of Japan Society of Mathematical Education*?217-222

Iijima, Y. (2010). On the ideas of solution of kafusoku-zan in "Kyusyou-Sanjyutu". *Journal of Japan Society of Mathematical Education*, 92(1), 2-11

Ishida, J. (1994)?A study on teaching of writing of mathematical expression based on meaning of words problems that including addition and subtraction. *Proceedings of the 27th annual conference of Japan Society of Mathematical Education* 395-400

Itagaki, M. (1997)?A study on lower secondary students' understanding of literal symbols: Focusing on the duality of literal symbols in the context of constructing of formula of getting area. *Proceedings of the 30th annual conference of Japan Society of Mathematical Education,* 235-240

Koiwa, D (2004). Developing task that probe students' understanding of literal symbols: Focusing on the process-product. *Proceedings of the 37th annual conference of Japan Society of Mathematical Education* 259-264

Koyama, M. (2008). Improving Teaching and Learning of the Number and Expression Strand in Lower Secondary School Mathematics. (Special Issue: Toward New Curriculum (III)). *Journal of Japan Society of Mathematical Education,* 90(9), 21-30

Kondo, Y. (2007). A study on lower secondary school students' ability of expressing words problems into mathematical expressions: Focusing on the words problems that particularly difficult for elementary school children. *Proceedings of the 407th annual conference of Japan Society of Mathematical Education* 355-360

Kunimune, S. & Kumakura, H. (1996). A study on levels of students' understanding of literal expressions. *Journal of Japan Society of Mathematical Education, Reports of Mathematical Education*, 65, 35-55

Kumakura, H. , & Kunimune, S. et al. (1993). Proofs by literal expressions (The 4th Report) : Proofs and variables. *Journal of Japan Society of Mathematical Education*, 75(7), 10-12

Kumakura, H. ,et al. (1994). Demonstration of letter symbols (Fifth Report). *Journal of Japan Society of Mathematical Education*, 76(7), 11-19

Kume,S., Matsuo,Y, Murakami, Y, & Takahashi,N. (1990). A study on children's understanding of literal symbols: Focusing on the results of written survey. *Tokyo Gakugei Journal of Mathematics Education*, 2?27-35

Kyougoku, K. (1999). Curriculum for teaching function based on correspondence and variability: Practice and evaluation in the area of numbers and algebraic expressions. *Journal of Japan Society of Mathematical Education*, 81(3), 69-77

Makino, M (1996). Cognitive gap in understanding of literal symbols. *Proceedings of the29th annual conference of Japan Society of Mathematical Education*?37-42

Matsubara, G (1977) . *Mathematical way of thinking*, Tokyo: Kokudo publishing company.

Mizoguchi, T (1998)?Constructing a frame work to grasp children's cognition of equal sign?*Proceedings of the31th annual conference of Japan Society of Mathematical Education*?75-80

Maruyama, T. (1978). On the Advisability of Using ?,?,etc.as Symbols : Centering around the Results of Fact-finding Investigations. *Journal of Japan Society of Mathematical Education, Reports of Mathematical Education*, 60(6), 6-9

Miwa, T?1996?Introduction of teaching of literal symbols, *Tsukuba Journal of Educational Study in Mathematics*. 15,1-14

Morimoto, A (1993). Hearing impairments children's leaning of literal symbols. *Proceedings of the26th annual conference of Japan Society of Mathematical Education*?459-465

Morita, T (2005). A study on transition from arithmetic to algebraic method of solving linear equation. *Proceedings of the38th annual conference of Japan Society of Mathematical Education*?325-330

Morota, T. (2006). A study on transition from arithmetic to algebraic method of applying linear equation. *Proceedings of the39th annual conference of Japan Society of Mathematical Education*?325-330

Morozumi, T (1993). Reading mathematical expressions in school mathematics? *Proceedings of the 26th annual conference of Japan Society of Mathematical Education*?133-138

Morozumi, T (1997). ?Reading mathematical expressions in school mathematics: Focusing on three years study with same students , *Proceedings of the30th annual conference of Japan Society of Mathematical Education* Reading mathematical expressions in school mathematics, 253-258

Naito, H. (1989). Teaching formulas that Employ Symbols and Letters Instead of Numbers. *Journal of Japan Society of Mathematical Education*, 71(10), 283-288

Nakamura, K. (1971). Commentary on Euclid's Elements (in Japanese). Tokyo: Kyoritsu.

Yokota, M(1994). On the duality of literal symbols *Proceedings of the27th annual conference of Japan Society of Mathematical Education*?47-52

Nakanishi, C & Suzuki, Y. et al. (1992). Proofs by literal expressions: A classroom study. *Journal of Japan Society of Mathematical Education*, 74(11), 2-12

Naomi, K.,et al. (1987). Teaching the way of solution in simultaneous equation: Solving problems through the specific situation and its evolutional treatment. *Journal of Japan Society of Mathematical Education*, 69(5), 147-157

Ohta, S. (1990). Cognitive development on a letter formula. *Journal of Japan Society of Mathematical Education*, 72(7), 242-251

Ohta, S. (1992). Cognition on literal expression of lower secondary students. *Journal of Japan Society of Mathematical Education*, 74(9), 275-283

Saito, Y. (1989). How equations be taught in elementary school: Special attention to developmental reasoning-. *Journal of Japan Society of Mathematical Education*, 71(4), 70-74

Sakamoto, Y. (2001). Teaching method of quadratic equations in the new Course of Study: A study based on number extension of the solution of quadratic equations. *Journal of Japan Society of Mathematical Education*, 83(5), 10-24

Sato, R. (1983). Solving some functional equations. *Journal of Japan Society of Mathematical Education*, 65(11), 286-289

Shimada, N. (1991). A study of developmental introductory assignment on quadratic equations. Journal of Japan Society of Mathematical Education, 73(5), 116-121

Shimizu, H. (1997)?Lower secondary students' understanding of literal symbols: Focusing on the way of seeing them from a structural perspective. *Proceedings of the 30th annual conference of Japan Society of Mathematical Education*, 247-252.

Soeda,Y.(1992)?Students' cognition of rhetoric of Mathematical notation: Based on the survey focusing on mathematical expressions. *Proceedings of the 25th annual conference of Japan Society of Mathematical Education*, 77-82

Sugiyama,Y.?1990??On reading of mathematical expressions?*Tokyo Gakugei Journal of Mathematics Education*, 2, 17-25

Suzuki,Y(2005)?Supporting students to express words problems of linear equation by using literal symbols: try and error, and quasi-variables. *Proceedings of the 38th annual conference of Japan Society of Mathematical Education*, 247-252

Suzuki, Y. (2006). Analysis of the Results of PISA2003 and TIMSS2003 : High School Mathematics. Journal of Japan Society of Mathematical Education, 88(1), 23-31

Suzuki,Y?1989??Analysis on high school students' strategies of solving inequalities: Focusing on the incorrect conception and its' relation to the strategy. *Proceedings of the 22th annual conference of Japan Society of Mathematical Education* 91-96

Takeshita, S (2009)?A study on the process of expressing situation into mathematical expression. *Proceedings of the 42th annual conference of Japan Society of Mathematical Education*?265-270

Takizawa, Y (2003)?A study on individual teaching method to become cognitively independent student in the context of learning of literal symbols? *Proceedings of the 36th annual conference of Japan Society of Mathematical Education*?145-150

Tanaka, F. (1978). On Recurring Inequalities. Journal of Japan Society of Mathematical Education, 60(5), 92-97

Tanaka,Y (2003). Teaching and learning of literal symbols from the pro-cept way of thinking in lower secondary school students. *Proceedings of the 36th annual conference of Japan Society of Mathematical Education*, 127-132

Tsukada,T (1992)?High school students' understanding of mathematical expression. *Proceedings of the 25th annual conference of Japan Society of Mathematical Education*?53-58

Tsunamoto,H(2008)??Teaching of literal symbols using the quasi-variables: Dveloping of teaching materials. *Proceedings of the 41th annual conference of Japan Society of Mathematical Education*?297-302

Tobioka, M. (1972). Changes on teaching method of quadratic equation. *Journal of Japan Society of Mathematical Education*, 54(9), 184-195

Yokoyama, S. (1957). Judgment and demonstration of the authenticity of inequality formed of polynomials of the first order on each letter. J*ournal of Japan Society of Mathematical Education, Reports of Mathematical Education*, 39(9), 145-151

Yoshino, K. (1993). A Consideration on Lesson for Introducing Letters as Variables. *Journal of Japan Society of Mathematical Education, Reports of Mathematical Education*, 75(10), 261-268

Yoshizawa, M. (2001). A suggestion for teaching algebraic equations. *Journal of Japan Society of Mathematical Education, Reports of Mathematical Education*, 83(7), 36-42

Yoshino, K. (1993). A consideration on lesson for introducing letters as variables. J*ournal of Japan Society of Mathematical Education, Reports of Mathematical Education*, 75(10), 261-268

CHAPTER 62

PROVING AS AN EXPLORATIVE ACTIVITY IN MATHEMATICS EDUCATION

New Trends in Japanese Research Into Proof

Mikio Miyazaki
Shinshu University, Japan

Taro Fujita
University of Exeter, United Kingdom

ABSTRACT

The aims of this chapter are to summarize the findings and outcomes of recent research into the teaching and learning of proof in Japan, and to specify their implications for future research. In section 1 we provide some brief general background information about the current teaching and learning of mathematics and specifically proof in Japan. Then, we classify research studies presented and published in the last 10 years into 'global studies' and 'local studies.' The former broadly examine general matters related to proof and proving such as the nature of proof, historical develop-

The First Sourcebook on Asian Research in Mathematics Education: China, Korea, Singapore, Japan, Malaysia, and India, pp. 1375–1407

ments, students' understanding of proof across the grades and so on. The latter examine more specific aspects of proof and proving. We then give a critical review of the studies in each category (sections 2 and 3), identifying key messages and conclusions. Finally, we outline the direction in which these suggest future research might be conducted in order to improve the current situation of the teaching and learning of proof in Japan (section 4).

Keywords: Proving, proof, explorative activity, vertical and horizontal-directions of research

1. PROOF AND PROVING IN JAPANESE MATHEMATICS EDUCATION

1.1 Proof and proving in the 'Course of Study'

In this section, we summarize the teaching and learning of proof in Japan as defined in the Japanese national curriculum, the 'Course of study'. In particular, we will describe the stage at which proof as a mathematical activity is introduced, what mathematical content is used to teach proof and so on. We also refer to results from recent national surveys in Japan to outline the current situation as regards to teachers' approach to teaching proof, students' understanding of proof, typical misconceptions and so on.

The Japanese 'Ministry of Education, Culture, Sports, Science and Technology in Japan' (MEXT) states that current Japanese school education is faced by a situation in which our society is becoming 'knowledge-based' and globalized, and as a consequence it is increasingly important for children and students to acquire and develop their 'Zest for life' through the harmony of a 'sound scholastic ability', a 'gentle mind, and a 'healthy body'. To cultivate their 'Zest for life', the educational laws which were prescribed 50 years ago were revised, and a new direction for education was proposed. In particular it is emphasized that education should aim for children and students systemically to develop the following three main areas of learning, i.e. 'basic and fundamental knowledge/skills', 'thinking/judgment/expression' and 'attitude for learning', within the balance between 'knowledge, morals and physical health'. In accordance with this reform, in 2008 the Japanese school curriculum was radically revised.

In this section, we shall give an overview of the teaching and learning of proof in the 'Course of Study'. Since 1958, the teaching and learning of proof has been taught to all students in Grades 8 and 9 of junior high schools (ages 13-15). In these grades, mathematical proofs are taught in two main topics 'Number and Algebraic expressions' and 'Geometry' as a means to study the properties of numbers and geometrical figures. While proof in 'Number and Algebraic expressions' is rather limited, proof is explicitly taught in 'Geometry'. Although there is no official teaching sequence and activity prescribed in the 'Course of Study', we can find the following progression in many textbooks and practice in schools:

- In Grade 7, students study properties of plane and solid figures informally but logically to establish the basis of the learning of proof;
- In Grade 8, students are introduced to formal proof through studying properties of angles and lines and congruent triangles, learn the structure of proofs (Figure 62.1, Okamoto et al., 2005, pp. 86-87) and how to construct proofs, and then explore and prove properties of triangles and quadrilaterals;
- In Grade 9, students study similar figures and properties of circles drawing on their consolidated ability to use proof.

The distinctive feature of the 2008 revision was the introduction of 'mathematical activities' as a key learning content across all of the four defined areas of mathematics 'Number and Algebra', 'Geometry', 'Function' and 'Making Use of Data'. These 'mathematical activities' are regarded as 'various activities related to mathematics, and in which students actively participate with a clear purpose', and include the following

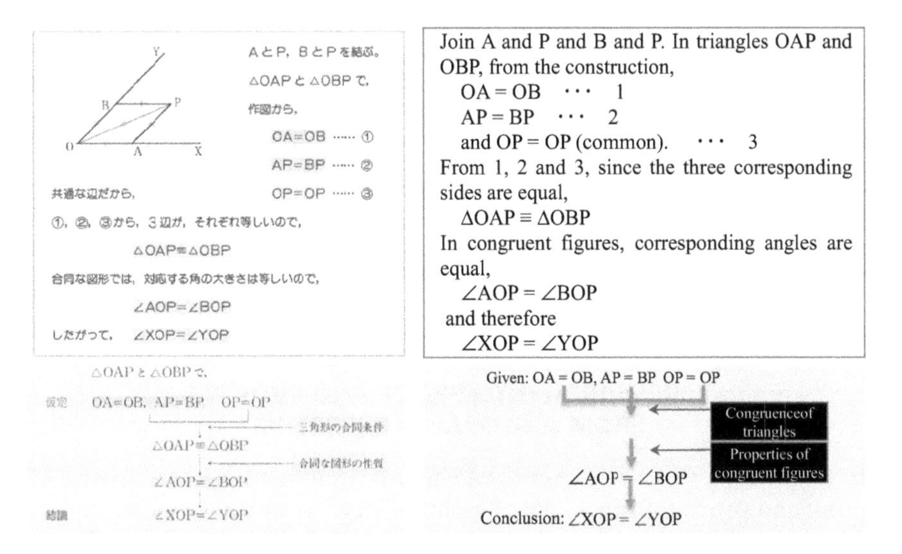

Figure 62.1. A proof and its structure in a mathematic textbook.

three kinds (MEXT, pp. 17-8): a) Activities for finding out the properties of numbers and geometrical figures based on previously learned mathematics; (b) Activities for making use of mathematics in daily life; (c) Activities for explaining logically and communicating with each other by clarifying grounds of reasoning with mathematical representations. Of these, the third point, i.e. formulating and communicating mathematical explanations, promotes learners' reflective thinking through interaction with others and supports the first two points. This aspect is emphasized in the teaching of mathematics in junior high school because of the particular emphasis on language activities, i.e. activities which promote use of language and discussion, which are considered to be vital pursuits to be introduced in all school subjects as one of the main themes in the revision of the 'Course of Study' (Nagata, 2008).

1.2. Students' Understanding of Proofs

While proof is taught to all junior high school students, do they really understand the concept of formal proofs? Evidence suggests the contrary. For example, according to findings in the National Assessment of Academic Ability over the whole of Japan, which has been undertaken by the Ministry of Education and National Institute for Educational Policy Research since 2007, many Grade 6 children in elementary schools have difficulties in explaining their mathematical thinking, and many Grade 9 students are unable to prove properties of numbers and geometrical figures. In particular, they struggle to distinguish between the roles of inductive and deductive explanations (the latter is proof), to plan and construct proofs (for more information, see the next section) and so on.

Against this backdrop, teachers and researchers have been trying hard to overcome various challenges with alternative approaches: some studies try to investigate the philosophical and epistemological nature of proofs and give historical accounts; others focus on various aspects of proof and proving, giving detailed accounts of children's understanding of proof, the proof construction process, and so on. In the next section, we shall describe some research results presenting a way forward to overcome these difficulties in proof and proving in school mathematics.

2. RESEARCH INTO PROOF AND PROVING FROM A GLOBAL PERSPECTIVE

In this section we review studies which have examined general aspects of proof and proving from a 'global' point of view, by which we mean studies of an overarching nature, not restricted to any particular stages / activities etc. In these studies, overall principles of the teaching of proofs and students' understanding of proof and proving across the grades are examined, and we focus on theoretical/historical studies (2.1) and empirical studies (2.1), and summarize their findings and implications (2.3).

2.1. Theoretical Studies Into the Teaching and Learning of Proof and Proving

We shall concentrate on two important studies here. One is Sugiyama's study (1985), which explores theoretically the teaching implications from a critical reflection of mathematical proofs. The other is an historical reflection of teaching and learning of proofs in Japan by Shimizu (1994, 1998).

2.1.1 Y. Sugiyama's Axiomatic Method

It is an important theme for mathematics education research to consider pedagogical implications and utilizations of genuine mathematical proof and proving, and Sugiyama (1985) focuses on 'axiomatic methods of mathematics', as the choice of axioms determines the systematic relationship between mathematical propositions and proofs. The study's main point is as follows: based on the views of axiomatic methods held by mathematicians such as van der Waerden (1967), he firstly distinguishes axiomatic methods into classical ones found in Euclid's Elements and modern ones used in Hilbert's Foundation of Geometry. Then, he summarizes the following two points as pedagogical principles of 'axiomatic methods' in school mathematics: 1) 'investigation of foundation of reasoning' and 2) 'thinking based on something assumed'. He considers the pedagogical value of the teaching of proof and proving based on this axiomatic method as follows: it promotes reflective thinking (Dewey) and strengthens 'deductive thinking based on premises' which is often lacking in Japanese people, and makes possible innovative and creative teaching and learning supported by the discipline of mathematics.

The teaching of mathematics based on the axiomatic method is that "students investigate the grounds of reasoning and seek sound knowledge and that students acquire new knowledge through solving challenging problems by using already established ideas and knowledge as assumptions" (p. 325), and this provides us with pedagogical principles for the teaching of proofs. For example, an introductory approach for the teaching of proof and an advanced approach in which students make use of the reflection of the already constructed proof are proposed. After studying a proof of the construction of an angle bisector, students should theoretically be encouraged to reflect on the construction process, and then should discover alternative methods to construct an angle bisector. It is also proposed that the axiomatic method is valuable to design effective teaching for not only proofs but other areas of mathematics, e.g. algebraic structure (Klein four group), multiplication of decimals, multiplication of positive and negative integers, inequality, square roots, division of fractions, areas and so on.

2.1.2 Historical Studies by S. Shimizu

To understand current mathematics education, it is essential to study its historical development. A comprehensive and detailed historical account of the development of the teaching of proof in Japan is given by Shizumi Shimizu (1994, 1997).

Accompanying the modernization in the mid 19th century which marked the end of the 'Samurai' era, 'Wasan ('Wa' and 'San' mean 'Japanese' and 'arithmetic' respectively)', i.e. Japanese traditional mathematics, was deemed outdated compared to western mathematics and was replaced by western-style mathematics. The first geometry textbooks Elementary Plane Geometry and Elementary Solid Geometry (1888 and 1889), which reflected the Euclidean style of rigid deductive proof and aimed to introduce such ways of thinking from the western world to Japanese people, were edited by Dairoku Kikuchi, who studied mathematics at St. John's College, Cambridge in England. Kikuchi excluded practical and experimental approaches in geometry, and treated the theory of proportion very rigorously. While Kikuchi successfully wrote a textbook whose quality was on a par with western textbooks, many people among both teachers and learners of geometry at that time found the content too difficult and demanding. Mathematics educators exchanged varied opinions as to how to tackle these issues over a long period. In particular the focus of debate was the tension between such difficulties and the educational value of following a rigorous way of the teaching of geometry, which it was considered would promote a more scientific attitude and so rectify Japanese people's defects in this area (Shimizu, 1998). During debates throughout the 1920's various reforms in western countries such as those triggered by the Perry movement and International Congresses of Mathematics Education (1908, 1911) were discussed, and in 1931 the syllabus for secondary mathematics teaching, which had been based on Kikuchi's ideas, was finally revised and these new ideas were introduced. In particular, "Kika-zukei (geometrical figures, 'Kika and Zukei' mean 'geometry' and 'figures' respectively)", comprising intuitive and experimental approaches such as observation, drawing, measurement, and constructions of figures as a preparation for deductive geometry, was introduced in the first grade in secondary schools (pupils aged 13). It included not only the study of plane figures, but also of solid figures, and developing students' concept of space was mentioned. Further, it was suggested that the teaching of proof should provide students with opportunities to discover mathematical concepts by themselves rather than giving already established content, to grasp the importance of observing dynamic aspects of geometrical figures, to accept intuitively trivial properties and relations and to use them as a foundation of their reasoning by easing rigorous axiomatic systems and so on (see also Yamamoto and Fujita, 2006).

After World War II, the school system was reformed and proof was introduced in the high school curriculum (15-18 yrs old) in 1951. A philosophical underpinning was the idea of relative views of truth. This is similar to what Fawcett (1938) pointed out, i.e. the discovery of non-Euclidean Geometry challenged the epistemological nature of mathematics truths, and the way of thinking that truths are absolute was replaced by one that they are in fact relative, and hence this idea should also be introduced in wider mathematics education. In the 'Course of Study' announced in 1958 all lower secondary school students were intended to study proof in geometry. The official teachers' instructional guidance (1959) emphasized the above idea of proof and proposed the teaching of proof by asking students 'to determine several starting points from already learnt contents and explore logical reasoning based on these' (Ministry of Education, 1958, p.93). Proofs in the overall discipline of mathematics and those in school mathematics were clearly differentiated, as it was suggested that too much emphasis on rigorous approaches would not motivate students' learning. The 'Course of Study' has been revised five times (1969, 1978, 1988 1998 and 2000). Although there have been small differences in the treatment of proof in each version, until the current 'Course of Study' announced in 2008, the fundamental ideas and principles of the teaching of proof have been consistent since the 1958 version.

In his historical study which reflects on the principles and philosophy of the teaching of proof in Japan, Shimizu (1994) reviews the critiques of Euclid by Arthur Schopenhauer, Ernst Mach's construction of deductive thoughts as an alternative to thought experiments, the difference between the process of discovering facts and systematizing these facts by Heath, and various other mathematicians' views regarding proofs. He then summarizes three perspectives about the teaching of proof: a) proof should be placed within creative activities in which students discover various mathematical facts and make conjectures about them; b) while ways of thinking in discovery/making conjecture should be clearly distinguished from systematic ways of thinking, the strengths of both approaches should be appreciated; c) 'analytical' thinking rather than 'synthetic' should be emphasized and utilized as the fundamental way of planning and constructing proofs.

2.2. Empirical Studies Into the Teaching and Learning of Proof and Proving

Several studies based on classroom practice or empirical data address students' understanding of generality and construction of geometrical proof across the grades (Koseki et al, 1978,1980, 1987; Kunimune, 1987). Large-scale surveys (National Institute for Educational Policy Research, 2009a, 2009b, 2009c) also provide insight into this.

2.2.1. Studies of the Significance of Geometrical Proof

In the study by Koseki, Kunimune et al (1978, 1980, 1987), they propose that it is necessary for students to understand at least the following five aspects to prove a statement in geometry: symbolization, proving, representative nature of figures, universal generalization, and significance of proof (1987, p. 22). These studies propose teaching strategies for each aspect based on relatively small empirical data from students in public junior high schools, with their main attention on the three aspects of symbolization, proving, and universal generalization.

Students' understanding of the significance of proof is further considered in terms of the following two aspects: 'Generality of proof' and 'Construction of proof' which students need to master. The first, an understanding of the 'generality of proof in geometry', means students need to grasp the universality and generality of geometrical theorems (proved statements), the roles of figures, the difference between formal proof and experimental verification and so on. At the same time, they also have to learn how to 'construct' deductive arguments in geometry; definitions, axioms, assumptions, proof, theorems, logical circularity, axiomatic systems and so on, i.e. 'Construction of proof in geometry'. In order to measure students' grasp of these two aspects, the following levels of understanding are proposed (for each level there are two sub-levels, a) and b) (Kunimune, Fujita and Jones, 2009a, p. 257):

- Level I: students consider experimental verifications are sufficient to demonstrate that geometrical statements are true (Level Ia: Students have not achieved both 'Generality of proof' and 'Construction of proof' and Level Ib: Students achieved 'Construction of proof' but not 'Generality of proof').
- Level II: students understand that proof is required to demonstrate that geometrical statements are true (Level IIa: Students have achieved 'Generality of proof', but have not understood logical circularity and Level IIb: Logical circularity is also understood).
- Level III: students can understand simple logical chains between theorems.

These levels are empirically validated by a survey conducted with students across the grades (G5-9). Together with these results and reflections from classroom practice, the instructional ideas which particularly encourage students to transfer from level I to II are proposed. For example, Grade 8 lessons should provide students with explicit opportunities to examine differences between experimental verifications and deductive proof using the topic of the sum of the inner angles of triangles (1987, p.156, see also Kunimune et al, 2009).

2.2.2. Large-scale Survey: National Assessment of Academic Ability

Since 2007 the MEXT and National Institute for Educational Policy Research has undertaken annually a large-scale survey, the National Assessment of Academic Ability in Japanese and Mathematics, with G6 children and G9 students. So far the entire population (2007-9) and a sample of 30% of students (2010) participated in this survey. For each subject, two categories of problems are given. The first category is aimed to establish the level of 'knowledge'. The questions in this category test children's and students' basic knowledge which is essential for their studies and to live everyday life. The other category tests 'application' and assesses children's and students' ability to use and apply their knowledge to various problems in everyday life, and their ability to make plans to solve a problem, solve it with the plans and then evaluate and improve the process and outcomes productively. With regard to the survey questions concerning proofs, the first category questions ('knowledge') mainly ask students the meanings and roles of deductive proofs, and the second category questions ('application') ask students to plan a proof analytically/synthetically, construct a proof based on already established plans, correct a proof with logical circularities, discover new properties based on already completed proofs, read proofs and extend them and so on (see also Chino, et al, 2010).

The results and findings of these large-scale surveys suggest useful information about the teaching and learning of geometrical proof. For example, from the 2008 survey, while only 44.1% of G9 students could analytically/synthetically plan and construct proofs, 67.0% could identify properties other than a conclusion, which can be deduced by using the conditions of congruent triangles used in the proof. The 2009 survey found that only 29.7% of G9 students could clearly distinguish the different roles between an inductive explanation and a deductive proof of the sum of the inner angles of triangles.[1] The summary reports by the Institute for Educational Policy Research suggest that it is important to make students appreciate deductive proofs by making explicit the difference between inductive and deductive reasoning, as the former cannot guarantee the generality of properties or relationships. (For further information, a teaching plan for improvement of lessons can be found on Institute for Educational Policy Research website (2009c).)

2.3. Implications of Research Findings

The main findings from global studies can be summarized as follows. From theoretical/historical studies (2.1), the pedagogical underpinning of proof and proving in schools can be regarded as 'axiomatic methods', and historical study identifies what proof and proving mean in school in Japanese school mathematics, i.e. 'proof should be placed within creative activities in which students discover various mathematical facts and make conjectures about them'. Also empirical studies (2.2) have revealed various aspects of students' understanding of proof and proving: significance of proof, planning and construction of proof (synthesis and analysis), discovery of new properties based on proved propositions, discrepancies between inductive explanations and deductive proofs.

These findings suggest that it is necessary for us to establish what it is about the teaching and learning of proof which we consider most valuable in school mathematics and can contribute most to life–long learning for independence. To achieve this, we should clarify the epistemological nature of proof and proving within Japanese historical and cultural contexts, and pursue ways to secure the position of proof and proving in the mathematics curriculum by working collaboratively and closely between researchers and teachers. This should help us to build a clear case for why the teaching and learning of proof is necessary and can compete for its place in the curriculum alongside other areas of mathematics and other subjects. Also, based on the findings from large-scale surveys and their teaching plans for improving everyday lessons, some research into administrative matters is necessary to consider how we can best disseminate these findings to local governments as well as head teachers and senior teachers who are responsible for improving local schools and hence are important decision makers we would most like to influence.

3. RESEARCH INTO THE TEACHING AND LEARNING OF PROOF AND PROVING FROM A LOCAL PERSPECTIVE

In this section we review studies which have examined closely and locally the specific aspects of proof and proving such as its emergence in children's learning, its development and valuable activities related to proof and proving in school mathematics since 2000 (before 2000, see Kunimune, 2010). These studies give a comprehensive account of several aspects of the teaching and learning of proof and proving which include the seeds (3.1), the developmental process (3.2), students' proving activities (3.3) concerning proof and proving, and curriculum development (3.4). In order to elucidate the key impacts of proof and proving on school mathematics it is necessary to clarify not only existing practice but also its sources and the developmental journey. Since this section 3 is long, we include discussion of the implications of research findings at the end of each subsection.

3.1. Research Into the Emergence of Ideas of Proof and Proving

It is important for educators to consider how the ideas of proof and proving are introduced to students, and recent studies have focused on informal proofs such as action proof and operative proof which can become the 'seeds' of proof and proving. These studies examined theoretically and empirically not only the significance and uses of informal proofs, but also how to promote children and students' mathematical thinking by taking such informal approaches in the teaching of proofs.

Komatsu (2009b) studies the interrelationship between enactive representations of objects (arrangement and manipulation) and symbolic representations such as positional notations, and finds the coordination between these two representations deepens students' mathematical understanding, and contributes to their ability to explain why (mathematically correct) conjectures they make are always true. Also, Komatsu (2010b) focuses on instructional ways to promote action proof within the context of mathematical inquiry in which students resolve their questions through interacting with others, make conjectures and prove them, and continue to investigate and extend what they have proved. In this study, he identifies several factors required for students to generate new mathematical statements: students are encouraged to perform new actions by manipulating objects and looking at them in a different way, to interpret the process and results of these new actions, to reflect on what conditions are generated by the actions, and so on. In another study (Komatsu, 2010a), he empirically shows that confronting counterexamples functions as a driving force for elementary school children to refine their conjectures and proofs. In his case study, children understood the reason why their conjecture was false through their analysis of its proof and therefore could improve their primitive conjecture. They also identified the part of the proof which was applicable to the counter-example, and this identification was essential for their invention of a more comprehensive conjecture.

Sasa and Yamamoto (2010) focus on the role of operative proof in the teaching and learning of mathematics, which originates from Piaget's genetic epistemology (Beth and Piaget, 1961) and Wittmann (1996)'s operative principle. Their study concludes that operative proof has the possibility to support the formal proofs at advanced levels and promote discovery of other mathematical patterns.

In summary, these studies mainly concentrate on algebraic content but have not yet investigated what action and operative proof approaches would be effective in other areas such as geometry and function. In addition, more theoretical reflection and frameworks are still necessary to firmly situate these approaches in elementary schools' mathematics curricula. Further, more study is required a) to investigate what we can identify as the 'seeds' of proof and proving to be 'planted' in children's minds, and b) to consider how these seeds can be nurtured and developed, not only in mathematics but also other subjects, i.e. research must be conducted from a cross-curricular perspective.

3.2. Research Into the Developmental Process of Proof and Proving

The studies which examine the developmental processes of proof and proving mainly focus on the transition processes between informal proof (triggering the emergence of insight into proof) and formal proof as the next step. By reflecting theoretically on the characteristics of algebraic and geometric proofs and the cognitive and social dimensions to learners' proving, these studies propose levels of understanding marking progress from informal to formal proofs and examine the validity of these levels and processes from evidence from classroom-based research.

Okazaki and Iwasaki (2003) first consider that it is necessary to bridge a 'gap' between the teaching and learning of geometry in elementary and junior high schools. In order to do this, they particularly focus on the part played by children's and students' learning about relationships between geometrical figures, and suggest ideas for activities such as utilizing shapes and properties as a means to investigate other shapes, constructing and determining shapes by combining several geometrical properties, considering the ordinal relationship between properties and scrutinizing the nature of assumptions. They then conduct further thorough research into geometrical constructions, selected as a topic which integrates these activities, and specify five levels and factors to facilitate the students' transition between them. Based on this theoretical study, Okazaki and

Komoto (2009a, 2009b) and Okazaki (2010) identify factors which advance students' understanding of proofs by considering inference, recognition of figures, social influences and the interrelationships among these three. For example, when students are tackling whether mathematical phenomena hold or not, if they investigate a range of propositions where the phenomena do hold, through making inductive explanations and/or finding counterexamples, then this might help them to progress their approach from an experimental to a more deductive level.

Miyazaki (2000) proposes six theoretical levels of proof in lower secondary school mathematics representing a transition process from an inductive argument to an algebraic proof on the basis of three axes, i.e. content of proof, representation of proof, and students' thinking. These levels and students' developmental shifts are examined in the context of one concrete example activity involving the properties of five consecutive numbers, and how ideas of more formal proof emerge when G7 students engage with action proofs.

In summary, the levels of attainment examined in these studies are derived from assessing students' mastering of the Japanese geometry/algebra curriculum. They are useful to describe Japanese students' understanding of proof and proving in the context of their own Japanese mathematics curriculum and to identify what is necessary to improve that understanding in the same context, but it is still uncertain what attainment as described by these levels would be deemed sufficient for the learning of proof and proving. In future research, we need to identify the sufficient conditions for the learning of proof and proving, and then to exemplify activities or teaching instructions based on these conditions, situate them within the mathematics curriculum, and examine how these approaches would be effective for the actual teaching and learning of algebraic and geometric proofs.

3.3. Research Into Students' Proving Activities

Intrinsically, proving activities are investigative in nature (3.3.1), and they can be characterized by at least the following three components i.e. producing propositions (3.3.2), producing proofs (3.3.3) and looking back at proving processes and proofs (3.3.4). It is important to study the roles and functions of these three components to promote students' understanding in the context of teaching and learning of proof and proving. Also it is necessary to develop and design tasks and a curriculum which enable both students and teachers to achieve proof by means of explorative activities (3.3.5). Research in this area is diverse and can be further categorized into both general and specific studies, i.e. some research deals with just one of the three components listed above, some with two and some with all three components and their contribution towards proving activities.

3.3.1. General Studies Into Proving Activities

Recent and ongoing studies have been focusing on the use of the rules of inference, the process of developing explanations in problem solving, cultural considerations in Japanese classrooms and so on. For example, Miyakawa (2004a, 2004b) investigates the way of using the rules of inference in the problem solving process of geometrical construction and proving problems related to reflective symmetry, and found by analyzing pairs of students' problem solving processes based on the cK¢ model that a) even if students can correctly construct a symmetrical figure and are conscious of the rules which are necessary for proof, they cannot always use the rule appropriately in proving, and b) even when the students do not have the appropriate rule for proving, they are still able to generate some of their own rules from geometrical properties already known.

Nunokawa (2010) examines how the experience of exploration and understanding influence the development of students' explanation-building processes and how the activities themselves affect students' exploration approaches and contribute to students' understanding by analyzing the problem solving process in examples from the USA mathematics Olympiad. In this study, implicit assumptions embedded in problems and expected explanations sometimes provide learners with appropriate clues as to the solutions or generate new directions of thoughts to be explored. Based on these findings, the interplay between understanding and exploration plays an important role in the process of learners' construction of full explanations arising out of local explanations to the example in their problem solving.

Sekiguchi (2002) studies the cultural aspects of proving and proofs in the Japanese classroom and found that these aspects correspond to what can be conceptualized as a 'group model' of sharing and reaching a consensus without challenge regarding proof and proving in Japan. He also found that Toulmin's model can be applied to reveal the structure of 'Discussion and sharing' which has traditionally been considered as an important process in Japanese mathematics classroom (though this tendency towards consensus approach is even easier to recognize in general styles of communication among Japanese people.) In addition, he concludes that "the idea of locating learning of mathematical proof in the context of argumentation is biased by Western culture", and notes that though it is culturally difficult to achieve argumentation and teach proof by means of argumentative activity because of the Japanese group model nature, the importance of argumentative activities should be recognized in our modern more internationalized society.

In summary, the studies in this section have revealed various aspects of proving activities: some students can use rules of inferences in geometrical constructions but not proofs; within the process of constructing explanations, the interrelationship between investigation and understanding plays an important role; the process of sharing proofs can be understood in terms of a 'group model' in which students respect each other's ideas, and so on. At the same time, it is still unknown how the interrelationship between investigation and understanding contributes to promote progress from the ability to explain local parts of a proof to being able to formulate a complete proof. Additionally, though the weakness itself is identified that argumentative activity is not culturally easily accepted for Japanese students, how we can achieve this in Japanese classrooms in our current internationalized society is not yet clear. Furthermore, the activities examined in these studies are limited to 'processes of construction of proofs of given propositions'. Meanwhile, in addition to these activities, the following activities and their interrelationships should be generally considered: producing propositions, producing proofs (planning and constructing) and looking back (evaluating, improving, discovering). We consider we should try to realize these activities and interrelationships in the classroom, and emphasize their importance in the teaching and learning of proof and proving. Future research should tackle the following research questions: 'How can we promote these activities in lessons?', 'What kinds of interrelationships among various activities can be considered?', 'How should learning activities be designed?' and so on. As we can see in 3.3.2-3.3.5, some researchers have started posing questions in order to try to bridge this gap.

3.3.2. Producing Propositions

Producing propositions is one of the most important proving activities as this is not only the starting point towards proving, but also a driving force behind the establishment of the universality of propositions. In the 1970s, Japanese researchers focused on children's logic, but recently they are studying children's understanding of statements in a wider context.

For example, Otsuka (2008) constructs a framework for capturing children's interpretation of propositions. This framework considers the subjects which children's proposition refers to and their perspectives how to see the subjects in the proposition. In addition it also examines the contexts where the children interpret the proposition. There might be three kinds of contexts as follows: the context in which children try to understand the proposition, the context in which children try to prove it and the context in which children try to use it as rules of reasoning. By using this framework he finds how children's varied interpretation of propositions leads them to identify counterexamples.

Studies in this area pay attention to children's recognition of the universality of propositions to examine how producing propositions can be a driving force towards learning how to construct proof. Miyazaki (2008) has constructed a conceptual framework to capture students' immature understanding regarding the establishment of the universality of propositions through experimentation/measurement. It is represented with a quadrangular-pyramid model comprised of five representative aspects as its vertices: idealism/teleology, pessimism, optimism, realism, and naïve pre-established harmony. In constructing the framework, he uses the following three viewpoints: (1) students take as a criterion of the universality of propositions whether the proposition matches the result of experimentation/measurement or not; (2) students assume that the result will always match the proposition; and (3) the students consider improving the method of experimentation/measurement. He suggests that students waver between these representative aspects, calling this their 'unstable perception', and identified through questionnaires the unstable perception of the universality of the sum of inner angles of a triangle.

Finally, how analogical reasoning is used to extend and develop propositions has also been studied. For example, Nakagawa (2006) defines 'two propositions are analogical' as showing 'correspondence of the relationship between elements of two propositions' based on Polya's idea, and applies this idea to two propositions proved by the mid point theorem of triangles. Then, he points out that if students observe what elements of propositions are similar or not, then it might be possible for them to examine whether these propositions are analogical or not, to generate propositions by examining the relationships of elements of 'analogical propositions' and general properties behind the propositions, and to organize the generated propositions. He designed 4 hours of a teaching experiment and 2 hours of a survey, which had the result that 12 very able students in a private cram school 'jyuku' could generate new propositions based on proofs and organize them by using analogical reasoning together with a teacher's appropriate teaching guidance.

In summary, in the process of producing propositions, frameworks are proposed to capture students' various interpretations of mathematical propositions. For effective planning and constructing of proofs, it is however necessary to study further the relationship between producing propositions, construction of proofs (planning and construction) and reflection, in particular asking questions such as 'How are propositions generated by students, and what properties and qualities do these generated propositions have/what should they have in terms of universality of propositions?', 'How does reflecting on steps taken to achieve proof contribute to generating propositions and what properties and qualities do these generated propositions have?' and so on.

3.3.3. Producing Proofs: Planning and Construction

We have already stated that the introduction of formal proof is located in G8 in the Course of Study. Following this most teachers hope that as many students as possible will be able to construct formal proof by themselves, and this goal is particularly emphasized in everyday mathematics lessons. However, a problem is that too much emphasis may be on just 'how to write' proofs and many students do not truly understand and so cannot use what they learn. Based on the research findings from large-scale surveys and global research projects, researchers have started addressing the importance of planning of proof.

For example, Tsujiyama (2011a) focuses on the role of looking back in proof-planning process from the view point of argumentation, and found three characteristics of the argument-production process based on Toulmin's mode of arguments: observing whether the conclusion holds or not, showing the argument's uncertainty and clarifying what condition is necessary to be added. Subsequently, he (2011b) examines the overall proof-planning process including the use of plausible reasoning and revised the above three characteristics into the following four points to capture the process with Toulmin's model: making arguments which might include a leap and/or uncertainty, removing the leap and/or uncertainty, clarifying conditions under which the arguments hold, and clarifying 'backings' for the 'warrant'. By considering the imaginary proof-planning process with the four points, clarifying conditions under which the arguments hold leads students to uncover the process where they can make use of the arguments in special cases, and asking students to clarify 'backings' for the 'warrant' reveals their thought process and assumptions.

Makino (2007) focuses on the relationship between planning and construction of proofs, and identifies empirically that the following three factors are necessary to effectively plan and construct proofs: (1) to change flexibly a prior proving plan when the proving process ends in deadlock through making a mistake; (2) in any state of planning and constructing a proof to check whether definition, hypotheses or conditions are enough to derive a conclusion; and (3) to examine the logical order of the elements of a proof in writing a proof.

Studies on the construction of proof examine various topics including curriculum development in the transition from elementary and secondary school mathematics, the use of metaphors in the teaching of proof, development of teaching units based on the structure of proofs, the use of flowchart proof at the introductory stage of the teaching of proof, the use of two-column proof in high school (Oguchi, 2008), teachers' effective interventions in the proof construction process in the Dynamic Geometry Environment (DGE) (Fukuzawa, 2002) and so on. Haneda and Shinba (2002) developed a geometry curriculum in G8 and implemented lessons and approaches to the teaching of deductive geometry based around a set of 'already-learnt' properties which are shared and discussed within the classroom. In these lessons students had the opportunity to conduct productive mathematical discussion and argument and were encouraged to think deductively to construct proofs. (See also Fujita, Jones and Kunimune (2010) which analyzes this teaching experiment in terms of cognitive unity.) In relation to the use of metaphors for teaching proofs, Sekiguchi (2000) employs a metaphor of 'Adventure' which encompasses the ideas of 'source', 'destination', 'path', and 'direction' and adds two more factors, 'tool' and 'ideas'. He concludes, based on a five-month teaching experiment, that the teaching of proof using these metaphors can be effective for many G8 students, and make it accessible and enjoyable for them.

In the development of teaching units based on the structure of proofs, Miyazaki and Yumoto (2009) found the teaching and learning of a proof as an 'object' in lower secondary school mathematics involves at least four factors: a) recognizing/constructing a proof roughly, b) making the distinction between universal propositions and singular propositions, c) recognizing/making the deductive relationship between a universal proposition and a singular proposition, and d) organizing all the deductive relationships between universal propositions and singular propositions. Similarly, Yumoto (2007) designed and implemented 7 lessons in which students are introduced to flow-chart proofs at the introductory stage of proof learning. During the first four lessons, through constructing open flow-chart proofs, students develop their holistic understanding of the structure of proofs which they define as the relational network via deductive reasoning that combines singular and universal propositions (see also Miyazaki and Fujita, 2010; Miyazaki, Fujita and Jones, 2011). Then in the last three lessons, after introduction of formal proofs, students spontaneously rewrote their flow-chart proofs into more formal ones, and used flow-chart proofs to check whether formal proofs which they constructed are correct or not.

In summary, one of the important implications from these studies is that it is possible to capture planning processes of proof by utilizing Toulmin's layout of arguments. For actual constructing proof processes, principles of curriculum development which bridge between elementary and junior high schools are proposed and established. Other studies examine various approaches based on ideas from metaphors, structure of proofs, flow-chart proofs, two-column proof, the teaching and learning with DGE and so on. Also, there are studies which focus on the relationship between planning and construction of proofs. For future research, it is necessary to develop more comprehensive frameworks to capture students' planning processes. Also we need to establish this process as one of the objectives of proof teaching within a curriculum, and develop the teaching and learning to promote students' synthesized and analytical thinking.

3.3.4. Looking Back of Proving: Interpretation, Evaluation, and Improvement

It is essential to utilize the functions of proofs to effectively look back at proving processes and learn from the experience. In research into this area, researchers have analyzed the functions of proofs, and examined how children utilize them, what factors are important to encourage children to use and apply these functions and so on. Also, some studies explore how students accept or do not accept logical circularities of proofs as a part of the looking back processes of proof and proving.

Makino (2003) defines an understanding of mathematics through proof as a "general-special" relationship which is made up of several already known facts, and shows that students are able to organize the relationship between propositions by reviewing an original proposition and its proof from "an alternative perspective". In particular, he indicates that such learning can generate "an alternative perspective" in the process of understanding through proof.

Miyazaki (2000) focuses on the discovery function of proof, i.e. that the process of mathematical proof provides us with new results and things to learn from including not only proposition and proof but also tacit assumptions, naïve concepts, counterexamples, refined definitions and so on. This study also found several conditions which trigger students to apply one or more functions in lower secondary school mathematics. Chino (2003) concentrates on the discovery function of proof, and finds that even if several students have constructed the same proof about a proposition, individually they might actually have imagined different propositions, and that some students could fully understand what they had proved for the first time only after having completed constructed a proof. Also Chino (2005) reports that, in theory, students should have an opportunity to generate a more sophisticated proposition based on already completed proofs when they are provided with a hint, for example the beginning words of proof problem 'see the figure on the right …'etc), but in practice they interpret the hint in many ways, and not many students manage to fully utilize the opportunity to generate a sophisticated proposition.

Komatsu (2009b) defines the concept of action proof and points out its three characteristic properties: embodiment, movability and genericity. This study finds two kinds of discovery functions of action proof, and the result indicates that action proof provides us with a productive learning opportunity at primary school level in which pupils invent new statements through previously established action proofs.

In order to evaluate students' understanding of the structure of proof more closely, Miyazaki and Fujita (2010) capture students' understanding of proof in terms of three levels: Pre-structural, Partial-structural (Elemental, Relational), and Holistic-structural. By analyzing an episode where students had begun to accept a proof with a logical circularity, the two aspects concerning universal instantiations and syllogism at the Relational level are considered to explain the reason why these students accepted the wrong proof. The study concludes that this is due to their lack of understanding of syllogism, i.e. they failed to make a logical connection between the assumptions of a proposition and its conclusion. In their recent studies, Fujita, Jones and Miyazaki (2011) studied this issue further, by utilising flow-chart proofs with open situations and web-based learning support system (flow-chart proof system see http://www.schoolmath.jp/flowchart_en/home.html). In open situations, students can construct different proofs by changing premises under certain given limitations. The case studies in this paper show that students who engaged proof problems in open situations could overcome the logical circularity gradually by considering possible combinations of assumptions and conclusion. Also feedback given by the flow-chart proof system enables learners to check whether their proof fell into logical circularity or not. They suggest that the kinds of activity available with the web-based flow-chart proof system are useful to understand the whole structural relationship between assumptions and conclusions more deeply, to encourage learners to shift the level of the understanding of proof structure, and that this may lead to them, in the end, overcoming the error of logical circularity.

In summary, it is evident in the above studies that the discovery function of proofs has been particularly studied, and the factors which promote this function are well identified in the teaching and learning of not only formal proofs but also informal proofs such as action proofs. Also, some students accept a proof which includes logical circularity during their evaluation processes, and the levels of understanding of the structure of proof are proposed and this framework is utilized to explain why these students accept the circularity. Based on these findings, future research should continue to further examine the process of reflection and looking back in proof and proving. For example, a spiral relationship is considered between producing propositions, producing proofs and looking back (see also 4.1) and it is interesting to examine how these aspects play various dynamic roles within the proving activities in the teaching and learning of proofs or within the curriculum as a whole. Also, producing new properties based on already completed proofs is now specified as one of the objectives in the Course of Study, and it is necessary to consider how this process, considered as a product of looking back, should be promoted in relation to the discovery function of proofs. Furthermore, there has not been much research into the other functions such as systematization (e.g. organizing already proved propositions) and communication (e.g. how to express proofs) and we need more consideration into these areas. From a life-long learning point of view, these perspectives are important not only for the teaching and learning of proofs but also, for more general human development as systematization and communication are essential skills for everyday life.

3.4. Curriculum and Task Development

It is essential to design a curriculum and tasks which enable both teachers and students to effectively generate propositions, plan and construct proofs and to evaluate and improve proofs, which are all important aspects of proof and proving. Traditionally, Japanese teachers and researchers have made a particular effort to develop such tasks based on interesting mathematics problems, for example nine-point circles, star-shapes, and so on. More recently, we can find studies which focus on the development and implementation of investigative proving tasks and lesson plans (e.g. Kyogoku, 2001), propose a sequence of learning content which should encourage students' logical thinking, and so on.

Other studies explore what activities should be designed for junior high school students based on the epistemological relationship between geometry and geometrical space. For example, as a task which encourages students' advanced mathematical thinking, Nakagawa examines the following, 'Consider a proof in which a bisector of inner angles is used. Replace the condition 'inner angles' to 'exterior angles' and examine dynamically what properties are changed/unchanged.' He concludes that this kind of investigation would encourage students to developmentally investigate propositions, see geometrical figures dynamically, appreciate proofs, and so on. As to a desired sequence of learning content, in order to enhance students' deductive and logical thinking in similar figures he (2010) first examines how properties of parallel lines and ratios are inductively and analogically derived from the generalization of mid points theorem, then proposes that it is important to start from the properties of parallel lines and ratios to prove other properties of similar figures. He designs a task in which students examine the relationship between the respective heights of a man and his shadow standing between two lamp posts which have the same height.

Miyazaki (2005) also points out that the following activities will be essential for junior high school students in the development of a geometry curriculum when we consider epistemological relationships between geometry and real space: 1) students should conjecture geometrical properties and relationships construct proofs, and then verify them by experiment and measurement and organizing the propositions deductively; 2) students should investigate, by reflecting on geometry and space surrounding us, the universality of propositions used as axioms in a local system of propositions.

In summary, the studies in this area propose principles to develop students' logical thinking, to promote developmental thinking based on already proved propositions, to give directions for activities based on epistemological relationships between geometry and space, and so on. One problem and challenge in this research area is the existence of the 'Course of Study' which has a legal power to restrict mathematical content to be taught in schools. Because of this, it is sometimes difficult to develop and design tasks and curriculum which are not intended in the 'Course of Study', and the design principles for ideal proof teaching have yet to be fully and theoretically established. For example, it is possible, theoretically, to design a curriculum which can achieve proofs in elementary number theory or systematization of transformation geometry, but because of curriculum restriction, such studies do not currently exist. In addition to examining whether it would be worth extending proof study to these branches of mathematics, for the next version of the Course of Study, it is essential to further cultivate our interest into various aspects of activities within proofs and their relationships, and it is necessary to develop a) a long-term spiral curriculum which has two axes, i.e. system of mathematics and activities of proofs, and b) an assessment scheme to evaluate student's understanding and achievement within this curriculum.

4. NEW PERSPECTIVES OF RESEARCH INTO PROOF AND PROVING

4.1. Proving as Explorative Activity

As we have stated in sections 1 and 2, the teaching and learning of proof in Japan has been emphasized in the mathematics curriculum since the late 19th century to 're-educate' the traditional Japanese mind-set by introducing a rigorous way of thinking, although there have been slight changes in accordance with educational, social and cultural demands and attitudes of the times. Thus the Course of Study requires students to study formal proof in mathematics, but the reality of the current situation, which is that many students have difficulty in studying proofs, is a disappointment for mathematics educators. Furthermore, because of the influence of the 100% paper-based high school entrance examinations, 'how to write proofs' is often over-emphasized in the teaching of proof in Japanese junior high schools. Under these circumstances, many students merely learn passively how to write a proof for a given proposition and do not consider what, why and how to prove in various (unfamiliar) situations, and hence it is very hard for them to think by themselves and express their own ideas in developing new proofs. For these students, the learning of proof is just an 'initiation process' or 'right of passage' for the entrance examinations. Consequently for many adults, their memory of learning proof is that it was dull and difficult. There have been efforts by many teachers and researchers to improve this situation, but we question whether these efforts are sufficient, and how we could establish through evidence whether the situation is genuinely improved. The teaching and learning of proof is now facing challenging issues and it is time to sincerely reflect on the past and consider what should be our new direction.

It is needless to say that in mathematics (discipline) proof and proving is not limited to just writing proofs. As we have seen some researchers particularly point out the importance of modern axiomatic methods and relative views of truths and have learned from the perspectives of heuristics and fallibilism, proving activities are flexible, dynamic and productive in nature, and various aspects of proving activities are interrelated and resonant with each other. Namely, we can see that proving activities 'breathe intellectual life' into mathematics, for example: producing propositions inductively/deductively/analogically, planning and constructing proofs for these produced propositions, and reflecting on and looking back at producing propositions, including planning and constructing proofs to overcome local and global counter examples, and then refining propositions and proofs. Mathematics as an activity is continuously developed by these processes which work dynamically together as 'intellectual gears', as if water wheels (proving activities) give power to propel a big steamboat (mathematics).

It is therefore our task to make the teaching and learning of proof truly valuable for all students and help their independence. To achieve this, we need to rid ourselves of the 'rote learning' and ritual culture attached to teaching and learning proof, and to bring intellectual breath of mathematics education as well as more general education.

As discussed in section 3.3, it is considered that there are at least three key components to proof and proving, i.e. producing propositions, producing proofs (planning and construction) and looking back (interpreting, evaluation and improving). This model (see Figure 62.2) will provide us with an idea of how to make the teaching of proof more valuable for teachers and students, i.e. 'proving as an explorative activity'. If students

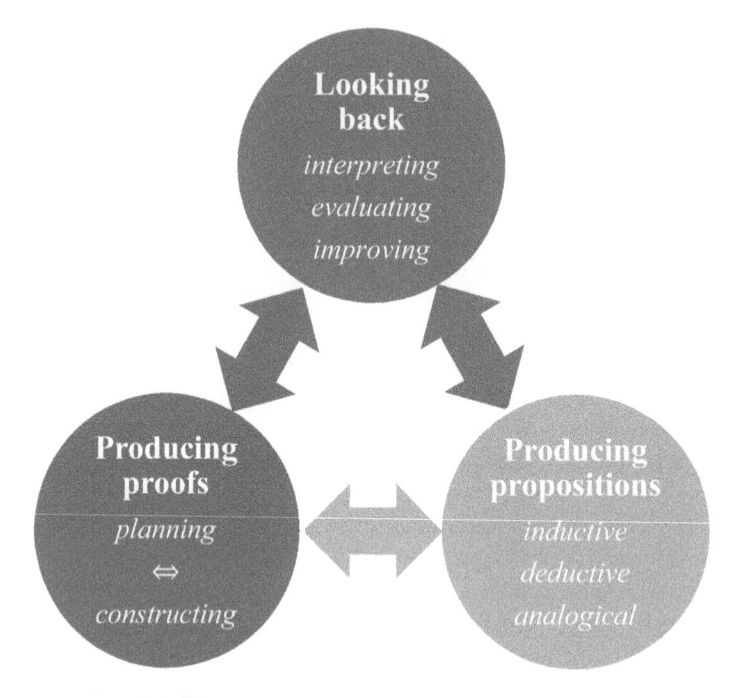

Figure 62.2. Proving as an explorative activity.

would be able to actively utilize the three components and understand their interrelationships, which should make proving activities more dynamic, then proving in school mathematics can be seen as a productive activity which is not only "transplanted" from mathematics but also should be a core feature of learning in general.

While research into 'proving as an explorative activity' has only recently started, a theoretical framework to evaluate it and the quality of learning which it offers has already been proposed by Chino et al (2010). In their paper, the framework is constructed based on the three components listed above and the responses to problems in large-scale surveys (as referred to in section 2.2.2) are analyzed to establish 'which component of proving does this survey problem assess?' As we have seen in section 3, there are several studies which examine each component individually, but research which recognizes these components combine to make 'proving as an explorative activity' is still limited. It is necessary to re-examine these existing studies from the perspective of 'proving as an explorative activity', to undertake further research into these components and their relationship as a whole, and establish theoretical underpinning and evidence from practice. For example, it is necessary to consider theoretically the value and functions of the relationships between the three components in relation to other disciplines such as philosophy of science and psychology, and then design tasks for students in which they could effectively utilize the values and functions by undertaking empirical studies in not only experimental but also classroom contexts. Based on these micro findings, in order to crystallize the idea of 'proving as an explorative activity' into the Course of Study for all students, we should investigating 'What principles of curriculum development will be necessary?', 'How do we develop the curriculum, and how would this curriculum be effective and limited?', and so on.

4.2. Vertical and Horizontal Directions of Research Into 'Proving as an Explorative Activity'

In order to achieve the learning of 'proving as an explorative activity' it is unwise to focus only on particular school-grades or age groups. Rather, research into the teaching and learning of proof should consider the long-term span of students' progress from elementary school to university, i.e. research of a 'vertical' nature. To achieve this, it is necessary to conduct new research based on theoretical foundations, 'seeds' (origins/ triggers), development of learning etc with regard to 'proving as an explorative activity' based on a thorough examination of the three components and their interrelationships.

Previous studies have been inclined to concentrate on formal proofs as the goal of learning, and have sought to establish what can be the foundations, seeds and development towards formal proof, concluding often that the answer lies in the disciplined nature of mathematics. But we should re-examine these themes of study in the light of 'proving as an explorative activity' as the new goal of learning. For example, we first of all should have a consensus about the following questions: what is, exactly, 'proving as an explorative activity', and why do we have to teach and learn this?, 'what activities can be worth being the seeds of 'proving as an explorative activity' to grow?', 'how should these seeds develop to more formal ways of thinking and approaches of proving, and what levels and stages of understanding could be necessary to develop?' Of course, we might utilize existing research findings, but we consider novel researches into 'proving as an explorative activity' should be undertaken from a completely new perspective because the nature and goal of the teaching and learning of 'explorative proving' (with its potential to make teaching richer and more productive) and conventional 'proof teaching' (which has been seen as 'dull' for many students) are totally different. Future knowledge based on collective research findings should then establish a firm basis and principles of 'proving as an explorative activity' and it is expected that we will be able to develop a curriculum in which a cyclical relationship between the three components we have outlined is embedded into all stages of the teaching and learning of proof and proving, and develops via an upward spiral from elementary school to university.

Fortunately in Japan, 'activities to explain' are envisaged in various contexts of the teaching of mathematics in the early stages of elementary school in the current 'Course of Study'. Core to 'proving as an explorative activity', is that explaining ideas and reasons is important and hence it is expected that this kind of learning opportunity will contribute to bridge the gap between 'activities to explain' and 'proving' in junior high schools, and will enable us to develop a consistent curriculum of explorative proving throughout compulsory education.

It is equally important to undertake new research to consider how to achieve the teaching and learning of explorative proving within all areas of mathematics, i.e. research of a 'horizontal' direction nature. As we have described, in Japan the teaching and learning of proof and proving is intended to be addressed mainly in the areas of 'Number and Algebra' and 'Geometry' through studying properties of number and figures (see section 1). At the moment less opportunities for proof and proving are provided for students in the other two areas of the Course of Study, 'Function' and 'Making Use of Data', and educators do not have a consensus about what 'proof' is in these areas, although students are still required to explain (prove) why facts are correct. Furthermore, little research has been carried out into proof and proving in these areas.

When considering the nature of 'proving as an explorative activity', we do not have to restrict its impact on proof and proving to the areas of school mathematics only; indeed it is theoretically possible to provide this in any area of school subjects (horizontal direction) because there should be no "walls" of how to learn for learners between these subjects and also between school life and everyday life. We also can and must consider the value of such an approach surrounding the teaching of proof and proving in other school subjects. In particular, if explorative learning is essential for students' independence and life-long learning, then it is necessary for us to examine the values of this approach from a wider perspective, and consider how to achieve explorative learning in each area and subject. A similar thing can be considered in terms of the horizontal direction, i.e. we do not have to restrict the teaching and learning of proof and proving to a specific age group of students, but it should be considered in any stage of education.

In summary, it is necessary to establish 'proving as an explorative activity' through the grades (horizontally) and across not only the areas of school mathematics but also subjects, even living areas (vertically). In

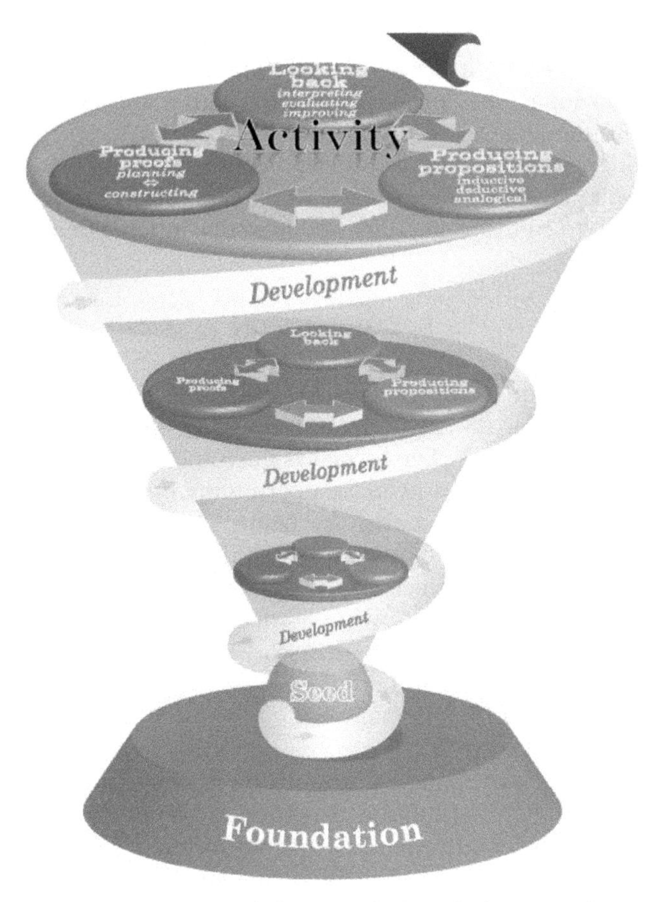

Figure 62.3. Research framework: Foundation × seed × development × activity.

order that we should undertake research studies into the learning of 'proving as an explorative activity' based on the nature of mathematical content, and researches into the relationship with other subjects and everyday life within the framework of the reciprocal relationships between the foundations, seeds, development, and activity of proving (Figure 62.3).

These research studies will contribute to make the uniqueness and necessity of learning proof in mathematics clearer by enriching the explorative aspects of proving and clarify the relations with the outside of mathematics. In other words, through such research, the specific and general purposes of proof and proving, how this relates to other subjects, and the processes involved in acquiring and mastering proof and proving will be clarified and widely accepted by not only mathematics educators but also very ordinary people. Proof and proving in Japanese school mathematics will no longer be "propagated" in isolation and become idiosyncratic like animals evolved in the Galapagos Islands. We can expect that it will not only be appreciated as a means to inherit mathematical culture but also it will gain a secured position in the National curriculm as a core content for all to cultivate students' ability to live independently.

5. CONCLUSION

Proof and proving in mathematics is a dialogue which bridges between two separate worlds, the metaphysical world of mathematics and the physical world of humans. The former is pure and logical and reflects an Ideal world, and there is no need for questioning or exploration; the latter is full of imperfection echoing the reality of our human lives, which include dynamic and chaotic processes, including identifying gaps, failed solutions, evaluation, recognizing mistakes, improvement and repetition etc. Through the dialogue of proving human beings can gradually access the former Idealistic and mathematical world, but it can also be said that no one can access the world without the process of dynamicproving activities in the physical world of humans.

The pedagogical idea of 'proving as an explorative activity' enables proving to become a dynamic and productive activity for many students. Furthermore, through explorative activities, students do not just learn proof 'from the top,' (authoritarianism), but by their own activities and engagement with the tasks, they keep an equal relationship between themselves and mathematics in the Ideal world, i.e. anti-authoritarianism (Hanna, 1995). We consider this to be one of the unique values of the explorative proving approach in mathematics education and that this gives weight to our claim that the learning and teaching of 'proving as an explorative activity' deserves to be firmly established. Our goal is that 'proving as an explorative activity' should become a powerful 'whirlpool' entangling both school education, with its stated aim to develop 'Zest for life', and general life-long learning, which should add value to all people's lives. The energy for this whirlpool is generated from learning with explorative activity in which children and students plan, implement and evaluate/improve their learning to solve problems and tasks. From this initial small whirlpool, momentum will be gathered from across a range of subjects and it will eventually become a maelstrom in school education, and even more broadly in the world of education. To achieve this, researchers coming with a theoretical perspective must establish a long-term relationship with teachers who have daily encounters with children and students, must identify and select truly important future tasks for mathematics education from issues identified in mathematics lessons, and must provide concrete and achievable plans to improve the status quo by dialectic approaches, i.e. critical interaction between theory and practice.

ACKNOWLEDGMENTS

The authors thank the following people: Chino Kimiho, Komatsu Kotaro, Makino Tomohiko, Miyakawa Takeshi, Mizutani Naoto, Nagata Junichiro, Nakagawa Hiroyuki, Okazaki Masakazu, Otsuka Shintaro, Sasa Hiroyuki, and Tsujiyama Yosuke. This research is supported by the Grant-in-Aid for Scientific Research (No. 18330187, 22330245, 23330255, 24243077), Ministry of Education, Culture, Sports, Science, and Technology, and the University of Plymouth, UK.

NOTE

1. Similar tendencies had been pointed out by other studies such as Kunimune, et al (1987, 2009).

REFERENCES

Beth, E. W. and Piaget, J. (1961). *Epistémologie mathématique et psychologie. Etudes d'Epistémologie Génétique XIV.* Presses Universitaires de France, Paris.

Chino, K. (2003). Rethinking "discovery" as a function of proof in school mathematics. *Annual Report of Graduate Studies in Education (Doctoral Program in Education, University of Tsukuba), 27,* 73-83. [In Japanese]

Chino, K. (2005). Students' interpretation of situation in a proved conjecture described in a particular/alphabetical statement fitted to a figure. *Tsukuba Journal of Educational Study in Mathematics, 24,* 31-38. [In Japanese]

Chino, K., Fujita, T., Komatsu, K., Makino, T., Miyakawa, T., Miyazaki, M., Mizutani, N., Nakagawa, H., Otsuka, S. & Tsujiyama, Y. (2010). An assessment framework for students' abilities/competencies in proving. In Y. Shimizu, Y. Sekiguchi and K. Hino (Eds.) *Proceedings of the 5th East Asia Regional Conference on Mathematics Education (EARCOME5)* (Vol. 2, pp. pp.416-423). Tokyo: Japan Society of Mathematical Education.

Fawcett, H.P. (1938). *The nature of proof: a description and evaluation of certain procedures used in a senior high school to develop an understanding of the nature of proof* (The National Council of Teachers of Mathematics, The Thirteenth Yearbook), AMS PRESS, New York.

Fukuzawa, T. (2002). On teacher support during proof activities with dynamic geometry software. *Journal of Japan Society of Mathematical Education (Mathematics Education), 84*(7), 10-18. [In Japanese]

Fujita, T. Jones, K., Kunimune, S., Kumakura, H. and Matsumoto, S. (2011). *Proof and Refutations in lower secondary school geometry*. Paper presented at 7th Congress of the European Society for Research in Mathematics Education, University of Rzeszów, Poland. Retrieved 20th December 2010 from www.cerme7.univ.rzeszow.pl/WG/4/WG4_Fujita.pdf

Fujita, T., Jones, K. & Kunimune, S. (2010). Students' geometrical constructions and proving activities: a case of cognitive unity? In M. M.F. Pinto & T. Kawasaki (Eds.) *Proceedings of the 34th Conference of the International Group for the Psychology of Mathematics Education* (Vol. 3, pp 9-16). Belo Horizonte, Brazil.

Fujita, T., Jones, K. and Miyazaki, M. (2011). Supporting students to overcome circular arguments in secondary school mathematics: the use of the flowchart proof learning platform. *Proceedings of the 35th Conference of the International Group for the Psychology of Mathematics Education* (Vol. 2, pp. 353-360). Ankara, Turkey: Middle East Technical University.

Hanna, G. (1995). Challenges to the importance of proof. *For the Learning of Mathematics, 15,* 42-49.

Haneda, A. and Shinba, S. (2002). Improving teaching geometry by focusing on the process of drawing figures to developing proof. *Journal of Japan Society of Mathematical Education (Mathematics Education), 84*(5), 20-28. [In Japanese]

Komatsu, K. (2009a). A study on function of "action proof" in school mathematics: Focusing on discovery. *Journal of Japan Society of Mathematical Education (Research Journal of Mathematical Education), 92,* 35-47. [In Japanese]

Komatsu, K. (2009b). Pupils' explaining process with manipulative objects. In M. Tzekaki, M. Kaldrimidou, & C. Sakonidis (Eds.), *Proceedings of the 33rd annual conference of the International Group for the Psychology of Mathematics Education* (Vol. 3, pp. 393-400). Aristotle University of Thessaloniki.

Komatsu, K. (2010a). Counter-examples for refinement of conjectures and proofs in primary school mathematics. *The Journal of Mathematical Behavior, 29*(1), 1-10.

Komatsu, K. (2010b). Facilitating pupils' utilization of "action proof" in mathematical inquiry: A case study. *Journal of Japan Society of Mathematical Education (Research Journal of Mathematical Education), 93,* 3-29. [In Japanese]

Koseki, K. et al. (1978). Teaching a demonstration in geometry: the first part. Journal of Japan Society of Mathematical Education (Mathematics Education), 60, 12–19. [In Japanese]

Koseki, K. et al. (1980). Teaching a demonstration in geometry: the third report (the first part). *Journal of Japan Society of Mathematical Education (Mathematics Education), 62,* 59–65. [In Japanese]

Kunimune, S. (1987). A developmental study on the understanding of "significance of demonstration", in K. Koseki (Ed.), *Teaching a demonstration in geometry* (pp. 129–158), Meijitosho, Tokyo. [In Japanese]

Kunimune, S. (2010). Demonstration in geometry, In Japan Society of Mathematical Education (Ed.), *Handbook of research in mathematics education of Japan* (pp.123-131). Tokyou: Touyoukan. [In Japanese]

Kunimune, S., Fujita, T. & Jones, K. (2009b) "Why do we have to prove this?" Fostering students' understanding of 'proof' in geometry in lower secondary school. Lin, F-L, Hsieh, F-J., Hanna, G. and de Villiers, M. (eds.), *Proof and proving in mathematics education ICMI study 19 conference proceeding* (Vol. 1, pp. 256-261). Taipei: National Taiwan Normal University.

Kunimune, S., Fujita, T., and Jones, K. (2009a). Students' understanding of 'proof' in geometry in lower secondary school, *Proceeding of the 6th Congress of the European Society for Research in Mathematics Education,* University of Lyon, France. Retrieved 20th December 2010 from http://www.inrp.fr/publications/edition-electronique/cerme6/wg5-09-kunimune-et-al.pdf

Kyogoku, K. (2001). On Students' Actual Situation about Application of Basis in Proof : Based on comparison between generalization and application of theorems. *Journal of Japan Society of Mathematical Education (Mathematics Education), 83*(3), 18-26. [In Japanese]

Makino, T. (2003). A study on the proof in junior high school mathematics: Focus on the promotion of understanding through proof, *Annual Report of Graduate Studies in Education (Doctoral Program in Education, University of Tsukuba), 27,* 97-106.

Makino, T. (2007). On proving process by junior high school students: Analysis of conditions to plan, construct and write a proof, *Tsukuba Journal of Educational Study in Mathematics, 26,* 39-46.

Ministry of Education (1959). *The Teaching Guide to the Course of Study for Lower Secondary School Mathematics.* Tokyo: Meijitosho. [in Japanese]

Miyakawa T. (2004a). Reflective symmetry in construction and proving. In M. J. Høines & A. B. Fuglestad (Eds.), *Proceedings of the 28th annual conference of the International Group for the Psychology of Mathematics Education* (vol. 3, pp.337-344). Bergen University College.

Miyakawa, T. (2004b, July). *The nature of students' rule of inference in proving: the case of reflective symmetry.* Paper presented at the International Congress on Mathematical Education (ICME-10), Copenhagen. Retrieved from http://www.icme-organisers.dk/tsg19/

Miyazaki, M. (2000). Levels of proof in lower secondary school mathematics : As steps from an inductive proof to an algebraic demonstration. *Educational Studies in Mathematics, 41,* 47–68.

Miyazaki, M. (2000). What are essential to apply the "discovery" function of proof in lower secondary school mathematics?. In T. Nakahara, & M. Koyama (Eds.), *Proceedings of the 24th annual conference of the International Group for the Psychology of Mathematics Education* (Vol. 4, pp. 1-8). Hiroshima University.

Miyazaki, M. (2005). What kinds of activity are necessary to develop the curriculum of lower secondary school mathematics with epistemological relations between geometry and space?. *Tsukuba Journal of Educational Study in Mathematics, 24,* 1-10. [In Japanese]

Miyazaki, M. (2008). Cognitive Incoherence of Students Regarding the Establishment of Universality of Propositions through Experimentation/Measurement. *International Journal of Science and Mathematics Education, 6*(3), 533-558.

Miyazaki, M. & Yumoto, T. (2009). Teaching and Learning a proof as an object in lower secondary school mathematics of Japan. In Lin, F. L., Hsieh, F. J., Hanna, G., & de Villiers, M. (Eds.), *Proceedings of the ICMI Study 19 Conference: Proof and Proving in Mathematics Education* (Vol. 2 , pp. 76-81). Taipei: National Taiwan Normal University.

Miyazaki, M. & Fujita, T. (2010). Students' understanding of the structure of proof: Why do students accept a proof with logical circularity? In Y. Shimizu, Y. Sekiguchi and K. Hino (Eds.) *Proceedings of the 5th East Asia Regional Conference on Mathematics Education* (Vol. 2, pp. 172-179). Tokyo: Japan Society of Mathematical Education.

Nagata, J. (2008). Mathematics education envisioned by the new course of study (Toward New Curriculum (I)). *Journal of Japan Society of Mathematical Education (Mathematics Education), 90* (5), 14-22.

Nakagawa, H. (2010). Relationship between properties of similarities and ratios ? parallel lines. *Journal of Japan Society of Mathematical Education (Mathematics Education), 92*(9), 27-34?

Nakagawa, H. (2007). On relationships between the bisectors of interior angles and exterior angles. *Journal of Japan Society of Mathematical Education (Mathematics Education), 89*(11), 19-25.

Nakagawa, H. (2006). On the activities of developing propositions based on the Relations called analogy. *Journal of Japan Society of Mathematical Education (Research Journal of Mathematical Education), 85,* 22-41.

National Institute for Educational Policy Research (NIER) (2009a). *Frameworks of National Assessment of Academic Ability over the whole of Japan,* Retrieved from http://www.nier.go.jp/09chousa/09kaisetsu_chuu_suugaku.pdf [In Japanese]

NIER (2009b). *Reports of National Assessment of Academic Ability over the whole of Japan,* Retrieved from http://www.nier.go.jp/09chousakekkahoukoku/03chuu_chousakekka_houkokusho.htm [In Japanese]

NIER (2009c). *Teaching ideas based on National Assessment of Academic Ability over the whole of Japan,* Retrieved from http://www.nier.go.jp/09jugyourei/21_chuu_jugyou_idea_houkoku.pdf [In Japanese]

Nunokawa. K. (2010). Proof, Mathematical Problem-Solving, and Explanation in Mathematics Teaching. In G. Hanna, H. N. Jahnke, & H. Pulte (Eds.), *Explanation and Proof in mathematics: Philosophical and Educational Perspectives* (pp. 223-236). New York: Springer.

Oguchi, Y. (2008). The effect of two-column format for writing a proof in plane geometry in upper secondary school. *Journal of Japan Society of Mathematical Education (Mathematics Education), 90*(7), 2-9?

Okamoto, K., Koseki, K., Morisugi, K. et al. (2005). *Mathematics for the future.* Osaka: Keirinkan.

Okazaki, M. & Iwasaki, H. (2003). Geometric construction as an educational material mediating between elementary and secondary school mathematics: Theory and practice to promote transformation from empirical to logical recognition, *Journal of Japan Society of Mathematical Education (Research Journal of Mathematical Education), 80,* 3-27.

Okazaki, M. & Komoto, S. (2009a). The transition process towards logical proofs through synthesizing geometric transformations and geometric constructions: A design experiment of 7th grade unit "Plane Geometry", *Journal of JASME (Research in Mathematics Education), 15*(2), 67-79.

Okazaki, M. & Komoto, S. (2009b). The recognition of geometric figures cultivated through geometric transformations: The design experiment aimed at the transition towards proof, *Journal of Japan Society of Mathematical Education (Mathematics Education), 91* (7), 2-11.

Okazaki, M. (2010)?Development of reasoning ability towards proof using seventh grade Plane Geometry? In Y. Shimizu, Y. Sekiguchi and K. Hino (Eds.), *Proceedings of the 5th East Asia Regional Conference on Mathematics Education (EARCOME5)* (Vol. 2, pp. 188-195). Tokyo: Japan Society of Mathematical Education.

Otsuka, S. (2008). Construction of the framework for seeing children's interpretations of proposition in school mathematics, *Tsukuba Journal of Educational Study in Mathematics, 27,* 21-30. [In Japanese]

Yamamoto, S. and Fujita, T. (2006). A brief history of the teaching of geometry in Japan, *Paedagogica Historica, 42*(4&5), 541-545.

Yumoto, T. (2007). Improving teaching and learning of proof by using a flow chart. *Journal of Japan Society of Mathematical Education (Mathematics Education), 89*(7), 22-28?

Sasa, H. & Yamamoto, S. (2010). A study on operative proof in mathematics education: Operative proof with counters and the place value table, *Journal of JASME (Research in Mathematics Education), 16*(2), 11-20.

Sekiguchi, Y. (2002). Mathematical proof, argumentation, and classroom communication: From a cultural perspective. *Tsukuba Journal of Educational Study in Mathematics, 21,* 11-20?

Sekiguchi, Y. (2000). A teaching experiment on mathematical proof: Roles of metaphor and externalization. *Proceedings of the 24th Conference of the International Group for the Psychology of Mathematics Education* (Vol. 4, pp. 129-136). Hiroshima University.

Shimizu, S. (1994). Demonstration. In S. Shimizu (Eds.), *CRECER Geometry and Demonstration* (Vol. 6, pp. 204-236). Tokyo: Nichibun. [In Japanese]

Shimizu, S. (1997). Views on scholastic ability by Dr. D. Kikuchi and Dr. R. Fujisawa. In Japan Society of Mathematical Education (Ed.), *Mathematics education in Japan 1996: Philosophies of mathematics education in the twentieth century* (pp. 17-28), Tokyo: Sangyotosho.

Sugiyama, Y. (1986). *Teaching mathematics based on axiomatic methods,* Tokyo: Touyoukan. (ISBN-13: 978-4491025452) [In Japanese]

Tsujiyama, Y. (2011a, February). *On the role of looking back at proving processes in school mathematics: focusing on argumentation.* Paper presented at 7th Congress of the European Society for Research in Mathematics Education, University of Rzeszów, Poland. Retrieved 20th December, 2010 from http://www.cerme7.univ.rzeszow.pl/WG/1/CERME7_WG1_Tsujiyama.pdf

Tsujiyama, Y. (2011b). Processes of devising a plan to prove in school mathematics: focusing on argumentation. *Bulletin of Institute of Education University of Tsukuba, 35,* 41-53. [In Japanese]

Waerden, B.L. van der (1967). Klassische und moderne Axiomatik, *Elemente der Mathematik : eine Zeitschrift der Schweizerischen Mathematischen Gesellschaft, 22,* 1-4.

Wittmann E. Ch. (1996). *Operative proofs in primary mathematics.* Paper presented at 8th International Congress of Mathematics Education (the Topic Study Groups 8 "Proofs and proving "Why, when, how?"), Seville, Spain.

CHAPTER 63

DEVELOPMENTS IN RESEARCH ON MATHEMATICAL PROBLEM SOLVING IN JAPAN

Kazuhiko Nunokawa
Joetsu University of Education

ABSTRACT

This chapter overviews the trends and issues in research on problem solving in Japan and illustrates that it has constantly been affected by its tradition which values mathematical nature of problem solving and mathematical richness of thinking, even after Japanese researchers learned much about problem solving research from the U.S. and European researchers. Giving the attention to those mathematical nature and richness in their research on problem solving, Japanese researchers have elaborated the theoretical components of problem solving processes to some extent, so that those components can reflect mathematical thinking more explicitly and work as facilitators of students' thinking in mathematics classes.

Keywords: Mathematical problem solving, mathematical thinking, sense-making, metacognition, Polya

The First Sourcebook on Asian Research in Mathematics Education: China, Korea, Singapore, Japan, Malaysia, and India, pp. 1409–1436

INTRODUCTION

This chapter overviews the research on mathematical problem solving which has been carried out in Japan to demonstrate the outline of changes in those research since 1950s. Problem solving approach is often used in Japanese classrooms to encourage our students to learn mathematics actively and understand mathematical knowledge deeply. Such teaching practices in Japan and their features were already reported and analyzed well by other researchers (e.g. Hino, 2007; Shimizu, 2009). Thus, this chapter focuses on the research rather than teaching practices implemented in Japanese classrooms.

Problem solving and mathematical thinking has been emphasized in Japanese mathematics education since 1930s. Even when the research on problem solving during 1950s, 1960s and 1970s focused on the word problems, the researchers intended to develop students' mathematical thinking and discussed teachers' interventions taking account of students' solving processes. At the end of 1970s and the beginning of 1980s, the reform with the emphasis on mathematical problem solving (NCTM, 1980) and the related research carried out in the U.S. and Europe was introduced to the mathematics education community in Japan. Those works, like Schoenfeld (1985) and Silver (1985), had a great impact on the research on problem solving in Japan. Thus, this chapter mainly deals with the research before this impact (from the mid of 1950s to 1970s) and after that (from 1980s to 2000s) and demonstrates that the developments in the research on problem solving after that impact can be considered a kind of integrations of the above-mentioned impacts and the intrinsic features of the research in Japanese mathematics education community. It also illustrates that such developments have enabled us to understand more deeply students' thinking in learning mathematics as well as teaching styles based on problem solving.

BEFORE THE IMPACT

When a new textbook series, usually called "Green-Cover Textbooks," was edited during 1930s, the disposition to think mathematically was introduced by Naomichi Shiono, the official chief editor of that textbook series, as a priority issue of mathematics education in elementary schools. He expected children to develop this disposition to find mathematical aspects in phenomena and think those phenomena mathematically, as well as to inquire mathematical phenomena and appreciate their beauty (*cf.* Oku, 1983). For example, at the end part of the textbook for 6th grade students, we find the following problem: "A tree is growing in at a certain place. This tree became 1 meter in height at the end of the first year. It grew by 50 cm next year, and it grew by 25cm in the third year. Like this, this tree grows by the half of the growth of the previous year. How tall can this tree become?" Because this problem was posed for elementary school children who had not learned a limit of series, students might be expected to find mathematical phenomenon in this situation, inquire a pattern in that phenomenon, and be impressed by the unexpected answer which could lead them to enjoying this inquiry. Shimada (1987) pointed out that developing such a disposition or scientific spirit is an appearance of the "tradition to seek a meta-physical or meta-technical attitude behind every kind of arts, techniques, or disciplines" (p. 10). He also suggested that this disposition and the underlying philosophy could be "interpreted as identifying the learning process with the process of problem solving in its broader sense" (p. 10). It is usually said that Shiono's emphasis on this disposition has been passed down and made an important aspect of mathematics education in Japan (*e.g.* Oku, 1982).

The influence of the emphasis on the disposition to think mathematically can be observed in mathematics education in 1950s. For example, Ohya & Fujiwara (1958) pointed out the necessity of developing the following aspects of students' thinking through mathematical problem solving: logical thinking, insights, competence in applying mathematical knowledge and skills, and thinking everyday phenomena mathematically. They also directed their attention to the problem solving process students should experience. This process is similar to 4 phases presented by Polya (1945) and consisted of the following four stages: (a) Understanding structures of problems; (b) Relating conditions and constructing necessary procedures; (c) Implementing necessary procedures; (d) Reflecting on the solution and searching for other solutions. Their recommendations suggest that in 1950s, appropriate problem solving processes, as well as developing mathematical thinking, was an important topic in mathematics education community.

Since the mid of 1950s, some researchers recommended the use of Kouzouzu (structure diagrams) for supporting appropriate problem solving processes and developing students' competencies to solve mathematical word problems. This diagram was similar to a semantic network for a problem in the schema theory (Greeno, 1987). The proponents of the use of this diagram tried, however, to introduce it even when students solved rather difficult and complicated problems. Figure 1 (a) (Ochi, 1961) and 1 (b) (Miki, 1954) show the Kouzouzu for solving the following problems respectively: (a) "A fruit shop bought in 1000 oranges whose unit price was 8 yen. During transporting them, 143 of them were damaged and they could not sell those oranges. They sold the remaining oranges at 10 yen each. How much profit did they make?"; (b) There is

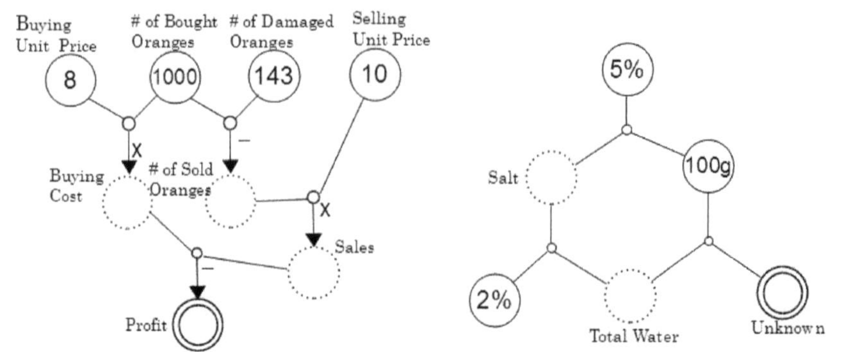

(a) Kouzouzu for Selling Orange Problem (b) Kouzouzu for Salt Water Problem

Figure 63.1.

100g of 5% salt water. You need to change it into 2% salt water by adding water. How much water do you need to add?" In other words, these researchers did not confine themselves to simple word problems in conducting their research so that they could aim at developing students' more complex thinking.

Some researchers attempted to show the positive effects of the use of this diagram on students' problem solving performances. For example, Toda (1954) reported the positive effect of the use of structure diagram on 6th grade students' performances of solving word problems. He also considered the use of structure diagrams appropriate for teaching because they could make students' inner representations of problems outer and help teachers observe and assess students' inner representations (p. 168). Structure diagrams found in Ochi (1961), for example Figure 63.1(a), became more appropriate for this purpose, because they showed structures of problems more explicitly than the diagrams in Toda (1954). The former ones included distinctions of known and unknown quantities, a name of object for each quantity, units of quantities, missing quantities to be required, and arithmetic operations relating known and unknown quantities. Ochi (1961) emphasized that when using structure diagrams, it was important for students to experience the processes in which they uncovered and made explicit the relationships in problems through completing structure diagrams.

According to Iida (2010), although the research on problem solving during 1960s mainly focused on the instruction of solutions of word problems, there were also some researchers who directed their attention to problem solving which integrated wider parts of mathematics and to developing students' mathematical thinking (pp. 225-226). Similar tendency seemed to remain during 1970s, especially in the classroom-practice-based research. Hirota (1980) summarized the trends in the problem solving section at the summer meeting of Japan Society of Mathematical Education and posed some issues to be explored further about problem solving in elementary school mathematics. He pointed out that most of the presentations in the problem solving section dealt with the instruction of solving word problems. And one of the issues he posed was to investigate what kinds of mathematical thinking could be developed in students through solving word problems and how we could evaluate their thinking.

In 1970s and the beginning of 1980s, some researchers attempted to develop the theoretical frameworks which focused on higher order thinking and students' constructions of problems, and also develop mathematical problems which fitted in those frameworks. Most of those attempts seem to originate with Shimda's (1977) project called "Open-End Approach," which started in 1971. Shimada (1977) intended to assess the "higher-order objectives" or "the objectives of higher-order thinking" (Hino, 2007). The project members thought that the higher-order thinking could be observed in the students' following behaviors: "In facing a problem situation, the students can mathematize the situation and deal with it" or "In analyzing a problem situation, the students bring forth an (important) aspect of the problem into their favored way of thinking by mobilizing their repertoire of learned mathematics, reinterpreting it to deal with the situation mathematically, and then applying their preferred technique" (Becker & Shimada, 1997, pp. 2-3; Shimada, 1977, p. 12). In other words, they attempted to observe how students drew on all of their mathematical knowledge to tackle unfamiliar problems. The project presented the diagram which showed the model of mathematical activities from their viewpoint (Becker & Shimada, 1997, p. 4; Shimada, 1977, p. 15). This diagram included the translation of the problem in the world of reality into mathematical language[1] as well as the development of a new theory of mathematics and the development of a general theory and algorithm.[2] We should also note that they insisted that "world of reality" in this model "may not necessarily be the empirical physical world but may be a conceptual world that is less abstract than ["world of mathematics" students are not yet familiar with]" (p. 5). Here, we can see a similarity to the above-mentioned disposition to think mathematically. In 1980s, "the idea to use some form of open-ended problems in the mathematics classroom spread all over the world, and research on its possibilities became very vivid in many countries" (Phekonen, 1997, p. 64).

Adoption of this viewpoint directed researchers' attention "toward the development of teaching materials and ways of organizing lessons using open-ended problems" (Hino, 2007, p. 508). The open-ended problems, a tool for assessing students' higher-order thinking, reflect the idea that finding out answers to simple word problems is not necessarily central to mathematical problem solving activities because "the situation [of a word problem] involves what students have already learned" and "it leaves too little room for their preferred way of thinking" (Becker & Shimada, 1997, p. 3). Thus, the project members developed many open-ended problems. In solving those problems, students can think more freely but need to integrate their mathematical knowledge to tackle them. In fact, some of the problems presented in Shimada (1977) are used, for example, in Lee *et al.* (2003) as a test tool for "mathematical creative problem solving ability." The Marble Problem, which is one of the popular problems Shimada's (1977) project developed and was used in Lee *et al.* (2003), shows that the open-ended problems are very different from word problems which can be solved with one or two operations: "Three students A, B, and C, throw five marbles that come to rest as in the figures above [Figure 63.2 (a)-(c)]. In this game, the student with the smallest scatter of marbles is the winner. To determine the winner, we will need to have some numerical way of measuring the scatter of the marbles. (a) Think about this situation from various points of view and write down different ways of indicating the degree of scattering. (b) Which way appeals to you?" (Hashimoto & Becker, 1999, p. 106; Becker & Shimada, 1997, p. 25). There are various possible mathematizations of this situation as follows: (a) Measure the area of a polygonal figure; (b) Measure the perimeter of a polygonal figure; (c) Measure the length of the longest segment connecting two points; (d) Sum the lengths of all segments connecting two points. Shimada (1977) pointed out that each method has its advantages and disadvantages and that it is important here for students to recognize possibilities and limitations of his/her method.

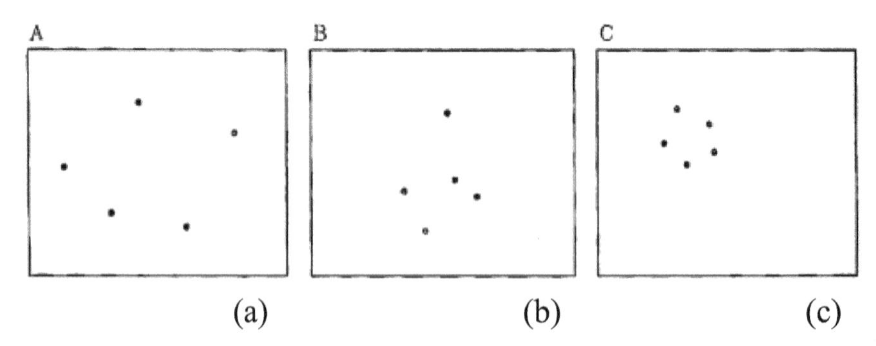

Figure 63.2.

Some of the project members thought that developing problems after solving the original ones and discussing solution methods is the important components of the mathematical thinking and the attitude supporting it. They called such an approach including problem development "Developmental Treatment" and recommended the classroom instruction based on this idea (Hashimoto & Sawada, 1983). The book by Takeuchi & Sawada (1984), the members of Shimada's (1977) project, can be taken as one of the outcomes inspired by this new idea. This book included many examples of the lessons based on the developmental treatment. In one example, the teacher posed the original problem as follows: "Are the quotients of 12÷4 and of (12x10)÷(4x10) the same?" (p. 95). When the 5th grade students were asked to develop new problems based on this original one, they first made problems by changing 12 and 4 in the original problem into other numbers including decimal numbers or by changing 10 in the original problem into other numbers including decimal numbers. Through this development of problems, the students noticed the more general principle: a÷b equals to (axc)÷(bxc). When the teacher asked the students to change the original problem in more various ways, some of the students developed the problems which included the other operations than division, like comparison of 12x4 and (12+1)x(4+1). Through this examination of wider cases, the students came to notice the more general rules which can be applied to certain cases. They also posed a further problem: What can we do with the case in which a dividend is indivisible by a divisor like (12÷3)÷(4÷3). Shimada's (1977) model of mathematical activities included learners' search for similar cases and their generalization/systematization activities based on those cases, which lead to a theory of mathematics. The instruction based on the developmental treatment seems to elaborate this part of the model. In fact, Hashimoto (1996) pointed out the importance of use of an open-ended problem and "the developmental treatment of a mathematical problem" in a lesson at all levels, and stated that the important thing is "to think generally about the problem after solving the problem" (p. 210) through the developmental treatment.

AFTER THE IMPACT

Encounter With U.S. and European Research

As mentioned in the introduction, research on problem solving in the U.S. and Europe was introduced to the Japanese mathematics education research community from the end of 1970s and they influenced the Japanese community (see Hino, 2007). The concepts of schemata, heuristics or problem solving strategies, metacognition, beliefs and problem posing, all of which were the themes actively discussed in the U.S. and European research at that time, became the issues the Japanese researchers dealt with in their works, although the ideas similar to some of them had been also discussed in the Japanese community without using those terms. While the Japanese researchers learned much from those U.S. and European research, however, they also attempted to examine the relationships between those new ideas on mathematical problem solving and the resources the Japanese community had accumulated and to develop those ideas further from the Japanese viewpoints. And through this examination, the important aspects of problem solving, which had been discussed in Japanese mathematics education community since 1930s, seemed to be highlighted again. Japanese researchers did not uncritically accept those ideas (Oku, 1982).

Kotoh (1983) examined the concept of problem solving NCTM recommended and the concept of mathematical thinking the Japanese mathematics education community had emphasized traditionally. He pointed out that the problem solving instruction recommended by NCTM seemed to be aimed at developing students' general problem solving abilities while the instruction attending to mathematical thinking aimed at development of students' thinking associated with creating and/or learning mathematics. Based on this difference, he recommended the followings: (a) It is important for students, especially younger students, to become able to think mathematically first rather than to master general problem solving; (b) problem solving strategies should keep mathematical natures and include ways of thinking using mathematical ideas like mathematical expressions and functions; (c) sequences of problems which have rich mathematical backgrounds should be developed. Mase (1983) examined the NCTM's recommendations using its PRISM report. While she pointed out based on her examination that Japanese mathematics educators tended to confine their discussion about problem solving to word problem solutions (see the above citation from Iida (2010)), Mase (1983) also pointed out the necessity to investigate the aspects of problem solving specific to mathematics.

Yamashita (1985) associated the developmental treatment of problems with Polya's four phases of problem solving to supplement this four-phase model so that it can develop students' "open mind" (NCTM, 1980, p. 3) more effectively. He recommended: (a) To adopt the developmental treatment of the previous problem before Understanding phase and use the problems made by students as problems in the next problem solving session in order to facilitate their understanding and planning as well as to make them more interested in the problems; (b) To adopt the developmental treatment at Looking-Back phase in order to make students' activities more creative. Yamashita et al. (1980) also pointed out that it is important for students to understand the problem situations per se to generate new problems after solving the original one. Recalling that generation of new problems facilitated students' understanding and planning, their recommendation seemed to suggest that understanding problem situations and finding mathematical aspects in them must play an important role in planning and implementing phases.[3] Nakajima (1985), who adopted the concept of mathematical thinking for the 1958 national curriculum, stated that developing students' mathematical thinking is the same as developing their problem solving abilities and that he had used the former in 1958 national curriculum to make explicit its relations with features specific to mathematics and higher-order objectives (p. 10). He mentioned that because the ways of thinking useful for problem solving were only treated in mathematics lessons in a rather implicit manner, the U.S. research on problem solving strategies should be considered a cue to investigate such useful ways of thinking more systematically in Japan.

Such efforts around 1980 might lead to the opportunity to interact with the US researchers. The seminar concerning problem solving in the U.S. and Japan was held by the researchers of both countries in 1986 (Becker & Miwa, 1987), which gave Japanese researchers a chance to reflect on the nature of problem solving and their research on it (Miwa, 1992). They asked the teachers in both countries to use the same problems in their lessons and compared their lessons and their students' solutions to highlight the characteristics in each country. Through such analyses, Miwa (1992) and his colleagues found, for instance, that the Japanese students were more likely to resort to number sentences or algebraic expressions than the U.S. counterparts and that this tendency was generated by the characteristics of the lessons in both countries: Japanese teachers required their students to use number sentences or algebraic expressions more often than the U.S. teachers. For example, Senuma (1992) compared the performances of Japanese and the U.S. 8th and 11th students in solving two Arithmogon problems and found that Japanese students tended to approach to them by equations and the U.S. students tended to approach them by (systematic) trial-and-error. She also pointed out that Japanese students examined whether their answers satisfied the given conditions more often than the U.S. counterparts. Sugiyama (1992) analyzed the mathematics lessons he observed in the U.S. focusing on their differences from Japanese ones. He pointed out that the U.S. lessons tended to put less emphasis on mathematical expressions than Japanese ones. Furthermore, the U.S. lessons did not direct students' attentions to the critical ideas for solving problems. In a 7th grade class he observed, the matchstick problem was used: "We will

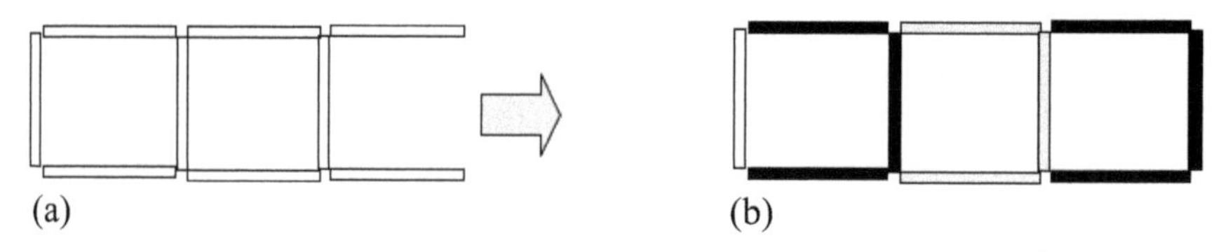

Figure 63.3.

arrange matchsticks like the figure below (Figure 63.3(a)). How many matchsticks do we need to make 5 squares?" In lessons in Japan, teachers usually direct students' attentions to the critical structures of this problem situation like Figure 63.3(b) and encouraged students to express the structures they found using mathematical expressions like 1+3x5 or $1+3n$. But Sugiyama (1992) did not find such teacher's behaviors in the U.S. lesson he observed. At least in 1980s, Japanese researchers considered such features about use of mathematical expressions and emphasis on relationships between critical structures and mathematical expressions to be an important difference between Japanese and the U.S. lessons. They also found that the emphasis on mathematical natures of problem solving, which had been observed in the Japanese research community, was reflected in the teaching practices in Japanese lessons, which in turn influenced Japanese students' thinking in their problem solving.

Hybrid of New and Traditional Perspectives

Majority of the research on problem solving in Japan which was carried out during 1990s and 2000s seemed to have a common characteristic: They kept mathematical natures of problem solving and the disposition to think mathematically in view and elaborated the above-mentioned concepts concerning problem solving from this viewpoint. Ishida (1997) designed the lessons where teachers taught the "set up an expression from a table" strategy to 5th grade students. Whereas in the experimental class, the students were encouraged to evaluate and improve their solutions from the viewpoint of far generalization and learned the above-mentioned strategy through this comparison, the students in the control class were not encouraged to do this. Analyzing the students' pre-, post-, and retention-tests performances, Ishida (1997) concluded that "the teaching strategy, focused on an evaluation and improvement activity, was effective to improve the students' performance on pattern finding problems" (p. 160). As he found that it was important for students to understand the advantages of the strategy when learning it, he further investigated whether students were aware of advantages of some solution methods, especially mathematical advantages. Ishida (2002) asked twelve 6th graders "which of their solution strategies for each question was better and whether it could be improved." He found that "students generally selected a strategy as being better because it was efficient, easy to use, or easy to understand. Mathematical values such as generality were rare" (p. 49). For example, some students "used the 'find a pattern' strategy but they were not aware of the possibility of improvement by setting up mathematical expressions" (p. 54). "Some students did not select what was regarded as the more mathematical solution method (find a mathematical expression) as the better solution method, although they had indeed found the mathematical expression" (p. 55). Based on this research, he pointed out "the need for further development of students' understanding of the mathematical value of different solutions" (p. 49). Ishida (2002) focused not only on the students' use and acquisition of problem solving strategies, but also on mathematical values which he expected their students to pay attention to in using problem solving strategies.

When he experimentally examined the relationship between students' abilities to use problem solving strategies and their problem solving performances, Norihiro Shimizu (1995) adopted the "quality of used strategy" as an important aspect to evaluate students' abilities to use strategies as well as "whether that student used any strategies or not." This aspect evaluated whether the strategy used by a student was general one in that the same strategy can be applied to the similar problems with larger numbers or higher dimensions. His statistical analyses of the data showed that the students' ability to use problem solving strategies evaluated in the above-mentioned manner had a greater impact on their problem solving than their metacognitive ability, while its impact was less than that of students' mathematical knowledge and skills. Shimizu's (1995) scoring criterion also reflected the mathematical value, generalizability of the solutions.

Shimizu & Yamada (2007) examined the effects of looking-back of Polya's four phases taking account of its mathematical values. When the students finished solving the first problem, Shimizu & Yamada (2007) encouraged the students to look back at their solutions in three different ways: (a) by asking them to check the correctness of their solutions; (b) by asking them to search for other solutions, especially better ones; (c) by asking them to examine the generalizability of their solutions. Then they asked the students to solve the second problem which was similar to the first one but included a larger number. Shimizu & Yamada (2007) investigated whether each way of looking-back could make the students' solutions to the second problem more sophisticated. Through their analysis of the data, Shimizu & Yamada (2007) found that looking-back for searching other solutions was effective for the students at all levels of ability and looking-back for examining generalizability was effective for students at lower level. Mathematical values were associated with the looking-back activity and were reflected in the criteria of effectiveness in their research framework.

Metacognition is one of the important factors which influence human problem solving (*e.g.* Schoenfeld, 1985). When he discussed students' metacognition in mathematical problem solving, Yoshinori Shimizu (1989) confined him to the metacognitive aspects which reflected the nature of mathematics subjectively or objectively perceived. He used the term 'metacognition' to refer to "what one believes about oneself as a doer of mathematics and about mathematics as a school subject and how one monitors and controls one's own behaviors" (p. 25). To focus on such metacognition that is mathematically significant, he paid attention to a kind of turning points in problem solving processes, called 'problem transformation,' in which "solvers transform the problem at hand to easier one." When he analyzed the problem solving behaviors of the 9th grade students solving a tough geometric construction problem, Shimizu (1989) attempted to identify "a series of 'problem transformations' in protocols" and attended to students' metacognition which occurred in such turning points. Because reducing a problem into easier problems is an important aspect of mathematics, attending to the metacognitions occurred in problem transformations enabled him to focus on the metacognitions that is mathematically significant and examine the roles of metacognitions critical for mathematical problem solving.

The conceptions of mathematical problem solving presented by some Japanese researchers also reflected the above-mentioned tradition. Nohda (1987a) already presented the following standpoint: Finding out underlying patterns in problems is critical for solving those problems and those patterns heavily influence students' modes of problem solving. For example, he showed Figure 63.4(a) to 2nd grade students, who learned only 1-digit-times-1-digit multiplications, and asked how many apples there are in this figure. The students could find various patterns, like Figure 63.4 (b) and (c), in this figure and those found patterns heavily influenced which operation students used or how they counted the apples. It should be noted here that the problem situation has no unique inherent pattern or structure and there are many ways in which the students understood this figure. Moreover, which pattern is useful depends on mathematical knowledge students can use. His conception of problem solving seemed to emphasize solvers' understanding of patterns in problem situations and activating his/her mathematical knowledge on the basis of those patterns. That is, what is required here is "to find

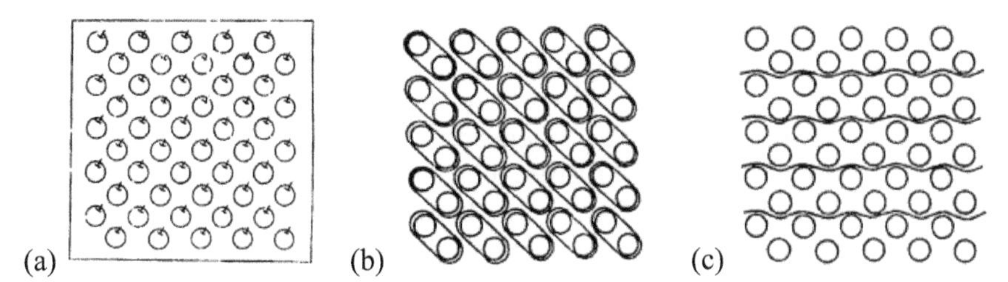

Figure 63.4.

mathematical aspects in phenomena and think phenomena mathematically" even though the problem is very elementary.

The conception of problem solving processes presented by Nunokawa (1994b, 2005) can be considered an expanded version of this idea. It characterized mathematical problem solving as follows: "Mathematical problem solving is a thinking process in which a solver tries to make sense of a problem situation using mathematical knowledge she/he has and attempts to obtain new information about that situation till she/he can resolve the tension or ambiguity" (Nunokawa, 2005). Instead of the concept of a fixed problem structure used in previous research on problem solving, he introduced the concept of "solver's structures of a problem situation" (Nunokawa, 1994b), which he thought a solver constructs and changes during his/her problem solving processes. Then, he paid attention to how solvers change their solvers' structures so that their mathematical knowledge can be associated with the problem situations at hand. Yamada (1998) attended to the changes of solver's goals in mathematical problem solving processes and stated that "the concept of 'transformation of solver's goal' is possible to become an effective viewpoint to analyze problem-solving processes from a context of control or strategy choice" (p. 373). As mentioned above, Shimizu (1989) considered the points where solvers transform the problem at hand to be turning points in mathematical problem solving and attempted to characterize problem solving processes as a series of 'problem transformations.' These efforts to perceive mathematical problem solving processes from other viewpoints than, for example, Polya's four phases can be seen as attempts to comprehend problem solving processes without loss of the nature of natural and authentic mathematical thinking.

When the conception of problem solving lays emphasis on thinking-mathematically aspects, the role of problem solving strategies becomes to promote students' mathematical thinking. Nunokawa (2000) recommended to consider problem solving strategies not only as useful ways of finding answers to problems, but also as tips for probing problem situations and understanding them more deeply so that solvers can find ways of making sense of those situations using their mathematical knowledge. For example, the systematic trial-and-error approach is often considered a way of finding answers to the problems to which students cannot easily find arithmetic operations, like the following problem: "There are some cranes and turtles. The total number of them is 20, and the total number of their legs is 52. How many cranes and how many turtles are there?" (Green-Cover textbook, 6th grade, vol. 2). If students examine the case of 19 cranes and 1 turtle, the case of 18 cranes and 2 turtles, and so on and check the total number of legs of each case, they can find the answer, 14 cranes and 6 turtles, even if they cannot find useful arithmetic operations. If they use the systematic trial-and-error approach as a tip for probing problem situation, they can find the answer in a way consistent with the disposition to find mathematical aspects in phenomena and think those phenomena mathematically. In checking some cases in this problem, students can notice that the total number of legs increases by 2 when one crane is changed into one turtle. In other words, the systematic trial-and-error approach can be used for finding patterns in this problem situation. Because in the case, for instance, of 20 cranes and 0 turtle, the total number of legs is 40, the number of necessary turtles can be found out by (52-40)÷2. The found pattern enables students to make sense of this situation using their mathematical knowledge, subtraction and division.

Based on the above-mentioned conception of problem solving processes, it sounds natural that diagrams as external representations of solvers' understanding can also change during problem solving processes. In other words, "solvers do not have to draw diagrams reflecting essential relationships or problem structures from the outset" and "it should be allowed that solvers improve their primitive diagrams toward more sophisticated ones in solving processes" (Nunokawa, 2000, p. 95). For example, when Hiroi (2001) asked some 5th grade students to solve the following problem in his interview research, the participants drew many diagrams during their problem solving processes: "There are a father lion and his child. The sum of their weights is 252kg, and the father weighs 3 times as much as his child. How much does each lion weigh?" Figure 63.5 demonstrated the first and last diagrams drawn by one pair of students during their problem solving process. In the first diagram (Figure 63.5(a)), the students drew pictures of the father lion, his child and two scales and wrote the given conditions, "252kg" and "x3," without relating those elements. In the last diagram (Figure 63.5(b)), which was very abstract, the students related the weight of the child indicated by smaller circles to the weight of the father indicated by an ellipse, and they noted that there are four small circles in the total weight, 252kg. It illustrated the critical structure of this problem situation to those students. Other

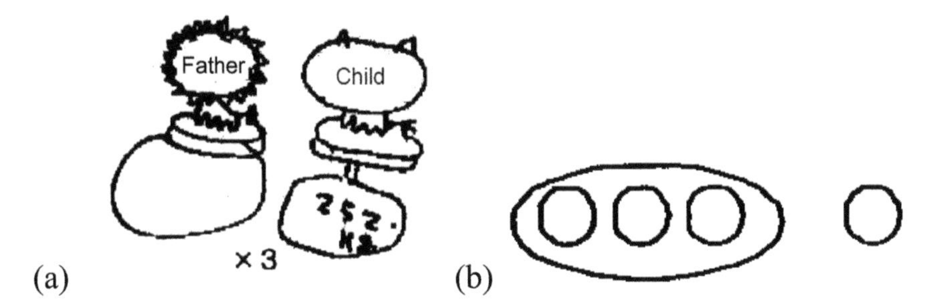

Figure 63.5.

diagrams they drew between the first and last ones helped the students notice new information about the problem situation as well as reflect on what they understood at that moment (Nunokawa & Hiroi, 2013). The key role of problem solving strategies from this viewpoint is to help solvers probe problem situations and find new information about them, rather than to lead them to the answers to those problems (Nunokawa, 2000).

Along this line of thought, Yuko Shimizu (2008) considered students' question-asking to be a critical activity for probing problem situations and attempted to develop their question-asking attitudes, a kind of scientific spirit, rather than to teach some of specific problem solving strategies. She designed a teaching experiment for developing students' question-asking in a scaffolding manner. She found that some students could develop their question-asking through this teaching and became more active problem solvers. Furthermore, comparing the learning processes of successful and unsuccessful students, she also found that successful students came to believe that it is important in mathematical problem solving to understand the problem situations as deeply as possible, whereas unsuccessful students kept focusing on getting the answers. This suggests that the effect of teaching about problem solving was influenced by students' internalization of mathematical values.

It can be easily assumed that trying to keep mathematical-thinking taste in research on problem solving enables us to apply the theoretical frameworks in those research to other research areas related to students' thinking. For example, Nunokawa (2010) reexamined proof construction processes from his viewpoint about problem solving processes. That is, he tried to see proof construction processes as the processes in which a solver tries to make sense of a problem situation using his/her mathematical knowledge so that he/she can resolve the tension or ambiguity about mathematical phenomena to be proved.

Problem Solving Components Observed in Lessons

It sounds natural that certain components of mathematical problem solving play important roles and are reflexively constructed in mathematics classrooms, because teaching and learning via problem solving is "the standard approach followed for teaching all mathematics content" (Shimizu, 2009, p. 100). Such an idea also reflects the philosophy underlying Green-Cover textbooks which can be "interpreted as identifying the learning process with the process of problem solving in its broader sense." Some researchers came to explore how such components work in students' learning or how such components are constructed through our teaching.

The role of metacognition in students' learning of mathematical ideas was one of the themes the researchers tried to pursue. Shimizu (1995, 1996) examined 6th and 7th grade students' understanding of division of fractions and analyzed their responses during "two-person problem solving" and interview sessions from the viewpoints of beliefs and metacognitive aspects, which he had focused on in analyzing students' problem solving activities. Based on the results of this analysis, Shimizu (1995) pointed out the importance of metacognitive aspects for flexibility in learning: The students who seemed to have difficulty in metacognitive aspects rejected the procedure presented by the teacher, which was nonstandard but correct one, without examining it using their mathematical knowledge. Shimizu (2000) implemented a teaching experiment in which students were encouraged to define unfamiliar quadrilaterals in problem solving settings. In this teaching experiment, the researcher worked as a facilitator of students' metacognition, as well as a facilitator of students' activities, by asking what they did or why they did so. Such interventions seemed to make the students imagine others who might criticize their ideas. The criticism assumed by the students became a cue for examining their ideas and evaluating their activities and a cue for revising their definitions to construct better ones. That is, the data showed the important role which students' metacognitive-level activities played in defining those figures and understanding the nature of definitions in mathematics.

Hirabayashi & Shigematsu (1986) laid more emphasis on how students' metacognition is constructed through our teaching activities. They pointed out that the results from the research at that time "have little implication useful to the practice of mathematics education" and attempted to define meta-cognition "in the teaching-learning context so as to be applicable to the practice of mathematics education" (p. 165). They attended to critical questions teachers ask during classes and stated that without the cognitive function corresponding to those questions, students cannot "develop his/her thinking by oneself even in the same situation" (p. 170). Thus, students need a teacher in themselves "who proposes an appropriate question and properly examines the answer to it." They introduced the concept of "Inner Teacher" to refer to this teacher in students themselves who play metacognitive roles in students' learning. This implies that quality of teachers' behaviours and utterances during their classes can influence their students' development of metacognitive aspects as well as their learning as problem solving processes. In fact, Hirabayashi & Shigematsu (1987) pointed out that rather negative metacognition of below-average students, like thinking "It's my weak point" when confronting a problem, was "originated from the teacher's attitude of teaching" (p. 248). The fact that Japanese teachers frequently said "Understand?" might be harmful, they thought, because "it's sounds something like authoritarianism to compel pupil to understand anything that teacher says as infallible" (p. 248). Shigematsu (1995) and his colleagues have continued their research along this line and attempted to investigate how students' metacognition is constructed in mathematics lessons and how we can make their metacognition more appropriate for learning mathematics.

Yamada & Shimizu (2002) examined the effect of teachers' interventions which they expected to trigger solvers' metacognition or "self-referential-activity," an elaborated version of metacognitive aspects in problem solving. They pointed out that asking "What are you doing now" (Schoenfeld, 1985) is too directive in that it directly suggests metacognitive actions, while it is too general to advise something helpful for solutions. Thus, in designing those interventions, they paid their attention to the condition that those interventions are not too general but not directive in that they don't directly invoke metacognitive actions. They chose the questions of "How did you locate point P?" or "Why did you locate point P at that place?" as one of such interventions during solving the problem which asked to determine point P so that given conditions are satisfied. Through their analysis of the data they obtained in the interview setting, Yamada & Shimizu (2002) concluded that such questions could elicit the solvers' self-referential-activities which made the solvers' solving processes turn for the better. In other words, the above questions can be examples of critical questions teachers should ask in order to help students and, at the same time, in order not to cut into students' mathematical thinking.

Concerning problem solving activities used for learning of new ideas in mathematics classes, Wada (2007) considered that devising-plan phase of problem solving processes is critical for students' problem solving because it determines the possibility of their solutions and then the possibility of their active participation in class discussions. He distinguished two types of relationships between problems students try to solve and mathematical knowledge they have learned, Extension Type and Generalization Type. He also adopted the scaffolding approach to set 5 levels of teachers' interventions. He used these dimensions as two axes and constructed a matrix which structured teachers' assistances in helping students devise their plans. He stated that when appropriate teachers' interventions are designed referring to this matrix and teachers make their students have their own plans and actively participate in class discussions, "those teachers could develop students' mathematical thinking through their experiences of elaborating their ideas in those class discussions" (p. 16). Wada (2007) emphasized the importance of changes in students' thoughts after their initial planning and expected students to exert their mathematical thinking for those changes in mathematics classrooms. He seemed to consider the planning phase to be a ticket to participation in a collaborative problem solving process. It can be also noted here that he focused on the changes in students' thinking, which have been emphasized in the conceptions of problem solving processes presented by Japanese researchers.

Applying the Products of Problem Solving Research to Exploration of Learning

If a learning process can be identified "with the process of problem solving in its broader sense," it may be expected that the use of the findings and the perspectives derived from problem solving research makes it possible to investigate students' learning processes in more detail, especially through focusing on their making-sense activities. Nunokawa & Kuwayama (2004) examined students' learning processes by paying attention to changes in their ways of drawing diagrams (Nunokawa, 1994a) during their learning. They implemented a series of mini-teaching, which consisted of one teacher and two students, and videotaped the whole learning processes of those students. Through their analysis referring to the above-mentioned changes, they found that the students made sense of the others' new ideas in the context of the old idea they had at that time and came to understand mathematical contents through construction of hybrids of old and new ideas. Their approach, which was inspired by the problem solving research, can be used for analyzing the processes in which students learn specific mathematical ideas.

Isono (2007) analyzed the students' learning of additions with carrying focusing on the subtle changes observed in the diagrams they drew. One student participating in her clinical interview drew a diagram shown in

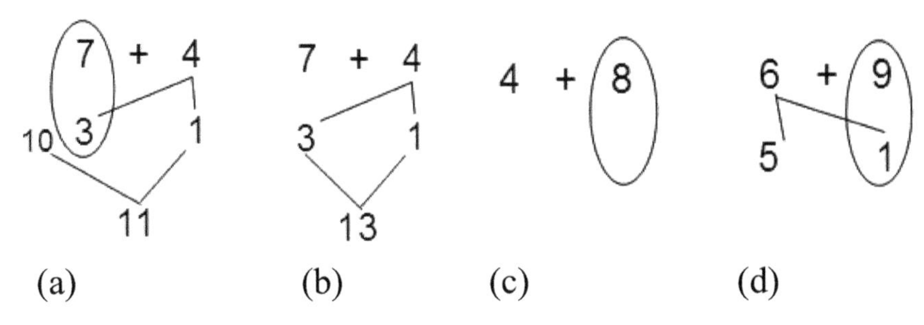

Figure 63.6.

Figure 63.6(b) instead of Figure 63.6(a) which illustrates the composition and the decomposition required for calculating 7+4, although this student could operate on base-10 blocks appropriately to calculate the same addition. Therefore, Isono (2007) encouraged this student to do live coverage of what she did in operating on blocks. The diagram became to play a "model-of" (Gravemeijer, 1997) her operations of moving blocks. The student came to draw an ellipse in the diagram at the earlier stage like Figure 63.6(c) and decomposed and composed the numbers to complete the diagram. Later, this student became able to find the answers of additions before completing the diagram (Figure 63.6(d)). In other words, the abbreviated version started to work. Here, the subtle changes in her ways of drawing diagrams seemed to reflect the progress of the student's understanding of additions with carrying, just as the changes in drawings observed by Hiroi (2001) reflected the progress of the students' understanding of the problem situation.

There were other researchers who tried to investigate processes in which students are learning specific mathematical ideas in authentic mathematics classrooms. They also paid attention to how students changed their diagrams and their sense-making of those diagrams during several lessons. Satoh (2007) investigated the students' learning processes in the lessons on proportional reasoning. He attended to how the students drew and used dual number lines in their learning. He checked the order in which the students added numbers to their dual number lines (Figure 63.7). Here, a number in a circle indicates the order in which each number was added, while the numbers without the circles indicate the numbers which the students deleted and the researcher retyped afterward for his analyses. Three diagrams in Figure 63.7 were drawn by the same student in solving the following three problems respectively: (a) "5m rubber hose costs 1400 yen. How much does this hose cost per meter? How about 7m hose?"; (b) "There is a 12m-long tape. If you divide this tape into 4/5m-long short tapes, how many short tapes can you

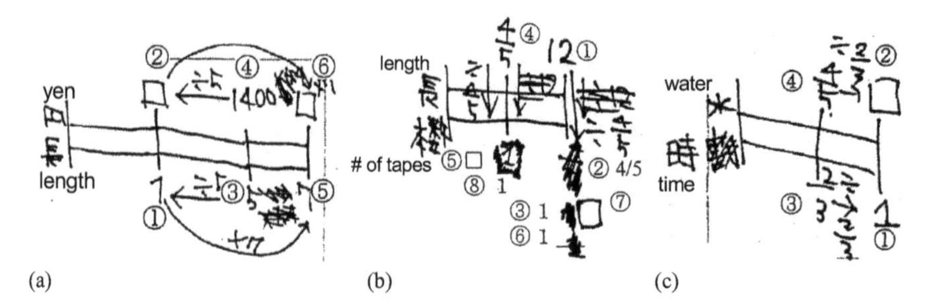

Figure 63.7.

get?"; (c) "4/5L water is poured during 2/3 minute. How many litters of water can be poured during one minute?" While, as Figure 63.7(a) shows, this student could build an appropriate dual line to represent such a multiplicative situation, she got confused in solving complicated problems including common fractions (Figure 63.7(b)). In the end of the post test, she could make an appropriate dual number line even when the problem included common fractions (Figure 63.7(c)). The orders illustrated in these diagrams implied that whether she added the number "1" first or later influenced the student's use of dual number lines and her comprehension of multiplicative situations.

In her detailed analysis of 5 students' learning of proportional reasoning in a 5th grade classroom, Hino (2003) attended to the role of inscriptions produced by learners and referred to the research on students' inscriptions in problem solving. Focusing on the students' uses of inscriptions, she tried to highlight students' sense-making of the mathematical notions taught in this class. She found that the students created their own ways of using mathematical notations which were slightly different from the ways introduced by the teacher and that the creation of those ways played important roles in the development of their proportional reasoning. She also pointed out that the students changed their sense-making of notations during the lessons and that those changes were facilitated through the interaction between public and private situations. On the basis of such findings, Hino (2003) stated that it was necessary to investigate how to facilitate students' reflection on their thoughts through their own uses of inscriptions.

In mathematics lessons, especially in the lessons based on problem solving approaches, students are often presented with problem situations which can lead them to new mathematical ideas. Therefore, we can expect that the research on problem solving processes can be used to analyze students' learning processes in detail and reveal how students change their understanding of those situations. Nunokawa (2008) attended to learners' attempts to make sense of a presented problem situation with their old mathematical knowledge and explored how they changed their sense-making and came to understand the situation in a new way which could lead them to a targeted new idea. He analyzed three 6th grade students learning processes in the lessons in which they were expected to learn relationships between scale factors of enlargement and the areas of enlarged figures. He found that the students unintentionally used their initial and naïve sense-making, making sense of this situation by linear relationship, in their later exploration of the situation (e.g. interpreting the data they gathered, setting an implicit assumption, interpreting the others' ideas) and these influences of initial sense-making prevented them from newly making sense of the situation and noticing the necessity of new mathematical ideas. Furthermore, he found that only when the students realized these influences of their initial sense-making on their explorations and thinking, they could look at the situation in a new way and began to search for new sense-making of the problem situation.

These research demonstrated that what is important in exploring students' learning is not whether students understand tasks and representations presented by teachers, but how students understand or make sense of them and how their understandings change during lessons.

CONCLUDING REMARKS

This chapter illustrated that research on problem solving in Japan has been constantly affected by its tradition which values mathematical nature of problem solving and mathematical richness of thinking, even after Japanese researchers learned much about problem solving research from the U.S. and European researchers. Giving attention to those mathematical nature and richness in their research on problem solving, the researchers elaborated the theoretical components concerning problem solving to some extent, so that those components can reflect mathematical thinking more explicitly and work as facilitators of students' mathematical thinking. Moreover, such an approach to the research on problem solving has made it easier to apply the viewpoints and the findings of the problem solving research to the research on students' learning processes in mathematics lessons and enabled us to investigate those learning processes more carefully.

Through his examination of previous works on problem solving in Japanese mathematics education community, Koyama (2009) presented three axes for discussing significances of problem solving in mathematics education: objectives-methods, mathematics-realities, and individuals-collectives. Concerning the second axis of mathematics-realities, he distinguished "solving mathematical problems" and "solving problems mathematically" and called the former the narrow sense of mathematical problem solving and the latter its broad sense. Recalling that both of these problem solving are included in the mathematics activities in Shimada's (1977, p. 14) diagram, it may be said that the conception of mathematical problem solving in Japan has kept a composite nature, which has helped researchers keep the mathematical nature in sight on the one hand and made ambiguous what each researcher means by the term "mathematical problem solving" on the other hand. Therefore, it is necessary to make explicit and clearer the relationship between the products of problem solving research and these narrow and broad senses (e.g. Can this product hold only in narrow sense or even in the broad sense? How should we modify that product for applying it to the broad sense?). It is also productive to promote an interaction between the research on mathematical problem solving and on students' learning processes in classrooms. It can be expected that our investigation of students' learning will be further developed using perspectives on problem solving, while frameworks of problem solving will be more elaborated by referring to students' learning processes so that those frameworks can deal with mathematical thinking more flexibly.

NOTES

1. Referring to Shimada (1977), Miwa (1987) pointed out that mathematical model-making "is often one of the more important issues stressed in Japanese school mathematics" (p. 95).
2. Nohda's (1987b, 1995) Open Approach seemed to put this diagram into the context of teaching and learning in classrooms. It attempted to foster both the creative activities of the students and the mathematics created through problem solving simultaneously so that both of them can be "carried out to the fullest extent" (Nohda, 1987b, p. 118). He presented three types of openness in it: Openness in students' activities, Openness in mathematical ideas, and Openness of interactive activities between students and mathematical ideas. While the first two types seem to correspond to certain parts of the processes in Shimda's (1977) diagram, third type seems to indicate the students-teachers interactions which are required to activate the first two.
3. Later, Yamashita (1994) used open-ended problems to incorporate mathematization phase into problem solving processes. He dealt with, for example, the following problem: "Three teams A, B and C participated in a game and got 45 points, 27 points and 18 points respectively. The prize of this game was 10 melons. How to divide these melons into three teams? Please develop various ways." According to what a kind of assumption they set (e.g. divide the melons in a manner proportional to the points or not),

students may mathematize this situation differently. On the other hand, Nunokawa (1995) examined problem solving processes from the viewpoint of mathematical modeling and insisted that the interpretation of mathematical results in problem solving processes should be considered sense-making of problem situations using those results, rather than sense-making of or interpreting those results referring to situations. In solving augmented-quotient division problems, for example, solvers should examine what quotients can tell her/him about situations rather than how quotient be adjusted. These works can be also considered attempts to reconcile 'problem solving' and 'thinking situations mathematically.'

REFERENCES

Becker, J. P. & Miwa, T. (Eds.). (1987). *Proceedings of the U.S.-Japan Seminar on mathematical problem solving*. Carbondale, IL: Board of Trustees of Southern Illinois University.

Becker, J. P. & Shimada, S. (1997). *The open-ended approach: A new proposal for teaching mathematics*. Reston, VA: National Council of Teachers of Mathematics.

Gravemeijer, K. (1997). Mediating between concrete and abstract. In T. Nunes & P. Bryant (Eds.), *Learning and teaching mathematics: An international perspective* (pp. 315-345). East Sussex, UK: Psychology Press.

Greeno, J. G. (1987). Instructional representations based on research about understanding. In A. H. Schoenfeld (Ed.), *Cognitive Science and Mathematics Education* (pp. 61-88). Hillsdale, NJ: Lawrence Erlbaum Associates.

Hashimoto, Y. (1996). Mathematical problem solving. In D. Zhang, T. Sawada, & J. P. Becker (Eds.), *Proceedings of the China-Japan-U.S. Seminar on Mathematical Education* (pp. 209-214). Carbondale, IL: Board of Trustees of Southern Illinois University.

Hashimoto, Y. (1997). The methods of fostering creativity through mathematical problem solving. *ZDM, 29* (3), 86-87.

Hashimoto, Y. & Becker, J. (1999). The open approach to teaching mathematics: Creating a culture of mathematics in the classroom: Japan. In L. J. Sheffield (Ed.), *Developing Mathematically Promising Students* (pp. 101-119). Reston, Va: National Council of Teachers of Mathematics.

Hashimoto, Y. & Sawada, T. (1983). Research on the mathematics teaching by developmental treatment of mathematical problems. In T. Kawaguchi (Ed.), *Proceedings of ICMI-JSME regional conference on mathematical education* (pp. 309-313). Tokyo.

Hino, K. (2003). *Developing a teaching unit for proportional reasoning from the viewpoint of students' learning processes in classrooms*. Technical Report for Grant-in-Aid for Scientific Research of Japan Society for the Promotion of Science (No. 12680178). [In Japanese]

Hino, K. (2007). Toward the problem-centered classroom: Trends in mathematical problem solving in Japan. *ZDM, 39*, 503-514.

Hirabayashi, I. & Shigematsu, K. (1986). Meta-cognition: The role of "Inner Teacher." In L. Burton & C. Hoyles (Eds.), *Proceedings of the 10th Annual Meeting of the International Group for the Psychology of Mathematics Education* (pp. 165-170). London.

Hirabayashi, I. & Shigematsu, K. (1987). Metacognition: The role of the "inner teacher" (2). In J.C. Bergeron, N. Herscovics & C. Kieran (Eds.), *Proceedings of the 11th Annual Meeting of the International Group for the Psychology of Mathematics Education, vol. II* (pp. 243-249). Montreal, Canada.

Hiroi, H. (2001). Students' comprehension of problems using diagrams in mathematical problem solving: The interview research with 5th grade students. In S. Itoh (Ed.), *Proceedings of the 34th Annual Meeting of Japan Society of Mathematical Education* (pp. 457-462). Tokyo. [In Japanese]

Hirota, K. (1980). Issues concerning mathematical problem solving. *Journal of Japan Society of Mathematical Education, 62* (2), 36-37. [In Japanese]

Iida, S. (2010). Mathematical problem solving. In Japan Society of Mathematical Education (Ed.), *Handbook of Research in Mathematics Education* (pp. 221-232). Tokyo: Touyoukan. [In Japanese]

Ishida, J. (1997). The teaching of general solution methods to pattern finding problems through focusing on an evaluation and improvement process. *School Science and Mathematics, 97* (3), 155-163.

Ishida, J. (2002). Students' evaluation of their strategies when they find several solution methods. *Journal of Mathematical Behavior, 21*, 49-56.

Isono, K. (2007). Effects of private-speech on the transition from operations on semi-concrete materials to operations on numbers: First grade students' learning of addition with carrying. In F. Ikeda (Ed.), *Proceedings of the 40th Annual Meeting of Japan Society of Mathematical Education* (pp. 397-402). Noda, Japan. [In Japanese]

Kotoh, S. (1983). Problem solving and mathematical thinking. *Tsukuba Journal of Educational Study in Mathematics, 3*, 1-8. [In Japanese]

Koyama, M. (2009). Significances of problem solving in mathematics education research and practices. In Kunimune (Ed.), *Proceedings of the 42nd Annual Meeting of Japan Society of Mathematical Education, vol. 2* (pp. 44-47). Shizuoka. [In Japanese]

Lee, K. S., Hwang, D.-J., & Seo, J. J. (2003). A development of the test for mathematical creative problem solving ability *Journal of the Korea Society of Mathematical Education Series D: Research in mathematical Education*, 7 (3), 163-189.

Mase, H. (1983). One perspective for discussing problem solving: Focusing on PRISM report. *Tsukuba Journal of Educational Study in Mathematics, 3*, 20-27. [In Japanese]

Miki, Y. (1954). Building problems by structural chart. *Journal of Japan Society of Mathematical Education, 3* (6), 16-17. [In Japanese]

Miwa, T. (1987). Mathematical model-making in problem solving: Japanese's pupil's performance and awareness of assumptions. *Tsukuba Journal of Educational Study in Mathematics, 6*, 93-105.

Miwa, T. (Ed.). (1992). *Teaching mathematical problem solving in Japan and the US*. Tokyo: Touyoukan. [In Japanese]

Nakajima, K. (1985). Developing mathematical thinking and instruction of problem solving. In K. Nakajima (Ed.), *Mathematical thinking and problem solving 1: Theoretical background* (pp. 1-16). Tokyo: Kaneko Shobo. [In Japanese]

National Council of Teachers of Mathematics. (1980). *An agenda for action: Recommendations for school mathematics of the 1980s*. Reston, VA: NCTM.

Nohda, N. (1983). The heart of 'Open-Approach' in mathematics teaching. In T. Kawaguchi (Ed.), *Proceedings of ICMI-JSME regional conference on mathematical education* (pp. 314-318). Tokyo.

Nohda, N. (1987a) Mathematical pattern-finding in problem solving: Mode and performance of Japanese students. In J. P. Becker & T. Miwa (Eds.), *Proceedings of the U.S.-Japan seminar of mathematical problem solving* (pp. 342-358). Honolulu, Hawaii.

Nohda, N. (1987b). Teaching and evaluation of problem solving using 'Open-Approach' in mathematics instruction. *Tsukuba Journal of Educational Study in Mathematics, 6*, 117-126.

Nohda, N. (1995). Teaching and evaluating using "open-ended problems" in classroom. *ZDM, 27* (2), 57-61.

Nunokawa, K. (1994a). Improving diagrams gradually: One approach to using diagrams in problem solving. *For the Learning of Mathematics, 14* (1), 34-38.

Nunokawa, K. (1994b). Solver's structures of a problem situation and their global restructuring. *Journal of Mathematical Behavior, 13* (3), 275-297.

Nunokawa, K. (1995). Problem solving as modelling: A case of augmented-quotient division problem. *International Journal of Mathematical Education in Science and Technology, 26* (5), 721-727.

Nunokawa, K. (2000). Heuristic strategies and probing problem situations. In J. Carrillo & L.C. Contreras (Eds.), *Problem Solving in the Beginning of the 21st Century* (pp. 81-117). Spain: Hergué.

Nunokawa, K. (2005). Mathematical problem solving and learning mathematics: What we expect students to obtain. *Journal of Mathematical Behavior, 24*, 325–340.

Nunokawa, K. (2008). On the 6th graders' problem solving processes in a mathematics lesson. *Reports of Mathematical Education: Journal of Japan Society of Mathematical Education, 90*, 19-39. [In Japanese with English summary]

Nunokawa, K. (2010). Proof, mathematical problem-solving, and explanation in mathematics teaching. In G. Hanna, H. N. Jahnke, & H. Pulte (Eds.), *Explanation and Proof in mathematics: Philosophical and Educational Perspectives* (pp. 223-236). New York: Springer.

Nunokawa, K. & Hiroi, H. (2013). Elementary school students' use of drawings and their problem solving. In S. Helie (Ed.), *Psychology of problem solving* (pp. 123-152). Hauppauge, NY: Nova Science Publishers.

Nunokawa, K. & Kuwayama, M. (2004). Students' appropriation process of mathematical ideas and their creation of hybrids of old and new ideas. *International Journal of Science and Mathematics Education, 1* (3), 283-309.

Ochi, M. (1961). Responses to the criticisms toward the use of structural charts. *Journal of the Mathematical Educational Society of Japan, 43* (6), 113-120. [In Japanese]

Ohya, S. & Fujiwara, Y. (1958). *Mathematical problem solving and its instruction*. Osaka, Keirinkan. [In Japanese]

Oku, S. (1982). Mathematical problem solving and mathematics education in Japan (1). *Tsukuba Journal of Educational Study in Mathematics, 1,* 27-36. [In Japanese]

Oku, S. (1983). A research for the goal of mathematics education in Japan: Study of Naomichi Shiono Theory of mathematics education. In T. Kawaguchi (Ed.), *Proceedings of ICMI-JSME regional conference on mathematical education* (pp. 407-410). Tokyo.

Pehkonen, E. (1997). The state-of-art in mathematical creativity. *ZDM, 29* (3), 63-67.

Polya, G. (1945/1973). *How to solve it: A new aspect of mathematical method*. Princeton, NJ: Princeton University Press.

Satoh, M. (2007). Development of proportional reasoning in the integration approach: Analysis of 6th grade students' learning of proportions and ratios. In F. Ikeda (Ed.), *Proceedings of the 40th Annual Meeting of Japan Society of Mathematical Education* (pp. 391-396). Noda, Japan. [In Japanese]

Schoenfeld, A. H. (1985). *Mathematical problem solving*. Orland, FL: Academic Press.

Senuma, H. (1992). A study on mathematical problem solving through the Japan-U.S. collaborative research: An analysis of the results of "Arithmogon" problem. In T. Miwa (Ed.), *Teaching mathematical problem solving in Japan and the US* (pp. 101-118). Tokyo: Touyoukan. [In Japanese]

Shigematsu, K. (1995). Metacognition in mathematics education. In Japan Society of Mathematical Education (Ed.), *Views of Mathematics Learning toward Theorizing: 1995 Yearbook* (pp. 237-249). Tokyo, Japan: Sangyo Tosho. [In Japanese with extended English summary]

Shimada, S. (Ed.). (1977). *Open-ended approach in arithmetic and mathematics*. Tokyo: Mizuumi Shobo. [In Japanese]

Shimada, S. (1987). Problem solving: The present state and historical background. In J. P. Becker & T. Miwa (Eds.), *Proceedings of the U.S.-Japan seminar of mathematical problem solving* (pp. 5-32). Honolulu, Hawaii.

Shimizu, N. (1995). A study on strategic ability in mathematical problem solving: An experimental study on the rate of contribution of strategic abilities toward problem solving ability. In T. Nakahara (Ed.), *Proceedings of the 28th Annual Meeting of Japan Society of Mathematical Education* (pp. 359-364). Hiroshima.

Shimizu, N. & Yamada, A. (2007). The development of solutions through looking-back activities after mathematical problem solving: A consideration in terms of potential ability for mathematics learning. *The Bulletin of Japanese Curriculum Research and Development, 30*(2), 1-8. [In Japanese]

Shimizu, Y. (1989). Metacognition in solution processes of junior high school students on a construction problem. *Reports of Mathematical Education: Journal of Japan Society of Mathematical Education, 52,* 3-25. [In Japanese with English summary]

Shimizu, Y. (1995). Students' thinking on division of fractions: The rigidity in their argument. *Reports of Mathematical Education: Journal of Japan Society of Mathematical Education, 63/64,* 3-26. [In Japanese with English summary]

Shimizu, Y. (1996). "High achievement" versus rigidity: Japanese students' thinking on division of fractions. In D. Zhang, T. Sawada, & J. P. Becker (Eds.), *Proceedings of the China-Japan-U.S. Seminar on Mathematical Education* (pp. 223-238). Carbondale, IL: Board of Trustees of Southern Illinois University.

Shimizu, Y. (2000). Understanding the role of mathematical definition through construction: A teaching experiment. *Reports of Mathematical Education: Journal of Japan Society of Mathematical Education, 73/74*, 3-26. [In Japanese with extended English summary]

Shimizu, Y. (2009). Japanese approach to teaching mathematics via problem solving. In B. Kaur, Y. B. Har, & M. Kapur (Eds.), *Mathematical problem solving: Yearbook 2009, Association of Mathematics Educators* (pp. 89-101). Toh Tuck, Singapore: World Scientific Publishing.

Shimizu, Yu. (2008). Teachers' supports promoting development of students' question-asking in mathematical problem solving: Adopting the idea of scaffolding. In S. Shimizu (Ed.), *Proceedings of the 41st Annual Meeting of Japan Society of Mathematical Education* (pp. 117-122). Tsukuba. [In Japanese]

Silver, E. A. (Ed.). (1985). *Teaching and learning mathematical problem solving: Multiple research perspectives*. Hillsdale, NJ: Lawrence Erlbaum Associates.

Sugiyama, Y. (1992). A comparison of Japanese and U.S. mathematics lessons. In T. Miwa (Ed.). *Teaching mathematical problem solving in Japan and the US* (pp. 172-187). Tokyo: Touyoukan. [In Japanese]

Takeuchi, Y. & Sawada, T. (1984). *From problems to problems: Improving mathematics lessons using the developmental treatment of mathematical problems*. Tokyo: Touyoukan. [In Japanese]

Toda, K. (1954). A focalization of the verbal problem teaching and utilizations of structural chart. *Journal of Japan Society of Mathematical Education, 3* (6), 8-10. [In Japanese]

Wada, S. (2007). Considering ways to support students during devising-plan phase of problem solving. *Journal of Japan Society of Mathematical Education, 89* (4), 11-17. [In Japanese]

Yamada, A. (1998). A study on cognitive process in mathematical problem solving: Segmentation of problem-solving process. In A. Yamashita (Ed.), *Proceedings of the 31st Annual Meeting of Japan Society of Mathematical Education* (pp. 371-376). Fukuoka.

Yamada, A. & Shimizu, N. (2002). Study on self-referential-activities in mathematical problem solving (VI): An analysis of self-referential-activities in solving "making-square-problem." *Journal of Japan Academic Society of Mathematics Education: Research in Mathematics Education, 8*, 95-107. [In Japanese with English summary]

Yamashita, A. (1985). On openness of problem solving in school mathematics. *Bulletin of Fukuoka University of Education: Part IV. Education and Psychology, 35*, 221-226. [In Japanese]

Yamashita, A. (1994). A study on open end problem. *Bulletin of Fukuoka University of Education: Part IV Education and Psychology, 43*, 509-515.

Yamashita, A., Kondoh, A., Kawaguchi, H., & Koga, M. (1980). On teaching arithmetic through a developmental treatment of problems. *Journal of Japan Society of Mathematical Education, 62* (10), 218-222. [In Japanese]

CHAPTER 64

TEACHING AND LEARNING MATHEMATICS WITH INFORMATION AND COMMUNICATION TECHNOLOGY IN JAPAN

The Case of Geometric Constructor

Yasuyuki Iijima
Aichi University of Education

ABSTRACT

This chapter illustrates aspects of research and practice of teaching and learning mathematics with Information and Communication Technology (ICT) in Japan. The term ICT here is mainly used to refer to computers and the Internet. The issues in the development of software are discussed with a particular focus on *Geometric Constructor*, GC in short, which is one of the popular dynamic geometry software in Japan. After a review of the historical development of computer use in mathematics classrooms in Japan, potentials and pitfalls are examined of the use of various devices in mathe-

The First Sourcebook on Asian Research in Mathematics Education: China, Korea, Singapore, Japan, Malaysia, and India, pp. 1437–1453

matics classrooms, as well as the Internet for providing a platform for discussing mathematical problems and students' difficulties among teachers. The current status of teaching and learning mathematics with ICT is illustrated with examples with a particular style of classroom practices. Implications for classroom instruction are discussed and needed directions for further studies are examined.

Keywords: ICT, computer, dynamic geometry software, the Internet, Geometric Constructor

A RETROSPECTIVE PERSPECTIVE ON THE DEVELOPMENT OF ICT IN JAPANESE CLASSROOMS

For more than a few decade, a variety of mathematical tool software is available in the classrooms in Japan. Such software can roughly be categorized into the following four groups; graphing software of function (e.g. Grapes), dynamic geometry software (e.g. GC, Cabri, Geometer's SketchPad, Cinderella, Geogebra), spreadsheet (Excel) and CAS (e.g. Mathematica, Risa/Asir). Since 1990, the style of educational practice was changing according to the principles of ministry of education about ICT use in education, course of studies, developments of the software and hardware and the Internet in Japan. To illustrate it, the case study about GC is suitable. In this section, we review the historical development of computer use in mathematics classrooms in Japan.

Emergence of Mathematical Tool Software: 1985 – 1996

In the era of BASIC (n88-basic, DOS-BASIC), we had made many programs written in BASIC for mathematical problem solving, but the size of them were very small and they were made individually for each problem. We used BASIC as a general tool to make programs for each mathematical problem, and didn't use it as a computer language to make mathematical tool software with which we could solve many mathematical problems. The emergence of Turbo Pascal, Quick BASIC had changed this situation. These languages had made us easily to develop mathematical tool software. In Japan, some free software have developed and been used in many schools. One of them was Grapes (GRAph Presentation & Experiment System), which is graphing software of function developed by Tomoda (1994, International Windows version 2003), and the other is Geometric Constructor which is dynamic geometry software developed by the author (in short GC/DOS: 1989). In this era, various types of software were developed such as Geometric Supposer and Cabri geometry (DOS

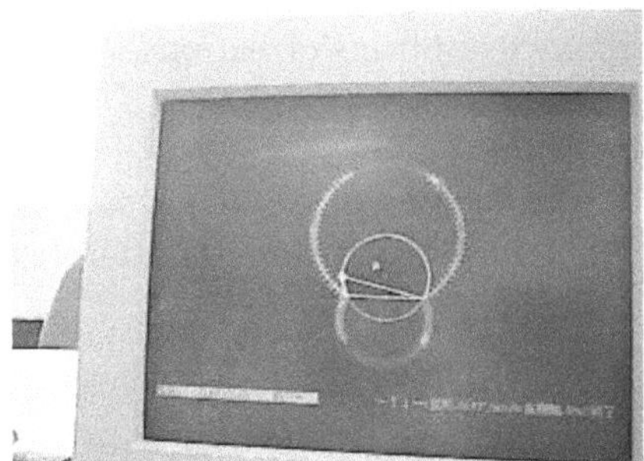

Figure 64.1. GC/DOS (1989).

version) as commercial software in abroad. Earlier GC/DOS was developed with Quick BASIC and later it was developed with Microsoft BASIC. It was one of the dynamic geometry software. We could make many geometric constructions and measurements, and we could drag some points (with mouse or keyboard) and observe what was changing or invariant. I distributed this software with floppies and printed manuals, and communicated with many teachers with telephone and fax.

In this era, each school had a computer room, in which they had 20 or 40 stand-alone computers. Educational practices with computers were done in a computer room. Each student or pair of students used a computer for their learning.

Some supervisors of prefecture's educational centers had many concern about this software (Tottori, Shimane, Aichi, Niigata, Nagano etc. and Kawasaki City). Some staffs in educational centers wrote papers, and made lectures to teachers in each prefectures.

In 1992-1993, we made lesson studies about GC/DOS with many teachers. Our slogan was "mathematical activity" which was able to do with GC/DOS. I (as the developer of GC/DOS) illustrated the concept of following mathematical activity and some problems to teachers, and teachers designed some lessons of the activity, and we discussed lesson plans. In many case, we had lessons in two hours. In first lesson, we posed the problem and investigated using computers, and in next lesson, we discussed the results and formulated the mathematical problems which were derived from the discussion and solved it with paper and pencils. Discussed mathematical activities are following; dragging and investigation of figures: finding special cases and common condition, proof of impossibility (absence of some special cases), changing of the conditions of problems, supposing with paper and pencils and testing it with GC, investigation of the change or invariance of measurements, searching the locations of point which satisfy some conditions, investigation of / with geometrical transformations,

Example 1

In the area of geometry, we have treated some problems dynamically. The following problem found in textbook is such an example.

1. ABCD is a quadrilateral. P, Q, R, S are midpoints of segments AB, BC, CD, DA. Prove that PQRS is a parallelogram.
2. What is PQRS, when ABCD is a rhombus.
3. What do you think about the relation of diagonals AC and BD, when PQRS is a rhombus (see Figure 64.2).

We intended to draw many cases with paper and pencil, observe them and investigate the relation of ABCD and PQRS. With GC, we could pose problem openly as following and make students to drag, observe figures and write the following worksheet and investigate the relation of ABCD and PQRS;

Problem posed to students with GC was as follows.

1. Construct a quadrilateral ABCD. Make midpoints of each segment and label them P, Q, R, S. Drag points A, B, C, D and observe the figure and write the result on the worksheet.
2. What do you think from the results?

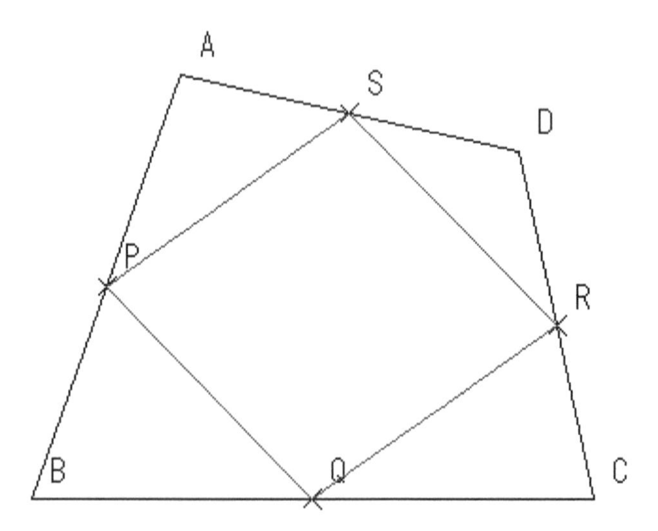

Figure 64.2. Problem in the textbook.

Table 64.1. The Relation of the Given Quadrilateral and Quadrilateral Made by Midpoints

ABCD	*PQRS*	*Sketch of the Figure*
Square	Square	
Rectangle	Rhombus	Students draw sketches in this place
Rhombus	Rectangle	
Parallelogram	Parallelogram	
Trapezoid	Parallelogram	
Isosceles trapezoid	Rhombus	
Quadrilateral	Parallelogram	

In the lesson, teacher showed how to write the worksheet and discuss which cases should be added to ABCD with students. Many students noticed the conjugate relation of rectangle and rhombus from above worksheet. Some students pointed out that if ABCD is isosceles trapezoid then PQRS is rhombus, and found that if the lengths of diagonals are same, then PQRS is rhombus.

Example 2

The above way of thinking is not only applied to above problem but also many problems. Following is the one of them.

The problem in the textbook is as follows:

ABCD is a parallelogram. P, Q, R, S are intersections of angle bisectors as follows.

Prove that PQRS is rectangle (see Figure 64.2a).

We revised the problem to next one.

Problem:

1. ABCD is a quadrilateral. P, Q, R, S are intersections of angle bisectors as follows. Drag points, observe the figure and Drag points and observe the figure and write on the worksheet.

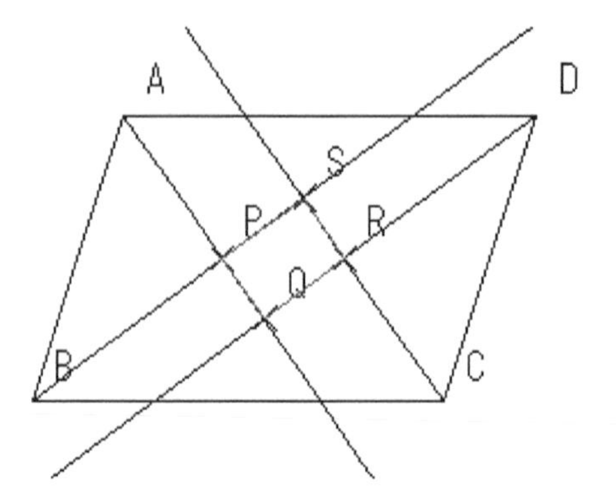

Figure 64.2a.

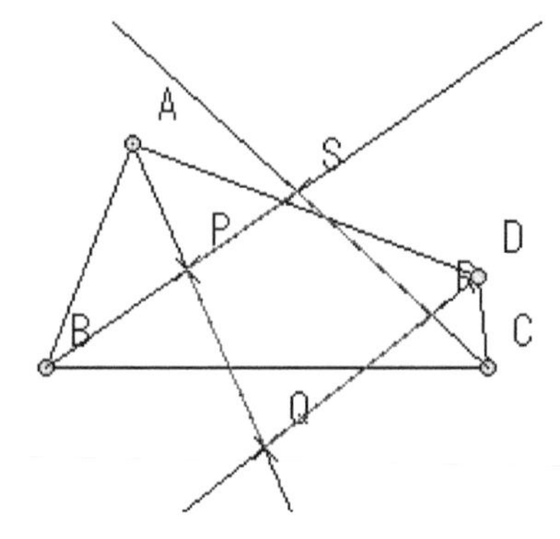

Figure 64.2b.

2. What do you think from the results?

Each student made sketches and wrote on the worksheets. Especially, the columns marked (*) are different and important for their discussion. In our lesson, one student noticed that there was not parallelogram in the column of PQRS. According to his notice, other student pointed out that there was not rhombus also. They supposed and proved that PQRS is cyclic quadrilateral, and proved the absence of parallelogram and rhombus in the column of PQRS.

While Iijima,Y. (1995, 1997) reported on the results of these lesson studies, other groups also implemented lesson studies with dynamic geometry software. One of the groups was consisted of researcher and graduate students of University of Tsukuba (Nohda, Kakihana, Shimizu, Harada, Tsuji, etc.). They researched the effect of Cabri. Shimizu & Kakihana (1999) reported their lesson studies. Other group was the teachers of the Nara Woman's University Secondary School (Yoshida, Ohnishi, Yamagami, etc.). They used Geometer's SketchPad (Mac version) in their lessons.

Figure 64.3. A lesson in the computer laboratory.

Table 64.2. The Relation of the Given Quadrilateral and Intersections of Angle Bisectors

ABCD	*PQRS*	*Sketch of the Figure*
Square	Point	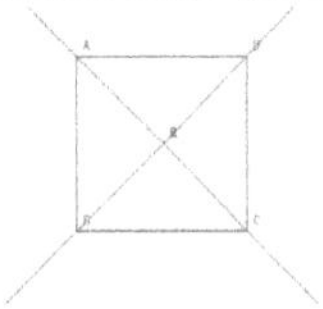
Rectangle	Square	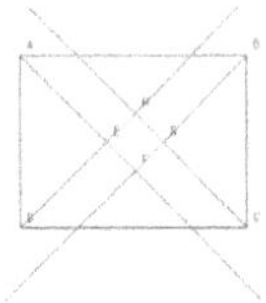
Rhombus	Point	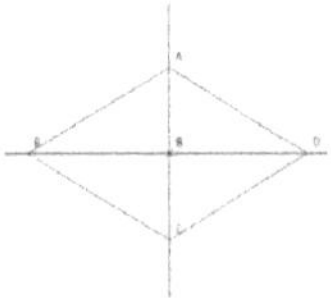
Parallelogram	Rectangle	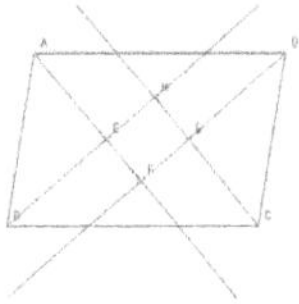
Trapezium	*quadrilateral with angle P, angle R are right angles.	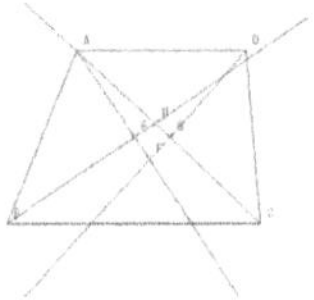
Kite	Point	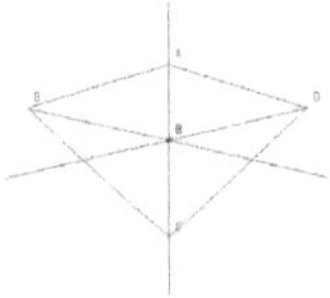
Convex quadrilateral	*cyclic quadrilateral	
Non-convex quadrilateral	*	

Figure 64.4. Forum of geometric constructor.

The Use of the Internet and Development of GC/Win

Since the emergence of windows3.1, various types of software has been developed based upon windows, including Cabri II, Geometer's Sketch-Pad (windows version). But in Japan, many schools used DOS-based computers and could not use windows application.

Since the emergence of windows95, many schools started to use the Internet. For the sake of distribution of software and manuals, we started the use of web server (Forum of Geometric Constructor: http://www.auemath.aichi-edu.ac.jp/teacher/iijima/index.htm) in 1995. A large numbers of teachers begun to download them, instead of coping and sending floppies and printed manuals.

According to the increase of the windows OS in the schools, GC/Win (windows version of GC) was developed from 1997. And we made the mailing list about GC to make the on-line community in 1997. This community helped the development of GC/Win.

When I make a new version of GC/Win, I upload it on the web and announce it with mailing list. Many teachers tested it and reported many things with mailing list almost every day. The experience of this cycle of software development made us recognize the importance and potentiality of the Internet.

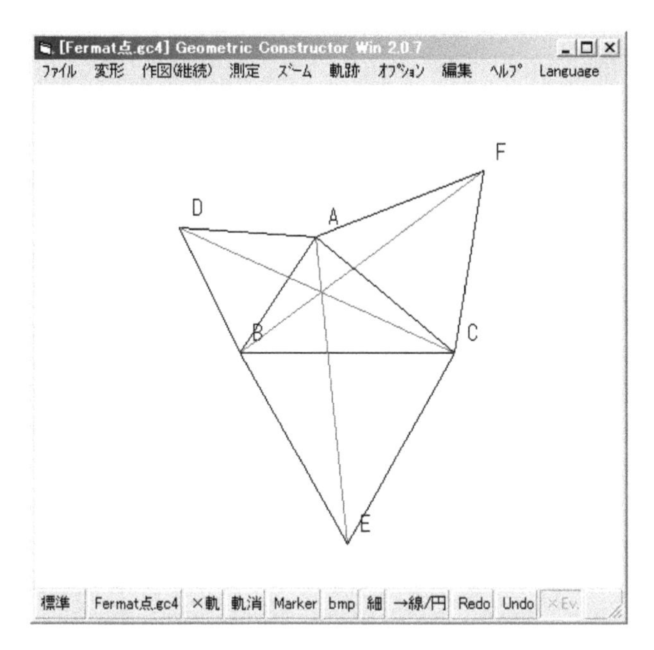

Figure 64.5. Fermat point constructed with GC/Win (1998).

Alignment With the "Millennium Project"

In 1999, Japanese government started "millennium project". According to it, Ministry of Education made new principles about ICT use for education, and settled the plans to achieve them by 2005. From the viewpoint of research and practice of mathematics education, it was important that (1) we can use computer and network in every classroom (not only computer room) (2) we can use a projector (or digital board) and a computer in each classroom for the purpose of teacher's presentation (not for students' individual investigation) (3) we can use digital textbook contents for teacher (developed by textbook publisher) in many subjects in primary and junior high school (4) we can use many educational resources from the Internet.

In some projects made by ministry of education for this plan, we have contributed with GC.

Digital Textbook Contents for Teachers With GC/Java

GC/Java was developed in 2000, with which we can use GC in the web page without installation of GC/Win, and we can make many digital contents with dynamic geometry software.

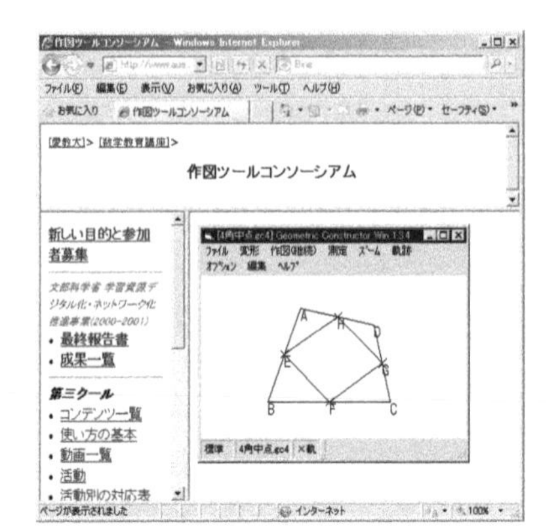

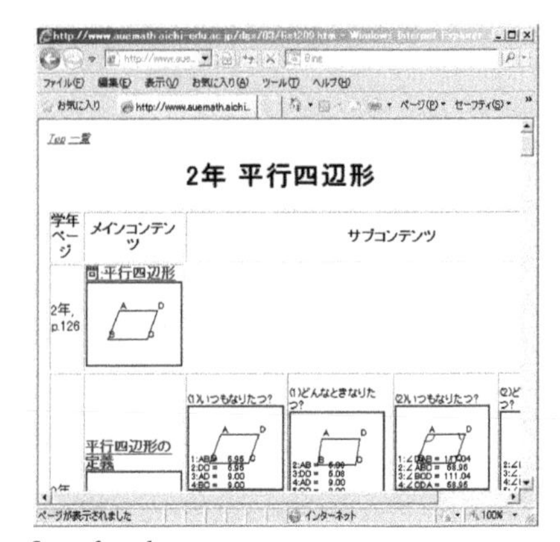

Figure 64.6. Top page and the list page of textbook contents.

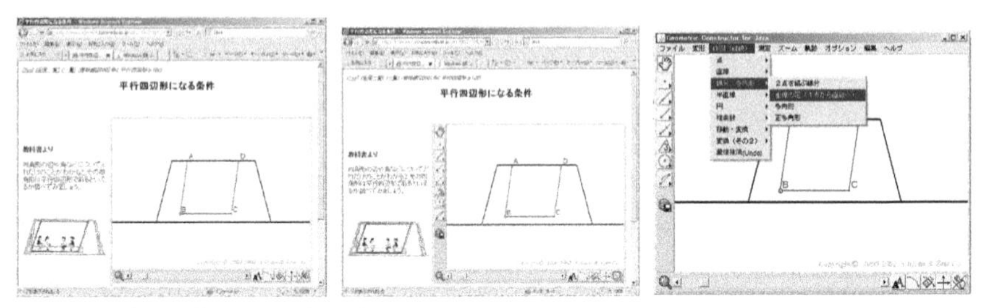

Figure 64.7. Three modes of GC/Java.

GC/Java has three modes; viewer mode, applet mode and window mode, which can be changed by pushing button. In viewer mode, it has only some functions (drag, trace, zoom bar, some buttons (with which we can change axis, width of lines, mode)), which can be used easily by beginners. In applet mode, we can use some typical geometrical construction with icon. In window mode, the function of GC/Java is almost same as that of GC/Win, it can save files to our server instead of a local computer.

In this era, we could use some Java applet of dynamic geometry software including CabriJava and JavaSketchPad. We could use these applet only as viewers for web contents. We could drag and investigate them but could not construct additional lines and circles to the given figures nor make a new figure. In the case of GC/Java, we can use almost all function of GC/Win including construction of additional objects, new construction and saving them.

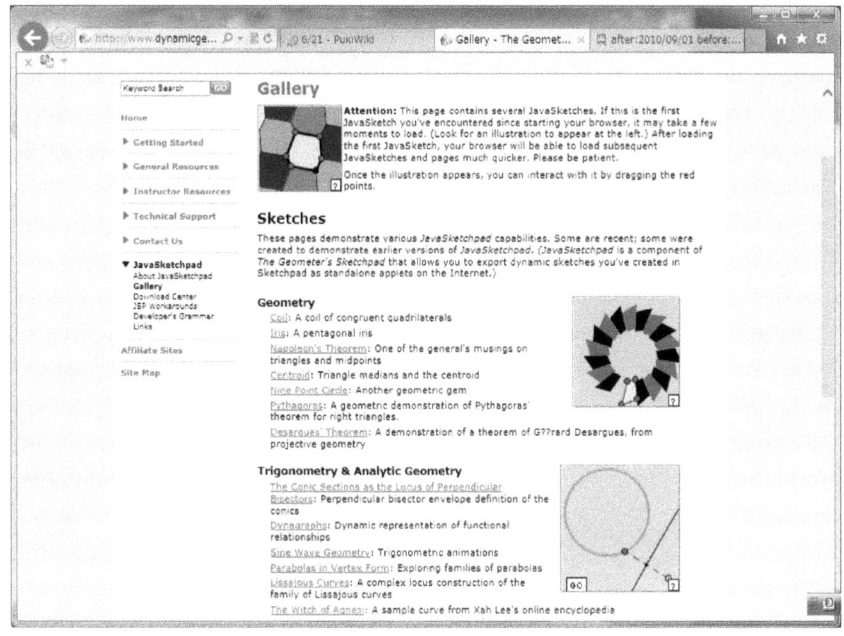

Figure 64.8. Gallery of java sketchpad.

Educational Practices With One Computer and Projector in a Regular Classroom

We have made many educational practices with above digital contents. In many case, teacher used GC/Java to pose problem dynamically, students thought about it with a paper and pencil, and discussed it with GC/Java projected to the blackboard in a usual classroom.

Discussion in the Community With the Internet

We have made a consortium of a developer, some researchers, a textbook maker and about 50 teachers of junior and senior high schools. And we had many discussions (about 1,200 mails during 18 months) with a mailing list referring to the resources in the Internet.

Development of Learning Environment With GC/Java

In above project, we used GC/Java as a contents viewer mainly. But, GC/Java could be used according to various aims. To realize it, we developed some server side software to make learning environment with GC/Java.

Figure 64.9. A lesson with GC projected to the blackboard.

One of them was "PukiWiki with GC". PukiWiki is one of the famous free CMS (Content Management System) written with PHP. Adding some plug-in software, server side software, we made it possible that we can use GC/Java in PukiWiki, and make new geometrical construction, and save it in our server, and make content with it. So we could use it as a web application of GC, with Java-installed computers connected to the Internet.

We have used it at undergraduate and graduate program of some university and at discussion of community of dynamic geometry software

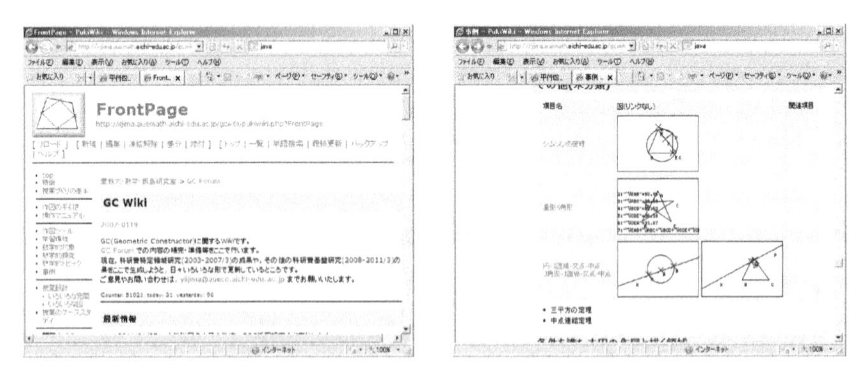

Figure 64.10. GCWiki and contents.

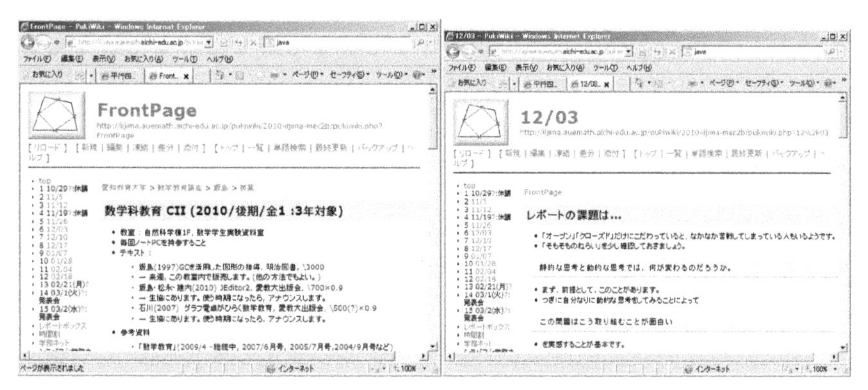

Figure 64.11. PukiWiki for the undergraduate course.

including GC, GeoGebra and Cinderella. It has been very useful for editing new educational materials to support educational research and practice. Using them, we could make some collaboration between university and educational practice in schools, easily.

Further Development: Emergence of a New Device to Be Used in a Regular Classroom

Since 2008, some new device has appeared and realized a new learning style. One of them is so-called netbook and the other is tablet device (tablet PC, iPad, Andoriod etc). We can use them in a usual classroom to make a group experiment and discussion or individual learning.

Since iPad doesn't support Java and Flash, we can't use GC/Java with iPad. So, we have started the development of GC/html5 from 2010. In 2012, the development of standard version of GC/html5 was finished, with which we can construct, measure and investigate figures with multi-touch, and we can save it to the server on the Internet. GC/html5 is written with JavaScript using html5, which make possible to use it with many device including PC, iPad and Andriod. We can use English and Japanese for the language of menu, and we can add the dictionary for the other language use.

Implications and Suggestions for Future Research

In this chapter, the author reviewed historical developments of research and practice of teaching and learning mathematics with ICT in Japan. We have learned some lessons in the course of development and

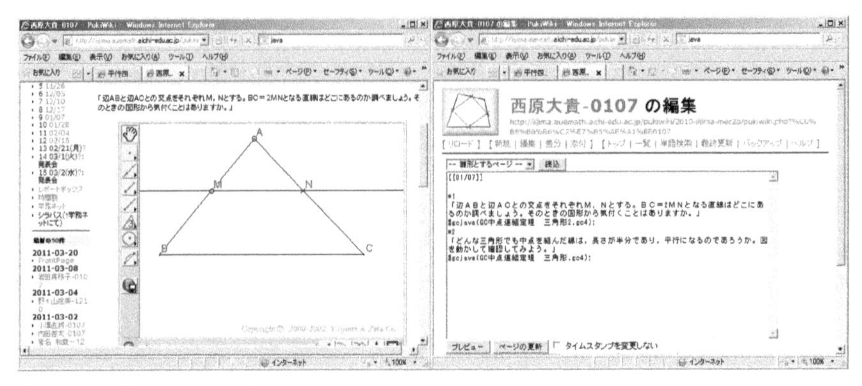

Figure 64.12. Student's works.

Figure 64.13. Problem posing with GC/Java projected to the blackboard.

use of Geometric Constructor. Implication for classroom practices and needed studies are as follows.

For the effective use of educational software, it is important for us to form a community of developer and teachers and to design a cycle of development; use – evaluation – revision with the discussion in the community. Also, for the use of educational software in various opportunities, it is useful to develop the software as web application. To achieve it, we can use such technologies as Java applet, Flash, html5 (and JavaScript) for client side software and php, asp.net, and so on for server side software.

In the current situation, digital contents are developed as digital textbook for teacher's presentation. For the regular use of ICT in many class-

Figure 64.14. Group experiment and discussion with notebook.

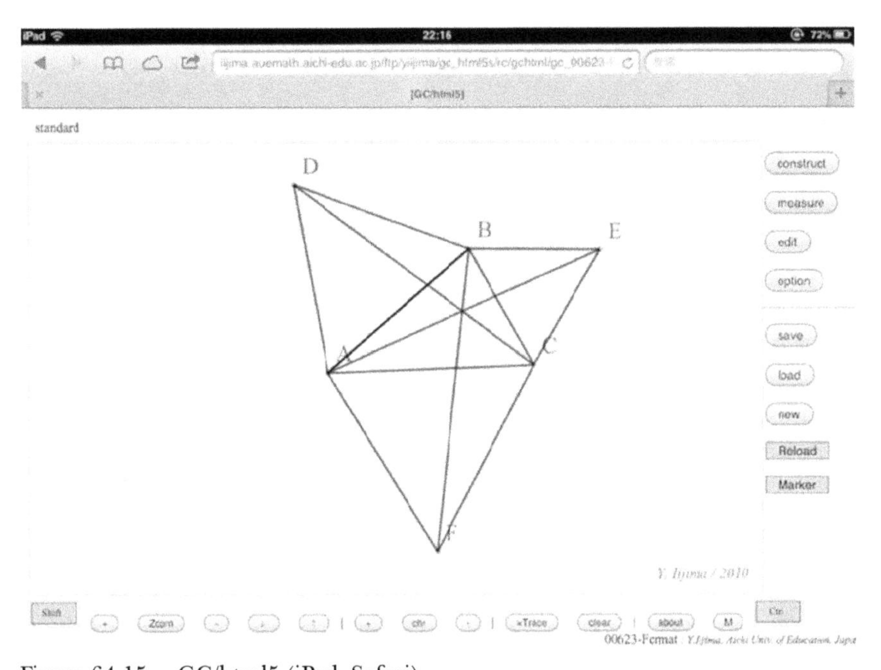

Figure 64.15. GC/html5 (iPad, Safari).

rooms, it is effective to make digital contents based on the textbooks for daily use. Digital textbook for teacher's presentation show the standard of educational practice using ICT. And we can use the Internet to share vari-

Figure 64.16. Problem posing with digital board

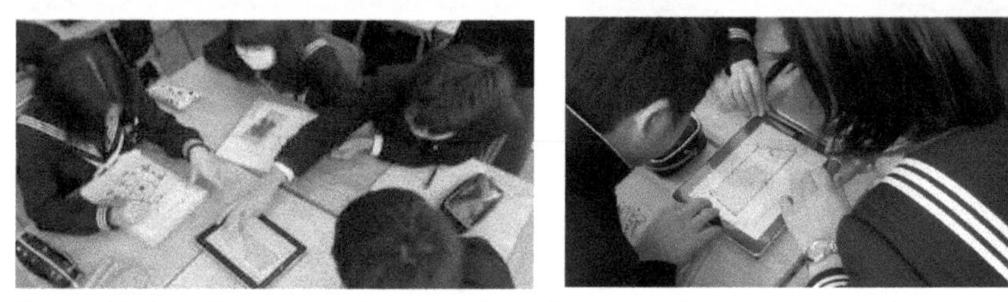

Figure 64.17. Group experiment with iPad and drawing to the OHP sheet on iPad.

Figure 64.18. OHP sheet and presentation of overlaid sheets on the LCD TV.

ous contents and ideas for planning and implementing lessons in our communities. Further research and practices are needed to utilize the possibility of tablets like iPad for individual and/or group mathematical investigations in regular classrooms.

REFERENCES

Harada, K. (1997) Using Cabri-Geometry of helping students in geometrical proof-problem solving: An analysis of students' inferences in geometrical construction. *Proceedings of 30th annual meeting of Japan Society of Mathematical Education*, 415-420, Tokyo: Japan Society of Mathematical Education. (in Japanese)

Japanese Government (1999) *Educational-use computer installation program: A millennium project.*

Iijima, Y. (1990) Basic consideration for development of Geometric Constructor: formulation of dynamic treatment of figure. *Tsukuba Journal of Educational Study in Mathematics, 9*, 105-117. Tsukuba, University of Tsukuba. (in Japanese)

Iijima, Y. (1995) *Change mathematics lessons with computer: using software GC.* Tokyo: Meiji Tosho (in Japanese)

Iijima, Y. (1997). *Lessons of geometry using dynamic geometry software GC.* Tokyo: Meiji Tosho (in Japanese)

Iijima, Y. (2001) Development of educational software and contents: report and proposal of the consortium of dynamic geometry software, *Journal of Japan Society of Mathematical Education, 93(12)*, 13-24, (in Japanese)

Iijima, Y. (2005) Development of various learning environments using GC/Java, *Journal of Science Education in Japan, 29(2)*, 110-119, (in Japanese)

Iijima, Y. (2010) Possibility of iPad as learning device for student and possibility of html5 to develop mathematics software and contents for education, *Proceedings of 41st annual meeting of Japan Society of Mathematical Education*, 741-746, Tokyo: Japan Society of Mathematical Education. (in Japanese)

Kakihana, K., Shimizu, K. & Nohda, N. (2000). A Study of students' and teachers' conception of the effects of dynamic geometry software, *Journal of Science Education in Japan, 24(3)*, 151-158

Shimizu, K., Kakihana, K. (1999). *Students' activity supported by computer – a new approach in geometry.* Tokyo: Meiji Tosho (in Japanese)

Tomoda, K. (1994), *Let's study high school mathematical graphics with computer,* Tokyo: Morikita Shuppan. (in Japanese)

Tomoda, K. (2003). GRAPES: GRAph Presentation & Experiment System. http://www.criced.tsukuba.ac.jp/grapes/ Retrieved on June 30, 2013.

Tsuji, H. (2003). A study on the development of the figural concepts by geometric construction in computer-based environments, *Journal of Science Education in Japan, 27(2)*, 85-93 (in Japanese)

Cabri http://www.cabri.com/

Cabrijava http://www.cabri.net/cabrijava/

Cinderella http://www.cinderella.de/

JavaSketchPad http://www.dynamicgeometry.com/JavaSketchpad.html

GC/html5(English) http://iijima.auemath.aichi-edu.ac.jp/ftp/yiijima/gc_html5e/

GeoGebra http://www.geogebra.org/

Geometer's Sketchpad Resource Center http://www.dynamicgeometry.com/

Mathematica http://www.wolfram.com/

Risa/Asir(Kobe Distribution) http://www.math.kobe-u.ac.jp/Asir/asir.html

CHAPTER 65

"INNER TEACHER"

The Role of Metacognition and Its Implication to Improving Classroom Practice

Keiichi Shigematsu
Nara University of Education

ABSTRACT

This chapter overviews research on metacognition in Japanese mathematics education and provides research findings that are related to the improvement of the teaching and learning mathematics in classrooms. A special attention is given to the original conceptualization of metacognition as "Inner Teacher" which assumes that a part of students' metacognition related learning mathematics derived from the interaction with their teacher in the classroom. Based on the review of the current status of research on metacognition as well as the development in the methodologies, issues for the further research are discussed.

Keywords: Metacognition, metacognitive skills, inner teacher, mathematical problem solving, questionnaire, journal writing,

The First Sourcebook on Asian Research in Mathematics Education: China, Korea, Singapore, Japan, Malaysia, and India, pp. 1455–1474

INTRODUCTION

Research on metacognition in the 1970's has focused mostly on reading and comprehension of texts. In the early 1980, pioneering researchers in mathematics education started to focus on meta-memory and metacognitive skills related to mathematics learning and mathematical problem solving. At that time, researchers in Japan, however, did not recognize the importance of metacognition in research and practice, and rarely discussed it at national conferences of mathematics education.

We are often inclined to emphasize pure mathematical knowledge only in education. Thus, finding it hard to impart to our students. As a result, they fail to solve mathematical problems and soon forget it after the paper and pencil test.

Researchers of mathematics and psychologists have come to notice metacognition as an important function of human cognitive activities. The ultimate goal of our research is to develop a clear concept about the nature of metacognition and applying this knowledge to improve the methods of teaching mathematics and teacher education.

We could regard metacognition as the knowledge and skills that activates the objective knowledge in one's thinking activity. Following the definition and suggestions by Flavell (1976), we categorize metacognitive knowledge into four divisions, and metacognitive skill into three divisions as follows.

(Metaknowledge)
1. the environment
2. the self
3. the task
4. the strategy

(Metaskill)
1. the monitor
2. the evaluation
3. the control

We start from a primitive view that teaching is a scenario where a teacher teaches students and each student learns from a teacher. In the process of teaching, a phenomenon remarkable from a psychological point of view will soon happen in the student's mind. We call this the splitting ego in a student or decentralization in a student, in a Piagetian terminology. Children according to Piaget are ego-centric by nature. In the earlier part of lower grades in elementary school, ego-centrism is gradually collapse and split into two egos. The acting ego and executive ego where the former is regarded as metacognition.

Recognition, in the narrow sense, but seems the same perception of mathematics learning is to calculate, measure, and plotting a graph as a scratch, knowledge and appeal of mathematical activity acts directly on the environment cognitive function, including the means and skills. On the other hand, has been successfully used and whether the knowledge and skills, and metacognitive awareness of action to adjust the action. Seen this way, the metacognitive rather than direct action on the environment, the perception that target cognitive function, and cognitive recognition of that, and say what happens in the mind.

This concept is derived from the study of memory development in mathematics education where solving problems with knowledge and skills.

Children poor in mathematics when guided through repeated teaching of the same content is an example. They often forgot the lesson after a week. In order for children to have good memory, it is important to exercise their meta-memory. In this case, lack of confidence in mathematics is due to lack of memory.

Let us consider another example, solving two problems with the same solution or two different solutions. Sometimes the other solution can not solve the problem. The difference of these solutions maybe cause by anything. At this point we can not attribute the differences in the knowledge of mathematics but on how one use his mathematical knowledge.

We can see that metacognition is a part of the problem-solution together with knowledge and skills a learner use.

RESEARCH ON METACOGNITION IN MATHEMATICS EDCUCATION IN JAPAN

Beginning of the Study on Metacognition

Definition of metacognition has long historical discussion before it is roughly defined as cognition about cognition. In the same line of thinking with Piaget and Vygotsky on reflective abstraction as how we intend to use the same word.

Flavell (1976) used the word meta-memory in a psychological memory research. Metacognition among other things refers to the active monitoring and consequent regulation and orchestration of these processes in relation to the cognitive objects or data on which they bear, usually in the service of some concrete goal or objective.

The difference in their interest and the definition of metacognition shows little difference, it captures the two sides of metacognition, metaknowledge and metaskills respectively.

> Metacognition refers, among other things, to the active monitoring and consequent regulation and orchestration of these processes in relation to the cognitive objects or data on which they bear, usually in the service of some concrete goal or objective.

Flavell who is also an avid researcher of Piaget, had a strong interest in metacognitive development of children. Following this, Brown (1978), based on the reading comprehension skills from the study of teaching retarded children to learn, have approached the problem of metacognition. The inclined foster enthusiasm for work by increasing the independent learner metacognitive ability of students.

Thus, the difference in their interest is not seeing much difference, but rather their definition of metacognition have a commonality that captures the two sides of metacognition. In other words, Flavell said, "knowledge about cognitive aspects of metacognitive knowledge," "adjustment control on cognitive aspects of metacognitive experience," which stipulates, for each of these. Brown et al., "Elements of knowledge" and "self-regulating element" that stipulates.

In psychology, depending on the interest on either side of the two sides in this, or depending on whether you emphasize both become different, the research idea is different as structured model and functional model for metacognition.

Thus, metacognitive research stemming from Flavell and Brown, was carried out in the research of cognitive psychology which focus on problem solving. As a result, in 1980, metacognition was added to the index at the world's largest bibliographic database ERIC (Educational Resources Information Center). These research tendency also affected mathematics education research which Silver say, its research topic is metamemory (1982). And Lester and Garofalo (1985) pointed out the need for systematic study of metacognition in solving mathematical problems, which began in earnest the study metacognition in mathematics education .

Research on Metacognition in Mathematics Education in Japan

Earlier research on metacognition in Japanese mathematics education community can be found in Takazawa (1986) who overviews studies of metacognition in mathematics education and presentationed his conceptulization based on the definition of metacognition of Flavell (1976). Shigematsu (1986) also examines the classroom activity in mathematics from the viewpoint of metacognition, such as "meta-understanding" and "metacognitive understanding governing the activities."

Hirabayashi (1987) speculates on the very relevant to Skemp's model of understanding research that explain Δ_1 understanding of intelligence involved in the objective world and Δ_2 intelligence to control it to the self-action (cognitive) and self-monitoring (metacognition). Thus, metacognitive studies in Japan begin with research such as understanding, and then expand the field to solve mathematical problems, which has been promoted as a unique study of mathematics education. At that time, the metacognition is divided into two such as "metacognitive skills" and "metacognitive knowledge" in cognitive psychology.

For example, Iwago (1990) thinks which metacognition is "knowledge about cognition" and classified its spects of metacognitive knowledge into three, "person", "task", and "strategy" following the idea of Flavell. In addition, he also think its "adjustment control on cognitive aspects" to "metacognitive skills," and, according to the provisions of the Brown classified into three, "monitor (monitoring)," "self-evaluation" and "control".

Shigematsu (1990) adds "environment" to three categories of Flavell which accumulated in the environment to determine the effect of cognitive status. In addition, he divided these metacognitive knowledge into a positive thought processes to function effectively and a negative inhibitory functioning in reverse. And, as Figure 65.1 shows the relationship of the following cognitive and metacognitive processes.

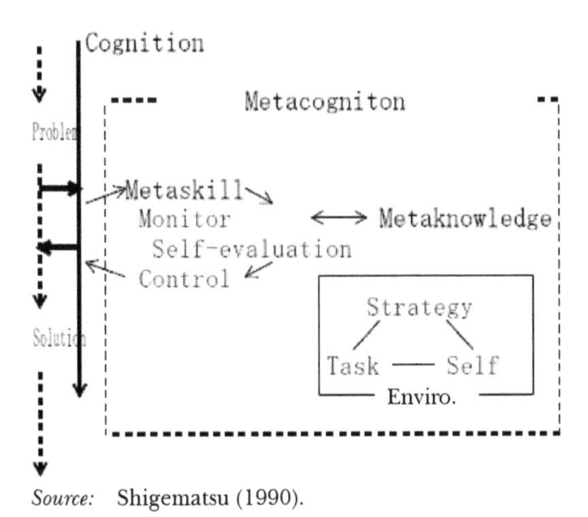

Source: Shigematsu (1990).

Figure 65.1. Relationship between cognition and metacognition.

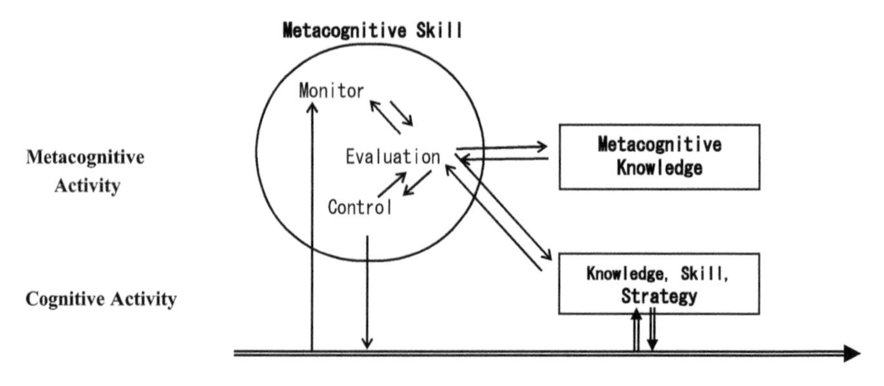

Source: Kato (2000).

Figure 65.2. Relationship between cognition and metacognition in mathematical problem solving.

The cognitive and metacognitive processes are explained that is the modus ponens syllogism logical relationship of metacognitive skills and metacognitive knowledge. In particular, the metacognitive process that occurs is explained as 'Monitoring → Self-evaluation → Control' model to the point that the cycle is also seen Figure 65.1 and is appropriate provisions of the metacognitive skills of each.

Kato (2000) creates a model that combines elements shown in Figure 65.2 for the relationships of "knowledge and skills" and "strategy".

This is suggested by Shimizu (1995) which mathematical problem solving relates to knowledge, understanding and skills, and more to strategies for application, associated with said metacognition.

These studies were cited in other studies, which metacognition is considered as an important driving force for solving mathematical problems. For example, Yamada and Shimizu (1997) regarded metacognition, self-evaluation, looking back and reflect as "self-referential activity" in general terms to refer to activities such as product or process of self-solving activities in mathematics learning. They refine the structure and role of them, and explore the implications to the teaching of problem solving based on them.

Methodologies for the Study of Metacognition

There are many difficult factors to capture metacognitive activities that actually happens. It is natural to capture its process in the process of problem solving, but metacognitive activity is less likely to be from the description of language. However, unlike research in psychology, we want to analyze the teaching situation that aims to develop students' metacognition in learning mathematics, and so we want to capture the metacognition that is active at learning mathematics. This is an overview of previous studies by the methodology.

Questionnaire

This is a way to capture the students with the metacognitive questionnaire.

Shigematsu (1994), under the hypothesis that the formation of student's metacognitive actions are internalized according to the teacher's utterances (as described, Explanation, Questions, Indication and Assessment) to collect and capture the metacognitive questionnaire was created for. Using the results of the questionnaires, the relationship between student 's metacognition and teacher's utterances is analyzed, and the transformation of metacognitive development of students is revealed.

Shimizu (1995), as well as metacognition, analyze the knowledge, understanding, skills and the use of element of strategy, based on quantitative data from questionnaires about the relationship between problem-solving skills and contribute to study the situation of solving mathematical problems.

These methods catch the actual condition of the learner with metacognitive knowledge, but have a certain effect, not intended for actual problem solving processes, it is difficult to capture directly to the metacognitive skill. Therefore, the process aimed at solving real problems, methods have been developed as follows.

Collaborative Problem Solving

Two or three students attack at one problem which will be resolved in cooperation with , or a way to analyze the thought processes of students there. By allowing for cooperative problem solving, increases the opportunity to target their own thinking, which is thought to promote further monitoring. The speech is to be represented by a metacognitive skills, it can be naturally described it as a protocol.

Using this method, Shimizu (1989) analyzed the process of problem solving the construction in junior high school based on the role of metacognition to problem-solving progress which investigates empirically. And in the process of "transformation of problem" as an analysis point of view, was to capture the occurrence of metacognition to identify major turning point. By using this analysis point of view, solving the problem in the process of drafting two junior high school students who have worked collaboratively to identify where it occurs using metacognition for considering the role of metacognition. Shimizu (2006) has implemented the data analysis focused on language at "Transformation of Problem" and obtained the results in this study focused on "Problem Solving" collaboratively as a set of learning environment. And more, he also analyzed the situation for the teaching of reasoning and the multiplication/division using fraction meta-analysis has also been thought of.

Yamaguchi (1992) analyzed the problem-solving process by using a pair of quadrangle drawing task for 3 and 4 elementary school graders and also analyze metacognitive activity occurring in the process. As a result, the identification of those six contents speechs was made, in which, in particular, "says denying the opinion of the peer," "presentation of questions for opinion of the peers," "opposing opinion to the opinion of the peers" are revealed as the induction of recognition for playing a major role. However, for problem solving in cooperation, that a framework is necessary to consider what influence the effects of metacognitive process of resolution of individuals and what form the two communication depending on whether progress has also been, are pointed out because the problems such as coming to a different impact on metacognitive processes occurs.

Stimulated-Recall Technique

Stimulated-recall technique is presented as a way to do research while stimulating the learning process after the resolution's own problem-solving. Therefore, this method has the advantage of reporting can be obtained when performing in line with thinking in problem-solving process.

Using this method which has demonstrated the process of internalization of metacognition in students, Shigematsu and Katsumi (1993), after solving the problem of elementary school fifth grader at the time of problem solving give a stimulus of VTR presented and students answer the questionnaire which plays stimulation of verbal utterances of teachers.

Kato (2000) develops a stimulating problem-solving method using the questionnaire about metacognition referring to students' answer for the problem at the end of problem-solving.

Using Student's Journal

This is the method to capture a metacognitive activity of learners using the learner's own written work.

Shigematsu and Katsumi et al. (2002) are analyzing the mathematics journal which students had written for a few minutes to take the time at the end of mathematics class. "The method of Math Jornal Writing" has been described as factors such as metacognitive knowledge and transformational change in the classroom. As a point of view to analyze their journal, they identify the five stages of knowledge transformation in the classroom, by focusing on metacognitive functions as a factor in their transformation, and model transformation has been proposed as a development of metacognition. This approach can also enable to present the development of metacognition and has become many examples can be seen in research practice.

DEVELOPMENT OF METACOGNITION

With the advancement of research to capture the aspect of metacognition, an issue of the mathematics education has been studied focusing on the development of metacognition in students better.

Metacognitive Framework of Development

In considering the development of metacognition, Iwago (1990) pointed out which metacognitive functioning and refer to the theory of Vygotsky's social interaction negotiation, individuals can not be resolved by, when the resolution will become available through involvement with others.

Hirabayashi and Shigematsu (1986) captures the process of developing metacognition, which has been expressed "Inner Teacher". In the definition of "Inner Teacher", they are pointing out the process of internalization of metacognition in seven steps. Their unique concept is that this "metacognition" is thought to originate from and internalized by the teacher him/herself. Teachers can not teach any knowledge per se directly to students but teach inevitably through their interaction with students in class. Their original conception is that this executive ego is really a substitute or a copy of the teacher from whom the student learns. The teacher, if he/she is a good teacher, should ultimately transfer some essential parts of his/her role to the executive ego of the student. In this context, they refer to the executive ego or "metacognition" as "Inner Teacher".

The advantage of this metaphor is that they could have the practical methodology to investigate the nature of metacognition; that is, they may collect varieties of teachers' behaviors and utterances in lessons and carefully examine and classify them into psychological view-points

Shigematsu and Katsumi et al. (1993), based on these considerations, the following seven steps of "internalization models of metacognition" to create, as shown in Figure 65.3 shows the process.

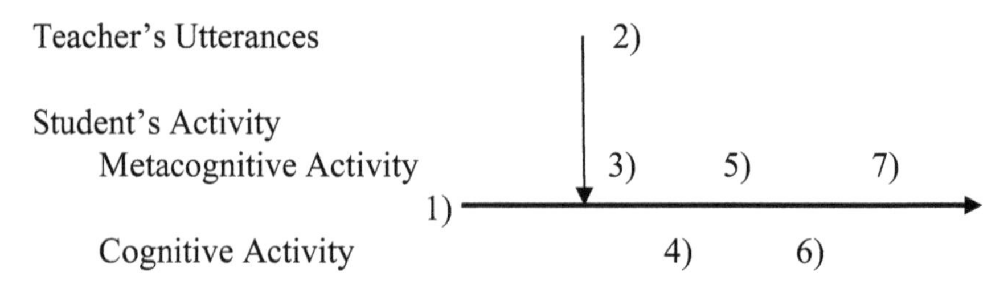

Figure 65.3. Internalization process model of metacognition in Shigematsu et al. (1993a).

First, we specify the process of internalization of metacognition (inner teacher) as follows:

1. A student is very much aware of the current problem-solving;
2. Before or while the student solves the problem, the teacher gives him/her a suitable metacognitive advice when he/she wants to get it;
3. The student remembers the teacher's metacognitive advice tentatively at this time;
4. The student can solve the problem referring to the teacher's metacognitive advice and has a good affective feeling;
5. The student wants to remember it permanently;
6. The student can solve similar problems referring to the teacher's metacognitive advice;
7. The student acquires the metacognition as the inner teacher.

We assume that the internalization of metacognition mainly originates from teaching-learning communication between a teacher and student in a classroom lesson as this hypothesis.

By these results, Kato (2000) is building a framework for teaching such as: Cultivating metacognitive ability.

(A) To resolve personal issues
(B) Understanding A, self-solving activities of others solving activities through negotiations
(C) Metacognitive activities carried out by A, a greater understanding of the metacognitive activities metacognitive activities of others and self through negotiations
(D) By applying the teaching of similar metacognitive activity, scenes that establish the usefulness of consciousness that work metacognitive activity

Then, by using this framework, work on the development of practical skills through teaching and tutoring metacognition aimed at fourth graders in the developing metacognition is that certain results obtained.

Ishida (2002) examine, utilizing these frameworks foster metacognition, incorporating mathematics lessons teaching of metacognition what the effects of students' problem-solving process. As a result, have found that is effective in problem solving processes and work out on their own students how to solve a variety of idea after teaching metacognition.

Metacognitive Support by Teachers

In order to foster the student's metacognition, teachers need appropriate support. About this, Shigematsu (1994) suggest the role of the teacher to develop metacognition as the next four mentions. The first, Schoenfeld (1987) cited, "a role as model" and emphasizing the work of the metacognitive problem-solving process is shown by the teacher. Moreover, as the one that will correspond three metacognitive skills described above.

Role as a monitor: between observation desk during tutoring and whole class discussions, to examine the studednts' problem-solving, the role of teacher to monitor on behalf of students, and advice.

Role as a evaluator: a direct evaluation of metacognitive knowledge against the results of student's problem-solving, the role of the teacher on behalf of student's self-esteem.

Role as a controller: self-evaluation based on the results of the student's teachers on behalf of control.

Kato (2000) explained the above Shigematsu's ideas about the role of the teacher to develop metacognition as the metacognitive support and surveyed the process of cognitive activity such as knowledge, skills and strategy and also the effect of metacognitive support for elementary school sixth grade students. As a result, the top group needs metacognitive support, and medium group needs cognitive support and also metacognitive support, finally subordinate position group needs effective cognitive support primarily be done.

Teaching/Learning to Develop Metacognition

Progress of research mentioned so far, it has become more common as research aimed at development of metacognition in teaching/learning prcesses.

For example, teaching students to recite what the teacher showing you a breeder of metacognition has been done before. Todome (1994) revised such a teaching statement which student's inner utterances occur. As a results, the improvement compared to previous guidance was shown held long, which also can be improved to ensure problem-solving scores.

In addition, as described in the previous section, teaching/learning processes are more likely to be practical and metacognitive by fostering and promoting the activities described in students.

For example, Kameoka (1996) use the students' note which are written so-called "Balloon" around the problem and he described metacognition. As a results, there are shows that can promote the metacognition.

Shigematsu and Katsumi et al. (2002) proposed that the practical continuous use of everyday mathematics teaching "Math Journal Writing Method" by the proposed step is divided into four specific teaching steps aim to help students in elementary school middle grades metacognitive that. And the possibility of making such a practice has been verified by a teacher writing mathematics teaching work for the first time. This method is putting teachers through continuous red pen has the characteristic that there are effects of teacher interaction with students. In other words, writing a single journal is or can be viewed through that continuously analyzes the changes in such students do not recognize and can return the student to the comments elicit metacognitive desirable by actively teachers you can be.

ENHANCING THE QUALITY OF STUDENTS' METACOGNITION

Teacher's Comments Using Red Pencil on Students' Journal Writing

Shigematsu and his colleagues (2002) started the study of metacognition from the early era and were interested on students' journal writing and teachers' utterances. They adopted metacognitive instructions with "using a red pencil", which refers to feedback written using red ink by the teacher on students' journal writing. It was essential leading to the changes in quantity and quality of metacognition. They proposed a scheme for effective feedback processes. Examples of analyzed cases are provided below.

Emphasizing discussions on the metacognitive instructions using the method of Math Jornal Writing are presented in Figure 65.4.

In this processes, we want to comment on some instructional points:

1. Supporting students' metacognitive activities with comments written in red pencil?

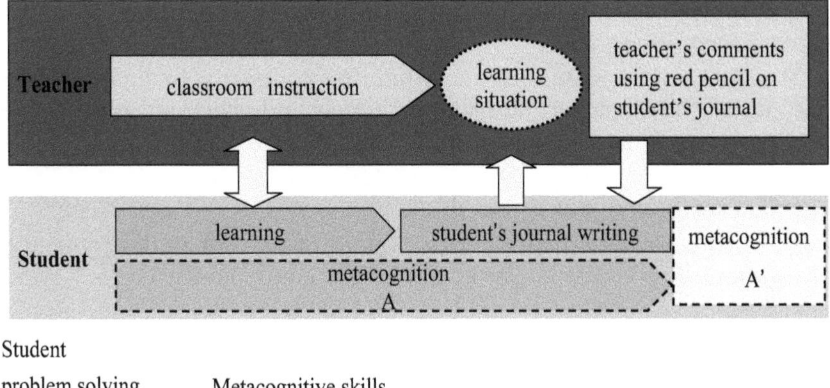

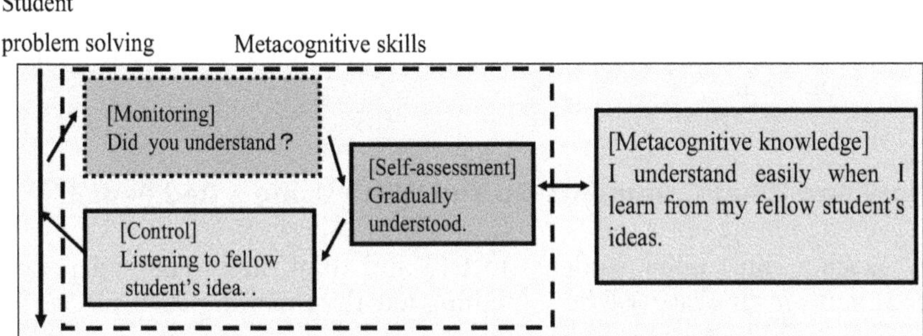

Figure 65.4. The internalization process of metacognition by student's journal writing.

2. The changing process in the development of students' journal.

For teachers who actually spend most lesson time for class instruction, journal writing is a practically helpful approach to check and foster metacognition in students. Focusing such function of writing as an instructive tool to activate metacognitive thinking, we have developed 'Math Journal Writing Method' based on teachers' metacognitive instruction in mathematics class discourse to foster "Inner Teacher". It is very important for a teacher to give some comments orally and using red pencil on students' journal writing.

At the end of the lesson, the teacher gives 5 minutes for students to write on their journal writing about the day's learning according to teacher's instruction as follows:

- "Please write on your journal about today's lesson, remember to emphasize on how you process the idea."

Afterwards the teacher collects the students' journal and gives comments using red pencil in students' journal writing sheet. Teacher then suits comments accordingly. We think this red pencil comments as a suitable metacognitive advice.

Students cannot write their metacognitive comments by themselves at first. At that time, teacher must support timely and give some appropriate comments as follows:

- "Please write freely as you think."
- "Please write something that you think is well understood and interesting."
- "Please write the next thing you want to do."
- "Please write on how you came up with this idea."
- "Please write something that you think that comes from behind your brain."

How and What Comments Do You Write Using a Red Pencil?

Teacher must allow students to feel and think first. The red pencil comments in this case will be something like the following example.

- "Well, you often thought."
- "You noticed a good place."
- "Great. That's true."
- "You've really struggled."
- "You often tried."
- "You noticed so many things to see like the chart."

The entire leadership is based on individual leadership not getting the first step. The students also will be writing the following comments.

- "Since my teacher praised me, I was happy."
- "I can not express my ideas in class, but in journal writing I can express it anytime with my teacher."

Changing the Teacher's Comments Referring to the Development of Students' Journal

With the changing content of student's journal writing, you may change the content of the teacher's comments using red pencil. Journal writing written by the first student is a lot like the following.

- "Today, I studied ...."
- "Today I made a mistake."
- "Today's study is interesting."
- "Today's study was good
- "Today's study was difficult."

In order to move away from just writing that describes the facts and impressions in this way, the teacher gives the following instructions all together.

- "Why did you got it wrong?"
- "How difficult was it?"
- "How did you know?"

However, it should be noted that their effects do not appear immediately. Therefore, for student to write more in detail, the teacher must occasionally give the following comments using red pencil.

- "What do you think?"
- "What makes you think you can not do it?"
- "Let's talk about the reason why"
- "How difficult was today's study?"
- "In what way did your friend explain it?"

In some cases, as the teacher returns a notebook to the child, it is also effective to say the following.

- "For the comments in red pencil, please feel free to write me back soon."

Finally, we suggest teacher's comments be changed from Step A to Step E according to student's metacognitive activity as shown in Table 65.1.

Typical Features of Comments Using Red Pencil in Response to Students' Attainment Level of achievement

The students in the group of upper-achieving rich in knowledge, and they often have successful problem-solving. Therefore, it is important that

Table 65.1. The Steps of Student's Metacognitive Development

Step	*Mathematical Description*	*Active Metacognition*
A	Feeling of attainment or failure, thoughts about tasks: I learned ...	metacognitive knowledge on strategy
B	Feeling of attainment or failure with reasoning, referring to other's opinions: Because ...	metacognitive knowledge on strategy
C	Statement of self-confidence, self-evaluation: I have become expert at(in) ...	metacognitive knowledge on self
D	Generalized questions, related to a similar problem: I try to advance ...	metacognitive knowledge on task, metacognitive knowledge on strategy
E	Inclusive statements on mathematics: I think math is ...	metacognitive knowledge on task, metacognitive knowledge on strategy

the teacher will answer questions on journal writing written by student. For example, in the following journals.

> "What will be the multiplication computation on paper of a 3-digit, 4-digit and 2-digit number? What will be the division computation on paper? Will the division computation on paper be like multiplication computation on paper? "

For this journal, in order to develop student's thinking, the teacher gives comments using red pencil in this manner.

> "Multiplication can be calculated as it is no matter how many digits it has. Division computation on paper is different than multiplication calculation method. We will study this soon, look forward for it."

The middle group of student often repeat the success or failure of a problem solving. So, for example, students wrote in their journal such as:

> "I could answer all of computational problems, so I feel confident with multiplication computation on paper."

For such journal, in giving student the following comments using red pencil, it is important that student become confident through it.

> "You know why it was wrong before, you were confident with problems of different number of digits."

In this way, if a student is successful the teacher acknowledges it, and if a student fails the teacher makes the student think of the reason for failure. This guidance will lead these students to success.

The lower group of student is often unsuccessful in problem solving. Therefore, teachers can help student solve the problem, and shall ensure that student get a little sense of solving it. For example, student wrote the following journal.

> "I did not quite understand the problems of the day. But, S's taught me so I could understand. I was happy."

This time the teacher gives the student the following comments using red pencil.

> "When you do not understand the question, it is better for you to ask your friends. This would be a good way to understand. "

Through this, the student may be aware of "Why I could solve the problem?" and "How could I solve the problem?"

Furthermore, the contents of the journal written by students will change. Repeating this experience, students will acquire the metacognition as the inner teacher.

NEEDED DIRECTIONS FOR FURTHER RESEARCH

Based on research reviewed in this chapter, the author raises issues and discusss needed directions for further research on metacognition in mathematics education. First, we need to have better understanding of the development of metacognitive learning during the time of problem solving by students their own in a class as well as those during "Neriage", which means in Japanese "polishing up" ideas students came up with.

In addition to explicit metacognitive training course content as the basis of previous research, it is desirable to demonstrate that the curriculum guidelines for mathematics education as one of the purposes of fostering metacognition. It is worthwhile for us to examine the curriculum framework for mathematics in Singapore, the "pentagon", which emphasizes metacognition as one of the key areas for learning mathematics.

Metacognition is an essential element to study the mechanism of students' learning of mathematics and mathematical problem solving. However, teaching mathematics in regular classrooms tends to focus more likely on the "visible" aspects of students' cognition. Identifying the relationship between metacognition and cognition is an important agenda for mathematics education research. These metacognitive studies for practice may also improve reform the student autonomous learning in mathematics education. Further research is needed which is closely related to the practice of fostering students' metacognition. Also, we need to connect different research findings from other research areas such as brain sciences, psychological research and other subject that relates to it. Furthermore, we need to analyze teaching and learning processes focusing on fostering students' metacognition.

REFERENCES

Brown, A. L. (1978). Knowing when, where, and how to remember: A problem of metacognition. In Glaser, R. (Ed.), *Advances in instructional psychology.* (pp. 75-165.) Hillsdale, NJ: Lawrence Erlbaum. (R. Yukawa & Y. Ishida (Trans.) *Metacognition: Knowledge about cognition.* Tokyo:Science Publishing. (in Japanese)

Flavell, J.H. (1976). Metacognitive aspects of problem solving. In L.B. Resnick, (Ed.) *The nature of intelligence.* Hillsdale, NJ: Lawrence Erlbaum, 231-235.

Fujita E. (1995). Learning and assessment in elementary school mathematics education: Self-assessment skills (metacognitive skill) and assessment. *Arithmetics Education. (Journal of JSME)*, 77 (2), 2-7. (in Japanese)

Garofalo, J. & Lester, F.K. (1985). Metacognition, cognitive monitoring, and mathematical performance. *Journal for Research in Mathematics Education*, 16, 163-176

Hirabayashi I. (1987). *Principles of activism in mathematics education.* Tokyo: Toyokan Publishing. (in Japanese)

Hirabayashi I. & Shigematsu, K. (1986). Meta-cognition: The role of the "Inner Teacher". *Proceedings of the 10th annual conference of the International Group for the Psychology of Mathematics Education*, 165-170.

Hirabayashi I. & Shigematsu, K. (1987). Meta-cognition: The role of the "Inner Teacher"(2). *Proceedings of the 11th annual conference of the International Group for the Psychology of Mathematics Education*, vol.2, 243-249.

Hirabayashi I. & Shigematsu, K. (1988). Meta-cognition: The role of the "Inner Teacher (3). *Proceedings of the 12th annual conference of the International Group for the Psychology of Mathematics Education*, vol.2, 410-416.

Hoshino M. (2000). A study on the role of mental models in acquisition and formation of mathematics knowledge : Focused on situation deriving procedures based on declarative knowledge. *Secondary Education (Journal of JSME)*, 82 (5), 3-12. (in Japanese)

Ishida, J. (2002). A study of the effects of teaching through meta-cognitive approach on sixth graders' problem solving process. *Research Journal of Mathematical Education (Journal of JSME)*,78, 3 -21. (in Japanese)

Iwago, K. (1990). Study of cognitive process on metacognition in mathematics education. *Scientific Research Grants(General Research C) Research Report* (63580233). (in Japanese)

Katsumi Y.(2007). A Study of Development and Analyzing on Curve for Elementary School Student's Mathematics Learning Situation. *Arithmetics Education (Journal of JSME)*, 89 (8), 10 -19. (in Japanese)
Katoh H. (2000). A study on the development and function of metacognition in mathematical problem solving. *Research Journal of Mathematical Education (Journal of JSME)*, 71/72, 21-27.(in Japanese)
Kameoka M. (1996). A Study on the Integration of Guidance with the "Balloon Method" and Evaluation. *Arithmetics Education(Journal of JSME)*, 78 (10) ,297 - 302. (in Japanese)
Maruno T. (Ed.) (2007). Special issue: Development of metacognition research and then. *Japanese Psychological Review*, 50 (3), 191-355.
Maruno T. (Ed.) (2008). *Metacognition as 'inner eye'. Modern Esprit 497*. Tokyo: Shibundo Publishing. (in Japanese)
Okamoto, M. (1999). A study of metacognition on solving elementary school mathematics word problems. Tokyo: Kazama Shobo.(in Japanese)
Sannomiya M. (ed.) (2008). *Metacognition, learning ability to support higher cognitive functions*. Kyoto: Kitaoji Shobo.(in Japanese)
Schoenfeld, A.H. (1987). What's all the fuss about metacognition? In A. H. Shoenfeld (Ed.). *Cognitive science and mathematics education* (pp. 189-215). Hillsdale, NJ: Lawrence Erlbaum.
Shigematsu, K. (1986). A study on understanding in mathematics education (2): Meta-understanding and understanding. *Research Report of Mathematics Education(JSME)* 19, 125-128. (in Japanese)
Shigematsu, K. (1990). Metacognition and mathematics education: The role of Inner Teacher. In Committee for the Commemorative Publication of Dr. Ichiei Hirabayashi (Ed.) *Perspectives on mathematics education* (pp.76-107). Tokyo: Seibunsya. (in Japanese)
Shigematsu, K., Katsumi, Y. & Ueda, Y. (1993). Research on metacognition in mathematics education (8): A survey of internalization of students' metacognition. *Research Report of Mathematics Education (JSME)* 26, 97-102. (in Japanese)
Shigematsu, K. (1994). A research on the development of a questionnaire to measure student's metacognition in solving mathematical problems that affect student, *Scientific Research Grants (General Research C) Research Report* (ID: 04680311). (in Japanese)
Shigematsu K., Katsumi Y., Katsui H., and Ikoma Y. (2002). Research on metacognitive support for students by using the method of math jornal writing for middle grades elementary mathematics. *Arithmetics Education (Journal of JSME)*, 84 (4), 10 -18. (in Japanese)
Shigematsu K. (1992). Metacognition: the Role of the "Inner Teacher"(4). *Proceedings of the 16th annual conference of the International Group for the Psychology of Mathematics Education*, vol.2, 322-329.
Shigematsu, K. & Katsumi, Y.(1993). Metacognition: The role of the 'Inner Teacher' (5). *Proceedings of the 17th annual conference of the International Group for the Psychology of Mathematics Education*, vol.2, 278-285.
Shigematsu, K. (1995). Metacognition in mathematics education. In Japan Society of Mathematical Education (Ed.). *Views of mathematics learning toward theoriz-*

ing: Mathematice education in Japan, 1995. (pp. 237-249). Tokyo: Sangyotosho. (in Japanese).

Shigematsu, K. & Katsumi, Y. (2000). Metacognition: The role of the 'Inner Teacher' (6). *Proceedings of the 24th annual conference of the International Group for the Psychology of Mathematics Education*, vol.4, 137-144.

Shigematsu, K. & Katsumi, Y. (2010). Metacognition. In Japan Society of Mathematical Education (Ed.) *Handbook of research in mathematics education.* (pp.310-317). Tokyo: Toyokan (in Japanese)

Shigematsu, K. & Katsumi, Y. (2010). Metacognition: The role of the "Inner Teacher" (8): Changes in quality of students' metacognition by teacher's comments using red pencil on students' journal writing. In Y. Shimizu, Y. Sekiguchi and K. Hino (Eds.), *Proceedings of the 5th East Asia Regional Conference on Mathematics Education (EARCOME5)* (Vol. 2, pp. 252-259). Tokyo: Japan Society of Mathematical Education.

Shimizu N. (1995). A research on strategy in solving mathematics problems: Developing questionnaire on metacognition. *Journal of JASME, Research in Mathematical Education 1*, 101-108 . (in Japanese)

Shimizu N. & Yamada A. (1997). A study on the activities of self-reference in solving mathematical problems (1): Identifying the activity of self-reference. *Journal of JASME, Research in Mathematical Education 3*, 47-58. (in Japanese)

Shimizu Y. (1989). Metacognition in solution processes of junior high school students on a construction problem. *Research Journal of Mathematical Education (Journal of JSME)*, 52, 3-25. (in Japanese with English abstract)

Shimizu, Y. (2006). An analysis and promotion of "meta-thinking" in learning mathematics. *Research Journal of Mathematical Education (Journal of JSME)*, 86, 5-11. (in Japanese)

Shimizu Y. (2008). *Teaching thinking in mathematics education.* Tokyo: Toyokan Publishing. (in Japanese)

Silver, E. A. (1982). Thinking about problem solving: Toward an understanding of metacognitive aspects of mathematical problem solving. *Paper presented at the conference on thinking*, Suba, Fiji, p.1

Takazawa S. (1986). A study of metacognition in mathematics problem solving. *Research Report of Mathematics Education(JSME)* 19, 9-12. (in Japanese)

Takazawa S. (1995). Research results and tasks about metacognition from practice point of view. *Research Report of Mathematics Education(JSME)* 28, 654. (in Japanese)

Todome M. (1994). A study of students' understanding through teacher's metacognitive instruction. *Research Report of Mathematics Education(JSME)* 27(23-28). (in Japanese)

Yamaguchi T. (1992). A preliminary study on metacognition in mathematics education: Impact of metacognition in problem solving and the concept of communication by a pair. *Research Journal of Mathematical Education (Journal of JSME)*, 25, 59-64. (in Japanese)

CHAPTER 66

CROSS-CULTURAL STUDIES OF MATHEMATICS CLASSROOM PRACTICES

Yoshinori Shimizu
University of Tsukuba

ABSTRACT

Mathematics education research in recent years tends to include more international endeavors than ever before. Research that crosses national boundaries provides new insights into the nature of teaching and learning of mathematics. This chapter reviews the selected findings of recent large-scale international studies of classroom practices in mathematics which include Japan as a participating country and discusses the uniqueness of how Japanese teachers structure and deliver their lessons and what Japanese teachers value in their instruction. Particular attention is given to the ways lessons are structured with an emphasis on presenting and discussing students' thinking on alternative solutions to the problem. While characteristics of Japanese mathematics classrooms are examined through the review of research findings, it is noted that in the recent development in mathematics education, research aims, technological advances, and methodological techniques have diversified, enabling more detailed analyses of learning taken place in mathematics classrooms.

The First Sourcebook on Asian Research in Mathematics Education: China, Korea, Singapore, Japan, Malaysia, and India, pp. 1475–1490

Keywords: Cross-cultural study, mathematics teaching, classroom practice, TIMSS Videotape Study, Learner's Perspective Study, qualitative methodology

INTRODUCTION

In 1980's a group of mathematics educators from Japan and the United States conducted U.S-Japan Collaborative Research on Mathematical Problem Solving (Becker & Miwa, 1987). The research project included a comparative study on classroom practices of mathematical problem solving as a major part of it (Miwa, 1991b; Fujii et al., 1998). In September 1988 the U.S. researchers visited Japanese schools to observe classroom practices in mathematics (Becker, Silver, Kantowski, Travers & Wilson, 1990). Then in November 1988 and April 1989, Japanese researchers visited the schools in the US to do the same (Miwa, 1991b; Sugiyama, 1991). The observations they made suggest a sharp contrast in classroom practices between two countries. For instance, one of the characteristics of mathematics lessons in Japanese elementary and lower secondary schools related to the frequent exposure of students to alternative solution methods to solve a problem. Also, in the Japanese classrooms students' solutions were mainly presented with mathematical expressions, while in the U.S. classrooms mathematical expressions were not commonly used.

Later, with powerful video technologies, large-scale international studies of mathematics classroom practices have been conducted (Hiebert, et al., 2003; Stigler & Hiebert, 1999). Complementary analyses of classroom practices have been conducted with a different research methodology (e.g. Clarke, Keitel & Shimizu, 2006; Clarke, Emanuelsson, Jablonka & Mok, 2006). The findings of these studies include aspects of instruction as identified with a resemblance among participating countries while instruction in Japan seemingly unique. Japanese mathematics teachers, for example, appeared to spend more time on the same task in one lesson than their counterparts in other countries by having students work on a challenging problem and discuss alternative solutions to it (Hiebert el al., 2003). Also, experienced teachers in Japan typically highlighted and summarized the main points at particular phases of lessons to have their students reflect on what they have learned (Shimizu, 2006b).

These striking characteristics can be regarded as indicating some indispensable elements of mathematics classroom instruction that are valued and emphasized by Japanese teachers. This chapter examines those aspects of mathematics classroom instruction that appear to make Japanese lessons different from the other countries and to explore key characteristics of mathematics instruction in Japanese classrooms.

"STRUCTURED PROBLEM-SOLVING" APPROACH TO TEACHING MATHEMATICS

Japanese teachers, in elementary and junior high schools, in particular, often organize an entire mathematics lesson around the multiple solutions to a single problem in a whole-class instructional mode (Shimizu, 1999; Fujii, 2010). A typical mathematics lesson in Japan, which lasts forty-five minutes in the elementary schools and fifty minutes in the lower secondary schools, has been observed as divided into several segments (e.g., Becker, Silver, Kantowski, Travers, & Wilson 1990, Stigler & Hiebert 1999). These segments serve as the "steps" or "stages" in both the teachers' planning and delivering actual teaching-learning processes in the classroom (Shimizu 1999):

- Posing a problem
- Students' problem solving on their own
- Whole-class discussion
- Summing up
- Exercises or extension (optional depending on time and how well students are able to solve the original problem.)

Lessons usually begin with a word problem in the textbook or a practical problem that is posed on the chalkboard by the teacher. After the problem is presented and read by the students, the teacher determines whether the students understand the problem well. If it appears that some students do not understand some aspect of the problem, the teacher may ask these students to read it again, or the teacher may ask questions to help clarify the problem. Also, in some cases, he or she may ask a few students to show their initial ideas of how to approach the problem or to make a guess at the answer. The intent of this initial stage is to help the students develop a clear understanding of what the problem is about and what certain unclear words or terms mean.

A certain amount of time (usually about 10-15 minutes) is assigned for the students to solve the problem on their own. Teachers often encourage their students to work together with classmates in pairs or in small groups. While students are working on the problem, the teacher moves about the classroom to observe the students as they work. The teacher gives suggestions or helps individually those students who are having difficulty in approaching the problem. He or she also looks for the students who have good ideas, with the intention of calling on them in a certain order during the subsequent whole-class discussion. If time allows, the students who have already gotten a solution are encouraged by the teacher to find an alternative method for solving to the problem.

When a whole-class discussion begins, students spend the majority of this time listening to the solutions that have been proposed by their classmates as well as presenting their own ideas. Finally, the teacher reviews and sums up the lesson and, if necessary and time allows, the he or she poses an exercise or an extension task that will apply what the students have just learned in the current lesson. In the development of a lesson, experienced teachers demonstrate sophisticated use of blackboard (Nakamura, 2010) and effective use of questioning (Koizumi, 2010).

In sum, from a teacher's perspective, Japanese lessons can be characterized as being structured with a set of segments that includes students' problem solving and a whole discussion as major parts. In this sense, Japanese lessons can be characterized as "structured problem solving."

LARGE-SCALE INTERNATIONAL STUDIES OF CLASSROOM PRACTICES

The video component of the Third International Mathematics and Science Study (TIMSS) was the first attempt ever made to collect and analyze videotapes from the classrooms of national probability samples of teacher at work (Stigler & Hiebert, 1999). Focusing on the actions of teachers, it has provided a rich source of information regarding what goes on inside eighth-grade mathematics classes in Germany, Japan and the United States with certain contrasts among three countries. One of the sharp contrasts between the lessons in Japan and those in the other two countries relates to how lessons were structured and delivered by the teacher.

Exploration of How Lessons Are Delivered

Focusing on the actions of teachers, TIMSS Video Study has provided a rich source of information regarding what goes on inside eighth-grade mathematics classes in Germany, Japan and the United States with certain contrasts among three countries. One of the sharp contrasts between the lessons in Japan and those in the other two countries relates to how lessons were structured and delivered by the teacher. The structure of Japanese lessons was characterized as "structured problem solving", here again, while a focus was on procedures in the characterizations of lessons in the other two countries.

The following sequence of five activities was described as the "Japanese pattern" (Table 66.1): reviewing the previous lesson; presenting the prob-

Table 66.1. The Japanese Lesson Pattern

Reviewing the previous lesson
Presenting the problems for the day
Students working individually or in groups
Discussing solution methods
Highlighting and summarizing the main point

Source: Stigler and Hiebert (1999, pp. 79-80).

lems for the day; students working individually or in groups; discussing solution methods; and highlighting and summarizing the main point.

In the lesson pattern, the discussion stage, in particular, depends on the solution methods that the students actually use. In order for making this lesson pattern to work effectively and naturally, teachers have to have not only a deep understanding of the mathematics content, but also a keen awareness of the possible solution methods their students will use. Having a very clear sense of the ways students are likely to think about and solve a problem prior to the start of a lesson makes it easier for teachers to know what to look for when they are observing students work on the problem. The pattern seems to be consistent with the description of mathematics lessons as problem solving in the previous section, though there are some differences between them as "reviewing the previous lessons" above and "exercises or extension" in the previous section.

Analyzing Classroom Lesson Events

Characterization of the practices of a nation's or a culture's mathematics classrooms with a single lesson pattern was, however, problematized by the results of the Learner's Perspective Study (LPS) (Clarke, Mesiti, O'Keefe, Jablonka, Mok & Shimizu, 2007). The analysis suggested that, in particular, the process of mathematics teaching and learning in Japanese classrooms could not be adequately represented by a single lesson pattern by, at least, the following two reasons. First, lesson pattern differs considerably within one teaching unit, which can be a topic or a series of topics, depending on the teacher's intentions through out the sequence of lessons. Second, elements in the pattern themselves can have different meanings and functions in the sequence of multiple lessons. Needless to say, it is an important aspect of teacher's work not only to implement a single lesson but also to weave multiple lessons that can stretch out over several days, or even a few weeks, into a coherent body of the unit. It would not be possible for us to capture the dynamic nature of activities in teaching and learning process if each lesson was analyzed as isolated.

An alternative approach was proposed to the international comparisons of lessons by the researchers in LPS team. That is, a postulated "lesson event" would be regarded to serve as the basis for comparisons of classroom practice internationally. In LPS, an analytical approach was taken to explore the form and functions of the particular lesson events such as "between desk instruction", "students at the front", and "highlighting and summarizing the main point" (Clarke, Emanuelsson, Jablonka & Mok, 2006).

In particular, the form and functions of the particular lesson event "highlighting and summarizing the main point", or "Matome" in Japanese, were analyzed in eighth-grade "well-taught" mathematics classrooms in Australia, Germany, Hong Kong, Japan, Mainland China (Shanghai), and the USA (Shimizu, 2006b). For the Japanese teachers, the event "Matome" appeared to have the following principal functions: (i) highlighting and summarizing the main point, (ii) promote students' reflection on what they have done, (iii) setting the context for introducing a new mathematical concept or term based on the previous experiences, and (iv) making connections between the current topic and previous one. For the teachers to be successful in maintaining these functions, the goals of lesson should be very clear to themselves, activities in the lesson as a whole need to be coherent, and students need to be involved deeply in the process of teaching and learning. The results suggest that clear goals of the lesson, a coherence of activities in the entire lesson, active students' involvement into the lesson, are all to be noted for the quality instruction in Japanese classrooms.

Supportive Relationship Between the Teacher and Students

In mathematics lessons, students engage in different activities such as review of the previous lesson, listening to the teacher's explanation, discussion about solution methods, practicing, and summarising. These activities usually take the form of whole-class interaction, small-group interaction or individual work. Hino (2006) reported analyses that focused on the seatwork activity that is commonly observed in mathematics lessons in different countries, and investigated its features and functions in the case of the Japanese mathematics lessons in the LPS study, from the perspective of both teacher and student. As a result, both similarities and differences were observed in the three sites (School J1, J2, J3). One similarity is that the seatwork activity is placed before the presentation of main content of the lesson. In classrooms J2 and J3, generally half of the problems in the seatwork activity related to the presentation of main content while in classroom J1, it was found that more than half of the problems were related. This result is consistent with the sequence of five activities described by TIMSS 1995 video study (Stigler & Hiebert, 1999, p. 79). Here, seatwork activity is seen in the activity of "Students working individually or in groups." In Germany and US on the other hand, seatwork activity usually followed the presentation of main content. In fact, as the time for seatwork, they identified the activity of 'Practicing' in the German pattern and 'Practicing' and 'Correcting Seatwork and Assigning Homework' in the US pattern. The three Japanese teachers observed in this study also spent the seatwork activity as the time for exercises. However, when presenting main content, they took the time to let students think about the content beforehand. Also in the TIMSS 1999 video study (Hiebert et al., 2003), based on the data of time spent on different purposes in the lesson, it was found that in Japanese mathematics lessons, introducing new content is emphasised more than in the other six countries. One reason for this, conjectured from this study, is that the teachers spent time that was allocated to seatwork activity on introducing new content.

The placement of seatwork activity before presenting main content gives a foundation for interpreting the teacher's actions during the seatwork activity. In all sites, during seatwork activity the teachers supported individual students by taking into consideration the development of the lesson after the activity. Here, another similarity is the emphasis on different ways of thinking. In all sites, observations were made that the teacher dealt with different solutions, ideas or opinions by the students or that the students sought for and found different solutions. Especially in the seatwork activity before the presentation of main content, it was a natural tendency that the students used somewhat naïve ideas and informal ways of thinking. This was considered to be good by the teachers. Rather than preventing these, the teachers in this study tried to elicit and make use of them when presenting the main content of the lesson.

Mathematical Norms in Classrooms

Sekiguchi (2006a, 2009) reported analyses of lessons taught by three eighth-grade Japanese and Australian teachers who participated in the LPS. The analysis focuses on mathematical norms introduced by the teacher. In Hiebert et al. (2003), the analysis of mathematics teaching focused on mathematical knowledge, procedures, and reasoning involved in the problems presented in the lessons. Teaching of mathematical norms was beyond their analysis. Though the mathematical norms are often not explicitly taught by teachers nor written in textbooks, they are crucial when the learning process of mathematics is conceived as mathematical activities.

Mathematical norms are knowledge "about" doing mathematics; therefore, they belong to the domain of meta-knowledge in mathematics. It is hypothesized that beginning teachers are often occupied with covering curriculum content, paying their attention to mathematical knowledge and skills: Competent teachers as selected in LPS by design would invest more time and effort in teaching of meta-knowledge.

Sekiguchi explored what mathematical norms would surface in the lessons and would the teacher introduce, negotiate or utilize those norms during the lessons. His analyses revealed that developing a norm that functions fully in various contexts cannot be done just one lesson; it requires a long patient effort. The use of previously developed norms, therefore, is a necessary process of regular lessons. The use of students' work seems very important. Since a norm is about how to work on mathematics, the use of mathematical work is natural for communicating a norm. Also, since students are familiar with their work, the use of students' work would facilitate students' understanding of the norm. Comparison of students' work would also be very helpful for students to produce clear understanding of the norm as well as their metacognition of their own work. Since pointing out students' violation of a norm may hurt their feeling, being considerate of those students who did not follow the norm seems a hallmark of "competent" teachers.

Studying norms requires understanding of relationships between various norms. A classroom in Japanese schools constitutes a community where a teacher and students stay together, negotiate meanings, share common goals, and shape their identities (Fujii, 2009). It forms a "community of practice" (Wenger, 1998). A community generates, maintains, modifies, or eliminates various kinds of patterns called norms, standards, obligations, rules, routines, and the like. Consider a mathematical norm that I identified above, "in mathematics you cannot write what you have not shown to be true yet." This is consistent with a general social norm: "You cannot write what you have not shown to be true yet," and more general moral "You should not tell a lie to people." Mathematical norms seem to be backed or authorized by social norms. This would make mathematical norms appear to be reasonable and no arbitrary rules.

DICHOTOMIES FOUND IN INTERNATIONAL COMPARATIVE STUDIES

International comparisons of classroom practices lead researchers to more explicit understanding of their own implicit theories about how teachers teach and how children learn mathematics in their local contexts as well as what is going on in school mathematics in other countries. While international studies provide new insights into the development and improvement of the teaching and learning of mathematics, there are many dichotomies evident in discussing on the findings; high-performing versus low performing, teacher-centered versus student-centered, and East versus West, among others.

In search of the identity of mathematics education in East Asia, Leung (2001) tries to describe distinctive feature of it by focusing on key differences between the East Asian and the Western traditions in mathematics education with six dichotomies; "product (content) versus process", "rote learning versus meaningful learning", "studying hard versus pleasurable learning", "extrinsic versus extrinsic motivation", "whole class teaching versus individualized learning", and "competence of teachers: subject matter versus pedagogy". In the first distinction 'product (content) versus process', for example, Leung (2001) describes East Asian mathematics classroom emphasizing mathematics content and procedures or skills putting basic knowledge and basic skills in the foreground, whereas Western education in the last decades tends to focus more on the process of doing mathematics.

The dichotomy "East versus West" has been foregrounded by international benchmark testing, and has led to a qualitative focus on learning in different geographical regions as a result. Accumulated research over the past decade has contributed to our understanding of similarities and differences in mathematics teaching and learning between East Asia and the West (e.g., Leung, Graf & Lopez-Real, 2006) or between Eastern and Western cultures. The discussion document for the ICMI study argued that "those based in East Asia and the West seem particularly promising for comparison." In this study a comparison was made between "Chinese/Confucian tradition on one side, and the Greek/Latin/Christian tradition on the other" (Leung, Graf & Lopez-Real, 2006).

Juxtaposing the two different cultures indicated that researchers wanted to examine teaching and learning in each cultural context by contrasting differences between them. The labels "East/Eastern" and "West/Western," however, could be problematic in several ways. First, the terms East and West literally mean geographical areas but not cultural regions. Needless to say, there are huge diversities in ethnicity, tools, and habits that are tied to the corresponding cultures. Further, Cobb and Hodge

(2011) argue that two different views of culture can be differentiated in the mathematics education literature on the issue of equity, and that both are relevant to the goal of ensuring that all students have access to significant mathematical ideas. "In one view, culture is treated as a characteristic of readily identified and thus circumscribable communities, whereas in the other view it is treated as a set of locally instantiated practices that are dynamic and improvisational" (p. 179). With the second view, in particular, it is problematic to specify different cultures based on geographical areas?

Second, it is possible to oversimplify and mislead the cultural influence on students' learning within each cultural tradition by using the same label for different communities. For example, there are studies that suggests much child education in Japan diverges from the Confucian approach in "East Asia" (Lewis, 1995). Also, in the special issue on exemplary mathematics instruction and its development in selected education systems in East Asia, it was manifested that there are variety of approaches to accomplish quality mathematics instruction in these different systems in East Asia (Li & Shimizu, 2009). Thus, any framework for differentiating cultural traditions runs the risk of oversimplifying the cultural interplay. In particular, there is a need to question whether polarizing descriptors such as "East" and "West," "Asian" and "European," are maximally useful. Perhaps we need more useful ways to examine differences, for the purposes of learning from each other and identifying ways to optimize learner practices.

The countries in East Asia, in the Confucian Heritage Culture (CHC) certainly share commonality and mathematics classroom practices in this area also have similarities in various aspects of teaching and learning. However, when we look into mathematics classrooms in local contexts in different countries, even within East Asian area, we immediately realize the diversity of practices in the teaching and learning. Teachers in different countries behave differently when teaching the same mathematical content and consequently students in each country learn the topic differently. Educational systems are embedded in their respective societies, with their particular cultural and historical backgrounds.

FINDING DIFFERENCES IN SIMILARITIES AND IDENTIFYING SIMILARITY IN DIFFERENCES

The mathematics education research community has recognized that mathematics classrooms need to be considered as cultural and social environments in which individuals participate, and that teaching and learning activities taking place in these environments should be studied as such (e.g. Cobb & Hodge, 2011). Teachers' actions that appear normal in a classroom reflect the social values, norms, or traditions that are prevalent outside it. These realizations have led to studies of differences between teaching behaviors and learning outcomes in different countries. Then, similarities and differences have been explored in topics such as exemplary mathematics classroom instruction and teachers' perspectives on effective mathematics teaching (Li & Shimizu, 2009).

One of the reasons for studying aspects of teaching and learning in classrooms across cultures is that teaching is a cultural activity (Stigler and Hiebert,1999), which takes place in particular cultural and social environments. Because cultural activities vary little within one community or society, they are often transparent and unnoticed. Cross-cultural comparison is a powerful approach to uncover unnoticed but ubiquitous practices, inviting examination of the things "taken for granted" in our teaching, as well as suggesting new approaches that have not evolved in our own society (Stigler et al., 2000). International comparative classroom research is viewed as the exploration of similarity and difference in order that our understanding of what is possible in mathematics classrooms can be expanded by consideration of what constitutes "good practice" in culturally diverse settings.

In sum, with the growing internationalization of education, and as the education community has given higher priority to international research in the last decade, it is timely to examine the insights from comparative analyses of aspects of teaching and learning of mathematics that are situated in very different cultures. The contrasts and unexpected similarities offered by cross-cultural studies reveal and challenge existing assumptions and theories (Clarke, 2003).

Story or Drama, as a Metaphor for Lessons

There seems to be supporting conditions and shared beliefs among the Japanese teachers for having "Matome" often at the end of the lessons or at the end of sub-units. Any lesson has parts of an opening, "core", and closing. This is particular the case for Japanese lessons which begins and ends by students' bowing. A lesson is regarded as a drama, which has a beginning and leads to a climax, by Japanese teachers. In fact, one of the characteristics of Japanese teachers' planning of lessons is the deliberate structuring of the lesson around a climax, "Yamaba" or "Miseba" in Japanese. Most teachers think that a lesson should have a highlight.

Stigler and Perry (1988) found *reflectivity* and *coherence* in Japanese mathematics classroom. The meaning they attached to *coherence* is similar to that used in the literature on story comprehension. Stigler and Perry (1988) noted as follows.

> A well-formed story, which also is the most easily comprehended, consist of a protaganist, a set of goals, and a sequence of event that are causally related to each other and to the eventual realization of the protaganist's goals. An ill-formed story, by contrast, might consist of a simple list of events strung together by phrases such as "and then...," but with no explicit reference to the relations among events....The analogy between a story and a mathematics classroom is not perfect, but it is close enough to be useful for thinking about the process by which children might construct meaning from their experience in mathematics class. A mathematics class, like a story, consists of sequences of events related to each other and, hopefully, to the goals of lesson. (p. 215)

Often mentioned idea of "Ki-Shou-Ten-Ketsu" by Japanese teachers in Lesson Study meetings (Lewis & Tsuchida, 1998), an idea originated in the Chinese poem, further suggests that Japanese lessons has a particular structure in which a flow is moving toward the end ("Ketsu", summary of the whole story).

The lesson event Matome appeared to serve for promoting the reflection by the teacher and the students. Stigler and Perry (1988) also found *reflectivity* in Japanese mathematics classroom. They pointed out that the Japanese teachers stress the process by which a problem is worked and exhort students to carry out procedure patiently, with care and precision. Given the fact that the schools are part of the larger society, it is worthwhile to look at how they fit into the society as a whole. The event type seems to rest on a tacit set of core beliefs about what should be valued and esteemed in the classroom. As Lewis noted, within Japanese schools, as within the larger Japanese culture, *Hansei*---self-critical reflection---is emphasized and esteemed (Lewis, 1995). Of special interest is in exploring a difference between cultures at this level.

FINAL REMARKS

Sekiguchi (1998) emphasized the importance of recognizing social and cultural situated-ness of mathematics education research.

> Research participants, settings, unit of data analysis, interpretation, educational implications are all socially and culturally constrained. The reliability and validity of research results are, therefore, also socially and culturally bounded. (p. 394)

We can learn something from other countries only if we have relevant information and interpret the information in a sensible way.

This chapter aimed to examine key characteristics of mathematics classroom instruction in Japan. The uniqueness of the way Japanese mathematics teachers structure and deliver their lessons was discussed in relation to what Japanese teachers value in their instruction by reviewing selected findings of large-scale international studies of classroom practices. Valuing students' thinking to be incorporated into the development of a lesson is discussed as key aspect of the approaches taken by Japanese teachers to develop and maintain quality mathematics instruction.

REFERENCES

Becker, J.P. & T. Miwa (1987). *Proceedings of the U.S.-Japan seminar on mathematical problem solving*. Columbus, OH: ERIC Clearinghouse for Science, Mathematics, and Environmental Education (ED 304315).

Becker, J.P., Silver, E.A., Kantowski, M.G., Travers, K.J. & and Wilson, J.W. (1990). Some observations of mathematics teaching in Japanese elementary and junior high schools. *Arithmetic Teacher*, 38, October, 12-21.

Clarke, D.J. (2003). International comparative studies in mathematics education. In A.J. Bishop, M.A. Clements, C. Keitel, J. Kilpatrick, and F.K.S. Leung (eds.) *Second international handbook of mathematics education* (pp. 145-186). Dordrecht: Kluwer Academic Publishers.

Clarke, D. J., Keitel, C., & Shimizu, Y. (eds.), (2006) *Mathematics classrooms in twelve countries: The insider's perspective*, Rotterdam: Sense Publishers.

Clarke, D. J., Shimizu, Y., Ulep, S., Gallos, F., Sethole, G., Adler, J., & Vithal, R., (2006) Cultural diversity and the learner's perspective: Attending to voice and context., In F. Leung, K. Graf & R. Lopez-Real (eds.) *Mathematics education in different cultural traditions: A comparative study of East Asian and the West.* (pp.353-380), New York, NY: Springer.

Clarke, D. J., Mesiti, C., O'Keefe, C. Jablonka, E., Mok, I. A. C & Shimizu, Y. (2007) Addressing the challenge of legitimate international comparisons of classroom practice. *International Journal of Educational Research, 46*, 280-293, 2007

Cobb, P., & Hodge, L. L (2011). Culture, identity, and equity in the mathematics classroom. In E. Yackel, K. Gravemeijer & A. Sfard (Eds.) *A journey in mathematics education research: Insights from the work of Paul Cobb* (pp. 179–195). New York, NY: Springer.

Fernandez, C. & Yoshida, M (2004). Lesson Study: A Japanese approach to improving mathematics teaching and learning. Mahwah, NJ: Lawrence Erlbaum Associates,

Fujii, T., Kumagai, K., Shimizu, Y. & Sugiyama, Y. (1998). A cross-cultural study of classroom practices based on a common topic. *Tsukuba Journal of Educational Study in Mathematics.*, 17. 185-194.

Fujii, T. (2009). A Japanese perspective on communities of inquiry: The meaning of leaning in whole-class lessons. In M. Tzekaki, M. Kaldeimidou & H. Sakonidis (eds.) *The Proceedings of 33rd annual conference of the International Group for the Psychology of Mathematics Education, Vol.1*, pp. 165-170.

Fujii, T. (2010) Aspects of "thinking in group" in a mathematics classroom: Is a group more than the sum of individuals? In Y. Shimizu (ed.) *Science of classroom practices: A new approach to mathematics classroom*. Tokyo: Gakubunsha (in Japanese).

Hino, K. (2006) The role of seatwork in three Japanese classrooms. In D. Clarke, C. Keitel, & Y. Shimizu (Eds.). Mathematics classrooms in twelve countries: The insider's perspective, (pp. 289-306), Rotterdam: Sense Publishers.

Hino, K. (2007). Studying lesson structure from the perspective of meaning construction: The case of two Japanese mathematics classrooms. *The proceedings of the 31st annual conference of the International Group for the Psychology of Mathematics Education, Vol. 3*, pp. 25-32.

Hino, K. (2009) Coherence in student's construction of mathematical meanings: Glimpses from three Japanese classrooms. *On-Line Proceedings of 3rd Redesigning Pedagogy International Conference*. Singapore, National Institute of Education.

Kaiser, G., Hino, K. & Knipping, C. (2006). Proposal for a framework to analyse mathematics education in Eastern and Western traditions. In F. Leung, K. Graf & R. Lopez-Real (eds.) *Mathematics education in different cultural traditions: A comparative study of East Asian and the West* (pp. 319-351). New York, NY: Springer.

Koizumi, Y. (2010) An investigation of teacher's questioning in the mathematics classrooms in Germany and Japan. In Y. Shimizu, Y. Sekiguchi & K. Hino (eds.) *Proceedings the 5th East Asia Regional Conference on Mathematics Education, Volume 1*, Tokyo: Japan Society of Mathematical Education

Lee, S.Y., Graham, T. & Stevenson, H.W. (1996). Teachers and teaching: Elementary schools in Japan and the United States. In T.P. Rohlen & G.K. Letendre (Eds.), *Teaching and learning in Japan*. New York: Cambridge University Press.

Leung, F.K.S. (2001). In search of an East Asian identity in mathematics education. *Educational Studies in Mathematics*, 47(1), 35-51.

Leung, F. K.S., Graf, K., & Lopez-Real, R. (Eds.) (2006). Mathematics education in different cultural traditions: A comparative study of East Asian and the West. New York, NY: Springer.

Lewis, C. (1995) *Educating hearts and minds: Reflections on Japanese preschool and elementary education*. New York: Cambridge University Press.

Lewis, C. & Tsuchida, I. (1998, Winter). A lesson is like a swiftly flowing river: How research lessons improve Japanese education. *American Educator. 22*(4). 12-17, 50-52.

Li, Y., & Shimizu, Y. (2009). Exemplary mathematics instruction and its development in East Asia. *ZDM – The International Journal on Mathematics Education, 41*(3), 257–262.

Miwa, T. (1991a). School mathematics in Japan and the U.S.: Focusing on recent trends in elementary and lower secondary school. In I. Wirszup & R. Streit (eds.) *Developments in school mathematics around the world*. Reston, VA: National Council of Teachers of Mathematics.

Miwa, T. (1991b). A comparative study on classroom practices of mathematical problem solving between Japan and the U.S. In T. Miwa (ed.) *Report of the Japan-U.S. collaborative research on mathematical problem solving*. Ibaraki: University of Tsukuba. (in Japanese with English summary).

Miwa, T. (ed.) (1992). *Teaching mathematical problem solving in Japan and the U.S.* Tokyo: Toyokan. (in Japanese)

Nagasaki, E. & Becker, J.P. (1993). Classroom assessment in Japanese mathematics education. In N. Webb (Ed.) *Assessment in the mathematics classroom*. Reston, VA: National Council of Teachers of Mathematics.

Nakamura, K (2010) The characteristics of board writing by an experienced teacher in mathematics: The structure of bansho. In Y. Shimizu (ed.) *Science of classroom practices: A new approach to mathematics classroom.* Tokyo: Gakubunsha (in Japanese).

Sekiguchi, Y (1998) Mathematics education research as socially and culturally situated. In J. Kilpatrick & A. Sierpinska (eds.) *Mathematics education as a research domain: A search for identity: An ICMI study*. Kluwer Academic Publishers.

Sekiguchi, Y. (2006a). Mathematical norms in Japanese mathematics lessons. In Clarke, D., Keitel, C., & Shimizu, Y. (eds). *Mathematics classrooms in twelve countries: The insider's perspective*. Rotterdam: Sense Publishers.

Sekiguchi, Y. (2006b) Coherence of mathematics lessons in Japanese eighth-grade classrooms. In *The proceedings of the 30th annual conference of the International Group for the Psychology of Mathematics Education, vol. 5*, pp. 81-88.

Sekiguchi, Y. (2008). Classroom mathematical norms in Australian lessons: Comparison with Japanese lessons. *The Proceedings of 32nd annual conference of the International Group for the Psychology of Mathematics Education, vol. 4*, pp. 241-248.

Shimizu, Y. (1999a) Aspects of mathematics teacher education in Japan: Focusing on teachers' roles, *Journal of Mathematics Teacher Education*, 2(1), 107-116.

Shimizu, Y. (1999b) Studying sample lessons rather than one excellent lesson: A Japanese perspective on the TIMSS Videotape Classroom Study. *Zentralblatt fur Didaktik der Mathematik. 99*(6), 191-195

Shimizu, Y. (2006a) Discrepancies in perceptions of mathematics lessons between teacher and the students in Japanese classrooms. In D. Clarke, C. Keitel & Y. Shimizu (eds.), *Mathematics classrooms in twelve countries: The insider's perspective*. Rotterdam: Sense Publishers.

Shimizu, Y. (2006b) How do you conclude today's lesson?: The form and functions of "matome" in mathematics lessons. (pp. 127-146) D. Clarke, J. Emanuelsson, E. Jablonka & I. Ah Chee Mok (eds.) *Making connections: Comparing mathematics classrooms around the world*. Rotterdam: Sense Publishers.

Shimizu, Y. (2009) Characterizing exemplary mathematics instruction in Japanese classrooms from the learner's perspective, *Zentralblatt für Didaktik der Mathematik, 41(3)*, 311-318.

Shimizu, Y. (2010). A task-specific analysis of explicit linking in the lesson sequence in three mathematics classrooms. In Shimizu, Y., Kaur, B., Huang,

R. & Clarke, D. (eds.) *Mathematical tasks in classrooms around the world.* Rotterdam: Sense Publishers.

Shimizu, Y., Kaur, B., Huang, R. & Clarke, D. (eds.) (2010) *Mathematical tasks in classrooms around the world.* Rotterdam: Sense Publishers.

Stevenson, H.W. & Stigler, J. W. (1992). *The learning gap: Why our schools are failing and what we can learn from Japanese and Chinese education.* NY: Summit Book.

Stigler, J. W., Gonzales, P., Kawanaka, T., Knoll, S. & Serrano, A. (1999) *The TIMSS Videotape Classroom Study: Methods and findings from an exploratory research project on eight-grade mathematics instruction in Germany, Japan, and the United States.* Washington, DC: U. S. Government Printing Office.

Stigler, J. W. & Hiebert, J. (1999) *The teaching gap: Best ideas from the world's teachers for improving education in the classroom.* New York: NY, Free Press.

Sugiyama, Y. (1991) Observation of mathematics classes in Japan and the U.S.-A comparative study. In T. Miwa (ed.) *Report of the Japan-U.S. collaborative research on mathematical problem solving.* Ibaraki: University of Tsukuba. (in Japanese with English summary).

CHAPTER 67

SYSTEMATIC SUPPORT OF LIFE-LONG PROFESSIONAL DEVELOPMENT FOR TEACHERS THROUGH LESSON STUDY

Akihiko Takahashi
DePaul University, USA

ABSTRACT

This chapter discusses how the Japanese systematic support for teacher professional learning has taken place and what Japanese teachers have learned though the process based on the results of the following two studies. The first study is about the forms of lesson study. Lesson study does not follow a uniform system or approach in Japan. It is a part of the Japanese teachers' cultural activity and it takes many different forms based on the purpose of the professional development goals and the vision of the members of the lesson study group. The second study is about the differences in knowledge and expertise that exist among prospective teachers, novice teachers, and experienced teachers. It discusses how the school system supports teachers to learn continuously through their careers and why it is important even after completing formal teacher training at universities. The study of the

The First Sourcebook on Asian Research in Mathematics Education: China, Korea, Singapore, Japan, Malaysia, and India, pp. 1491–1503

implementation of Japanese professional development structure in different cultures is discussed as a possible future research.

Keywords: Professional development, lesson study, mathematics teacher

INTRODUCTION

For more than a decade, educators and researchers especially in the field of mathematics education have been interested in lesson study as a promising source of ideas for improving education (Stigler & Hiebert, 1999). Although Japanese teachers and educators have been deeply involved in lesson study for more than a century, this might be the first time for educators and researchers around the world to become interested in leaning about the Japanese professional development approach. Moreover, schools and teachers in different countries have been trying to implement lesson study into their own education systems.

At the beginning of this movement, a number of school districts in North America have attempted to use lesson study in order to change their practices and impact student learning (Council for Basic Education, 2000; Germain-McCarthy, 2001; Catherine Lewis, Perry, Hurd, & O'Connell, 2006; Research for Better Schools Currents Newsletter, 2000; Stepanek, 2001; Weeks, 2001). In 2003, at least 29 states, 140 lesson study clusters/groups, 245 schools, 80 school districts, and 1100 teachers across the United States were involved in lesson study (Lesson Study Research Group). Although these are the only numbers that we have regarding how widely lesson study has been implemented, more and more teachers and educators have become familiar with lesson study and have participated in lesson study after this data was collected. In fact, many U.S. federal and state grant programs that focus on teacher development include some aspects of lesson study.

The lesson study programs in the U.S. have been involving not only various grade levels in mathematics including kindergarten to high school, but also other subject areas such as English Language Arts and American History. Since this is only the tip of the iceberg, I believe that lesson study can be implemented in many different subject areas and many different schools and districts throughout the United States and Canada.

In contrast with the vast interest of lesson study around the world from the late 90's, most Japanese educators and teachers did not know why researchers and educators around the world became interested in lesson study. Furthermore, they did not even know that the term *lesson study* existed as it was coined by a U.S. researcher who had studied the process of *jyugyo kenkyu*.

Because lesson study has been a major professional development activity in the Japanese school system as both pre-service and in-service, few Japanese researchers in mathematics education have studied about lesson study itself. There are some articles discussing the issues around professional development, which includes lesson study, improvements for teaching and learning mathematics, but most are available only in Japanese.

Concerning the above, this chapter will discuss the research on the Japanese approach of professional development for mathematics teachers with the focus of lesson study by reviewing the research conducted by not only Japanese researchers but also U.S. researchers who have carefully looked at lesson study.

LITERATURE REVIEW

Lesson Study Research in Japan

The recent movement of lesson study in mathematics education outside Japan has affected the Japanese mathematics education community. Immediately after the ICME-9 in Tokyo in 2000, U.S. and Japanese mathematics education researchers who participated in this event extended their stay in Tokyo to learn and discuss about the Japanese professional development approach that focuses on lesson study. The workshop, "*Studying classroom teaching as a medium for professional development*", was organized by the National Research Council and the United States National Commission on Mathematics Instruction. Another workshop was organized by the National Council of Teaches of Mathematics. The report of the former workshop was published in 2002 (National Research Council). These workshops and meetings attract a strong interest of lessons study among Japanese mathematics education researchers and practitioners. The background of the lesson study movement outside of Japan and the discussions during the above two workshops have been reported in the Journal of Japan Society of Mathematical Education (Takahashi, 2000). Also, *The Teaching Gap* (Stigler & Hiebert, 1999) was translated into Japanese by Saburo Minato in 2002. More and more Japanese educators have realized that lesson study is a unique professional development approach that many western countries do not have in their school systems, and many researchers outside of Japan have become interested in learning about lesson study.

A frequently asked question by researchers concerns the origin of lesson study and how long Japanese mathematics educators practice lesson study. Makinae (2010) argues that the origin of Japanese lesson study, *jyugyo kenkyu*, was influenced by U.S. books for educators to introduce new approaches to teach during the late 1880's. He pointed out that one of these books by Sheldon (1862) describes methods to learn about new teaching approaches called *criticism lesson and model lesson,* may be the beginning of Japanese lesson study. In fact, Inagaki (1995) argues that one of these methods, *criticism lesson* was already practiced among elementary schools affiliated to the normal schools in Japan as early as the late 1890s. According to Makinae (2010), teacher conferences utilizing *criticism lessons* were conducted by local school districts in the early 1900s. Some of these conferences were already called lesson study conferences, *Jugyo-kenkyu-kai* in Japanese.

Despite the long history of lesson study in Japan, Japanese mathematics researchers and other researchers have not been interested in studying lesson study itself until recently. In fact, the recent publication by the Japan Society of Mathematics Education, Handbook of Research in Mathematics Education (2010) does not include any research that focuses on mathematics lesson study under the section of mathematics teacher education/professional development.

Lesson Study Research and Practice in the U.S.

After researchers in the U.S. introduced lesson study to the mathematics education community during late 90s, the term *lesson study* spread among researchers and educators in the U.S. and later around the world. One of the most influential books that discusses about lesson study is *The Teaching Gap* (Stigler & Hiebert, 1999). Some U.S. school districts give this book to each teacher in their schools and professional development programs use the book as a major resource to discuss. As a result, lesson study attracts many mathematics teachers and school districts to try and replicate its success in changing mathematics teaching and learning.

The rapid increase in popularity of lesson study, especially in North America, has been spearheaded by several key researchers, practitioners, and organizations whom support teacher professional development. At the beginning of lesson study endeavours in the late 1990s, a research project was conducted by Catherine Lewis (C. Lewis, 2000; C. Lewis & Tsuchida, 1998), who is fluent in Japanese and has observed various lesson study sessions in Japan, and Makoto Yoshida (Yoshida, 1999a, 1999b), who was born and raised in Japan. An intensive case study was conducted, analysing an elementary school teachers' lesson study process, drew interest among several schools and practitioners. Working with a classroom teacher and a math coach at the San Mateo School District in California,

Catherine Lewis supports the teachers in the San Francisco Bay Area in practicing lesson study. This project was supported by The Noyce Foundation through The Silicon Valley Mathematics Initiative[1] and by several federal grants led by Catherine Lewis. For establishing lesson study in California's Bay Area, Lewis invited several Japanese researchers of mathematics education and Japanese lesson study practitioners in order to learn the mechanisms of lesson study in order to implement lesson study in U.S. schools. The results of this process was carefully studied by researchers and published (Catherine Lewis, Perry, & Hurd, 2009; Catherine Lewis, et al., 2006). The authors reported significant changes were observed not only among participating teachers in the lesson study but also within the students who were taught mathematics by the teachers of the lesson study. This is one of a few research projects that examine the process of the systematic efforts of lesson study involving entire faculties in one U.S. public school. Since full implementation of lesson study requires collaboration among researchers who fully understand the lesson study process, school administrators who are knowledgeable about research regarding effective professional development, and teachers who are willing to take a risk in order to improve their classroom instruction, it is extremely time consuming for researchers to conduct a study because they must support schools in implementing lesson study and carefully observe both teachers and students changes over long periods of time.

At the same time on the East Coast, Makoto Yoshida and Clea Fernandez supported the teachers at Paterson School Number 2 in New Jersey to start lesson study. This NSF funded project enabled the teachers at Paterson School Number 2 to work with teachers at the Greenwich Japanese School in Connecticut to learn how to plan, observe, and reflect on teaching through lesson study during the years 2000 and 2002. Even after the funding ended, teachers at both schools continued their collaboration and conducted lesson study open houses in their schools with the support of Makoto Yoshida at the Global Education Resources and Patsy Wang Iverson at Research for Better Schools.

In addition to these two major lesson-study projects, several lesson-study initiatives were conducted by professional organizations. On the west coast, the Northwest Regional Educational Laboratory supported local public schools to explore lesson study, and Sonoma County Office of Education initiated lesson study among their schools. On the east cost, Research for Better Schools supported schools that include the Paterson School Number 2 to support the implementation of lesson study. National organizations such as The National Council of Teachers of Mathematics (NCTM) and The American Federation of Teachers (AFT) also supported lesson study. NCTM offered online workshops for teachers to learn about lesson study. AFT offered nationwide support for schools and teachers to learn and try lesson study as professional development. These schools include those in Rochester, New York and Volusia County, Florida.

Looking at the successes of the pioneers' work above, many researchers and practitioners applied for state and federal professional development grants to try implementing lesson study to support their teachers in developing knowledge and skills to improve teaching and learning. As a result, many universities and school districts have received a large amount of grant money to initiate lesson study in schools throughout the U.S. These lesson study initiatives include the program led by Fresno Unify School District, California, New Mexico State University, New Mexico, Albuquerque Public Schools, New Mexico, Little Rock Public Schools, Arkansas, Loras College, Iowa, Eastern Michigan University in Michigan, and Lancaster School District, in Pennsylvania. The research conducted by U.S. researchers in various lesson study programs in the U.S. has been published by Hart et al. (2011). Although enormous amount of effort and money has been spent for establishing lesson study in various school settings, it is still too early to see the full implementation of lesson study in many cases among U.S. schools (Takahashi, 2011).

Unlike other grant funded projects, the Chicago Lesson Study Group originated as a volunteer teacher group for exploring the possibilities of replicating the success of Japanese lesson study in a U.S. setting. The Chicago Lesson Study Group was launched in November of 2002. The Chicago Lesson Study Group has become well known among lesson study researchers and practitioners as one of the few groups that conduct public research lessons. The first public research lesson was conducted by the Chicago Lesson Study Group in May of 2003. This was the beginning of the Annual Chicago Lesson Study Conference. The Chicago Lesson Study Group has conducted this conference every year for 10 years. A unique feature of the conference is that the participants can observe a live research lesson, in addition to presentations given by leading researchers and practitioners involved in lesson study. In each conference, teachers and educators from not only the Chicago area, but also from other states have been invited to discuss how to create student-centred classrooms in mathematics. Around one hundred participants from various U.S. states and around Canada have attended the conferences each year and have participated in discussions on how to help students develop algebraic thinking skills through problem solving.

Although the most popular form of lesson study in Japan takes place within a single school as a school-based professional development program (Yoshida, 1999), the Chicago Lesson Study Group has adopted a cross-school volunteer model for its lesson study group from the beginning. The reasons for this adaptation are twofold. First, an effective model of lesson study is often one that is started as a grassroots movement of enthusiastic teachers rather than as a top-down formation (Lewis, 2002; Takahashi & Yoshida, 2004; Yoshida, 1999). For this reason, starting a lesson study group as a cross-school volunteer group was thought to be appropriate. Furthermore, it is sometimes difficult to establish a school-based lesson-study group in the U.S. because many teachers do not have experience working with other teachers in the same school to accomplish a shared goal. Secondly, in order to have a sufficient number of enthusiastic elementary and middle school teachers who are interested in lesson study focusing on mathematics, a cross-school model was found to be more appropriate in the U.S. setting.

After several years, the teachers and the educators who have been a part of the Chicago Lesson Study Group started establishing their own lesson study teams. Now various schools and teachers are conducting lesson study at their schools. Since 2002 more than 20 Chicago area schools have been voluntary practicing lesson study and the number is growing year after year.

DISCUSSION

An enormous interest in lesson study from other countries made Japanese mathematics education researchers realize that lesson study is a unique approach to teacher professional development in mathematics education. At the same time, many Japanese mathematics education researchers have come to understand the need for in-depth research on lesson study in mathematics. For example, even U.S. researchers have pointed out that Japanese mathematics classrooms have changed over time by implementing ideas from research using lesson study (Lewis & Tsuchida, 1998; Stigler & Hiebert, 1999; Yoshida, 1999b). How lesson study actually contributes to the changes in teaching and in student learning are still not clear. Some other U.S. researchers have asked Japanese mathematics education researchers what the definition of lesson study is, but Japanese researchers have had a hard time clearly defining it because lesson study can be seen in many different settings, and in different forms and scales.

As a result, several studies about the different forms and scales of lesson study in mathematics were conducted and published during the 2000s. Using some of these studies, the author will discuss different purposes and different forms of lesson study to describe a framework of Japanese mathematics teacher professional development that include both pre-service and in-service as well as the role of lesson study in different forms and purposes.

First, lesson study does not follow a uniform system or approach in Japan. It is a part of the Japanese teachers' cultural activity and it takes many different forms based on the purpose. Hirabayashi (2002) argues that there are two major functions of lesson study. One functions as a method of research and another functions as a place for presenting new findings. The former function of lesson study is for teachers to discuss issues in teaching and learning by looking at actual classrooms. This function might be suitable for professional development for teachers to study the effectiveness of mathematics teaching and learning in their own classrooms. In the early stages of Japanese lesson study in the early 1890s, *criticism lesson* (Sheldon, 1862) utilized this function to carefully examine the effectiveness of teaching and publicly discuss ways to improve teaching and learning. The name, research lesson, *kenkyu-jyugyo*, might come from this function of lesson study. On the other hand, *model lesson* (Sheldon, 1862) utilized the latter function of lesson study as demonstration lessons for showcasing exemplary lessons or presenting new approaches for teaching. For this purpose, the lesson should be thoughtfully prepared and carefully based on research conducted by a teacher researcher or a group of researchers so that the participants can see actual lessons instead of simply reading papers to describe the results of the study.

Considering the origin of lesson study, *criticism lesson* and *model lesson*, and the two functions of lesson study, a variety of lesson study sessions practiced among schools in Japan can be found in-between the *criticism lesson* and *model lesson* as shown in Figure 67.1.

From this view, the public research lessons at public open houses or nation-wide conferences hosted by schools affiliated with universities or volunteer teacher organizations share the presentation demonstration function with the *model lesson*. On the other hand, the type of lesson study conducted by a group of public school teachers such as the ones described by Lewis and Tsuchida (1998), and Yoshida (1999b) share the method of research function with the *criticism lesson*. Thus, it can be said that the various forms of lesson study currently conducted in Japan may fall in-between the *criticism lesson* and *model lesson*.

Several researchers have tried to sort out the forms of mathematics lesson study in Japan. Using several criteria, these researchers have sorted out lesson study in Japan into 2 to 7 forms. These differences are based

Figure 67.1. Japanese lesson study.

on how a researcher looks at each form of lesson study. For example, Lewis categorizes lesson study into two forms, lesson study for in-service professional development and lesson study as a part of public open house in 1998 (C. Lewis & Tsuchida), and later she categorized them into seven forms based on who organizes the lesson study (C. Lewis, 2002b). Takahashi (2006) sorted out lesson study by practicing teachers in Japan into three major forms, school-based lesson study, district-wide lesson study, and cross-district lesson study, and examined how each form was received by the participant differently. Each form has a different format and has a different focus. For example, the school-based lesson study mainly focuses on developing a common view to educate all the students in the school. Knowing each student and adopting the Course of Study, the national standards, and government-authorized textbooks to maximize their students learning is the primary goal. The district-wide lesson study is often planned and conducted using district professional development days. The major purpose of this form is to discuss issues that are common among schools in the district. For example, teachers can discuss about the potential benefits and challenges of using a new technology tool that the district is considering using to equip their schools by looking at actual classroom instruction using the tool.

Cross-district lesson study is usually conducted by national schools who are affiliated with universities and national professional organizations holding particular interests. This form of lesson study is often conducted during regional or national teacher conferences. Leading researchers and practitioners form an expert panel in order to observe and discuss the research lesson. Japanese teachers recognize that these three forms of lesson study have different emphases and they value each of these forms of lesson study. By using the chart in Figure 1, each of these three forms of lesson study in professional development can be put in a different place. The school-based lesson study might be the closest to the criticize lesson; and the cross-district lesson study might be the closest to the model lesson. The district-wide lesson study might have both research and presentation functions as shown in Figure 67.2.

In addition to the lesson study in in-service professional development, lesson study plays an important role in pre-service education as well. In

Figure 67.2. Japanese lesson study.

teacher preparation programs, prospective students in teacher training institutes usually have both opportunities: observing cooperating teachers' teaching, and teaching lessons under the guidance of cooperating teachers. Like any teacher training programs, Japanese university students also have the same kind of opportunities using lesson study.

A recent study discusses the differences in knowledge and expertise that exists among prospective teachers, novice teachers, and experienced teachers in Japan (Takahashi, 2011). It also touches upon how the school system uses lesson study to support teachers to learn continuously through their careers and why it is important even after completing formal teacher training at the universities. As a way to uncover the differences in teacher expertise, three levels of teaching, Level 1, Level 2, and Level 3 are used in the study.

Japanese mathematics educators and teachers identify three levels of expertise to teach mathematics; the teachers who can teach important basic ideas of mathematics such as facts, concepts, and procedures (Level 1), the teachers who can explain meanings and reasons for important basic ideas of mathematics in order for students to understand them (Level 2), and teachers who can provide students opportunities to understand these basic ideas, and support their learning in order for students to become independent learners (Level 3) (Sugiyama, 2008).

An empirical study was conducted with selected prospective and practicing teachers in Japan to see how they would plan a lesson from the same page of a mathematics textbook. Level 1, Level 2, and Level 3 teachers were asked to complete a questionnaire designed to elicit how they use their knowledge of mathematics and pedagogy to plan a lesson based on a textbook. Teachers worked from one textbook page from the mathematics textbook series most widely used in Japanese public elementary schools. From the results of this study, two areas of expertise can be identified as important for using textbooks effectively in mathematics teaching: expertise in structured problem solving, and expertise in anticipating student responses.

In both pre-service and in-service teacher development programs, lesson study plays an important role in supporting teacher growth. The author argues that lecture and workshop type of professional development and lesson study should complement each other to support teachers in developing knowledge and expertise for teaching mathematics based on the results of the study. Lecture and workshop type of professional development, known as phase 1 professional development, could be effective in increasing a teacher's knowledge for teaching mathematics. This includes content knowledge of mathematics, pedagogical content knowledge for teaching mathematics, curricular knowledge for designing lessons, and general pedagogical knowledge (Shulman, 1986). In order for teachers to develop such knowledge, phase 1 professional development usually provides opportunities to learn through reading books and articles, listening to lectures, and watching videos or demonstration lessons. Most university coursework falls into this category. On the other hand, lesson study, known as phase 2 professional development, focuses on developing expertise for teaching mathematics. This includes the expertise needed to develop lessons for particular students, to use various questioning techniques, to design and implement formative assessments, to anticipate student responses to questions and tasks, and to make purposeful observations during class. For teachers to develop such expertise, they need opportunities to plan lessons carefully, to teach the lesson based on the plan, and to reflect upon the teaching and learning based on careful observation. Japanese teachers and educators obtain these experiences through lesson study (Lewis & Tsuchida, 1998; Stigler & Hiebert, 1999; Takahashi & Yoshida, 2004; Yoshida, 1999).

Implications and Suggestions for Future Research

Since lesson study is a unique feature of Japanese teacher professional development, the readers will be able to see how the school systems and universities work together to support teachers to become life-long learners. Like some Asian countries, the culture among Japanese teachers and schools expect a teacher to be a learner who can be a model of life-long learning. Lesson study plays an important role for Japanese teachers to continuously learn through out their carrier, which sometimes is as long as nearly 40 years, to improve teaching by working with colleagues. Those countries where a teacher's carrier is much shorter and that do not have schools or a society that expects teachers to learn while they teach their students, it will be important to establish a supporting system. It is essential for teachers to continuously learn to be effective because by just going through teacher preparation programs may not be enough for a majority of teachers trying to develop knowledge and expertise to become effective mathematics teachers. This could bring new insight for researches working with various school systems in the world. Possible future research related to this lesson study would be the study of the implementation of not only lesson study but also a professional development structure like the Japanese one in different cultures.

NOTE

1. http://www.noycefdn.org/svmi.php

REFERENCES

Council for Basic Education (2000, September 24-27). *The eye of the storm: Improving teaching practices to achieve higher standards.* Paper presented at the Wingspread Conference, Racine, Wisconsin.

Germain-McCarthy, Y. (2001). *Bringing the NCTM standards to life: Exemplary practices for middle schools*. Larchmont, NY: Eye on Education.

Hart, L. C., Alston, A., & Murata, A. (Eds.). (2011). *Lesson Study Research and Practice in Mathematics Education*. Now York: Springer.

Hirabayashi, I. (2002). Lesson as a drama and lesson as another form of thesis presentation. In H. Bass, Z. P. Usiskin & G. Burrill (Eds.), *Studying classroom teaching as a medium for professional development. Proceedings of a U.S. - Japan workshop*. Washington, DC: National Academy Press.

Inagaki, T. (1995). *Meiji Kyouju Rironshi Kenkyu [A Historical Research on Teaching Theory in Meiji-Era]*. Tokyo: Hyuuron-Sya. (in Japanese)

Lewis, C. (2000, April 2000). *Lesson Study: The core of Japanese professional development.* Paper presented at the AERA annual meeting.

Lewis, C. (2002b). *Lesson study: A handbook of teacher-led instructional improvement*. Philadelphia: Research for Better Schools.

Lewis, C., Perry, R., & Hurd, J. (2009). Improving mathematics instruction through lesson study: a theoretical model and North American case. *Journal of Mathematics Teacher Education, Volume 12*(Number 4), 285-304.

Lewis, C., Perry, R., Hurd, J., & O'Connell, M. P. (2006). Lesson study comes of age in north America. *Phi Delta Kappan, 88*(04), 273-281.

Lewis, C., & Tsuchida, I. (1998). A lesson like a swiftly flowing river: Research lessons and the improvement of Japanese education. *American Educator, 22*(4). 12-17, 50-52.

Makinae, N. (2010). The Origin of Lesson Study in Japan. In Y. Shimizu, Y. Sekiguchi, & K. Hino (eds.) *In Search of Excellence in Mathematics Education: The Proceedings of The 5th East Asia Regional Conference on Mathematics Education, Vol.2*. (pp. 140-147). Tokyo: Japan Society of Mathematical Education.

National Research Council (Ed.). (2002). *Studying classroom teaching as a medium for professional development. Proceedings of a U.S. - Japan workshop.* Makuhari, Japan: National Academy Press.

Research for Better Schools Currents Newsletter (2000). Against the odds, America's lesson study laboratoy emerges. *Research for Better Schools, 4.1*.

Sheldon, E. A. (1862). *Object-teaching*. New York: Charles Scribner.

Stepanek, J. (2001). A new view of professional development. *Northwest Teacher, 2*(2), 2-5.

Stigler, J., & Hiebert, J. (1999). *The teaching gap: Best ideas from the world's teachers for improving education in the classroom*. New York: Free Press.

Sugiyama, Y. (2008). *Introduction to elementary mathematics education*. Tokyo: Toyokan publishing Co. (in Japanese)

Takahashi, A. (2000). Current trends and issues in lesson study in Japan and the United States. *Journal of Japan Society of Mathematical Education, 82*(12), 15-21.

Takahashi, A. (2006). Types of Elementary Mathematics Lesson Study in Japan: Analysis of Features and Characteristics. *Journal of Japan Society of Mathematical Education, Volume LXXXVIII*, 15-21.

Takahashi, A. (2011). Jumping into Lesson Study—Inservice Mathematics Teacher Education. In L. C. Hart, A. Alston & A. Murata (Eds.), *Lesson Study Research and Practice in Mathematics Education*. New York: Springer.

Takahashi, A. (2011). The Japanese approach to developing expertise in using the textbook to teach mathematics rather than teaching the textbook. In Y. Li & G. Kaiser (Eds.), *Expertise in Mathematics Instruction: An international perspective*. New York: Springer.

Weeks, D. J. (2001). Creating happy teachers. *Northwest Teacher.*

Yoshida, M. (1999a, April). *Lesson Study [jugyokenkyu] in elementary school mathematics in Japan: A case study.* Paper presented at the American Educational Research Association Annual Meeting, Montreal, Canada.

Yoshida, M. (1999b). *Lesson study: A case study of a Japanese approach to improving instruction through school-based teacher development.* Unpublished Dissertation, University of Chicago, Chicago.

INDIA

CHAPTER 68

INTRODUCTION TO THE INDIA SECTION

K. Subramaniam
Homi Bhabha Centre for Science Education (TIFR), Mumbai

Two facts need to be kept in mind as one reads the chapters in the India section in this volume. First, that India is a country of diversity and large numbers, and second, that education in India has witnessed major initiatives in the last decade. In 2010-2011, over 15 million students were in Class X (i.e., Grade 10), which is the year of completion of secondary school. About 8 million young Indians from this age cohort were not in school (corresponding to a Gross Enrollment ratio of 65%). In the same year, there were over 19.5 million children in Class VIII, which is the year of completion of the compulsory stage of schooling. The number is much larger because of the higher Gross Enrollment Ratio of 85% (Ministry of Human Resource Development, 2012).

These numbers are likely to have increased in the last two years due to strong legislative action in 2010 by the central (federal) government in the form of the Right to Education Act (frequently referred to as the "RTE" Act). The Act makes it mandatory on the part of the Government to provide school education to every child from the age of 6 to 14, amounting to eight years of school education. It specifies certain parameters of quality for school education which include adopting the National

The First Sourcebook on Asian Research in Mathematics Education: China, Korea, Singapore, Japan, Malaysia, and India, pp. 1507–1513

Curriculum Framework as broadly defining the curricular aspect of quality education.

The framers of the constitution of independent India had, in 1950, expected school education to become universal within a decade. It is nearly 60 years later that the Government enacted legislation to realize this stubbornly elusive goal. Implementing the RTE Act requires navigation of the complex federal governance structure in India. There are 28 states in India, many with a separate dominant language, representing a diversity of cultures. All the states have their own elected legislatures and governments. In the mid-1970s, education, which till then was the domain of the state governments, was made a "concurrent subject", that is, a joint domain of responsibility and action by the Central and State governments. In recent decades, the central government has played an increasingly pro-active role, using its leverage to align the policies of the state governments with those of the Center.

The Central Government called for the creation of a new National Curriculum Framework in 2005, often referred to as "NCF 2005". The relation of the NCF 2005 to the history of education policy making in India, and its perspective on mathematics education are discussed in detail in Khan (this volume). Nearly every chapter in this section takes account of the background of the new curriculum framework and locates the discussion in the context of educational reform that the NCF 2005 represents. The creation of NCF 2005 was a mammoth exercise carried out by the National Council of Educational Research and Training (NCERT), drawing on the educational expertise available in the country. There were 21 focus groups composed of experts drawn from institutions across the country that formulated policies and guidelines on various aspects of school education – one of the focus groups was on the teaching of mathematics in school. The reports of these groups formed the input for new framework. NCF 2005 gave primacy to placing the child at the center, to moving away from rote learning and stereotypical assessment, to linking school learning with the child's lived experience, and to building a democratic ethos. All the documents that comprise the NCF 2005 are available on the webpages of the NCERT: http://www.ncert.nic.in.

Although the NCF 2005 is the primary policy document for the school curriculum at the national level, its direct impact is limited to schools that are regulated by the central government. These are the schools affiliated to the Central Board of Secondary Education, which follow the textbooks published by the NCERT. In order to impact the curricula and textbooks in the various states of India, the effect of NCF 2005 must percolate through the complex federal structure, a process that is still underway. Each state has its own regulatory body in the form of a secondary education board. Table 68.1 gives an idea of the relative sizes of some of the

Table 68.1. Number of Students (in Millions) Who Appeared for the 2012 Class X Examination of Various School Boards

School Board	*Number of Students Who Appeared for Class X Exam in 2012 (in millions)*
Central Board of Secondary Education (country wide)	1.18
Council for the Indian School Certificate Examinations (country wide)	0.13
National Institute of Open Schooling (country wide)	0.53
Maharastra State Secondary Education Board	1.35
Tamil Nadu State Secondary Education Board	1.05
Gujarat State Secondary Education Board	0.93

Source: Compiled from various news reports announcing the results of the school board examinations.

major school boards in the country. The first three school boards in the table operate across the country, while the others operate within their respective states.

The vast majority of schools located in rural areas and schools catering to low socio-economic groups are affiliated to the state boards. In these schools, the medium of education is typically the language of the state. For example, for the states in the last three rows of Table 68.1, the languages are Marathi, Tamil and Gujarati respectively. In contrast, the schools affiliated to the Council for the Indian School Certificate Examinations are typically urban, English medium schools that cater to middle or higher income families. Differences in the availability of resources, distance from urban centers, access to and aspiration for the English language, and position in the social, especially caste hierarchy, are some of the critical factors in determining the outcome of education in India. An indicator of the range of differences is the stark contrast between the performance of the country's top education and research institutions in science and technology and the poor quality of education in the vast majority of its institutions. These factors need to be kept in the background while discussing problems of mathematics education in the Indian context. The larger context of education and its relation to mathematics education have also been discussed in several articles in a recent volume (Ramanujam & Subramaniam, 2012).

The chapters related to India in the sourcebook deal with the issues of K-12 mathematics education, or from primary to senior secondary stages, as they are called in India. The opening chapter by Farida Khan discusses the perspective on mathematics education in NCF 2005. She sets this in the context of how earlier policy documents viewed the place of mathematics in the school curriculum. Although many of the recommendations of NCF 2005 echo those made in earlier policy documents, she notes that there is a shift from viewing mathematics as a static, unquestionable body of knowledge to emphasizing mathematics as a way of thinking and reasoning. The curriculum framework and the textbooks that followed did bring in a focus on the learner that is new in the Indian context. She cautions, however that the resources that a majority of schools are currently able to muster may prove inadequate to realize the vision of mathematics education that is envisioned in NCF 2005.

Amitabha Mukherjee and Vijaya Varma describe an important effort aimed at reforming the primary mathematics curriculum. The effort is one among several efforts outside the formal education sector that were led by non-profit voluntary organizations (commonly referred to as "NGOs" or "Non-Governmental Organizations"). The significance of the School Mathematics Project lies in the fact that it was an effort initiated by University professors of science to address problems in primary mathematics education, but one in which primary teachers played a central role. The Project had an impact beyond its field of work in influencing first, the curriculum and textbooks of the Delhi state and later, the post NCF 2005 national curriculum and textbooks of the NCERT. The chapter thus documents an important episode in the history of reform efforts in primary mathematics education in India.

The work of Usha Menon is another example of the voluntary efforts to improve primary mathematics education, one that is informed by the perspective of Realistic Mathematics Education. The chapter explores the nuances of early number learning focusing on the ordinal and cardinal meanings of number. It analyses the role of pedagogical supports, especially the 100-bead string and the empty number line in developing number sense among young children. The author attempts to develop empirical indicators of number sense in terms of the diversity of solution strategies that children use in solving problems. In relating theory to practice, the author draws on her own extensive teaching experience and her interactions with teachers.

Jayasree Subramanian, Mohammed Umar and Sunil Verma summarize some interventions aimed at developing new approaches to the teaching of core topics in middle school mathematics. The authors set this in the context of the work of the NGO Eklavya, which is widely known for its seminal contributions to innovations in science and social science education at the school level. Efforts at innovating curricular approaches are set against the often harsh realities of Indian schools, especially those that serve students from poor families. The authors raise sharp questions about whose interests the current curriculum really serves and if the curriculum can be re-visioned to make it more relevant to the lives of children from less affluent backgrounds.

Rakhi Banerjee draws on her previous research on approaches to teaching beginning algebra and analyzes the impact of the new curricula and textbooks on classroom teaching at the middle school level. Although the textbooks adopt a more student-friendly approach to introduce symbolic algebra than earlier textbooks, she argues that they do not prepare students adequately nor take advantage of their knowledge of arithmetic. She makes a case for the textbook approaches to be better informed by the vast literature on algebra education, and to focus on algebraic thinking as a central goal of algebra education.

The chapter by Ruchi Kumar, Subramaniam and Shweta Naik provides a brief overview of issues related to the development of in-service mathematics teachers in India. The authors' focus is workshops for mathematics teachers, which is the main mode of in-service teacher development in India. They suggest a broad framework consisting of goals, principles and themes for analyzing the design and enactment aspects of a workshop, and illustrate its application through the analysis of the components of a workshop and its enactment. Their analysis reveals how the agencies of the teacher and the teacher educators shape the interaction in the light of the goals and the principles.

Assessment of mathematical learning is one of the areas that is recognized as needing urgent and comprehensive reform. The two chapters on assessment address this problem from two different perspectives. The chapter by Aaloka Kahnere, Anupriya Gupta and Maulik Shah discuss insights about student errors gained from large scale diagnostic tests conducted across many states in India. The large scale data allows them to draw what may prove to be robust insights about Indian students' mathematical conceptions. They also draw on interviews of groups of students conducted in a classroom setting and recorded on video, to probe misconceptions in greater detail.

Shailesh Shirali's chapter on assessment discusses test items from high stakes examinations such as the public "board" exams at the end of Class 10 and 12, and the highly competitive entrance examinations to prestigious higher education institutes. Such examinations cast a long shadow on education, and drive the major effort and spending by families on education. Shirali sketches some of the damaging effects of a competitive, high stakes examination culture. He calls for research on innovations in assessment including the use of technology and investigative projects in mathematics, remarking that there is better acceptance of the latter in science than in mathematics education.

Jonaki Ghosh explores the use of technological tools such as computer algebra systems, dynamic geometry, spreadsheets and graphics calculators for teaching mathematics to higher secondary school students. Drawing on her own experience of teaching secondary school students over several years using a project based approach combined with technological tools, she describes several examples of investigative projects. The chapter highlights the mathematical ideas that have engaged students as they worked on the projects. Ghosh locates her work in the context of the challenges in using technology for teaching mathematics in the mostly resource-poor schools in India.

The last two chapters illuminate mathematics education and school mathematics from a historical perspective. The chapter by Senthil Babu explores the curriculum of indigenous schools in South India in the pre-British period. Based on a study of manuscripts that functioned as arithmetic manuals in the *tiṇṇai* or veranda schools, he argues that memory played a basic role in learning and in the organization of knowledge. The manuals contained arithmetic tables of various kinds which were committed to memory and recalled in the context of solving a problem. The numbers, units and problem contexts show the rich links that the elementary mathematics curriculum had with practical problems in a local context.

Unlike Babu's study of a local, vernacular form of mathematics close to the domain of application, the chapter by Raja Sridharan and Subramaniam is closer the grand narrative of the history of Indian mathematics. Marking a departure from the domains of astronomy, geometry and algebra, they discuss the mathematics of combinatorial problems associated with prosody and music. The discussion is located in the context of contemporary Indian music and Sanskrit prosody. They explore the occurrence in Indian mathematical texts of binary arithmetic, of "Fibonacci numbers", of factorial and other representations for positive integers. They argue that the mathematics explored in this chapter has relevance at the school level in terms of the novel ideas and surprising connections.

ACKNOWLEDGMENTS

I thank the authors and editors of the India section for their contributions and for bearing with repeated and at times unreasonable demands. The reviews by the editors have contributed greatly to enhancing the quality of the chapters. I thank Bharath Sriraman for his support at several stages of the project. My colleagues and students at the Homi Bhabha Centre for Science Education have supported and encouraged me. I especially thank Shweta Naik, Ruchi Kumar, Arindam Bose and Jeenath Rahman, who pitched in enthusiastically, helping with last minute proof reading and corrections.

REFERENCES

Ministry of Human Resource and Development. [MHRD] (2012). Statistics of school education 2010-2011. New Delhi: Ministry of Human Resource and Development, Government of India. Available at http://mhrd.gov.in/sites/upload_files/mhrd/files/SES-School_201011_0.pdf.

Ramanujam, R. & Subramaniam, K. (Eds.) (2012). *Mathematics education in India: Status and outlook*, Mumbai: Homi Bhabha Centre for Science Education (TIFR). Available at http://nime.hbcse.tifr.res.in/articles/INPBook.pdf

CHAPTER 69

EVOLVING CONCERNS AROUND MATHEMATICS AS A SCHOOL DISCIPLINE

Curricular Vision, Educational Policy and the National Curriculum Framework (2005)

Farida Abdulla Khan
Jamia Millia Islamia, New Delhi

ABSTRACT

This chapter traces the trajectory of mathematics education in the Indian school curriculum and makes the argument that it was shaped and driven by the developmental trajectory that India adopted after independence, wherein the importance of mathematics was attributed primarily to its usefulness as a foundational requirement for training in technology and the sciences. The contribution of the National Curriculum Framework 2005 has been largely in shifting the focus away from the needs of the discipline to the learning-teaching process, to the child and to the doing of mathematics as a meaningful and rewarding activity. In conclusion, attention is drawn to concerns regarding the implementation of the new curriculum, given the

The First Sourcebook on Asian Research in Mathematics Education: China, Korea, Singapore, Japan, Malaysia, and India, pp. 1515–1537

complex nature of the social negotiations of curriculum transaction within a deeply unequal and discriminatory system of schooling.

Keywords: Mathematics curriculum; national policies on education; social concerns and mathematics education; national curriculum framework

The history of schooling in independent India, the developmental trajectory that was adopted by a newly independent Indian nation, and the ensuing national and educational policies have all contributed to a heavily content-loaded curriculum for mathematics at the school level. The subject forms an integral and highly valued component of the school curriculum and is a mandatory requirement up to Class X, the completion of which marks an important milestone in Indian schooling.

The two major initiatives on school education in independent India – the Education Commission of 1964-1966, and the National Policy on Education (NPE) 1986 – gave central status to the development of science and technology and underlined the critical role of mathematics in achieving this objective. Both documents refer to mathematics as a foundational discipline and emphasize its contribution to the development of science and technology. In the present context of economic globalization, advances in science and technology and the domination of multinational business corporations, mathematical knowledge and expertise represents a critical asset and an essential component of training in the sciences, business and information processing. As a school subject it becomes indispensable for higher education in the sciences, technology, economics and business. A wider appeal for retaining mathematics as a core school subject is made on the basis of its usefulness in dealing with the widespread use and increased dependence on technology.

By the 1980s however, concerns regarding the school curriculum, its disconnect from lived experience and its routinized transaction in the classroom were being expressed from within and outside the educational establishment. The Yash Pal Committee Report (Government of India [GoI],1993) entitled "Learning without Burden" was a response to the extremely stressful learning routines and an equally stressful assessment and examination system that characterizes contemporary Indian schooling. It was a well grounded critique that focused attention on some major problems of school education and touched a chord in the general population as well as the educational establishment. These concerns were addressed by the National Council for Educational Research and Training (NCERT) when it launched the important exercise of rethinking the objectives of schooling and the school curriculum. This led to the preparation of a new curricular document, the National Curriculum Framework

(NCF) in 2005, and a reworking of the entire school syllabus followed by a major revision of school textbooks.

This chapter traces the development of mathematics as a school subject in India and situates it in the economic and political contexts of a post colonial society and polity, grappling with issues of nationhood and development. With mathematics as its focus, it traces the impact that the agenda for development has had on educational policy, school curriculum and syllabi and how this has shaped the place and value and subsequently the content of the discipline of mathematics in Indian schools. It situates the NCF 2005 as the latest effort in a series of reforms within an educational system that, along with interventions from the Government, non-governmental organizations and international agencies, has also witnessed interesting innovations, important critiques and serious debates especially in the last few decades.[1]

The chapter goes on to provide an overview of the NCF 2005 (National Council of Educational Research and Training [NCERT], 2005) as a significant change in perspective and the emphasis on a socially "constructed" view of knowledge, along with a renewed focus on the child, the classroom and the teaching-learning process. It then discusses the position paper of the focus group on Mathematics (NCERT, 2006a) that calls for a curriculum which is ambitious and teaches important mathematics and at the same time allows children to enjoy the subject, and to find meaning within it.

Although the document and its vision have been generally well received, there are apprehensions about the modalities of its translation into practice within the classrooms and the limits of the infrastructure – both social and physical, that is currently available for doing this. In conclusion, attention is drawn to some issues of concern regarding the implementation of the new curriculum, given the complex nature of the social negotiations of curriculum transaction within a deeply unequal and discriminatory system of schooling.

THE VISION OF EDUCATION FOR A NEW INDEPENDENT NATION

The formal and modern system of Indian education as it exists today, evolved during the colonial period of India's history. Put into place by a regime with its own colonial agenda, there were nevertheless concerns about the population that it was meant to serve. There were debates and differences over several issues but the two that gained most prominence were to do with the medium of instruction, and what knowledge was to be the focus of teaching in school. Once the model of English education had been accepted and the fact that it was to be a vehicle of modernization was established, the teaching of the three R's (reading, writing and arithmetic) at the primary level was put into place. The need for numerical skills was universally recognized and there was never any challenge to the inclusion of mathematics as a core school subject.

When India gained independence in 1947, its major priorities were to build a self-sufficient economy and a cohesive social order, given the backdrop of a long period of colonial rule and the social upheaval of the partition of the country into India and Pakistan. The preferred model of development that was adopted after early deliberations and debates was that of centralized planning through the five-year plans. Education was to play a major role in the modernizing project – in matters economic as well as social – but most importantly, at this juncture, in the creation of a cohesive nation and a modern and prosperous nation state. The constitution of independent India reflects a commitment to creating a democratic republic based on the principles of justice, liberty, equality and fraternity. The references in the constitution to education are aimed primarily at fulfilling the vision of a just and equitable society. The constitution issues directives to the state to make education universally accessible and also to safeguard social, cultural and political rights.

Article 45 of the Indian constitution directs the state to make free and compulsory education available to all children up to the age of 14 years. Although this was to be accomplished within ten years of the commencement of the constitution in 1950, it is only in 2009 that an act of parliament making education free and compulsory for children between 6 to 14 years, was passed. Other articles of the constitution protect the rights of minorities, women and children and provide safeguards for instruction in the mother tongue for all languages and linguistic minorities. The constitution therefore acknowledges an important role for education in forging the newly envisaged republic and specifies the major objectives that it is expected to fulfill.

EDUCATION COMMISSIONS AND CONCERNS OF THE POST-INDEPENDENCE PERIOD

The rhetoric of social justice, equity and inclusion figure prominently in all policy documents related to education, but scientific and technological aspirations leading eventually to economic development seem to have been the dominant concern for planning and policy making for the first few decades after independence. Although individual and social development are important considerations, education is ultimately expected to harness the human resource pool needed for the development of the "nation." Commitment to universal primary education figured prominently in the constitution but the immediate focus of both policy and planning in the early years was on higher education, and on the development of the sciences and technology.

An analysis of these initiatives and a careful reading of the documents is presented for understanding the involvement of the state in the project of education; the priorities that influenced the ways in which the educational system evolved; and how this shaped the Indian school system and the content and processes of school education.

The first major review of education in independent India came in the form of the University Education Commission with a mandate to "report on Indian university education and suggest improvements and extensions as they may be desirable to suit present and future requirements of the country" (GoI, 1949, p. 1). The report reflects the euphoria of a newly formed nation and the excitement of creating a new social order. The tone is lofty and projects both the idealism and the concerns of the constitution, exhorting universities to "...provide leadership in politics and administration, the professions, industry and commerce. They have to meet the increasing demand for every type of higher education, literary and scientific, technical and professional. They must enable the country to attain, in as short a time as possible, freedom from want, disease and ignorance, by the application and development of scientific and technical knowledge" (GoI, 1949, p. 33). It suggests courses in general education that would include sciences, humanities and the arts. Mathematics figures as a component of all three streams, and although several disciplines are discussed in detail, there are no recommendations specifically related to the teaching of mathematics or to mathematics as a discipline. The importance as well as the subject matter of mathematics seems to require neither advocacy nor elaboration.

In 1952, the Government of India appointed the Secondary Education Commission to examine the prevailing system of secondary education and to suggest measures for its reorganization and improvement. Its task was to reconsider the prevailing system of secondary education "which is unilateral and predominantly academic in nature" and to explore possibilities of changing to a system which would cater to different aptitudes and interests (GoI, 1956). This commission lists three important objectives of education – to cultivate in its citizens a broad, national and secular outlook; to improve productive efficiency to increase the national wealth and thereby to raise appreciably the standard of living of the people; and lastly, to recognize the need for orienting the educational system in such a way that it will stimulate a cultural renaissance (GoI, 1956, p. 23).

The critique of the education system elaborated in this document in the form of the specific problems mentioned here, was to surface in almost identical form in many subsequent documents including the NCF 2005:

- education given in schools is isolated from life and the communities that children live in;
- it is narrowly academic and does not cater to the broader personality of the child – emotional, aesthetic needs, aptitudes, etc.;
- importance given to English disadvantages a large section of the student population;
- methods of teaching do not engage students and fail to inspire them;
- large numbers of children in each class prevents personal contact and attention; and
- an inappropriate examination system that relies on memorization of facts curbs all creativity and experimentation both in the teachers and in the students (GoI, 1956, p. 21-22).

The concerns are almost identical to those stated in the University Commission as are the sources which it draws upon, to describe what secondary education is to accomplish. The vision of a new social order through the long struggle for independence is evoked, as are the reference points and motivating spirit of the constitution. A spirit of democratic functioning within a national and secular outlook is a dominant concern in all policy documents of the era, as is the concern with economic development. It refers specifically to tackling poverty and to raising the standard of living of the people. The broad characteristics of secondary education are to be "…the training of character to fit the students to participate creatively as citizens in the emerging democratic social order; the improvement of their practical and vocational efficiency so that they may play their part in building up the economic prosperity of their country; and the development of their literary, artistic and cultural interests, which are necessary for self-expression and for the full development of the human personality, without which a living national culture cannot come into being" (GoI, 1956, p. 23).

Mathematics, although mentioned as a core subject both for a general and a technical degree, is never discussed in any detail, whereas the sciences, social sciences and languages are all adequately elaborated. It is however included in the category of subjects that are "to introduce the pupil, in a general way to the significant departments of human knowledge and activity. These will naturally and obviously include language and literature, social studies, natural sciences, and mathematics which have always formed part of every secondary school curriculum" (GoI, 1956, p. 82).

A NATIONAL SYSTEM OF EDUCATION: THE EDUCATION COMMISSION 1966

The first major and comprehensive exercise of reviewing school education in independent India came in the wake of the third five-year plan (1961-66) and debates around the developmental trajectory that the centralized project of planning was looking to adopt. Given the tensions that existed around a range of issues such as poverty and land reforms in the context of a food crisis, the existing economic and social inequality of the Indian population, linguistic and caste conflicts, the cornerstones of the new model of development were science, technology, industrialization and modernization. Within this model, education was seen as a powerful instrument of social and economic change that would inculcate rationality and modernity and work towards improving the lives of people. The Education Commission of 1964-66 (also known as the Kothari Commission) was constituted at this juncture and it was assigned the task of recommending a national system of education from the pre-primary stage to professional and higher education. The Commission produced a landmark document that was to shape the future of education and educational debates in India. It also became a reference point later for the NPE (GoI, 1986/1992).

The report of the Education Commission opens with the following statement: "The destiny of India is now being shaped in her classrooms. This, we believe, is no mere rhetoric. In a world based on science and technology, it is education that determines the level of prosperity, welfare and security of the people" (GoI, 1986/1992, section 1.01). It aims

- to raise the standard of living of our people,
- to achieve social and national integration and create an integrated and egalitarian society, and
- to strengthen democracy and make it permanently viable.

A chapter of the Education Commission Report, dealing extensively with the curriculum, highlights the importance of science and mathematics and defines the parameters for the mathematics curriculum. As in all prior documents on education, mathematics is considered an integral part of "general" education (i.e., education for all students prior to streaming). All references to it are in conjunction with sciences and technology and it is seen as a foundational aspect of these disciplines. There is "great emphasis on making science an important element in the school curriculum" along with the recommendation that "science and mathematics be taught on a compulsory basis to all pupils as a part of general education during the first ten years" (GoI, 1966, section 8.50). Elaborating on the role of mathematics, the document states,

> One of the outstanding characteristics of scientific culture is quantification. Mathematics therefore assumes a prominent position in modern education. Apart from its role in the growth of the physical sciences, it is now playing an increasingly important part in the development of the biological sciences. The advent of automation and cybernetics in this century marks the beginning of the new scientific industrial revolution and makes it all the more imperative to devote special attention to the study of mathematics. Proper foundation for the knowledge of the subject should be laid at school. (GoI, 1966, section 8.62)

These assumptions about the importance of mathematics, and of its critical role in creating a scientific and technological workforce, were to become increasingly entrenched and accepted. It was this understanding of mathematics that was to shape not only the mathematics curriculum and pedagogy but also to uncritically accord to it a gate-keeping function for entry into courses in the sciences, technology, economics and business.

There are specific suggestions for the teaching of science and mathematics:

- investigatory instead of lecture methods to be adopted;
- close connection between science, industry and agriculture to be stressed reflecting the industrial and agricultural interests of the local communities;
- strengthening of laboratory work and investigatory experiments under supervision by teachers;
- understanding of basic principles rather than mechanical teaching of computation; and
- flexibility to be provided to cater to the varying needs of gifted students.

The recommendations of the Education Commission were translated into a document entitled "The Curriculum for the Ten Year School – A Framework" (NCERT, 1975) and syllabus outlines were drawn up for use in schools. It reiterates the objectives of "providing a broad base of general education to all pupils in the first ten years of schooling", and mentions a list of the proposed curricular areas with the sciences and mathematics heading this list. Other areas include work experience;

social sciences; languages; art, music and other aesthetic activities; and health and physical education with the caveat that this list is not to be considered exhaustive (NCERT, 1975, p. 14).

Specific objectives of mathematical education at the school level are specified in the 1975 document as:

- enabling the students to cultivate a mathematical way of thinking,
- enabling students to quantify their own experiences in the world,
- enabling students to learn basic structures of mathematics through unifying concepts, and
- to stimulate students to study mathematics on their own, to develop a feel and an interest in mathematics.

As is evident, deliberations on the mathematics syllabus were aware of both the changing pedagogies as they were of the need for students to comprehend and engage with mathematical content.

THE NATIONAL POLICY ON EDUCATION 1986

The exhaustive Education Commission report of 1966 and its recommendations were an important reference point for the formulation of the NPE in 1986. Its policy implications as elaborated in the "Plan of Action", covered a variety of areas, of which some important ones are listed below:

- access to education of a comparable quality for all,
- introduction of minimum levels of learning and provision of threshold facilities,
- articulation of a national system of education,
- improvement of teaching and learning through examination reform,
- development of culture-specific curricula and instructional materials for the disadvantaged,
- overhauling of the teacher education system, and
- use of modern communication technology in education, training and awareness creation.

Science and mathematics teaching is seen as extremely important and as in earlier documents is to form a part of the core and compulsory syllabus up to Class X.

Following the National Policy on Education, the NCERT produced the second National Curricular Framework for School Education (NCERT, 1988) and despite the many references to a need for reforming the curriculum, pedagogy and examination system, the articulation of the framework through courses and textbooks resulted in an increase in curricular content at all levels (NCERT, 2005, p. 4). This "load" of textbooks became a matter of serious concern within the larger community of parents, teachers, educationists and the general public to such an extent that the Yash Pal Committee was set up to look into this issue which prepared a report entitled "Learning without Burden" (GoI, 1993). The committee highlighted the inadequacies of the school curriculum and hit out scathingly at a system of education that rewarded students on their ability for rote memorization of vast amounts of information in order to pass examinations.

THE NATIONAL CURRICULUM FRAMEWORK 2005

A review of the Indian school curriculum was conducted in 2000 and a curriculum framework was brought out by the NCERT, but not only did this framework fail to tackle the problem of the information-packed curriculum, it also became mired in controversies at both the academic and the political level.[2] It was finally in 2005 that the NCF 2005 was prepared by the NCERT to address issues raised by the Education Commission of 1966, the National Policy on Education of 1986 and the Yash Pal Committee Report of 1993.

The NCF 2005 did not limit itself to preparing a curriculum framework document on the basis of the policy guidelines alone, but initiated a comprehensive critique of the educational system and set itself the task of rethinking and reorganizing classroom practice through a new curricular framework. The exercise began with the formation of 21 focus groups to deliberate upon and to produce position papers on what were considered critical areas in contemporary school education. One of these focus groups was on the teaching of mathematics. We discuss the position paper of this focus group later in this chapter. Subsequent to the NCF 2005, committees for re-working syllabus outlines and guidelines for textbook writing for each subject were constituted that re-worked the syllabi for each discipline. This was followed by re-writing all textbooks from Classes I to XII. These were not treated as secondary outcomes of the deliberations of the Curriculum Framework, but were conceptualized as critical to the process of transforming a vision into a concrete reality in the classrooms. The textbook writing was a large-scale exercise at the national level carried out by a core of academics, activists, teachers and other practitioners whose inputs had shaped the framework document and some of whom had been part of the focus groups and therefore shared and endorsed its theoretical and philosophical perspective.

The following quote from the foreword of the NCF 2005 highlights some of its essential features.

> There is much analysis and a lot of advice. All this is accompanied by frequent reminders that specifics matter, that the mother tongue is a critical conduit, that social, economic and ethnic backgrounds are important for enabling children to construct their own knowledge. Media and educational technologies are recognized as significant but the teacher remains central. Diversities are emphasized but never viewed as problems. There is a continuing recognition that societal learning is an asset and that the formal curriculum will be greatly enriched by integrating with that. There is a celebration of plurality and an understanding that within a broad framework plural approaches will lead to enhanced creativity.... Since children usually perceive and observe more than grown-ups, their potential role as knowledge creators needs to be appreciated (NCERT, 2005, p. iii).

The NCF 2005 spells out an ambitious agenda and lists the following core concerns:

- commitment to Universal Elementary Education, not only representing cultural diversity but also ensuring that children from different social and economic backgrounds are able to learn in schools and to addressing disadvantages arising from inequalities of gender, caste, language, culture, religion or disabilities;
- to broaden the scope of the curriculum to include the rich inheritance of different traditions of knowledge, work and crafts;
- the development of self esteem and ethics and the need to cultivate children's creativity along with the need to respect a child's native wisdom and imagination;
- making children sensitive to the environment and the need to protect it;
- creating a citizenry conscious of their rights and duties and commitment to the principles embodied in the Constitution (NCERT, 2005, pp. 5-7).

It says, "that learning has become a source of burden and stress on children and their parents is an evidence of a deep distortion in educational aims and quality" (NCERT, 2005, p. iv). To address this, and other problems associated with schooling and learning, the NCF 2005 proposes the following guiding principles for curriculum development:

- connecting knowledge to life outside school,
- ensuring that learning is shifted away from rote methods,
- enriching the curriculum to provide for overall development of children rather than remain textbook centric,
- making examinations more flexible and integrated into classroom life, and
- nurturing an over-riding identity informed by caring concerns within the democratic polity of the country (NCERT, 2005, p. 5).

The changes proposed by the NCF 2005 have been overdue and largely well received. A major contribution of the new curriculum is the critique of knowledge systems and the power hierarchies that sustain them. An understanding of how knowledge is selected or excluded in the curriculum has allowed the textbook writers to select content with a much needed sensitivity to the many levels of exclusion that operate within a highly divided Indian social structure. These divisions span class and gender as much as they do caste, religious community, ethnicity, region and ability, both physical and psychological. The NCF 2005 demonstrates sensitivity to these issues and a sincere effort to formulate a curriculum and to produce textbooks that are inclusive, child friendly and meaningful. Within a highly unequal system that privileges children of an elite minority, an effort has also been made to make learning materials and texts accessible in a variety of ways by incorporating diverse social registers, and respecting linguistic, social and cultural variations.

CHARACTERIZING THE SHIFT IN PERSPECTIVE OF THE NCF 2005

The framing of the NCF 2005 was an unusual event in the recent history of Indian education in that a national curriculum framework was deliberated upon as an independent exercise and not merely a follow-up exercise of a major educational review or policy as was the case with earlier curricular formulations. It was able to draw upon the benefit of hindsight based on research, discussion and debates around the Indian education system and evidence of its successes and failures over several decades of initiative and reform. The twenty odd years between the formulation of the NPE 1986 and the NCF 2005 were a period of important changes in India's economic and political landscape. The fiscal crisis of the early 1990s, and the subsequent opening up and liberalization of the economy were to have a major impact on the educational sector in India.

Significant attempts were made during the 1990s to achieve universal elementary education and substantial progress was made in universalizing enrollment at the primary level. Following the NPE 1986, several large scale initiatives were launched at the national as well as state levels, especially after the entry of international donor agencies into the Indian education sector. The initiatives were at both national and local levels and by both governmental as well as non-governmental agencies (see Mukherjee & Varma, this volume). This inevitably led to a widened interest in education at all levels and sparked off several debates over issues of quality. The NCF 2005 came at a moment when there was an urgently felt need to channel these educational debates and reform proposals into action.

The review of educational documents presented earlier in this chapter marks the similarities of educational aims and objectives between these and the NCF 2005. There is a common thread of commitment to democratic values and social justice and to the social, personal and academic concerns of the child. In fact, the core concerns regarding the general state of education and educational practice are so similar that little seems to have changed from the time of the University Commission in 1949 to the NCF 2005. Both the Education Commission of 1966 and especially the NPE 1986 elaborate upon the importance of making learning joyful and activity oriented, adopting methods of teaching that allow for understanding and comprehension rather than meaningless memorizing, supporting and encouraging children's curiosity, and helping children to become critical thinkers. Notwithstanding these concerns about curriculum and pedagogy, it was the more pressing issues of making education accessible, of improving infrastructure and of reducing the glaring disparities at a macro level that took precedence at the level of planning and execution in the wake of the earlier documents. Why these goals were never quite achieved would need discussion that is beyond the scope of this chapter.

A marked shift of the NCF 2005 is its emphasis on the teaching-learning process and classroom practice with the learner as its central concern. Strongly espousing a social constructivist view of knowledge and a personal constructivist view of learning, there is a clear shift from "curriculum as fact" to "curriculum as practice."[3] A great deal of reflection and analysis has been devoted to thinking through the educational process at the level of the school and the classroom. The process of intervention is woven into the selection of content along with its presentation through the writing of the textbooks and this is intended to be a major tool for altering pedagogical practice.

Although adhering to the general principles and objectives of the NPE 1986, and of earlier curricular documents, there are important features that distinguish the NCF 2005 and the shift in practice that it aims for. *Firstly,* while mindful of and sensitive to larger social and national objectives, the immediate focus is the child, the classroom, and the teaching-learning process. The syllabi and textbooks are especially targeted towards initiating a major pedagogical shift. *Secondly*, there is serious reconsideration of the role of social sciences and their content in creating sensitivity and critical thinking with reference to the problems of contemporary Indian society. And *lastly*, and as a corollary of these, there is a shift in conceptualizing the disciplines of the physical sciences and mathematics, not solely as abstract disciplinary knowledge that can serve the cause of a modern industrialized and technologically driven economy but as tools of rational and critical thinking and enquiry, to serve individual curiosity and a quest for knowledge. This aspect of the natural sciences and mathematics had not been ignored by the earlier commissions or by the NPE 1986, but the NCF 2005 sought to translate it into practice by its serious commitment to the exercise of textbook writing which it considers essential and critical to changes in pedagogy and classroom practice.

While the NCF 2005 is fully aware that the school is not an isolated institution and that it exists and functions within larger social and political structures, it nevertheless reposes great faith in the power of the curriculum and the changes that are expected to result from this. It proposes and attempts to initiate a fundamental shift in conceptualizing the learner and the learning-teaching environment with the expectation that this will carry over into the larger social and political world outside the school.

MATHEMATICS AND THE NCF 2005: CURRICULAR VISION AND OBJECTIVES

Prior to the NCF 2005, all policy statements and educational documents ascribe special importance to mathematics as a foundation for courses in science, technology and economics and business. As a result of this the syllabi and content of school textbooks was determined largely by domain experts in the field. A primary concern seems to have been to keep up with the considerable advances in the discipline of mathematics and to prepare students for higher education in science and technology. This history of educational planning and practice, and the status of mathematics within it ensured that with each textbook revision, the content of school textbooks was updated, accelerated and increased.

Although the "loaded" school curriculum and the examination oriented system (where the volume of memorized information represents success and failure) has had an adverse effect on every school subject, the discipline of mathematics has been particularly affected. It needs to be recognized that within the repertoire of school subjects, the subject matter of mathematics is unique. The levels of complexity and abstraction that characterize the formal discipline of mathematics are impossible to master without a reasonable amount of comprehensive engagement. Memorization of formulas and algorithms, although important, leaves children perplexed and very often leads to fear and eventually dislike of the subject. The subject matter of the other core school disciplines such as language or the natural and social environment, all form a part of the child's lived experience to a much greater extent than does the subject matter of formal mathematics. These subjects, especially at the primary school level, have some quantum of "meaning" for the child in her immediate and everyday world. Mathematics and the formal mathematics curriculum, beyond counting, operations in their simplest forms and a sense of shapes and patterns does not relate easily to the child's lived experience. The move to the symbolic and the abstract has to be painstakingly guided and supported through appropriate pedagogies, with regard to which traditional learning techniques have not met with any great success.

Mathematics has been instrumental in inculcating a fear of schooling and of classrooms and continues to be a contributing factor in drop-out rates, especially for children who lack academic support systems outside of school. Mathematics classrooms epitomize all that is wrong with the Indian school system and have been an area of special concern to those working with the educational system. The focus group on mathematics, which was one of the 21 focus groups that supported the formulation of NCF 2005, was well aware of these concerns and took them into serious consideration while making its recommendations.

The vision of mathematics and mathematics education of the NCF 2005 is elaborated in the position paper of the focus group on mathematics education (NCERT, 2006a). "The main goal of mathematics education" it states, "is the mathematization of the child's thinking. Clarity of thought and pursuing assumptions to logical conclusions is central to the mathematical enterprise. There are many ways of thinking and the kind of thinking one learns in mathematics is an ability to handle abstractions, and an approach to problem solving" (NCERT, 2006a, p. 5). Mathematics, according to the NCF 2005 is an important tool for rational thinking and an integral part of a child's education but it is no longer projected as the very foundation of national development. It is seen as one discipline among many and the endeavour within the classroom is to make it interesting and meaningful so that children learn to engage with it and to understand it.

Supporting the right of every child to have access to quality mathematics education, the position paper identifies the current problems of school mathematics as (a) a sense of fear and failure regarding mathematics among a majority of children, (b) a curriculum that disappoints both a talented minority as well as the non-participating majority at the same time,

(c) crude methods of assessment that encourage perception of mathematics as mechanical computation, and (d) lack of teacher preparation and support in the teaching of mathematics (NCERT, 2006a, p. 4). It sets out a vision of school mathematics as a space for learning where

- children learn to enjoy mathematics;
- children learn important mathematics;
- mathematics is a part of children's life experience which they talk about;
- children pose and solve meaningful problems;
- children use abstractions to perceive relationships and structure;
- children understand the basic structure of mathematics; and
- teachers expect to engage every child in class (NCERT, 2006a, pp. 2-3).

It recognizes the need for children to enjoy mathematics and to be meaningfully engaged in the learning process, but at the same time emphasizes the importance and seriousness of the subject matter of the discipline and recommends:

- shifting the focus of mathematical education from achieving 'narrow' goals to 'higher' goals;[4]
- engaging every student with a sense of success, while at the same time offering conceptual challenges to the emerging mathematician;
- changing modes of assessment to examine students' mathematization abilities rather than procedural knowledge; and
- enhancing teachers with a variety of mathematical resources (NCERT, 2006a, p. 8).

The recommendations for the new curriculum are based on two premises: that *all students can learn mathematics* and *all students need to learn mathematics*. The position paper reflects sensitivity to the needs and capacities of a child in the classroom, and also awareness of the conceptual depth that mathematical knowledge and mathematization entails. It recognizes core content areas and the need for a solid foundation, for factual knowledge, procedural fluency and conceptual understanding. It rejects unthinking memorization of formulas and facts and what it characterizes as "trivial" mathematics (NCERT, 2006a, p. 8). It expects mathematics classrooms to "invite participation, engage children and offer a sense of success" (NCERT, 2006a, p. 9). It also emphasizes the use of precise and unambiguous language for mathematical communication and the need for children to appreciate the use of a deliberate, conscious and stylized mathematical discourse and to feel comfortable using it.

MATHEMATICS AND THE NCF 2005: SYLLABUS AND TEXTBOOKS

As indicated earlier, although most of these objectives and concerns were a part of educational discourse and documents prior to the NCF 2005, it did not result in changes at the level of the teaching and learning nor did it improve the educational experience of the child, the teacher or pedagogical practice. An awareness of the problems faced by children in the classrooms was not entirely missing prior to the NCF 2005, nor was the awareness of changing educational practices and pedagogies and the need for students to comprehend and engage with mathematical content. The relationship to experience and real life was also mentioned as was the need for children to be actively involved. For example, the "Curriculum for the Ten Year School: A Framework" 1975, states that

> The teaching of science and mathematics will have to be upgraded and the curriculum continually renewed in order to give our children modern knowledge, develop their curiosity, teach them the scientific method of inquiry and prepare them for competent participation in a changing society and culture, increasingly dependent on a rational outlook leading to better utilization of science and technology (NCERT, 1975, section 2.3).

Despite recommendations calling for pedagogies that value inquiry, investigation and problem solving, mathematics curricula in Indian schools continued to rely on what has been termed an absolutist philosophy of mathematics (Ernest, 1991), one that relies on a static, unquestionable body of accumulated knowledge. Strengthening the national economy on the basis of scientific and technological advances was the overarching concern of educational discourse and deliberations prior to and following the NPE 1986. The school mathematics curriculum, syllabi and textbooks were dominated by the need to keep pace with the advances in mathematics, science and technology worldwide.

The NCF 2005 reversed this focus and re-defined the aims of learning mathematics. Although aware of the importance of the discipline, the focus of teaching and learning shifted to the child. The emphasis of the syllabus documents and the textbooks that followed the NCF 2005, is predominantly on the meaningfulness of the content materials, its connection to children's lived experiences, accessible language of the texts and the pedagogical shift that it is expected to foster. The syllabus outline for mathematics that appeared as a corollary to the NCF 2005, lays down two sets of guidelines, one for designing of textbooks for the primary (Classes I to V) and a second one for the upper primary level (Classes VI to VIII) (NCERT, 2006b). The introductory comments spell out a perspective of what and how mathematics is to be understood and transacted within the classroom. Mathematics is defined as a "certain way of thinking and reasoning .... which should be reflected in the way materials are written and exercises treated" (NCERT, 2006b, p. 68).

Great significance is attached to activities and exercises built around children's real life experiences and to learning outside the classroom. In the initial years the effort is to consolidate mathematical concepts that children already possess and use of language that is comprehensible to them. There are recommendations to make such materials available that would allow children to make generalizations and find solutions to problems on their own initiative; to allow for problems with multiple solutions and to encourage children to think through their problem solving. Definitions are not to be given, but to be arrived at, in the course of generalizations and observation of patterns. Thus, "the course would de-emphasize algorithms and remembering of facts, and would emphasize the ability to follow logical steps, develop and understand arguments as well" (NCERT, 2006b, p. 80).

The following guidelines for the textbook revision are an illustration of the manner in which the framework intends to translate the perspective on mathematics into practice.

- To encourage development of learning outside the classroom.
- Overlapping themes within mathematics and with other subjects are to be encouraged.
- Study materials to be developed with focus on activites and exercises that are built around the child's real-life experiences. Enough space to be given for children's current local interests.
- To allow children to think and reason in creative ways and to reflect upon their reasons for doing things in certain ways.
- Language used in books for classes II to V should be what children use and understand.
- The sequencing of concepts to be spiral rather than linear.
- An effort to include problems with multiple solutions should be made.
- Textbooks to be made visually attractive with illustrations and cartoons if possible.
- Problems to be introduced in a manner that allows children to generalize.

- Definitions should not be given but arrived at collectively.
- Allow children to formulate math problems (NCERT, 2006b, pp. 68-69).

A similar set of guidelines for the upper primary stage reflect the major concerns around language, understanding and engagement.

- Use of language that the child can understand so that the teacher is needed only to provide support and facilitation.
- Textbook materials to emerge from the contexts of children who will be expected to verbalize their understanding, generalizations and formulations of concepts and propose and improve definitions.
- Need to be aware of multiple solutions and alternative algorithms for solving problems.
- An awareness of differences between verification and proof.
- Mathematics should emerge as a subject of exploration and creation rather than finding old answers to old problems.
- Textbooks should be attractive, teachers must formulate and create contextual problems and anticipation for later topics should be set up (NCERT, 2006b, p. 81).

IMPLICATIONS OF THE NEW CURRICULUM AND ITS TRANSLATION INTO PRACTICE – A NOTE OF CAUTION

The NCF 2005 came as a timely response to the general disillusionment with the Indian school system and the critiques that it had generated. Although there has been no dearth of national and local level initiatives, both at governmental as well as non-governmental levels, the educational system is a major area of discontent at a variety of levels. Disparities and inequities continue to flourish and the quality of learning leaves much to be desired, even as children reel under the tyranny of qualifying examinations and assessments that determine selection into higher learning, better employment opportunities and ultimately entry into high prestige occupations.

The classroom however, is not a decontextualized space, but functions within larger social, political and economic structures. The education system remains highly inequitable and wide disparities in terms of class, caste, gender and community amongst others, exist at every level. The challenges to creating a truly equitable system however are enormous and although the new curricular framework addresses them, it can hardly be expected to resolve them. Although it is too early for an extensive evaluation, some preliminary studies with teachers and in classrooms are raising questions about the effectiveness of the translation of syllabi into classroom practice (Banerjee, this volume; Minocha, 2013). In this context a note of caution, especially with reference to the teaching of mathematics, seems necessary, and therefore the following observations are made.

First, although the new textbooks have been altered not only in content but in ways that should radically alter pedagogy, they rest upon assumptions about teaching, learning, education and knowledge that is neither practiced nor encouraged within the contemporary structure of schooling. Unless teachers in the classrooms are guided through these ways of thinking about subject matter and re-conceptualizing the learner and the content of learning as well as the processes of learning and pedagogy, the task of translating the texts into meaningful learning routines is going to be at best difficult and at worst, chaotic.

Second, the NCF 2005 is firmly based on a theory of social construction of knowledge and makes its critiques of the curricular content on this basis. In the writing of the textbooks there seems to be a conflation between the social construction, and individual constructivist views of learning and knowing. That knowledge systems are socially constructed is a widely accepted assumption in mathematics, and this sets the parameters for entry into the system. That children as individuals need to actively understand these systems and to internalize the knowledge in meaningful ways is also largely accepted by psychologists. However, one has to make a distinction between the process of understanding and internalizing knowledge and the process of creating it. Children's spontaneous knowledge therefore may not always correspond to social knowledge systems and although "rational" it may not always be "mathematical". There is therefore a need to socialize children into a formal system of knowledge in the mathematical classroom, and into a knowledge system that is validated by a social community of mathematics experts for them to become active participants in the creation of knowledge. Confounding the two can only become a source of confusion for both teachers and students.

Third, concomitant to the last point, is the need to recognize the power of hierarchical systems of knowledge even as one critiques them, especially in an unequal system of social structures of the kind that exist in India. Nor can one deny the resistance that privileged groups are likely to exert to maintain their privilege, and the status accruing to it. The phenomenal growth and influence of information technology, coupled with a globalized economy and growth and spread of big business has only led to heightened anxieties about keeping up with scientific and technological knowledge and added to the lure of disciplines associated with economics, commerce and finance. Mathematics as a school discipline is deeply implicated within these relationships and how and what is to be learned is often determined by them. If school systems are to ensure equal and democratic access of "powerful mathematical ideas" and "important" mathematics to every child then this access has to be made available in the classroom and within the school time-table. At the present moment, given the state of our schools and classrooms, the resources available in them along with the quality of teaching, this hardly seems possible.

CONCLUSION

The NCF 2005 is a welcome and much needed intervention in the educational process in Indian schools. It has addressed both the issue of knowledge and content and aims at radically changing the culture of learning and pedagogy in the classrooms. The position papers of the several focus groups are an extremely useful compendium of documentation that encompass both critiques and constructive suggestions regarding the diverse areas that were covered in their preparation. The curriculum framework itself and the syllabus revision and textbooks that have followed in its wake, are beginning to challenge accepted hierarchies of knowledge as well as the disconnectedness of school and classroom learning from lived experience. All this has been achieved with serious sensitivity to the burden of learning that the present classroom imposes on the child and also to the many social and economic inequalities which form the contexts of the children's lived realities.

Learning important mathematics requires resources which a majority of our schools at the present moment, are not able to provide, especially to children that need them most. Not the least of these are adequately qualified and trained teachers and well equipped classrooms and schools. Given the importance and status of mathematics and the power that it wields as a mark of accreditation, the mathematics curriculum and its transaction need to be carefully considered. The challenges to creating a truly equitable educational system are enormous and inherently implicated within larger societal, political and economic structures. A school curriculum can begin to address these issues, how effectively it can challenge structures and bring about change is a complex and continuing debate.

NOTES

1. Amongst government initiatives there was Operation Blackboard, 1987; the National Literacy Mission, 1988; the District Primary Education Programme (DPEP), that came in the wake of foreign investment in the Edu-

cation sector and shifted the parameters of debates on education for a variety of reasons; and the much later Sarva Shiksha Abhiyan; all of which have raised important issues about provision, access and quality. Among non-governmental initiatives, the Hoshangabad Science Teaching Programme, in Madhya Pradesh, 1972; The Kerala Sastra Sahitya Parishad, Kerala, 1987; the Lok Jumbish in Rajasthan, 1992 (among others) had a major impact in terms of action and of imagination.

2. See Sahmat & Sabrang.Com (eds.), (2001) *Against Communalization of Education*, New Delhi, for a range of critiques of the framework and also the journal *Seminar*, New Delhi, on Redesigning Curricula (no. 493), September 2000.
3. These terms are used by Maxine Greene (1971) and correspond roughly with what has been a traditional view of the curriculum vs. one that considers the curriculum as socially constructed that had a powerful influence on the sociology of education in the 1970s.
4. This is with reference to 'narrow aim' vs. the 'higher aim' for school education proposed by George Polya and is elaborated in Section I of the position paper of the focus group on mathematics education, NCF 2005 (NCERT, 2006a).

REFERENCES

Ernest, P. (1991). *The Philosophy of Mathematics Education*. London: Falmer Press.

Government of India. (1949). *Report of the University Education Commission* (1948-1949) New Delhi: Ministry of Education.

Government of India. (1956). *Report of the Secondary Education Commission.* New Delhi: Ministry of Education.

Government of India. (1966). *Report of the Education Commission (1964-66): Education and National Development.* New Delhi: Ministry of Education.

Government of India. (1968). *National Policy on Education 1968.* New Delhi: Ministry of Education.

Government of India. (1986/1992). *National Policy on Education 1986 (with modifications undertaken in 1992)* New Delhi: Department of Education, Ministry of Human Resource.

Government of India. (1993). *Learning without Burden (Yash Pal Committee Report).* New Delhi. Available at http://www.teindia.nic.in/Files/Reports/CCR/Yash%20Pal_committe_report_lwb.pdf

Greene, M. (1971). Curriculum and Consciousness, *Teachers College Record. 73*, (2), 253-269.

Minocha, D. (2013) *Understanding teachers' perceptions of mathematics NCERT textbooks of grades 3, 4 &5.* Unpublished field attachment report. Mumbai: Tata Institute of Social Sciences.

National Council of Educational Research and Training. (1975). *The Curriculum for the Ten-Year School: A Framework.* New Delhi: NCERT.

National Council of Educational Research and Training. (1988). *National Curriculum for Elementary and Secondary Education: A Framework.* New Delhi: NCERT.

National Council of Educational Research and Training. (2005). *National Curricular Framework*. New Delhi: NCERT.
National Council of Educational Research and Training. (2006a). *Position Paper, National focus group on teaching of Mathematics*. New Delhi: NCERT.
National Council of Educational Research and Training. (2006b). *Syllabus, Volume 1 (Elementary Level)*. New Delhi: NCERT.

CHAPTER 70

THE SCHOOL MATHEMATICS PROJECT

An Account of a Mathematics Curriculum Development Project

Amitabha Mukherjee
University of Delhi

Vijaya S. Varma
Ambedkar University, New Delhi

ABSTRACT

The School Mathematics Project was started in 1992 at the University of Delhi with the aim to develop a curriculum to address the fear of mathematics in school children. Its core group included school teachers, university and college teachers and researchers. The project experience of over eight years demonstrated that it was indeed possible in the Indian context to create and implement a curriculum for primary school mathematics that leads to freedom from fear, and for teachers to not only teach in a different way, but to create new materials for their own classroom use and to reflect on and critique their own practice. The work of the project had an impact on the

The First Sourcebook on Asian Research in Mathematics Education: China, Korea, Singapore, Japan, Malaysia, and India, pp. 1539–1558

revision of the curriculum and learning materials first in the state of Delhi and later at the National level.

Keywords: Primary mathematics, curriculum development, mathematics curriculum in India, school mathematics project

The School Mathematics Project (SMP) is an important link in a chain of programs of curricular reform and material development in elementary schools in India in recent times, starting from the All India Science Teachers' Association effort to develop a middle school physics teaching program in elite public schools in north India under the strong influence of the Nuffield program in the mid 1960s. This was followed by the Hoshangabad Science Teaching Program (HSTP) in the middle 1970s, the Social Science program and the Prashika programs of Eklavya in the 1980s, and then SMP in the mid 1990s. Although developed in different locations and different circumstances, these programs were all characterised by a common approach which stands in stark contrast to the paradigm of curriculum development prevailing in India during this period. The prevalent paradigm can be characterised by the feature that all decisions emanate from some central controlling authority, supplemented by experts drawn from select institutions of higher learning with little or no experience of actual school teaching and with a small representation of school teachers from some specially selected schools. There is no trialling of teaching materials in schools before actual implementation, and therefore no possibility of incorporating feedback from classroom experiences in the final set of materials.

In contrast, all the five programs listed above can be characterised by local decentralised groups empowered to take decisions, with representations of university academics together with practising school teachers, in which curricula are jointly developed, with extensive trialling with school teachers and their students and feedback from actual classroom experiences playing an integral and critical role in the finalisation of the curriculum and the teaching materials. This was particularly true in SMP, in which school teachers from many schools played a seminal role in the formulation of policy and the development of the curriculum, the pedagogy and the teaching materials.

An important aspect worth emphasising at this stage is the serious and sustained involvement in all the programs mentioned of University academics in school education with a sustained engagement over a period of time. HSTP was able to persuade the University Grants Commission (the national institution charged with the responsibility of overseeing the functioning of all universities in the country) to allow the University of Delhi to second its faculty to the program on a long-term basis. Some half a dozen university teachers went from Delhi and spent six months at a time in the field becoming familiar with the conditions prevailing in the program schools in the villages of Hoshangabad district of the central Indian state of Madhya Pradesh. The same philosophy underlay the setting up of the Centre for Science Education and Communication (CSEC) as a constituent centre of the University of Delhi in the late 1980s, with a charter to work for the improvement of science education at both school and university levels.

Another reason for documenting SMP is that, as described later, three subsequent developments – the primary mathematics course of the Indira Gandhi National Open University, the Delhi state textbooks and the curriculum and textbooks of the National Council of Educational Research and Training (NCERT) at the elementary level – drew substantially from the SMP experience, as well as on the human resources created by the project.

ANTECEDENTS

SMP started in 1992, initially as a discussion group based at the CSEC University of Delhi. However, its roots went back almost two decades, to the HSTP a program of intervention in science teaching in the middle grades in government schools in the state of Madhya Pradesh in India. Here we can give only a brief description of HSTP; for details the reader is referred to Mukherjee, Sadgopal, Srivastava and Varma (1999) and Eklavya (2013). A more detailed discussion of the program is available (Joshi, 2014).

HSTP began in 1972 as a pilot project in 16 schools in the Hoshangabad district of Madhya Pradesh. The project was initiated by two non-governmental organizations (NGOs) located in the district, and received academic support from a core group drawn largely (though not exclusively) from the University of Delhi. It evolved an approach to the teaching of science in middle school that was inquiry-based and participatory. A complete curricular package was developed, including a workbook for each class, a kit of experimental equipment to be made available to schools so that all children working in groups of four could perform all the experiments themselves, and assessment based on open book examinations and experimentation. All participating teachers received intensive training for three weeks during the summer for three summers, supplemented by school-based support by members of the core group. In the year 1978, in a major development, the Madhya Pradesh state government adopted HSTP as a state program in all the middle schools (then over 200 in number) of Hoshangabad district. This led to the creation of new structures for school-based support within the government department of education, as well as to the formation of an extended academic support group.[1]

Following the adoption of HSTP by the state government, the need was felt for an independent institution which would work to consolidate HSTP and work for innovation in school education on a wider canvas. Eklavya (named after a character in the epic *Mahabharata*) was set up in 1982, after which the academic coordination of HSTP was entrusted to it by the Madhya Pradesh government.

In the second half of the 1980s, Eklavya launched two more programs for curriculum development and intervention: the Social Science Programme, and Prashika — a program for primary education. The latter, which had a direct influence on SMP, was initially conceptualised as an integrated program of study for Classes I to V (Agnihotri, Khanna & Shukla, 1994). A single textbook, called *Khushi-Khushi*, was created for each class (Prashika Group, 1988-1992). Although Prashika, as a program, remained at the pilot stage, it had an influence on many groups working at the primary level, including the SMP group.

PHASE I: FORMATION OF THE CORE GROUP AND INITIAL ACTIVITIES

The experiences of HSTP as well as Prashika had indicated that an intensive effort for reforms in school curricula was required, with a specific focus on mathematics. SMP was conceptualised as a program of intervention in mathematics in elementary schools to try and understand why so many students were failing mathematics in school. The program was located in the CSEC, an independent centre of the University of Delhi, whose charter included serious engagement with school science and mathematics education.

The project received financial support, one year at a time,[2] from the Ministry of Human Resource Development, Government of India. Initial exploratory meetings led to the formation of a core group that included school teachers, university and college teachers, researchers and others interested in the teaching of mathematics in schools. The stated aim of the group was to try and develop a program that would address the fear of mathematics in school children, and in that way improve their performance in mathematics and reduce the failure rate in examinations. From the outset, the group was interested in working at the primary level. It was felt that it would not be difficult to exploit the freedom in the early years of the school curriculum to develop our own materials and teaching methods. It may be added here that in India the public examinations cast a very long shadow leading to a rigid school system. The higher the school classes, the greater the resistance to any innovation in curriculum and pedagogy.

The first phase of activities consisted of a series of discussion meetings, a national conference and two studies. The discussion meetings, held roughly every fortnight, were attended by members of the core group and occasionally by members of Eklavya and others who were interested in issues related to school mathematics. The discussions covered a wide range, from research findings on children's learning of mathematics to the classroom situation in government schools in Delhi. The focus was on evolving a common approach which could form the basis for curriculum development.

During the summer of 1993, the group decided to carry out a study on teachers' attitudes towards children and mathematics. A questionnaire was designed for the purpose, piloted and finally administered by core group members to 31 teachers in municipal schools in Delhi. A description of the survey items and a brief summary of the findings are given in Appendix A. The survey pinpointed some difficult areas in primary mathematics, such as the concept of place value and the algorithms, especially those for multiplication and division. It also revealed that most teachers did not think textbooks or teaching methods were responsible for the difficulties. The findings of the survey played an important role in evolving the group's strategy.

The group also planned a street survey on the extent ordinary people use mathematics in daily life, to see whether or not this could be used as a basis for designing the school curriculum in mathematics. Although carried out by college students without any special training, the study threw up some interesting findings. Details are presented in Appendix B. This survey can be considered as complementary to the work of Khan (2004), who carried out a survey of the mathematics usage among working-class children in Delhi, in that it did not attempt any in-depth study but instead focused on the usage of mathematics in the daily lives of a much broader category of people. It was quite clear from the survey, however, that the level of usage of mathematics among common people engaged in their livelihood tasks was too elementary to form the basis of a school curriculum in mathematics.

In October 1993, CSEC organised a conference on 'Primary School Mathematics: The Basis for Curricular Choices', which was attended by a number of participants both from academic institutions and NGOs active in educational reform, from all over the country. The background paper to the conference said "The development of a curriculum based on principles that are tied to the structure of the discipline, to the modalities of classroom teaching, and to a model of how children learn mathematics is a matter of some priority." The conference was an attempt to address a very real problem the group was then facing – what basis could one use in deciding what should or should not be included in a curriculum for school mathematics, particularly in the early years? Although the conference did not yield straightforward answers, it succeeded in establishing contact between members of the group and others working in the field of school mathematics. There were presentations by members of the SMP group, especially some school teachers, including one on the teachers' survey.

In early 1994, a baseline survey was designed to ascertain the mathematical readiness of the students who had just been admitted to Class I. This was based on a test initially designed at the Department of Education, University of Delhi, and adapted by the group in the light of its understanding. The survey was in the form of 11 test items, mostly in the form of oral questions, some with concrete materials and one or two involving pencil and paper. In April-May 1994, during the first weeks of the new school session, the test was administered to 108 children in five Government schools. The test revealed that whereas 83% of children could add small numbers in the concrete mode, only 28% and 21%, respectively, could add the same numbers in the contextual and abstract modes. This may be compared to the findings of Hughes (1986). More surprisingly perhaps, while many children could rattle off number names in sequence, sometimes up to 100 or more, 59% could not count correctly as few as seven small objects placed before them. Moreover, only 5% of the children could consistently conserve number when given a classic Piagetian task. The results were an eye-opener for those administering the test as well as the teachers who were to teach the Class I students.

These initial interactions and studies convinced the core group that any meaningful intervention in school mathematics had not only to begin from the primary classes, but indeed from Class I. Starting later would require undoing the damage that would have otherwise been done to students in the lower classes.

PHASE II: INTERACTION WITH A LARGER GROUP AND ARTICULATION OF THE APPROACH

In the second phase, there was interaction with a larger group in Delhi. To identify the intervention schools and initiate a dialogue with teachers, a series of meetings called *Ganit par Gupshup* (Chatting about Mathematics) was started. These went on over a period of several months. The meetings were given some structure by posing questions to which teachers responded on the basis of their experiences, and also by reading out excerpts from books and writings on mathematics and inviting discussion. The aim of identifying the intervention schools was not fully realised. Nevertheless, the Ganit par Gupshup teachers appear to have found the sessions useful. This emerged from a number of chance meetings with individual teachers that different core group members had over the next few years.

In the winter and spring of 1994-1995, three material creation workshops were organised. The aim was to create material that was consistent with the group's approach and could be used in Class I in the intervention schools. The topics chosen were Number, Addition and Subtraction, and Shapes, Size and Patterns. Before each workshop, one of the core group members prepared and circulated a background note. The workshops were attended by around 50 people from schools, academic institutions and NGOs. The output of the workshops was subsequently edited and compiled into a volume entitled *Games and Activities for Class I* (School Mathematics Project, 1996).

During this period, and in parallel with the above activities, the group was in the process of articulating its approach. In doing so, the group was influenced by the work of Piaget and Vygotsky, and contemporary researchers such as Kamii (1985), Hughes (1986), Liebeck (1990) and Holt (1982). In March 1995 a formal document entitled *The SMP Approach* was ready, and was subsequently included as an appendix in the book of games and activities (School Mathematics Project, 1996). The key points of the document are given below, with some explanatory comments incorporated.

On Children

- Children are not blank slates when they enter school. They come equipped with a certain awareness of number and operations ("initial mathematics") that is independent of formal instruction. Disregard of this leads to the growth of fear of mathematics.
- The classroom process should not be viewed as a one-way transfer of "knowledge" from the teacher to the taught. The emphasis should be on elucidation rather than on instruction.
- Children are individuals with their own pace and often their own strategies of learning. The curriculum should provide room for them to remain different. One method, one activity, one technique cannot provide for all children. There is a natural pace at which each child picks up new concepts and skills in mathematics. Riding roughshod over this in an attempt to maintain a pace of learning

dictated solely by an externally imposed pre-determined curriculum leads to the development of fear of mathematics.

On Mathematics

- Mathematics is more than number, operations and algorithms. It encompasses shape and space, patterns, structures, data handling and measurement.
- Mathematics is inherently beautiful and a potential source of joy – but only if the teacher feels this herself can she communicate it to children.
- Facility comes naturally when there is a meaningful context for mathematics.

On Teachers and Teaching

- If the teacher is not convinced of the need for change, no curricular change will work. Teacher training is not just a matter of training teachers in new concepts and techniques, but of changing their attitude to mathematics and to teaching, especially in their relationship with children. This cannot be done by imposition and may be possible only through involvement and association.
- Ultimately the teacher has to transact the curriculum in the classroom. It is neither possible nor desirable to spell out exactly how everything should be done. It is nevertheless necessary to provide the teacher all possible support.
- Symbolic notation is a powerful tool for computation as well as a means of recording the results of computation. However, familiarity with the symbol for something does not imply facility with what the symbol stands for.

One implication, drawn for Class I but also applicable beyond it, was that the progression in learning any new concept should be from the concrete to the oral-contextual mode to the symbolic and abstract. In particular, the teaching of algorithms should be deferred as long as possible. This was also advocated by Kamii and Domenick (1998) in their influential paper published three years after the SMP Approach document.

In addition to those listed above, there was agreement in the group on a number of other points relating to curriculum and pedagogy, some explicitly stated in the background notes mentioned above and in other documents, others implicit in the group's work.

- Children are basically creative and eager to learn. They learn when they are actively engaged in what they are doing. The curriculum should have activities that engage children and it is important that these have a sense of play and of challenge.
- Children should be encouraged to talk in the mathematics classroom. They should explain aloud in class the way in which they solve simple problems, and to listen carefully to such articulation by other children. This will help bring out individual methods and also develop an awareness of other ways of solving problems. Moreover, they should be encouraged to work in groups and to learn from each other.
- Making up problems is a kind of learning task that can be usefully employed in the classroom.
- Children should be encouraged to work on projects in groups as this leads to better learning, greater enjoyment and increased self-confidence. The challenge is to think of projects that are simple enough for children to undertake, yet interesting enough to keep them engrossed.
- Discussion of errors and alternative strategies in problem solving, practising estimation and developing a feeling for the magnitude of the correct answer in relation to numerical work should all form an integral part of the primary mathematics curriculum.
- Learning does not take place abruptly. It is a slow process involving many loops and feedback cycles. Thus the progression should not be linear, but should involve spiralling, so ideas and concepts are not merely introduced once but are revisited again and again. Given the hierarchical nature of the discipline, unless we ensure that the foundations are well laid, future progress becomes impossible for a child.
- The curriculum must allow for horizontal elaboration. Children must have opportunities to use emerging concepts in a variety of diverse situations. The school environment should provide a rich experience base for the child to make this possible.
- We must try and promote understanding and not merely accuracy in computation. Nothing is so mind numbing and deadening as mere repetition of exercise after exercise in an attempt to build accuracy, competence and speed.

- We must use games, puzzles and riddles to attract and maintain the interest of children in mathematics and to excite their imagination. These also provide non-didactic feedback to the child.

The most important implication of all is that such a curriculum should only be developed in conjunction with and in continuous interaction with teachers who are actually engaged in teaching primary school mathematics, with feedback from the classroom constituting an integral part of such development.

PHASE III: INTERVENTION

The third phase, starting April 1995, consisted of intervention in five schools, representing a mix of school types. Two of these were regular government schools (Sarvodaya Schools), while one was a school managed by the University of Delhi and two were private schools managed by the armed forces.[3]

Teachers who were to teach Class I in the five schools carried out a baseline survey on students newly admitted to Class I. Subsequently, they went through an intensive orientation program before embarking on teaching according to the new curriculum. A draft version of *Games and Activities for Class I* served as a manual for the teachers (School Mathematics Project, 1996). Over the rest of the academic year the program unfolded in the target schools. Classroom teaching was supplemented by regular feedback meetings, which were held by turn in the five schools, and by follow-up school visits by core group members. In two of the schools, regular classroom observation was carried out by a research associate, who maintained a detailed diary.

In March 1996, when the program was about to complete the first year of work in the classroom, financial support was abruptly withdrawn by the Ministry of Human Resource Development.[4] The project continued on the ground, though with some difficulty. While a lot of the material in the Class I book could be used directly for Class II, teachers created other materials as and when needed. Although feedback meetings were no longer regular, they were still frequent enough at this point to think of the material as being evolved collectively. At some point during this year, one of the Sarvodaya Schools dropped out of the program.

PHASE IV: STRUGGLES ON THE GROUND

During the years 1997-1999, the intervention moved to Classes III and IV. Meetings were held largely on a felt-need basis. However, while materials used in the classroom were pooled together for Class III, this was never done for Classes IV and V. Thus the classroom transaction was based completely on individual teachers' interpretation of the group's approach. During this time, the second Sarvodaya School also dropped out of the program, while the participation of the Delhi University school became marginal.

In January 1999, CSEC organised and hosted a Seminar on "Aspects of Teaching and Learning of Mathematics in the Primary School." This was attended among others by participants from the Homi Bhabha Centre for Science Education (Mumbai), Vidya Bhawan Society (Udaipur), Tamil Nadu Science Forum (Chennai), Digantar (Jaipur), Eklavya (Bhopal) as well as various academic institutions and NGOs in Delhi. A number of papers were presented, based on the wide experience of the participants. The SMP group also presented its concerns and ongoing work, and it is worth noting that these presentations were not only made by members of the original core group, but also by some of the teachers in program schools, reflecting their growing confidence.

In 1999-2000 the program got some fresh financial support, as it came under the ambit of CSEC's project "Elementary Education Teachers' Research Network" – a collaborative program between Homerton College of the University of Cambridge and CSEC. The project envisaged setting up a network of practising teachers across the country. It was envisaged as a network of networks with CSEC acting as a central node which was connected to nine other nodes which were themselves networks of school teachers spread across five states of the country. The intention was to facilitate the participating teachers in reflecting on and carrying out research related to their own teaching practices, and enabling them to share their findings electronically over the Internet. This led to SMP teachers exchanging and sharing their experiences with teachers working in very different settings across the country. In particular, the interactions with teachers working in non-formal education centres run by Ankur were both interesting and fruitful. In some respects, however, the program was ahead of its time, because the ICT infrastructure in India, particularly connectivity, had not by then developed sufficiently so as not to become a bottleneck to the flow of information between the participating nodes of the network.

In April 2000, the children who had been in the first intervention batch of SMP entered middle school (Class VI). An evaluation study of the five years of the intervention was carried out later that year. This consisted of interviews with teachers and children as well as a written test administered to children in two program schools and to a comparable group of children in a non-program school. Some of the key findings include an increase in the self-confidence of the teachers and a greater ability on their part to articulate their understanding of how mathematics should be taught in the lower classes. The interviews with children showed that they were not afraid of mathematics. However, the written test, carried out after more than six months of conventional instruction in Class VI, did not show a large difference between the SMP children and the control group children. Perhaps there is a lesson in this.

According to some teachers, SMP has never been closed down, and continues to live in their classroom practice. However, formally, no specific activities were undertaken after the evaluation study, except that the Class II materials were edited and brought out in the form of a book for teachers and a workbook in 2002 (School Mathematics Project, 2002a; 2002b).

LESSONS LEARNT

The SMP experience of over eight years taught participants many things which are not easy to summarise. Nevertheless, one can abstract some lessons learned from it which could be useful for any future program of intervention in school mathematics.

The first lesson is that it is indeed possible to create and implement a curriculum for primary school mathematics that leads to freedom from fear. However, to keep it going needs constant work, and teachers need all the support they can get. In this, the role of the core group is crucial.

The second lesson is that it is possible for teachers to not only teach in a different way, but to create new materials for their own classroom use and to reflect on and critique their own practice (See Appendix C for an example). This assumes, of course, appropriate orientation and continuing support. It is worth noting that SMP went beyond HSTP and Prashika in one respect: it was truly driven by the teachers. Except for the first year, classroom activities were not only carried out by the teachers, but conceptualised by them as well. Indeed, it would be fair to say that the interaction between the core group and the program teachers was one of equals, each learning from the other.

The third lesson which emerged is that we need to work with a cross-section of mainstream schools. However, we have to acknowledge that the Government schools are hard to retain, and need special attention. This, of course, is not something specific to interventions in mathematics.

In the group's early interactions with teachers, it was often reported that, at entry to Class I, many children actually loved mathematics. Their fear of the subject developed only as they progressed to higher classes, probably arising from not enough time being spent on individual concepts and possibly from poor teaching approaches. This often led to frustration and a freezing of their desire to engage with mathematics later.

The hierarchical nature of the discipline meant that incomprehensibility was likely to grow in subsequent years if the foundations were not carefully established.

Too much emphasis on arithmetic in the early years at the expense of shapes, patterns, symmetries, etc. tends to turn many children away from the subject. Teachers, parents and curriculum designers also need to realise that it is highly unrealistic to expect that concepts, which had historically taken hundreds of years to develop across cultures and civilizations, can be acquired by young children in a matter of a few years. Such an expectation is likely to do great damage to the psyche of a young child. A corollary to this is that it is unrealistic to expect that children will grasp a concept at first exposure. The curricular design must be such that it acknowledges this and revisits a given concept repeatedly over a period of time.

Major sources of difficulty in the early years are place value and the four arithmetic operations which are called by the same name and yet are apparently associated with different rules when performed on whole numbers, fractions or decimal numbers – something which most young children find completely baffling.

EPILOGUE

As an epilogue, we may note that SMP appears to have had an influence that goes beyond its small scale, its brief intervention time span and its difficulties on the ground. Some SMP ideas have been absorbed into Indian mainstream curriculum design.

In 1999-2000, the Indira Gandhi National Open University decided to create a course on Learning Mathematics as part of its Certificate Programme in the Teaching of Primary Mathematics. Several members of the SMP core group were invited to be members of the course design team, and to develop materials for the course. Some of the other course writers had also had contact with SMP. It is not surprising then that the influence of the SMP approach is visible in the materials (Indira Gandhi National Open University, 2000; 2001).

In 2003, the Delhi State Council of Educational Research and Training undertook the task of writing textbooks for children in Government schools. The resulting effort was a significant event in the recent history of Indian education, and has been documented by Agnihotri, Khanna and Rajan (2008). The Director of the Council, who was aware of SMP, approached CSEC for help in creating mathematics textbooks. Although the time frame was totally unrealistic, CSEC coordinated the creation of not only the primary mathematics books, but the primary and middle science books as well. The resulting textbooks (State Council of Educational Research and Training Delhi, 2004; 2005), were radically different from books used earlier. The writing team included several people who had been in contact with SMP, and they made an effort to incorporate spiralling and looping in the structure of the books. Moreover, algorithms were deferred, and adequate space was given to shapes, spatial relationships, patterns and data handling.

When NCERT embarked on the process of creating the National Curriculum Framework 2005, many people who had been involved with the Delhi textbooks in all subjects were involved in various capacities. Thus the experiences of the Delhi effort also contributed in part to the NCERT framework. The Position Paper of the National Focus Group on the Teaching of Mathematics (National Council of Educational Research and Training [NCERT], 2006a) shows the influence of SMP in its approach to the curriculum at the primary level (see Khan, this volume). The SMP emphasis that non-number areas of mathematics – Shapes and Space, Patterns, Measurement and Data Handling – should get their due place in the primary classes finds an echo in the position paper. It is also reflected in the syllabus for Primary Mathematics (NCERT, 2006b). Moreover, the present authors had advocated in the 1999 Seminar that operations on fractions should be postponed to higher classes when children are older, more mature and better able to handle the new concepts involved. The new syllabus has done precisely that. The current NCERT textbooks *Math-Magic* (NCERT, 2006c; 2007; 2008), although informed by other experiences too, show the clear imprint of SMP. Incidentally, the team of authors of these books includes one of the first program teachers of SMP.

ACKNOWLEDGMENTS

The authors would like to thank all core group members, program teachers and volunteers of SMP for the long association on which this article is based. In particular, we have drawn on unpublished papers and reports authored by Sandhya Kumar, Harsh Kumari, Jasbir Malik, Nargis Panchapakesan, Dharam Prakash and Padma Sarangapani. We wish to acknowledge the contributions of others who attended our meetings, seminars and workshops, and who enriched our understanding with their experiences. We would also like to thank Pramod Srivastava, who was Director CSEC during the period covered in this account, for constant support and intellectual engagement of the highest quality, not only with SMP but also with other programs.

NOTES

1. In fact, when the expansion of HSTP to the district and the state level was under intense internal debate, the plan proposed by the University group was that such expansion should be done, not with one single group acting as the co-ordinating agency, but by seeding colleges in each district to take responsibility for such expansion and being academically responsible for all schools in their catchment areas. These would share a common philosophy and approach to the school curriculum but would develop their own teaching materials in tune with the conditions obtaining in their areas.
2. The nascent SMP group, in its naïveté, asked for funding only for a year-long project to explore the feasibility of undertaking a program of intervention in mathematics. By the time the group had convinced itself that indeed such a program was feasible, the official in the Ministry of Human Resource Development who had supported the idea had been transferred and the Ministry was no longer interested in funding the program. The lesson to be learnt from this experience is that policies in government can change not only when governments change or when ministers of education change, but also when relatively minor officials in the ministry change. So if funding is available, it is much better to accept the challenge and try for the change one wants rather than ask for short-term funding at the start for fear of being too ambitious.
3. The original plan was to work in seven schools, including two municipal schools which had been identified through the NGO Ankur, but eventually they did not join the program.
4. This step did not appear to be based on any assessment of the work carried out under the project, and the reasons for the withdrawal of funds were never communicated to the group. At an emergency meeting, the program teachers expressed their resolve to carry on. The program thus continued with skeletal support from the core funds of CSEC. However, the absence of funds did have an impact on the program. Photocopying of materials had to be restricted. The feedback meetings became irregular, as did school visits. Classroom observation had to be abandoned altogether, since the salary of the researcher could not be paid.

APPENDIX A: THE TEACHERS' SURVEY

A questionnaire was developed to explore the attitudes of teachers to mathematics teaching and learning. Based on a preliminary trial, the final version of the questionnaire consisted of twenty-three items. It was administered to 31 teachers, teaching Class III in primary schools run by the Municipal Corporation of Delhi. The data were collected through personal interviews, each interview lasting about an hour and a half.

The first part of the survey consisted of questions related to children taught by the respondents, and their problems in learning mathematics.

Contrary to the group's preconceived notions, teachers did not rate their own students as inferior. The majority (24/31) felt that the socio-economic background of students is not a determining factor in their ability to learn mathematics. However, when asked why children found mathematics difficult, a majority chose reasons related to children themselves (e.g. inability to concentrate) or to lack of parental support. Only a minority thought curricular factors were responsible for mathematics being seen as difficult (12/31) or boring (6/31). About half the respondents felt that children brought some mathematical notions with them when they entered school.

The second part of the survey was designed to elicit the views of teachers on the practices used by them in mathematics classes. A majority (20/31) said that they followed the sequence of topics as given in the textbook. On the issue of rote-learning, about half of the teachers felt that students need to memorise multiplication tables, formulae, etc., while all but one said that children had already memorised multiplication tables in lower classes. Why do some children still have difficulties with multiplication in spite of having memorised tables? The answers to this were confusing. A minority (6/31) felt that children needed to understand multiplication and apply it in different contexts. The final question in this section sought to elicit information on the kinds of activity which provided an opportunity for children to think, understand, explore and experience mathematical notions by themselves. Some of the choices offered were: activities which engaged the interest of students, consulting each other and group discussions. The teacher's responses, however, did not show any clear pattern. Some typical responses were, "Small children don't have the understanding to discuss things with each other," "I divide children into groups and ask them to check whose answer is right," "Maths is not a subject where you can have group work. The intelligent children will tell the others," "There are no such activities in maths ... may be in class I, counting with marbles." The responses suggest that the respondent teachers did not recognise children's ability to reconstruct their experiences or make mathematical sense of them.

The third part of the survey was concerned with teachers' opinions about some of the content that was then being taught in Class III. Questions pertained to place value, algorithms for various operations on whole numbers, properties of operations and word problems. The responses to this part indicated that the concept of place value and the algorithms, especially those for multiplication and division, are problem areas for children and need to be looked at in greater detail.

Overall, the survey was an attempt to investigate the perceptions of teachers about children, classroom practices and mathematics content in the existing curriculum. In spite of the small sample size, the findings were quite revealing, and constituted an important input for the SMP group in devising its strategy for developing an alternate curriculum for primary mathematics.

APPENDIX B: THE STREET MATHEMATICS SURVEY

The survey was carried out by two young previously untrained field workers who interviewed about fifty persons like bus drivers, bus conductors, taxi drivers, small shop keepers, vegetable sellers, street vendors, peons and even some minor functionaries in the university. No tests were administered but records were kept of the identity of the person being interviewed and the results of the interviews. An analysis of the responses collected showed that, whereas most people knew how to add and frequently used addition with whole numbers although usually not beyond 100, very few carried out subtraction by the method they were taught in school (if they had gone to school). When required, they did subtraction by adding on. Thus a fruit seller when offered say Rs 100 in payment for goods worth Rs 37 never actually subtracted 37 from 100 to calculate the change. Instead she would count on from 37 first with one-rupee coins: 38, 39, 40, then continue with a ten-rupee note: 50, and then give a fifty-rupee note and say 100 to signal the end of the transaction.

There were a few occasions on which multiplication was used but people seldom remembered all the tables they had learnt. Usually they got by by remembering their tables till 5 and of course the 10 times table. Thus if the price of 7 kilos of something were to be calculated, the price per kilo would be multiplied first by 5 and then by 2 and the two results added together.

Nobody actually used division. Intrigued how fruit sellers could figure out what price to charge for their goods while ensuring they did not make a loss, it was fascinating to find out that they did it by a process of multiplication – an extension of the technique they used for subtraction. Suppose they had bought 40 kilos of fruit for Rs 650 and had to pay an additional cartage of Rs 40 to transport the goods from the wholesale market to the point of sale. They would seldom add up their costs (Rs 690) and divide it by 40 to fix the sale price. Instead they would use a process of hit and trial. If they were to sell the fruit at Rs 15 a kilo, they would figure in their minds, they would recover only Rs 600 which would cause a loss, on the other hand if they sold the fruit at Rs 20 a kilo they would recover Rs 800. Then they would have to decide if that was enough profit and match it against what the consumers would be willing to pay. In any case the price was always negotiable, varied with the time of the day and how much stock remained unsold at the end of the day and whether the fruits would be in a condition fit for sale the next day.

Almost none of the people interviewed professed to use any other mathematical operation. When asked didn't they need to know how to use percentages, say to check the interest that their bank accounts earned, they would laugh and say they had so little money in the bank that it didn't really matter and in any case the banks must be doing it right.

Even a preliminary assessment of the findings showed that the level of use of mathematics in everyday lives of common people could not possibly be used as a basis for designing the school mathematics curriculum. The exercise provoked a debate within the group. On the one hand, it was argued that designing the curriculum bearing in mind only the requirements of those who would study mathematics beyond their school years was unfair and potentially damaging to the vast majority of children who would live ordinary lives. On the other, it would be disastrous to ignore completely the requirements of teaching children in a manner that they could become future mathematicians, and there was also the argument that all students should get a chance to appreciate the beauty of "higher" mathematics. Perhaps, some members felt, there was a case for streaming in mathematics beyond the middle school. It is interesting to note that this debate within the SMP group, which was shared with others interested in school mathematics, foreshadowed a similar debate that was to take place over a decade later at the national level (NCERT, 2006a).

APPENDIX C: LONG DIVISION—AN EXAMPLE OF ALTERNATE STRATEGIES

In the course of the intervention period 1995-2000, a host of alternate pedagogical strategies were developed by the program teachers, through a dialectical process of classroom trial and discussion in the group. The teaching of long division is taken here as an example.

As described in the main body of the paper, there was a consensus on postponing the teaching of algorithms as far as possible. Thus division was introduced contextually, and children went through many exercises in doing division without introducing the algorithm. Eventually, it was felt that children (in Class IV) were ready for the traditional long division algorithm. However, this was quickly proved wrong. Teachers reported that, in spite of the contextual exercises, children found it no easier to follow the algorithm than with the conventional approach followed earlier. At group meetings, several suggestions were made to modify the algorithm so that it was made somewhat more transparent. Some were tried out and rejected on the basis of classroom trials.

Finally, teachers reported that the way of doing long division given below worked the best:

```
                                  3
        3                        40
       40                       100
      500                       400
     1000  1543                1000  1543
 5 | 7715                  5 | 7715
     5000                      5000
     ----                      ----
     2715                      2715
     2500                      2000
     ----                      ----
      215                       715
      200                       500
     ----                      ----
       15                       215
       15                       200
     ----                      ----
        0                        15
     ----                        15
                               ----
                                  0
                               ----
```

This will work even if the division is not optimal at every stage, as can be seen in the reworking of the same division on the right hand side.

REFERENCES

Agnihotri, R. K., Khanna, A. L., & Shukla, S. (1994). *Prashika, Eklavya's Innovative Experiment in Primary Education.* New Delhi: Ratna Sagar.

Agnihotri, R.K., Khanna, A.L., & Rajan, J. (2008). *Indradhanush Elementary Textbook Series: From Creation to Reception.* Delhi: State Council of Educational Research and Training.

Eklavya. (2013). *Hoshangabad Science Teaching Program.* Retrieved from http://eklavya.in/go/index.php?option=com_content&task=category§ionid=12&id=52&Itemid=74

Holt, J. (1982). *How children fail* (2nd ed.). New York: Delacorte Press/Seymour Lawrence.

Hughes, M. (1986). *Children and Number: Difficulties in learning Mathematics.* Oxford: Blackwell Publishing.

Joshi, S. (2014). *Never a dull moment* (Translated from Hindi by Rex D'Rosario). Bhopal: Eklavya.

Kamii, C. (1985). *Young children reinvent arithmetic.* New York: Teachers College Press.

Kamii, C., & Domenick, A. (1998). The Harmful Effects of Algorithms in Grades 1–4. In L. J. Morrow & M. J. Kenney (Eds.), *The teaching and learning of algorithms in school mathematics: 1998 NCTM yearbook* (pp. 130-140). Reston, VA: National Council of Teachers of Mathematics.

Khan, F.A. (2004). Living, learning and doing mathematics: a study of working class children in Delhi. *Contemporary Education Dialogue, 1,* 199-227.

Liebeck, P. (1990). *How children learn mathematics*. London: Penguin.

Indira Gandhi National Open University. (2000, 2001). *Learning mathematics (LMT-01: Blocks 1-6)*. New Delhi: Indira Gandhi National Open University.

Mukherjee, A., Sadgopal, A., Srivastava, P.K., & Varma, V.S. (1999). The Hoshangabad Science Teaching Programme, presented at the *South Asian Conference on Education*, Delhi. Available at http://www.cisl.columbia.edu/grads/presi/EKLAVYA/Anil_5fAmitabh_5fArticle.html

National Council of Educational Research and Training. (2006a). *Position Paper: National Focus Group on the Teaching of Mathematics*. New Delhi: National Council of Educational Research and Training. Available at http://www.ncert.nic.in/new_ncert/ncert/rightside/links/pdf/focus_group/math.pdf

National Council of Educational Research and Training. (2006b). *Syllabus Volume I: Elementary level*. New Delhi: National Council of Educational Research and Training.

National Council of Educational Research and Training. (2006c, 2007, 2008). *Math-Magic 2, 3, 4, 5*. New Delhi: National Council of Educational Research and Training.

Prashika Group (1988-1992). *Khushi-Khushi 1-5* (in Hindi): a series of textbooks for the primary classes. Bhopal: Eklavya.

School Mathematics Project (1996). *Games and Activities for Class I*. Delhi: Centre for Science Education and Communication.

School Mathematics Project (2002a). *Games and Activities for Class II*. Delhi: Centre for Science Education and Communication.

School Mathematics Project (2002b). *Worksheets for Class II*. Delhi: Centre for Science Education and Communication.

State Council of Educational Research and Training Delhi (2004-2005). *Mathematics* 1-5: a series of textbooks for primary mathematics. Delhi: Delhi Bureau of Textbooks.

CHAPTER 71

INTERVENING FOR NUMBER SENSE IN PRIMARY MATHEMATICS

Usha Menon
Jodo Gyan, New Delhi

ABSTRACT

The Empty Number Line is being increasingly recognized as a tool that can support calculation with number sense in the early years. To ensure the possibility of using this tool flexibly, it is necessary that it should be preceded by activities with the bead string as was originally envisaged. This lays the basis for connecting the cardinal and ordinal aspects of number through the work on the bead string (*Ganit mala*). The empirical basis for the arguments made in this paper draws on an intervention for teaching number sense using the Empty Number Line in India and the modifications made to strengthen the cardinality aspect of the Ganit Mala.

Keywords: Early number-sense, empty number line, addition-subtraction strategies, arithmetic in the early grades

The First Sourcebook on Asian Research in Mathematics Education: China, Korea, Singapore, Japan, Malaysia, and India, pp. 1559–1582

India is part of the reform movement in school education which has been developing across the world, especially since the introduction of the National Curriculum Framework, 2005 (Khan, this volume). The spirit of the Framework was anticipated and preceded by various initiatives especially from non-governmental agencies involving actual intervention in the classrooms. Many of these interventions began as informal small scale interventions and eventually had an impact on initiatives in the government sector. In the two decades or so before the National Curriculum Framework, many organizations and individuals made a passionate appeal to change the formal approach to the teaching of mathematics and to introduce activities: to go from the 'concrete to the abstract', to link to the 'everyday life of children', to introduce 'learning with understanding' (Mukherjee & Varma, this volume). These wide-ranging efforts at intervention aimed at introducing activities into the classroom, drew their inspiration and perspectives from different sources, which, as discussed below, can be classified as either structuralist or empiricist (Treffers, 1991a). Jodo Gyan is a non-profit social enterprise that joined this effort towards the end of the nineties. A main focus of the work of Jodo Gyan has been the development and communication of activities to support the development of number sense in the primary grades.

A source of inspiration for this effort was Freudenthal's perspective of "mathematics as a human activity" and the work that has come to be known as Realistic Mathematics Education (RME) (Treffers, 1993). Number sense, which is considered "difficult to define but easy to recognize" has been a major theme in the last few decades and there is a general consensus that it involves an intuitive understanding of numbers, their magnitudes, positions and relationships and how they are affected by operations. It is recognized that it cannot be taught as a set of fixed procedures but has to emerge from a deeper understanding of numbers (Greeno, 1991).

The development of the Empty Number Line (ENL) and bead string combination from the school of RME in the nineties was a major contribution for supporting the development of early number sense (Klein, Beishuizen & Treffers, 1998; Van den Heuvel-Panhuizen, 2008). It developed a tool that children could use flexibly to deal with numbers, especially for addition and subtraction. The present paper considers the modifications related to the use of the ENL introduced by Jodo Gyan and analyses them in terms of the relationship between the quantity and order aspects of number. The approach to these two aspects is related to the different perspectives on the role of contexts, concrete materials and activities in mathematics education.

APPROACHES TO CONTEXTS AND ACTIVITIES

In the case of early numbers, the reform movements have sought to move away from the formal methods of teaching the algorithms that had been the foundation of the traditional approach. Attempts to develop activities in this area can be broadly divided into two different types. The structuralist approach was to find objects that can have a structure similar to the structure of the operations that are to be carried out. In the case of addition and subtractions with numbers up to 100 with carry-over and borrowing (trading), the attempt was to find objects that can exemplify the relationships between the different place-value columns. The aim was not only to show the relationship between the place values but also to develop activities that were isomorphic to the actions carried out while solving mathematical problems. A good example of this is the prescriptive use of Dienes blocks in which first the smaller blocks are added as is done in the written algorithm without considering the spontaneous tendency to first add the larger units. Although less alienating than the earlier formal approaches, the problems with such structuralist approaches have been brought out by many authors (Gravemeijer, 1994; Treffers, 1991a). In fact, attempts to develop concrete models to represent the laws of arithmetic precede the current reform movements (Courant & Robbins, 2007/1941).

The empiricist approach for reform sought to look for contexts and situations in which children experienced meanings that were analogous to the mathematical ideas that were to be learnt: quantity, increase, take way and so forth. Within this group at present we can discern two major streams regarding the nature of the relationship between the familiar meaningful context of the child and the concept to be learnt. The dominant stream takes to exposing the child to many different everyday contexts to support concept formation through similarity recognition (Mitchelmore & White, 2004; National Council of Educational Research and Training, 2007). The other stream mainly represented by RME, considers a special well-designed context as the basis for developing understanding. Through a series of level-raisings with increasing abstraction a (mature) understanding of the concept is reached (Treffers, 1991a). The algorithms and rules for operations are also reached through a series of stages with each stage forming the basis for the next stage of level-raising.

These two approaches can again be seen to be related to different understandings about the nature of a concept. The first assumes a concept to be a generalisation from a large number of examples, which is the classical idea of a concept. The second considers model building to be the main activity in the development of a concept and the model to emerge from a typical context.

> In the opinion of the realists, on the other hand, it is especially by means of strong models that children are given the opportunity to bridge the gap between informal, context-bound work and the formal, standardised manner of operation, through the constructive contribution of the children themselves. In short they take the position that assistance can be given via the presentation of fitting models and schemes, that direction can be given to learning, that something essential can be offered from "outside" (Treffers, 1991a, p. 33).

RME in this sense is different from constructivist or socio-constructivist approaches. In the RME tradition this model building activity has been put forward as a series of steps. Both in the RME and Vygotskian perspectives a paradigmatic situation forms the basis for the development of a concept. In Jodo Gyan also we have followed the approach of focusing on a paradigmatic context which we call as the anchoring context and have been exploring the different ways of modeling with that context.

NUMBER, QUANTITY AND ORDER

The question of how the concept of number is understood has deep implications for classroom practices. Number can be seen purely formally in terms of the laws of arithmetic and the successor function, or it can be seen in terms of the meanings that are attached to the concept of number with connections to the formal aspects. Historically number has been thought in terms of quantity: as discrete quantities (multitudes) or continuous quantities (magnitudes). The number sense literature has for long stressed the importance of quantities. For example, Greeno (1991) characterizes number sense as 'knowing in an environment of numbers and quantities' so as to be able to create appropriate mental models for flexible functioning. We can consider that it was the great contribution of the reform movement with its emphasis on number sense that they again directed attention to the quantity aspect which had virtually disappeared from the school curriculum with its focus on the teaching of formal algorithms.

Quantity as such can be grasped without counting through means of one-to-one correspondence, but with counting new dimensions emerge. Both Freudenthal and Vygotsky noted the important role played by counting. Freudenthal argued that numerosity was a primitive number concept that was soon superseded in human history by more refined concepts as humans learnt to count (Freudenthal, 1973, p. 188). Vygotsky discussed how counting on the fingers served as a bridge between natural arithmetic (what we would today call subitizing/parallel individuation) and cultural arithmetic. Vygotsky argued that "the quantitative characteristic of any group of objects is perceived initially as one of the qualitative characteristics. The matter changes as soon as man, in reacting to the quantitative aspect of any situation, resorts to his fingers as a tool to aid in carrying out the counting operation" (Vygotsky, 1987, p. 52).

With the emergence of counting to deal with quantities, the order principle also emerges. We can therefore consider number to have both the quantity and order aspects in its constitution – or the cardinality and ordinality aspects. These two aspects, together with one-to-one correspondence form the three 'how-to-count' principles put forward by Gelman and Gallistel. The ENL and the bead string combination designed almost two decades ago can in fact be seen to integrate both quantity and order aspects. This approach gains more significance in the context of the recent discovery of the wide presence across different cultures and age groups of number-space mappings, to begin with, on a logarithmic scale and later on a linear scale (Dehaene, Izard, Spelke & Pica, 2008; Siegler & Booth, 2004).

THE INTERVENTIONS

The modifications in the uses of ENL introduced by Jodo Gyan were made as part of a larger intervention to improve the quality of primary education in Delhi especially in the fields of mathematics and science. The effort took formal shape in 1999 with the setting of a non-funded, not-for-profit social enterprise called Jodo Gyan to work in the area of primary education. The initial focus of Jodo Gyan was on disseminating existing best practices, but it was forced to innovate when confronted with the paucity of implementable solutions. The innovations were made by problem solving in the classroom within a broad Vygotskian framework of meaning emerging from purposeful joint activity. In retrospect the work can be considered as "theory informed bricolage" rather than "theory guided bricolage" (Gravemeijer, 1994, pp. 181-186). In the method that was followed, I took the try-out lessons both in our own little experimental school and in the schools where we started working intensively. Observation of these lessons and discussion thereafter became the method to share the insights and orient others in Jodo Gyan and develop a new community of practice (Menon, 2004). The curriculum work as it progressed meant the development of a complete curriculum for mathematics from kindergarten to Class VI. It was developed mainly grade by grade over the years reaching Class VI in 2009-10. The innovations reported in this paper were developed as a part of this on-going work of intervention and not as stand-alone innovations and therefore the documentation was done only as needed by the intervention.

The interaction of Jodo Gyan with the school community, while developing and sharing its curricular innovations has involved varying levels of intensity. Some interventions were very intensive and transformative in character and were of long-term duration. These involved school visits once or twice a week with the holding of demonstration lessons followed by reflections on the activities conducted. Conducting these demonstration lessons and teachers observing them is considered necessary as the curriculum proposed has meant a paradigm shift for teachers in terms of the prevalent classroom norms and practices. Monthly workshops are conducted in these schools to discuss topic wise trajectories and the broad approaches and philosophy of the program with the teachers. It has been our experience that the first year is needed to prepare a common ground for communication with the teachers and only from the second year does the program actually take-off. For a successful partnership with a school it is seen that both formal and informal interactions dealing with the belief systems of the teachers about the nature of mathematics as well as about the identity of a teacher as a professional are crucial. Over the years there have been a little over a hundred teachers who share the common approach and values with the potential of forming a 'community of practice'. About 2000 children are following the curriculum in four schools in Delhi and its neighborhood.

Apart from this intensive work, Jodo Gyan has been also trying to disseminate the approach to mathematics education through workshops (of one to 10 days in duration). These workshops have been held with diverse groups of teachers – teachers from metropolitan upper end private schools to rural schools run by NGOs, teachers from the local community as well as government school teachers. A more diffuse but important space for reaching out to the school community has been the teaching learning materials, mainly produced by Jodo Gyan and used in the concept and introductory workshops.

NUMBER SENSE WITH THE EMPTY NUMBER LINE —DOMAIN OF NUMBERS UP TO 100

The use of the ENL can be located within the stream of realistic approaches in the math teaching reform movement: of developing activities that are close to the world of children rather than starting from the structure of the subject matter. The recognition that children's informal strategies for addition and subtraction often follow a stringing strategy rather than the taught algorithms was a reason for researchers to consider the use of the number line in the early years (Beishuizen, 1993). Yet the incorporation of the numbered number line into the classroom was ineffective due to the fact that it called forth counting and reading-off responses from children and not higher order structuring responses (Gravemeijer, 2002).

The situation changed when following an idea of the American mathematician Hassler Whitney, Treffers introduced the ENL (Whitney, 1985; Treffers & de Moor, 1990). The ENL was chosen because it provided a tool to do addition and subtraction using the 'stringing' method. The stringing method not only connected with the forward counting strategy that children naturally use to do addition (Treffers, 1991b), but also was seen to avoid the pitfalls associated with the splitting strategy, especially in subtraction (Beishuizen, 1993).

The ENL can be considered to be one of the most significant contributions of RME for the primary classes, especially for addition and subtraction. Following its development in the Netherlands, (Treffers, van den Heuvel-Panhuizen, & Buys, 1999), the ENL has been used in other countries such as the U.K., Australia and New Zealand (Department for Education, U.K., 2010; Department of Education, Tasmania, n.d.; New Zealand Maths - Ministry of Education, n. d.). I started experimenting with the use of ENL in India after a meeting with Prof. Treffers in 2000.

The Role of the Bead String (*Ganit mala*)[1]

The original formulation of the empty number line by Hassler Whitney (1985) as well as by Adri Treffers (Treffers & de Moor, 1990) linked it to the use of a ten structured bead string for learning to count numbers up to 100 as well as to locate them. In this process it was recognised that there is an important process taking place in which the ordinal and cardinal aspects of number were coming together - an aspect that attracted Treffers to the idea of Whitney.

Treffers suggested that children needed to have three important abilities to be able to effectively use the empty number line for addition and subtraction in a flexible way. These were (1) the ability to count by tens synchronizing with the showing on the ten structured bead string or on the number line, (2) the ability to quickly locate quantities on the bead string and numbers on the empty number line, and (3) the ability to move forward and backward by making jumps of ten from any random number on the number line (Treffers & de Moor, 1990, pp. 52-53). Following these early formulations, the ideas were field tested with positive outcomes (Klein, 1998; Klein, Beishuizen & Treffers, 1998). Julie Menne in her thesis added the need to develop mastery in the basic number facts up to 10 or 20 (Menne, 2001). Her methods for 'productive practice' in the development of the basic skills needed to add and subtract on the empty number line have been very useful.

ENL Without the Bead String

In the extension of the use of the ENL to other countries the use of the bead string appears to have been given a lesser role. In the England and Wales National Numeracy Strategy, the Empty Number Line seems to be preceded by a marked number line rather than the bead string. Bobis (2007) from Australia informs us that the ENL was not introduced through a bead string in her daughter's class but with a marked number line. Unfortunately the strength of the ENL, which is to support mental calculation through its flexibile use, seems to be getting compromised in the Netherlands itself where structured work on the marked number line is incorporated in many textbooks (Van den Heuvel-Panhuizen, 2008). In countries such as Greece, the number line has been introduced in the primary classes as a marked number line and not as an empty number line. It has been reported that the marked number lines did not support the number operation, and in fact, was more of an obstacle (Skoumpourdi, 2010).

The numbered number line in fact supports the ordinal aspect of number and not the quantitative or cardinal aspect. Using it to do addition and subtraction can give rise to the familiar mistake that children make of counting on from the first number – in adding 4 and 3, counting 3 further from 4 rather than from 5 and therefore reaching 6 instead of 7. There are also differences in the cognitive processes involved in adding 4 and 3 mentally and in adding by counting on the numbered number line, although both might be considered as counting-on. While adding mentally we count "four…five, six, seven" and while adding on the numbered number line we count "one, two, three" from 4 and read off 7 (Thompson, 2003, p. 10). Therefore the numbered number line cannot be considered as a prelude to the ENL.

WORK IN INDIA—DEEPENING AND EXTENDING THE USE OF THE GANIT MALA

The trajectory that was developed by us through working with children in different schools in India has broadly followed the direction set by Treffers, Menne and Klein and Beishuizen. Some innovations have been also introduced in the course of practice.

One major difference in our approach is that much more time is spent with the concrete material, that is, the Ganit Mala than in the original trajectory. Table 71.1 gives the flow of the main activities done with Ganit Mala in the trajectory as it has developed over the years. The items that are not numbered such as skip counting and guess my number do not fall into a strict order. Similarly the two activities of "jumps of ten" and "one number to another" intertwine. The table includes the modifications made to the "Treffers" trajectory (extracted from Klein, 1998; Menne, 2001; Menne, 2008/09; Treffers et al., 1990; Treffers & Buys, 2001). The

Table 71.1. Jodo Gyan Trajectory of ENL/Ganit Mala

No.	*Activities in Jodo Gyan Trajectory*
1.	Clip 1: (Teacher puts the clip) "How many beads from the beginning till the butterfly (clip)?"
2.	Clip 2: (Children put the clip on a number) (number context) (forward) "Can you come and put a clip on 25?"
2a.	Clip 2a: After Diwali (light festival) "Can you show 25 lamps on the Ganit mala?" (concrete context) (Done only in schools where found necessary for learners who do not seem to be ready to think about bare numbers)
3.	Clip 3: (*maan* card/ Gattegno card) Teacher puts a clip on the Ganit Mala and children make that number using Gattegno cards. Later, variations in which children put the clip are introduced.
4.	Hanging the Number card on the Ganit Mala
5.	Introduction of 0 and forward counting by counting on the **position.** Till this counting of the beads is done, but after introduction of 0 the position becomes the basis for counting.(See next section for details).
6.	Backward counting from **position** 100 to 0
6a.	Location of numbers on the GM by reverse counting
7.	Location of numbers on the empty number line
8.	Jumping from one number to another on GM (forward)
9.	Jumping from one number to another (backward)
	Skip counting by 2s
11.	Jumps of ten on the Ganitmala (forward and backward)
12.	Jumps of ten on the empty number line
13.	Jumping from one number to another on the empty number line
14.	Addition and subtraction on the empty number line
15.	Transition to the arrow notation
16.	Skip counting by other numbers up to 10
	Complementing to 100

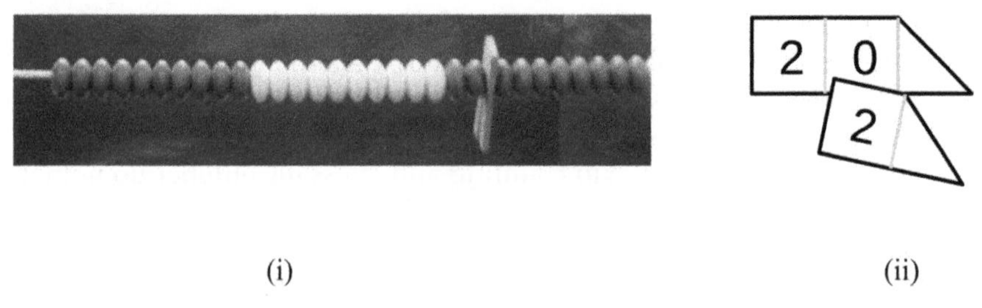

(i) (ii)

Figure 71.1. (i) Clip placed on the Ganit Mala and (ii) Gattegno cards.

main difference is related to the activity called "Clip 1", which has functioned to support the ordinal-cardinal switch and whose details are given separately. Another minor innovation which has found a very good response from children and teachers has been the use of Gattegno cards (also called arrow cards or place value cards, see Figure 71.1) along with the Ganit Mala ("Clip 3" in Table 71.1). In this activity the teacher hangs a clip on the Ganit Mala and each child in the class makes the number shown by the clip using Gattegno cards. This has functioned as a useful prelude to the use of number cards and seems to deal with the problem of children seeing two digit numbers as mere concatenated digits. The joint use of the Ganit Mala and Gattegno cards appears to support the development of "quantity value" since children see for example 85 as 80 and 5 (Thompson, 2000). In the Jodo Gyan trajectory, when children are able to comfortably jump from one number to another on the Ganit Mala, which might take 4-6 months, the Ganit Mala is set aside and full-fledged activities are started with the ENL. This activity although mentioned in Klein (1998) has not found a place in the detailed description of activities in Menne (2001). In our practice this functions as a key point in the transition from Ganit Mala to ENL. The ability of children to go from any random number to another random number by structuring seems to indicate a higher level of understanding of number relationships than the ability to go to any random number from 0. As in the original Treffers trajectory, in Jodo Gyan trajectory also, children work on a smaller Ganit Mala (with two children sharing one) along with the whole class Ganit Mala.

Innovating to Support Ordinal-Cardinal Switch

The relationship between the ordinal and cardinal aspects in the use of the Ganit Mala can be understood by studying the process of locating numbers on the bead string. When beads are just counted the ordinal aspect of number is foregrounded. When a clip is put after counting that many beads to locate a number, the cardinality aspect emerges where the clip represents the position where there are that many beads from the beginning till that point.

Among the changes that have been introduced in the original trajectory is an activity called Clip 1. Bright coloured rubber clips are used along with the Ganit Mala and it is thought as a butterfly, a parrot or a monkey as the children choose. In this activity, teacher puts the clip somewhere on the bead string and asks, "How many beads are there from the *beginning* till the 'monkey'?" Children come forward to count and find the answer.

This simple modification has different functions:

- to 'slip in' the meaning of the 'origin' or 'beginning' for the counting process,
- to establish the left to right convention for working on the Ganit Mala, and
- to link the cardinal and ordinal aspects of the Ganit Mala.

This connection between the cardinal and ordinal aspect needs some clarification. A simple counting of the beads would in fact only refer to the ordinal aspect of number. It is when a clip is put on the string that the cardinal or quantity aspect emerges. But this cardinal meaning needs more support. The position of a number on the Ganit Mala had to be established as the position after counting that many beads while counting forward and not as being on the bead itself. It was in order to foreground this aspect that the extra activity of clip 1 had to be introduced.

Gravemeijer has reported about the problems that were faced when the empty number line was introduced without working earlier with the bead string partly because it was felt that the children were responding well to the empty number line. But it was seen not to be so easy for the children to see whether going from 90 to 88 represented a reduction of two or three. A child counted as 90, 89, 88 and argued that 3 were taken away. Without the experience of the bead string it was not easy for the children to differentiate between the two positions (Gravemeijer, 2002). Gravemeijer underlines the fact that in the bead string, the "beads themselves form a clearly visible *quantity*, while the clip demarcates this amount and at the same time marks a *position* in the counting row" (Gravemeijer, 2002, p. 17. Translation mine). In a footnote to this discussion Gravemeijer remarks that "however, the clip in fact marks the position *next* to the last counted bead, while in general the counting number indicates the last counted bead." (Gravemeijer, 2002, p. 22. Translation mine).

It is this crucial marking of the *position* that is the subject matter of Clip 1 activity (see Figure 71.2). In order to remove any ambiguity another

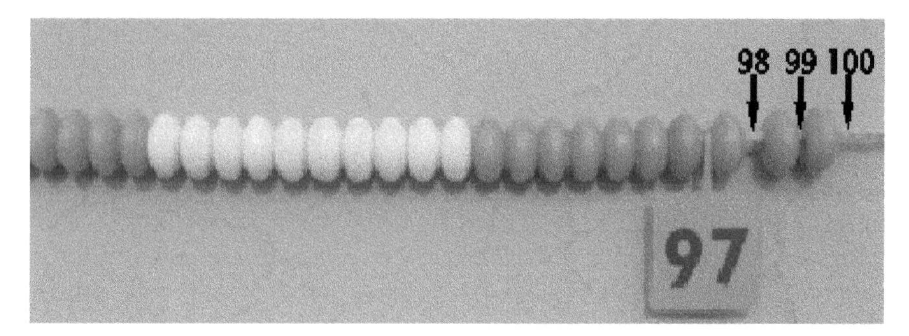

Figure 71.2. Reverse counting by position to locate 97.

innovation was introduced: going over from counting by the beads to counting by the position. This counting by the position happens with the introduction of 0. When children are very comfortable with location of numbers on the Ganit mala, then the 0 number card is introduced. After the positioning of the 0 number card, children count from 0 to 100. This naturally requires that the positions on the string are counted and not the beads. The meaning of the location of numbers is further underlined – that there are that many beads from the beginning till that point. Once children count from 0 to 100, then they also count from 100 to 0 on the Ganit Mala by holding the position.

I became aware of the different aspects related to the positioning of the clip when a teacher spoke about the difficulties with reverse counting on the Ganit Rack (Counting Rack up to 20) in the first year of our intervention in 2004. When reverse counting on the Ganit Rack or Ganit Mala, a problem would emerge if the same convention is followed as when counting forward of putting the clip after the last counted bead. By reflecting on this problem clip 1 and the activity of counting from 0 by counting the positions were developed.

In fact many consider that the clip should be put on the bead and not on the string. In India we normally come across high school mathematics teachers who insist that the clip or the card should be on the bead itself. Here we have a question of the exclusive perception of the ordinal aspect of the Ganitmala. This is an aspect that seems to be dominant in the introduction of ENL in other countries. The U.K. Standards has a fair amount of detail about the use of ENL in the classroom, but except for a picture of the bead string (apparently) to build a model of the marked and then empty number line there is no mention of the work that needs to be done with the bead string (Department for Education, U.K., 2010, p. 18).

Some Glimpses From the Classroom

The impact of the new practices using the ENL on the conceptual understanding of the children would need a separate study. This paper presents number sense related aspects of the classroom practices as reflected in the written answers of children as they solve word problems. The details are taken from the responses of children from two schools where we have been working since 2005 and 2006. The data is for 2009-10 and was collected as a part of the routine of the intervention program and are the results of assessments conducted at the end of the academic year of children of Class II (8+ years old). In these assessments, children were given sheets with questions and the questions were also read out to them. Apart from explaining the question (including in Hindi), no other help was given to the children. In School 1 the assessment was done a month after children had gone over to the next class and in School 2 it was done just before the academic year ended. Therefore at the time of assessment children would have had about three months to four months of time in which they used the ENL (and in School 1 the arrow notation also) to solve addition and subtraction problems in their classes.

Meaningful Operations—Multi-Step Word Problems

The ability of children to correctly solve multistep word problems can be considered to be an indication of their number sense. This indicates not only computational abilities but also the ability to choose the correct operation, indicating a sense of the meaning of the operations. Table 71.2 shows the number and percentage of children who gave correct answers to two multistep word problems. The two word problems were as follows.

- Q1 (gujias): For *Holi* celebrations Varun's parents made 36 *gujias* (special sweets made during the festival of *Holi* occurring around the time when the assessment was done). They ate 11 of them. And then Nani and Nana brought 15 more. How many *gujias* do they still have in the house?
- Q2 (birds leaving): At the Sultanpur National park there were 74 white storks in the beginning of February. As it started getting hot they started leaving the park. 13 white storks left the park in the beginning of March and 22 left at the end of March. How many white storks were still there in the park?

The numbers in the table do not include children who have comprehended the problem but have made errors in calculation since our focus is

Table 71.2. Number of Children With Correct Answers for Multi-Step Word Problems

School (Total No. of Children)	*Q1 (Gujias)*	*Q2 (Birds Leaving)*
School 1 (148)	118 (80.3 %)	106 (72.1%)
School 2 (163)	131 (80.4 %)	131 (80.4 %)

on the development of calculation skills along with understanding of the operations in the number range up to 100 emerging from an ENL program.

Since these assessments are done as part of an on-going intervention program there are no control groups to report from. Children from a school with comparative socio-economic background and a traditional curriculum performed at a significantly lower level for these same questions. But the validity of such comparison can be questioned since there is a close link between the pedagogy and the assessment: multistep word problems are not included in the normal Class II curriculum (Van den Heuvel–Panhuizen, 2001).

Flexibility—Strategies and Procedures

The ability of children to flexibly choose the most efficient strategy to solve a problem has been widely considered as a desirable outcome of schooling. This ability is connected to the knowledge of different strategies as well as to the ability to choose the most efficient strategy appropriate for a particular context. As far as addition subtraction problems are concerned we can distinguish between solution strategies and computational procedures (Beishuizen, 1997). A problem such as the following can be solved by different strategies.

- Q3 (pages read): Sohail is reading a book of 92 pages. He has read 85 pages. How many pages does he need to read to complete the book?

It can be done by direct subtraction of 85 from 92 (SUB as per the code for strategy used by Beishuizen, 1997) (Figure 71.3) or by indirect subtraction by going from 92 till 85 is reached (TAT – Taking-away-to). It can also be found by indirect addition (AOT - Adding on-to). In the examples given, children follow the convention of showing addition above ENL and subtraction below the ENL. In both SUB and TAT, the operation starts from the larger number while in AOT the operation starts from the

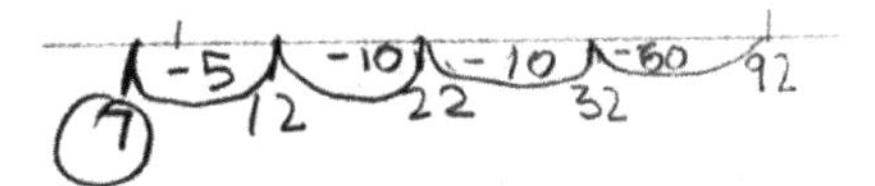

strategy: SUB, procedure N10

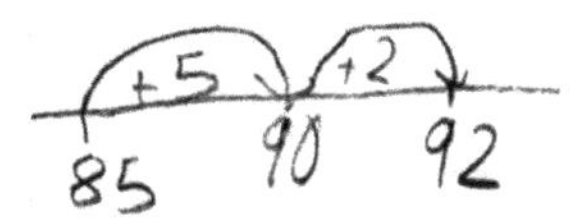

strategy: AOT, procedure A10

strategy: TAT, procedure A10

strategy: AOT, procedure N10c

Figure 71.3. Examples of strategies and procedures used by students to solve "92–85".

smaller number, but in both TAT and AOT the difference between the two numbers is in focus.

Each of these solution strategies can be solved through different computation procedures. The focus of the discussion in the literature in the last two decades has been mainly about the distinction between splitting or 10-10 procedure and the sequential or N10 (Number ± 10) procedure (Beishuizen, 1997a). A more advanced sequential procedure involving compensation (N10c) was also differentiated and became the basis of classroom instruction. Later Beishuizen and colleagues have noticed another variant of the stringing procedure called A10 ('Adding on to the next round ten') which the better pupils preferably use (Beishuizen, 1997, p. 91). These give rise to different options and the question of the most efficient combination of strategies and procedures for different semantic contexts. Thus for example, indirect addition along with A10 is an efficient strategy to solve Equalization problems or Change problems in which an addend is missing.

Table 71.3 shows the spread of solution strategies and computation procedures used by the children in the two schools. Both the strategy of indirect addition (AOT) and the small size of the number to be added could have contributed to the high correlation between AOT and A10 shown in solving Q3 (pages read).

Flexibility Across Problems

Flexibility in the use of solution strategies across different problems can be explored by comparing the strategies used by children for Q3 (pages) with that used to solve Q2 (birds leaving). Since Q2 is a multistep

Table 71.3. Distribution of Solution Strategies and Computation Procedures for Q3 (Pages Read)

		Procedure					
School	*Strategy*	*All N10*	*A10*	*Single Jump*	*Others*	*Unclear/ No Answer*	*Total*
School1	AOT/TAT	8 (9%)	53 (58%)	31 (34%)	0 (0%)	0 (0%)	92 (100%)
	SUB	31 (66%)	3 (6%)	5 (11%)	8 (17%)	0 (0%)	47 (100%)
	Total*	39 (26%)	56 (38%)	36 (24%)	13 (9%)	4 (3%)	148 (100%)
School 2	AOT/TAT	8 (7%)	84 (78%)	10 (9%)	5 (5%)	1 (1%)	108 (100%)
	SUB	37 (73%)	12 (24%)	1 (2%)	0 (0%)	1 (2%)	51 (100%)
	Total*	46 (29%)	96 (60%)	11 (7%)	5 (3%)	3(2%)	161 (100%)
Total	AOT/TAT	16 (8%)	137 (69%)	41 (21%)	5 (3%)	1 (1%)	200 (100%)
	SUB	68 (69%)	15 (15%)	6 (6%)	8 (8%)	1 (1%)	98 (100%)
	Total*	85 (28%)	152 (49%)	47 (15%)	18 (6%)	7(2%)	309 (100%)

Note: *In some cases the total number is greater than the sum of the procedures since "Direct answer" and "Unclear/No answer" in strategy have not been shown.

Table 71.4. School Wise Distribution of Solution Steps for Q2 (Birds Leaving)

	Solution Steps				
School	*Subtract One by One*	*First Add*	*Direct Answer*	*Unclear/ No Answer*	*Total*
School1	109 (74%)	25 (17%)	6 (4%)	8 (5%)	148 (100%)
School 2	132 (82%)	24 (15%)	0 (0%)	5 (3%)	161 (100%)
Total	241 (78%)	49 (16 %)	6 (2%)	13 (4%)	309 (100%)

problem children need to first make the strategic choice of whether to deal with the birds leaving sequentially or together and then choose the operation. Most children follow the logic of the question to repeatedly subtract the number of birds leaving, while a more advanced group is able to transform the question and solve the problem by first adding together the number of birds leaving. These children show a meta-awareness of the operations involved.

The strategy and computation procedures used by the 49 children who first added and then subtracted are given below in Table 71.5. As can be seen, six among them did indirect addition.

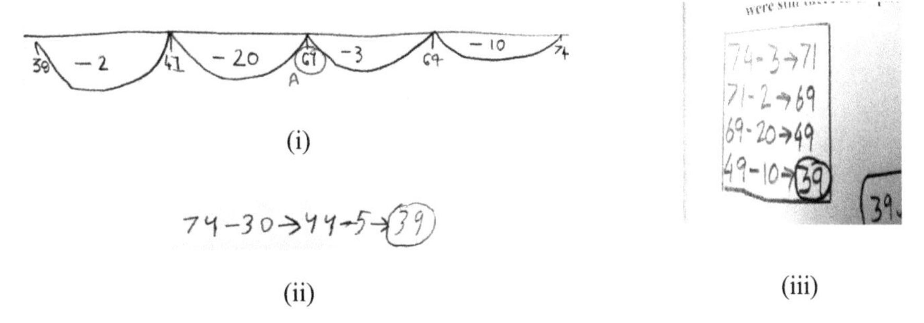

Figure 71.4. Examples of students' solutions to Q2 (birds leaving): (i) Subtracting one by one using procedure N10, (ii) and (iii) are variations of procedures using the arrow notation.

Table 71.5. Q2 (Birds Leaving)—Students' Solution Strategies and Procedures

	Q2 Procedure				
Q2 Strategy	*N10*	*A10*	*Single Jump/ Direct Answer*	*Other*	*Total*
Subtraction/SUB	31 (72%)	2 (5%)	7 (16%)	3 (7%)	43 (100%)
Indirect addition/AOT	4 (67%)	2 (33%)	0 (0%)	0 (0%)	6 (100%)
Total	35 (71%)	4 (8%)	7 (14%)	3 (6%)	49 (100%)

Variability in the Class

Research has shown that children are able to model different actions and relationships with concrete materials and use these different strategies to solve different semantic categories of problems – an ability they seem to lose once they start schooling (Carpenter, Hiebert & Moser, 1983). This can be due to the fact that children are not able to relate what they learn as addition and subtraction in school to the modeling operations they would do to solve the problems. One of the posited strengths of ENL is its ability to provide a tool that can bridge this gap – to model different operations such as addition, indirect addition, subtraction and indirect subtraction, without having to transform them into whole/part relationships and then use direct addition and subtraction.

The ENL also provides the opportunity to do computation at different levels apart from the differentiation into procedures such as N10 and A10 which we have discussed. Within each of these also there are possibilities of level raisings as children use more anticipatory and consolidated procedures. The minimum level expected in the classroom is that of jumps of 10 and one and as children become more competent, variability increases as children start taking consolidated jumps. The sequential procedures afforded by ENL are followed by the arrow notation (Figure 71.4) and the possibility of different levels within a procedure exists with the use of arrow notation also.

A few of the students in the class go on to developing even more advanced procedures such as those involving compensation such as N10c (Figure 71.3). Therefore some amount of variability can be considered as a characteristic of a classroom where the methods are being used with a sense of what is being done rather than as a mechanical execution of a single taught procedure. Children in Table 71.6 are in the early stages of the use of the ENL and an insightful use can be expected to involve wide variability.

This aspect is explored by taking the individual classroom as a unit. Class II shows the variability in each class as children answer the multistep word problem Q2 (birds leaving). Only the children who had answered correctly are included in the analysis. In this analysis each individual variation – of steps of solution (of doing two subtractions or one subtraction after adding the number of birds leaving), of solution strategy or computation procedure or levels within a computation procedure is counted as separate. In this analysis both the notation of ENL and arrow are also coded as separate. In fact School 2 used exclusively the ENL notation although normally before the end of Class II the arrow notation is introduced. The data shows a long tailed structure with a few children using one or two of the most efficient ways and the others using different variations.

As can be seen there is a wide variation even among the children who have answered correctly. In some of the classes, if we exclude the top one or two groups it can be seen that almost all the children have used a method of their choice, indicating an active differentiated class.

Usually the variation is expected to reduce as more and more children become comfortable with the most efficient strategy. But even among proficient users some variation is visible as can be seen from the examples from another set of users of the sequential (ENL/arrow) approach. Thus for example in the subtraction of "936 – 597" two strategies are used: adding on from 597 and subtracting 597. But in adding on itself there are at least two different variations of A10 procedure that competent users of the sequencing approach can employ as seen in Figure 71.5. While one child adds 300 and then 36 another adds them together.

In the Jodo Gyan trajectory as given earlier, children work with Ganit mala for an extended period of about 4-5 months with various activities including learning number names, spend about a month with ENL doing

Table 71.6. Variability in Solving Multi-Step Word Problem Q2 (Birds Leaving)

School	*Section*	*No. of Children Who Answered Correctly*	*No. of Procedures*	*No. of Children Using the Most 'Popular' Procedure*
1	A	22	14	5
1	B	27	17	4 each for two procedures
1	C	22	14	6
1	D	23	14	6
1	E	18	6	5 each for two procedures
2	A	24	9	8 and 7
2	B	24	13	9
2	C	25	8	14
2	D	23	10	9
2	E	21	9	7
2	F	19	11	4

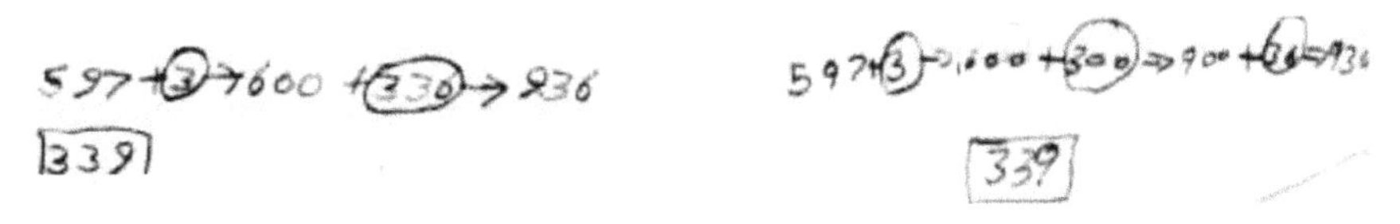

Figure 71.5. Variations of the A10 procedure used to solve "936–597".

activities such as jumps of ten and jumping from one number to another using different contexts and then go on to do actual addition and subtraction on the ENL. Through this process children appear to have developed number sense, a structured quantity sense that supports them to use procedures and methods as per the numbers and contexts. Given the positive response of the children a Ganit Mala with 1000 beads was also introduced for activities before using ENL and arrow language for addition and subtraction of numbers up to 1000. These glimpses from the classroom indicate that the use of Ganitmala and then ENL can create spaces in which 'pupils are working in an active and differentiated way within vertically planned teaching/learning process' (Treffers, 1978).

INTERVENING WITH THE TEACHERS

Developing an innovation and sharing the innovation with the school community are two different processes with the second being the less explored one. In presenting the results of the Wiskobas innovative curriculum, Treffers (1978) put forward the importance of a 'three dimensional' description rather than one involving only the activities and the content goals. The importance of presentation of the didactical context, with the leading questions asked, the didactical hints given and so on are considered necessary if the communication has to be effective. In our experience also we have found the crucial significance of the didactical context in sharing the experience with the teachers.

Reflecting on our experiences of interacting with the teachers over the years, we can see that different levels have emerged in these interactions. A major factor affecting the response of the teachers is the fact almost all the schools and textbooks use a place-value based vertical algorithm for teaching addition and subtraction and therefore the stringing approach of a Ganit Mala/ENL combination would mean a paradigm shift. In spite of this there has been a good response and teachers have started incorporating the trajectory to varying extent. Accordingly, the interactions with teachers have as targets different levels to which teachers may incorporate the learning trajectories as below:

1. Intensive Trajectory Program: Sharing the complete trajectory including the use of the ENL for doing addition and subtraction – done through a 2 year program involving biweekly visits tapering off after the two years and finally stabilizing at half yearly assessments and workshops.
2. Introductory Trajectory Program: Sharing all the Ganit Mala activities and only the activity of locating numbers on ENL – usually done during 3-10 day workshops on primary mathematics. New forms of these programs involving bimonthly school visits have also been designed.
3. General Introduction: Sharing only the activities that involve forward jumps such as Clip 1, Clip 2 and so on including from one number to another and no marking on the ENL – shared during introductory workshops of one day or shorter.

We find that many schools who continue to use the vertical algorithm for teaching addition and subtraction up to 100, have included the use of the Ganit Mala in their classroom practices. The response of the children appears to have played a role in this decision. The fact that Jodo Gyan has sold more than 13,000 Ganit Malas (about 12,000 in the last four years) indicates the extent of interest that is developing. Further some other organisations working with teachers have also since included the Ganit Mala in their repertoire although they have not included a stringing approach to operations and continue to focus on the place value based algorithm.

The use of ENL provides a number-sense based approach involving quantitative reasoning, counting-on and flexible methods to do addition and subtraction of numbers. Our experiences indicate that Ganit Mala can lay a basis for the development of number sense: an intuition from the field that has been strongly supported by recent research. Siegler and colleagues have investigated the internal representation of children across different ages through estimation tasks on a number line and have found that estimation progressed from logarithmic to linear scales. Most interestingly they have shown that children with a more linear mapping of numbers have a better knowledge of numbers (Siegler & Booth, 2004) and that playing numerical board games (numbers from 1 to 10) can improve not only number line estimation, numeral recognition and counting but also magnitude comparison as compared to children who played a colour-based board game (Ramani & Siegler, 2008). Even more significantly, later studies showed similar effects for numerical magnitude and the learning of addition and subtraction sums for the linear numerical games and not for circular numerical games (Siegler & Ramani, 2009). These researchers have argued that these results could be attributed to the greater transparency of mapping between the physical material and the internal number line and "that board games provide children with strongly correlated spatial, temporal, kinesthetic and verbal or auditory cues to numerical magnitude' and constitute an 'ideal support for constructing a linear representation of numerical magnitude" (Siegler & Booth, 2004, p. 441). This perspective coming from a very different school of research can be considered as a major support for the approach presented here. Ganit Mala has all the aspects of the linear board game and one might add, more transparency, as far as quantity and therefore counting-on are concerned.

SUMMING UP

The reform movement in mathematics education had foregrounded the importance of number sense. The ENL/ bead string combination from the Netherlands introduced a number-sense based method of doing flexible addition and subtraction. In its adoption in other countries the ENL is often introduced without the bead string, while in the intervention in India, the role of the bead string has been further strengthened. The ability of the bead string/Ganit Mala to support the cardinal/ordinal switch and thus base further developments on the quantitative intuitions of children justifies its use prior to introducing the ENL. Recent research on linear board games supports this observation. The strong component of the Ganit Mala activities in the Jodo Gyan curriculum has created a further reach-out for the number sense approach in India than would have been possible with only the ENL.

NOTE

1. The terms bead string and Ganit Mala are used inter-changeably, using Ganit Mala in the Indian Context.

REFERENCES

Beishuizen, M. (1993). Mental Strategies and Materials or models for addition and subtraction up to 100 in Dutch second grades. *Journal for Research in Mathematics Education*,294-323.

Beishuizen, M. (1997). Development of Mathematical Strategies and Procedures up to 100. In M. Beishuizen, K. Gravemeijer, & E. van Lieshout (Eds.), *The Role of Contexts and Models in the Development of Mathematical Strategies and Procedures* (pp. 127-162). Utrecht.

Bobis, J. (2007). Empty Number Line: A useful tool or just another procedure?. *Teaching Children Mathematics, 13*(8), 410-413.

Carpenter, T. P., Hiebert, J., & Moser, J. M. (1983). The effect of instruction on children's solutions of addition and subtraction word problems. *Educational Studies in Mathematics,14*(1), 55-72.

Courant, R., & Robbins, H. (2007/1941). *What is Mathematics?* Delhi: Oxford University Press.

Dehaene, S. Izard, V., Spelke, E., & Pica, P. (2008). Log or Linear? Distinct Intuitions of the Number Scale in Western and Amazonian Indigene Cultures. *Science,320*(5880), 1217-1220.

Department for Education, U.K. (2010). *Teaching Children to Calculate Mentally.* Retrieved from http://dera.ioe.ac.uk/778/1/735bbb0036bed2dcdb32de11c7435b55.pdf

Department of Education, Tasmania. (n.d.). *Tasmanian Curriculum - Mathematics-Numeracy.* Retrieved on 2011 from http://www.education.tas.gov.au/curriculum/standards/maths/syl-mnall.pdf

Freudenthal, H. (1973). *Mathematics as an Educational Task.* Dordrecht : D.Reidel

Gravemeijer, K. (1994). *Developing Realistic Mathematics Education.* Utrecht: CD Beta Press.

Gravemeijer, K. (2002). Didactisch gebruik van de lege getallenlijn - een persoonlijk perpsectief. (Didactical use of the empty numberline - a personal perspective) . *Panama-Post, Tijschrift voor nascholing en onderzoek van het reken-wiskunde onderwijs*, 11-23.

Greeno, J. G. (1991). Number Sense as Situated Knowing in a Conceptual Domain. *Journal for Research in Mathematics Education*, 170-218.

Klein, A. S., & de Klein, T. (1998). *Flexibilization of mental arithmetic strategies on a different knowledge base: the empty number line in a realistic versus gradual program design*. Leiden University.

Klein, A. S., Beishuizen, M., & Treffers, A. (1998). The empty Number Line in Dutch Second Grades: Realistic versus Gradual Program Design. *Journal for Research in Mathematics Education*, 443-464.

Menne, J. (2008/09). Kralenketting (Bead string). *Volgens Bartjens*, 12-15.

Menne, J. (2001). *Met Sprongen vooruit (Jumping Ahead)*. Utrecht: Freudenthal Institute.

Menon, U. (2004). Transformative Links and the Training Strategies in Mass Educational Movements: Some Conceptual Issues. In K. Chanana (Ed.), *Transformative Links between Higher and Basic Education* (pp. 319-337). New Delhi: Sage Publications.

Mitchelmore, M. & White, P. (2004). Abstraction in Mathematics and mathematics learning. In *Proceedings of the 28th Conference of the International Group for the Psychology of Mathematics Education* (Vol 3, pp. 329–336).

National Council of Educational Research and Training. (2007). *Math Magic. Class II*. New Delhi: NCERT.

New Zealand Maths - Ministry of Education. (n.d.). *Jumping the Number Line*. Retrieved from: http://nzmaths.co.nz/resource/jumping-number-line

Ramani, G.B. & Siegler, R.S. (2008). Promoting Broad and Stable Improvements in Low-Income Children's Numerical Knowledge Through Playing Number Board Games. *Child Development*, *79*(2), 375-394.

Siegler, R.S. & Booth, J. (2004). Development of Numerical Estimation in Young Children. *Child Development, 75*(2), 428-444.

Siegler, R. S. & Ramani, G.B. (2009). Playing Linear Board Games - But Not Circular Ones - Improves Low-Income Preschoolers' Numerical Understanding. *Journal of Educational Psychology, 101*(3), 545-560.

Skoumpourdi, C. (2010). The Number Line: An Auxiliary Means or an obstacle? *International Journal for Mathematics Teaching and Learning*. http://www.cimt.plymouth.ac.uk/journal/skoumpourdi.pdf

Thompson, I (2000). Teaching place value in the UK: time for a reappraisal?. *Educational Review, 52*(3), 291-298.

Thompson, I. (2003). On the right track? *Primary Mathematics*, Autumn,. 9-11.

Treffers, A. (1991a). Didactical background of a mathematics program for primary education. In L. Streefland (Ed.), *Realistic Mathematics Education in Primary School* (pp. 21-57). Utrecht, The Netherlands: CDß Press.

Treffers, A. (1991b). Meeting Innumeracy at Primary School. *Educational Studies in Mathematics*, *22*(4), pp. 333-352.

Treffers, A. (1993). Wiskobas and Freudenthal Realistic Mathematics Education. *Educational Studies in Mathematics*, *25*(1-2), 89-108.

Treffers, A. (1978). *Wiskobas doelgericht (Wiskobas to the point)*. Instituut voor Ontwikkeling van het Wiskunde Onderwijs.

Treffers, A., & Buys, K. (2001). Grade 2 (and 3) – Calculation up to 100. In M. Van den Heuvel-Panhuizen (Ed.), *Children Learn Mathematics* (pp. 61-88). Utrecht, The Netherlands: Cdß Press.

Treffers, A., & de Moor, E. (1990). *Prove van een nationaal programma voor het reken-wiskunde onderwijs op de basisschool Deel 2 [Specimen of a National Program for Primary Mathematics Education]*. Tilburg: Zwijsen.

Treffers, A., Van den Heuvel-Panhuizen, M., & Buys, K. (1999). *Jonge Kinderen Leren Rekenen [Young Children Learn Calculation]*. Groningen: Wolters-Noorhoff.

Van den Heuvel-Panhuizen, M. (2001). *Towards a didactics based model for assessment design*. Retrieved October 16, 2009, from http://www.math.ntnu.edu.tw/~cyc/_private/mathedu/me1/me1_2001/mhp2.doc

Van den Heuvel-Panhuizen, M. (2008). Learning from Didactikids: An Impetus for Revisiting the empty Number Line. *Mathematics Education Research Journal, 20*(3), 6-31.

Vygotsky, L. S. (1987/1934). Thinking and Speech. In R. W. Rieber, & A. S. Carton (Eds, translated by Minick N.), *The Collected Works of L.S.Vygotsky - Volume 1* (pp. 39-285). New York: Plenum.

Whitney, H. (1985). Taking Responsibility in School Mathematics Education. In *Proceedings of the ninth international conference on the psychology of mathematics education* (pp. 123-141). Noordwijkerhout: PME.

CHAPTER 72

SOME ETHICAL CONCERNS IN DESIGNING THE UPPER PRIMARY MATHEMATICS CURRICULUM

A Report From the Field

Jayasree Subramanian
Tata Institute of Social Sciences, Hyderabad

Mohammad Umar
Azim Premji Foundation, Chittorgarh

Sunil Verma
Eklavya, Hoshangabad

ABSTRACT

This paper gives a brief description of the upper primary mathematics curriculum in India, the aims and objective of the curriculum as stated in the National Curriculum Framework 2005 and discusses the implication of the curriculum for the socio-economically marginalized students. Through a detailed description of the learners' socio-economic and educational back-

The First Sourcebook on Asian Research in Mathematics Education: China, Korea, Singapore, Japan, Malaysia, and India, pp. 1583–1606

ground, the physical space of the government and private schools where the authors intervened and carried out design experiments, the kind of teaching learning activities that usually take place in these classrooms, the learning levels of the students at upper primary level and by citing studies on education of the marginalized and students' achievement, the paper argues that the students from socio-economically marginalized sections are in no position to cope with the prescribed curriculum at the upper primary level and will in fact fail to access the benefits of education because of failure in mathematics. Questioning what necessitates or justifies a curriculum that cannot be transacted in large number of classrooms and whose interest such a curriculum serves, the paper argues that designing a curriculum involves ethical considerations and calls for a Re-Visioning of the upper primary mathematics curriculum that offers the (marginalized) learners scope for a critical understanding of the social reality in which they are placed.

Keywords: Ethics in curriculum development; design experiments in mathematics; curriculum in India; research and development in mathematics education

INTRODUCTION

Curriculum development, it is often said, is a negotiated activity particularly when it is intended for large scale implementation. It needs to take into account concerns raised by educationists, those brought in by the domain experts, teachers' knowledge and their understanding of the discipline, their experiences of teaching the content, the kind of training and support that they would require, perceived needs of the learners, public opinion, socio-political concerns and many more. The nature and context of how this negotiation is to be carried out raises several questions: Is it possible that such a negotiation be done in some objective way resulting in an optimal curriculum that meets the interests of all those who have a stake in it? If not, whose needs and interests does a curriculum represent and in the process who does it marginalize? When education departments of governments decide curriculum for the state or for the whole country, what is the vision that drives such an exercise and what kinds of justification are offered for curricular decisions? How do popular notions and expert perceptions about the nature and history of a discipline impact curriculum design and how does school curriculum in turn help to reinforce popular notions about a discipline and its importance? What is the relation between curriculum, school, and the social reality of a learner? Should curriculum development be informed by studies on students' achievements and preparedness for what they are going to be taught? Are there ethical concerns involved in curriculum development?

Curriculum development in India is done largely at two levels. At the national level, National Council for Education Research and Training (NCERT) develops the curriculum which is followed by all the central schools and schools affiliated to the Central Board of Secondary Education (CBSE) across the country. At the state level, twenty four of the twenty eight states have their own state boards of education and develop a curriculum that is followed in all the government schools in the state and also in the private schools affiliated to the state board. Though the National curriculum and the textbooks brought out by NCERT function as a reference point for the state level curriculum, the state boards are not bound to conform to it. Unlike history and other social sciences, disciplines like mathematics and science, are viewed as neutral and objective, impervious to the socio-political values of those in power and also significant in deciding the learners' career opportunities and economic status.[1] There is often, a large overlap in the curricula prescribed by different boards, across states and internationally. At the primary school level, the focus of the curriculum is on ensuring that students acquire basic competencies in numbers and operations on them, basic ideas in spatial understanding, measurement and their application in real life situations. At the upper primary level students are expected to move beyond everyday mathematics and engage with abstraction, acquire logical thinking, ease with symbolic representation and competence to do mathematics at the higher classes (NCERT, 2006a). (In most states in India, "upper primary" refers to Grades 6 to 8; in some it may refer to Grades 5 to 8 or 5 to 7). Typically the upper primary curriculum deals with integers, rational numbers and their properties, algebra, geometry, data handling and "commercial mathematics." This article will discuss the context in which explorations and the design experiments were conducted by the authors, which included government schools in rural India and some private urban schools catering to low income groups. It will describe the socio-economic background of the students, their competence at the upper primary level and the implication of what is taught at the upper primary level for these students. In keeping with the critical approach to mathematics education, we believe that curriculum design is an ethical issue and based on evidence presented here we argue in this article that curricular choices made by the boards of education, rather than representing the needs and interests of these children, function to further marginalize them. We stress the need for an alternate vision of upper primary mathematics curriculum that serves primarily the interests of the majority of the children rather than the interest of the discipline.

The structure of the paper is as follows. The first section gives a brief description of the existing upper primary mathematics curriculum and makes a few comments on design experiments and curricular research at the upper primary level in India. The next section outlines Eklavya's (the organization that the authors belonged to at the time of the design experiments) earlier engagement in primary mathematics curriculum development and discusses in some detail its experience in trying to evolve alternative resource material to teach upper primary mathematics. The next section situates our experience in the context of literature on education in India and on social justice issues in mathematics education. The final section concludes by raising ethical concerns related to curriculum design and the need for research.

THE UPPER PRIMARY MATHEMATICS CURRICULUM IN INDIA

A significant step in curriculum development in mathematics in recent times in India is the stated shift in focus in the national curriculum "from mathematical content to mathematical learning environments, where a whole range of processes take precedence: formal problem solving, use of heuristics, estimation and approximation, optimization, use of patterns, visualization" (NCERT, 2006a, p. v)

The Position Paper contends that, "making mathematics a part of children's life experience is the best mathematics education possible" (NCERT, 2006a, p. 2) and lists inculcating logical thinking, helping children understand the basic structure of mathematics and mathematization of the child's thought as part of its vision and goal. The syllabus for upper primary mathematics states that the emphasis is on "the need to look at the upper primary stage as the stage of transition towards greater abstraction, where the child will move from using concrete materials and experiences to deal with abstract notions" (NCERT, 2006b, p. 80). Consistent with these, and in a major departure from the traditional textbooks in India, the colorful NCERT primary mathematics textbooks with pictures and photographs, situate learning of mathematics in a range of contexts that include everyday experiences, puzzles, games and stories and invest in developing ideas by inviting children to explore, experiment and think on their own. In contrast the textbooks of the Madhya Pradesh (M.P.) state government (to which the schools that we work with are affiliated), impart content largely in the form of rules and algorithms and encourage drill and practice, even though they claim that the national curriculum framework has been taken into consideration while developing them. A feature common to the curricula of both the Center and the M.P. state is that they take a particular view of the discipline as given and center the curriculum on it. This is not singular to the Indian context as one can see by glancing through the upper primary curriculum in other countries. However, within the community of mathematics educators at least, the question of what constitutes as mathematics has been problematised for sometime now. From the point of view of school education, academic or research mathematics, which constitutes the discipline of mathematics, is referred to as "mathematicians' mathematics" and is seen as one of the many kinds of mathematics (Ernest, n.d.).

Prevalent Curricular Content and Approaches

In spite of the very significant difference between their approaches, there is a major overlap between the two boards in the content of the upper primary curriculum. For the strand on numbers and number systems, beginning with natural numbers, integers and their properties in Grade 6, the syllabus moves on to introduce decimals, rational numbers, their representation on the number line, factors, multiples, exponential notation, prime factorization of numbers, ratio, proportion, percentages and finishes with finding square roots and cube roots, including the algorithm for finding square roots in both the boards. In addition, the state board introduces fractional exponents and uses the algorithm to find the approximate values of square roots of 2, 3 and 5.

Algebra begins in Grade 6 with introduction of symbols to stand for numbers. While the NCERT textbook limits itself to writing linear polynomial expressions and equations in one variable at the Grade 6 level, the state board introduces higher degree polynomial expressions in two variables and the operations of addition and subtraction on them, ending with solving linear equations in one variable in Grade 6. But by Grade 8, both the boards cover the three operations on higher degree polynomials in two variables, factorization (NCERT) or division (State board) and algebraic identities.

In geometry and measurement, the content begins by giving some idea of the meaning of geometrical objects such as points, lines and line segments, then move on to polygonal figures, the notion of angle, using the geometry kit to measure angles, construct triangles, quadrilateral, perpendiculars and angle bisectors, properties of triangles and quadrilaterals, circles, three dimensional objects namely sphere, cylinder, cone, pyramids. Formulas for finding the perimeter, area and volume are either arrived at or given straightaway depending on the board. Apart from these three major themes, namely, numbers and number systems, algebra and geometry, there is some exposure to data handling (pictographs, frequency tables, chance and probability/ measures of central tendency, graphs) and commercial mathematics (profit and loss, simple and compound interest).

Design Experiments in Upper Primary Mathematics

In India, investment for research in mathematics, though very selective, is many orders of magnitude more than investment for research in mathematics education. In fact, systematic curricular research and material development in mathematics education is still an emerging area in India. This is partly because education as an academic domain has not received the kind of attention it deserves from the state. Education departments have largely focused on developing and implementing preservice teacher training programs. In fact, barring a few institutions that are engaged in education and curricular research, most of the experimentation and innovation in education have come from the Non-Governmental Organizations (NGO) which often function to address the needs of the marginalized sections and have to work against a range of constraints. Often these innovations are not documented and hence do not inform curriculum development. Research publication from the country therefore does not match the efforts that have gone into curricular intervention in education.

At the upper primary level, Homi Bhabha Center for Science Education at Tata Institute of Fundamental Research (HBCSE) and Eklavya have been actively engaged in design experiments. In algebra, the design experiment by the HBCSE team, working with students from low and mixed income groups in a metropolitan city, uses Grade 6 students' knowledge of arithmetic as a starting point to build a bridge between arithmetic and symbolic algebra. Using "terms" as objects that combine with other "terms", the experiment exposes children to a new way of looking at arithmetical expressions, operating on them and comparing them (Banerjee & Subramaniam, 2004; Banerjee & Subramaniam, 2005). This experience is then used in contexts such as generalizing, predicting, and proving, all of which require symbolic algebra. The "terms approach" enables children to acquire structural understanding alongside procedural understanding by requiring children to initially decompose an arithmetic and later with that experience an algebraic expression into signed "product terms" which are first evaluated (in the arithmetic context) and then "combined" (meaning, added) (Banerjee & Subramaniam, 2008). In a recent paper, Subramaniam and Banerjee (2010) argue that historically algebra has been seen as foundation to arithmetic in the Indian tradition and that the "terms approach" could be seen as drawing on that tradition at least to clarify ideas. They use the term "operational composition" to denote information contained in an expression and argue that a refined understanding of operational composition includes judgments about relational and transformational aspects and it may play a role in developing understanding of functions and thus point to the importance of emphasizing structural understanding of numerical expressions as a beginning for symbolic algebra.

In fractions, a collaborative design experiment at HBCSE and Eklavya combines the share and measure interpretation of fractions to facilitate students' reasoning in making sense of the fraction symbol and in comparison tasks (Subramaniam & Naik, 2007; Naik & Subramaniam, 2008). The alternative approach developed by the two teams share a basic conviction that share and measure meaning of fractions enable children in acquiring a situated and conceptual understanding of the notion of fraction. Classroom trials in four different sites demonstrate that children see the connection between these two interpretations of fractions and use them flexibly to reason in comparison tasks (Subramanian, Subramaniam, Naik & Verma, 2008; Subramanian & Verma, 2009). It is hoped that these experiments would contribute to both curriculum development and teacher education.

EKLAVYA'S ENGAGEMENT WITH UPPER PRIMARY MATHEMATICS

Eklavya, an NGO in central India, was set up in 1984 to intervene in education with the aim of developing alternative curricula that engages the child's mind. A rare opportunity given by the state government made it possible for Eklavya to intervene in the state schools in some of the backward districts of the state and to implement its alternate curriculum in science (known as the Hoshangabad Science Teaching Program) and social science at the upper primary level. Important features of the interventions include the involvement of teachers in curriculum development, use of locally available material to teach science, and sustained support to teachers throughout the academic year (Agnihotri, Khanna & Shukla, 1994; Batra, 2010). Its primary school programme, *Prashika*, developed between 1983 and 1987, was able to innovatively combine language and mathematics learning at the primary school level. *Prashika* is an acronym for **Pra**thamik **Shi**ksha **Ka**ryakram (Agnihotri et al. 1994). For an engaging description of *Prashika* curriculum and its textbooks "*Khushi Khushi*" see Khan (2005). The government first implemented *Prashika* in a few schools in 1987 and extended it to all schools in two districts. The government closed the *Prashika* programme in 2000 and the other Eklavya programs by 2002. Subsequently, Eklavya's efforts have been directed towards providing "outside school" support at the primary school level to first generation learners in about 200 centers and developing alternative curricular material as alternative resource material for teachers and curriculum developers. These are in the form of modules at the upper primary level in mathematics and at the high school level in science.

Upper Primary Mathematics—A Report From the Classrooms

Our explorations and sustained experiments in curriculum development have been going on for a little over six years and have used three different settings: (i) four private, English medium schools (three urban and one rural) catering to children coming from the low income group for a period of one to two years, (ii) Muskaan, a school run for scrap pickers in an urban location for less than a year, and (iii) three government run schools in a rural location for more than three years, of which one is a primary school for girls where we worked with the same set of children from Grades 3 to 5 and the other two upper primary schools. In most of the schools we worked with, we directly taught the students. In one school we worked with the same set of students for three years. In all other school we focused our attention on Grade 6 students. Only in two private schools our engagement was limited to providing support to the teachers.

Government schools in India cater to the poorest sections of society. This also means most of them come from marginalized castes or tribes. In one government school 68% of the children belonged to "Other Backward Castes", 17% were Dalits (Scheduled castes) and 10% were *Adivasi* (Scheduled Tribes). (All the three categories have low social status, and the government provides a variety of quotas in education and jobs to support individuals from these groups.) In another the percentages were 57, 37 and 4 respectively. Most of the children going to a government school do not have geometrical instrument boxes and carry a single notebook with them in which they copy everything. The government provides one meal a day, a set of school uniforms and the required textbooks to the children. However, there is a lot of resentment from some of the teachers (majority of whom belong to the "dominant castes") and it is not uncommon to hear comments like "these children are not capable of learning. They come to school for the free meal". Parents of the students are either illiterate or barely literate. Therefore children do not get any support from the parents in their studies nor can they afford paid tutors; all the teaching they receive is from the school.

Student absenteeism was a common feature in both the government schools that we worked with. There are several reasons for student absenteeism (Pratham Organisation, 2012; The Probe team, 1998; Sinha & Reddy, 2011). Part of the reason for absenteeism lies in the fact that for the large majority of children, school offers nothing interesting to engage them and make them feel welcome. During our informal interaction, students sometimes reported to us that they do not attend school regularly because they are bored at school as very little teaching happens. Our regular visits to the school only confirmed this, as it was common for teachers to leave the classroom assigning the students some writing work or asking them to memorize the tables. While part of this could be attributed to the bias teachers carry against the caste/class background of the students, a significant part of it also has to do with the lack of training and continued support for the teachers to cope with teaching first generation learners. On several occasions we have witnessed students being beaten up in spite of our protest. Witnessing and being subjected to such severe punishments demotivate students from attending school. Other reasons for staying away from school are poverty that forces them to give a helping hand to the family in both government schools and private schools: they stay back to look after younger siblings and attend to domestic work when the parents go to work in the fields during winter and rainy seasons, and to help the adults during harvest.

Parents of the children studying in the private schools we worked with were literate or semi literate. In most of the cases, fathers had completed school education and in a few cases mothers had done so, many having dropped out of school at some point. The parental ambitions for their children could be considered high as they had enrolled their children in private schools in which the medium of instruction was English. (The so called "English medium" schools often may not even have teachers who are comfortable with the language.) However it would be difficult to believe that the children got direct help from their parents to support or supplement what they were taught at school. We did not witness teachers beating up students in the private schools as often and as indiscriminately as we did in government schools. It was rare to find a classroom without a teacher. However, the teachers' own subject matter knowledge and pedagogic content knowledge, their comfort level with the language of communication and instruction in the classroom, the number of classes and hours they teach in a day and the class size were major limitations. All the four private schools associated with Eklavya had some commitment to provide good education to the socio-economically marginalized and attempted to do so within the many constraints in which they were functioning. So, in that sense, these private schools should be considered as better than typical private schools catering to the poor.

Description of the Intervention Classrooms

The physical space. Referring to the physical space of the school and the classroom reported about by Bopape in South Africa, Skovsmose (2007) says,

> How could it be that this hole in the roof has not been seriously addressed by mainstream research in mathematics education? Learning obstacles can be looked for in the actual situation of the children and with respect to the opportunities which society makes available for the children. The actual distribution of wealth and poverty includes a distribution of learning possibilities and learning obstacles. This distribution is a political act. Paying attention to this means re-establishing the politics of learning obstacles (p. 84).

Let us get a picture of the classrooms in the government and private schools with which we have been interacting.

The physical ambiance and the amenities of classrooms are marked by the socio-economic class of the children who come to study there. The classrooms in government schools have no furniture for children. They usually sit on *tat pattis* (mat rolls made of jute) or *dhurries* (thick woven material like carpets) spread on the floor. Often the *tat pattis* and *dhurries* are dirty and torn, having been in use for years. During winter when the floors are too cold for them to sit on, or if there is a shortage of *tat pattis*, they are forced to huddle together. The job of cleaning the classrooms is left to the students. As the classrooms double up as dining halls where they are served the mid-day meals, the students have the choice of sitting in classrooms with little bit of food lying around all over or sweeping the classroom for the second time in a day. Typically the government schools have no power supply. The classrooms in private schools catering to low income groups are small in size packed with benches and desks leaving very little room for the teacher to walk around. The situation could become worse when it rains as rain water could leak from the rooftop and force a seating rearrangement making the already crammed classroom more crowded. Sometimes one large hall was used for two classes, which meant there was continued noise and distraction from one class disturbing the other. In contrast, Navodaya schools and central schools run by central government and private schools catering to economically better off children have much better facilities with spacious classrooms, with enough furniture for children, built in shelves, and electricity. The differences in physical amenities are important because they make a difference to what is possible for teaching in a classroom. On a rainy day, with children sitting on dirty and wet floors, in winter with children too poor to afford warm clothing crowding around on the torn *tat pattis* and *dhurries*, in classrooms where the traffic noise can drown the din of children, well designed curriculum, good textbooks, well worked out lesson plans and the beauty of mathematics seem the last thing on anyone's mind.

Teaching learning activities in the classrooms. What could teaching and learning mean in such classrooms and for these children? On a typical day when one enters the classroom it is very likely that one finds the children busy writing. The teacher is at the board or has left, having worked out a problem on the board which the children are copying into their notebooks, or, there is no teacher and the children are copying something from commercially available keys in which problems are already worked out. Having graduated from copying three digit subtraction problem with "borrowing" in Grade 3 to copying division of a four digit number by a two digit number in Grade 5, they could be copying how to simplify expressions like $(118 - \{121 + (11\times11) - (-4) - (+3 - 7)\})$ and $4x^2 - [9x^2 - \{-5x^3 - (2 - 7x^2) -6x\}]$ or if the statement "The difference between 6^5 and 5^6 is zero" is true or false or how to "subtract 13x – 4y from the sum of 6x–4y and –4x–9y" (These are problems given in the Grade 6 state board textbook.)

Children going to private schools buy their own books and notebooks and have a separate notebook for mathematics. (Since our work has only been with private schools catering to lower middle class, the term "private school" in the article refers to only those catering to low income group.) Teachers may assign them homework and the notebooks may contain signs of having been checked (in the form of tick or cross mark or an occasional remark) by the teacher. However in the government schools, there is no evidence of any feedback being given for written work that children hand in.

Given that this is what serious (and understandably tedious and boring) learning means, any attempt to engage children in discussions to understand and solve problems seem like "*khel*" (play) for them. For Grade 6 children, an activity like writing four digit numbers or their successors, doing mental addition of two digit numbers or making sense of fractions by paper folding seem not like doing mathematics but whiling away time pleasantly. In the private schools where "covering the syllabus" is important, children themselves sometimes express anxiety over losing precious time in "*khel*". In the government schools children on the other hand welcomed such "*khel*" and looked forward to our visits.

Learning levels and Oral vs written mathematical skills. In our interaction with Grade 6 children from government school we found that many of them help their parents at the vegetable market or shops and nearly half of them could solve simple arithmetic problems orally but when asked to write the same systematically, they had difficulty. As the numbers become larger oral skills become less effective. Inability to read and write numbers clearly make it impossible for them to use algorithmic approaches. In a Grade 6 class, children could not compute $95 \div 5$ using the standard algorithm, but when asked "if 95 toffees are equally shared among 5 how much would each get" many of them could figure out that each child would get 19 using distribution schemes of their own.

We have been interacting with Grade 6 children over the last few years in an attempt to figure out if we could introduce algebra and negative numbers in some way. Each year, before we began explorations, we tried to assess the comfort level of the students in number representation and operations on them. In one year we found only 4 out of 31 students in Grade 6 could write numbers (two, three and four digit numbers) correctly on hearing the number words, the following year we found only about 70% of the children in Grade 6 could do three digit addition without error, only about 55% of the children could subtract a three digit number from 1000 or multiply a three digit number by 3. Many had no quantity sense of four digit numbers. On another occasion out of the 25 Grade 6 children tested individually, we found 2 children still could not write two digit numbers without errors, 9 could not write 3 digit numbers without error and 19 children could not write four digit numbers without error. For most of the children there was no concept of writing down the steps in a systematic manner. They would do some rough calculations and write the answer. Our experience with Grade 3 children was also similar – most of them could not write two digit numbers. This is also because number words for two digit numbers in Hindi do not follow the order in which they are written. The Hindi equivalent of thirty six for example would be something like "six, thirty" (Khan 2008). Children going to private school are better – but we found that there were quite a few students in Grades 6 and 7 who would write 3110 as a successor of 319 and 70036 or 700306 for seven hundred and thirty six.

Similarly, most of the children in Grade 6 and 7 cannot link the fraction words they know with the fraction symbols, think 1/3 is bigger than ½ even though the primary curriculum introduces fraction in Grade 3 and finishes all operations on fractions by Grade 5. Some of the private school children may be able to use standard algorithm for comparison, addition and subtraction of fractions mechanically, even though the fraction symbol may not mean anything to them.

Design experiments and findings. As our objective was to evolve alternative approaches to teach mathematics at the upper primary level, we began our explorations with integers for Grade 6 students and rational numbers for Grade 7 in one of the private schools. Soon we realized that we need to develop an alternative approach for teaching fractions right from the primary school level, which we did, continuing alongside an interaction with Grade 6 students on numerical representations and division algorithm, algebra, measurement and geometry. A brief report from these interventions and explorations follows.

An alternative approach to teach fractions. From our initial engagements at Grade 7 level in private schools, it became clear to us that if we want to transact percentages, ratio, proportion and rational numbers at the upper primary level we need to intervene at the primary level and try alternative approaches to teach fractions. Encouraged by the results of our collaborative work with the HBCSE team reported earlier, we took up a longitudinal study beginning from Grade 3 and worked with the same set of children up to Grade 5 in a government primary school for girls. We drew heavily from the work of Streefland (1993) in designing the alternative.

We introduce fraction as the share that a child gets when, for example, 12 *rotis* (circular homemade bread, the Indian equivalent of pizza for teaching fractions and familiar to all children in central and north India) are equally shared among 3 children; the fractional share and the symbol to denote it, are introduced beginning from unit fractions. The symbol 7/8 represents a share situation in which 7 *rotis* are to be equally shared among 8 children. Children draw circles on top to represent *rotis* and stick figures to represent children and divide and distribute writing the share of each child below. Some distribution scheme might result in the relation 7/8 =1/2+1/8+1/8+1/8 = 1/2 +3/8, another might result in 7/8=1/2+1/4+1/8 (Umar, 2010; Verma, 2010). With this experience they know how to compare unit fractions. Over a period of time they also learn the equivalence of different division schemes and can move from one to another with ease. Children carried out measurement activities using (a large enough) unit scale and subunits from ½ to 1/10. Children know for example that 7 pieces of size 1/7 make 1 unit. In the third year of the intervention (i.e., in Grade 5) we introduced the number line, representing fractions on the number line, equivalence of fractions and gave an idea of how to use equivalence to compare or add fractions.

Some instances of students' reasoning schemes that emerged spontaneously: Nikita (Grade 5) compares 4/5 and 7/8 as follows. First she divides each of the 4 *rotis* each into 5 equal parts and gives each child 4/5. Then she shares out 7 *rotis* among 8 as 1/2 +3/8, quickly changes her mind about 4/5 and writes it as ½+1/10+2/10 =1/2+3/10 and concludes 4/5 which is equal to 1/2+3/10 is less than 7/8 that is equal to ½+3/8 because 1/10 is less than 1/8. Bharati (Grade 5) shows where 52+3/7 and 52+4/7 would lie on the number line and argues that 52½ is greater than 52+3/7 because the space between 52 and 53 is divided in to 7 equal parts and 52½ would be in the middle with three parts to its left and three to its right and so 52½ = 52+3/7+1/14. There are several such instances of spontaneous reasoning that children come up with because for them over a period of three years, the fraction symbol means many things.

Though the experiment convinced us that children can be taught fractions in a meaningful way from Grade 3, it also gave us an opportunity understand the limitations in a government school context. Most of the children had difficulty reading and so arguments that children came up with were oral, which the teacher documented after the class, though, children certainly could write equations like 7/8=1/2+1/4+1/8.

Student absenteeism was a major factor that we had to deal with. On a given day out of the 57 students only 30 or 35 of them would be present. Since it is not the same 35 who were present everyday, we had to spend time to repeat what has been already taught. Students who were detained in Grade 4 and Grade 5 joined in the second and third year respectively which meant some part of what was taught in the previous years had to be repeated. Eventually only 30 out of the 57 children we started with stayed with us through all the three years and among them 15 got the benefit of the experiment, as the rest were either detained in the successive years or they did not attend classes regularly. So at the end of three years, we had children who were at different levels.

In a trial with Grade 6 children in a government school, we found that they were able to hold on to the meaning of the fraction symbol as a share and model many "word problems" in the language of sharing and answer them correctly. Also children used proportional reasoning to help them when faced with "large numbers". To answer "which *bhartha* will be spicier – the one using 300 g of chillies in 5 kg of eggplant or 200 g of chillies in 3 kg of eggplant" children used 3 and 2 to represent 300 g and 200 g respectively and compared 3/5 and 2/3 (Umar, 2010). Similarly, Pooja in Grade 5, who was part of the longitudinal study showed 1000+300/700 on the number line by declaring "no one can divide the gap between 1000 and 1001 into 700 equal parts. So we will divide into 7 equal parts and assume that each part represents 100" though she was not sure how to write one thousand and one and hesitantly wrote 10001.

Partial Quotients Method for division. In continuation of the work in fractions with the same students, now at Grade 6 level, we wanted the students to use the standard algorithmic approach for comparison and addition of fractions. This led us to dealing with factors and multiples and eventually to division as most of the students could not carry out the standard division algorithm as it made no sense to them. We introduced the Partial Quotients method currently used in the NCERT Grade 5 text book that works with the whole number rather than individual digits as the standard division algorithm does. Some of the Grade 6 students needed the support of materials such as match sticks to distribute and record the results while many others could carry out division using the Partial Quotients method on paper and pencil; more importantly, almost all the students were able to relate to division in a meaningful way (Khemani & Subramanian, 2012). Our attempts at maximizing quotients and shifting to the standard algorithm failed, indicating that the students do not see a ready link between the two approaches.

Explorations in Negative numbers. We came across several students at upper primary and secondary level, who could not carry out addition and subtraction of integers and so we attempted to introduce negative numbers through games and meaningful situations to Grade 6 and 7 students in two of the private schools we were working with. We found that while students could carry out addition of integers, subtraction and ordering posed major challenges; also students did not relate to negative numbers in any meaningful way, though with practice they could carry out addition. We did not consider it possible to introduce negative numbers for Grade 6 students in the government schools given that they have not received appropriate training in number representation and basic arithmetic operations.

Explorations in algebra. In our effort to assess what levels of abstraction children can engage with, we tried introducing algebra to Grade 6 children in a government school. Any trial in algebra is very challenging because children cannot write simple arithmetic expressions or even numbers properly. We adopted a procedural approach and began by making children compute solutions for verbally stated arithmetic expressions and we wrote down on the board what they said and used these as basis to introduce algebraic expressions. We also tried to introduce algebra without introducing students to negative numbers and hence ensured that the expressions would not result in a negative number for any choice of number that students make. For example, they could be asked "think of a number, double it, add 5 to it, subtract the number you thought, add another 5 subtract again the number you thought and tell me what you got". Each child was asked to carry this out by picking a number of her choice and report how she calculated. The whole equation was recorded by the teacher for the benefit of all to see. For example, if a child said 'I thought the number 3' and proceeded with the calculation, it would be recorded as: $(2x3)+5-3+5-3=6+5-3+5-3= 10$. After a few examples they realize that everyone gets the same answer and at that level they were asked to explain why everyone gets the same answer. Most often they were able to notice that the number they doubled got subtracted twice and so only 10 will remain. After much discussion and reasoning we also arrived at the equation $2x+5-x+5-x=10$ by saying 'x represents the number in someone's mind which we do not know'.

We also wrote on the board an expression such as $3x-2x$ or $2x-1+2$, explaining what $3x$ and $2x$ meant and ask them to find out what would be the result. Barring a couple of students, they attempted such questions by picking a number and substituting for x. Almost always they were able to notice the pattern, say what would be the result and why, though it would never be x or $2x+1$. For example they may say that in the expression $2x-1+2$, the number gets doubled and 1 gets added to it. In one rare instance a student even made a purely mathematical remark by saying 'the result of 2x-5+3-1 would always be an odd number' meaning if we substitute natural numbers for x we will only get odd numbers. Although the expression 2x-5+3-1 would give a negative number if x equals 1 or 0, students did not choose these values for x.

While students participated in these exercises eagerly, it was clear that barring three or four students in the class, the rest would not be able to write these expressions on their own. Also, when asked to compute 2x+x, most of them gave numerical answers even though they were told to write the answer that will be correct for all possible numbers in the place of x. In other words our attempt to make them see algebra as generalized arithmetic and work with symbols rather than numbers did not succeed. However, it provided them an opportunity to do mental arithmetic and also play around with numbers. In the process they were also testing out if 0 and "two and a half" could be a legitimate choice of numbers (even though he/she may not know how to write two and a half numerically).

DISCIPLINE, CURRICULUM AND LEARNERS: A CALL FOR RE-VISION

The above report from the classrooms may not come as a surprise to anyone who has worked with children of the marginalized groups in India. The classrooms we encounter are not exceptions but representative of a large number of classrooms in the country. Though there is a paucity of literature on mathematics education, there are several studies and a large body of literature in the country that address the issue of access to education, poverty, social status and educational achievements. Govinda (2002; 2011) are rich sources to get a sense of who has access to education, how it is related to caste, class, gender and social location, what have been the interventions by some states and by NGOs, how curriculum can be made responsive to the context and what are challenges in addressing the issue of quality and teacher development. Ramachandran (2003) and Kumar and Sarangapani (2005), for example, provide an account of innovative interventions outside and inside the school system. Leclercq (2002) describes in detail the Education Guarantee Scheme and its impact on two blocks in central India, with some information on students' achievement in mathematics. The Annual Status of Education Report (Pratham organisation, 2012) gives a state-wise picture of educational achievements of children enrolled in government and private schools. According to the report 50% of children in Grade 7 cannot divide a 3 digit number by a single digit number.

It is also well known that children from marginalized backgrounds, who are engaged in economic activity to supplement family income, bring oral arithmetic skills to the classroom from their work place. Studies on children from India engaged in economic and other activities describe the kind of oral and informal mathematical computations that children can perform (Khan, 2004; Subramaniam, 1998). Based on their systematic study in Brazil, Nunes, Schliemann and Carraher (1993) argue that oral mathematics calls for abstraction, and that oral and written arithmetic are based on the same implicit logico-mathematical principles namely the associative, commutative and distributive properties. They argue further that, oral approaches preserve meaning and relative values of the numbers under consideration unlike written algorithms that focus on individual digits. Also as the direction of the calculations is from the larger units to the smaller units, the size of errors tend to be smaller in comparison with written procedures (Nunes et al., 1993). A quick look at the upper primary syllabus would convince us that in the present upper primary curriculum there is very little place for the competencies that such children bring with them. Integers and algebra call for working with signed numbers and abstract symbols and manipulating numerical and algebraic expressions. How can these be taught if the students cannot make sense of simple written arithmetic sentences? If primary schools have not been able to provide the basis for students to cope with the upper primary mathematics, how could we expect that the teachers will be able to provide the basis at the upper primary level and also transact the upper primary content? Is there any evidence to show that children in that age group can take the load of what they missed out at the primary level and what is prescribed for them at the upper primary level? Is it a fair demand on students who are already faced with several odds against them? It is the near impossibility of transacting a certain curriculum for children from certain socio-economic background that forces us to question the relevance of the curriculum not only for these children but for all. Why is it that there is very little in the upper primary mathematics that prepares the learners for critically understanding the socio-economic conditions in which they live? Why is it that practical mathematics that enables one to read and make sense of simple data that one finds in the newspaper, not part of the upper primary curriculum?

There are necessarily several assumptions that curriculum design begins with. These include certain normative notions about childhood, assumptions about the nature of the disciplinary knowledge that it plans to impart and its relevance to the learner, about the learners' prior knowledge, cognitive capabilities, their interest in what the curriculum proposes to deliver, about teachers' competence and conviction to teach the content. Since a curriculum does not exit in a vacuum but is realized in practice, one way to know what these assumptions are would be to inspect who it turns out as successful at the end and what purpose the curriculum serves them. If the upper primary mathematics curriculum turns out predominantly, urban and semi urban middle class children as successful, then it would be fair to say that the curriculum is premised on their prior knowledge and support structure and that it functions to meet their aspirations.[2] In other words, the curriculum would not thrust on the "normative" students, the kind of mathematics that they cannot cope with as it does with children from the marginalized backgrounds. On the other hand if the curriculum had a vision for those from the marginalized who constitute the bulk of students and who drop out from school at various stages, then it would attempt to incorporate mathematics that will add value to their life. If a curriculum that advances the pursuit of mathematics as an end in itself or as a sophisticated instrument that shapes the technologies of the globalized world, then it would assume as potential learners, those for whom such a pursuit is tangible. Such a curriculum would turn right away a whole lot of children out of focus. Conversely, it is impossible, for example, to conceive of an upper primary mathematics curriculum that seeks to engage the learners (whatever their socio-cultural and economic background might be) with operations on polynomial expressions if the learners we have in mind are whom we described in the previous section: children who have the same reasoning skills but have not been trained in written mathematics. Oral arithmetic competence, while necessary and important, cannot serve as a sufficient platform to launch teaching of algebra as generalized arithmetic but the cognitive and computational skills involved in oral arithmetic can be productively used in mathematical projects designed to understand some of the socio-economic realities. Children who do not have notebooks, pencils and geometry boxes cannot be taught geometric constructions and measurement of angles. Children for whom school learning is synonymous with copying meaningless symbols and whose reading, writing and arithmetic skills at Grade 6 level are at the level of what is expected of a Grade 2 child, cannot be the children the upper primary curriculum has in mind if the content of the curriculum were to remain what it is. It would instead factor in the oral competencies they bring in and incorporate a curriculum content that builds on it. Curriculum design, like any other human endeavor, involves ethical considerations. How is it ethical to design a curriculum ignoring the fact that a large number of students do not have the academic basis and the material resources that they require to learn the prescribed content? How is it ethical or even possible to expect that a large number of these children should or would carry the dual load of the prescribed curriculum and what they missed when they reach the upper primary stage?

Curriculum viewed from a certain point can appear to be an equalizer. It can be viewed as an invitation to stake one's claim in our shared heritage, namely the disciplinary knowledge. Viewed thus, it would seem only fair that, even though the society at the present moment is not ready to share its tangible, finite, material resources in an equitable way, it should allow the best knowledge products to trickle down to the last child in line. As a knowledge product to be bequeathed equitably to all, what can be better than mathematics with its long history across cultures and civilizations, levels of abstraction promising immense intellectual fulfillment, commitment to logical thinking, long list of problems and puzzles to challenge and tease one's mind and its centrality to the globalized world we are living in? (See the introduction in Gellert and Jablonka (2007) and Skovsmose (2007) for mathematization in the present world.) And it even evokes compelling images of a future society in which, empowered with mathematical knowledge, the last child in line (along with others in front) is in a position to stake claims for material wellbeing. Yet, the real picture we have in front of us is that children refuse our offer and worse still, mathematics figures as a major cause for failure in school education thus closing options for higher studies or gainful employment for them.[3]

One way to account for failure could be to blame it on a range of intervening factors: inappropriate pedagogy, lack of subject matter knowledge or pedagogic content knowledge on the part of the teachers, their unwillingness to teach the children, their notions about the learners socio-cultural background and their ability to learn, the teacher-pupil ratio in the classroom, students' background and parental ambitions for their children, the opportunity they have to study after school hours and so on. One can argue that our effort should be to set right the intervening factors. It is known from several small experiments in the country and outside that if they are set right a large majority of students who cannot access higher education because of failure in mathematics will be able to do so. In a country where the state does not have the vision or will to provide quality education for all, where education is getting more and more privatized, it would be naïve to wish that the conditions would change in the immediate future. And it is this reality that provides us an opportunity (perversely) and forces us to take a critical look at the curriculum and ask what is necessary about it and if an alternative could be visioned.

The view of curriculum as an equalizer and the accompanying arguments are premised on an assumption that the nature of mathematical knowledge is given. It assumes that, asking questions about numbers and space as abstract entities, defined linguistically in very specific ways in order to enable unambiguous logical inferences using certain well defined rules of inference, is not just one way to make mathematical knowledge but the right way to make it and that this knowledge is of relevance to all. Indeed, perceived from within the discipline such a view may seem perfectly valid. However, this view of mathematics has been challenged by many from within mathematics education (Poovey, 2002; Ernest,1998; Mukherjee, 2010; Skovsmose, n.d). According to them, starting from its many cultural roots and historical moorings the discipline of mathematics as construed today, is one of the many ways in which the discipline could have emerged. To privilege the interest of the discipline over the interests of the learners is bound to fail those from the marginalized sections. To design a curriculum knowing that it cannot reach large categories of learners raises ethical questions and calls for imaginative resolutions. Tony Cotton says, "the idea that we can achieve equality in mathematics education by developing a curriculum that is identical for all actually goes against the idea of social justice. A social justice model would be based on the assumption that people are alike in some ways and yet different in some ways" (Cotton, 2001). There is a strong resistance against differentiated curriculum, from progressive thinkers in India, who fear that it would amount to caste based education and division of labor. But is it not obvious that in an unequal society, with increasing privatization of education, uniform curriculum would only marginalize the already marginalized? If the existing curriculum serves the interest of the discipline and the middle class, treating children from marginalized sections and any concern for their future as disposable, it becomes imperative that curriculum designers arrive at an alternate vision for upper primary mathematics curriculum that acknowledges the differences among the learners, goes beyond achieving numeracy and includes use of mathematics for critical understanding.

SOME QUESTIONS TO BEGIN ENQUIRY

Curriculum development warrants alternate, critical conceptions of the nature of mathematics that may depart from the notion that practicing mathematicians hold about the nature of the discipline. It demands engagement with alternate approaches to designing curriculum so that it can do more than imparting minimal levels of numeracy to the socio-economically marginalized or discovering and celebrating mathematics embedded in their cultural practices (Vithal & Skovsmose, 1997). To do so, the vision of mathematics curriculum developers should expand to include aspects of mathematical knowledge that contribute to critical understanding of one's social reality and a scope for empowerment. It should explore ways to organize learning of mathematics so that, it is not focused on individual achievements but on encouraging groups of learners who work together to acquire knowledge that they can share. It must demand not just the ability to solve problems but also that of inquiring into situations in which the knowledge can be productively applied, and the ability to communicate the knowledge to fellow learners. In chapters 1, 2 and 4 of his book *An Invitation to Critical Mathematical Education,* Skovmose (2011) gives some examples how this can be done even in a small way. Such approaches to mathematics curriculum development would have something to offer to those who are visible in the classroom only as disposable learners but who nevertheless bring into the classroom their ability to reason and a desire that sitting in the classroom for long hours each day would amount to something other than destroying their sense of self, dulling their minds and a training in subordination to authority. In fact such a curriculum would add value to the quality of mathematics education for all students in some very important ways.

Curriculum development has far reaching consequences particularly for those belong to the margins of the society and therefore has ethical commitments towards all the learners. Curriculum development should involve considerations which would term it unethical to design a curriculum that has very little to offer a large number of students with significant part of them belonging to social margins by virtue of their economic status, caste, gender, religion or geographical location. If the curriculum is to evolve in truly inclusive and socially responsive ways it will have to undertake academic inquiry and to engage with the questions that have been raised here.

ACKNOWLEDGMENT

The authors wish to thank their colleagues Rashmi Paliwal, Anjali Noronha, Arvind Sardana, C. N. Subramaniam, Rajesh Kindri, and Maxine Berntsen for discussions and inputs.

NOTES

1. However it should be noted that, there is a renewed interest in understanding the history of mathematics in India with the potential to point to alternate traditions of doing mathematics and there is also a parallel attempt by the jingoistic right wing to incorporate in textbooks some algorithms in the name of "Vedic Mathematics". See for example http://mptbc.nic.in/class%209%20to%2012.htm
2. Direct data on what percentage of children from rural and urban background finish school is difficult to come by. However, from studies on dropout rates we can infer that children from rural, poor and marginalized caste background get pushed out of the education system. Shanta Sinha and Anugula Reddy (2011) list inappropriate curriculum, poor quality and irrelevance of education as some of the reasons for school dropout.

3. While the relevance and meaningfulness of the curriculum could be critiqued on similar ground for other subjects as well (See Batra, 2011), in mathematics a student cannot secure minimum marks to pass the examination merely on the strength of one's memory while one can do so in other subjects.

REFERENCES

Agnihotri, R., Khanna, A. L. & Shukla, S. (1994). *Prashika: Eklavya's innovative experiment in primary education.* New Delhi: Ratna Sagar.

Banerjee, R. & Subramaniam, K. (2004). 'Term' as a bridge concept between arithmetic and algebra. In J. Ramdas & S. Chunawala (Eds.), *Proceedings of Episteme-1* (pp.76-77). Mumbai: Homi Bhabha Centre for Science Education.

Banerjee, R. & Subramaniam, K. (2005). Developing procedure and structure sense of arithmetic expressions. In H. L. Chick & J. L. Vincent (Eds.), *Proceedings of the 29th conference of the International Group of the Psychology of Mathematics Education.* Melbourne, Australia.

Banerjee, R. & Subramaniam, K. (2008) Bridging arithmetic and algebra: Evolution of a teaching sequence. In O.Figueras, JL Cortina, S. Alatorre, T. Rojano and A. Sepúlveda (Eds.), *International group of the psychology of mathematics education: Proceedings of the Joint Meeting of PME32 and PME-NA XXX (PME29),* Vol 2, 121-128, Morelia, Mexico

Batra, P. (Ed) (2010). *Social Science Learning in Schools: Perspective and Challenges.* New Delhi: Sage Publication.

Cotton, T. (2001). Mathematics teaching in the real world. In Peter Gates (Ed.), *Issues in Mathematics Teaching* (pp 23-37). Rouledge Falmer.

Ernest, P. (n.d), *Why teach mathematics.* Retrieved May 31, 2013, from http://people.exeter.ac.uk/PErnest/

Ernest, P. (1998) The nature of the mathematics classroom and the relations between personal and public knowledge: An epistemological perspective. In F. Seeger, J. Voigt & U. Waschescio (Eds.), *The culture of mathematics classroom* (pp. 245-268). UK: Cambridge University Press.

Gellert, U. & Jablonka, E. (Eds.) (2007). *Mathematisation and demathematisation: Social, philosophical and educational ramifications.* Rotterdam: Sense Publishers.

Govinda, R. (Ed.) (2002). *Indian education report: A profile of basic education.* National Institute of Education Planning and Administration & UNESCO. New Delhi: Oxford University Press.

Govinda, R. (2011). *Indian education report: A profile of basic education.* National Institute of Education Plannig and Administration & UNESCO. New Delhi: Oxford University Press.

EdCil (India) Limited. (2009). *Teachers' absence in primary and upper primary schools (Synthesis report of study conducted in Andra Pradesh, Madhya Pradesh and Uttar Pradesh).* Research, Evaluation and Studies Unit, Technical Support Group. Retrieved May 31, 2013, from http://ssa.nic.in/page_portletlinks?foldername=research-studies

Khan, F. A. (2004). Living, learning and doing mathematics: A study of working class children in Delhi. *Contemporary Education Dialogue*, Vol 1:2 pp 199-227.

Khan, F.A. (2005). Placing the text in context. In M. Kumar. & P. Sarangapani (Eds.), *Improving government schools: What has been tried and what works* (pp 41-48). Bangalore: Books for Change.

Khan, F. A. (2008). Formal Number Systems in the Context of Early Schooling. *Contemporary Education Dialogue*, Vol 6:1 pp 5-24.

Khemani, S. & Subramanian, J. (2012). Tackling the division algorithm. *TSG-7 International Conference of Mathematics Education: ICME-12*, Seoul, South Korea. Retreived 31, May 2013, from http://nime.hbcse.tifr.res.in/indian-participants-at-icme-2012

Kumar, M. & Sarangapani, P. (2005). *Improving government schools: What has been tried and what works.* Bangalore: Books for Change.

Leclercq, F. (2002). *The impact of education policy reforms on the school system: A field study of EGS and other primary schools in Madhya Pradesh.* New Delhi: Centre De Sciences Humaines.

Mukerjee, A. (2010). The Nature of mathematics and its relation to school education. *Learning Curve Issue XIV. Special Issue on School Mathematics.* Retrieved May 31, 2013, from http://www.azimpremjifoundation.org/pdf/LCIssue14.pdf

Naik, S. & Subramaniam, K. (2008). Integrating the measure and quotient interpretation of fractions. In O.Figueras, JL Cortina, S. Alatorre, T. Rojano and A. Sepúlveda (Eds.). *International Group of the Psychology of Mathematics Education: Proceedings of the Joint Meeting of PME 32 and PME-NA XXX (PME29)*, Vol 4, 17-24, Morelia, Mexico.

Nawani, D. (2002). Role and contribution of non governmental organizations in basic education. In R. Govinda (Ed.), *Indian Education Report: A Profile of Basic Education* (pp. 121-130). National Institute of Education Plannig and Administration & UNESCO. New Delhi: Oxford University Press.

National Council of Educational Research and Training (2005). *National curriculum framework*. New Delhi: NCERT.. Retrieved May 31, 2013, from http://www.ncert.nic.in/html/pdf/schoolcurriculum/framework05/nf2005.pdf.

National Council of Educational Research and Training. (2006a). *National focus group on teaching of mathematics report*. New Delhi: NCERT. Retrived May 31, 2013, from http://www.ncert.nic.in/rightside/links/focus_group.html#

National Council of Educational Research and Training (2006b). Syllabus for classes at elementary level. New Delhi: NCERT. Retrieved May 31, 2013 from: http://www.ncert.nic.in/rightside/links/syllabus.html

Nunes,T., Schliemann, A.D. & Carraher, D.W. (1993). *Street mathematics and school mathematics.* Cambridge University Press.

Poovey, H. (2002). Promoting social justice in and through the mathematics curriculum: Exploring the connections with epistemologies of mathematics. *Mathematics Education Research Journal*, Vol. 14, No. 3, 190-201.

Pratham organisation. (2012). "Annual Status of Education Report 2012". Mumbai, India. Pratham resource center. retrieved from: http://www.asercentre.org/education/India/status/p/143.html

The Probe Team (1998). *Public report on basic education in India.* Oxford University Press: New Delhi Retrived May 31, 2013 from http://www.undp.org/content/

india/en/home/library/hdr/thematic-reading-resources/education-/public-report-on-basic-education-in-india.html

Ramachandran, V. (Ed.) (2003). *Getting children back to school: Case Studies in primary education.* New Delhi: Sage Publication.

Sinha, S. & Reddy, A. N. (2011). School dropouts or "Pushouts"? In R.Govinda (Ed.), *Who goes to school?: Exploring exclusion in Indian education* (pp. 166-204). New Delhi: Oxford University Press.

Skovsmose, O. (2007). Students' foregrounds and the politics of learning obstacles. In U. Gellert & E. Jablonka (Eds.), *Mathematisation and demathematisation: Social, philosophical and educational ramifications* (pp 81-94). Rotterdam: Sense Publishers

Skovsmose, O. (2011). *An invitation to critical mathematics education.* Rotterdam: Sense Publishers

Skovsmose, O. (n.d). Mathematics: A critical rationality. Retrieved May 31, 2013 from http://people.exeter.ac.uk/PErnest/pome25/index.html

Streefland, L. (1993). Fractions: A realistic approach. In T.Carpenter, E.Fennema & T.A. Romberg (Eds.), *Rational numbers: An integration of research* (pp 289-325), Lawrence Erlbaum Associates Publishers

Subramaniam, K. & Naik, S. (2007). Extending the meaning of the fraction notation. In C. Natarajan & B. Choksi (Eds.), *Proceedings of Episteme-2,* 223-227, Mumbai: Homi Bhabha Centre for Science Education.

Subramaniam, K. & Banerjee, R. (2010). The arithmetic-algebra connection: A historical-pedagogical perspective. In J. Cai & E. Knuth (Eds.), *Early algebraization: A global dialogue from multiple perspectives* (pp. 87-107). Heidelberg: Springer.

Subramanian, J., Subramaniam, K., Naik, S. & Verma, B. (2008). Combining share and measure meaning of fractions to facilitate students' reasoning. *International Conference of Mathematics Education: ICME-11,* Monterrey, Mexico. Reterieved May 31, 2013, from http://tsg.icme11.org/document/get/823

Subramanian, J & Verma, B. (2009). Introducing fractions using share and measure Interpretation: A report from classroom trials. In K. Subramaniam & A. Majumdar (Eds.), *Proceedings of Episteme-3*, Mumbai: Homi Bhabha Centre for Science Education.

Subramaniam, K. B. (1998). *A study of informal acquisition of some elementary mathematical skills by uneducated.* Regional Institute of Education, NCERT. New Delhi

Umar, M. (2010). From fraction to proportion. *Seshnik Sandharb.* Vol 2, No. 11 (Original No. 68). Retrieved May 31, 2013, from http://www.eklavya.in/pdfs/Sandarbh_68/45-57_From_Fraction_to_Proportion.pdf

Verma, S. (2010). One hour of my class. Seshnik Sandharb Vol 2, No. 12 (Original No. 69). Retrieved May 31, 2013, from http://www.eklavya.in/pdfs/Sandarbh_69/57-64_One_Hour_Of_My_Class.pdf

Vithal, R. & Skovmose, O. (1997). The end of innocence: A critique of "ethnomathematics". *Educational Studies in Mathematics* 34: 131-158.

CHAPTER 73

STUDENTS' UNDERSTANDING OF ALGEBRA AND CURRICULUM REFORM

Rakhi Banerjee
Azim Premji University, Bangalore

ABSTRACT

This chapter reports on a study conducted in one school with a group of Grade 6 and 7 students and teachers to understand the impact of the new curriculum framework and the new textbooks on the teaching and learning of mathematics. The study highlights the complexity of any curriculum development process. It brings forth the need to address issues of teacher professional development focused on content enrichment as well as issues of their beliefs and values simultaneously in order to bring a substantial change in the teaching learning process.

Keywords: Algebra, curriculum, textbooks, students' understanding of algebra

Mathematics teaching-learning has been an area of concern for many teachers, parents, educators, and researchers due to constant failure of

The First Sourcebook on Asian Research in Mathematics Education: China, Korea, Singapore, Japan, Malaysia, and India, pp. 1607–1630

students in the subject and has been a reason for students dropping out of school. People do not largely question the importance of mathematics and validate its existence in the school curriculum due to its "supposed" utility in everyday life. For some others, it is the gateway to access higher education in prestigious disciplines or professions.

Even though one tries to make some connection with students' daily lives in the primary grades, it is perhaps hard to do the same in the middle school level. The concepts/ definitions and ideas grow in abstraction and the rules and the procedures gradually become delinked from concrete experiences, leading to a sense of arbitrariness in the symbol manipulation process. Studies in the past, such as those by Hart, Brown, Kuchemann, Kerslake, Ruddock and McCartney (1981) and Booth (1984), have shown students' difficulties in understanding the symbols and procedures in various areas of mathematics. There is no such systematic research in the Indian context, but the findings are likely to be true for India as well. To be successful in mathematics in the middle and secondary grades, it is essential for students to make the shift from mathematics based on concrete everyday experiences to one which is abstract, rule governed and symbolic. A quick look at mathematics taught in the middle school level reflects this abstraction with emphasis on definitions and procedures and building the capacity to use conventional symbols.

In the recent times, mathematics teaching that is only aimed at symbolic competence and ability to use rules and procedures in various problems (which leads to the exclusion of many from this strand of schooling) has been questioned. Like many other countries, India too responded to the challenge by engaging in a country wide deliberation resulting in a reformed curriculum document based on the constructivist philosophy, the National Curriculum Framework (NCF) 2005. (The different states of the country have been subsequently engaged in revising and drafting state level curriculum frameworks.). NCF 2005 strongly recommends the need to move away from the narrow aim of "utility" of mathematics to focus on higher aims of developing "child's resources to think and reason mathematically, to pursue assumptions to their logical conclusion and to handle abstraction. It includes ways of doing things, and the ability and the attitude to formulate and solve problems" (NCERT, 2006a, p. 42). The document further recommends the creation of a curriculum which teaches important and meaningful mathematics, which is coherent and engages children in communicating with mathematics and posing and solving problems. There is an underlying expectation of change in teacher attitude and action. They need to be well prepared to deal with the content and its pedagogic aspects as well as understand the motivation for the change. The classroom interactions need to be redefined to be able to engage students in the "(re-)creation" of mathematics. Thus, curriculum reform has multiple dimensions and there are many determinants: textbooks, teachers, classroom, and assessment.

These deliberations and ideas led to the drafting of a new set of textbooks developed by the National Council for Educational Research and Training (NCERT) that are now being used by schools which are affiliated to the Central Board of Secondary Education since 2006. Compared to the older textbooks, the presentation in the new books is more lucid, addresses the students directly, there is an effort to lend meaning to the concepts and procedures and there is a change in the way the content is organized and divided across grades (NCERT, 2006b; 2006c; 2006d). The assumption behind the change is to include more students in the teaching-learning process and achieve the "higher aims" of mathematics learning in the elementary school. Since textbooks are the major resource for most teachers (and for some the only resource), they carry most of the burden of conveying the principles and the philosophy of the new curriculum.

It is in this backdrop that a study was conducted in 2009 to understand the impact of the new curriculum framework and the new textbooks on the teaching and learning of mathematics in classrooms. The study was restricted to the area of algebra for Grades 6 and 7, where one finds a gradual introduction to the use of conventions and dealing with an abstract symbolism to communicate one's understanding. In this chapter, I will explore what sense students are making of introductory algebra which is a part of the new textbooks. In the process, I will try to evaluate the extent to which the visions of NCF 2005 are being fulfilled in terms of the content being taught, teacher preparation and classroom transaction of the content.

BACKGROUND

As indicated in the previous section, mathematics education does not have a distinct disciplinary standing in this country. Also, there does not exist any systematic body of research in the area. However, post NCF 2005, there has been some writing on the philosophy behind the new curriculum framework, its impact on how knowledge would be constructed in classrooms in different subject areas and issues which are likely to arise as a result of its implementation (e.g. Batra, 2006; Paliwal and Subramaniam, 2006; Saxena, 2006;). None of these address the area of mathematics. The new mathematics textbooks were reviewed, though not systematically analyzed, by Tripathi (2006) where she notes the marked improvement in the books for primary and secondary/ senior secondary but not at the middle school level where a large amount of the content receives the same treatment as earlier.

One can also draw from studies in some other countries in the past that have shown the difficulties in the implementation of reforms. Such difficulties arise due to lack of adequate content knowledge among teachers, inability to reach all sectors to train teachers, unavailability of textbooks and resource materials, teachers' differential perception of the curriculum with respect to the load, the content covered, different interpretations of the textbooks and the materials, unawareness of new techniques of assessment and the motivations for change. (Research Advisory Committee of the NCTM, 1988; Remillard, 1999; Swarts, 1998; Bulut, 2007). The use of curricular materials involves "interpreting the meanings and intents of these resources" and "the enacted curriculum is more than what is captured in the official document" (Remillard, 1999, p. 317). Several studies have also indicated the influence of teachers' consciously or unconsciously held beliefs, notions and preferences on the success of curricular reforms (Thompson, 1988; Handal & Herrington, 2003; Yates, 2006; Choksi, 2007). Teachers thus respond differently to curricular changes in varying situations – some resisting it, some changing their practices superficially to match the new expectations without changing their fundamental beliefs about the content and nature of mathematics and teaching and learning process and some changing their practices for the better.

Some scepticism for any reform comes mainly due to the lack of sufficient research base (empirical or theoretical) for many of the recommendations (Research Advisory Committee of the NCTM, 1988). In the context of this study, it is important to know that there is no consensus on the "best approach to introducing algebra" (for different approaches, see Mason, Graham, Pimm & Gowar, 1985). Each approach brings with it some solutions and some difficulties and they are all important to grasp the content and purpose of algebra. It is therefore important to know the difficulties in teaching and learning of algebra as well as in the use of different approaches to introduce algebra while writing the textbooks. Past studies have shown the demands that each of these approaches puts on the students and the core ideas which students must grasp in order to understand algebra (e.g. Kuchemann, 1981; Booth, 1984; MacGregor & Stacey, 1997; Stacey & MacGregor, 2001; Banerjee, 2011).

One of the popular alternatives to introduce algebra is through generalization of patterns of shapes and figures. This is the approach which the new textbooks use for introducing algebra in Grade 6. This context for algebra indeed affords many possibilities in the classroom, in that it provides a situation which can be seen in multiple ways by the students and thus arrive at seemingly different relations and their representations. It is important to focus on multiple ways of arriving at a functional relation to describe the pattern. Verbalizing and communicating the structures and relationships that students observe within the shapes/ figures in the pattern is an important aspect in the whole process. It is equally important to discuss appropriateness of symbols and equivalence of different rules/ generalizations. Thus a complementary set of skills and understanding must be developed to manipulate symbols with understanding. Students must develop a sound understanding of procedures of transforming expressions and its relation to the structure of expressions (having a sense of the composition of the expression, how the components are related to the value of the expression and their relation among each other) (Subramaniam & Banerjee, 2011). They must also understand that any valid transformation of the expression would leave its value invariant. The evolution of a teaching approach based on some of these above ideas, to bridge the transition from arithmetic to algebra for Grade 6 students, can be found in Banerjee (2008).

Broadly speaking, the goal of the teaching and learning of algebra is the ability to think algebraically. Algebraic thinking is the kind of thinking involved in analysing relationships between quantities, noticing structure, studying change, generalizing, problem solving, modelling, justifying, proving and predicting (Bell, 1995; Kieran, 2006). It is possible to engage in this form of thinking without working on the letter and the conventions of algebra. This idea of algebraic thinking would also be consonant with the vision of NCF 2005 to move away from narrow aims to higher goals of teaching mathematics.

METHODOLOGY

The study being reported in this chapter was carried out in an English medium school belonging to a nationwide chain of schools run by the Government of India, largely for government employees. It caters to a heterogeneous population with students from different backgrounds – parents workings as officers as well as lower cadres in the defence sector, employees in the school, low and medium income government/ public servants and parents working in the private sector. The school had three mathematics teachers who were teaching at the middle school/ secondary level. This school appeared to be progressive in its vision of education, teaching and learning and designed ways and means to support students' learning of mathematics. It also engaged teachers in discussions about NCF 2005 and content of school mathematics at various levels, with opportunities to participate in workshops and training sessions for enrichment.

The aim of the study was not to carry out a survey of different schools to understand the impact of the new NCF but to study in-depth one school's efforts to make sense of it and change its practices. The study tried to analyze the content of algebra which is part of the textbooks, explored students' understanding of various symbols, concepts and procedures, ways in which teachers' knowledge of the content of algebra influenced students' understanding or meaning making processes, and explored the same teachers' understanding of the NCF and the new textbooks. It aimed to understand the extent to which the vision of NCF was being implemented in the context of the middle school mathematics teaching and learning and the extent of teachers' preparation for it. It also sought to evaluate students' learning in this context.

A combination of tools were used to collect the data – classroom observations, informal discussion with the teachers after the class and semi-structured interviews with them, written test for students followed by interviews with some selected students. Classrooms of two teachers (henceforth referred to as T1 and T2), both of whom taught one section each of Grades 6 and 7 (referred to as 6A, 6B, 7A, 7B), were observed as they taught algebra. In all, 20 lessons were observed, 14 for T1 and 6 for T2. Semi-structured interviews were conducted with these two teachers at the end of the observations as well as with the school principal in order to get an overall sense of the school, its vision, its initiatives, its policies. All students in Classes 6A, 6B, 7A, 7B were administered a common written test. In all, 85 students from Grades 6 and 7 each took the written test (total 170 students). Subsequently, 12 students from Grades 6 and 7 each were interviewed (6 from each of the four sections, total 24 students). The students were selected by the researcher on the basis of their performance in the test. Two students each were randomly selected from below the class average, average and above average in the test.

ALGEBRA IN GRADES 6 AND 7 TEXTBOOKS

The new Grade 6 textbook contains one chapter of algebra and introduces it as generalization of simple patterns (see Figure 73.1). It is done in the context of growing patterns of simple shapes, like A, L, V, thus using only one operation while expressing the generalization (in the case of the above examples, all expressions will be of the form $a \times x$). Other situations include writing an expression for recording rules and formulas, like area, perimeter, relations between cost per item and number of items, properties of operations (the generalized expressions are of the form $a \pm x$). The fact that a letter stands for a number and it acts like

11.4 More Matchstick Patterns

Ameena and Sarita have become quite interested in matchstick patterns. They now want to try a pattern of the letter C. To make one C, they use three matchsticks as shown in Fig. 11.2(a).

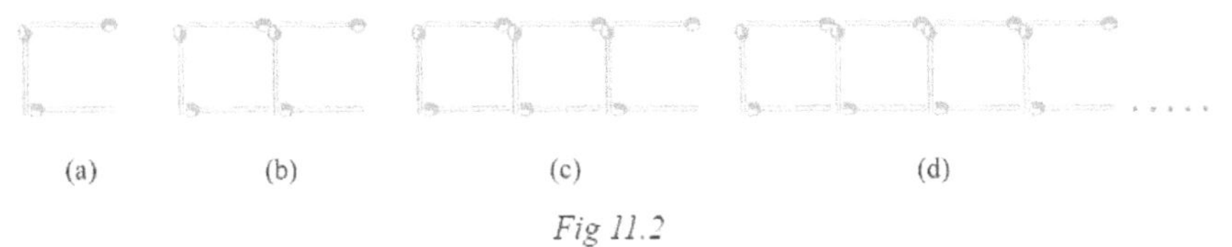

Fig 11.2

Table 2 gives the number of matchsticks required to make a pattern of Cs.

Table 2

Number of Cs formed	1	2	3	4	5	6	7	8	....	....	....
Number of matchsticks required	3	6	9	12	15	18	21	24	....	....	....

Figure 73.1. A sample pattern generalization exercise from the mathematics textbook of Grade 6.

a variable is repeatedly stressed throughout the chapter. Towards the end of the chapter a flavor for its use as an unknown in equation solving, again in the context of pattern generalization activity is provided. However, the chapter does not introduce formal ways of solving an equation but restricts it to the use of tables for identifying the solution. A small section is devoted to the discussion about arithmetic and algebraic expressions. Brackets are used to denote precedence operations like $3+(4\times5)$ whereas, an algebraic expression like $4x+5$ is itself an answer, and can be further computed only when the value of the letter is given. In this way it highlights the procedural/ operational aspect of expressions and symbols. Other situations like translating algebraic expressions into natural language and vice-versa also emphasise the procedural/ operational understanding. For example, $4x+5$ means "multiply a number by 4 and add 5 to it". It does not discuss this expression as a relation (an "object" meaning) expressing a number: "5 more than the product of 4 and any number".

There are two chapters in the Grade 7 book devoted to algebra: simple equations and algebraic expressions. The chapters are spaced wide apart in the book, equations chapter to be covered in the beginning of the year and the expressions chapter towards the end of the academic year. However, the classroom observation was restricted to the algebraic expressions chapter.

Equations are introduced as a condition on the variable, and then solved first by trial and error and then by the "balance model". Finally, the "transposition rule" is stated, followed by examples demonstrating how it works. Generally, simple linear equations with only one variable on one side of the "=" sign are discussed. All solutions are verified in an effort to underline the significance of the "=" sign. The chapter on algebraic expressions introduces algebraic expressions as a result of a combination of operations on numbers and letters (see Figure 73.2). The chapter includes linear as well as non-linear expressions with one or more variables. The rest of the chapter deals with identification of different components of the expression (terms, factors, coefficients, etc.), like and unlike terms, different types of polynomials and addition and subtraction of polynomials using the distributive property. Finally, the chapter introduces evaluating algebraic expressions for a given value of the letter and using letters to record rules and formulas. The last section focuses on pattern generalization which are slightly more complex than what was done in Grade 6 (of the form $ax + b$).

12.2 How are Expressions Formed?

We now know very well what a variable is. We use letters x, y, l, m, ... etc. to denote variables. A **variable** can take various values. Its value is not fixed. On the other hand, a **constant** has a fixed value. Examples of constants are: 4, 100, –17, etc.

We combine variables and constants to make algebraic expressions. For this, we use the operations of addition, subtraction, multiplication and division. We have already come across expressions like $4x + 5$, $10y - 20$. The expression $4x + 5$ is obtained from the variable x, first by multiplying x by the constant 4 and then adding the constant 5 to the product. Similarly, $10y - 20$ is obtained by first multiplying y by 10 and then subtracting 20 from the product.

The above expressions were obtained by combining variables with constants. We can also obtain expressions by combining variables with themselves or with other variables.

Look at how the following expressions are obtained:

$$x^2,\ 2y^2,\ 3x^2 - 5,\ xy,\ 4xy + 7$$

(i) The expression x^2 is obtained by multiplying the variable x by itself;

$$x \times x = x^2$$

Just as 4×4 is written as 4^2, we write $x \times x = x^2$. It is commonly read as x squared.

Figure 73.2. An introduction to algebraic expressions in Grade 7 mathematics textbook.

Analysis of the Algebra Content

A careful look at these chapters reveal that although the treatment of algebra in Grade 6 is very different from the older books but the difference in new and the old Grade 7 textbooks is less evident. No doubt that the presentation is more lucid and effort is being made to communicate to the student, but there is no radical conceptual departure from the earlier one. There are many evident difficulties in the way the algebra content in both grades is presented. The introduction to algebra in Grade 6 does not follow the formal and symbolic transformation approach but rather situates it in the context of pattern generalization. The change is a result of the view that rule-based syntactic transformations do not carry any meaning for students while pattern generalization provides students a space to construct meaning for the symbols in algebra. The simplicity of the tasks may help students to engage with algebra and may be found less daunting. However, the situations used are too simplistic (of the form a $\times$ x or a $\pm$ x), especially for Grade 6, to convey the power of algebra as well as convincingly demonstrate the need for letters and their use as variables. Pattern generalization is not an activity only to record a rule using algebraic expressions but is also supposed to inculcate algebraic thinking – explore multiple ways of seeing the same pattern, analyse the relationships between various components, especially find different strategies to count, use this strategy in order to continue the pattern, generalize the strategy and express it using algebra. Further, it should lead to discussions about equivalence of two or more such generalizations/ rules for any given pattern. The task taken in its entirety is not simple and potentially poses many challenges for the students. But it could also create a forum for fruitful discussions about notations and conventions, writing expressions using the letter, transformations of algebraic expressions, ideas of equality, equivalence; these being integral aspects of learning algebra. A consequence of using only very simple patterns, which would lead to rules of the form a$\times x$ or a$\pm x$, is that one is more likely to miss out on the opportunities that the generalization activity can provide for learning a substantial portion of algebra.

Moreover, the chapters in both the grades introduce students to arithmetic and algebraic expressions only as encoding procedures. This reinforces a procedural understanding of expressions rather than helping students appreciate the duality – that expressions encode operations (procedural / operational) but also are the products of those operations ("object"). This is important for making sense of symbolic algebraic expressions and manipulating them. In Grade 6, the algebra chapter introduces both arithmetic and algebraic expressions, without any background of evaluation of arithmetic expressions with multiple operations.

There is no discussion of precedence of operations (or order of operations) in the context of arithmetic expressions and brackets are used to signify precedence of certain operations. One may wonder about the extent to which this strategy is useful in making sense of algebraic expressions and transformations on them. A large part of the algebra chapter in Grade 7 textbook emphasises manipulation of algebraic expressions and only a small section is devoted to pattern generalization (which again does not highlight the processes involved in generalization). The difficulty for the student lies in the fact that enough groundwork has not been done either in Grade 6 or in the earlier chapters of Grade 7 to deal with such abstract symbolism and operations on them. In the context of algebra, the precedence of operations needs to be internalized and transformations on expressions are governed by properties of operations, including understanding of bracket opening rules. Transformational activities in algebra must lead to an understanding of properties of operations, discussions about equivalence of expressions and equations (value of an expression, operations or transformations which leave the value invariant), and the relation between transformational rules in arithmetic and algebra.

Overall, the chapters in both grades do not seem to adequately take into consideration the difficulties which students face in learning algebra and the prerequisites for learning them and the affordances and limitations of any approach to introducing algebra. The chapters do not convey clearly the purpose of algebra and the various tasks, its content and the processes which are important to learn algebra. In the remaining part of the chapter, I will discuss students' understanding of beginning algebra and the different aspects they have learnt in Grades 6 and 7 together with teachers' own understanding of the new curriculum and the content and purpose of algebra.

TEACHERS AND THE TRANSACTION OF ALGEBRA IN THE CLASSROOM

The way a curriculum or a textbook is transacted depends a lot on the teacher's own understanding of the subject matter and his/ her beliefs about teaching and learning of mathematics and the students who study that mathematics. Of the two teachers whose classrooms were observed, one had a Master's degree in mathematics and the other had a Bachelor's degree. Both of them had a degree which certified them as teachers. Teacher T1 had 23 years of experience in teaching mathematics and teacher T2 had an experience of 33 years of teaching mathematics, out of which she had spent 12 years in teaching in the primary school. They had been working in this particular school for more than 5 years. Teacher T1 had a slower pace than teacher T2, who finished the chapters in around half the number of lessons taken by teacher T1. Moreover, teacher T2 could not be observed by the researcher (for reasons beyond her control) while teaching ideas of pattern generalization to any of the classes.

Both the teachers held beliefs about teaching and learning of mathematics which can be said to be largely traditional. They considered teaching of concepts, systematic writing of procedures and adequate practice to be important. According to them, students' own motivation, their family backgrounds (parental education, outside help like private tutors), their language skills (especially in English), acted as factors influencing their success in the classroom or in the tests. Even when they appreciated the changes brought forth by NCF 2005 in terms of child centeredness and activity approach to teaching and learning, they were unable to see the larger aims of teaching mathematics professed by the document. This of course influenced their own teaching in the classroom (as will be discussed below), together with paucity of time to do things in a way which would be more aligned with the philosophy of NCF 2005 and the new textbooks.

The teachers also had a narrow understanding of algebra. Teacher T1 described algebra "as a branch of mathematics which deals with operations on constants and variables". On the other hand, teacher T2 considered algebra to be "like a puzzle", "where one learns to solve problems". Their descriptions did not highlight any processes like generalization or thinking/ reasoning with symbols as a characteristic of learning algebra. Even when the teachers saw some connection of algebra with other branches of mathematics (like arithmetic for teacher T1 and geometry for teacher T2), they did not use these ideas to teach it in the classroom. Teacher T1 was able to involve her students in the teaching learning process more often than teacher T2, by eliciting more examples and information about the concept or task. Teacher T2 tried to do so sometimes by seeking help from students to complete the solution to a routine transformation exercise on the blackboard. However, her fast pace did not give enough time to most students to think and complete the solutions. On top of it, her emphasis on systematic writing of solutions seemed daunting for many, more so, when they did not understand the reason for the solution process. This made the classroom somewhat monotonous and quiet.

Both teachers experienced difficulties in explaining some of the ideas in the class. This was a result of their own experiences of having learnt and taught mathematics for a long time in a particular way as well as inadequate understanding of the pedagogical content knowledge required to teach the content. Both the teachers spent a lot time discussing and recording rules and procedures. Even though teacher T1 actively involved her students in solving various tasks/ problems, she treated problems in a narrow way. She too focused on arriving at an answer and not on the processes and some of the discussion was extraneous to the problem itself. The pattern generalization activity was dealt in a manner as if the goal was to find *the* rule for counting matchsticks in various patterns of shapes/ figures. The impression most students would have gathered would be that it has one fixed method and one fixed solution. When faced with a situation like "what is the result of $x+5$" in both Grades 6 and 7, she found it difficult to find an explanation suitable for the students in these grades. The students did not have any knowledge of order of operations but had some knowledge of distributive property in the context of numbers. However, the teacher used the same kind of explanation in both places – 5p + 5p = 10p as 5 pencils and 5 pencils are 10 pencils but 8p + 5r are 8 pencils and 5 rubbers which cannot be written as 13 pencils and rubbers. This explanation has obvious difficulties and students did not seem to be very convinced by this. She had no alternative but to rely on her past experience of having used it in combination with distributive property and students' understanding of order of operations. Teacher T2 also faced similar difficulties while students simplified but not being fully aware of the sources of such difficulties, she tended to repeat the standard rule and procedures as explanations. In fact during the interview, T2 commented that "what we have studied indirectly gets introduced to the child irrespective of them having included or not included in the textbook. I always teach them the way I have learnt. ...What we have studied is basic; they cannot go out of the system even if the book does not mention it". She was referring to the BODMAS rule in this context.

In the rush to finish the course content and other administrative responsibilities that teachers often carry, the higher aims of mathematics, which is algebraic thinking in this case, got compromised. The interviews with the teachers failed to elicit responses which could indicate their awareness of the larger goals of teaching mathematics – preparing to think mathematically, reason, argue, understand and skilfully use symbols, and especially algebra; or reasons for doing particular tasks, like patterns or symbol manipulation, as they appear in the textbook. Need for practice, entry level preparedness to deal with mathematics content seemed to override the need for learning to think, reason mathematically. They were candid about the difficulties they face, given the fact that curriculum development is a top-down process in this country and on the whole teachers do not get sufficient training or help in understanding or interpreting the changes. Moreover, they felt that the primary school does not prepare students enough in mathematics concepts and procedures (though it instils confidence in them) that they can cope with the demands of Grade 6.

STUDENTS' UNDERSTANDING OF ALGEBRA

Students' understanding of algebra was revealed through the written test and interviews conducted with six students from each of the four sections. In Grade 6, the algebra items mainly consisted of pattern generalizing task (Repeating H and Triangles, see Figure 73.3) which involved identifying number of matchsticks for given figures (GF), for next few figures (near generalization "NG", e.g. number of matchsticks for 4th, 6th figure), for figures distant from the given ones (far generalization "FG", e.g. number of matchsticks for 15th, 38th figure) and a general rule (GR). Other items involved equation solving through trial-and-error (ES) and translating sentences in natural language to algebraic expressions (AE) and vice-versa (NL). Many other items checked for their understanding of arithmetic expressions and multiple operations on them and integer operations.

Q7. How many matchsticks will be required for continuing this pattern?

(1) (2) (3) ....

Number of H	1	3	4	6	8	...	15	...	38	n
Number of matchsticks										

Q8. How many matchsticks will be required for continuing this pattern of triangles?

(1) (2) (3) ....

Number of triangles	1	2	3	4	5	...	10	...	22	n
Number of matchsticks										

Figure 73.3. Pattern generalization task for Grade 6 students: Repeating H and Triangle.

Table 73.1. Performance of Grade 6 Students on Algebra Items (n = 85)

Item	*Repeating H*				*Triangle*				*ES*	*AE*	*NL*
	GF (2)	NG (3)	FG (2)	GR	GF (3)	NG (2)	FG (2)	GR	x-4 = 12	(4)	(4)
% success	71	54	36	25	58	24	9	1	10	49	40

Note: The numbers in the bracket indicate the number of items in each category.

Grade 6 Students' Understanding of Algebra

The Table 73.1 gives the performance of students in Grade 6 on the various algebra items.

Students' understanding of pattern generalization. The table indicates that they have very poor understanding of the process of generalization. This was expected, largely because of the way this was dealt in the classroom, and its presentation in the textbook. It was fairly difficult for a few of them to even count the number of matchsticks for the given figures and think systematically about extending the pattern and find a general rule. This difficulty increased with more complex figures (e.g. Triangle), where most students could not identify the number of matchsticks for figures near or far from the given figures or come up with a generalized rule. Responses of a large number of students to most of the items/ positions are unrelated numbers or a combination of numbers and letters; there seems to be hardly any pattern or underlying logic for these responses. In spite of errors in specific numerical positions, a few students (5) managed to write the correct general rule in the "Repeating H" pattern. This could be a result of recognizing a known pattern rather than understand what it means to generalize patterns. Quite a few others (~25%) found the number of matchsticks for specific numerical positions, both near and far from the given figures, but failed to write the correct rule for the same pattern.

The interviews with students further substantiated the results of the written test. Students who had performed reasonably well in the test (above class average) were the ones who attempted to answer the questions during the interview. Only a few students could explain their response to the pattern generalization task, others were not able to articulate what the task meant. These responses were largely received for the "Repeating H" task, the "Triangle" task was much harder for majority of the students. One of the students explained that "*to make 1 H, 5 matchsticks are needed. So for 3 H, I have multiplied 3 by 5 and got 15*" (VU). When the interviewer presented this student with another rule for the same pattern

$4n+n$, he could also see how counting the matchsticks in the pattern could lead to this rule "*4n would mean the matchsticks on top and the bottom and n could mean the matchstick in the middle*". Although only a couple of students could respond to this task in this manner, it provided an opportunity to at least these students to connect ways of counting with pattern generalization, which otherwise seemed to them to be a mechanical activity of finding *the* correct answer. The others were intrigued by the presence of the letter "*n*" and gave random explanation for the general rule, like "*we have to add this, write the answer and then write n*".

Students' understanding of equations. Equation solving as a concept and solving it using trial and error were also poorly understood. Most of the students (see Table 73.1) could not find the value of the letter in an equation with a single variable through a process of trial-and-error. Similarly, only a few could state what an equation $x - 4 = 12$ meant: "*a number from which if we subtract 4, we will get 12*" (KS). Most of them thought that "=" "*tells us the answer*". Some of them in further attempts could state that $2 + 5 = 7 + 1$ is a wrong statement but $2 + 5 = 10 - 3$ is a correct statement, demonstrating a correct understanding of "=". Moreover, most of the students who managed to solve the equation also had a correct conception of the letter as an unknown number and considered an equation like $m - 4 = 12$ to have the same solution as $x - 4 = 12$ because: "*no matter what variable we write, we will get the same answer when we subtract 4 from 12*" (MS). Students who did not understand the meaning of equations also revealed a poor understanding of arithmetic/ numerical equality.

Students' understanding of simple algebraic expressions. Students did not do too well in translating algebraic expressions into natural language and vice-versa (see Table 73.1). Students performed slightly better in translating a sentence in natural language into algebraic expression. However, students displayed the common errors of conjoining (i.e. writing $2+x$ as $2x$ or $x2$), or literal translation (translating "8 subtracted from b" as $8\text{-}b$) or conjoining to write $-8b$ or $b8$. In a more complex item like "z is multiplied by -2 and 5 is added to the result", responses consisted of all combination of letters and numbers (e.g. $-2z+5=2$, $z+2$, $z\times\text{-}2=2z=2z+5=7z$, $z\text{-}2+5$), giving a sense that the students hardly understood expressions either as encoding procedures or as stating relationships. Thus, even though these questions did not require any manipulation of the symbols but only representation using the letter, students on their own made these transformations on the symbols to arrive at a "closed" answer. Many of these errors are standard and have been documented elsewhere. It is likely that they were using their understanding of numbers and operations and the little exposure to letters that they had in order to make sense of these tasks and the idea of expression. One gets enough evidence of the fact that they do not understand the meaning of the letter and its use very well. A doubt also surfaces about the reasonableness of dissociating manipulation of symbolic expressions from using the symbol for representation, as in these questions.

Students' understanding of operations on arithmetic expressions and its relation to algebraic expressions. The items which required them to solve a few arithmetic expressions using properties of operation like commutativity, associativity or distributivity showed their inability to deviate from sequential operations on the expression (that is, moving from left to right) and many were seen to perform operations by vertically arranging the numbers. The performance dipped in the expressions with multiple operations (e.g. 1343+125-343 or 215×14+215×16) compared to those which had a single operation of addition or multiplication (e.g. 243+169+257 or 165×5×20).

Students' performance in the different tasks in algebra was not very satisfactory. The nature of students' responses in the arithmetic tasks indicate that for them operation signs still indicate an action of computation making it harder for them to understand algebraic expression as relations and representations. This perhaps is leading to a poor grasp of introductory algebra in Grade 6, where the letter has been introduced in contexts of representation of relationships. Although, the letter has not been imbued with the operational properties yet, students' intuitive sense-making of symbols is happening through operations, which is what they have experienced all through the primary grades.

Grade 7 Students' Understanding of Algebra

In Grade 7, the algebra items tested for pattern generalization (repeating and overlapping H, see Figure 73.4), solving equations (ES) and framing and solving equations in one variable (FSE), simplifying algebraic expressions (SAE, which included items with only lettered terms "LT" and items with numbers and lettered terms "NLT") and finding the value of an expression given the value of the letter (Value). Like in Grade 6, there were a few items testing students' understanding of relevant arithmetic expressions and their evaluation using properties of operations and integer operations.

We can see once again that students' performance by and large was not very satisfactory, especially in the more complex tasks (see Table 73.2).

Students' understanding of pattern generalization. Students' responses to the pattern generalization task again revealed that many did not appreciate generalization as a process; a few continued to write unrelated numbers. Some others (~20%) now tried to use recursive relations to predict the number of matchsticks but this worked well only for the

Q5. How many matchsticks will be required for continuing this pattern?

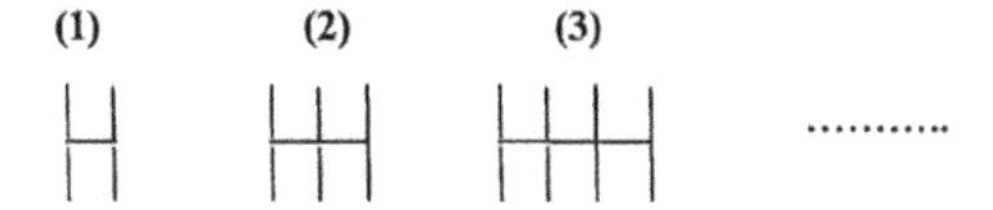

Number of H	1	2	3	4	5	...	15	...	38	n	n+2
Number of matchsticks											

Figure 73.4. Pattern generalization task for Grade 7: Repeating and overlapping H.

Table 73.2. Performance of Grade 7 Students on Algebra Items (n = 85)

Item	*Repeating/Overlapping H*				*ES*	*FSE*	*SAE*		*Value*
	GF (3)	NG (2)	FG (2)	GR	(4)	(2)	LT (2)	NLT (3)	(3)
% success	75	62	28	1	48	51	81	44	70

Note: The numbers in the bracket indicate the number of items in each category.

immediate few positions after the given ones (near generalization). They made errors in trying to generalize this pattern. Quite a few students (~15%) actually were able to find the answers to specific numerical positions but failed to generalize their procedure into a rule. However, these students did not show any working to support their answer and therefore it is hard to infer the reasoning used. During the interviews, of the five students who had attempted the pattern generalization task beyond the given figures, only one of them managed to complete it by systematically generalizing his counting action. The rest said that they had added 3 repeatedly or multiplied by 3 but could not write an expression to indicate what they had done and thus could not generalize their rule.

Students' understanding of symbol manipulation in the context of equations and simplifying algebraic expressions. The overall performance in solving equations is less than 50% (see Table 73.2). There were in all four items, of which only one contained a variable on both sides of the "=" sign, solved by only eight students. The rest were of the form $ax \pm b = c$. In each of the items, a few students (~6) tried to use an arithmetic way of solving the equation (that is, solving $5m + 9 = 24$ as $5\times3 + 9 = 24$). Majority of the students used the standard algebraic way of solving the equations and made numerous errors in symbol manipulation, including sign errors. For example, the above equation was solved as: 24 – 5m + 9 = 24 – 14 = 10 => 5 + 10 + 9 => m = 10 or 5m + 9 = 24 – 5 => m + 9 = 9. This shows a poor understanding of both the letter and the "=" sign.

Students performed similarly in the task that required them to first form an equation for the given situation and then solve them. A few students (6-8) could formulate the equation but could not solve it correctly. Many others (~30%) either formulated a wrong equation or solved it incorrectly. Random symbol manipulation was also seen in the context of simplification of algebraic expressions. Students conjoined terms, ignored the letter, ignored brackets and made sign errors. For example, students added $5y + 10$ and $6y + 4$ as $15y - 10y = 5y$ or subtracted $12m - 2$ from $14m - 2$ as $12m - 2 - 14m - 2$ or simplified $2x + 5y - 8 - 3y + 6x - 5$ as $8x + 8y - 13 = 16y - 13 = 3y$. Students were more comfortable adding and subtracting expressions with only two like terms (e.g. $2y + 6y$).

In the interview, most students repeated rules that they had learnt to explain their solution for simple equations or simplification of algebraic expressions. For example, one student (MA) who had used an arithmetic procedure in the written test to find the solution for $5m + 9 = 24$ showed the algebraic solution and said that "*the signs change when numbers are moved to the other side*". All students were quite sure that changing the variable did not impact the solution. However, they were quite unsure of the impact of adding/ subtracting the same number (say, 5) from both sides of the equation ($5m + 9 = 24$) or adding a number (say, 9) to one side and subtracting the same number from the other side. In most cases, their guess was that the solution will change. A few students correctly thought that the value will not change in the first situation but would change in the second but could not articulate the reason well. For the first situation one student (UC) said that "*-5 and +5 will get cancelled*" ($5m + 9 + 5 = 24 + 5$). Another student (AA) said for the latter case that "*-9 and +9 will get cancelled. On the other side, 9 will be added to 24, dividing the result by 5 would give a different value of m*"($5m + 9 - 9 = 24 + 9$). Some others thought that it would be different in both cases as "*the numbers are different*" or that it would be different in the first situation and same in the second one due to obvious reasons – "*I would take the 5 and -5, and 5, 5 would make this 34 and then 34 – 9*", and that "*+9 and –9 would become 0*". Most students were sure that one has to add like terms to simplify algebraic expressions and that one cannot add unlike terms. However, they had little understanding about the reason for doing so or of the fact that transformation of expressions leaves their value same or that they are equivalent expressions. Evaluating the algebraic expressions (the starting expression and the simplified one) for a value of the letter did not convince most of them of the equality of the two expressions.

Students' understanding of evaluating algebraic expressions. The question on finding the value of an algebraic expression given the value of the letter was interesting. More students were able to perform the task successfully (see Table 73.2). A small percentage of students (~13%) did not show any understanding of substitution or subsequent simplification (e.g. $3q - 7$ for $q = 5$ was solved as $3q - 5 = 29 - 7 = 5q$ or as $35 - 7$). Quite a few students substituted correctly and simplified the resulting arithmetic expression correctly, following the correct rules of evaluation. There were some students who could substitute correctly but failed to simplify the more complicated arithmetic expressions, like those resulting from substituting the value of the letter in $5q + 13 - 2q$. This indicates that students by the end of Grade 7 understand algebraic expressions procedurally and are able to think of the letter as standing for a number. However, they have poor understanding of rules of evaluating arithmetic expressions, making it harder to use them consistently to simplify algebraic expressions. They may, in fact, be finding the rules of transforming algebraic expressions arbitrary.

Students' understanding of arithmetic expressions and its relation to their understanding of algebraic expression. Students' responses to evaluation of arithmetic expressions show that most of them evaluated expressions in the left to right order, solving each operation sequentially and some still aligned them vertically. Very few could find easy ways of evaluating expressions like $1343 + 125 - 343$. Like for Grade 6 students, it was slightly better in the case of expressions containing only multiplication sign (e.g. $15\times75\times2$) where they could identify easy ways of evaluation. They also showed very poor understanding of the distributive property or order of operations (especially when expressions involved both "+" and "×" sign).

To be able to parse expressions with multiple operations into distinct component "terms", combine these components flexibly using relevant properties of operations are some skills and concepts that are used while transforming algebraic expressions and are easier to learn in the arithmetic context. Since most students have not abstracted these rules, it was hard for them to understand manipulation of algebraic expressions or equations. Students' written answers and interview responses indicate that they are largely bound by procedures and rules, without adequate understanding of concepts and reasons for the procedures. They also displayed a poor understanding of symbols like "=". In most situations, even after two years of learning algebra, students seemed to be doing algebra on/with symbols but the actions were devoid of any meaning.

DISCUSSION AND CONCLUSION

The analysis of the textbook material and the data collected from classroom observations, teacher interviews and students indicate that the nature of curriculum reform is fairly complex. The textbook analysis reveals the kind of demands the new textbooks place on students in order to understand algebra. The new textbooks made a significant advance over the earlier ones by introducing algebra using a new approach, that of pattern generalization (numerical as well as shapes of figures), to students in Grade 6 in very simple situations. However, this approach failed to instill among the students a way of thinking crucial to algebra – algebraic thinking, due to several interrelated reasons seen in this specific situation. Using the data analysed in the previous sections of the chapter, I will try to highlight these in the next few paragraphs.

The books do not adequately highlight the processes involved in pattern generalization. This is largely a result of the very contrived nature of the patterns taken for discussion, which are restricted to a$\pm x$ and a$\times x$, and in turn lead to very fixed ways of looking at the pattern and recording rules. It also overemphasizes a procedural understanding of symbols and expressions, rather than helping the students to deal with the complex process-product ("object") duality. Moreover, algebraic transformations, which were the highlight of the old textbooks, have not been totally abandoned but pushed to the next grade (Grade 7). Unfortunately, the preparation required for understanding such transformation on algebraic symbols is missing from both Grades 6 and 7. They do appear in discrete places but not connected enough with algebraic symbol manipulation, with the result that students on their own do not draw these connections. An earlier study (Banerjee & Subramanian, 2012) had shown the necessity of drawing explicit connections between arithmetic and algebra and substantial discussion about rules, concepts and procedures in both contexts for students to use them consistently across different situations and develop an understanding of them. Suddenly, there is a huge cognitive demand placed on students to understand multiple meanings of the letter and other symbols, notations and conventions and accept the expression, rather than a number, as an answer ("lack of closure"). The textbooks do make the effort of conveying these to the students as cautionary notes; however, there is no opportunity created for students to internalize these by being convinced or by developing justifications. Similar difficulties are observed while teachers transact the content of the textbook. More research within the country on different areas of school curriculum would provide resources and alternative approaches for the textbooks and may serve the purpose of making the books better for everyone.

The responses of students to various questions in the written test and interview display their poor understanding of almost all aspects of algebra that have been taught to them in the two years. They did not seem to understand the process of generalization or a way of thinking involved with generalization. Most students understood it as having a fixed correct answer, that is, a single rule, emanating out of a standard procedure, just like all other tasks in mathematics. Similarly, they did not think about underlying reasons for transforming algebraic expressions or equations and thereby missed crucial ideas. They based their responses (written or interview) on mere rules, which were sometimes remembered correctly and at other times incorrectly. They had not grasped the important idea of equality/ equivalence of expressions or equations which underlies all such transformations. Many of them were not even sure of arithmetic (numeric) equality. A positive point was that all students interviewed said that the letter denoted a number. But they failed to endow the letter with operational properties of numbers, which is important for manipulating expressions with letter. Students probably understood algebraic expressions as procedures for computation, an interpretation highlighted by the textbook and also used by the teachers in the classroom. However, they had no means to understand the computation of arithmetic expressions with multiple operation signs and the fact that an expression has a *unique* value. Moreover, Grade 6 students on their own seemed to manipulate algebraic expressions (that is, looked for closing the expressions to a single termed answer) when they had not yet been introduced to it formally. This is not very surprising given the emphasis on a procedural understanding of expressions, without highlighting the fact that they also express relations and manipulation must follow certain properties of operations.

In this context, it is important to underline the classroom processes that play a major role in students' construction of new ideas. A framework document and rewriting of the textbooks is not sufficient to change the teaching-learning process or radically improve students' understanding. Meaning does not reside in the activities themselves; one has to construct meaning out of activities and teachers play an important role in this process. Thus, there is a need for professional development of teachers, engaging them in the reform process - allowing them to build an understanding of the motivations behind the reform, the philosophy guiding the reform and the nature of change in the content, its implications for classroom practice. The classroom observations of these four classes did not indicate any significant change from classrooms before NCF 2005. They continued to be unidirectional from the teacher to the student and do not incorporate the kind of discussions required for transacting the textbook material. The new material has brought in a few crucial ideas in the classroom (especially the idea of generalization) but the transaction remaining more or less the same, it did not impact significantly students' understanding. Teachers being a central force in any curriculum reform also need to be supported to re-visit and enrich their own content knowledge, pedagogical practices as well as explicate their own beliefs about teaching, learning and students studying mathematics. There does not exist any document which informs teachers of the changes made in the textbooks and the way they need to approach the teaching and learning of it. In the absence of a teachers' handbook and adequately developed anchoring ideas and concepts among students, they are often not in a position to teach the ideas in the way given in the new textbook. They may take recourse to older ways of teaching them, which they know and have developed and used over the years, or feel constrained and helpless in not being able to explain things. It may be important that textbooks be able to convey some of the principles of teaching the content, as for many this is the only resource for organizing teaching.

On the whole, NCF 2005 brings with it many progressive ideas which were not heard in the public discourse earlier. The fact that students bring with them mathematical ideas to the classroom and can construct mathematics by making connections (within mathematics and with the outside world) would have been hard for most people to accept some time back. Whether all those who are involved in the process of education (teachers, educators) understand the essence and implication of this statement is not clear, and there are possibilities of a naïve interpretation. This small study revealed the many gaps which make it difficult to realize the vision of the document. Due to the many factors which impinge on the translation of a curriculum in the school system, it may fall short of achieving the "higher aims" of teaching mathematics. It is necessary that many such studies are conducted, some on a larger scale with a broader focus, involving other content areas so as to assess the impact of the envisioned curricular reforms.

REFERENCES

Banerjee, R. (2011). Is arithmetic useful for the teaching and learning of algebra? *Contemporary education dialogue*, *8* (2), 137–159.

Banerjee, R. & Subramaniam, K. (2012). Evolution of a teaching approach for beginning algebra. *Educational studies in mathematics*, *81*(2), 351-367.

Banerjee, R. (2008). *Developing a learning sequence for transiting from arithmetic to elementary algebra* (Unpublished doctoral dissertation). Homi Bhabha Centre for Science Education, T. I. F. R., Mumbai.

Batra, P. (2006). Building on the National Curriculum Framework to Enable the Agency of Teachers. *Contemporary education dialogue*, *4*(1), 88-118.

Bell, A. (1995). Purpose in school algebra. *Journal of Mathematical Behavior, 14*(1), 41-73.

Booth, L. R. (1984). *Algebra: Children's Strategies and Errors.* Windsor, UK: NFER-Nelson.

Bulut, M. (2007). Curriculum reform in Turkey: a case of primary school mathematics curriculum. *Eurasia Journal of Mathematics, Science and technology education, 3*(3), 203-212.

Choksi, B. (2007). *Evaluating the Homi Bhabha Curriculum for Primary Science: In situ. Technical report no. I (07-08),* Homi Bhabha Centre for Science Education: Mumbai.

Handal, B. & Herrington, A. (2003). Mathematics teachers' beliefs and curriculum reform. *Mathematics Education Research Journal, 15*(1), 59-69.

Hart, K. M., Brown, M. L., Kuchemann, D. E., Kerslake, D., Ruddock, G., & McCartney, M. (Eds.). (1981). *Children's Understanding of Mathematics: 11-16.* London: John Murray.

Kieran, C. (2006). Research on the learning and teaching of algebra: a broad sources of meaning. In A. Gutierrez and P. Boero (Eds.), *Handbook of Research on the Psychology of Mathematics Education: Past, Present and Future* (pp. 11-49). Rotterdam, The Netherlands: Sense Publishers.

Kuchemann, D. E. (1981). Algebra. In K. M. Hart, M. L. Brown, D. E. Kuchemann, D. Kerslake, G. Ruddock, M. McCartney (Eds.), *Children's Understanding of Mathematics: 11-16* (pp. 102-119). London: John Murray.

MacGregor, M. & Stacey, K. (1997). Students' understanding of algebraic notation:11-15. *Educational Studies in Mathematics, 33*(1), 1-19.

Mason, J., Graham, A., Pimm, D. & Gowar, N. (1985). *Routes to/Roots of Algebra.* Walton Hall, Milton Keynes: The Open University Press.

National Council for Educational Research and Training (2006a). T*he national focus group on the teaching of mathematics: Focus Group Paper.* Delhi: NCERT.

National Council for Educational Research and Training (2006b). *Syllabus: Elementary level (Vol I).* Delhi: NCERT.

National Council for Educational Research and Training (2006c). *Math-Magic: Textbook for Class I-V.* Delhi: NCERT.

National Council for Educational Research and Training (2006d). *Mathematics: Textbook for Class VI-VIII.* Delhi: NCERT.

Paliwal, R. & Subramaniam, C. N. (2006). *Contextualising the Curriculum of the Poor: Concerns Raised by the Blocked Chimney Theory. Contemporary education dialogue, 4*(1), 10-24.

Remillard, J. T. (1999). Curriculum materials in mathematics education reform: a framework for examining teachers' curriculum development. *Curriculum Inquiry, 29*(3), 315-342.

Research Advisory Committee of the National Council of Teachers of Mathematics (1988). NCTM curriculum and evaluation standards for school mathematics: responses from the research community. *Journal for Research in Mathematics Education, 19*(4), 338-344.

Saxena, S. (2006). Questions of Epistemology: Re-evaluating Constructivism and the NCF 2005. *Contemporary education dialogue, 4*(1), 25-51.

Stacey, K. & MacGregor, M. (2001). Curriculum reform and approaches to algebra. In R. Sutherland, T. Rojano, A. Bell and R. Lins (Eds.), *Perspectives on School Algebra* (pp. 141-153). Dordrecht, The Netherlands: Kluwer Academic Publishers.

Subramaniam, K. & Banerjee, R. (2011). The arithmetic-algebra connection: A historical pedagogical perspective. In J. Cai and E. Knuth (Eds.), *Early Algebraization: A Global Dialogue from Multiple Perspectives* (pp. 87-105). Springer.

Swarts, P. (1998). *Evaluation and monitoring exercise of the mathematics curriculum. Executive summary of the report.* National Institute for Educational Devlopment: Namibia.

Thompson, A. G. (1988). The relationship of teachers' conceptions of mathematics and mathematics teaching to instructional practice. *Educational Studies in Mathematics, 15*(2), 105-127.

Yates, S. M. (2006) Primary teachers' mathematics beliefs, teaching practices and curriculum reform experiences. In Australian Association for Research in Education Conference. AARE: Australia.

CHAPTER 74

PROFESSIONAL DEVELOPMENT WORKSHOPS FOR IN-SERVICE MATHEMATICS TEACHERS IN INDIA

Ruchi S. Kumar, K. Subramaniam, Shweta Naik
Homi Bhabha Centre for Science Education (TIFR), Mumbai

ABSTRACT

In this chapter we present the background of teacher professional development workshops conducted in the Indian context with the need for rethinking the goals of the workshops in light of new policy initiatives. We elaborate on the goals, principles and the framework adopted for design and enactment of the workshops conducted at Homi Bhabha Centre for Science Education, Mumbai along with description and a few examples of the different types of tasks and sessions planned for the workshop. We describe three principles which guided the design and enactment of the workshop- (i) situatedness in the work of teaching, (ii) offering challenges to teachers' to revisit their knowledge and beliefs, and (iii) developing a sense of belonging to a professional community. Subsequently we present analysis of two episodes from a ten day professional development workshop to illustrate – (a) how the three principles and goals of the workshop design shaped the tasks

The First Sourcebook on Asian Research in Mathematics Education: China, Korea, Singapore, Japan, Malaysia, and India, pp. 1631–1654

and enactments of those tasks, (b) how authenticity of the tasks and enactment lead to agency of teachers and teacher educators and (c) how teachers demonstrated sensitivity towards students' thinking.

Keywords: In-service teacher professional development, professional development in India, professional community for teachers, framework for design and enactment of workshops, teacher agency, teacher beliefs, teacher knowledge

Teachers are central to any education system. There is growing realization across the world that reform in education cannot be brought about without adequately addressing teachers' role in it. In India, two major policy initiatives aimed at providing quality elementary education to all children have been launched in the last few years: the National Curriculum Framework for school education 2005 and the Right to Education Act 2009 (see Khan, this volume). As India struggles to implement these two initiatives, teachers face challenges like providing children quality education through student centered pedagogy, assessing students comprehensively and continuously, and relating school subjects to the daily lives of children. The National Curriculum Framework has been criticized for being silent on how teachers are supposed to bring about the change in their classroom and for not addressing the much needed teacher development to support curriculum renewal (Batra, 2005). Efforts undertaken thus far like changing textbooks and issuing directives to schools and teachers, sidestep the issue of developing adequate knowledge and enabling beliefs among teachers, which is needed to realize the vision conveyed in the new curriculum framework.

Several national committees have over the years recognized the need for continuing professional education of teachers and recommended "at least two or three months of in-service education in every five years of service" (Government of India [GoI], 1966, section 4.56). The New Education Policy of 1986 recommended a rapid expansion of the infrastructure for education of teachers at the elementary level through the setting up of institutions at the district and block levels, which would deal with both pre-service and in-service teacher education (GoI, 1986). While the teacher education infrastructure has indeed expanded vastly, issues of poor quality and low relevance of teacher preparation remain (Sharma & Ramachandran, 2009). Further, teacher education institutions have tended to focus more on pre-service education leading to the neglect of in-service education (Ministry of Human Resources Development [MHRD], 2009). A renewed attempt to address the problems of pre-service and in-service teacher education is made by the new National Curriculum Framework for Teacher Education (NCFTE), which re-affirms the importance of in-service teacher development, and puts forth several principles that should govern the design of in-service teacher education programs (National Council of Teacher Education [NCTE], 2009). Of these, we highlight the following three principles:

- designing programs with clarity about aims and strategies for achieving these aims,
- allowing teachers to relate the content of the program to their experiences and also to find opportunities to reflect on their experiences, and
- need to respect the professional identity and knowledge of a teacher and to work with and from it (NCTE, 2009, pp. 66-67).

Besides these, the principles advocate creating spaces for sharing of teachers' experiences, addressing teachers' needs and extended interaction with a group of teachers. They caution against compromising interactivity especially through the use of electronic media, aiming at quick fixes, over-training, and routinised and superficial training. The principles highlighted above are especially important for the focus of this article on in-service teacher professional development (TPD) workshops.

IN-SERVICE TPD PROGRAMS IN INDIA

As emphasized by the NCFTE, there is a need to develop a clear vision of the goals that programs must achieve and the means by which they can be achieved. Most in-service TPD programs in India are designed in response to the need of curriculum reform and view teachers as agents of the state, who implement the reforms rather than as participants in the process of reconstruction of the curriculum. Underlying this is the assumption that teaching can be changed by directing changes in the content or structure of interactions in classrooms while not directly addressing the teacher's own conceptions of teaching, learning and mathematics. In-service TPD is seen as training for content or pedagogy, mostly revolving around the changed curriculum, but not necessarily as important for continuous teacher development. Content-focused interventions often consist of lectures delivered by "experts" and the mathematical content is typically divorced from the context of teaching and learning. Another common focus of TPD programs is "how to teach a particular topic". This may appear to be close to the work of teaching and hence directly relevant to teachers. However, there is a large variation in the contexts and life experiences that students bring to the classroom and teachers need to be flexible and adaptive in addressing the needs of a student (NCTE, 2009). Instruction to teachers are guided by a *transmission model*, where recommendations on how to teach a topic tend to be recipe-like. The effectiveness of such an approach is limited and is not consistent with the vision articulated in the NCFTE. Teachers need to develop their own vision of the changed goals of instruction and adapt their teaching in self-determined ways to meet these changed goals.

In India, workshops are an important component of TPD programs on which the greatest time, effort and resources of the state are spent. In our experience, and as reported elsewhere, TPD workshops are often organized in an ad hoc manner on the basis of expediency, sometimes driven by the need to utilise funds (MHRD, 2009, pp. 2, 15-16). There is no clear consensus about what needs to be done in these workshops and how it is to be done. In structured large-scale programs, TPD is sought to be achieved through the "cascade model" of training (MHRD, 2009, p.15), where master resource teachers are trained first, who in turn train other teachers. The design and content of the modules, which are used repeatedly at each tier of the cascade training, is generally not research-based. The vision underlying most of these programs restrict teachers' agency to implementing a new textbook, a pre-designed pedagogy or a prescribed assessment technique. In our view, TPD programs however need to have a broader vision of the needs of a teacher as a developing professional, view the teacher as an 'active learner', and must address issues of knowledge, beliefs, attitudes and practices in a comprehensive manner, rather than in the narrow context of a particular reform.

The new curriculum, arguably expects from the teacher a deeper understanding of subject matter as well as the teaching learning process, rather than merely adopting new techniques. Teachers in elementary and middle grades are expected to not only make their students fluent in computational mathematics but to also address process goals in the learning of mathematics, such as reasoning, using multiple ways to solve problems, justifying their solution, making generalizations and conjectures, and analyzing the mathematical work of others (National Council of Educational Research and Training [NCERT], 2006). However, there have been few TPD programs in India, which have focused on the skills and knowledge required to facilitate this kind of teaching. Research studies of teachers' knowledge in other countries have pointed to the importance of knowledge that integrates subject matter and pedagogy. Although pedagogical content knowledge and subject matter knowledge have been considered as useful constructs to describe essential knowledge for teaching (Shulman, 1986; Ma, 1999), it is rarely the central focus of any phase of teacher education in India (Naik, 2008; Kumar, Dewan & Subramaniam, 2012). Thus, we consider providing opportunities for deepening teachers' knowledge of mathematics and of pedagogy revolving around mathematical practices to be one of the central goals for TPD programs.

Bringing about change in teachers' knowledge of mathematics relevant to teaching is clearly a challenging task, but only partly addresses the TPD need. Studies have shown that teachers' beliefs also strongly influence teaching practice and determine what teachers notice in the classroom (Thompson, 1992; Philipp, 2007). In the Indian context, commonly held views include the belief that mathematics is a body of knowledge consisting of known solutions to a well defined set of problems and that not all children are capable of learning mathematics (Kumar & Subramaniam, 2013). A study by Dewan (2009) indicates that such beliefs, which stand in contrast to the ones envisioned in the National Curriculum Framework (NCERT, 2005), are held by not only teachers but even administrators, faculty members and directors of teacher education institutions, thereby indicating the extent of challenge to implement the new framework. This points to the need to create spaces where teachers articulate and reflect on the beliefs that they hold while respecting the identity of the teacher. Teachers need to not only experience alternative ways of doing mathematics, but also to build an awareness of and sensitivity to students' mathematical thinking.

Research studies have illustrated how the development of professional learning communities contribute to teachers' professional growth, by providing a site for articulation and reflection on the beliefs, for sharing the knowledge held and practice adopted by the teachers. (Kazemi & Franke, 2004; Jaworski, 2008; Brodie & Shalem, 2011).

Thus, in the Indian context as elsewhere, the goals that TPD programs need to focus on include:

- enabling teachers to develop a vision for the changed goals of instruction and become "active learners,"
- providing opportunities to make teachers' knowledge and beliefs explicit,
- strengthening teachers' knowledge integrating content and pedagogy,
- revisiting beliefs through reflection and engagement, and
- fostering professional communities as spaces for developing shared understanding about teaching and learning of mathematics.

In this chapter, our focus is how the components and interaction in a teacher professional workshop can be shaped to address these goals. The design as well as the enactment of the workshop contribute towards meeting these goals. Hence we develop a framework that illuminates both these aspects. The framework is drawn from our own experience of TPD, from the literature on teacher development and from guiding policy documents in the Indian context such as the NCFTE. We analyze two interaction episodes from the workshop and illustrate how the framework illuminates the task design and the agency of the participating teachers and teacher educators in addressing the workshop goals. Our purpose is to illuminate critical aspects in a particular professional development workshop for mathematics teachers through an analysis of sessions enacted in it.

FRAMEWORK FOR WORKSHOP DESIGN AND ANALYSIS

In designing TPD workshops to address the goals described above, we consider three guiding principles as essential. These are drawn mainly from our own practice, but are related to the theoretical perspectives of situated learning theory (Lave & Wenger, 1991) and communities of practice (Wenger, 1998). These principles are also related to those from the NCFTE highlighted previously. The three principles, which, in our view, must inform the design and conduct of TPD workshops through all its activities in a comprehensive manner are:

- situatedness in the work of teaching,
- offering challenges to teachers' to revisit their knowledge and beliefs, and
- developing a sense of belonging to a professional community.

The aspect of situatedness is addressed through the choice of tasks as well as the mode of presentation of the task. The use of artifacts like students' errors, examples from textbooks, or examples emerging from live or video records of classroom teaching with questions, prompts and examples used in the interaction can recall the context of teaching and learning. It is this aspect that allows teachers to make strong connections with their own practice thereby providing a stimulus for participation and reflection. Moreover, the use of artifacts from the daily activity of teaching has been emphasised by practice based professional development (Ball & Cohen, 1999).

The second principle of challenging teachers' beliefs and knowledge needs to be built into the tasks chosen for the sessions and reflected in the actions by teacher educators such as re-voicing individual teachers' views for consideration by the participants, and providing counterarguments, explanations and questions to help teachers think about the tacit aspects of teaching and mathematical content. In the TPD workshops, such responses were made not only by teacher educators, but teachers on their own also reacted to their colleagues' articulations by making conjectures, arguments, assertions, counterarguments, explanations and reflective remarks.

The third principle of building a sense of a professional learning community acknowledges that teaching is a cultural activity, and the development of a teacher is not to be viewed in individual terms, but in the setting of a community. We adopt a broad view of community as encompassing teacher educators, researchers and teachers, all of whom are engaged in the enterprise of analyzing teaching. We provided opportunities for discussion, sharing and inter-animation of ideas to enable members of the community to share their histories as a resource for participation (Wenger, 1998). In the workshops organized by the authors, this aspect was addressed by posing tasks and questions for the whole group rather than to individual teachers. The teacher educators attempted to situate themselves as members of the larger teaching community by using "we" in their language as well as drawing on their own teaching experiences with students in the course of their research work. Teacher educators adopted several words and categories commonly used by teachers and also elicited and acknowledged teachers' knowledge about students and teaching gained through years of experience.

The three principles described above of challenge, situatedness and community building are interrelated. Focusing on the work of teaching helped in fostering the solidarity among teachers, who were regarded as knowledgeable members of community as they are engaged in the work of teaching and thus are entitled to have and voice their views. Belonging to a community entails the work of making claims and conjectures, making arguments or counterarguments to support one's claims drawing on the knowledge gained from experience, and supporting the growth of knowledge in a community. Thus challenging beliefs and knowledge was an integral aspect of community building as much as situatedness in the work of teaching.

Understanding the role that interventions such as workshops play in the professional development of teachers requires consideration of not only design aspects, but also of enactment aspects. The affordances of the task that participants work on, and the interaction among the participants determine whether the workshop addresses the goals adequately. The framework outlined thus far includes goals and principles for the design and conduct of the workshop. To facilitate the analysis of the enactment of the workshop, we add further elements to this framework relevant to key features of the interaction during a workshop. We draw these elements from the notion of the teacher education triangle adapted from the didactic triangle (Goodchild & Sriraman, 2012) as shown in Figure 74.1. The interaction during a TPD session can be conceptualized as an interaction between the three elements of the task, the teachers and the teacher edu-

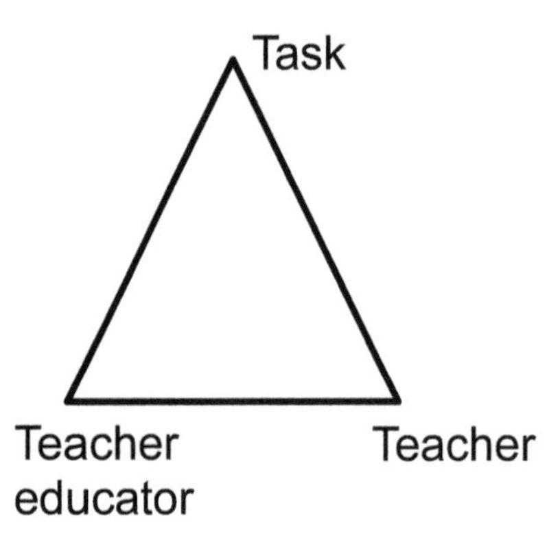

Figure 74.1. The teacher education triangle.

cators. We focus on the affordances of the tasks and on the agency of both teachers and teacher educators. Rather than viewing agency as "associated with the individual subject as a self-standing entity," we describe how this "arises out of engagement" (Wenger, 1998, p. 15). The engagement that we focus on is with colleagues and teacher educators who share the common enterprise of improving mathematics education in schools.

In the next section, we briefly describe the components of the workshop and how they relate to the goals and design principles in the framework. In the subsequent section, using the framework consisting of goals, design principles and interaction elements (see Table 74.1), we provide an analysis of the interaction during two episodes from the TPD workshop under sections dealing with *the nature of the tasks, the agency of the teachers and the agency of the teacher educators*. The choice of the task, the communicative devices used by the teacher educators and the efforts to shape the interaction, all reveal the importance and inseparability of the aspects of situatedness and challenge. We describe the efforts made to situate the discussion of teaching and learning both in the context of the work of classroom teaching and within the community of teachers. The questions that we specifically address are the following:

- How is the teachers' repertoire of knowledge and beliefs gained from experience brought into play in their engagement in the workshop?
- How are learning opportunities or opportunities for reflection on beliefs created in the course of the teachers' engagement?
- What aspects of teacher educators' enactment of the task and interaction with the teachers facilitated the engagement of teachers?

Table 74.1. Framework for Analysing Design and Enactment of a TPD Workshop

Workshop Goals	*Principles for Designing Components and Tasks*	*Interaction Aspects*
• Strengthen teachers' knowledge relevant to mathematics teaching • Provide opportunities to articulate and reflect on beliefs relevant to teaching mathematics • Foster the development of professional communities of learning	• Situatedness • Challenge • Community building	• Task structure • Teachers' agency • Teacher educators' agency

COMPONENTS OF THE TPD WORKSHOP

The design of the TPD workshops held at Homi Bhabha Centre is informed by the goals of providing opportunities to teachers to strengthen their knowledge for teaching mathematics, to reflect on their beliefs, and to foster the building of professional learning communities. The principles of situatedness, challenge and community building guide the design of the workshop components, which include sessions involving the study of classroom teaching to learn about content rooted in pedagogy, learning about students' thinking from students' responses, working on mathematical problems and understanding relevant research on teacher learning and the work of teaching. Every experience with teachers has helped us modify our thinking about what form of the content is most relevant and functional for teachers' reflection and development. Our aim here is not to present ideal designs for workshops, but to highlight the opportunities for the development of teachers' knowledge and reflection on beliefs that emerged in the workshops. To provide concrete details of the design, we focus on a residential workshop that was held for ten days, describing the various components and their functions. Following this description, we discuss in detail two episodes from this workshop, focusing on the enactment aspects.

The majority of teachers who participated in this workshop belonged to a nationwide government school system catering to children from a range of socio-economic backgrounds. The school system had taken various steps to implement reform based on National Curriculum Framework mainly through teacher training and issuing notices and circulars. Of the 20 teacher participants, 12 were from this school system: four middle school teachers from outside Mumbai, four primary and four middle school teachers from Mumbai. All the eight local teachers were engaged in follow up activity after the workshop. Among these, four teachers also received support in their classroom teaching by one of the authors. The follow-up phase of the project is not reported in this chapter.

The following is a description of the components in the workshop, with an elaboration of their role in the development of mathematical knowledge required for teaching and in reflecting on beliefs and attitudes related to teaching and learning of mathematics.

Observing classroom teaching. Teachers viewed and participated in three consecutive days of live teaching during the workshop. The involvement of the teachers was in contributing to the plan for the class, observing the lesson and then reflecting on it. The teaching observed was atypical, in its focus on eliciting students' ideas and building on them and thus was intended to be a source of reflection and challenge for teachers. These sessions provided teachers a context for making their situated knowledge about pedagogical approaches and students' capability explicit. Prompts to elicit reflection on the lesson included inviting teachers to make conjectures about the intentions of the teacher in making specific moves, what children were thinking and what alternative pathways could have been taken at critical points in the lesson.

Learning through problems. Teachers worked on mathematical problems during these sessions, which were posed in contexts close to either their teaching practice or daily life. The problems were content specific and therefore separate sessions were conducted for different content topics such as number sense, fractions and ratios, and algebra. The main objective was to create distractions in a mathematics problem based on familiar alternative conceptions that teachers or students have, leading to cognitive conflict and eventually to reflective learning. The problem presented in Table 74.2 is an example.

These sessions occasionally led to deep exploration of mathematical concepts and making connections between various mathematical constructs, and providing a space for teachers to reflect and build upon their mathematical knowledge for teaching.

Table 74.2. Example of a Workshop Problem Task

A student in the class had added fractions like this: 3/7 + 2/3 = 5/10. Why do you think students add in this way? When the teacher asked the student why she had done it in this way, the student said that her father had taught her. The teacher explained that this method was wrong. On the following day there was a complaint from the father. He pointed out that the teacher had added exactly like his method. This was his example, Marks in history: 35/50. Marks in geography: 24/50. Total marks in social studies: 35/50 + 24/50 = 59/100. How would you respond to the parent's criticism?

Working on students' thinking. These sessions included working on students' errors, uncovering students' thinking by analyzing strategies, and inferring potential misunderstanding underlying these errors. These led to discussions on issues such as – which questions are efficient in evaluating understanding of a specific concept, what do these errors tell us about students' and teachers' own conceptions including their beliefs about the nature of mathematics, and what do students' errors imply in terms of shaping the instruction. In the next section, we discuss in detail the interaction in two sessions of the workshop that involved working on students' errors.

Reading literature based on research. In these sessions, teachers in groups of three to four studied a research article from the field of mathematics education and made presentations to colleagues. The sessions were found to be valuable in fostering the sense that a teacher is a part of a community that systematically studies content and pedagogy with the goal of improving teaching and learning. The readings stretched the boundaries of the participant teacher community from the immediate peer group to the professional community of mathematics educators including researchers.

Analysing curriculum material. These sessions were included to add connection, coherence and depth to teachers' comprehension of textbooks so as to use them efficiently. Teachers in groups of two to three analysed textbooks from Grades 3 to 6 for a specific mathematical topic for the following prompts – what is the hierarchical development of the topic, how are context and real life connections brought about, what is the role of the various examples provided, and how are representations used in the textbook. Although teachers use these textbooks on an everyday basis, these sessions provided an opportunity to distance themselves from the sole purpose of teaching and look at the textbook critically.

Expressing beliefs about teaching, students and mathematics. In this session held at the beginning of the workshop teachers completed a 6-part questionnaire based on the Likert-type scale, which provided them an opportunity to reflect about their own beliefs about teaching, students, self and mathematics. Teachers worked individually on these questionnaires. The questionnaire items at times framed the discussions in subsequent sessions. Teachers also reported that the statements in the questionnaire made them think about issues that they had never thought of. At the end of the workshop, teachers were given parts of the belief questionnaire on mathematics, mathematics teaching and preferred practices and were asked to record the changes in their views. Table 74.3 presents a sample of the items used in the questionnaire.

Table 74.3. Sample Items Used in the Questionnaire for Teachers

Domain	*Sample Item*	*5 Point Likert-Type Scale*
Practice	I ask students to practice problems very similar to the one done in the class as homework.	Almost never to almost always
Belief about mathematics	Being good at mathematics means being able to perform calculation quickly and accurately.	Strongly agree to strongly disagree
Belief about teaching mathematics	Listening carefully to the teacher explain the mathematics lesson is the most effective way to learn mathematics.	Strongly agree to strongly disagree
Belief about self	If something is not clear in the textbook I am confident that I can work it out on my own.	Strongly agree to strongly disagree
Belief about students	Students from poor homes tend to struggle in mathematics.	Strongly agree to strongly disagree

WORKSHOP ENACTMENT

The sessions of the workshop were generally characterized by high levels of interaction and participation by teachers. The two episodes of the workshop that have been selected for analysis here, deal with errors made by students and the learning for teachers from these errors. The two sessions were led by different teacher educators, and were structured differently. Transcripts of the two episodes were prepared from the video records of the sessions. The coding process was adapted from Miles and Huberman (1994). The transcripts were coded broadly into three categories: the task design features, the interaction features and teachers' explorations and reflections. Dimensions within these categories were identified and consensus about the coding was established by discussion among the coders (authors). After the initial coding, the themes explicated in this chapter were arrived at keeping in view the framework presented in Table 74.1.

The Task as a Resource for Teacher Education

In Episode 1 (Day 2, about 30 min), the task was to describe and explain student errors from looking at their responses to seven test items on the topics of number, place value and fractions (see Table 74.4 for sam-

Table 74.4. Example of Test Items and Student Errors Shown to Teachers in Episode 1

Questions	*Student Error 1*	*Student Error 2*
Write next three numbers: 3097, 3098, ------, ------, -------	3097, 3098, 3099, 30910, 30911	3097, 3098, 3099, 30100,30101
14 tens + 23 ones	1423	14023
Draw 7/4	Drawing representing 4/7	Drawing representing 7/11

Table 74.5. Task Given to Teachers in Episode 2

Mohsin is in class 5. He helps his father, who is a vegetable seller, with home deliveries. He can find the total amount a customer has to pay and often does the addition mentally. He also knows a lot about how much things cost: Televisions, cycles, two wheelers, washing machines, etc. But when his teacher asked him to write 'rupees two thousand twenty five' in numerals, he wrote 'Rs 200025'.

Think about what Mohsin's problem is and how his teacher can help him. How can the teacher make use of what he already knows so that he can learn something he doesn't know?

ple items). At the beginning of the session, the teacher educator provided a set of prompts to guide the discussion for each question: identify competencies being tested, find all possible correct answers and understand what caused the student errors.

In Episode 2 (Day 1, about 30 min) teachers were given a handout that described a student Mohsin's difficulty in writing numerals despite his familiarity with numbers (see Table 74.5). The teachers discussed in groups and presented their suggestions about how Mohsin could be helped.

Both the tasks are invoked together with supports drawn from teaching practice itself to give authenticity to the task and to invite deeper engagement on the part of the teachers. In the first episode, the student errors were drawn from a pre-test of students participating in the vacation course, the same group of students whom the teachers knew they would be observing later. In the second episode, the teacher educator informs teachers that Mohsin is a real boy whom he teaches, and furnishes details about Mohsin's responses to other tasks in the course of the discussion. Teachers identified the students' responses from both the episodes as "common" among their students. However it required further probing on the part of teacher educators to make teachers think about students' thinking underlying the errors and the sources of the errors, beyond identifying them just as "common" errors.

The tasks chosen for the two episodes are situated in the context of teaching while challenge is introduced by requiring the teachers to think beyond the normal requirements of everyday teaching. In the first episode, this is achieved through the three prompts inviting teachers to uncover a deeper layer of students' thinking that can explain their responses. Teachers at first thought that the errors surfaced because of the non-typical questions asked in the test. For example, to explain why the student incorrectly showed 4/7 instead of 7/4, teachers said "because numerator is greater than the denominator in the given fraction." Several teachers thought that it was not possible to represent improper fractions. The teachers interpreted fractions in terms of the "part-whole" meaning as number of parts out of total parts, which made representation of improper fraction impossible. The task thus challenged them to re-consider the meaning attributed to fractions. We will return to the discussion of this example later.

In the second episode, the teachers were asked to consider both what the child does and does not know so as to induce a sense of conflict by juxtaposing these together. For e.g., teachers identified that Mohsin knows numbers in the thousands, can add mentally and can read the price of a bicycle but cannot write the amount 2025 correctly. What accounts for the child's capability in the context of everyday calculations, and his profound lack of understanding of a related part of school mathematics? This tension sets a dialectic in motion allowing the teachers to revisit the relatively hidden and unspoken aspects of their everyday teaching.

Both the tasks worked as a vehicle for reflection and engagement on the part of teachers by articulating their beliefs and knowledge. The discussion moved beyond the immediate demands of the tasks to broader concerns like connecting teaching of mathematics with out-of-school experiences or considering students' thinking underlying their mistakes as a resource for teaching mathematics. Thus the discussion of the tasks needs to shift focus from the tasks themselves to opening up a space for deeper engagement, where teachers can share and critically reflect on what they know, understand, believe and practice. The role of teachers' and teacher educators' agency in engaging with the task and the emergent issues will be discussed in the sections below.

Teachers' Agency in Engaging With Knowledge and Beliefs

An important part of understanding the work of teaching as a profession is a shared agreement about the specialized knowledge and expertise that informs the work of teaching. Professional development programs need to elicit and build on such knowledge, much as teachers elicit and build on students' knowledge in the classroom. This process also allows the community of teachers and teacher educators to develop a shared view of the contours of such knowledge. The process of eliciting teachers' knowledge and beliefs can create opportunities for the expression of teachers' agency. We understand teachers' agency as initiatives and autonomy expressed by teachers during the course of interaction to assert and justify their beliefs. In this section, we explore how teachers' agency was expressed in the course of the interaction during the selected episodes.

Anticipating students' responses. Requests to teachers to anticipate and predict student responses were either built into the task itself or were made by the teacher educator in the course of discussion. This aspect is embedded in teachers' everyday work of teaching. Over the years teachers develop an implicit knowledge about typical and atypical student responses. In the TPD context, making this knowledge explicit works as a resource in building the shared knowledge between teachers and teacher educators and providing ways to discuss students' thinking. In Episode 1, teachers were able to anticipate some student errors for the questions (see Table 74.4) which paved the way for discussing student thinking. In episode 2, for example, the teacher educator asked teachers to predict Mohsin's strategy to find the cost of 10 kg of potatoes given the cost of 1 kg. The teachers anticipated that Mohsin would repeatedly add the unit cost to arrive at the cost of 10 kg, which the teacher educator confirmed was what Mohsin actually did. These questions were significant as they directed teachers' attention towards what the student did know at a point when they were focusing only on his incapability. Sharing the anticipations paved the way to a discussion about the differences between mathematics students learn outside school and in school and the need to bridge the gap between the two.

Identifying "key knowledge pieces". At times, teachers contributed centrally to the goal of building mathematical knowledge for teaching. In episode 2, while trying to elaborate why Mohsin made the error of writing two thousand twenty five as 200025 a teacher explained that understanding the "meaning of zero" involves understanding how it changes value with position – it has no value when written in the leftmost position of the numeral and in other positions it determines the place value of the other digits in the numeral. Her explanation of the concept of the position of zero can be characterized as a "key knowledge piece" (Ma, 1999) that is important in understanding place value of a number. Her intervention led other teachers to also identify the conceptual gap in the student's thinking and a teacher asserted, "he knows 2000 and he knows 25 but how to write [2025] he doesn't know".

Conjecturing underlying causes. In episode 1, the discussion on understanding student errors led to discussing underlying causes of the errors. For the question of drawing a representation of the fraction 7/4, some students had drawn part-whole representations of 4/7 or 7/11. Teachers tried to explain the thinking that might underlie these responses. As discussed earlier, initially teachers identified the cause of the error as unfamiliarity with question. In the course of the discussion, a teacher put forth an alternative explanation of why students drew 7/11 to show the fraction 7/4. The student, he argued, may have interpreted 7/4 to mean "7 shaded and 4 unshaded parts" thus making a total of 11 parts of which 7 were shaded. Thus teachers began to engage with the reasoning that the student must have applied to create such representations. The discussion moved to how counting of shaded parts (using whole numbers) was generally over-emphasized while teaching fractions. Thus, teachers also reflected on their own teaching as a possible cause for student difficulties in learning. To cite another example, when a question was raised in episode 2 about why students are not able to learn mathematics even after five years of schooling while they learn quickly outside the school, a teacher observed that "we do not correlate mathematics taught in school with everyday life".

Articulating and contesting beliefs. We consider the occasions when teachers' beliefs were explicitly articulated as important moments in the workshop. At times this took the form of reflection suggesting a revisiting of beliefs. For e.g., in episode 1, after discussing student errors and students' thinking underlying these, a teacher reflected "in fact we know their mistakes but we don't really see into their thinking". Teachers contested and challenged views articulated not only by their colleagues, but also by teacher educators. For e.g., in episode 1, the question, "Add 337 + 33700" was discussed, where students had made an error in vertically aligning the digits. A teacher reacted to the example and said "Addition questions should not be given in horizontal manner as it will lead to error" thus indicating her belief that errors should be avoided during instruction. Similarly, during a discussion of teaching aids for teaching place value, a teacher voiced his opinion that using teaching aids will cause lot of confusion and it would be better if students are told the rules and asked to practice. Another teacher contested this by saying that they themselves (teachers) had learnt mathematics by rote when they studied in school but it is important now to emphasize understanding concepts. At a point in the discussion, a teacher asserted that the abacus was easier for students to learn place value. Another teacher responded by explaining how stick bundles representing different units (tens and hundreds) can build understanding of place value better. Thus voicing of assertions led to sharing of alternative viewpoints, which created a need for justification, thus deepening the engagement in the workshop. In this instance, the difference between the stick bundles and the abacus led to exploring the difference between the grouping principle and the positional value principle discussed later.

These articulations were important points in the sessions, which provided a window into teachers' thinking as well as created a space for revisiting and reflecting on beliefs relevant to teaching and learning. Teachers also assessed their own learning in the workshop as reflected in an appreciative comment by a teacher at the end of episode 1. The active interventions by the teachers described above were indicative of teachers' agency as they were not merely involved in affirming or contesting what the teacher educator or other teachers were saying but were engaging in their own sense making about the aspects discussed related to the task.

Agency of the Teacher Educator: Inter-Animation, Knowledge and Beliefs

In a TPD context to what extent teachers' knowledge and understanding are elicited, what aspects of knowledge are negotiated and in what direction the discussion moves during a session depends critically on the interventions made by the teacher educator. Not only is the participants' engagement crucial, but also the degree of inter-animation of ideas. Mortimer and Scott (2003) have used Bakhtin's idea of inter-animation to illustrate how an interaction in the classroom is "functionally dialogic" when more than one point of view about an issue is represented as well as explored. Mortimer and Scott define low inter-animation as just listing of the varied responses shared in the group while high inter-animation means that there is an engagement with the different views expressed by the participants as a group. In the context of TPD too, we found that the aspect of inter-animation of teachers' responses was crucial in how teachers perceived their roles in the session. In episode 1, inter-animation was low partly because the interaction was structured in such a manner that there was one correct answer for the questions posed and even when different opinions were voiced, the teacher educator responded to the individual teacher by elaborating and emphasizing the interpretation that she thought was correct. On the other hand in episode 2, the teacher educator considered different conjectures voiced and posed the differences in opinions as questions to be considered by the whole group. Not only did teachers contest the views of other participants but they also contested views expressed by teacher educator leading to discussion. The moves that led to high inter-animation in episode 2 included inviting teachers to respond to each other, problematizing teachers' responses to be discussed by the group and the use of open-ended questions.

The teacher educators' interventions during the course of the interaction in turn are guided by their own beliefs and the knowledge that they bring to bear on the discussions. In this section, we briefly discuss first the moves made by the teacher educators that reflect the goals and the principles outlined in the workshop design framework and second, the interaction between beliefs and knowledge as reflected in the teacher educators' interventions.

Teacher educators' moves in alignment with the goals of TPD. The teacher educators frequently invited teachers to respond to the views expressed by their colleagues communicating that the teacher educator is not solely responsible for evaluating teachers' responses but that it has to be decided by the deliberation of the whole group. This led to situations where the teacher educators had to handle disagreement and conflict of views. Teacher educators usually welcomed disagreement and considered it a healthy sign of engagement, which allowed teachers to articulate their beliefs, which could then be taken up for discussion. The teacher educator sometimes restated a view in more general terms by placing it in the broader educational context. In episode 2, when there was disagreement over whether it is better to teach students rules for algorithms, the teacher educator contrasted learning inside and outside school (translated and summarised: "if students hold strongly what they learn from outside the school, but they are not able to hold on to what is learned at school, why it is not held we must think and talk about it"). This can be interpreted as an attempt to build a sense of community by inviting teachers to reflect on their beliefs in the context of larger educational debates.

The use of open-ended or probing questions by the teacher educator elicited more and more varied responses from the teachers leading to richer discussion strands. There were several examples of this in episode 2. The teacher educator asked teachers to suggest a variety of learning aids that could be useful for Mohsin. He then invited them to think about which teaching aids are better and why. Teachers' responses to such questions were often elaborate with some teachers recounting their own teaching experiences. Another significant move by the teacher educator was asking for clarification of the meaning of terms used by teachers and moving towards shared meaning and vocabulary. For e.g., in episode 2, the teacher educator asked for a deeper probing of the meaning of "place value", a term that occurred frequently in the discussion.

Interaction between teacher educators' beliefs and knowledge. The teacher educators held and acted on beliefs that were at times different from those held by teachers. Since the teacher educators' actions were guided by an expectation that teachers accept these beliefs, they could be considered to be belief goals for the TPD program. Some of the beliefs had to do with the emphasis or value ascribed to elements in the teaching-learning context. For example, the teacher educators believed that what a student knows is more important or at least as important as what he or she does not know. The emphasis placed on this was prominent in episode 2, when the facilitator repeatedly brought teachers' focus back to what the student (Mohsin) knew when the discussion turned to what he did not know. In episode 1, the following student error was discussed. In response to the question "show the number made of 14 tens and 23 ones," a student wrote "14023". Teachers thought that this student did not have a concept of place value. A teacher educator present in the audience pointed out that in fact it did show a partial understanding of place value since the student knew that 14 tens can be written as 140 and 23 ones as 23.

The teacher educators' emphasis on what the student knows in contrast to what she does not know is consistent with the related belief of ascribing value to students' thinking as a resource for teaching. This belief interacts with knowledge about students' ways of thinking in enhancing teachers' awareness and sensitivity. The teacher educators attempted at times to lead the discussion into understanding students' responses more deeply. In the example discussed earlier of why some students incorrectly represented the fraction 7/4 by drawing a picture for 7/11, a teacher educator conjectured that it could be due to excessive emphasis on the part whole representation of fractions by making as many parts as the denominator and shading as many parts as the numerator. Thus instruction treats numerator and denominator as separate whole numbers. Taking the discussion further, a teacher educator in the audience proposed an analysis of the fraction concept, by distinguishing between counting and measuring contexts. He suggested that measuring rather than counting is a better context to understand a fraction as a single number indicating a particular quantity. Counting contexts tend to reinforce the idea that a fraction is made up of two numbers. These interventions not only led to a deeper probing of students' thinking, but also to an understanding of how the choice of a teaching approach may play an important role in students developing certain conceptions.

Another belief held by the teacher educators, which is related to valuing students' thinking, is the belief in the efficacy of using students' previous knowledge (especially knowledge acquired from the everyday life/culture) as a resource for teaching. This was foregrounded in episode 2 by asking teachers to anticipate Mohsin's responses while using information about the daily life activities in which he engages and raising the question about the need to bridge the gap between the mathematics learned outside and inside the school.

The teacher educators attempted to communicate that umbrella concepts like "place value" need to be understood in detail, in terms of how they play a role in specific contexts of learning related concepts, of problem solving, or of understanding an algorithm. To become more useful, they need to be decomposed into sub-concepts like "grouping principle" and "positional value principle". The grouping principle determines that in the decimal number system, 10 units form the next higher unit in the sequence of units, tens, hundreds and so on. Number words encode the grouping principle by naming the different powers of ten: "four thousand", "six hundred", etc. The positional principle in contrast determines that in the written numeral the value of a "digit" depends on its position. This principle is essential to understand written numerals. It includes the understanding that when a zero appears at a certain position, it indicates that there are no units corresponding to that position. Both the grouping and the positional value principles help to reconstruct the number from the written numeral. This distinction was important in episode 2 in understanding what Mohsin knew (prices of articles and composing money in terms of currency units) and what he did not know (writing a number). The distinction between the grouping and positional value principles was new to many teachers, at least in an explicit sense.

The teacher educators believed that the usefulness of a "teaching aid" depends on the context and specific needs/difficulties faced by students. In episode 2, to help address Mohsin's difficulty, teachers had selected aids embodying both grouping principle as well as the positional principle. This did not take account of the fact that while Mohsin had a weak understanding of the positional principle, he had a strong grasp of grouping of multi-units because of his familiarity with money. The teacher educator was able to use teachers' responses to explicate how understanding of place value embodies both principles and how the abacus specifically caters to developing understanding of the positional principle.

DISCUSSION AND CONCLUSION

The effectiveness of TPD workshops depend both on design and enactment aspects. We have attempted here to present a framework that can aid in the understanding of both aspects and to illustrate how the framework may be applied through an analysis of the components of a workshop and two interaction episodes. The framework assumes that the central goals are to address teachers' knowledge and beliefs relevant to mathematics teaching. The framework does not describe what constitutes knowledge for teaching mathematics, nor does it elaborate on the nature of beliefs conducive to teaching for understanding. A framework that elaborates on the specifics of knowledge and beliefs relevant to teaching mathematics will need to be contextualized with regard to topics and to teacher communities. The framework presented here, in contrast, identifies certain principles that are important for the design of tasks and their enactment in workshop sessions. The study reported here does not aim to provide evidence for the effectiveness of a TPD intervention. The framework proposed here, we believe, is useful in identifying and providing rich descriptions of elements that are important in a TPD intervention.

The principles we consider important are situatedness, challenge and community building. The components of the workshop and the tasks worked on were chosen and designed to embody these principles. As the analysis of interactions in the two episodes shows, not only is the design of tasks important, but also how interactions between teachers and teacher educators are shaped to support teacher learning. We used the teacher education triangle having the three corners as task, teachers' agency and teacher educators' agency, as a framework to analyse the interaction aspects. The task incorporated contexts and artifacts that are situated in the work of teaching thus facilitating the involvement of teachers by identifying elements common with their teaching practice and engaging in deeper exploration of the contexts and artifacts. In the episodes discussed above, a prominent artifact was students' errors or responses. Discussion centered on these led teachers to analyse conceptual gaps, alternative explanation of errors and develop a perspective of explaining student errors by thinking about students' sense making efforts.

The evidences of the types of teacher engagement that occurred during the episodes throw light on the kind of opportunities that arise for teacher learning. Teachers' engagement took the form of anticipating and predicting students' responses, identifying key knowledge pieces, conjecturing underlying causes, articulating and contesting beliefs and assessing a teaching resource or a teaching approach. Such engagement was crucial in building shared understanding not only among teachers, who rarely get opportunities to reflect collectively about teaching in their schools and professional development contexts, but also for teacher educators by providing windows into teacher thinking. Teachers' assertions, counterarguments, alternative explanations and assessments were also a resource, which deepened fellow teachers' and teacher educators' understanding about mathematics teaching as it takes place in classrooms. The key knowledge piece of the meaning of zero, identified by an expert teacher, was important in deepening the participants' understanding of the conceptual gap that needs to be addressed to help Mohsin in writing numbers correctly.

We have elaborated on the principles of situatedness of tasks, challenges and development of community as guiding the design and enactment of the sessions. These aspects inform the decisions of the teacher educators about how interventions are to be made in sessions to facilitate active learning of the teacher. The belief goals of the teacher educators helped in guiding what interventions are to be made in terms of prompts presented to the teachers and in identifying aspects of teachers' responses that could be problematized. The agency exercised by the teacher educator is important to not only actualize the opportunities afforded by tasks, but also in guiding discussions beyond the resolution of the tasks in order to relate to the broader goals of teaching mathematics. The actions of the teacher educator like inviting teachers to respond to each other, handling disagreement and conflicts by posing issues as more general questions, use of open ended questions and building shared vocabulary paved the way for building a sense of community while challenging teachers to explain and justify their thinking.

We claim that while designing workshops for teachers it is essential to not only focus on the aspects that need to be discussed but also how the session needs to be enacted to allow teachers to exercise their agency and take ownership of their own learning rather than looking for answers from outside. This is important because we need to provide ways through which teachers can build on the knowledge of students and mathematics teaching that they already have rather than merely providing knowledge, which teachers may or may not find useful in their own classroom contexts. This point is important for designing workshops for in-service teachers as they have already developed identity as well as situated knowledge of students and teaching which must be respected and built upon.

REFERENCES

Batra, P. (2005).Voice and Agency of Teachers: The missing link in the National Curriculum Framework 2005, *Economic and Political Weekly, 40*(36), 4347-4356.

Ball, D., & Cohen, D. (1999). Toward a Practice-Based Theory of Professional Education. *Teaching as the Learning Profession. San Francisco: Jossey-Bass.*

Brodie, K., & Shalem, Y. (2011). Accountability conversations: Mathematics teachers' learning through challenge and solidarity. *Journal of Mathematics Teacher Education, 14*(6), 419-439.

Dewan, H. K. (2009) Teaching and Learning: The Practices. In R. Sharma & V. Ramachandran (Eds.). *The elementary education system in India: Exploring institutional structures, processes and dynamics.* New Delhi: Routledge.

Goodchild, S., & Sriraman, B. (2012). Revisiting the didactic triangle: from the particular to the general. *ZDM, 44*(5), 581-585.

Government of India. (1966). *Education and National Development: Report of the Education Commission* (Vol 2, pp. 1-89). New Delhi. Ministry of Education, Government of India. (Reprint by the National Council of Educational Research and Training, March 1971)

Government of India. (1986). *National policy on education*. New Delhi: Ministry of Human Resource and Development.

Jaworski, B. (2008). Building and sustaining inquiry communities in mathematics teaching development: teachers and didacticians in collaboration. In K. Krainer and T. Wood (Eds.). *The International Handbook of Mathematics Teacher Education volume 3: Participants in Mathematics Teacher Education: Individuals, Teams, Communities and Networks* (pp.309-330). Rotterdam: Sense Publishers.

Kazemi, E., & Franke, M. L. (2004). Teacher learning in mathematics: Using student work to promote collective inquiry. *Journal of Mathematics Teacher Education*, 7(3), 203-235.

Kumar, R. S., Dewan, H. & Subramaniam, K. (2012). The preparation and professional development of mathematics teachers. In R. Ramanujam & K. Subramaniam (Eds.). *Mathematics education in India: Status and outlook*. Mumbai: Homi Bhabha Centre for Science Education.

Kumar, R, S. & Subramaniam, K. (2013). Elementary teachers' beliefs and practices for teaching of mathematics. In Nagarjuna, G., A. Jamakhandi & E. M. Sam (Eds.). *Proceedings of epiSTEME-5: International conference to review research on Science, TEchnology and Mathematics Education*, pp. 247-254, Margao, India: Cinnamon Teal Publishing.

Lave, J., & Wenger, E. (1991). *Situated learning: Legitimate peripheral participation*. Cambridge, UK: Cambridge University Press.

Ma, L. (1999). *Knowing and teaching elementary mathematics.* Mahwah, NJ: Lawrence Erlbaum associates.

Ministry of Human Resource and Development. (2009). *Proceedings of the International Conference in Teacher Development and Management, Discussions and Suggestions for Policy and Practice*. New Delhi: Ministry of Human Resource and Development.

Miles, M. B., & Huberman, A. M. (1994). *Qualitative data analysis: An expanded sourcebook*. London: Sage Publications.

Mortimer, E. F., & Scott, P. H. (2003). *Meaning making in secondary science classrooms.* Maidenhead, UK: Open University Press.

National Council of Educational Research and Training (2005). National curriculum framework. New Delhi: NCERT.

National Council of Educational Research and Training. (2006). *Position paper: National Focus Group on Teaching of Mathematics*. New Delhi: NCERT.

National Council for Teacher Education. (2009). *National curriculum framework for teacher education: Towards preparing professional and humane teacher.* New Delhi: NCTE.

Philipp, R. A. (2007). Mathematics teachers' beliefs and affect. In F. Lester (Ed.)., *Second handbook of research on mathematics teaching and learning* (pp. 257-315). Reston, VA: NCTM.

Sharma, R., & Ramachandaran, V. (Eds.). (2009). *The elementary education system in India: Exploring institutional structures, processes and dynamics.* New Delhi: Routledge.

Shulman, L. S. (1986). Those who understand: Knowledge growth in teaching. *Educational Researcher* Feb. 4-14.

Thompson, A. (1992). Teachers' beliefs and conceptions: A synthesis of the research. In D. A. Grouws (Ed.), *Second handbook of research on mathematics teaching and learning* (pp. 127-146). New York: Macmillan.

Wenger, E. (1998). *Communities of practice: Learning, meaning, and identity.* Cambridge, UK: Cambridge University Press.

CHAPTER 75

INSIGHTS INTO STUDENTS' ERRORS BASED ON DATA FROM LARGE SCALE ASSESSMENTS

Aaloka Kanhere
Eklavya, Hoshangabad

Anupriya Gupta and Maulik Shah
Educational Initiatives, Ahmedabad

ABSTRACT

In this chapter we share insights on student errors based on item response data of large scale assessments carried out in India. We use response data on items from diagnostic tests and an intelligent tutoring system available across classes to examine the nature of students' errors in detail. We also do a comparative study across different student groups to further enhance our understanding of some student errors. We believe that such an analysis of student errors and underlying incorrect notions can contribute to enhancing teachers' pedagogical content knowledge and would be useful to curriculum developers. The approach of studying the nature of student errors can be further used in mathematics education research to throw light on how or why children develop incorrect notions.

The First Sourcebook on Asian Research in Mathematics Education: China, Korea, Singapore, Japan, Malaysia, and India, pp. 1655–1684

Keywords: Student errors, large scale assessments, student interviews, diagnostic tests

We will begin by describing one of the many instances from interactions with school students in India which provide the motivation for writing this chapter. The episode occurred in Goa, which is the smallest state in India located on its Western coast. Goa has the highest GDP per capita among all states, two and a half times that of the country average. It has one of the fastest growth rates: 8.23% (yearly average 1990–2000) ("Goa", n.d.). Gaurav (name changed) is a Class 4 boy from one of the better private English medium schools in Goa. His first language is English. He is comfortable in English but when asked a simple mathematics word problem in English, Gaurav follows rules rather than try to understand the problem and choose an appropriate arithmetic operation. The problem given to him was, "Ram had 187 marbles. Shyam had 245 marbles. How many more marbles did Shyam have than Ram?"

Gaurav's answer was 432. When asked how he got this, Gaurav didn't have to think twice. Confidently he answered, "When '*less*' appears in the question one needs to subtract and when the word '*more*' appears the numbers are added." Whether Gaurav picked this up in school or in private tuition classes at home is not clear. Such simplistic algorithms usually fail when there are more than three numbers and students are left to make guesses.

In many Indian schools, the approach of teaching operations of whole numbers usually lays stress on attaining computational skills in the four operations. The school curriculum and the teacher usually emphasize learning procedures dealing with the four operations separately and then "reinforce" the procedure for that particular operation through a large number of word problems on that operation. Even though a student is able to add or subtract two 3-digit numbers, he or she may not know whether to add or to subtract in a given word problem. He may have only understood the algorithm which is to be carried out with the symbol "+" or "–". This is borne out by student response data on items from large scale assessments conducted in English medium private schools in India. Given below is one such item with the student response data.

Item 1. Ravi has 93 stamps. He has 17 stamps more than Sid. How many stamps does Sid have?

(A) 76 (B) 84 (C) 86 (D) 110

In Table 75.1, the most commonly chosen incorrect option(s) is(are) in bold and the correct answer option is underlined. We follow the same

Table 75.1 Responses of Class 4 Students (N = 14106) From Private English Medium Schools (Category Explained in the Methodology Section) to Item 1

	Option A	*Option B*	*Option C*	*Option D*
% students	61.90%	5.50%	6.40%	**23.80%**

convention for the rest of the chapter. As the table shows, about 24% of the students may have, like Gaurav, added because the word "more" appears in the question.

After the interview with Gaurav, it was clear that difficulty with the English language was definitely not the reason for getting an incorrect answer. He might be unsure of the applications of operations on numbers and hence preferred to follow the "*rule*" given in his class. He has been encouraged to memorize a seemingly correct rule: "*more*" in the question indicates that one must perform "*addition*".

Rote-learning or memorization of rules and procedures often leads to mistakes. Mistakes can be either slips or errors. Slips are wrong answers which are not systematic and are one-off. They result from carelessness and can be done even by experts. An example of a common slip would be the following: an engineering student in a prestigious technology institute in the country, mistakenly calculated 7^{-2} as –49 in an exam. He quickly realized his mistake corrected his answer to 1/49. Errors are more systematic than slips. They are indicators of "*errors*" in the underlying concepts of a child, which are hurdles for the child and prevent her from developing a clear understanding (Olivier, 1989).

In this chapter, we look at a broad category of errors and wrong notions, but exclude slips or careless errors which any student, or even experts, might commit without any rationale. An in-depth understanding of errors and the underlying student ideas, their prevalence across grades, and the likelihood of their occurrence in different groups of students is a critical part of a teacher's pedagogical content knowledge (Shulman, 1987).

In India, research on student errors so far has been largely based on classroom observations and interviews with individual students. With the improvement in data collection technologies in the last two decades, it has been possible to find students' responses to a large number of questions across the grades. Benchmarking studies on student learning standards and diagnostic tests in each subject are also being carried out globally (Loveless, 2008). Student errors and incorrect notions are captured in test items with highly attractive multiple choice distracters (Sadler, 1998), designed on the basis of prior research about student errors. The data from such large scale studies administered across populations and different student groups serve as a repository for finding common patterns in errors that students make, and provide granular information on student knowledge. Data on student errors is also collected by Intelligent Tutoring Systems (ITS) and used as formative assessment tools to give records of student learning processes (Pellegrino, Chudowsky & Glaser, 2001). An advantage of using an ITS is that it provides data across grades for a large number of test items designed to test researchers' hypotheses about students' incorrect notions. This allows researchers and trainers to get insights about students' conceptions across grades.

In India, the research in mathematics education on student errors is mostly done in the context of interventions to improve classroom instruction. For instance, Menon's study of the difficulties faced by students when they are first introduced to the topic of angles in school, such as associating the magnitude of an angle with the length of its arms (Menon, 2009), is a part of a project on the development of an alternative trajectory to teach geometry. The study of alternate teaching-learning practices is carried out by various groups working with children (see Mukherjee & Varma; Menon, this volume). Pradhan and Mavlankar (1994) provide a limited compilation of student errors. These studies need to be supplemented with studies on the prevalence of such errors over large student populations. It is also important to know if different student-populations show different error-patterns. Information about the persistence of errors across learning levels is also useful for mathematics educators.

METHODOLOGY OF TEST PREPARATION, ADMINISTRATION AND STUDENT INTERVIEW

For the purpose of this chapter, the student response data from the following sources have been used:

- ASSET, developed by Educational Initiatives Pvt. Ltd. (EI), is a diagnostic test taken by about 450,000 students every year across different classes of private English medium schools in India in five major subjects. All the students of the participating schools take the test and hence the tests are taken by different ability groups. Students from Indian schools located outside India also take this test. The data presented in this chapter is taken from the tests taken by students in Classes 3 to 9 in the last 5 years.
- Mindspark (ITS) – developed by EI, is a computer based adaptive self-learning program utilized by private English medium schools in India.

- Data from benchmarking studies conducted by EI over large groups of students such as Student Learning Study 2009 (assessment of student achievement in Classes 4, 6 and 8 in 18 states in 15 languages), Student Learning In Metros Study study 2006 (assessment of student learning among 30,000 students from "top" 40 schools in five leading metro cities) and Wipro Quality Education Study 2011(a research study in five metro cities and a few schools with a different learning environment).

Most of the assessment items used in the tests are multiple choice questions with four choices and only one correct option. The test items are carefully chosen and created by a team of specialist test developers in elementary school mathematics based on the curriculum followed by the schools and educational boards in India. The test items are at times created using the student response data on items administered earlier in similar and different populations. The student interviews and current pedagogical research are also taken into consideration while choosing distracters to capture the errors in students' thinking. The data already collected is used to create similar items to test the existence of errors across different populations and grades.

The aim of the selected items is to find learning gaps in a child's mathematical understanding. The test items go through various levels of checks including statistical tests, within the team both before and after the administration of tests. Test papers are sent to schools and test takers take the tests in the presence of invigilators who are school teachers in the case of ASSET. All the tests for the other studies mentioned were administered by trained representatives of EI to ensure that the test administration processes were standardized. Test takers filled their responses in OMR sheets in all the studies except the Student Learning Study 2009, where trained evaluators filled OMR sheets based on student responses on answer sheets.

Besides the data collected from the tests as described above, we also draw on data from interviews with groups of students focused on questions, often taken from the written tests. In an interview with students, generally a question is projected on an overhead display and the entire class is asked to be ready with their answers. Once students have chosen their answer option, a few of them are asked to share their answers and reasons for their choice one by one. The discussion happens around their responses (without the interviewers sharing the right answer or their views). Rather, the interviewers probe the thinking of students and often many incorrect notions are revealed. Many times even a student, who has the correct answer, may have reasoned wrongly. These instances provide valuable insights to teachers and researchers. Some of the student interviews are video recorded. The analysis of the data from the sources mentioned can confirm or contradict the findings from small-scale studies, and lead to more fine-grained insights about student errors especially in the Indian context.

In this chapter, we have also quoted content from the textbooks prescribed by National Council of Educational Research and Training (NCERT), which are commonly used across the country. We have attempted to point to some implications for writing textbooks and teaching through the analysis of student errors.

COMMON STUDENT ERRORS AND THEIR PREVALENCE AMONG INDIAN STUDENTS

In this section, we discuss student errors indicated by the student response data on assessment items with discerning distracters in large scale diagnostic tests. In the subsections to follow, we focus on incorrect notions commonly found in primary and middle classes in the topics of measurement of length and area, whole number arithmetic, ordering of decimal numbers, fractions, geometry and algebra. In a few cases, item response curves (IRC) which show the extent of the error in students of different ability groups have been included.

Measurement

Measuring length using a marked ruler. Using a ruler to measure length is a practical skill which is expected to be developed by Class 3 or Class 4. The concept of length measure as obtained by iterating a unit of standard length is crucial for building the concept of measurement. This concept is extended in higher classes to measurement of the area of a closed shape and the volume of a solid. To test this concept, the question shown in Figure 75.1 was asked to 1551 students of Class 4.

The question in Figure 75.1 was given to students from private schools who typically come from high income families with professionally employed parents. Table 75.2 shows that only 15.5% of students could correctly measure the length of a pencil using a ruler. There were two common wrong answers, options C and D. To understand these student errors, interviews were conducted by the team who were conducting this study.

In the student interviews conducted subsequently, it was found that there are two ways in which students answer this question incorrectly. (The recordings of these student interviews are available and can by obtained

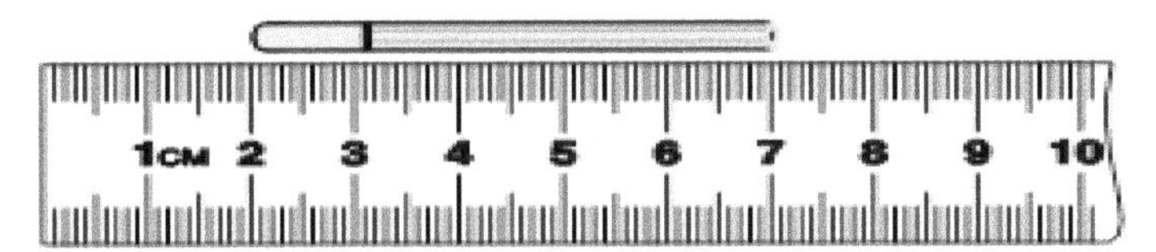

A. 4 cm
B. 5 cm
C. 6 cm
D. 7 cm

Figure 75.1. Testing understanding of length measurement.

Table 75.2. Performance on Length Measurement Question in Figure 75.1 of 1554 Class 4 Students From English Medium Schools in Metro Cities

	Option A	*Option B*	*Option C*	*Option D*
% students	10.80%	15.50%	**31.20%**	**42.50%**

by writing to the authors.) In the first interpretation, students look for the mark on the ruler where the object to be measured ends. They don't take into account the fact that the starting point of the object is not the "0" mark on the ruler. We will call this the "*endpoint rule*". In the second interpretation, students count the marks on the ruler between the two end points of the object which is to be measured. By using this method, they would count starting from 1 to 6, to arrive at 6 cm as the answer to this question. We call such an error-pattern as "*counting division marks rule*".

In both these cases, students don't understand how the measure of "length" is obtained by iterating a unit length. Student interviews suggest that many students seem to not know that the length between the 1 cm and 2 cm marks on the ruler is 1 cm and do not apply the notion of unit iteration to measure the length. Most often students are found to be following a procedure without understanding how and why it works as evident from student interviews. In the later sections in this paper, we will discuss the prevalence of these two different errors over classes.

Perimeter as a measure of length. Students are usually confused about the concept of perimeter as a measure of length. This confusion seems to arise due to incomplete understanding of area and perimeter. Students are often found to be struggling when asked to apply their conceptual understanding of area and perimeter as seen from the response data on

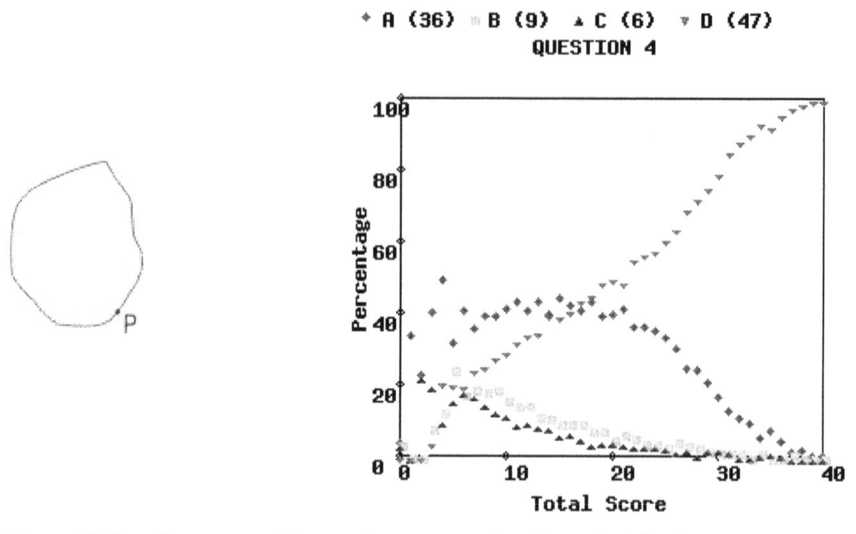

Figure 75.2. Curve traced for perimeter question (Item 2). The item response curve (Percentage of correct responses vs total score on the test) is shown on the right.

the Items 2 and 3. Item 2 requires the students to understand the concept of perimeter as a measure of length.

> **Item 2.** Anu draws the following shape on paper (see Figure 75.2). She starts drawing from P and ends at the same point as shown below. She does not take the pen off the paper from start to finish nor go back any time. The length of the line traced by her pen is the
>
> (A) area of the shape. (B) width of the shape.
>
> (C) height of the shape. (D) perimeter of the shape.

Large numbers of students confuse between perimeter and area. The IRC for Item 2 in Figure 75.2 shows that almost 40% of the students with scores between 0 and 20 on a 40 question test have a wrong notion that the length of the boundary of a closed shape is the area of the shape. Even a few high performing students are found to have such a wrong notion as seen in the IRC in Figure 75.2.

Item 3 from another test asked for the area of the shape formed by sticks shown in Figure 75.3. About 55% of the Class 6 students of English medium private schools across India gave the perimeter of the shape as the answer.

Table 75.3. Responses of Class 6 Students (*N* = 25710) From Various English Medium Private Schools Across India to Item 2

	Option A	*Option B*	*Option C*	*Option D*
% students	**36.3%**	8.5%	6.3%	47.1%

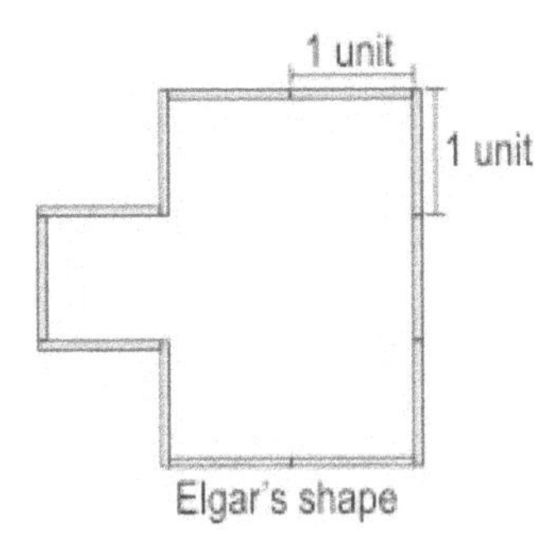

Figure 75.3. Area of shape with matchsticks (Item 3).

Table 75.4. Responses to Item 3 of Class Students (*N* = 27834) of English Medium Private Schools Across India

	Option A	*Option B*	*Option C*	*Option D*
% students	11.7%	21.7%	10.6%	**54.2%**

Item 3. Elgar made a shape using sticks, 1 unit long, as shown here (see Figure 75.3). What is the AREA of the shape?

(A) 6 sq units (B) 7 sq units (C) 10 sq units (D) 12 sq units

Area as a measure of space. Students often do not realize that area is an attribute of any simple closed shape. There is an emphasis while teaching on computing using formulas areas of shapes which are squares or rectangles or polygons which can be divided into rectangles or squares. Inadequate exposure to estimating areas of irregular shapes may lead to formation of wrong notions among students such as area can exist only when it can be calculated (using a formula) or when the shape is a polygon. Item 4 was developed to test such conceptions. The distracters were chosen so as to capture the incorrect notion that "only closed shapes with straight sides have area or only shapes for which there is a formula known to compute its area have area".

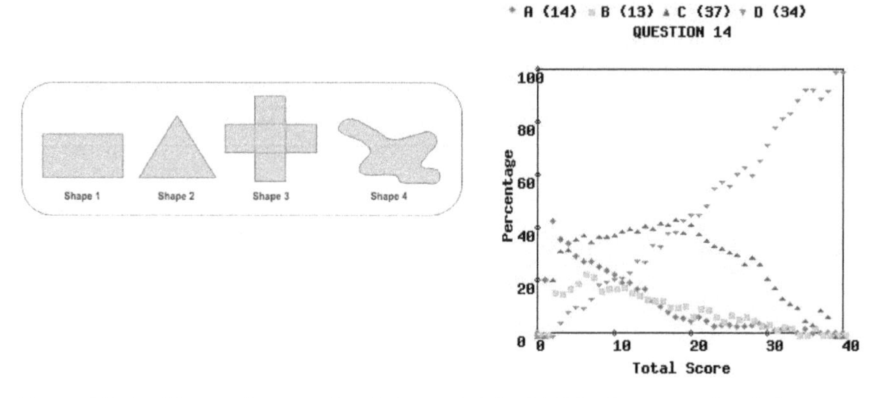

Figure 75.4. Shapes for question in Item 4. The IRC is shown on the right.

Item 4. Which of these shapes (shown in Figure 75.4) has/have area?

(A) only shapes 1 and 2 because the area can be calculated using formulae
(B) only shapes 1 and 3 as they are the only ones which can be covered completely by unit squares
(C) only shapes 1, 2, and 3 as they have straight sides
(D) all four shapes - 1, 2, 3 and 4

The response pattern in Table 75.5 indicates that many Indian students of Class 7 have an incorrect notion that only closed shapes with straight sides have area, and that a closed shape with irregular boundary does not have area. As per the guidelines of National Curriculum Framework 2005 (NCF) the concept of area is introduced informally in Class 5 and more systematically in Class 6. Textbooks prescribed by National Council of Educational Research and Training (NCERT) and the commonly used textbooks in India introduce students to the idea of approximating area of an irregular shape using a graph paper. However, the student response data (Figure 75.4) suggest that even in Class 7, about 30% of the students across ability levels (with total score on the test in the

Table 75.5. Responses to Item 4 of Class 7 Students (*N* = 25371) From English Medium Private Schools Across India

	Option A	*Option B*	*Option C*	*Option D*
% students	13.6%	13.0%	**37.1%**	34.8%

range of 0 – 30) have this notion. It can be seen that about 20% of the students, who answered three-fourth of all 40 questions in the test correctly, have this incorrect notion. This wrong notion exists even among few students answering more than 30 questions on a 40 question test correctly.

Numbers and Number Systems

Subtracting two 4-digit whole numbers. The National Curriculum Framework 2005 in India recommends 4-digit arithmetic in Class 5 or 6. However, in most of the states the curriculum introduces it in earlier classes. A question which required students to subtract two 4-digit numbers (Item 5) was asked in a test given to about 35000 Class 8 students of government schools across the country.

Item 5. 6000 – 2369 = ____

(A) 4369 (B) 3742 (C) 3631 (D) 3531

As seen from Figure 75.5, only 59.1% of Class 8 students answered this question correctly. Moreover 17.4% of students gave the answer as 4369 which may indicate either the well-known "Smaller-From-Larger" error (Resnick, 1982) or the '0 - N = N' error.

Ordering decimal numbers. Research studies have shown that students extend the ordering of whole numbers inappropriately to order

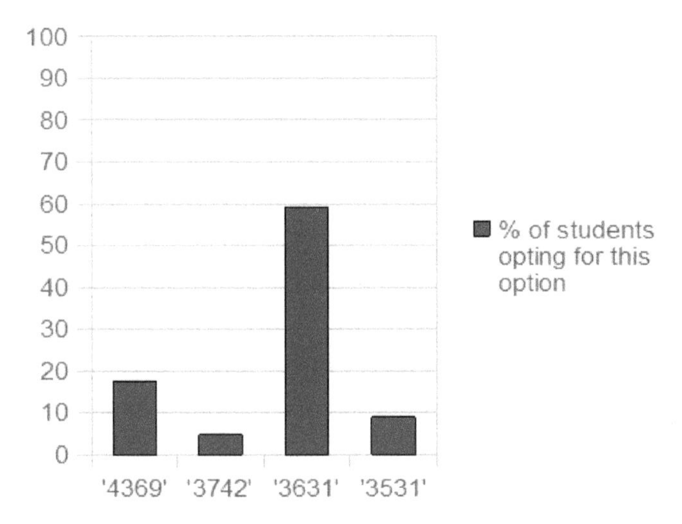

Figure 75.5. Graph showing student response data.

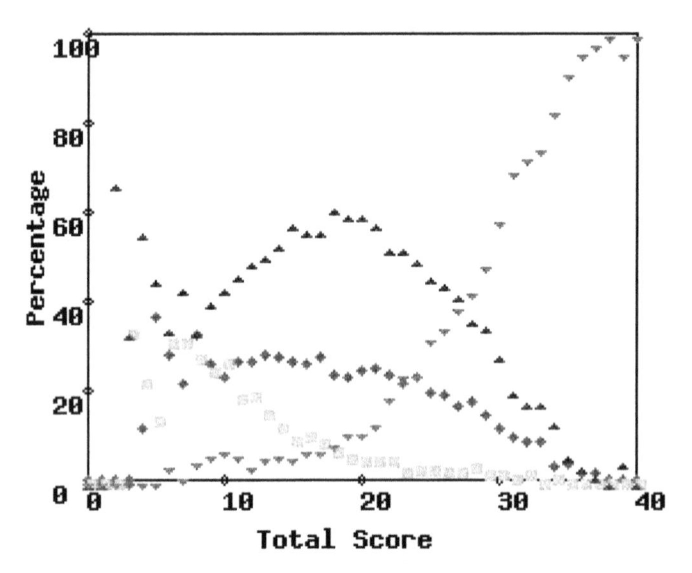

Figure 75.6. IRC for Item 6.

Table 75.6. Responses to Item 6 of Class 6 Students (*N* = 14327) From English Medium Private Schools

	Option A	*Option B*	*Option C*	*Option D*
% students	21.8%	8.1%	**47.7%**	<u>21.3%</u>

decimal fractions. The response data of Items 6 and 7 shed light on students' notions about ordering decimal fractions.

Item 6. Which of these lies between 6.3 and 6.6?

(A) 6.2 (B) 6.9 (C) 6.05 (D) 6.41

Table 75.6 shows that only about 21% of the students have ordered decimals correctly. Research studies reveal that students in general adopt one of the following reasoning to order decimal fractions incorrectly.

String length thinking. These students judge only on the basis of the length of the number string in a decimal. They treat a decimal as two whole numbers separated by a marker with longer numbers being larger

(Steinle & Stacey, 1998). So, 1.65 is judged to be larger than 1.9. Similarly 6.05 is taken to be larger than 6.5.

In another variant of "string length thinking", students ignore the decimal point, treat the number string in the decimal fraction as a whole number and order. Such students would take 5.3 as 53 and 5.25 as 525 to conclude 5.25 is bigger than 5.3.

Numerator focused thinking. These students ignore "0" after the decimal point even if non-zero digits follow "0" and then order decimal fractions (Steinle & Stacey, 1998). Just as 05 is same as 5 in the case of whole numbers, these students treat 6.05 same as 6.5. Students compare digits after the decimal point as whole numbers to compare decimal fractions. For the Item 6, students may have ignored the "0" in 6.05 to treat it as 6.5 to say that 6.5 lies between 6.3 and 6.6.

About 48% of 14327 Class 6 students seem to have used "numerator focused thinking". Additionally it is worth mentioning that, about 20% of the students who scored about 30 in a 40 question test, with each question adding either 0 or 1 to the score, were also applying the above erroneous strategy as seen from the IRC (Figure 75.6).

The following item (Item 7) provides additional data that throws light on the reasoning used by students to order decimal fractions.

Item 7. In which of the following are the decimal fractions arranged from the smallest to the largest?

(A) 2.5, 2.03, 2.07 (B) 2.03, 2.5, 2.07

(C) 2.03, 2.07, 2.5 (D) 2.07, 2.03, 2.5

About 30% of the students seem to have applied the *numerator focused thinking*. About 22% of the students seem to have applied the *string length thinking* judging 2.5 to be less than 2.03.

Fractions

In India, students are introduced to fractions by Class 4 in most of the curricula. We discuss students' errors related to fractions below.

Table 75.7. Responses to Item 7 of Class 6 Students (*N* = 8617) From English Medium Private Schools Across India and a Few Abroad

	Option A	*Option B*	*Option C*	*Option D*
% students	**22%**	**29.7%**	<u>37.1%</u>	9.9%

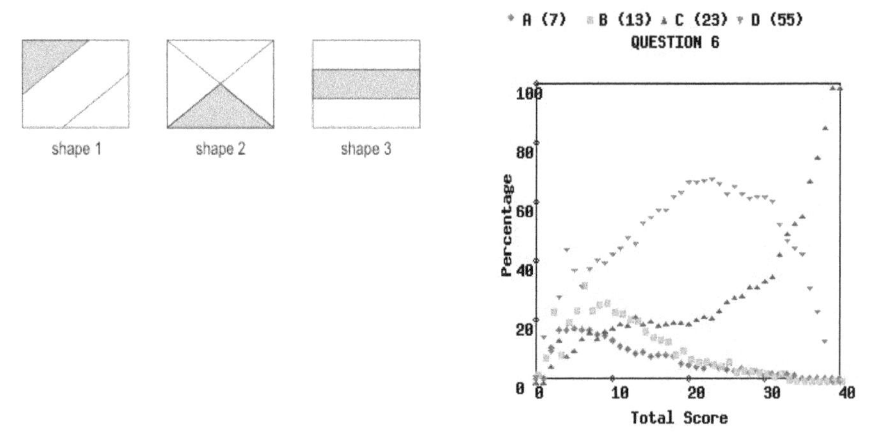

Figure 75.7. Fraction of area shaded (Item 8). The IRC is shown on the right.

Table 75.8. Responses to Item 8 of Class 4 Students (*N* = 26636) of English Medium Private Schools Across India

	Option A	*Option B*	*Option C*	*Option D*
% students	6.9%	13.0%	23.5%	**55.0%**

Representing fractions using the area model. Students learn to represent fractions using area model. Item 8 checks for ignoring the "equal parts" requirement in part-whole representation of fractions as well as for the error that the fraction shows a "part-part" relation rather than the "part-whole" relation.

Item 8. Which of the shapes (see Figure 75.7) is/are 1/3 shaded?

(A) only shape 1 (B) only shape 2

(C) only shape 3 (D) only shapes 1 and 3

As seen from the student response data, about 62% of the students choosing option A or D think that shape 1 is 1/3 shaded ignoring the requirement of equal area. As seen from the IRC (Figure 75.7), this notion prevails in more than 60% of the students scoring between 20 to 30 on a 40 question test each of 1 mark.

Figure 75.8 shows excerpts from the chapter "Halves and Quarters" of a Class 4 NCERT textbook. The chapter introduces the fractions half and

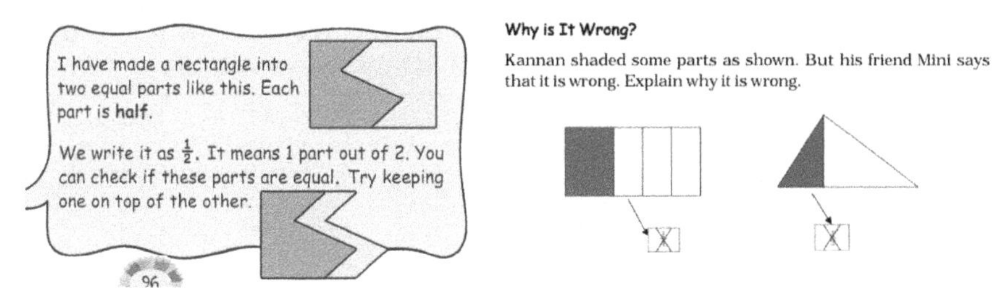

Figure 75.8. Excerpts from the Class 4 NCERT textbook.

quarter using different contexts and emphasizes that parts must be equal in a part-whole representation of fractions.

The notion that parts of the whole need to be equal is stressed in the chapter in many other ways. Large numbers of the students taking the test are from schools that follow the NCERT textbook. The data shows that the students have not understood the equal parts requirement.

In an item in a different test (Item 9), about 58% of the 27863 Class 5 students of English medium private school said that ¾ of the rectangle shown in Option C (Figure 75.9) is shaded. So the wrong notion continues to exist even among Class 5 students, and as other data shows, even in Classes 6 and 7.

Item 9. Which of these figures (shown in Figure 75.9) is ¾ shaded?

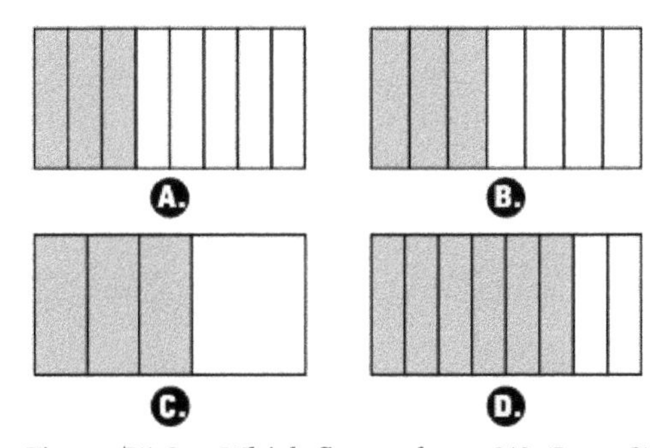

Figure 75.9. Which figure shows ¾? (Item 9).

Table 75.9. Responses to Item 9 of Students of Class 5 (N = 27863) of English Medium Private Schools Across India a Few Indian Schools Abroad

	Option A	*Option B*	*Option C*	*Option D*
% students	1.6%	**27.6%**	**58.2%**	11.0%

Table 75.10. Responses to Item 10 of Class 6 Students (N = 15543) From English Medium Private Schools Across India

	Option A	*Option B*	*Option C*	*Option D*
% students	14.7%	8.10%	**42.60%**	**32.90%**

Ordering a proper fraction. By Class 6, students are expected to develop a sense of the magnitude of fractional numbers. They are introduced to ordering fractions in Class 5. Item 10 deals with ordering fractions and specifically focuses on the incorrect notion that a proper fraction m/n is a number that lies between its numerator m and denominator n.

Item 10. $\frac{110}{570}$ lies between ____________.

(A) 0 and 1 (B) 1 and 2 (C) 11 and 57 (D)110 and 570

The data in Table 75.10 shows that about 76% students of Class 6 (i.e. those choosing options C and D) think that a fraction is a number that lies between its numerator and denominator. Students answering C are likely to have reduced $\frac{110}{570}$ to $\frac{11}{57}$ first and concluded it lies between 11 and 57. Less than 15% of the entire Class 6 population demonstrated the understanding that $\frac{110}{570}$ is less than 1.

Interpreting mixed fraction notation. As per the NCERT textbooks, students are introduced to mixed fractions informally in Class 5 and in detail in Class 6. By Class 7 students are expected to work with improper and mixed fractions. Many Class 7 students however, struggle with ordering improper and mixed fractions and determining between which two whole numbers the given improper or mixed fraction lies.

Item 11. Which of these is the same as $23\frac{1}{19}$?

(A) $23 + \frac{1}{19}$ (B) $23 - \frac{1}{19}$ (C) $23 \times \frac{1}{19}$ (D) $23 \div \frac{1}{19}$

As seen from the student response data (Table 75.11) only about 31% of Class 7 students understand the key idea that $23\frac{1}{19}$ the same as $23 + \frac{1}{19}$. More than half of the students scoring less than 28 marks on a 40 mark test think that $23\frac{1}{19}$ is equal to $23 \times \frac{1}{19}$ as seen from the IRC in Figure 75.10.

Table 75.11. Responses to Item 11 of Class 7 students (N = 14245) of English Medium Private Schools Across India

	Option A	*Option B*	*Option C*	*Option D*
% students	30.7%	3.8%	**56.4%**	8.1%

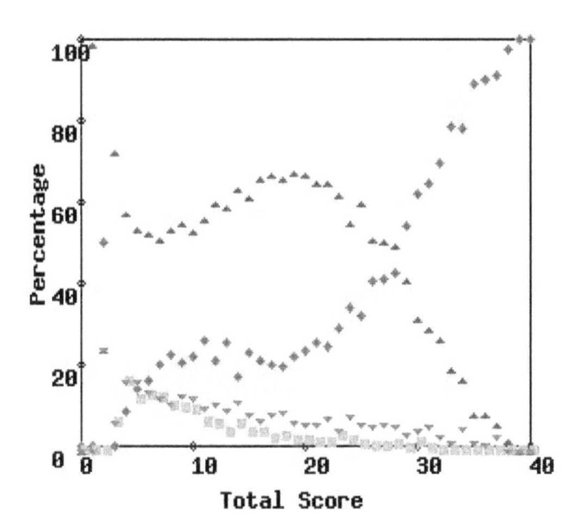

Figure 75.10. IRC for Item 11.

The student response data however indicate that many students may not understand the mixed fraction notation. It is possible that algebraic notation where $2x = 2 \times x$ (not $2 + x$), a new learning for students may be coming in their way of understanding the notation for mixed fractions. However, the response data for Item 12 below suggests that students more commonly interpret $2x$ as $2 + x$.

Item 12. Which of these expressions gives the value 6 when $x = 3$.

(A) $3x$ (B) $x - 3$ (C) $x + 3$ (D) $2x - 6$

Table 75.12. Responses to Item 12 of Class 6 Students (N = 1095) of English Medium Private Schools Across India

	Option A	*Option B*	*Option C*	*Option D*
% students	**21.6%**	11.5%	57.9%	8.4%

Geometry

Understanding angle as a measure of turn. Many students think that measure of an angle depends on the length of its arms or area enclosed between its arms, as the student response data for Item 13 indicates.

> **Item 13.** Which angle (of those shown in Figure 75.11) has the greatest degree measure?

It is possible that students develop such a wrong notion when an angle is introduced as the union of two rays sharing a vertex, rather than as a

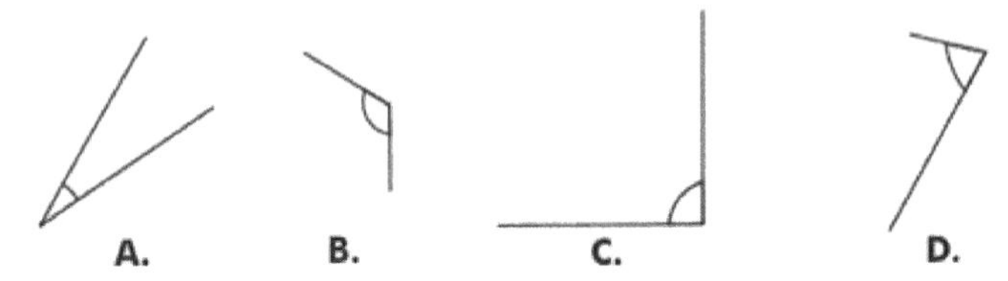

Figure 75.11. Which angles has the greatest degree measure? (Item 13).

Table 75.13. Responses to Item 13 of Class Students (*N* = 1871) of English Medium Private Schools Across India

	Option A	*Option B*	*Option C*	*Option D*
% students	4.2%	39.7%	**45.1%**	10.6%

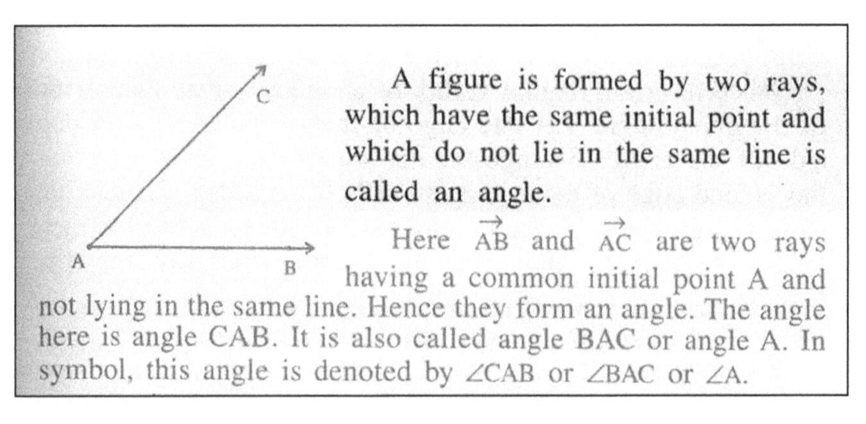

A figure is formed by two rays, which have the same initial point and which do not lie in the same line is called an angle.

Here $\overrightarrow{AB}$ and $\overrightarrow{AC}$ are two rays having a common initial point A and not lying in the same line. Hence they form an angle. The angle here is angle CAB. It is also called angle BAC or angle A. In symbol, this angle is denoted by $\angle CAB$ or $\angle BAC$ or $\angle A$.

Figure 75.12. Excerpt from a textbook showing definition of an angle.

"turn". Figure 75.12 shows an excerpt from a textbook, where angle is defined as a figure formed by two rays hence students might be tempted to think that the length of those rays determines the "size" of an angle.

Understanding the angle sum property of triangles. Many students have a wrong notion that the sum of the angles in an obtuse-angled triangle is greater than the sum of the angles in an acute-angled triangle.

Item 14. In which of these triangles (shown in Figure 75.13) is the sum of the measures of the three angles the greatest?

(A) Triangle I (B) Triangle II
(C) Triangle III (D) (It is equal for all 3 triangles.)

The response data in Table 75.14 indicates about 45% of Class 8 students do not know that the angle sum is equal for all triangles. About 33% of the students think that the sum of the angles of an obtuse-angled triangle will be greater than that in an acute-angled triangle. The extent of this error decreases steadily with the increase in mathematical ability as measured by the total score seen from the IRC in Figure 75.13.

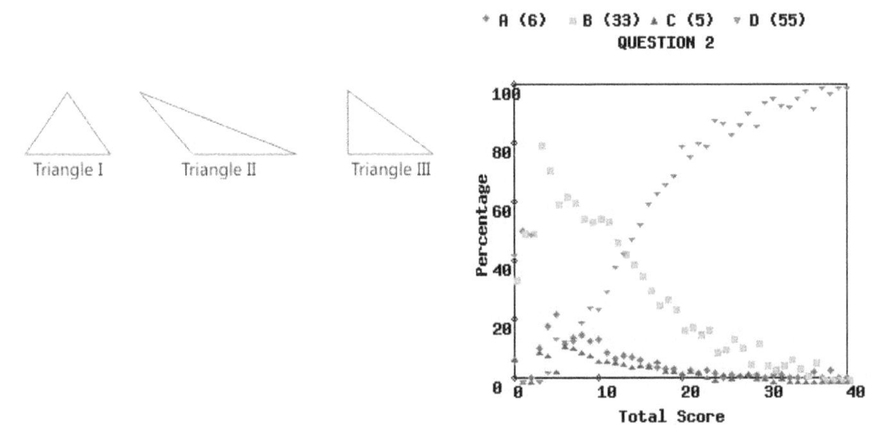

Figure 75.13. In which triangle is the sum of angles the greatest? (Item 14). The IRC is shown on the right.

Table 75.14. Responses to Item 14 of 12414 Students of Class 8 (*N* = 12414) of English Medium Private School Across and Few Indian Schools Abroad

	Option A	*Option B*	*Option C*	*Option D*
% students	5.7%	**33.4%**	4.9%	55.2%

Table 75.15. Responses to Item 15 of Class 8 students (N = 22581) of English Medium Private Schools Across India and Few Indian Schools Abroad

	Option A	*Option B*	*Option C*	*Option D*
% students	**42.7%**	34.9%	3.8%	16.9%

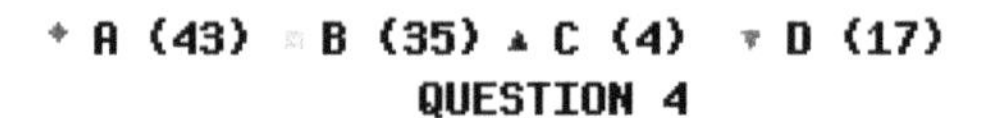

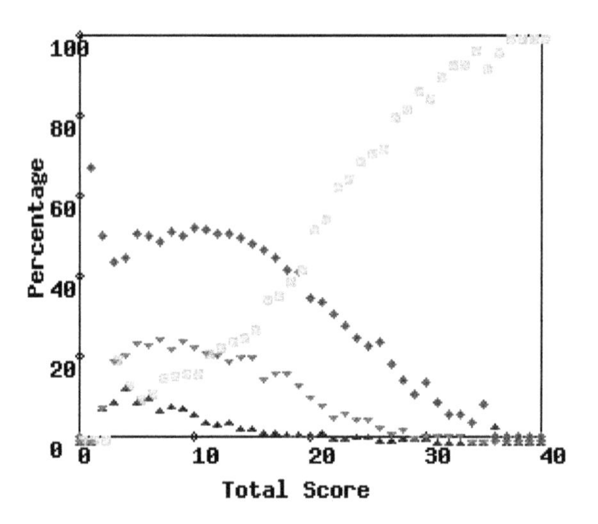

Figure 75.14. IRC for Item 15.

Understanding the Definition of an Acute Angled Triangle

Item 15. Two of the angles of a triangle are of measures 30° and 35° each. Then the triangle is

(A) an acute-angled triangle. (B) an obtuse-angled triangle. (C) a right angled triangle. (D) (Cannot be determined unless measures of all the three angles are known.)

As seen from Table 75.15, about 43% of the students were unable to apply the angle sum property of a triangle and deduce that the given triangle is an obtuse angled triangle. A large number of students think that a triangle with two acute angles is always an acute angled triangle. This includes about 40% of the students from the lower ability group (students scoring less than 18 mark on a 40 mark test of 40 questions) and a few students from the higher ability group (answering more than 30 questions correctly in the same test) as seen from the IRC in Figure 75.14. A few students think that an obtuse-angled triangle is a triangle with all its angles as obtuse.

We present an excerpt below from a student interview of Ankit (name changed), a Class 8 student in an English medium school of Ahmedabad, about his response to Item 15.

Interviewer: What is the answer to the question [in Item 15]?
Ankit: The answer is D.
Interviewer: Why?
Ankit: (Reads the line in distracter D.)
Interviewer: Since two of the angles are given can you not find the measure of the third one?
Ankit: Yes, I can find using angle sum property.
Interviewer: Then D cannot be the answer, right?
Ankit: Yes.
Interviewer: So what could be the answer?
Ankit: It is neither an acute angled triangle nor obtuse-angled triangle.
Interviewer: Why is it not an obtuse-angled triangle?
Ankit: *Because ALL the 3 angles of the triangle are not obtuse.*
Interviewer: Is it possible to have a triangle with all its angles obtuse?
Interviewer: (helping Ankit, who looked confused) Is it possible to have a triangle with all 3 of its angles greater than 90°?
Interviewer: (Ankit was still looking confused) Is it possible to have a triangle with 3 of its angles as 100°?
Ankit: No.
Interviewer: Why?
Ankit: Because sum of its angles is exceeding 180°.
Interviewer: Is it possible that two of the angles are greater than 90°? (Ankit was further asked to consider both the angles of 100° each and check if the angle sum property was violated.)
Interviewer: So, an obtuse angled triangle will have only one of its angles as more than 90°, right?
Ankit: Ya.
Interviewer: Can you draw an obtuse angled triangle?
(Ankit draws an obtuse angled triangle and shows the obtuse angle.)

Algebra

Understanding variables as negative numbers. When an unknown number represents a negative number students tend to represent it by $-x$ and not x failing to appreciate that a variable like "x" can be any number and can stand for even a negative number. The student response data on the following Item 16 throws light on the same.

Item 16. y is a negative number. Which of the following is 2 less than y?

(A) $-y - 2$ (B) $-y + 2$ (C) $y + 2$ (D) $y - 2$

About 36% of Class 8 students think that 2 less than y is $-y - 2$ when y is a negative number. In most of the examples discussed in the commonly used NCERT Class 6 textbook, a variable always stands for a natural number. This may be one of the factors influencing the students' response. In Class 7, students solve linear equations in one variable which could be a negative integer or a fraction. As shown by the Class 8 student response data, a large number of students however continue to think that a variable stands only for a positive number.

Understanding Special identities

Item 17. Which of these is the same as $(a + b)^4$?

(A) $a^4 + b^4$ (B) $(a + b)(a + b)^3$ (C) $(a^2 + b^2)(a^2 + b^2)$ (D) $a^4 + 2a^2b^2 + b^4$

Only 31.5% of Class 9 students were clear about the exponent notation involving algebraic expressions as can be seen from Table 75.17. A large number of students are aware of the identity $(a + b)^2 = a^2 + 2ab + b^2$,

Table 75.16. Responses to Item 16 of Students of Class 8 (*N* = 21994) of English Medium Private Schools

	Option A	*Option B*	*Option C*	*Option D*
% students	**36.4%**	29.3%	5.3%	27.6%

Table 75.17. Responses to Item 17 of Students of Class 9 (*N* = 3823) of English Medium Private Schools Across India

	Option A	*Option B*	*Option C*	*Option D*
% students	15.4%	31.5%	9.1%	**43.4%**

which they generalize inappropriately and think that $(a + b)^4 = a^4 + 2a^2b^2 + b^4$.

COMPARISON OF STUDENT ERRORS ACROSS CLASSES

Multiple choice items with strong distracters can be helpful to unearth wrong notions students have (Sadler, 1998). Although multiple choice questions have their own limitations, especially if the distracters are not chosen carefully, student response data on a large scale assessment test, as we have seen in the previous section, can provide an accurate picture of the extent of wrong notions among different ability groups of the student population. Making this knowledge available to teachers in professional development programs can be very useful. In many cases, making students aware of their own wrong notions and then building appropriate concepts can be a part of an effective teaching practice. Teachers can also use these items as formative assessment tools. In many of the cases, teachers may not be aware of wrong notions nurtured by their students. In this section we look at patterns of errors across classes, noting errors which decrease in incidence and those that tend to persist.

Errors in subtraction of whole numbers across student groups and classes. Item 5 discussed earlier ("6000 – 2369") also appeared in TIMSS 1999 for both Grade 4 and Grade 8. The TIMSS average international average for Grade 4 was 71%, which is better than the performance of Indian Class 8 students (59.1%) from Government schools across the country. The TIMSS average for Class 8 was 88% for this item.

When the same set of students from India were asked to compute "70 – 43", about 80% of the students answered correctly and only about 4% students showed the "Smaller-from-Larger" or the "0 - N = N" error. The difference in percentages of correct responses for these two questions indicates that Class 8 students from government schools in the country still have difficulty with operations on 4-digit numbers.

Error pattern across classes in measuring angles. Research studies show that when students are asked to compare the measures of angles, they compare the lengths of the arms (Menon, 2009). We have already seen many students make this error in the discussion of Item 13. We use the data from Mindspark (an ITS) to know about the extent of this error across classes. Figure 75.15 shows the question and the incidence of the "larger angle has longer arms" misconception across Classes 4 to 7.

When the question in Figure 75.15 was asked to Class 4 students of English medium private schools, 18% of students chose option B, the angle with the longest arms of the angles. In Class 7, only 6% of students

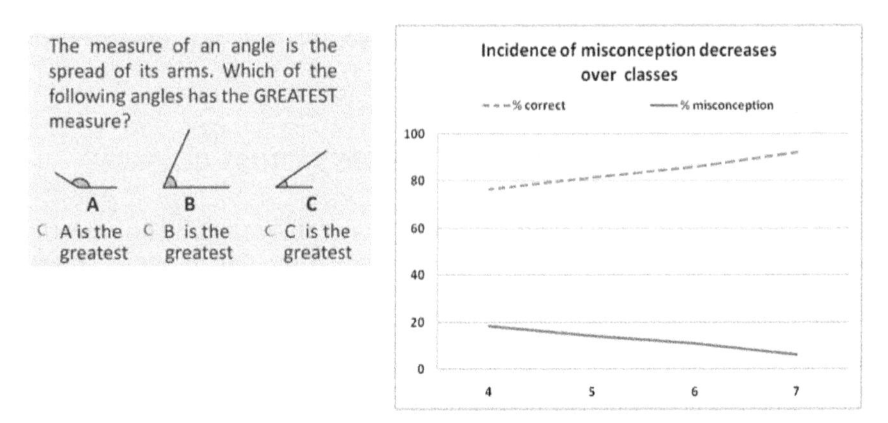

Figure 75.15. Question on largest angle and incidence of misconception across classes.

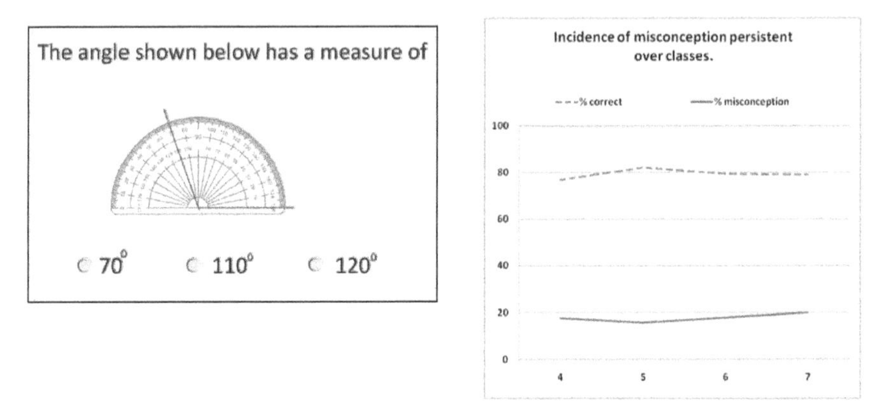

Figure 75.16. Incorrect reading of protractor: Incidence across classes.

chose that option. This suggests that the understanding of the measure of an angle improves with class levels as desired.

The skill of visually estimating the measure of an angle helps students in measuring an angle using a protractor. When checked with other examples it is evident that even though students learn to compare the sizes of angles fairly well as they move to higher classes, they are still not able to measure the angle correctly using a protractor. Figure 75.16 shows the angle to be measured (110°) and the incidence of incorrect reading of the protractor across Classes 4 to 7. The question is from the Mindspark ITS and the data is of students from English medium private schools across India.

The student response data for the question in Figure 75.16 suggests that many students have difficulty in correctly reading the scale on the protractor. Student interviews suggest that many students fail to notice that the angle shown is greater than 90°. In Figure 75.16, the angle being measured is in the standard orientation. From other items on angle measurement, we found that the incidence of error is higher when students are asked to measure angles using a protractor with angles in non-standard orientations.

Errors in measuring length using a marked ruler across classes. The common errors that students make while measuring length of a pencil using a marked ruler were discussed earlier. We examine the student response data from across classes for the same item discussed earlier (see Figure 75.1). Table 75.18 shows the error patterns for students of Classes

Table 75.18. Responses of Students From Classes 4, 6 and 8 to Question on Measuring the Length of a Pencil (see Figure 75.1)

Class	*Number of Respondents (N)*	*% Students Who Answered as 5 cm*	*% Students Who Answered as 6 cm (Counting Division Marks Rule)*	*% Students Who Answered as 7 cm (Endpoint Rule)*
4	1556	15.5%	**31.2%**	**42.5%**
6	1816	38.3%	**36.6%**	**19.9%**
8	1559	63.6%	**28.9%**	**4.6%**

Note: This data from students of "top" English medium private schools of 5 metro cities of India.

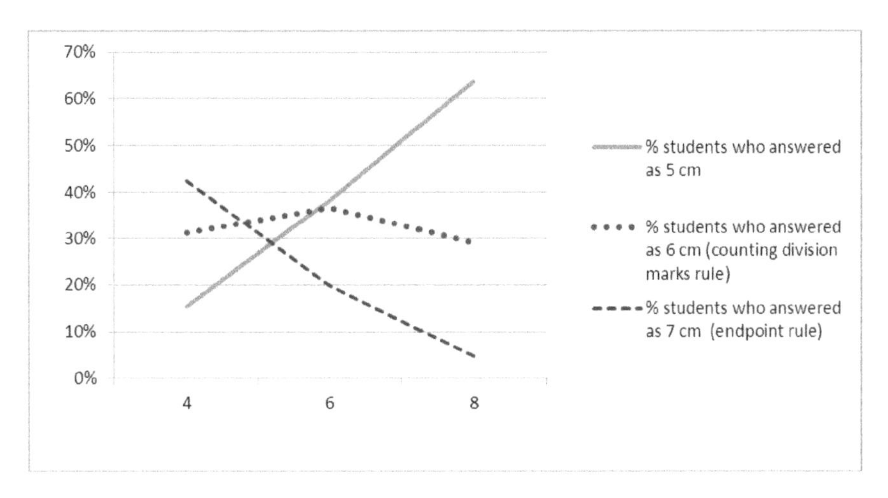

Figure 75.17. Graph showing the extent of the 2 error types across Classes 4, 6 and 8.

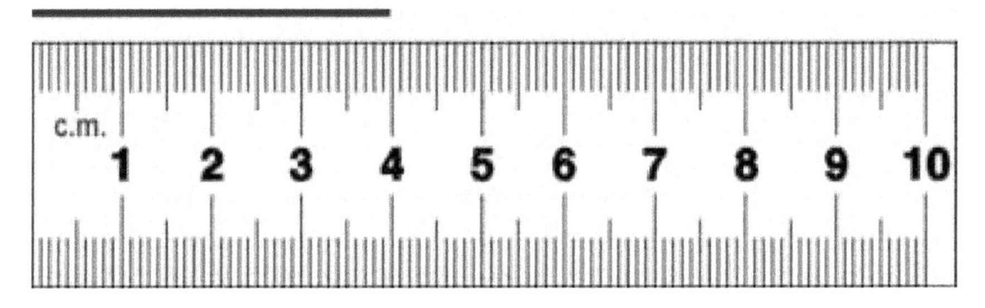

Figure 75.18. "Length of the line is 4 cm". (Item 18)

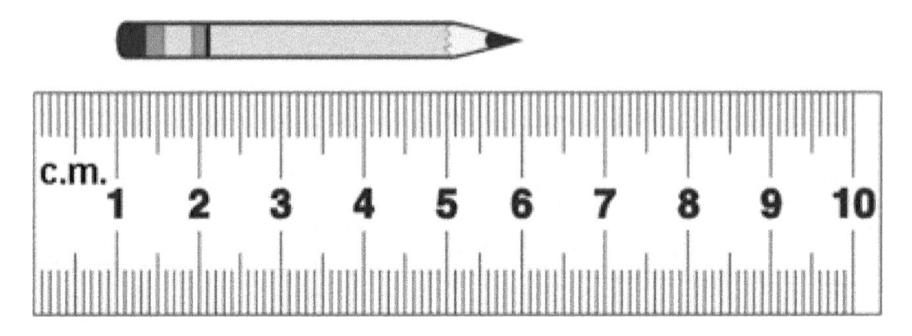

Figure 75.19. "How long is the pencil?". (Item 18)

4, 6 and 8 for the question on measuring length using a marked ruler (Figure 75.1). Figure 75.18 shows the trend of the two types of errors as we move from Class 4 to Class 8.

The response data shows a sharp decline in the end point error from Class 4 to Class 8. However the counting division marks error is persistent across classes 4, 6 and 8. About 30% of the student population across classes 4 to 8 makes the counting division marks error. The response to another test item, Item 18, given to a larger number of students confirmed this trend. 29513 students from Class 4, 35604 students from Class 6 and 35967 students from Class 8 from various government funded schools in 18 of the 25 states and a union territory of India were asked to respond to Item 18.

Item 18. The length of the line in the figure (Figure 75.18) above is 4 cm. How long is the pencil shown in the picture (Figure 75.19)? (Use the ruler shown in the picture.)

Answer: ________________ cm

Table 75.19. Responses of Students of Classes 4, 6 and 8 to Item 18

Class	*No. of Students (N)*	*% Students Whose Answer Was 5 cm*	*% Students Whose Answer as 6 cm*
4	29513	23.0%	**46.0%**
6	35604	22.1%	**41.7%**
8	35967	34.7%	**38.8%**

Note: The student response data is of a representative sample of students from government funded schools in 18 of the 25 states and a union territory of India.

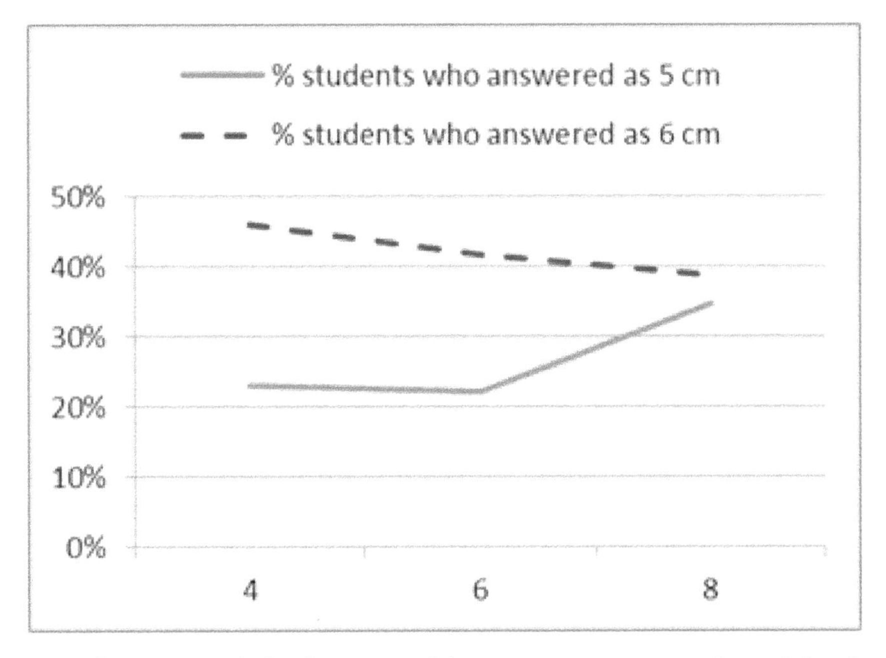

Figure 75.20. Graph showing extent of the two error types across classes 4, 6 and 8 for Item 18.

This was a free response item with the correct answer as 5 cm. The most common wrong answer was "6 cm". Even though the percentage of students answering "6 cm" reduced slightly over classes, it remained greater than the percentage of students answering correctly, as seen from the graph in Figure 75.20.

Error in decimal ordering in Class 8. In continuation to the earlier discussion on students' understanding of decimal number ordering, now we present another item which was administered in Class 8 in the same category of schools (English medium privately owned). The student response data on Item 19 shows the persistence of error in understanding decimal numbers in Class 8 as well.

Item 19. Manshakti watches the Olympic events. He is cheering for the Indian runners in the men's 100 m final. After seeing the nail-biting finish, he finds the following timings on the large display board of the stadium.

Runner	*Finishing Time (in seconds)*
Carl Joe	9.69
Baba Adams	9.069
Jay Armstrong	9.78
Balwan Pahelvan	9.078
Abbas Ali	11.407
Chin-Chin Chow	10.95
Huan Ki	11.013

The one who finishes the race in the shortest time among all the participants wins the race. So who won the race?

(A) Carl Joe (B) Baba Adams (C) Jay Armstrong (D) Balwan Pahelvan

In Item 19 while comparing decimal fractions like 9.069 and 9.69, where the numbers after the decimal point are "069" and "69", students seemed to be applying whole number ordering. Some students may have judged that 9.069 is larger than 9.69 because 9069 is larger than 969 (string length thinking, described earlier). These students would choose option A. Students applying numerator focused thinking may choose either option A or option B.

About 32% of the 9179 students of Class 8 felt that 9.069 is greater than 9.69 (Table 75.20). These errors are seen to steadily decrease with increase in the mathematical ability of students (Figure 75.21).

Table 75.20. Responses to Item 19 of Students of Class 8 (*N* = 9179) of English Medium Private Schools Across India

	Option A	*Option B*	*Option C*	*Option D*
% students	**32%**	60.6%	3.1%	3.9%

Note: This data is of Class 8 students.

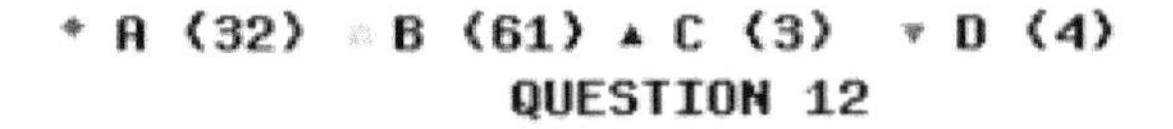

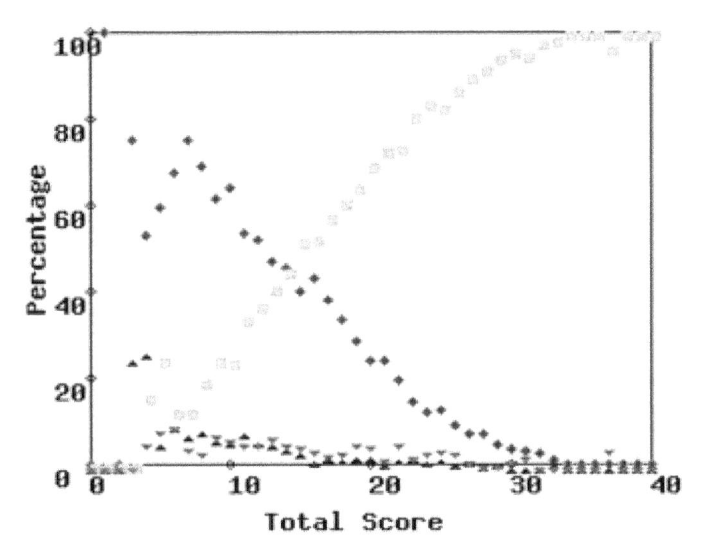

Figure 75.21. IRC for Item 19.

CONCLUSIONS AND IMPLICATIONS FOR RESEARCH

Errors and misconceptions are the natural result of children's efforts to construct their own knowledge (Olivier, 1989). Though we cannot avoid errors, a teacher can create a classroom where errors are looked at as opportunities to learn. But this is only possible if a teacher is aware of the misconceptions that are prevalent in the topic and at the level of the class. Hence a detailed compilation of errors made by children can prove to be a useful resource for teachers. It has been our experience that many teachers are surprised to see the extent of error in their students. On watching one of the videos of a student interview, recording an error in understanding of triangles, teachers were surprised that their students did not know what triangles were. Such a compilation can also be a useful tool in teacher development in enhancing teachers' pedagogical content knowledge.

Resnick (1982) found a relation between curricular sequences and the errors that students make. Hence, a comparison of errors of students following different curricula or different school boards can also help in designing curricula. One can also investigate possible reasons for the student errors and incorrect notions. Changes in pedagogical approach or changes in curriculum may reduce the extent of such errors among learners. One can study how one incorrect notion may lead to to errors or incorrect notions in another concept. A comparative study of the errors prevalent among Indian students and students from other Asian countries may suggest valuable insights.

AUTHORS' NOTE

A. Kanhere, A. Gupta and M. Shah have contributed equally to this work.

REFERENCES

Goa. (2013, June 11). In *Wikipedia, The Free Encyclopedia.* Retrieved from http://en.wikipedia.org/w/index.php?title=Goa&oldid=559382112

Loveless, T. (2008). What can TIMMS Surveys Tell us about Mathematics Reforms in the United States during the 1990s. In *Lessons Learned: What International Assessments Tell Us About Math Achievement* (pp. 127-174). Washington, DC: Brookings Institution Press.

Menon, U. (2009). The Introduction of Angles. In K. Subramaniam & A. Mujumdar (Eds.), *Proceedings of epiSTEME-3: International Conference to Review Research on Science, Technology and Mathematics Education* (pp. 133-138). Mumbai: Macmillan.

Olivier, A. (1989). Handling pupils' misconceptions. *Pythagoras, 21*, 10-19.

Pellegrino, J. W., Chudowsky, N., & Glaser, R. (2001). *Knowing what Students Know The Science and Design of Educational Assessment.* Washington: National Academy Press.

Pradhan, H. C., & Mavalankar, A. T. (1994). *Compendium of Errors in Middle School Mathematics.* Mumbai: Homi Bhabha Centre for Science Education.

Resnick, L. (1982). Syntax and semantics in learning to subtract. In T. Carpenter, J. Moser and T. Romberg (Eds.), *Addition and subtraction: A cognitive perspective* (pp. 136-155). Hillsdale, NJ: Lawrence Erlbaum Assoc.

Sadler, P. M., (1998). Psychometric Models of Student Conceptions in Science: Reconciling Qualitative Studies and Distracter-Driven Assessment Instruments, *Journal of Research in Science Teaching, 35*(3), p265–296

Shulman, L. S., (1987). Knowledge and teaching: Foundations of the new reform. *Harvard educational review, 57*(1), 1-23.

Steinle, V., & Stacey, K. (1998). The incidence of misconceptions of decimal notation amongst students in grades 5 to 10. In *Teaching Mathematics in New Times: Proceedings of the 21st Annual Conference of MERGA* (Vol. 2, pp. 548-555).

CHAPTER 76

ASSESSMENT OF MATHEMATICAL LEARNING

Issues and Challenges

Shailesh Shirali
Sahyadri School (KFI) and Community Mathematics Center, Rishi Valley School (KFI)

ABSTRACT

Though it is axiomatic that assessment should be integrally linked with learning and should feed into the learning process, the reality in Indian schools is very different. Across the board, assessment tends to be high-stakes and summative, driven by considerations of convenience and practicality rather than educational pedagogy or child psychology. In this paper we look at the assessment scenario in India with reference to public and competitive examinations, and at the Continuous and Comprehensive Evaluation system recently introduced by the Central Board for Secondary Education. Though there is scope for improving the variety of questions asked in these examinations, we suggest that this matter is secondary in comparison with the matter of quality of nurture that ought to be the driving force behind assessment systems. But for such reform to happen, adequate teacher preparation is essential.

The First Sourcebook on Asian Research in Mathematics Education: China, Korea, Singapore, Japan, Malaysia, and India, pp. 1685–1702

***Keywords*:** Assessment, summative assessment, Indian school education

Assessment in many ways constitutes the "bottom line" in a school education system; it is here that one can look back at the system and see whether it has achieved something of significance. It is also a "finish line" and this analogy is closer than one may imagine at first. A system of assessment provides an assessment not just of the learning of pupils, but also of the system itself; it assesses how well the system has done what it set out to do. If we look at assessment in this light, then what we find is not very encouraging.

It is axiomatic in education that assessment should be integrally linked with learning, that it not be merely a device to measure cumulative learning at the end of a teaching unit or end of an academic year. As a farmer pithily remarked at a public hearing in USA, "You can't fatten a hog by weighing it" (National Research Council, 1993); in our context, we may say that summative assessment does not add significant educational value, and that its benefit (if any) lies in its lateral action.

The sad reality about the Indian school education system is that "assessment" is just that; it is limited almost entirely to the summative variety, and is for the most part a device to measure cumulative learning: a device used to help teachers write reports and to help make pass/fail decisions. Thus there is little or no "feedback" into the learning process.

In few countries is it as perniciously true as in India that summative assessment holds the key to one's future, in the sense of opening or closing doors of opportunity; in the latter case, closing them very tightly indeed. The problem is of sufficient gravity that every single year there are suicides associated with it: children unable to cope with the disappointment and shame of failure, or with the fear of condemnation. Inevitably the specter of such assessment exercises a significant influence on the ambient educational culture, inviting poor educational practices and the creation of a powerful parallel education system called "coaching". Indeed, it invites criminal activity as well, through the leakage and sale of examination papers. The last implication is a significant area of concern in Indian education today.

Schoenfeld (2007) offers a framework for studying what different stakeholders want from mathematics assessment. We have adapted this list to our purpose here:

- Parents, who want indicators that will help them understand their children's knowledge and progress; they want: (a) performance statistics that place the student on a continuum (a grade or percentile score); and (b) information about what their child is doing well and doing poorly, so that they can arrange for help in areas where it is needed (most likely through evening tuition). Accurate assessment clearly has high priority for this group.

- Teachers, who want indicators on what the student knows and how well the student knows it, to thus know the areas where the student needs improvement. A priority requirement for teachers is that assessment tasks should have curricular value, else they take away valuable time from the job of teaching.
- Students, who want indicators that will reveal what they know and what they do not know. Tests that only return a number, especially tests that return results weeks or months later, are of little use to students. Tests also need to be fair.
- Schools, which want indicators that will help them in making decisions in areas such as: passing students to a higher grade; streaming according to ability.
- Examination boards and government education departments, which want indicators on how well the system is doing and how the country is performing in comparison with other countries.

Schoenfeld (2007) also lists some unintended consequences of assessment. Once again we have adapted the list to our purpose:

- test score inflation and the "illusion of competence",
- curriculum deformation (WYTIWYG) – "What You Test Is What You Get,"
- stifling innovation (requiring answers to be in the exact format prescribed by the teacher),
- language related issues (related to the fact that many students do not study the subject in their mother tongue), and
- negative effects on self-esteem, with possible serious physical harm involved.

It will be of value to keep these lists in mind during our study, and we shall return to it later.

ASSESSMENT AND EDUCATION IN INDIA

Significant efforts have been initiated in recent years by the National Council for Educational Research and Training (NCERT) through the revision of the National Curriculum Framework (NCF 2005, see Khan, this volume). Textbooks have appeared in nearly all subjects through its efforts, and they are of a significantly higher quality than its earlier books. The NCF deals with teaching methodology, styles of assessment and a great many educational issues. While it has had an effect on textbook writing, the effect on assessment culture and teaching methodology has been minimal. This is related to the fact that school education is a state subject in India, and a central body like the NCERT can only make recommendations, which are not binding on individual states.

School education in India follows a ten plus two system: ten years of compulsory schooling in which all students follow the same stream, followed by two years in which one chooses a set of optional subjects. These are grouped into streams: Mathematics, Physics, Chemistry (commonly known as “MPC”), Biology, Physics, Chemistry (“BiPC”) and so on.

The Central Board of Secondary Education (CBSE) and the Council for the Indian School Certificate Examinations (CISCE) are national examination boards, and many of the better known schools in the country are affiliated with one of them (see Subramaniam, this volume, for a discussion of school boards in India). CBSE follows the syllabus set by NCERT and uses NCERT textbooks, whereas CISCE sets its own syllabus, at both the 10th standard and 12th standard levels and does not prescribe textbooks; schools are free to use any textbook of their choice. The academic standards of the two boards are roughly comparable. The major difference is that CBSE has done away with the 10th standard examination, and has substituted it with the system of continuous and comprehensive evaluation, which will be discussed later. However, the 12th standard examination continues in its original form.

Entry into colleges is decided either on the basis of the marks secured in the 12th standard or entrance examinations conducted by the respective colleges. Population pressures mean that admissions are a highly competitive process, particularly for prestigious colleges like the Indian Institutes of Technology or the All India Institute of Medical Sciences. This single fact has had a great influence on secondary school education—unfortunately, not a positive one; indeed, one that trickles down to the primary level. The entrance examinations of a few institutes have now become bench marks in the country. We shall look at the style of a few of these examinations later in this essay.

A situation peculiar to this country is the phenomenon of tutorial colleges (“coaching centers”) which seek to prepare students for entrance to highly sought-after institutions. Some of these colleges are themselves quite highly sought after, and they have their own selection examinations, a situation which invites the possibility of an infinite iterative loop! One could laugh in good humor at the situation if it was not so wasteful of human energy, and some of its implications were not so grim. The methods used by these colleges amount to all-out drill, mastery of pattern recognition through analysis of past papers ("reverse engineering" set in an educational context), and reliance on huge memory banks. Over the last several decades these practices have been absorbed into the ambient educational culture of the country.

Still more far-reaching in its impact is the phenomenon of evening tuition. This is a practice which has now become ubiquitous; even children in the junior-most classes are not spared. For the most part the approach followed is all-out drill, done under close supervision of the tutor. For individual children the hours spent can be long, with tuition in mathematics and typically one or two other subjects.

ASSESSMENT STUDIES IN INDIA

In this section we briefly present views on assessment in the Indian context from a few key published documents. This will set the context for a closer look at the content of assessment in mathematics in some high stakes examinations.

One of the areas of concern listed by the NCF 2005 is the "crude methods of assessment that encourage a perception of mathematics as mechanical computation" (NCERT, 2006, p. 4). Amplifying on this point, it says, "While what happens in class may alienate, it never evokes panic, as does the examination. Most of the problems cited relate to the tyranny of procedure and memorization of formulas, and the central reason for the ascendancy of procedure is the nature of assessment. Tests are designed to assess knowledge of procedure and memory of formulas. Concept learning is replaced by procedural memory. Such antiquated and crude methods of assessment have to be thoroughly overhauled" (NCERT, 2006, p. 6). It recommends the following: (a) shift the focus of mathematics education from achieving narrow goals to higher goals; (b) engage every student with a sense of success, and at the same time offer challenges to the emerging mathematician; (c) change modes of assessment to examine mathematisation abilities rather than procedural knowledge; (d) enrich teachers with a variety of mathematical resources.

The third and fourth points are of relevance here. With regard to the third point it adds: "Since the Board examination for Class X is for a certificate given by the State, implications of certified failure must be considered seriously. Given the reality of the educational scenario, the fact that Class X is a terminal point for many is relevant; applying the same standard of assessment for these students as well as for rendering eligibility for the higher secondary stage seems indefensible. [Given] the high failure rate in mathematics, we suggest that the Board examinations be restructured. They must ensure that all numerate citizens become eligible for a State certificate. Nearly half the content of the examination may be geared towards this. However, the rest of the examination needs to challenge students far more than it does now, emphasizing competence and expertise rather than memory. Evaluating conceptual understanding rather than fast computational ability in the Board examinations will send a signal of intent to the entire system, and over a period of time, cause a shift in pedagogy as well. These remarks pertain to all forms of summative examinations. Multiple modes of assessment need to be encouraged. This calls for research and a wide variety of assessment models to be created and widely disseminated" (NCERT, 2006, p. 13) These are stirring and important words, and we hope they will be visited repeatedly by examination boards and state education departments in the years to come, because the lack of alternate assessment models and resources accessible to teachers has been and continues to be one of the central obstacles in this matter.

International assessments such as the PISA are "designed to assess the extent to which students are prepared to meet challenges of the modern world [and to] measure proficiencies that reflect the skills needed by mathematically literate individuals in a modern society" (Turner, 2010, p. 85). Some of the skills tested for by the PISA are:

1. the ability to recognize the existence of mathematical elements within a situation;
2. the ability to pare [a] situation down to its elements;
3. the ability to call on relevant mathematical knowledge and to apply [it];
4. the ability to relate that solution back to the original situation, [and] to recognize the extent and limitations of the solution;
5. the ability to communicate the outcomes to others; and
6. the ability to step outside the process and [to] exercise control mechanisms that help direct thought and action to achieve the desired outcome.

Turner's remarks made in an article published in an Indian periodical on education are relevant here:

> Being mathematically literate in the modern world means more than memorizing rules and formulae and mastering a set of procedures. It also means being able to think creatively about challenges. How much teaching and learning time [are] devoted to presenting students with problems set in authentic contexts? To what extent does India's education system value the investment of effort by students? (Turner, 2010, p. 85)

We also quote below from a report published by Education Initiatives, an organization that carries out large testing programs in the country (see Kanhere, Gupta & Shah, this volume). The report is of a set of diagnostic tests that Educational Initiatives conducted in 2006 for about 30,000 students in schools in many metropolitan centers of the country.

> [Our findings] do not present a happy picture of the state even in "top" schools. Students are able to answer questions based on recall or standard procedures quite well. However, their performance on questions testing understanding or application is far below what we consider to be acceptable levels. [They] are unable to tackle questions that appear a little different from what they find in textbooks or in the class. Their ability to apply what they have learnt to new, unfamiliar problems is not very high. [Students] slot learning into artificial compartments. They may learn something, but are able to answer it only in the context in which the learning first occurred. They may be using an aspect of what they have learnt in their [daily] lives, but be unaware of that connection. Another finding is that students [are] weak in certain real-life competencies like measurements and problem solving, which should be developed through the formal curriculum. [These] findings were corroborated through the study in which learning levels across classes were compared. While learning clearly improved from class 4 to 6 to 8, a number of students seem to be learning class 3 and 4 concepts only around class 6 or later. (Educational Initiatives, 2006, p. iv)

HOW EXAMINATION BOARDS HANDLE ASSESSMENT: CASE STUDIES

In this section we study the way the all India CISCE board deals with some selected topics and with problem solving in general. The following questions are sampled from public examinations for Class 10.

"Value Added Tax" computation. A manufacturer marks an article for Rs 5000. He sells it to a wholesaler at a discount of 25% on the marked price and the wholesaler sells it to a retailer at a discount of 15% on the marked price. The retailer sells it to a consumer at the marked price; at each stage the VAT is 8%. Calculate the VAT received by the Government from: (a) the wholesaler, (b) the retailer.

Trigonometry. Without using trigonometric tables, find the value of $\dfrac{\sin 35^\circ \cos 55^\circ + \cos 35^\circ \sin 55^\circ}{\csc^2 10^\circ - \tan^2 80^\circ}$.

Algebra. If x, y, z are in continued proportion, prove that $\dfrac{(x+y)^2}{(y+z)^2} = \dfrac{x}{z}$.

Algebra. Use the remainder theorem to factorize the following expression: $2x^3 + x^2 - 13x + 6$.

Trigonometry. From the top of a light house 100 m high, the angles of depression of two ships on opposite sides of it are 48° and 36° respectively. Find the distance between the two ships to the nearest meter. (The accompanying diagram has not been displayed here.)

Descriptive statistics: mean, mode, median. The distribution given below shows the marks obtained by 25 students in an aptitude test. Find the mean, median and mode of the distribution.

Marks	5	6	7	8	9	10
Students	3	9	6	4	2	1

Probability. Cards marked with numbers 1, 2, 3, 4, ..., 20 are well shuffled and a card is drawn at random. Find the probability that the number on the card is: (a) a prime number; (b) divisible by 3; (c) a perfect square.

Comments. One would be hard put to fault any of these questions, except for the problem in data analysis: one would want questions that consider the significance of what is computed rather than merely ask for some statistics to be computed. However, data analysis suffers from the syndrome of number-crunching-with-little-or-no-interpretation all through the school system in India. (The teaching of data analysis at the school level remains at a rather primitive level in the country, as does assessment of this topic. If we place these observations against the fact that so many great statisticians have come from this country, the situation seems a little ironic.)

Note that most of the questions are of an "all or nothing" kind, where potentially the grader can award a mark by looking at just the last line of the pupil's response. This is clearly not desirable. We need to move towards questions that have a more discriminatory dialogue between examiner and pupil, in which smaller, more searching questions are asked. In short, a continuum model is required, as different from a discrete, binary model. We also need at least a small component of questions in which the focus is not a single numerical answer but on descriptive and interpretative aspects of the situation at hand.

Related to the point made in the preceding sentence is the fact that assessment in mathematics in India tends to be totally in the problem solving mode, with little or no scope for open ended exploration (which would need to be assessed through project work; it would be meaningless and unfair to attempt to assess open-ended work through time bound tests). Most such problems tend to be of the "standalone" kind, with no accompanying context or background, so they do not help nurture in the student an appreciation of a larger picture. Solving such problems does not contribute to lateral links being developed. Thus the learning associated with such activity is highly limited. Indeed, students may form a skewed and rather inaccurate picture of the subject, particularly if they do well in these kinds of examinations. As a result, such forms of assessment contribute to what Schoenfeld (2007) has described as the "illusion of competence". At a deeper level, a real issue with this form of assessment is that it carries a high stake and has—potentially—an important bearing on one's future.

The following questions are sampled from the public examinations of the CISCE board for Class 12.

Matrix algebra. If the matrix $\begin{bmatrix} 6 & x & 2 \\ 2 & -1 & 2 \\ -10 & 5 & 2 \end{bmatrix}$ is singular, find the value of x.

Coordinate geometry. Show that the line $y = x + \sqrt{7}$ touches the hyperbola $9x^2 - 16y^2 = 144$.

Calculus, differentiation. Using a variable substitution, find the derivative of $\tan^{-1}\sqrt{\frac{a-x}{a+x}}$ with respect to x.

Calculus, integration. Evaluate the following integral: $\int_{-3}^{3} |x+2|\,dx$.

Calculus, integration. Evaluate the following integral: $\int\left(\frac{x\,e^x}{(x+1)^2}\right)dx$.

Calculus, differential equations. Solve the following differential equation: $\csc^3 x\,dy - \csc y\,dx = 0$.

Boolean algebra. x, y, z represent three switches in an "ON" position, and x', y', z' represent the same three switches in an "OFF" position. Construct a switching circuit representing the polynomial $(x + y)(x' + z) + y(y' + z')$. Using the laws of Boolean algebra, show that the above polynomial is equivalent to $xz + y$, and construct an equivalent switching circuit.

Descriptive statistics: correlation. Calculate Spearman's coefficient of rank correlation from the following data and interpret the result:

x	16	19	22	28	25	31	27	40	43	49
y	25	25	27	31	27	33	35	41	45	41

Descriptive statistics: regression. Find the equations of the two lines of regression for the following data: (3,6), (4,5), (5,4), (6,3), (7,2). Hence find an estimate of y for $x = 25$.

Descriptive statistics (data analysis): moving averages. Consider the following data.

Dates	12	13	14	15	16	17	18	19	20	21	22	23	24	25
Sales	2	5	0	12	13	25	45	13	31	18	11	2	3	1

Calculate three-day moving averages and display these and the original figures on the same graph.

Mean value theorem: verification, interpretation, significance. Using Rolle's theorem, find a point on the curve $y = \sin x + \cos x - 1$, with x in $[0, \pi/2]$, at which the tangent is parallel to the x-axis.

L'Hôpital's rule for limits of indeterminate forms. Evaluate:

$$\lim_{x \to \pi/2}\left[x \tan x - \frac{\pi}{2} \sec x\right]$$

Comments. All the comments made earlier apply here too, particularly those pertaining to Data Analysis, and to the "all or nothing" nature of the questions. I would strongly query the wisdom of asking for the indefinite integral of $x\, e^x/(x + 1)^2$ in a public examination; this is rather of a "trick integral" and it is not clear whether such questions are appropriate in a high stakes time-bound examination. (We must note also that Rolle's Theorem is not needed to "find a point on the curve $y = \sin x + \cos x - 1$, x in $[0, \pi/2]$, at which the tangent is parallel to the x-axis".)

Competitive Examinations: The AIEEE and JEE-IIT

Here we study some questions asked in post school entrance examinations. We limit ourselves to two such exams: the All India Engineering Entrance Examination (AIEEE) and the Joint Entrance Examination (JEE) for the Indian Institutes of Technology. A glance of the questions reveals the high level of preparation needed to do well in the exams. The situation is complicated by the fierce competition and very large number of candidates, which imply that a difference of just one mark may account for many hundreds of candidates. In response, students use strategies based on pattern recognition and memorization of large numbers of solved problems. One can well imagine the effects of this kind of high intensity input when it is continued for a year or longer: the effects on conceptual understanding, and on the mindsets of individuals.

Here are the test paper formats:

- AIEEE: 30 Multiple Choice Questions each in physics, chemistry, and mathematics (duration 3 hours).
- JEE: Paper I: 28 questions each in physics, chemistry, and mathematics; 8 multiple choice questions with single correct choice; 5

multiple choice questions with one or more correct choices; 5 comprehension type multiple choice questions; 10 numerical answer (with a two digit answer); 84 questions in all (duration 3 hours).

- JEE Paper II: 19 questions each in physics, chemistry, and mathematics; 6 multiple choice questions, single correct choice; 5 numerical answer (with a single digit answer); 6 comprehension type multiple choice questions; 2 matrix column matching type; 57 questions in all (duration 3 hours).

We now take a look at some problems from AIEEE and JEE papers.

Calculus, differentiation. Three normals are drawn from the point $(c, 0)$ to the curve $y^2 = x$. Show that $c \geq ½$.

Calculus, integration. Let I and J be defined thus:

$$I = \int_0^1 \frac{\sin x}{\sqrt{x}} dx, J = \int_0^1 \frac{\cos x}{\sqrt{x}} dx.$$

Then: (a) $I < 2/3, J < 2$ (b) $I < 2/3, J > 2$ (c) $I > 2/3, J < 2$ (d) $I > 2/3, J > 2$.

Calculus, limits. The value of $\lim_{x \to 0} \frac{1}{x^3} \int_0^x \frac{t \ln(1+t)}{t^4+4} dt$ is equal to:

(a) 0 (b) 1/12 (c) 1/24 (d) 1/64.

Combinatorics. Statement 1: $\sum_{r=0}^{n} (r+1)\binom{n}{r} = (n+2)2^{n-1}$

a) Statement 2: $\sum_{r=0}^{n} (r+1)\binom{n}{r} x^r = (1+x)^n + n\,x(1+x)^{n-1}$

Then: (a) Statement 1 is true, statement 2 is false; (b) Statement 2 is true, statement 1 is false; (c) Statement 1 is true, statement 2 is true, and statement 2 is a correct explanation of statement 1; (d) Statement 1 is true, statement 2 is true, and statement 2 is not a correct explanation of statement 1.

Inequalities. Statement 1: For positive integers $n > 1$, $\sum_1^n \frac{1}{\sqrt{k}} > n+1$.

Statement 2: For positive integers $n > 1$, $\sqrt{n(n+1)} < n+1$.

Then: (choices a, b, c, d exactly as above).

Comments. The questions are mathematically challenging, sophisticated and elegant; one would be hard put to fault them, and may even find them quite appealing. But we frequently find "trick questions" asked in such examinations: those in which it is possible to examine the options given and eliminate all but one by an elementary argument; or those solvable by highly specialized tricks. The following question from the AIEEE exam of 2012 is an example of this kind.

If n is a positive integer then $(\sqrt{3} + 1)^{2n} - (\sqrt{3} - 1)^{2n}$ is:

(A) An irrational number (B) An odd positive integer

(C) An even positive integer (D) A rational number other than a positive integer

The substitution $n = 1$ yields the quantity $4\sqrt{3}$. This instantly eliminates alternatives (B), (C), (D). So the answer "must" be (A). The use of these tricks appears to be restricted solely to examinations. (We recall an aphorism stated by Pólya and Szegö: "An idea which can be used only once is a trick. If you can use it more than once it becomes a method.") Thus, such questions test mastery of such tricks more than anything else.

The relevant question yet again lies in the social setting that lies behind these entrance examinations—the coaching culture, which is numbing of human initiative, and the intense expectations placed on pupils by parents. The primary purpose behind these highly sophisticated examinations is—surely—to filter out and exclude, rather than to include and nurture.

NEW INITIATIVES AND LIMITATIONS

Recently, the Central Board for Secondary Education has taken steps to bring in alternate assessment systems and has introduced a "Continuous Comprehensive Evaluation" system (CCE) (CBSE, 2011b). It has prepared elaborate manuals for teachers on how CCE is to be transacted, and has held workshops on CCE methodology. The CCE scheme offers the possibility of allowing the teacher to introduce topics with significant mathematical content but which are inappropriate for the closed book examination mode; for example, mathematical modeling or guided investigation. It is clearly an idea with promise, and one can hope that it will take root. However, without the necessary teacher preparation to back it up, its future remains in doubt.

Some examination boards (notably the CISCE) have made progress regarding projects and have a component for investigatory project work, done under a teacher's supervision. The marks allocation is modest: 10% of the total. But strangely, this provision is limited to the Sciences (Physics, Chemistry and Biology) and does not extend to Mathematics.

One of the prestigious national scholarships is the Kishore Vaigyanik Protsahan Yojana (KVPY, "Young scientists support plan"). The scholarship is given to meritorious students for undergraduate and graduate studies in pure sciences and mathematics, and is awarded on the basis of an examination of a high level of difficulty. Till 2011, it was possible for students to get the scholarship on the basis of original individual project work in the sciences or mathematics. The organizers had (in writing) recognized that the examination mode discriminates against a certain type of student, and hence that a dual approach is needed at a national level for the promotion of the sciences and mathematics. But in 2012 the KVPY disbanded the option of project work without giving any explanation.

Gilderdale, C. (2007) points to the need for bringing higher order thinking skills ("HOTS") into the educational mainstream. The notion has a pithy catch phrase: "HOTS, not MOTS!" where "MOTS" stands for "more of the same". The notion is attractive and easy to describe, but not so easy to carry out in practice, because an unavoidable prerequisite is a right mindset on the part of the teacher, and the preparation required to create such a mindset does not exist in the country at present. Help sites have sprung up in response to this need (CBSE, 2011a).

THE QUESTION OF INTEGRITY OF ASSESSMENT

A major issue connected with assessment in the Indian context is that of integrity. There is no guarantee that an internally awarded grade carries any intrinsic value, for its basis may well have little or nothing to do with academic worth. In many parts of the country at present, there is the widespread phenomenon of cheating in public examinations, in some cases through the ingenious use of modern technology.

Another area of concern is the fact that in several states, the style of the paper encourages memorization of answers to "standard" questions asked in earlier years (which tend to be repeated; there is thus the factor of predictability: reverse engineering once again). Added to this is the phenomenon of grade inflation—of state boards competing against the two national boards (CBSE and CISCE) by inflating their scores so that their students secure an added advantage in college admissions. This is surely a classic example of a "lose-lose" situation because it is the students who in the end are the real losers.

WHAT IS BEING ASSESSED?

The reality in India, it would seem, is that tests that only return a number and nothing else are the default mode all across the country. The various stakeholders involved, listed earlier adapting the framework by Schoenfeld (2007) seem to be quite comfortable with the status quo. Parents typically decide on the basis of the percentage score whether or not to engage the services of an evening tutor; they may even base childrens' career related decisions on percentage scores. Schools typically make pass/fail decisions and streaming decisions on the basis of a percentage score alone. Even students, who surely are best placed to know that "Marks don't tell the full story", do not demand any credible alternative; though, of course, their voice might not get heard even if they did make such a demand.

In short, assessment of a more searching and sensitive nature, in which the examiner does not merely focus on the last line of a solution but examines the intermediate steps in an effort to ascertain what might be going on in the students' mind, is hardly ever to be seen.

With regard to the unintended consequences too, there are hard realities to be faced. In the public examinations today there is a significant degree of grade inflation, with individual state level examination boards seeing themselves in competition with national boards and other state boards. It has become routine nowadays to see news reports of "toppers" who score an average of 98% or more. At the college and university level, however, there is typically a considerable degree of skepticism about such percentages, which is precisely why college entrance examinations have assumed such prominence. The JEE and AIEEE discussed earlier are merely two such examinations in a very large list.

So what Schoenfeld refers to as Curriculum Deformation and WYTIWYG ("what you test is what you get") is very much a part of the Indian educational scene. Likewise for the malaise of requiring answers to be in the exact format prescribed by the teacher, and the deadly and tragic phenomenon of examination related suicides. Some of these statements may seem like extreme caricaturing, but unfortunately their reality cannot be denied.

GUIDING PRINCIPLES FOR ASSESSMENT

Some points stated below have been adapted from the recommendations of the National Research Council (1993).

1. *Assessment has tended to be driven by practical concerns rather than by educational priorities.* This clearly *must* change.
2. *Assessment must be tailored to fit legitimate educational goals, and not the other way round. It must also test for "Relational Understanding" and not confine itself to the much more easily measurable category of "Instrumental Understanding"* (Skemp, 1989). In other words, it must look closely not only at the "how" but also the "why".
3. *Assessment must offer authentic feedback to students so that they become aware of their individual areas of strength and weakness, and do not*

develop an "illusion of competence". In particular, they—and their teachers—need to know whether they are operating purely at the instrumental level, or have understood the "why".

4. *Assessment must involve a component of articulation, in which students reflect on their own understanding and their attitudes to the subject.* This is an important way in which assessment feeds back into and connects with the learning process.
5. *Assessment must tell teachers something about learner attitudes and predispositions to the subject.*
6. *Assessment must be based on what is transacted in the classroom on a daily basis and not merely on end-of-term or end-of-year events.* Ways need to be found of translating the everyday transactions between teacher and student and between students into opportunities for assessment and learning. We mean here the observations made by the teacher on the attitudes and style of working of the student; the misconceptions visible as the teacher walks past the student's work table, errors that reveal themselves when the student works out a problem on the board or in single conversation with the teacher; and so on. Observations made in such a context are likely to be authentic and reliable, but they demand some skill and care on the part of the teacher; for example, in ensuring that the atmosphere is non-threatening. There is considerable scope in all such forms of informal assessment.
7. *Research is needed to explore how assessment can be facilitated using technology.* This includes self-assessment, in which repetitive trials are possible, offering an environment in which self-corrective action can take place, neutralizing to some extent the high-stakes nature of assessment and the stigmatic nature of remediation. (This has particular value for the testing of procedural skills. The underlying principle is the NCF 2005 recommendation: "Engaging every student with a sense of success, [and] offering conceptual challenges to the emerging mathematician.")
8. *Teachers are a fundamental key to assessment reform.* As evaluation of student achievement moves away from short-answer recall of facts and algorithms, teachers will have to become skilled in using and interpreting new forms of assessment which are not of an all-or-nothing nature. Professional development of teachers at both the pre-service and in-service levels will become increasingly important.
9. *Assessment can be the engine that propels educational reform forward*—but only if the driving force is a deep concern for children's well-being, and not measurement.

For these to actually happen (particularly the first six points), deep changes are needed in the mindsets of teacher and teacher educators, and therefore in teacher education programmes as well. At present most such programmes fall woefully short of the mark.

IMPLICATIONS FOR FUTURE RESEARCH

The NCF 2005 document states:

> While there has been a great deal of research in mathematics education and some of it has led to changes in pedagogy and curriculum ..., the area that has seen little change ... over a hundred years is evaluation procedures in mathematics. It is not accidental that even a quarterly examination in Class VII is not very different in style from a Board examination in Class X, and the same pattern dominates even the end-of chapter exercises given in textbooks. It is always application of some piece of information given in the text to solve a specific problem that tests use of formalism. Such antiquated and crude methods of assessment have to be thoroughly overhauled if any basic change is to be brought about (NCERT, 2006, p. 6).

Whatever is the driving ideology or philosophy behind an education system, the manner in which it actually is transacted depends wholly on a few crucial media:

- the culture and mindset of the teacher community;
- the culture of the assessment system, and in particular the examination system;
- the style and approach followed by the textbooks; and
- the curriculum itself.

With these remarks as backdrop, here are some directions we envisage (and recommend) for assessment related research in the country, in the coming years:

1. *Empirical studies on diagnostic and normative assessment.* In the mainstream of the country, there is low awareness of the potential of diagnostic assessment. We need to take steps which will demonstrate its potential and thus bring it into the mainstream. There is similarly great resistance to normative testing; currently, TIMMS and PISA are almost unknown in the country. A study of their pedagogical utility would surely have value. (This is not to say that we need to embrace such tests with open arms; far from it, for it is

within the bounds of possibility that this too would lead to "teaching to the test" in the same way as for so many other tests.)

2. *Scope and effects of computerized testing, and the possibility it holds for diagnosing and identifying gaps in understanding.* Included here would be the matter of gender bias involved in multiple choice testing. Also, whether students with difficulties in the subject benefit from low stakes computerized testing. However we need to proceed with caution. It is important that such mechanisms are used for diagnostic purposes only. Else they result in an unseemly race for the top positions—a state of affairs that is all too familiar.
3. *Scope of investigative projects, and more generally of open-ended work*—what would be a fair way of assessing such work? How can concerns for integrity be addressed?
4. Effects on conceptual understanding, originality and creativity in mathematics as a result of high intensity examination coaching during the formative school years.
5. Incidence of dyscalculia in urban and rural India and how examinations affect those with such difficulties.
6. Effects on mathematical abilities brought about by the use of calculators in school, including graphics calculators.

CLOSING REMARKS

Educational assessment is an exceedingly complex matter, for the issues it touches are so many and of so varied a nature—from educational pedagogy to inherited societal problems such as caste barriers with which India is struggling. (These have held back the movement in society of whole communities within the country.) Yet, it is not an "open" problem in the sense that a mathematical problem may be described as "open". The primary difficulty—we feel—is a lack of clarity of educational vision. What is needed is not so much the subtlety of mathematical questions (not that this is not relevant, but it is far from being a primary issue) but the matter of quality of nurture and care that should inform and be the ultimate driving force behind assessment.

Many centuries back, the great Indian poet-philosopher Kabir Das wrote these words: *Guru kumbhar ek saman, ghad ghad mare ghot. Pat andhar se hat dhare aur bahar mare chot.* He is telling us that a true teacher is like the potter, who while fashioning a pot may beat it from the outside even while holding it from the inside, supporting and nurturing it. I think we must see assessment in this metaphor—as an instrument for nurture, for support, and for growth. Though the current reality would seem to be opposite in both direction and spirit, yet we express hope for the future, because more and more people are beginning to ask pertinent and vital questions.

REFERENCES

Central Board of Secondary Education (CBSE). (2011a). *CBSE HOTS questions.* Retrieved from http://www.icbse.com/2009/cbse-hots-question-paper-class-10-12/

Central Board of Secondary Education (CBSE). (2011b). *Continuous and Comprehensive Evaluation.* Retrieved from http://www.cbse.nic.in/cce/index.html

Education in India. (n.d.). Retrieved on 11 March 2011 from http://en.wikipedia.org/wiki/Education_in_India

Educational Initiatives. (2006). *How well are our students learning?* Retrieved from http://www.ei-ndia.com/full-report.pdf

Gilderdale, C. (2007). Lower and Higher Order Thinking. Retrieved from http://nrich.maths.org/5795

National Research Council. (1993). *Measuring What Counts: A Conceptual Guide for Mathematics Assessment*. Washington, DC: The National Academies Press.

National Council of Educational Research and Training (2005). *National curriculum framework.* New Delhi: NCERT.

National Council of Educational Research and Training. (2006). *National focus group on teaching of mathematics report.* New Delhi: NCERT.

Schoenfeld, A. (2007). Issues and Tensions in the Assessment of Mathematical Proficiency. In Schoenfeld, A. (Ed.) *Assessing Mathematical Proficiency* (pp. 3-15). Cambridge: Cambridge University Press.

Skemp, R. (1989). "Relational and Instrumental Understanding", *Mathematics in the Primary School,* 1–17. (Routledge Falmer London). Retrieved from http://www.grahamtall.co.uk/skemp/pdfs/instrumental-relational.pdf

Turner, R. (2010). Lessons from the International PISA Project. *Learning Curve* XIV (Azim Premji Foundation). 84–87.

CHAPTER 77

OPPORTUNITIES FOR USING TECHNOLOGY IN MATHEMATICS CLASSROOMS

Jonaki B Ghosh
Lady Shri Ram College, Delhi

ABSTRACT

This chapter highlights the role of technology in mathematics education and focuses on the pedagogical affordances of tools such as computer algebra systems, dynamic geometry software, spreadsheets and graphics calculators in learning mathematics in secondary and higher secondary school. Through examples of studies conducted with students, the chapter illustrates that the considered use of technology can lead to a rich and motivating environment for exploring mathematical ideas. The chapter also points at the challenges of integrating technology in school mathematics education in India and goes on to suggest that mathematics laboratories can be a medium to provide access to technology enabled explorations and mathematical modeling activities.

The First Sourcebook on Asian Research in Mathematics Education: China, Korea, Singapore, Japan, Malaysia, and India, pp. 1703–1726

Keywords: Computer algebra system, dynamic geometry software, spreadsheets, graphics calculators, mathematical modeling, mathematics laboratories.

Mathematics has for years been the common language for classification, representation and analysis. Learning mathematics forms an integral part of a child's education. Yet, it is also the subject, which has traditionally been perceived as difficult. The primary reason for this state of mathematics learning today is the significant gap between content and pedagogy. The last few decades have witnessed serious experimentation and research in mathematics education all over the world and there has been a shift of paradigm as far as mathematics teaching and learning is concerned. Mathematics education is being revolutionized with the advent of new and powerful technological tools in the form of dynamic geometry software, computer algebra systems, spreadsheets and graphic calculators which enable students to focus on exploring, conjecturing, reasoning and problem solving and not be weighed down by rote memorization of procedures, computational algorithms, paper-pencil-drills and symbol manipulation. Some research studies have shown that the appropriate use of technology can lead to better conceptual understanding (Heid, 1988), and make higher level processes accessible to students without compromising on paper-pencil skills (Stacey, 2001). While many countries are increasingly using technology for mathematics instruction, in India mathematics continues to be taught in the traditional "chalk and board" manner. This chapter highlights the many challenges facing school mathematics education in India and elucidates some of the reasons which are a hindrance to large scale integration of technology in mathematics education. The focus of the chapter is however on how technology can be integrated into the secondary and senior secondary mathematics curriculum. The author suggests the use of mathematics laboratories as a way of integrating technology and describes some projects conducted by her with students.

CHALLENGES OF SCHOOL MATHEMATICS EDUCATION IN INDIA

This section will highlight some of the challenges facing school mathematics education in India. One of the challenges is to do with the transaction of the curriculum. Mathematics is largely taught as an abstract subject without any substantial reference to applications of the various topics to real life problems, thus making the subject pedantic. Traditionally, the emphasis has been on the development of manipulative skills and the school assessment also tests the same. There is an apparent lack of encouragement for creative and original thinking. The teaching and learning of mathematics, especially at the senior secondary school, is driven by preparation for the school leaving examinations (at the end of year 12) which often determines a student's future. The position paper on teaching of mathematics of the National Curriculum Framework, begins by stating that the primary goal of mathematics education is the "mathematization of the child's thought processes" (National Council for Educational Research and Training [NCERT], 2006, p. 1). It recommends that mathematics teaching at all levels be made more "activity oriented" and student centered, so that students understand the basic structure of mathematics, learn to think mathematically and relate mathematics to life experiences. The document suggests that the emphasis should be on the processes of learning mathematics such as visualization, estimation, approximation, use of heuristics, reasoning, proof and problem solving. While these recommendations have been the primary driving force for revisiting and revamping the elementary school mathematics curriculum, their impact on the senior secondary curriculum has been somewhat limited.

The curriculum that exists today is thus disappointing for the talented minority as well as the average or below average majority. Over the years curriculum coverage has been enhanced by adding new topics or removing certain topics but there has been no major change in the approach to dealing with these topics. Text books are often unimaginative, examination centered and far removed from the everyday experiences of students.

The second challenge is inappropriate assessment. Assessment is largely of the summative variety and is usually driven by preparation for the school leaving examinations (at the end of year 12). The school mathematics curriculum suffers from the ailment of obtaining "the one right answer". Exactness is overemphasized and problems posed in examinations test the ability to obtain the "right" answer. Assessment focuses on computational and manipulative aspects of the various topics. Examinations conducted by the central and state boards, which are high stakes examinations, end with the awarding of marks or grades to the student by which she tends to get branded for life.

The third challenge is that of teacher preparation. Teachers are indeed the most critical resource in the system. In India however a vast majority of mathematics teachers lack the opportunity to equip themselves with modern means to teach the subject imaginatively due to lack of proper training and development. Teachers who have taught for several years find little opportunity for professional development and upgradation. There is a strong need for sustainable and structured professional development opportunities for teachers.

Each of these challenges can be significantly impacted by the appropriate use of technology. For example, technology can aid in the visualization of concepts, in exploration and discovery, in bringing the experimental approach into mathematics, in focusing on applications, in redefining the teacher's role, in helping sustain students' interest, in individualized grading and assessment, and in teacher outreach.

Implementation of technology in the curriculum, however, poses many challenges, the greatest being the socio-economic challenge. The priority of government is to reach education to a large population of students. Technology must therefore be cost effective and easy to deploy. The last few years have witnessed extensive use of computer technology in various schools across the country. In some states there have been attempts to integrate open source software in classroom teaching. However, by and large, in the vast majority of Indian classrooms, mathematics teaching continues in the traditional "chalk and board" manner. Technology, if used for teaching mathematics, is primarily for demonstration purposes and does not involve the student actively.

To successfully face the challenges in implementing technology in the Indian context, pre-service teacher education programs must be designed where student teachers are taught mathematics using various technological tools. This will help future mathematics teachers develop new perspectives on integrating technology in their teaching-learning process. In-service teacher training programs must focus on changing the practising teacher's mindset towards technology and in developing their pedagogical content knowledge. Such professional development would require collaboration with technology solution providers who can provide ongoing support for the use of the technological tools in the schools. Schools will have to ensure that students have adequate access to technology on a daily basis. Further, involvement of teachers on a large scale will require fundamental changes in teaching practices.

Mathematics laboratories can be the platform through which students are given access to technology to explore and visualize mathematical concepts and ideas. The potentialities of this medium needs to be explored to its fullest extent.

ROLE OF TECHNOLOGY IN MATHEMATICS LEARNING

This section will highlight the findings of some research studies based on the use of various technology tools in mathematics teaching and learning. These include computer software such as computer algebra systems (CAS), dynamic geometry software (DGS), spreadsheets and graphic calculators. One of the first leading studies in the use of technology in mathematics education is the one reported by Heid (1988) in which computer algebra and some graphing tools were integrated into a first year university calculus course. The technology provided scope for a "concept first" approach, in which, calculus concepts were extensively taught using the tools and computational aspects were dealt with at the end of the course. The results showed that the experimental group, which attended the technology intensive course, outperformed the control group which attended the traditional course in conceptual tasks, and did as well as the traditional group in computational tasks. The study suggests that technology helped to develop conceptual understanding without hampering procedural skills. Herwaarden and Gielden (2002) conducted a study in which an attempt was made to integrate paper and pencil techniques with computer algebra in first year university calculus and linear algebra courses. This integration helped to enhance conceptual understanding as students were required to solve their assignments in two different ways: first using paper and pencil and then using computer algebra. The paper-pencil assignments focused only on developing the basic concepts while the computer algebra assignments were carefully designed to create a connection between the paper-pencil methods and computer manipulations.

Arnold (2004) reports that research on the classroom use of computer algebra strongly points to better understanding of concepts and does not lead to any loss of algebraic manipulations on the part of students. According to him, CAS gives students more control over the mathematics they are learning and the ways in which they learn it. He describes CAS as "the ultimate mathematical investigative assistant" which allows the student to engage in "purposeful and strategic investigation of problems" (p. 21). Lagrange (1999) suggests that CAS helps to set a balance between skills and understanding. It lightens the technical work thus allowing students to focus on concepts and applications. According to him, students using CAS can build deep links between their enactive knowledge and computable representations. Lindsay (1995) highlights the ability of CAS to extend student's understanding through transfer between algebraic and graphical representation of problems and in enabling visualization as a facilitator in the learning process. In another study, Heid (2001) describes CAS as a cognitive technology which makes higher level mathematical processes accessible to students. On the one hand, CAS plays the role of an "amplifier" by making it possible to generate a larger number and greater range of examples and thus can be used to extend the curriculum. On the other hand it serves the role of a "reorganizer" by changing the fundamental nature and arrangement of the curriculum. This refers to a resequencing of concepts and procedures, reallocation of time traditionally spent on the refinement of paper and pencil methods to the interpretation of symbolic results and application of concepts. Thomas and Hong (2004) describe how some students in their study used CAS to check their procedural work while others used it for performing a procedure within a complex process to reduce cognitive load.

Kissane (2008) describes some "pedagogical affordances" of the graphics calculator in visualizing and exploring concepts in calculus, such as limits, derivatives, discontinuity, convergence of series and asymptotic behaviour of functions. According to him "procedural competence can be best developed when students understand the underlying ideas well" (p. 16). Kutzler (1999) describes how technology such as algebraic calculators (graphics calculators which are enabled by computer algebra features) can play a critical role in trivialization of computations, thus enabling us to include with more complex and realistic problems in mathematics teaching. These tools can enhance visualization which can be symbolic, numeric or graphic. He further suggests that technology can be a scaffolding tool since it helps "to break down the learning process into smaller, easier digestible pieces" (p. 13). It helps to "compensate any weakness with the lower level skills" (p. 12) so that the student can work on the higher level skills.

Laborde (2002) suggests that learning geometry with a Dynamic Geometry Environment (DGE) might offer students possibilities to construct and manipulate geometrical figures and do empirical investigations. These activities are hard to conduct in a static geometry environment. A DGE allows the student to drag parts of a dynamic figure thereby changing its properties. Leung (2012) describes that "a key feature of DGE is its ability to visually represent geometrical invariants amidst simultaneous variations induced by dragging activities" (p. 2). Within these simultaneous variations induced by dragging activities, emerges an invariant structure. Leung (2010) describes this as "the discernment of invariant patterns and/or shapes and the re-production or re-presentation of that pattern" (p. 2). The emergence of these invariant patterns leads students to generalize patterns and make conjectures. There are a plethora of similar research studies which have focused on the pedagogical opportunities offered by technology and their learning outcomes.

The next section of this chapter will focus on some technology enabled modules conducted by the author in a mathematics laboratory. All the modules discussed draw heavily from the studies mentioned in this section.

MATHEMATICS LABORATORIES AND INTEGRATION OF TECHNOLOGY

This section will focus on the concept of mathematics laboratories and how technology enabled explorations can be integrated into the curriculum through such laboratories. Typically, in a mathematics laboratory, the student should be given the opportunity to explore and visualize mathematical ideas and concepts by participating in activities which enhance their understanding of the subject as taught in the classroom and also provide a glimpse of what is beyond. The activities may comprise of projects, experiments and modelling exercises based on the mathematics taught in the curriculum. The projects or exercises performed in a mathematics laboratory may be designed to fulfil one or more of the following criteria. The exercise should

- highlight some known concept based on a well known mathematical theory,
- shed new light on some aspect of the topic being studied,
- lead to some original discovery on the part of the student, and
- focus on some interesting application of mathematics to a real life problem.

In the laboratory, the teacher's role is primarily to facilitate students' explorations and lead them to develop mathematical concepts. Technological tools such as computer algebra systems, dynamic geometry software, spreadsheets and graphics calculators can play a critical role in the activities of a mathematics laboratory. The remaining part of this section will focus on specific activities which were conducted by the author with students in a laboratory setting where various technological tools enabled students to explore, visualize and compute. The activities conducted by the author may be categorized under the following heads:

1. Visualization and exploration of concepts using various technological tools.
2. Exploring geometrical ideas using dynamic geometry software.
3. Simulation of problems in probability using spreadsheets.
4. Investigatory projects based on mathematical modeling and applications of topics taught in the curriculum.

Visualization and Exploration of Concepts Using Various Technological Tools

Exploration of concepts in the mathematics laboratory was primarily done through carefully designed modules. Each module was designed to enable the student to explore concepts in a given topic, through exercises and problems, which were given in the form of worksheets. The exercises were sequenced so as to enable the students to explore the concepts in a step by step manner. The author made extensive use of some leading brands of graphics calculators, CAS enabled calculators, Mathematica (a computer algebra system) and Geogebra (an open source dynamic geometry software) in enabling students' explorations.

Graphics calculators proved to be extremely handy for performing mathematical modelling activities in the laboratory. Students felt comfortable using graphics calculators especially since they are hand held and portable. As a tool the graphics calculator has significant mathematical capabilities. Matrix manipulation, operations on complex numbers, symbolic differentiation and integration, simple regression analysis and graphical as well as numerical exploration of functions are some of the features of the calculators. The advantages of the graphics calculator in enabling students' explorations were as follows:

(a) The power of visualization was brought right into the palm of the hand. The graphing capability of the calculator could be suitably exploited to lend a deeper insight into problems and concepts. The calculator also served as a handy demonstration tool since it could be connected with an overhead projector or the software version could be displayed on a white board. For example, in a laboratory module on calculus (Ghosh, 2005) students explored the concept of the derivative at a point by drawing tangents to the curve at various points. They also explored the notion of "local linearity" by repeatedly "zooming" around a point.
(b) The graphics calculator helped to illustrate mathematical ideas and applications in pedagogically powerful geometric settings. For example, students were able to visualize the relationship between the unit circle and the trigonometric functions very effectively using the parametric mode of the graphics calculator.
(c) Graphical and numerical explorations could be performed simultaneously. The calculator allowed the students to view functions graphically, symbolically as well as numerically. Figure 77.1 shows how students used a CAS enabled calculator to explore the limit of a function.
(d) Simple regression analysis could be done using graphics calculators. The graphics calculator allows the fitting of curves to data, which proved to be extremely useful for performing simple modelling activities in the laboratory. Apart from the linear fit one could obtain a quadratic, power, exponential, logarithmic or even logistic fit to a given data. The equation of the quadratic fit was used to predict the value of the dependent variable for given values of the

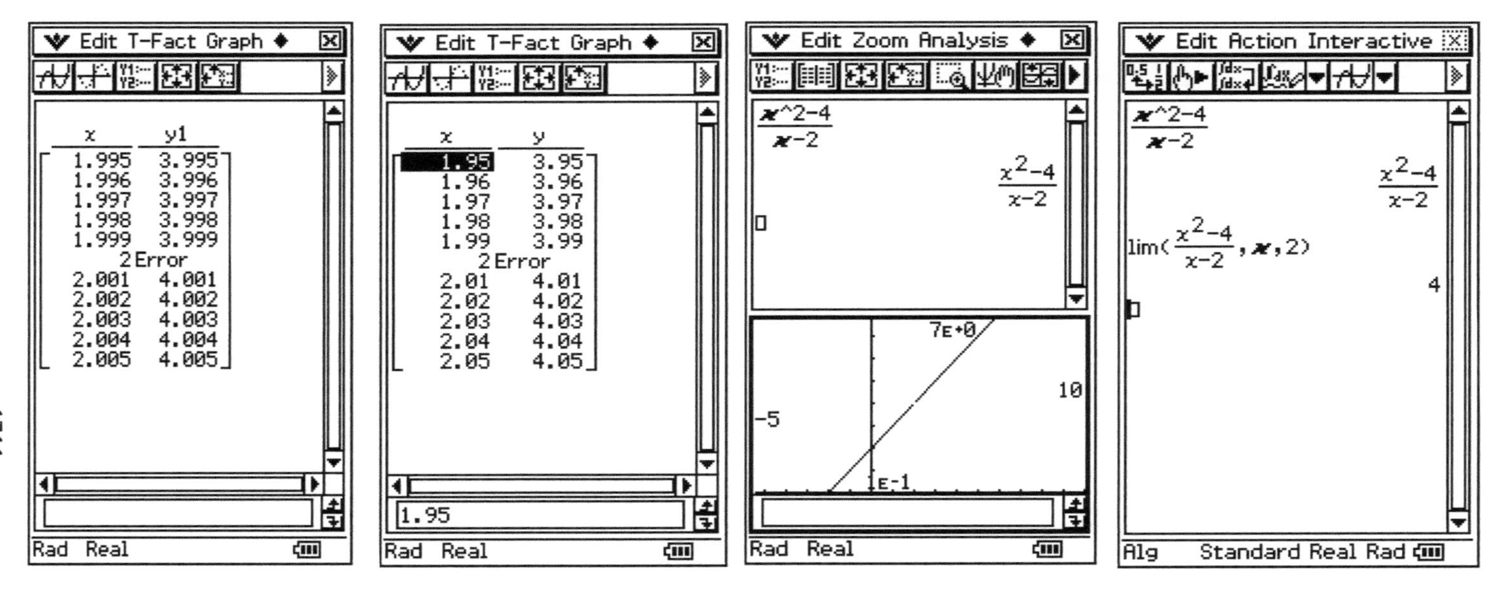

(i) (ii) (iii) (iv)

Figure 77.1. Screen shots of a CAS enabled calculator for visualizing the limit $\lim_{x \to 2} \frac{x^2 - 4}{x - 2} = 4$. The graphical output (iii) shows a gap while the tabular representations (i) and (ii) display “Error” at x = 2 indicating that the function is not defined at x = 2. Tracing the cursor along the graph and the table of values indicate that the function approaches 4 from either side as x approaches 2 and the built-in lim function (iv) confirms that the limit is 4.

Table 77.1. Solving Five Systems of Equations in Three Unknowns

(i)	*(ii)*	*(iii)*	*(iv)*	*(v)*
$x + y + z = 2$	$x + y + z = 1$	$x + y + z = 1$	$x - 2y + 3z = 2$	$x + y + z = 1$
$x + 3y - z = 1$	$x + 4y + 9z = 3$	$x + y + z = 7$	$2x - y + 2z = 3$	$x + y + z = 12$
$-x + 4y + 9z = 3$	$2x - y - 6z = 0$	$2x + 2y + 2z = 25$	$x + y - z = 4$	$8x + y - 6z = 0$

independent variable. Also the quadratic regression was compared with the linear regression for the same data.

In some of the laboratory modules, Mathematica was used as the vehicle for exploration. In these modules students were encouraged to work in pairs or in groups of three. In one particular module, students of Grade 12 visualized the solutions of systems of equations in three unknowns by plotting the planes representing the equations (Ghosh, 2003). In the worksheet five systems of equations in the unknowns x, y and z were given (Table 77.1).

The students were instructed to solve these by hand by reducing the systems to echelon form using elementary row operations. The solution of (i) led to the unique solution $x = 3/2, y = 0$, and $z = 1/2$. The students then verified their manual solution using Mathematica's **Solve** command. This was repeated for all five systems of equations. The system of equations (ii) had infinitely many solutions and systems (iii), (iv) and (v) were inconsistent. After finding the nature of solution for each system of equations, students were required to use Mathematica's **Plot3D** command to plot the planes. The plot for (i) showed three planes meeting at a point indicating that the three equations have a unique solution. The output for (ii) revealed three planes meeting in a line. Since any point on the line is a solution of the system, this was identified as a case of infinitely many solutions.

The Mathematica outputs for (iii), (iv) and (v) revealed three different situations of inconsistency (Figure 77.2). In (iii) the planes were parallel. Since the planes do not meet, students concluded that the system was inconsistent. The plot of (iv) revealed that the *three planes intersected pair wise in three non-planar parallel lines*. Since all three planes do not intersect in a single point or line, students concluded that the three equations are inconsistent. For (v) students concluded that *two parallel planes were intersected in two parallel lines by a third non-parallel plane*. Since all three planes do not intersect in a single point or line the three equations are inconsistent. Visualizing the different forms of inconsistency led to a graphical insight which would not have been possible without the 3-D graphing feature of Mathematica.

The primary objective of this module was to help students visualize solutions of equations in three unknowns and interpret them graphically.

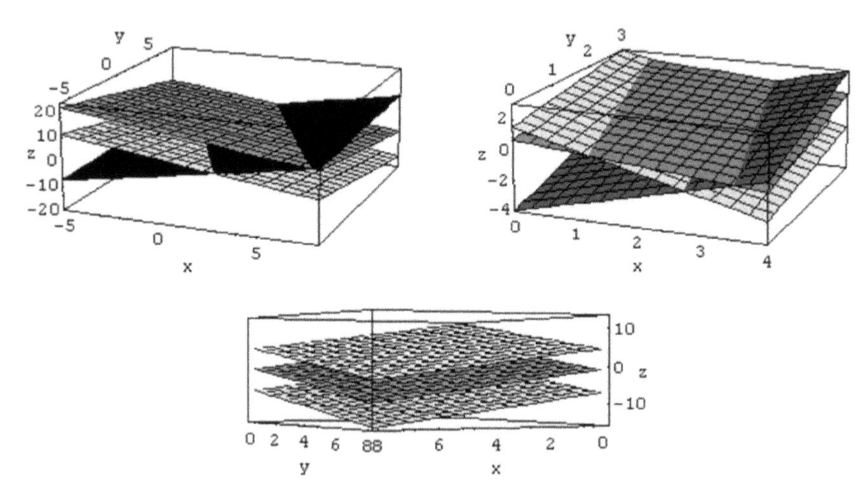

Figure 77.2. Mathematica plots of systems of equations in three unknowns show three different situations of inconsistency.

An attempt was made to ensure that there was no compromise on "by hand" skills as students were also required to work out the solutions manually using the matrix inverse method or by reducing the system of equations to triangular form. A comprehensive feedback questionnaire which students were required to complete at the end of the module included the following comments.

> I could solve the equations manually before going through the labs but Mathematica helped me to understand what they represent physically.

> I enjoyed arriving at the results on my own. The exercises in the worksheet were interesting because we could try the problems by hand and then ask Mathematica for the answer.

The students' responses revealed that Mathematica enabled them to explore concepts and arrive at results on their own or with some guidance from their teacher. This gave them a sense of discovery, which would be difficult to achieve in a traditional classroom. While Mathematica was used in this module, it could perhaps be readily adapted to free and open-source platforms like Maxima and Sage.

Exploring Geometrical Ideas Using Dynamic Geometry Software

Some exploratory activities using Geogebra, an open source dynamic geometry software, were also conducted by the author in the mathematics laboratory. Through these activities the author was specifically interested in observing whether a dynamic geometry environment helped to facilitate the process of making conjectures.

In a particular study, 22 Grade 9 students explored the centers of a triangle by constructing the circumcenter, centroid and orthocenter, first through paper folding on a paper triangle and then using Geogebra. The objective was to study if the explorations led students to make a conjecture regarding the collinearity of the centers leading to the construction of the Euler's line. This was to be followed by a discussion of the analytical proof of collinearity. Students were divided into pairs, each pair was given a paper cut out of a triangle and was required to identify the three centers through paper folding. This exercise was done so that they could later compare the paper folding method of construction with the GeoGebra construction. Initially students needed some facilitation to help them identify the folds which would lead to the perpendicular bisectors, altitudes and medians of the paper triangle. After identifying the three centers on their respective triangles (Each pair of students had a different triangle. Some were acute angled and scalene, some were obtuse angled, some right angled while others were isosceles or equilateral), students were asked to make observations regarding the positions of the three centers. The collinearity of the three centers was not evident to many simply by looking at their own triangles, so they resorted to looking at their neighbor's triangles. This activity led to interesting discussions but no consensus was reached regarding the centers. Questions such as "what kind of triangles have circumcenters or orthocenters lying inside or outside" were asked. Some students who had made accurate folds guessed that the three centers lie on a straight line.

After the paper folding task, students were given a worksheet which described the construction of the three centers stepwise in Geogebra. At the end of the session, students were asked to compare their Geogebra explorations with paper and pencil or paper folding constructions.

A majority of the students felt that GeoGebra constructions were easier, more accurate (than paper and pencil constructions) and were time saving. Out of 22 students, 18 concluded that the three perpendicular bisectors were concurrent, 16 concluded that the altitudes were concurrent and all students were convinced about the concurrency of the three medians. 21 out of 22 students were able to conjecture that the circumcenter lies inside, outside or on the hypotenuse of a triangle if the triangle was acute angled, obtuse angled or right angled respectively. Similarly 16 out of 22 students were able to make similar conjectures regarding the orthocenter. These conjectures were not made by students as easily during the paper folding exercises. However after dragging the vertices of the triangle on Geogebra, nearly all students arrived at the conjecture that the three centers are

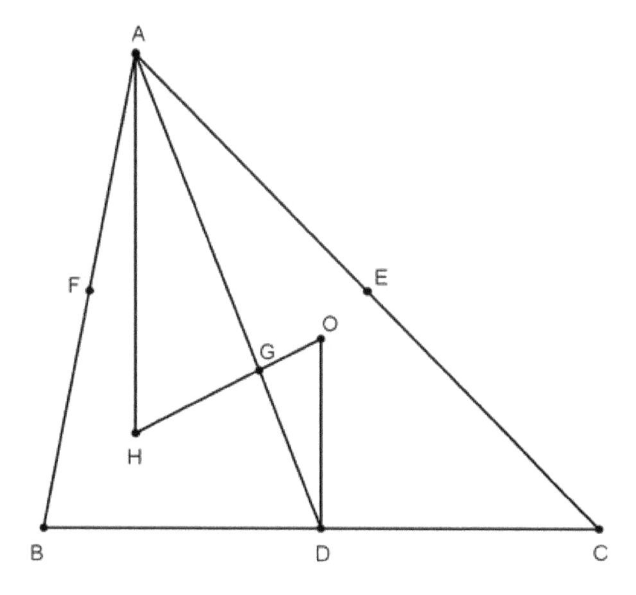

Figure 77.3. GeoGebra figure used by Indian 9th Graders to prove collinearity of O,G and H.

collinear. Also 16 students were able to describe the relationship among the three centers as HG = 2 OG (where H is the orthocenter, O the circumcenter and G the centroid). They were unable to make these conjectures using paper folding because of the inaccuracy of measurements and folds. The comments of the students revealed that students had a very positive impression of exploring the constructions in GeoGebra. At the end of the module students were required to write the proof of the collinearity of O, H and G. It was emphasized that the GeoGebra only helped to verify the collinearity of the three centers and that this was not a proof.

The first step in the exploration process was to construct the circumcenter O and the centroid G (Figure 77.3). It was then suggested that OG be extended to a point H such that HG = 2 OG. It was now required to be proved that H is indeed the orthocenter. After some discussions and argumentation students decided that they should focus on triangles AHG and OGD. They proceeded to prove the similarity of these two triangles. They further concluded that angles AHG and ODG are equal (being angles of similar triangles). This implied that AH is parallel to OD. Since OD is perpendicular to BC (being the perpendicular bisector of BC), it implied that AH, when extended is also perpendicular to BC. Students then concluded that AH is an altitude. Using similar arguments students proved that BH and CH when extended are also altitudes which was further used to conclude that H is indeed the orthocenter. Thus Geogebra provided a rich and motivating environment for exploring the ideas which led students to construct the proof of collinearity of the three triangle centers.

Simulation of Problems in Probability Using Spreadsheets

One of the fundamental concepts in statistics is the concept of probability and randomness and this forms an integral part of most mathematics curricula at the high school level. The topic of probability has many interesting problems which can be explored by students using technology tools. Simulation can be an effective modelling tool for conducting experiments such as the throw of a dice or coin, or exploring problems such as the Birthday Paradox. Most technological tools are equipped with random number generators which enable students to generate data, explore the data meaningfully and grasp important probability concepts.

One of the introductory modules on probability (Ghosh, 2004) conducted by the author focused on the concept of randomness and required students to perform experiments which would highlight the relationship between the empirical probability and the theoretical probability of an event. This particular module was implemented with Grade 9 students and the primary tool used for exploration was a graphics calculator. The **Ran#** function was used to generate a random number between 0 and 1. Students estimated the probability of an event, (such as the probability of obtaining a 6 in the throw of a die), by simulating the experiment a number of times (say 100) by entering the command **Int(6*Ran# + 1)** on the calculator. They also simulated the experiment of throwing a pair of dice and estimating the probability of obtaining the sums 2, 3, 4 up to 12. This was followed by the students calculating the theoretical probabilities of the events using the definition of probability.

In a subsequent module students explored the birthday problem using the spreadsheet feature of a CAS enabled graphics calculator. The random number generator of the tool, **rand( )**, was used to randomly generate birthdays in the spreadsheet application. The birthday problem, more popularly referred to as the birthday paradox, asks the following: *How many people do you need in a group so that there are at least two people who share the same birthday?*

The common response from most students is 366. However it can be shown that the answer is surprisingly smaller than that. In fact in a group size of about 50 we can be almost certain to find at least one birthday match and in group sizes of 24, the chances of finding a match is around half. The explanation for this can lead to an interesting classroom discussion where basic concepts of probability play an important role.

It can be an interesting, although tedious, exercise to actually verify the claim empirically by randomly collecting birthdays, dividing them into groups of 50 and checking if each group has a match. Another way of conducting the experiment is to ask each student to note the birthdays of persons known to her on slips of paper, fold them and put them in a box. After shaking the box, each student is asked to select a slip randomly and report the date, which is then marked off on a calendar. The box is circulated till a date is repeated and number of dates marked before finding the match is noted. After performing this experiment several times the average number of dates required to find a match is calculated. Suppose 10 sets of 24 slips each are created from the contents of the same box, then students can verify that very often 5 out of the 10 sets will contain a match while the other 5 will not have a match. This helps to convince them that the probability of a match among 24 randomly selected persons is around half. While the exercise is exciting it can be very time consuming.

The birthday problem was simulated on the spreadsheet application of a graphics calculator by generating 50 random birthdays. This was done by generating random integers from 1 to 12 (to indicate the month) in column A, and integers from 1 to 31 (indicating the day of the month) in column B. In column C the formula **=100*A1+B1** was entered. All the "dates" were converted to three or four digit numbers in which the first one or two digits indicated the month and the last two digits indicated the day of that month. For example, 225 indicated 25th of February and 1019 indicates 19th of October. The entries in column C were then sorted to check for a match and entered in cell D1. The experiment was run about 10 times to confirm that in each simulation of 50 birthdays (representing the birthdays of 50 randomly selected people) there is at least one match.

This module highlighted the fact that hand-held calculators enabled by spreadsheet capabilities can enable students to visualize, explore and discover important concepts without getting into the rigor of mathematical derivations. The problem can be simulated on any spreadsheet such as Excel, using the same steps. The Birthday problem can be conducted with students of Grades 9 and 10 without getting into the mathematical derivations. However, in Grades 11 and 12 the spreadsheet verification of the problem can be followed by an analysis of the underlying concepts which are rooted in probability theory.

Investigatory Projects Based on Mathematical Modelling and Applications

The position paper on the teaching of mathematics of the National Curriculum Framework 2005 suggests that mathematical modelling should be included in the curriculum especially at the secondary and senior secondary stage (NCERT, 2006). Mathematical modelling activities, in the form of problems and exploratory exercises, which enable the students to see the practical relevance of topics taught in the curriculum, should form an integral part of a mathematics laboratory. The remaining part of this section will focus on two laboratory modules in which students of Grades 11 and 12 used technological tools to explore ideas and visualize concepts. One module was based on exploring Fourier series and Gibbs phenomenon and the other was an exploration of autosomal inheritance in genetics.

The rationale for including Fourier series is that they are used to model various types of problems in physics, engineering and biology and arise in many practical applications such as modelling air flow in lungs, electric sources that generate wave forms that are periodic and frequency analysis of signals. In contrast to Taylor series which can be used only to approximate functions that have many derivatives, Fourier series can be used to represent functions that are continuous as well as discontinuous. The partial sums of the series, approximates the function at each point and this approximation improves as the number of terms are increased. However, if the function to be approximated is discontinuous, the graph of the Fourier series partial sums exhibits oscillations whose value overshoots the value of the function. These oscillations do not disappear even as the terms are increased. This phenomenon is referred to as Gibbs phenomenon (Libii, 2005). The topic of Fourier series is an integral part of mathematics courses at most undergraduate programs in engineering and science. However CAS enables students to visualize the series and perform computations quite easily and making it accessible to students of Grades 11 and 12. Since most students who participated in this module were aspiring for engineering courses for their higher studies, this module seemed relevant for them. Genetics, as a subject appeals to most students at the Grade 11 and 12 level. The module on autosomal inheritance was introduced as an application of matrices which is a topic in Grade 12.

Exploring Fourier series and Gibbs phenomenon. In this laboratory module 32 Grade 12 students explored Fourier series and Gibbs phenomenon over five one hour sessions. Students' explorations were guided by worksheets consisting of various tasks and Mathematica codes. The worksheets required each student to show their paper and pencil calculations as well as their observations from the Mathematica outputs. At the end of the module students were required to respond to a short questionnaire and give a written feedback describing their experience in the module. This module was the part of a course titled *Applicable Mathematics* which was designed by the author to enable students to visualize important concepts in Calculus, Linear Algebra and Probability and Statistics through the use of technology. The course was optional and was specially designed for students who planned to pursue an engineering discipline or a bachelor degree in mathematics for their undergraduate studies. The 32 Grade 12 students who were a part of the study, had chosen *Applicable Mathematics*, which required them to complete five exploratory laboratory modules based on real world applications. These students were also undertaking a Calculus course as a part of the regular curriculum which was taught in a traditional manner, without any technology. However while studying *Applicable Mathematics*, the same students were given access to technology and had undertaken a few calculus laboratory sessions in which technology played a vital role.

The laboratory module on exploration of Fourier series and Gibbs phenomenon was one of the five modules which the students were required to complete. In the first session, students were required to evaluate the trigonometric integrals such as

$$\int_0^{\pi} \sin^2 nt\, dt = \frac{\pi}{2} = \int_0^{\pi} \cos^2 nt\, dt \quad \text{for } n = 1,2,3..., \quad \int_0^{\pi} \cos nt \cos mt\, dt = 0$$

and $\int_0^{\pi} \sin nt \sin mt\, dt = 0$ for $n = 1,2,3...$, and for $n \neq \pm m$ by hand and then verify their answers using Mathematica. In some cases the Mathematica output was different from their paper and pencil solution and they had to manipulate the output using commands such as **Simplify** to obtain their solution. In the second session, after being introduced to the basics of Fourier series, the students had to manually evaluate the Fourier series of $f(t) = 1, t, t^2$ and t^3. This was followed by plotting the Fourier series partial sums using Mathematica. In subsequent sessions, students used Mathematica extensively for plotting partial sums, calculating function values at the peaks and finding values of the overshoots. Mathematica facilitated the computational process by helping to quickly generate the table of values of the partial sums without which it would not have been possible to observe Gibbs phenomenon. Thus paper and pencil methods helped students to understand the computations while Mathematica gave meaning to the computations. This supports Herwaarden's study (2001) which concluded that the integration of CAS helps to create a connection between the paper-pencil methods and computer manipulations. The extensive use of Mathematica in the last three sessions helped to lighten the technical work so that students could focus on making observations from the graphical and numerical outputs. Thus, as suggested by Lagrange (1999), Mathematica helped to balance "by hand" calculations and conceptual understanding. Student feedback taken at the end of the modules revealed that they began to perceive the paper and pencil tasks (evaluation of integrals, in this case) as more meaningful after using Mathematica. Mathematica helped to illustrate concepts and processes, which would be difficult to explain using only chalk and board. For example, in this module it would be impossible to visualize the Fourier Series Partial Sums of the functions without graphing them (see Figure 77.4). Students created their own Mathematica codes to produce the desired outputs and in creating these codes they used their mathematical understanding of Fourier series. In fact Mathematica served the purpose of a "mathematical investigation assistant" as proposed by Arnold (2004) and gave students control over what they were learning. Initially the Fourier series partial sum was only a symbolic expression but after graphing, it became a physical entity which could be modified and manipulated. The integration of Mathematica permitted students to understand the concepts through the three modes of representation, namely, symbolic, graphic and numeric and also understand the concept of Fourier series and Gibbs phenomenon which is far beyond the scope of the regular curriculum. This supports Heid's (2001) theory that CAS facilitates a multi-representational approach to learning mathematical concepts and also acts like an "amplifier" giving students access to higher level mathematical concepts. Some of the students' comments were as follows:

> Although we were familiar with trigonometric integrals (from our regular class), in the module we had to work out the Fourier coefficients and Fourier series expansions of some simple functions. All of this did not make much sense until we plotted the partial sums using Mathematica.

> Mathematica made the Fourier series come alive ... although writing the programs (codes) took some getting used to. Finally we could actually see Gibbs phenomenon.

Application of matrices to genetics: Autosomal inheritance. In this module, students of Grade 11 used matrix models to explore how genetic traits can be transmitted from one generation to the next in a plant population under autosomal inheritance. In autosomal inheritance, the inherited trait under consideration (say petal color) is assumed to be governed by a set of two genes, denoted by **A** (red color) and **a** (white color). The three possible genotypes are **AA**, **Aa** and **aa** where **AA** produces red flowers, **Aa** produces pink flowers and **aa** produces white flowers. Every individual inherits one gene from each parent plant with equal probability. Thus, if the parent pairing is **AA-Aa**, the offspring will inherit an "**A**" gene from the first parent and either an "**A**" or an "**a**" (with equal probability) from the second parent. Thus the offspring is likely to inherit a genotype of **AA** or **Aa** with equal probability.

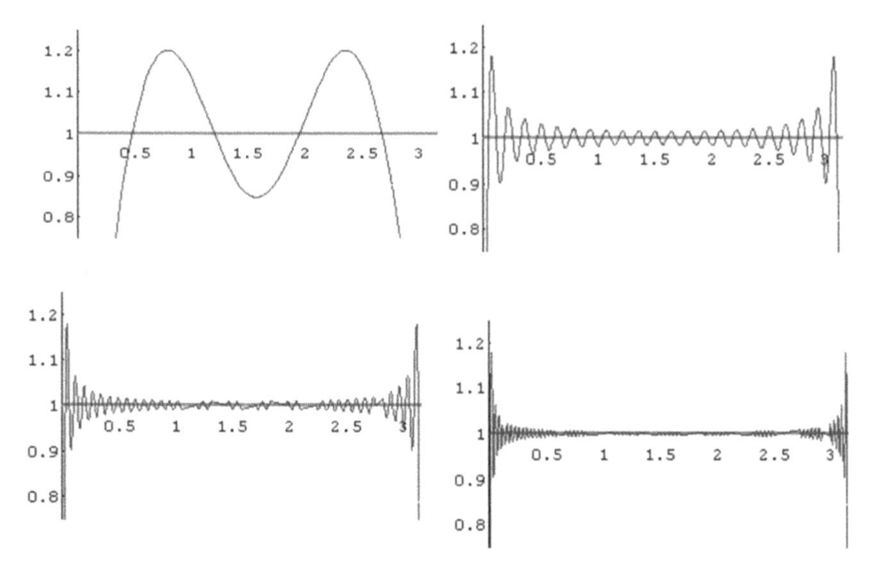

Figure 77.4. Mathematica plots of the Fourier Series Partial Sums for the function $f(t) = 1$.

Students were asked to list all possible parent pairings along with the probabilities of the resulting offspring combinations which resulted in the *genotype probability matrix* shown in Table 77.2.

The objective was to create a mathematical model which could predict the genotype distribution of the plant population after a certain number of generations under specific breeding programs (for example, where each plant of the population is fertilized with a plant of a particular genotype). Students were familiar with the elementary matrix theory and they were required to formulate the problem using matrices. For formulating the problem it was assumed that a_n, b_n and c_n are the fraction of plants of genotype **AA**, **Aa** and **aa** in the nth generation where $n = 0,1,2,...$ with subscript "0" denoting the initial fractions and $a_n+b_n+c_n=1$ for all n.

Students were first asked to consider the case when all the plants are fertilized with type **AA**. This led to the equations $a_n = a_{n-1} + \frac{1}{2} b_{n-1}$, $b_n = \frac{1}{2} b_{n-1} + c_{n-1}$ and $c_n = 0$.

In matrix notation this was expressed as $\mathbf{X^{(n)}} = \mathbf{MX^{(n-1)}}$, $n = 1,2,......$

where $\mathbf{X^{(n)}}$ is the column vector $[\, a_n, b_n, c_n \,]$ in transposed form and $\mathbf{M}$

$$= \begin{bmatrix} 1 & 1/2 & 0 \\ 0 & 1/2 & 1 \\ 0 & 0 & 0 \end{bmatrix}.$$

Table 77.2. Genotype Probability Matrix for the Six Possible Parent Pairings

Parent Pairings		*AA - AA*	*AA-Aa*	*AA - aa*	*Aa - Aa*	*Aa - aa*	*aa - aa*
Offspring outcomes	AA	1	1/2	0	1/4	0	0
	Aa	0	1/2	1	1/2	1/2	0
	aa	0	0	0	1/4	1/2	1

Students observed that **M** is the sub-matrix of the genotype probability matrix comprising of the first three columns. Following this, they used the equation $\mathbf{X}^{(\mathbf{n})} = \mathbf{M}\mathbf{X}^{(\mathbf{n-1})}$ to verify that $\mathbf{X}^{(\mathbf{n})} = \mathbf{M}^{\mathbf{n}}\mathbf{X}^{(\mathbf{0})}$ where $\mathbf{X}^{(\mathbf{0})}$ is the column vector of the initial population fractions. This equation helped them to find the genotype distribution of any generation of plants given the initial distribution $\mathbf{X}^{(\mathbf{0})}$. An exercise was given where the students were required to calculate the genotype distribution of the plant population in the first, second, third, fourth and fifth generations for different initial distributions. The graphics calculator proved to be a useful tool since they could manipulate the matrix equations in the **Mat** mode of the calculator.

Using the output matrix students made the observation that if all plants are fertilized with plants of red petal color then in the first generation 35% of the plants will have red petal color, 65% will have pink petal color and there will be no plants with white flower petals. For the second generation, the students obtained the fractions 0.675, 0.325 and 0 for the red, pink and white petal plants respectively. Some students calculated for the successive generations and concluded that the distribution becomes steady after a few iterations. The steady state distribution is [1, 0, 0] (in transposed form), which led to the interpretation that in the long run all plants will have red flowers. Further, students changed matrix **M** to explore the case when each plant is fertilized with type Aa. Here the steady state distribution was obtained as [0.25, 0.5, 0.25] (in transposed form). Thus the calculator took over the computational part of the modeling process and enabled the students to focus on the interpretation of the results. Students then used the following Mathematica code to graph the genotype distributions which helped to obtain a graphical insight into the problem.

```
M={{1,0.5,0},{0,0.5,1},{0,0,0}};
x[0]={{.2},{.3},{.5}};
F[n_]:=MatrixPower[M,n];
AA=Table[F[n][[1]].x[0],{n,0,20}];
Aa=Table[F[n][[2]].x[0],{n,0,20}];
```

```
aa=Table[F[n][[3]].x[0],{n,0,20}];
dataAA=Flatten[AA];
dataAa=Flatten[Aa];
dataaa=Flatten[aa];
plot1=ListPlot[dataAA,PlotJoined->True,PlotStyle->
{RGBColor[1,0,0],Thickness[0.02]},
AxesLabel->{"t","AA"},PlotRange->{{0,20},{0,1}}]
plot2=ListPlot[dataAa,PlotJoined->True,PlotStyle->
{CMYKColor[0,1,0,0],Thickness[0.02]},
AxesLabel->{"t","Aa"},PlotRange->{{0,20},{0,1}}]
plot3=ListPlot[dataaa,PlotJoined->True,PlotStyle->
{CMYKColor[0,0,1,0],Thickness[0.02]},
AxesLabel->{"t","aa"},PlotRange->{{0,20},{0,1}}]
Show[plot1,plot2,plot3,AxesLabel->None]
```

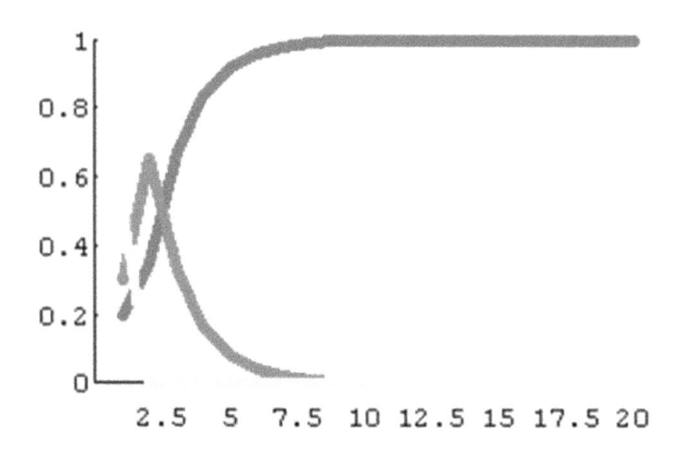

Figure 77.5. Mathematica plot indicating that all plants will have genotype AA in the long run (when fertilized with AA genotype).The graphics output helped students to conclude that when all plants are fertilized with genotype AA, in the long run all plants will have red flowers. In a similar manner, they explored steady state distributions after considering several other breeding schemes. Each time the graphics calculator was used for matrix computations while the Mathematica code was used to generate graphical outputs. In this module technology helped to increase the range of investigations.

CONCLUSION

This chapter highlights the role of technology in mathematics education and focuses on some pedagogical affordances of different technology tools, namely computer algebra systems, graphics calculators, dynamic geometry software and spreadsheets. In the second section, the discussion focuses on the challenges of school mathematics education in India. A disappointing curriculum in terms of transaction of content, inappropriate assessment and teacher preparation are some of the challenges. Integrating technology in the school mathematics curriculum is also fraught with numerous challenges. Technology must be cost effective and easy to deploy in order to achieve large scale integration. The author suggests that mathematics laboratories be used as a medium to provide students access to technology enabled explorations. However the greatest challenge is teacher preparation, that is, of developing sustainable professional development programs for teachers. The third section describes the role of technology in mathematics learning by briefly summarizing the findings of some key research studies conducted by researchers across the world. In the next section, the idea of including mathematical modeling activities in the curriculum through a mathematics laboratory has been suggested. Excerpts of some modules conducted by the author with students from Grades 9 to 12 in a laboratory setting have been described. These have been categorized under four heads, namely, those related to visualization and exploration of concepts using some form of technology, simulation of probability problems using spreadsheets, use of dynamic geometry software in exploring geometrical ideas, and investigatory projects based on mathematical modeling and applications of topics taught in the curriculum. In most of the modules, graphics calculators, spreadsheets, computer algebra systems or dynamic geometry software were the vehicles of exploration and students' investigations were guided by carefully designed worksheets which enabled them to explore the concepts in a step by step manner. Students had to provide manual solutions in some parts of the worksheets which helped to maintain a balance between by hand skills and use of technology. Feedback taken from the students at the end of the modules revealed that they preferred the laboratory modules to their traditional classes and that their levels of interest in the topic could be sustained over longer periods of time. The author emphasizes that in some modules technology served the purpose of a "mathematical investigation assistant" by taking over tedious computations, allowing students to focus on insights and concepts and by giving them more control over what they were learning. In other modules technology served the purpose of an "amplifier" by increasing the range of students' explorations and by giving them students access to higher level concepts which would have been difficult to achieve in a traditional teaching environment. The role of the teacher was more of a facilitator guiding and scaffolding students' explorations. In most cases the integration of technology provided a rich and motivating environment for exploring and developing mathematical concepts. It enriched mathematics teaching and learning and opened up possibilities for both students and teacher alike.

REFERENCES

Arnold, S. (2004). Classroom computer algebra: some issues and approaches. *Australian Mathematics Teacher, 60*(2), 17-21.

Ghosh, J. (2003). Visualizing solutions of systems of equations using Mathematica. *Australian Senior Mathematics Journal, 17*(2), 13-28.

Ghosh, J. (2004). Exploring concepts in probability using graphics calculators. *Australian Mathematics Teacher, 60*(3), 25-31.

Ghosh, J. (2005). Visualizing and exploring concepts in calculus using hand-held technology. *Proceedings of the 10th Asian Technology Conference in Mathematics (ATCM)*. Retrieved from http://www.atcminc.com/mPublications/EP/EPATCM2005/enter.shtml

Heid, M. K. (1988). Resequencing skills and concepts in applied calculus using the computer as a tool. *Journal for Research in Mathematics Education, 19*(1), 3-25.

Heid, M. K. (2001). Theories that inform the use of CAS in the teaching and learning of mathematics. Plenary paper presented at *Computer Algebra in Mathematics Education (CAME) 2001 symposium*. Retrieved from http://www.lkl.ac.uk/research/came/events/freudenthal/3-Presentation-Heid.pdf

Herwaarden, O. V., & Gielden, J. (2002). Linking computer algebra systems and paper-and-pencil techniques to support the teaching of mathematics. *International Journal of Computer Algebra in Mathematics Education, 9*(2), 139-154.

Kissane, B. (2008). Some calculus affordances of a graphics calculator. *Australian Senior Mathematics Journal, 22*(2), 15-27.

Kutzler, B. (1999). The algebraic calculator as a pedagogical tool for teaching mathematics. *International Journal of Computer Algebra in Mathematics Education,* 7(1), 5-23.

Laborde, C. (2002). Integration of technology in the design of geometry tasks with Cabri-Geometry. *International Journal of Computers for Mathematical Learning, 6*(3), 283-317.

Lagrange, J. B. (1999). A didactic approach of the use of computer algebra systems to learn mathematics. Paper presented at *Computer Algebra in Mathematics Education workshop, Weizmann Institute, Israel*. Retrieved from http://www.lkl.ac.uk/research/came/events/Weizmann/CAME-Forum1.pdf

Leung, A. (2012). *Discernment and reasoning in dynamic geometry environments*. Regular lecture delivered at 12th International Congress on Mathematical Education (ICME-12), Seoul, Korea.

Leung, A. (2010). Empowering learning with rich mathematical experience: Reflections on a primary lesson on area and perimeter. *International Journal for Mathematics Teaching and Learning* [e-Journal]. Retrieved April 1, 2010, from, http://www.cimt.plymouth.ac.uk/journal/leung.pdf

Libii, J. N. (2005). Gibbs Phenomenon and its applications in science and engineering. *Proceedings of the 2005 American Society for Engineering Education (ASEE) Annual Conference and Exposition, Portland.*

Lindsay, M. (1995, December). Computer algebra systems: sophisticated 'number crunchers' or an educational tool for learning to think mathematically? Paper presented at *Annual Conference of the Australasian Society for Computers in Learning in Tertiary Education (ASCILITE), Melbourne, Australia.*

National Council for Educational Research and Training (2005). *Position paper of National Focus Group on Teaching of Mathematics*. Retrieved from http://www.ncert.nic.in/rightside/links/pdf/framework/nf2005.pdf

Stacey, K. (2001). *Teaching with CAS in a time of transition.* Plenary paper presented at the *Computer Algebra in Mathematics Education (CAME) 2001 symposium.* Retrieved from http://www.lkl.ac.uk/research/came/events/freudenthal/2-Presentation-Stacey.pdf

Thomas, M. O. J., & Hong, Y. Y. (2004). Integrating CAS calculators into mathematics learning: Partnership issues. In *Proceedings of the 28th Conference of the International Group for the Psychology of Mathematics Education, 4,* 297-304.

CHAPTER 78

MATHEMATICS EDUCATION IN PRECOLONIAL AND COLONIAL SOUTH INDIA

Senthil Babu D.
Homi Bhabha Centre for Science Education (TIFR), Mumbai

ABSTRACT

The focus of this chapter is the broad area of indigenous and regional traditions of mathematical practices in India, in contrast to the image of a pan-Indian and monolithic Indian tradition. The paper describes and analyses arithmetic manuals, which exist as palm leaf manuscripts, that were used in the indigenous *tiṇṇai* or verandah schools, specifically, the *poṉṉilakkam, nellilakkam, eṇcuvaṭi*, which form part of the *eṇcuvaṭi* corpus. The contents of these are described in detail and the distinctive features of the corpus are outlined. This is followed by a discussion of the social setting of the tiṇṇai pedagogy. The paper concludes with a brief description of the transition from the tiṇṇai schools to the modern (colonial) school system.

Keywords: History of mathematics education in India, tiṇṇai schools, indigenous schools, arithmetic manuals

The First Sourcebook on Asian Research in Mathematics Education: China, Korea, Singapore, Japan, Malaysia, and India, pp. 1727–1748

This chapter may be viewed as a part of a larger project, whose aim is to reconstruct regional traditions of mathematical practice as a contrast to the commonly assumed image of a mathematical tradition that is pan-Indian and monolithic. Here, we introduce the nature and characteristics of elementary mathematics education in South India during the eighteenth and the nineteenth centuries with particular reference to the Tamil region. This is part of an attempt to probe the possibilities of different regional mathematical traditions, which have implications for contemporary mathematics education, besides relevance to the study of history of mathematics in the country. However, this chapter will confine itself to the historical context, the nature of the institutions, the curriculum, and aspects of its pedagogy in relation to one particular set of manuscripts, which also constitutes the primary evidence for the reconstruction of the history of mathematics education in the region. The chapter would also briefly outline the nature of the changes that the teaching and learning of mathematics went through during the course of the nineteenth century, against the background of the larger shifts in the colonial policies towards indigenous education.

There are several kinds of manuscripts that could provide a clue to the nature of the mathematical tradition in pre-colonial India. In the case of the Tamil region, these include

- the accounts or revenue manuscripts, which use various systems of measures and accounting procedures;
- the numeracy primers or the *eṇcuvaṭis*, which include the *poṉṉilakkam, nellilakkam, eṇcuvaṭi* and *kulimāttu**; and
- mathematical treatises or the *kaṇakkatikāram* manuscripts which are of a different order and involve a systematic exposition of mathematical engagement, beyond the elementary level.

In this paper, I will mostly talk about the second set of manuscripts – the numeracy primers – *poṉṉilakkam, nellilakkam, eṇcuvaṭi* and the *kulimāttu*. The *poṉṉilakkam* is the most basic numeracy primer; it introduces numbers and notation. The *nellilakkam* is the primer that deals with volumetric or cubic measures. The *eṇcuvaṭi* is the quintessential Tamil table book. The *kulimāttu* is the table book of squares.

Before going into the specificities of the number formation in these texts, it will be relevant here to provide context to these texts. This involves three different lines of historical reconstruction as follows:

- history of arithmetic practice in the non-institutional context,

---

*First occurrences of Tamil words are in italics. Subsequent occurrences are not italicized (except for names of texts) to enhance readability.

- history of elementary schooling institutions called the *tiṇṇai* schools, and
- history of these texts in relation to these institutions.

For the purpose of this chapter, we will dwell mainly on the second in order to contextualize the elementary number primers mentioned above. These primers were the product of the teaching and learning processes of indigenous elementary educational institutions called the 'tiṇṇai' schools or veranda schools. The primary historical sources related to tiṇṇai schools by themselves are not sufficient to provide us with a holistic picture. We have the texts in Tamil, which were products of the curriculum practiced in these schools, very few biographical accounts that mention them, a lot of oral lore that are often disjoint memories of these institutions, and the British educational records, most of which are statistical. We know nothing of the antiquity of the tiṇṇai schools in the Tamil region. No inscriptions available in Tamil seem to mention the tiṇṇai schools, which of course does not mean that they were absent during those periods. However, they were not unique to the Tamil country and seem to have been very common institutions almost all over the country. Some European travel accounts in the seventeenth century have accounts of such institutions, and the only detailed engagement of their presence was documented by the British Educational surveys in the beginning of the nineteenth century in Bengal and Madras and later in Bombay and Punjab.

Thomas Munro, as Governor-General of Madras Presidency ordered a detailed survey of indigenous schools in the year 1822 that culminated in the year 1826. He wrote a Minute, based on the survey, which became the basis for his initiatives to improve indigenous education. Dharampal (1983) used these surveys, along with the surveys of W. Adam in Bengal, Elphinstone in Bombay and G. Leitner in Punjab in his project to document the nature of indigenous education in the pre-colonial period. His aim was to find support for the views of Mahatma Gandhi, who in the course of a debate with Philip Hartog claimed that British rule uprooted indigenous education in India and hence literacy actually declined. This debate brought the British surveys of indigenous education out into the public. Dharampal reproduced these surveys to show how the British were responsible for erasing out a well established indigenous system of education by letting them "stagnate and die", through fiscal measures that affected the village society on the one hand, and by sheer "ridicule" and insensitivity towards that tradition on the other (Dharampal, 1983). B.S. Baliga, an archivist, engaging with the same debate used the annual returns of the Director of Public Instruction to show that the number of children in the school going age had actually increased and that more number of elementary schools were under the supervision of the Directorate of Public Instruction (Baliga, 1960). In that sense, in the second half of the nineteenth century, for him, literacy did not decline and the British cannot be accused of eroding literacy, but they succeeded in assimilating and fostering the tiṇṇai schools, resulting in better coverage. P. Radhakrishnan's project was to review the same set of British surveys; to bring out aspects of caste based discrimination in these schools (Radhakrishnan, 1990). His concern was to look at "differentiation in the participation of education and the role of the caste system in working out such a differentiation", because according to him, "from Vedic to village education is involved a process of Brahminic ascendancy in bureaucratic hierarchy ... [which] subsequently [worked] to restrict literacy ...[for the] lower classes, to guard their [i.e., brahmins'] interests'.

A more recent study of the Bengali *pathsalas* by Poromesh Acharya (1996), profiles the Bengali variant of the tiṇṇai schools. This account is different from the above works, in the sense that it tries to address the nature and orientation of the *pathsala* curriculum in relation to the society which supported the teacher and the schools. Calling these the "three – R" schools, Acharya says that these institutions flourished during the sixteenth to the eighteenth centuries. They were widespread, decentralized, with a curriculum that was oriented towards "practical" competence by following a "rote method" in the teaching of reading, writing and arithmetic, a curriculum that was sustained by the community and the teacher, providing them a sense of "spontaneity", which the British failed to appreciate.

All the above works are useful in providing us with important information related to the nature and extent of the tiṇṇai schools. Most of them rely on British accounts, when it comes to the task of understanding the curriculum and pedagogy related issues of these schools. Such perceptions were conditioned by particular understanding of what constituted "learning" at the elementary levels of a scheme of education. The tiṇṇai curriculum had to be perceived in relation to such frames of understanding, perpetuating a stereotype, characterizing them as "practical" or "reading, writing and arithmetic" schools. But the task is to attempt a reconstruction of the tiṇṇai mode, now that we know more about the content of the curriculum, nature and organization of the texts that were products of this curriculum, accompanied by distinct modes of pedagogic practice that was central to their functioning.

Elsewhere, I have discussed in detail the nature of curriculum and routine of these elementary learning institutions along with a brief discussion on the learning of mathematics using the primers, modes of problem solving and the social context in which such curriculum was sustained (Babu, 2007). The central place that memory as a mode of learning occupied in the learning processes is outlined in that article. Here, distinct features of this mode of memory are discussed in relation to the set of primers that I refer to as the *eṇcuvaṭi* corpus.

NUMBERS IN THE *PO<u>NN</u>ILAKKAM*

Children in these institutions began their lessons in mathematics with the learning of the Tamil numerals. The process began with the monitor pronouncing the number, which the student would recite, following after the monitor. The monitor then introduced the graphic symbol for that number in the Tamil notation, which the students had to recognize, associate with the number name, and while still reciting, write it by themselves on sand, all in unison. Often the monitor would also hold the hands of the beginners, assisting them in the process. This would continue till the student acquired familiarity with each number, with their name and symbol, stepping into the first process of committing himself to know numerical notation. Each number would be made familiar in an ordered pattern, as recorded in the text, called the *po<u>nn</u>ilakkam*, which is the elementary number primer in Tamil. This is not a textbook in the modern sense, but functions like a manual, which wasn't given to the students beforehand. However, each student, would create his own manual, when he gained sufficient skill to be able to write confidently on a palm-leaf, acquiring a book for himself in the process, his first *ēṭu* or *cuvaṭi*, as known in Tamil. When he reached that stage, the student was proficient in the Tamil number system, called as *muntiri ilakkam*.

The Tamil number system has three layers.

- Numbers from one and above, up to a crore are grouped as *pērilakkam* or *pēreṇ* (literally "large number").
- The second layer is called the middle number group *iṭai eṇ* or *kī<u>l</u>vāy ilakkam*, also known as *mēlvāy ci<u>r</u><u>r</u>ilakkam* (meaning middle numbers or small numbers, in the literal sense) which comprise the fractions from 1/320 leading up to one. The significant units that occur in this additive series are the following: *muntiri* (1/320); *araikkāṇi* (1/160); *kāṇi* (1/80); *araimā* (1/40); *mukkāṇi* (3/80), *mā* (1/20); *mākāṇi* (1/20 + 1/80 or 1/16); *irumā* (2/20); *araikkāl* (1/8); *mummā* (3/20); *mummākāṇi* (3/20 + 3/80 or 3/16); *nālumā* (4/20); *kāl* (1/4); *arai* (1/2); *mukkāl* (3/4) and *o<u>n</u><u>r</u>u* (1). Each of these numbers have standardized notation in Tamil. While the units muntiri, kāṇi and mā are "basic", the others are derived from these units: for example, "araikkāṇi" means "half of a kāṇi".

- The third group called the small number group – *ciṟṟeṇ* or *kīḻvāy ciṟṟilakkam* comprise fractions from the number (1/320 x 1/320) up to 1/320. This is muntiri = 1/320 is taken as one, and further divided into 320 parts, and called *kīḻ muntiri* (kīḻ meaning below) and the series proceeds as kīḻ muntiri (1/320 x 1/320); *kīḻ araikkāṇi* (1/320 x 1/160); *kīḻ* kāṇi (1/320 x 1/80) and so on till *kīḻ mukkāl* (1/320 x ¾).

This order of numbers is structured as an additive series, with respect to all the three layers. The significant numbers in the series become standard numbers and were simultaneously units of measures. For example, muntiri, kāṇi and mā are all units of land measures. In fact all the standard fractions in the middle series and the small number series stood for standard measures of land, except for mā which was also a unit for gold. The notations then are semantically loaded, not mere numbers, and would immediately strike a chord of familiarity to the beginner in the tiṇṇai school.

The second aspect important in *poṉṉilakkam* is that all the standard fractions can be expressed in terms of each other. This is called the *varicai* or series in Tamil, and they are muntiri varicai, kāṇi varicai and mā var-

Figure 78.1. A palm leaf from *Poṉṉilakkam* showing tables in Tamil numerical notation.

icai. This helps in learning fractions – as numbers, as quantities, as familiar units – all of which can be represented in relation to each other, but are distinguished by distinct notational forms.

What adds significance to the representation of the number series, along with the fact that they are additive and expressed as iterations, is that they are also represented as tables, in rows and columns. This becomes a structural device in representing the numbers and has significance in the memory mode of learning. The two-dimensional representation of numbers in a tabular form, immediately situates itself as memory images, visualized in a system of medians and reference points. A closer look at the structural organization of the *poṉṉilakkam* would reveal this feature. For example, the middle series begins with muntiri. Then,

muntiri + muntiri = araikkāṇi (Note that there is no symbol for "+". One needs to infer the addition operation from the space between the addends and the result in the right hand column).
araikkāṇi + muntiri = kāṇiyē araikkāṇiyē muntiri
kāṇiyē araikkāṇiyē muntiri + muntiri = kāṇi
kāṇi + muntiri = kāṇiyē muntiri
kāṇiyē muntiri + muntiri = kāṇiyē araikkāṇi

---
till you reach the next standard fraction, which is araimā, then
araimā + muntiri = araimā muntiri

---
till you reach the next standard fraction, which is mukkāṇi, then
mukkāṇi + muntiri = mukkāṇiyē muntiri

--- = mā (1/20)

Then mā proceeds in an iteration, when it gets added to all the above numbers in a series, till the next standard fraction is reached, which is mākāṇi (1/16), which will then add itself to all the previous combinations to reach irumā (2/20), then the series proceeds to reach kālaraikkāl, then till mummā (3/20), then till mummākāṇi (3/16), then till nālumā (4/20), then till kāl (1/4), then arai (1/2), then mukkāl (3/4), and finally one. After one, one gets added to kāl, to yield oṇṇēkāl, then to arai, which is oṇṇarai, then with mukkāl to yield oṇṇēmukkāl, then two, and so on till the standard integers occur to begin the series of large numbers, or the *pērilakkam*. This will constitute the process of learning the muntiri ilakkam.

The mā varicai, or the mā series on the other hand, will begin with mā (1/20), and after successive additions with mā, till one is reached, the same set of fractions are obtained in the iteration. Again mā, which is one whole divided into twenty parts, can be represented as a series involving the standard fractions. Twenty mā is equal to one vēli in a land measure, thus that series becomes immediately familiar in a context of practical measure. Likewise the kāṇi series would begin with 1/80, and proceed with successive additions of kāṇi till one is reached. Standard fractions occur in the series, which would again be represented as combinations of the fractions in addition. The idea of fractions as parts of whole – or parts of parts in a whole can be well established in this form of representation. To establish this aspect, there is another table, which had to be recited in a rhyming fashion:

> In one, mukkāl (3/4) is three parts in four
> In one, kāl (1/4) is one part in four
> In one, arai (1/2) is one part in two
> ---
> ---
> In one, mākāṇi (1/20 + 1/80) is one part in sixteen
> ---
> ---
> In one, muntiri (1/320) is one part in three hundred and twenty
> ---
> ---
> In one, kīḻ araikkāl (1/320 x 1/8), is one part in two thousand five hundred and sixty
> ---
> ---
> In one, kīḻ muntiri (1/320 x 1/320) is one part in one lakh and two thousand four hundred.

This marked the end of fractions, and the beginning of whole numbers in the order of the *poṉṉilakkam*. Each and every number in all the series mentioned above would be memorized in that particular order, in the same pattern of integrating the sound of the number name, visual recognition of the symbol, loud recital and writing, with concurrent testing at each level by the monitor or the teacher. Such an extensive system of fractions when represented as addition-based iterations, becomes the organizing basis to learn numbers in the memory mode of learning. There were also separate sessions as mentioned above, where the children would stand up and recite the entire series in unison, loudly in front of the teacher, one series after the other, repeatedly, day after day till the logic of addition as the basis of number organization can be cognitively internalized along with the process of building memory registers for the numbers in a particular order. This will finalize the memory learning of the elementary number series in Tamil, which is called the *poṉṉilakkam*, (poṉ = gold; ilakkam = number place, in the literal sense), denoting a particular order of numbers, as quantities.

Next in line is the *nellilakkam* (nel = paddy; ilakkam = number place) which is a number series that takes the units of Tamil volumetric measures as numbers and proceeds along similar lines as that of the *poṉṉilakkam*. Here, the standard numbers that occur in the series are the standard units for grain measures in Tamil. This series was also organized on the principle of iterations of addition, where the basic unit of grain measure, the *ceviṭu* becomes the number to be added repeatedly till the highest unit, the *kalam* is reached. The standard units of the grain measure, that occur on the way from ceviṭu to kalam are ceviṭu, *āḻākku*, *uḻakku*, *uri*, *nāḻi*, *kuṟuṇi*, *patakku*, *tūṇi* and kalam. In a similar pattern, all these units that occur below the unit kalam, and would be represented in combinations with each other, paired by addition. For example,

Oru ceviṭu + oru ceviṭu = iru ceviṭu
Iru ceviṭu + oru ceviṭu = mucceviṭu
mucceviṭu + oru ceviṭu = nālu ceviṭu
nālu ceviṭu + oru ceviṭu = oru āḻākku (hence, one āḻākku = five ceviṭu)
āḻākku + oru ceviṭu = āḻākkē oru ceviṭu

āḻākkē nālu ceviṭu + one ceviṭu = uḻakku (hence, two āḻākku = one uḻakku or ten ceviṭu = one uḻakku)
uḻakku + one ceviṭu = ulakkē oru ceviṭu

The table proceeds till one reaches the next unit, the uri which is five uḻakku or fifty ceviṭu. The next higher units keep occurring as representations of additions of ceviṭu, āḻākku, uḻakku, till nāḻi is reached, which will then be represented as additions of ceviṭu, āḻākku, uḻakku, and uri;

then occurs kuṟuṇi, which will be represented as additions of all the preceding units, then patakku, mukkuṟuṇi, tūṇi and finally kalam. After kalam, it still proceeds as kalam in addition with fractions until ten kalams, after which it becomes a regular whole number series when the standard kalam remains till one thousand kalam is reached, which marks the culmination of the *nellilakkam* series. It should be remembered that each of these units had distinct notational forms and were written as such, while here for the sake of clarity, their names are written and not the symbols. Although there is no information for the time taken by a student to become proficient in *poṉṉilakkam* and *nellilakkam*, it seems that this alone took about two years to complete (Cettiar, 1989).This also marks a process by which natural memory ability was trained into a cultivated memory, where reading and writing were only incidental to the learning process, not ends in themselves (Carruthers, 1990, p. 70).

LEARNING *EṆCUVAṬI*

The *eṇcuvaṭi* was the quintessential Tamil multiplication table book. They were so central to the life of the tiṇṇai school rhythm that they prompted an early observer to call them as "multiplication schools" (Dharampal, 1983, p. 261). The multiplication tables were followed by the learning of squares, called the *kulimāttu*. The *eṇcuvaṭi* is a compilation of several kinds of multiplication tables. All the numbers learnt during the course of *poṉṉilakkam* and *nellilakkam* would be subjected to multiplication with each other, to yield an entire set of tables, that were to be committed to memory. The organizing basis of the *eṇcuvaṭi* was multiplication, represented in a tabular format to secure an order, helping recollective memory. There are several layers of multiplication tables involved.

The first kind of tables in the *eṇcuvaṭi* are multiplication tables of whole numbers, beginning with one, and usually proceeding up to ten. Each table concludes with a line of verse that provides the sum of all the products and functions as a mnemonic. For example, the first table will appear as follows:

1	1	1
10	1	10
2	1	2
20	1	20
3	1	3
30	1	30

90	1	90
100	1	100

"five jasmine buds bloomed into 90 jasmines, shared by five *alakunilai* 595" ("595" is the sum of all the products in the table.)

Next, the *eṇcuvaṭi* presents multiplication tables of whole numbers with fractions, that is, each standard fraction in the middle series of the *poṉṉilakkam* will be multiplied with whole numbers from one to ten. For example, the muntiri table will be as follows:

1	muntiri	muntiri
10	muntiri	araimā araikkāṇi
100	muntiri	kālē mākāṇi
2	muntiri	araikkāṇi
20	muntiri	mākāṇi
200	muntiri	araiyē araikkāl

900	muntiri	iraṇṭē mukkālē mākāṇi
1000	muntiri	muṉṟē araikkāl

a verse "alakunilai patinette arai nālumā
araimā kāṇiyē araikkāṇiyē muntiri"
(= 18 + ½ + 4/20 + 1/40 + 1/160 + 1/320)

Third, we find multiplication tables of fractions with fractions: that is, each standard fraction in the middle series of the *poṉṉilakkam* will by multiplied with other fractions and represented in tables. For example, the mukkāl (3/4) table will be as follows:

mukkāl	mukkāl	araiyē mākāṇi (1/2 + 1/20 + 1/80)
arai (1/2)	mukkāl	kālē kālaraikkāl (1/4 + 1/8)
kāl (1/4)	-	mummā *mukkāṇi* (3/20 + 3/80)
nālumā (4/20)	-	mummā (3/20)
mukkāṇi	-	irumā *mukkāṇiyē* muntiri (1/20 + 3/80 + 1/320)

---		
araikkāṇi	mukkāl	muntiriyē kīḻ arai (1/320 + [1/320 x ½])
muntiri	mukkāl	kil mukkāl (1/320 x ¾)

Fourth, multiplication tables of fractions in the grain measures series as discussed in the *nellilakkam* series would be multiplied with whole numbers from one to ten and represented as tables. For example, the table for the first unit ceviṭu will be:

1	ceviṭu	ceviṭu
10	ceviṭu	oru uḻakku
100	ceviṭu	iru nāḻiyē ōruri (2 nāḻi and one uri)
2	ceviṭu	iru ceviṭu (2 ceviṭu)
20	-	ōruri(one uri)
200	-	ain良ḻi(five nāḻi)

9	-	oru āḻākkē nāṟceviṭu (one āḻākku and four ceviṭu)
90	-	irunāḻiyē ōruḻakku(two nāḻi and one uḻakku)
900	-	orupatakkē orunāḻiyē ōruri (one patakku, six nāḻi and one uri)
1000	ceviṭu	mukkuṟuṇiyē orunāḻi (one mukkuṟuṇi and one nāḻi)

"alakunilai oru kalamē nānku patakkē ainnāḻiyē oru mūvuḻakkē orāḻākku" (which is one kalam, four patakku, five nāḻi, one mūvuḻakku and one āḻākku)

Fifth, we find multiplication tables involving fractions in the middle series of the *poṉṉilakkam* with that of the fractions in the *nellilakkam* series, represented as tables. For example, the table for the unit nāḻi, will be:

one	nāḻi	nāḻi
mukkāl (3/4)	nāḻi	oru mūvuḻakku
arai (1/2)	nāḻi	ōruri (one uri)
kāl (1/4)	-	oruḻakku (one uḻakku)
nālumā (4/20)	-	ōruḻakkē mucceviṭu (one uḻākku and three ceviṭu)

---		
araikkāṇi (1/160)	-	kāl ceviṭu (1/4th of a ceviṭu)
muntiri (1/320)	nāḻi	kālaraikkāl ceviṭu (1/8th of a ceviṭu)

Below each table, there is a Tamil verse in a prosodic form, that functions as a mnemonic to remember a number, which is the sum of the products of that particular table. For example, in the first table given, the verse using jasmine buds, is to denote the number 595, which is the sum of the products of the first table. When this verse is sung aloud, it has a rhyme that aids quick retention in memory. Here is an instance, where we see that in a memory based learning system, a prosodic verse acts as a mnemonic to remember an entire table. Also, such verses would lend meaning to the mnemonic only in a context. It is called as alakunilai, in Tamil meaning "position that points" (alaku = pointer and nilai = place or position). The mnemonic verse also serves an additional function of verification in the course of solving a problem or executing algorithms. As Carruthers (1990) says, "accuracy comes about through the act of recreating in memory the complete occasion of which the accurate quotation is a part" (p. 61).

If the entire table set had to be memorized there was another very crucial resource, that had to play a inevitable role – language and rhyme. All the tables in the act of memorization went through the process of a rhythmic singing aloud, with the monitor setting the tune in the beginning, but later on almost universally followed by all the students of the tiṇṇai schools. For example, the first table will be sung as

ōroṇṇu oṇṇu (one one [is] one),
paittoṇṇu pattu (ten ones [are] ten),
ıroṇṇu iranṭu (two ones [are] two),
irupatoṇṇu irupatu (twenty ones [are] twenty) and so on till
nūṟoṇṇu nūṟu (hundred ones [are] hundred),
followed by the reciting of the mnemonic verse, which is sung as
mallikai aintu malarnta pū tonnūṟu
kolluvār aivar parittu" alakunilai 595.

Though the entire task of memorizing all the tables appear as a monumental task, we should remember that in the given context, even language learning proceeded in gradual stages, moving towards composition of verses as a desired objective. Learning language and learning mathematics were dependent on prosody and meter, elements that were essential for building Tamil verses. All the learning, including literature, grammar, styles – all were learnt through forms of prosody, and a learned man was one, who would immediately recognize the form of meter used in a particular verse and compose by himself, another in response to it. Texts like the *nannūl*, for instance, were manuals, which were not guidebooks to attain mastery in the art of composition but practical manuals, which guided one to write properly. This association between language learning and the learning of numbers was integral to the tiṇṇai pedagogy; it was a natural resource put to good advantage.

The process of committing entire tables to memory would proceed step by step. Practicing to write on sand would also occur either simultaneously or in the time allotted for that purpose. A Portuguese traveler, Peter Della Valle, traveling in the Malabar area of south India in the year 1623 described a typical *eṇcuvaṭi*, which testifies to the mutual mode of instruction practiced in this mode of simultaneous recital and writing, both acting as aids to memory (Dharampal, 1983, p. 260).

In the memory mode of learning, repetition in a context of mutual instruction proceeded through recognition of the sound of a number by hearing its name, followed by loud reciting, followed by writing, thus cognitively associating the sound with a symbol, in association with two numbers in a relationship, say multiplication. Again, vocalization and visualization are central. Each table committed to memory becomes a distinct lesson. For instance, if one table was memorized in a day, through repeated practice, the next day morning, that particular table had to be recited as the first thing in the morning, or else in the forenoon sessions, when the teacher himself would verify, when students take turns, to recite that table. When memorization proceeded in distinct parts, in a well-designed order, it also helped avoiding the problem of overloading of memory. These sessions of reciting the previous day's lessons, (called as *muṟai collutal* – meaning reciting in order, in the literal sense) included not just the language lessons based on the *ariccuvaṭi*, but also the tables of the *eṇcuvaṭi*. It is not surprising then that the learning of the *eṇcuvaṭi* was one continuous process throughout the course of instruction of the tiṇṇai schools, as Subbiah Cettiar (1989) testifies. Of course, there would have been plenty of occasions to commit error, but those were occasions, marked by moments of recollective loss, that could have been due to "improper imprinting in the first instance". Therefore the option was to carefully imprint, and by repetition and practice ensure that they are in the "long-term memory". The ethos of this mode of learning was not to "tire memory" by trying to memorize too much at a time, or too quickly.

> however large the number of things one has to remember, because, all are linked to one another, all join with what precedes to what follows, no trouble is required except the preliminary labour of memorizing (Carruthers, 1990, pp. 61-62).

The tables are memorized and stored in long term memory, assisted by the rhythms of a language, which helps in associating facts with words. Moreover the words for numbers had concrete meaning in the context of various kinds of transaction within the community. All of this aided in strengthening recollection. Representations in tabular order further assist recollection, allowing the possibility of identifying a median (one easy number in the middle of a table, say five, fifty, five hundred) so that both sides from that point could be remembered and recollected. Even though logical constructions involving numbers of this order are universal, (in contrast to words, that require habit and repeated practice for recollective memory), in the pedagogic practice of the *eṇcuvaṭi*, we find a situation where language plays a central role, integrating itself strongly to number learning, when not a single number name would appear strange to a child growing up in a community, that laboured and lived by measuring in its everyday life.

Towards the end, the students also commit to memory the various conversion tables involving measures of weight. Reduction of measures again proceeded through repeated exercises of memorizing the conversion tables. Some of the nineteenth century *eṇcuvaṭis* published in the region, had to devote more and more pages to these conversion tables because new sets of units were introduced and the need for conversion between the local and the newly introduced set of units in all realms kept expanding. To summarize, the following distinct features of the *eṇcuvaṭi* corpus in their pedagogic context can be discerned.

1. They are dependent on memory as a way of learning rather than as an aid or a skill. This is evident in the extensive system of mnemonics that are not only the constitutive elements of the structure of *eṇcuvaṭi* but often these mnemonics themselves are algorithms, involving transformations of specific quantities.
2. Such quantities are structured as sequences, addition centered iterations, marked by strong presence of mnemonics, which is both a pedagogic mode and an organizing basis.
3. In memory as a mode of learning, memorization, internalization and recollection – all will have to happen in a continuum.
4. But how can this be accomplished pedagogically? By loud recital, reciting after the teacher or the monitor, writing while reciting loudly and as externalized representations as symbolic objects.
5. In such a memory mode of learning, writing using particular notations helps the process of visualization that in turn aids associative memory, crucial in arithmetic operations. Notation and number name, together contribute to numeracy.
6. The arithmetic operations are in strong relationship with associative memory. For example, addition of fractions would imply both multiplication as an operation as well as factorization of fixed quantities or numbers. But there is no explicit mathematical representation of this process, and no references to the relatedness of arithmetic operations are found.
7. In a way, it is tempting to stress the absence of explicit engagements with transformation and concomitant representations. Interestingly, there are no names given to the concepts, which would help distinguish between processes and products. But curiously enough, language here becomes crucially important. For

addition based operations, *ye* sound becomes the mark (kāṇi + muntiri = kāṇiyē muntiri) and when multiplication is involved, kāṇi × muntiri = kāṇi muntiri, the multiplicands are named without the additional sound.

8. There is another instance where language learning and number learning go together. The fixed quantities like kāṇi, mā are semantically loaded since they are units that occur in everyday life. The learning of such concepts is not divorced from the ordinary day to day use of language of the people at large which has pedagogic significance. This can provide us with important insights into the relationship between the number idea and the development of notation and names.

TIṆṆAI PEDAGOGY AND SOCIETY

What would be the social context, as far as one can reconstruct, which sustained such a mode of learning of mathematics? How can the texts and the practice that they signify, help us to look at the relationship between arithmetic practice, curriculum, pedagogy and the society?

The orientation of the curriculum is thoroughly local. In an internalist sense, the characteristics of arithmetic expressions at the level of the *eṇcuvaṭi*, is socially sufficient, in the context of localized socio-economic transactions. The pedagogic strategies within the tiṇṇai institution worked to ensure skill and functionality, in a primarily agrarian and mercantile social order. But from the perspective of the history of mathematics, what were the possibilities of these arithmetic means and expressions to evolve into higher order representations? Can such goals or possibilities be discerned internally? Are they evident in the institutional rhythm of the tiṇṇai schools? The transition from *eṇcuvaṭi* to the higher treatises like *kaṇakkatikāram* seems quite plausible in terms of a shift in the cognitive levels/skills and goals of learning.

Even in the higher order of arithmetic engagement, as represented in the *kaṇakkatikāram* corpus, functionality and the local world of transactions remain as the focus, though the handling of arithmetic means belonged to a different plane of engagement – in the enumeration of "rule of three", magic squares, exhaustion problems, recursions and partitions. In a typical *kaṇakkatikāram* text we will have at least six distinct sections, classified according to the objects of computation. These texts primarily set out the rules of computation using different techniques. Normally, they are found to have sixty types or '*inam*' (in the sense of a 'genre') in as many verses. These verses enumerate techniques involving various kinds of measures related to land, gold, grain, solid stones, volumetric measures and a general section. For example, the section on land would deal with various ways to measure area of land of different dimensions, in both whole and fractional magnitude; estimation of total produce from a given area of land; assessment of yields and profit and so on. The section on gold would deal with computations related to estimation of quality of gold, calculation of price, and combinations of mixture in the making of particular grades of gold. Since gold was also a unit of money, usually this section would deal with computations involved in transactions of money in different situations. Verses dealing with grains for example would deal with techniques of conversion of measures, profit and loss calculations and so on (Kamesvaran, 1998). In a social sense, all such arithmetic representation with embedded cognitive aspirations, are characterized by a yearning to enable a person to be in "control" of a situation, to "plan", to "anticipate" and to recognize patterns. Yet, the occasions were the "normal" day to day socio-economic transactions.

The *eṇcuvaṭi* learning sessions involved regular afternoon sessions of problem posing and solving in the tiṇṇai schools. Here, recollective memory would have to score well in a problem solving context, which might involve more than one arithmetic operation at the same time. The students would carry back problems to their homes, where problem solving would happen in a non-institutional context, entirely orally and subsequently for the student the algorithm and the result of the problem, should ideally cement together, in the process strengthening the cognitive apparatus of associative memory. The next morning, the results are collected and discussed. The school and its pedagogic strategies was immersed in the cultural context, imbued with shared learning, where creativity in a child is associated with a whole set of agents outside the institution. The exact nature of the problems posed in these sessions are not evident from the *eṇcuvaṭi*s that are now available. These were manuals created by students themselves as a mark of the completion of the learning of the *eṇcuvaṭi*. However, there are some interesting sources that provide us with an idea of the nature of problems solved. I reproduce below an excerpt from a satirical play of unknown origin, reproduced in a published collection. The excerpt is an imaginary conversation between a tiṇṇai teacher and his students in the setting of a tiṇṇai school (Deivanayakam, 1986, pp.1-8).

The play begins with the lament of the teacher who gives vent to his plight and misery, yet takes pride in leading a virtuous life. As if to substantiate his claims over virtuosity, the teacher tries to demonstrate the ability of his students in the skills that he trained them in.

Teacher (calling out the name of a student):

"Hey, *muttukkumarā pāvātē rācāli*, read!!
Student: What to read, sir?
Teacher: Read the *kālaraikkāl* lesson (*kālaraikkāl*, that is the table for the fraction 1/8)
Student (the comedian's version to the audience): All lessons sound like that *araikkā* only. Quarter lesson, half lesson, three quarter of a lesson, anyway, there will never be a full lesson, to hear from you, right?
Teacher: Read, you little boy!
Student: What to read, *ātticcūti*, this and that?
Teacher: Great, how is it that you are telling me names of words (*muṟai collutal*)
Student: No, I am reading the book sir, not telling you words
Teacher (to the audience): See, how this guy fares. (points to another) Why don't you read?
Student: What to read sir? *ampikai mālai* or *mōtira muppatu*?
Teacher: This boy is keen only to tell me words, not to read. (looks at another student and says) What was the name of the flower I told you yesterday?
Student: *Makiḻampū māmpū mallikaippū* (names of flowers)
Teacher (to another student): What was the name of the flower for you?
Student: *Tāmaraippū tāḻampū, tāmarattāmpū* (names of flowers)
Teacher (to another student): Hey, what was the problem given yesterday?
Student: One *kuli* for one, ten and a quarter for one; cancel the eight, add two, mix, stir and multiply in the mouth to give it to eleven people.
Teacher: What was the problem for you? (to another student)
Student: *Nāle kālaraikkāl* is one banana. *kālē kālaraikkāl* is one banana. How much is the money?
Teacher: Hey, you there, give me the rod, at this rate, you all will read fifty books in fifty years.

Though the above scenario was meant to make fun of the plight of the tiṇṇai teacher and the way he handles his institution, what is significant to us is the information it gives us about the practice of taking tables as lessons – when the teacher asks the student to tell the "lesson" of kālaraikkāl (1/8) table. Subsequently, we also get to know a bit about the way problem solving sessions would have been conducted in these institutions. The problems that occur in the script above was to incite humor, and do not make any sense. But it shows how problems were posed – demanding more than one arithmetic operation at the same time, involving more than one variable. Also in the second problem, the banana problem, we get to see, how the recollective memory based on tables would have been tested in an algorithmic context. Here, the student had to associate the particular table, in that context, identify the product and proceed to compute. In the process, a common pool of problems and algorithms could have been collected over time. They belong to the institutions as texts and still remain as riddles and aphorisms outside the textbooks. Therefore, the memory mode of learning of numbers combined with the training in problem solving could prepare a tiṇṇai student to move on to look for a suitable teacher who could teach him higher mathematics, provided the family of the student could afford the costs. It is with such teachers that the mathematics as represented in texts like the *kaṇakkatikāram* was learnt.

Internally, within the tiṇṇai rhythm, we can see glimpses of the possibility of the given set of arithmetic means to develop into higher orders of engagement. But externally, did the tiṇṇai school as an institution offer such a possibility? This brings us to the question of the relationship between cognitive universals and social distribution of ability, strongly associated with questions of access and denial.

The tiṇṇai school needed strong monetary contributions, and had to operate in a caste society. The other important dimension however was that the curriculum was continuously monitored and validated by the community. There was a notion of sufficiency of the curriculum that relegated learning to functionality and socially credible notions of capability. The students were often evaluated in public/social spheres bringing in the teacher's capability at stake. (At times, poor performance appears to have led to consensual decisions to mete out severe punishment to the children.) If concepts, skills, reasoning, ways of abstraction were all immersed in the tiṇṇai mode, then it is interesting to study how the "modern" internally sealed off functionality, privileging reasoning, situated in the liberal individualism of early nineteenth century Europe wrought by the colonial encounter.

TIṆṆAI SCHOOLS IN TRANSITION

The story of elementary education in nineteenth century Madras is essentially a story of the tiṇṇai schools in their encounter with an ever persistent colonial state, which was bent on subjecting them to highly bureaucratic processes, based on alternating phases of contempt, reconciliation and accommodation. But through all these, the tiṇṇai schools were unrelenting, and they survived well into the early decades of the twentieth century.

The history of nineteenth century elementary education would appear quite different if the tiṇṇai schools were recognized as the most widespread and popular institutions of learning in comparison with the tiny number of institutions started at the behest of the various British policies. The story associated with the process of transition is a long one. To cut it short, poorly paid school inspectors and their assistants went on long tours, trying to gain the sympathy of the tiṇṇai school masters and the village elite, to shift to the modern/new curriculum, demanding that the teachers use the modern textbooks, send results and reports to them, asking them to get trained in the modern curriculum, asking them to learn what was then called school management techniques. But every time, after their long and tiring tours, they came back, and produced copious pages of unconvincing reports about the prospects of such a shift actually happening. It was a story of two curriculum structures, perceived and played out differently, marked by an idea of "relevance", not to mention questions of ideologies that involved struggles between utilitarian, liberal and continental experiences of learning and teaching among the Europeans themselves (Evidence taken before the Madras Provincial Committee of Education, 1882). This arguably resulted in a curiously mixed bag of what came to be called the techno-economic complex of a colonial state machinery, that had to perpetually contend with local traditions of institutionalized learning, rooted in the sphere of practice and whose orientation was thoroughly local. Extensive financial incentive based schemes were repeatedly worked out by the ever persistent state, to woo the unrelenting tiṇṇai school masters and students, who were at the same time, compelled to partake in the locally sanctioned educational processes with goals of credible learning, based on memory and functionality.

These dynamics persisted with concomitant processes of state making, when circulation of local skills was repeatedly sought to be incorporated into the revenue administrative mechanisms, spinning off processes of social exclusion into the networks of publicly legitimate ideas of good life, whose new parameter then was state recruitment.

What happened to the *eṇcuvaṭi* in this process? They were deeply mired in the encounter, and were also the first to face it, among the other texts like *kaṇakkatikāram*. The novel demands of a formal system of elementary education, organized through processes initiated by the company state and missionaries, almost independent of each other subjected the *eṇcuvaṭi* to continuous re-organization. The terms of this reorganization were set by the conception of the modern textbooks, the mainstay of European efforts at intervening in education of the natives. This process of reorganization was marked by the incorporation of the *eṇcuvaṭi*, in parts, into the modern textbooks of mathematics. They were in that sense, not integrated into the system of math teaching and learning, that was explicitly based on reasoning and abstraction, as its primary goals. Rather, they either stood alone as separate textbooks or remained as annexures to the modern textbooks. However they earned themselves a different name – *mental arithmetic or bazaar mathematics*. It is useful here to mention that this section gradually was removed out of the purview of official evaluation based on public examinations, when the *eṇcuvaṭi* remained as mere annexure, occupying the last few pages.

However, the *eṇcuvaṭi* remained vibrant in the nascent publishing market in the provincial centers. One can discern two independent and parallel streams in this regard. The *eṇcuvaṭis* with Tamil numerals and Tamil system of measures and tables were published by regional publishing houses, unedited and almost verbatim from the palm leaf manuscripts, through the course of the nineteenth century and well into the first three decades of the twentieth century. The earliest of these were the books published by local presses in Tanjore, Tiruvallur, Vedaranyam, Kumbakonam and Chidambaram in the 1830s. These were sponsored by the traders, who wanted them to persist as reckoners, sometimes with conversion tables for the old and the new system of measures. The Saivite mutt and the newly articulate social groups who established independent press houses on the other hand also published *eṇcuvaṭi*, in print, probably anticipating the tiṇṇai schools to shift to print, yet remain with the *eṇcuvaṭi*.

The other stream was more dynamic, organized by a new set of players of the emerging textbook market, the new class of petty traders in textbooks, or more generally in the print market. At this time producing for a newfound market for textbooks with a relatively new form of technology, the printing press must have been an exciting prospect for the polymaths of this class. Their aspirations were based on the continuous compulsions of the colonial state to accommodate the *eṇcuvaṭi*, even as annexure in the modern textbooks. Of course, how would this dampen the entrepreneurial spirit of the traders of print? They went about their business and kept printing both the modern table books and the *eṇcuvaṭi* for the market. These two streams of engagement with the *eṇcuvaṭi* remained almost parallel through the nineteenth century.

There was an interesting exception to this story though. In the early nineteenth century, Vedanayakam Sastri educated along with a Maratha prince under the guidance of German missionaries, thought it worthwhile to actually reorganize the *eṇcuvaṭi*, which he did. While most thought it better not to bother about the details beyond merely changing the Tamil notation into the modern, he reorganized the *eṇcuvaṭi*, premised on the perceived importance of conceptual understanding of the number and its transformations, wherein the processes of transformations of numbers became the guiding principle of reorganization, enabling him to call his work – *en-vilakkam*, "number explanation", as opposed to *en-cuvati* – the fixed, static "number text". However, this manuscript did not reach both the petit bourgeoisie's regional press houses or the presidency town of Madras and remained confined to a private archive, along with thousands of *eṇcuvaṭi* in various repositories whose only value was archaic (Sastri, 1807). Thus the *eṇcuvaṭi* remained separate from modern school mathematics textbooks till eventually they faded away.

REFERENCES

Acharya, P. (1996). Indigenous Education and Brahminical Hegemony in Bengal. In Nigel Crook (Ed.), *The Transmission of Knowledge in South Asia: Essays on Education, Religion, History and Politics* (pp. 98-118). Delhi. Oxford University Press.

Babu, S. D. (2007). Memory and Mathematics in the Tamil tiṇṇai Schools of South India in the eighteenth and nineteenth centuries, *International Journal for the History of Mathematics Education, 2*(1), 15–37.

Baliga, B. S. (1960). Literacy in Madras 1822-1931. In Baliga, B. S. (Ed.), *Studies in Madras Administration, II.* Government of Madras.

Carruthers, J. M. (1990). *The Book of Memory: A Study of Memory in Medieval Culture.* New York: Cambridge University Press.

Caturveta Siddhanta Sabha. (1845). *Eṇcuvaṭi.* Madras: Caturveta Siddhanta Sabha.

Caturveta Siddhanta Sabha. (1845). *Poṉṉilakkam.* Madras: Caturveta Siddhanta Sabha.

Cettiar, S. (1989). Interview with Prof. Y. Subbarayulu; âttankuti.

Deivanayakam. C. K. (1986). *Palajatika Vikatam.* Tanjore: Tanjore Sarasvati Mahal. Library Publications.

Dharampal. B. (1983). *The Beautiful Tree: Indigenous Education in the Eighteenth Century.* Biblia Impex.

Evidence taken before the Madras Provincial Committee of Education. (1882). Madras: Government Press.

Kamesvaran, K. S. (Ed.). (1998). *Kaṇakkatikāram – Tokuppu Nū l.* Tanjore: Sarasvati Mahal. Publication Series No. 388.

Radhakrishnan, P. (1986). *Caste Discriminations in Indigenous Indian Education – I: Nature and Extent of Education in Early 19th century British India.* Working Paper No. 63. Madras: Madras Institute of Development Studies.

Radhakrishnan, P. (1990). Indigenous Education in British India: A Profile. *Contributions to Indian Sociology,* 24 (1), 1-27.

Report of the Committee for the Revision of English, Telugu and Tamil School Books in the Madras Presidency. (1874). Madras.

Sastri, V. (1807). *Eṇvilakkam.*(Manuscript)

CHAPTER 79

REPRESENTATIONS OF NUMBERS AND THE INDIAN MATHEMATICAL TRADITION OF COMBINATORIAL PROBLEMS

Raja Sridharan
Tata institute of Fundamental Research, Mumbai

K. Subramaniam
Homi Bhabha Centre for Science Education (TIFR), Mumbai

ABSTRACT

This chapter provides an introduction to the mathematics associated with combinatorial problems that have their origin in music and prosody, which were studied by Indian mathematicians over the centuries starting from around the third century BC. Large parts of this mathematics are accessible without a knowledge of advanced mathematics, and there are several connections with what is learned in school or in early university education. The chapter presents expositions of such connections with, for example, binary arithmetic and Fibonacci numbers. In solving some of the problems, Indian mathematicians worked implicitly with the idea that all positive integers can be represented uniquely as sums of specific kinds of numbers such as the powers of 2, Fibonacci numbers and factorial numbers. These ideas are interesting, both in themselves and for the connections they make with aspects of culture, and hold promise for mathematics education and the popularization of mathematics.

Keywords: binary arithmetic, combinatorial problems, Fibonacci numbers, Indian mathematics, mathematics and music, mathematics and prosody

The First Sourcebook on Asian Research in Mathematics Education: China, Korea, Singapore, Japan, Malaysia, and India, pp. 1749–1767

REPRESENTATIONS OF NUMBERS AND THE INDIAN MATHEMATICAL TRADITION OF COMBINATORIAL PROBLEMS

The history of Indian mathematics has been an area of exciting new discoveries in recent decades. Fresh insights into the contributions of the Kerala mathematicians from the fourteenth to the seventeenth centuries CE are among the better known discoveries. Mathematical work from earlier periods too have been more thoroughly studied and better understood. Plofker (2009) provides a recent overview of the history of Indian mathematics. Several recent anthologies convey the excitement of current work in the field (see e.g., Emch, Srinivas, & Sridharan, 2005; Seshadri, 2010).

Our purpose in this chapter is to explore aspects of the history of Indian mathematics that may be of interest to the mathematics education community. Specifically we explore the work on combinatorial problems beginning from around the third century BC and continuing till the fourteenth century CE. The problems and the mathematical ideas developed by this tradition need only a level of mathematical knowledge available to many secondary school students. The ideas have interesting connections with Indian cultural forms, both living and historical and hence may appeal to a wider audience than those with a taste for mathematics. The material in this chapter is largely expository and draws heavily on the recent historical work and textual interpretations of among others, R. Sridharan, of whom the first author of this chapter is a collaborator (Sridharan, 2005, 2006; Sridharan, Sridharan, & Srinivas, 2010).

The development of numeral notation and forms in India provides a backdrop for the discussion of the connections of combinatorial ideas with number representations in this chapter. It is fairly well known that the decimal numeral system currently used had its origins in India and was transmitted to the West through contact with Arab culture. A decimal system of number names with Sanskrit names for the numbers from 1 to 9, and for powers of 10 up to a trillion was already developed in the second millennium BC and appears in the vedas, the oldest extant literature from India (Plofker, 2009). Large numbers were denoted by compounding names for 1 to 9 with names for powers of 10, much like in present day English. Besides these, once also finds in the vedic literature "concrete" number names, which are salient cardinalities (e.g., "moon" = 1, "eyes" = 2, "sages" = 7 from the well known *saptarṣi** or seven ancient sages). The earliest inscriptions in which a *positional* decimal numeral system is used, date to the second half of the first millennium CE. However

---

*First occurrences of Sanskrit or Tamil words are in italics. Subsequent occurrences are not italicized (except for names of texts) to enhance readability.

much earlier evidence for positional value exists in the form textual references. For example, the year in which a work was authored is mentioned as "*Viṣṇu* hook-sign moon." These are concrete number names where "*Viṣṇu*" (a leading diety) stands for one, "hook-sign" for nine (from the shape of the written numeral) and "moon" for one. Thus the number translates to "191" using decimal positional notation, a year measured in the Śaka era, which corresponds to the year 269 or 270 CE (Plofker, 2009). The order is actually right to left, which does not matter here since "191" is a palindrome.

It must be noted that there were other numeral systems in use through the centuries. Some of these were not based on positional value, like the alphanumeric system used by Āryabhaṭa, the author of the foundational astronomical-mathematical work *Āryabhaṭīya* written in 499 CE. In this system, consonants of the Sanskrit alphabet had specific numerical values depending on their position in the alphabet, while vowels indicated powers of 10 (Plofker, 2009). For example, the consonant "kh" had a value of 2, while the vowels "i" and "u" had respectively values of 10^2 and 10^4. Thus the syllable "khi" would mean 200, while "khu" would mean 20000. The syllable "ni" would denote 2000: "n" = 20 and "i" = 10^2. It was possible to denote the large numbers that are needed for astronomical calculations in a compact manner using Āryabhaṭa's notation, but the syllable-words that were produced were difficult to pronounce.

A more popular number system was the *positional value* based *kaṭapayādi* system in which consonants took numerical values from 0 to 9, depending on their order in the Sanskrit alphabet. The first ten consonants in the first two rows of consonants in the Sanskrit alphabet ("*k*" to "*ñ*") denote, in order, the digits "1" to "0". The next two rows also denote the same digits. So this system had redundancies—three or four consonants denoted the same digit, and vowels did not have numerical value. The redundancies allowed flexibility in the choice of a syllable combination to denote a number—often an actual Sanskrit word could be used to denote a number. Thus the word "*dhīra*" (meaning resolute or courageous) would denote 29, since "*dh*" denotes the digit "9" and "*r*" denotes "2". (Note that the order is right to left.) Of all these systems, the positional value based concrete number system described above was the most widely used in mathematical texts. For more details about the various number systems, see Plofker (2009).

COMBINATORICS IN MUSIC AND PROSODY

A rich tradition of combinatorial problems associated with the enumeration of symbol strings and mathematical techniques to solve them has existed in Indian mathematics for over two millennia. These problems have their origin not in a branch of science or technology, but in the arts—in prosody and in music. However, the mathematical ideas were pursued for their own sake as a distinct "mathematical" tradition, beyond the practical needs in poetry or music. While Sanskrit poetry largely belongs to the past or is pursued by specialized groups, Indian classical music is a living and vibrant aspect of contemporary Indian culture. (It must be noted that prosody in modern Indian languages is heavily influenced by Sanskrit prosody.) A penchant for classification and systematic organization is reflected in the classical musical traditions of India. These typically take the form of specifying an underlying basic structure or alphabet and enumerating melodic or rhythmic possibilities emerging from the basic structure. We will first look at this aspect of classical music, both with regard to melody and rhythm and then provide a brief introduction to these aspects in Sanskrit prosody. This introduction to cultural aspects provides a background for better appreciation of the mathematical discussion that follows. However, the mathematical sections can be understood independent of this background.

Combinatorics in Karnāṭak music. The two great streams of classical music in contemporary India are Karnāṭak and Hindustāni music. They share many characteristics and similarities, although the musical compositions, musicians and serious audiences are largely separate groups. We give a brief introduction to the Karnāṭak musical tradition, whose geographical center is in South India, to highlight the role of combinatorial structure in its melodic and rhythmic forms.

The melodic forms that provide the basis for both Karnāṭak and Hindustāni music are called *rāgas*, which are roughly analogous to scales in Western music. The basic specification of a rāga is in terms of the ascending and descending sequence of notes (*svaras*) in the scale. The notes are expressed using the seven "solfege" syllables of Indian classical music, which are pronounced as "Sa, Re, Ga, Ma, Pa, Dha, Ni", and commonly notated in writing using the first letter. These are short forms for the names of the notes: *Śaḍja, Riṣaba, Gāndhāra, Madhyama, Pancama, Dhaivata* and *Niṣada*. The tonic Sa (or Śaḍja) is fixed arbitrarily, and the remaining notes have a specific tonal relation to the tonic. The notes have higher and lower tonal values, which are shown in Table 79.1. However, the seven notes correspond only to 12 distinct tone positions because of overlaps. There are four pairs of duplicate names for the same position: R2=G1, R3=G2, D2=N1, D3=N2. It must be noted however that in Karnāṭak music as it is actually performed, the tonal values of the svaras are flexible (Krishnaswamy, 2003).

Table 79.1. Notes (Svaras) and Tonal Values (*Śrutis*) in Karnāṭak Music

Note	*Tonal Values*
Sa (tonic) & Pa (fifth)	Fixed
Ma (fourth)	Higher, lower (M1, M2)
Re (second), Ga (third), Dha (sixth) & Ni (seventh)	Higher, middle, lower (R1, R2, R3, G1, G2, G3, D1, D2, D3, N1, N2, N3)

The basic rāgas of Karnāṭak music, called the "*meḷakarta*" rāgas, always have the seven notes in the correct order in the ascending and descending sequences. The meḷakarta rāgas are the "mother" rāgas, from which other rāgas are derived ("born") by omitting some notes, varying the sequence of the notes, or by interpolating notes from other rāgas. By taking different combinations of the *distinct* tonal values of the seven notes, a total of 72 meḷakarta rāgas are obtained in the following manner: 6 (number of possibilities for R-G combinations) × 6 (number of possibilities for D-N combinations) × 2 (number of possibilities for M) = 72. It is interesting to note that the enumeration of the meḷakarta rāgas has a fixed order determined by the sequence in which the tonal values are varied. For example, the well known rāga "Śankarabharanam", which is analogous to the major scale in Western music, is number 29 in the meḷakarta sequence and has the following notes: S, R2, G3, M1, P, D2, N3.

The enumeration of the meḷakarta combinations provides a convenient organization of the alphabet and vocabulary of Karnāṭak music. A useful mnemonic system exists to identify the sequence number of a meḷakarta rāga. The rāgas have formal names (sometimes different from the common names) where the first two syllables in the name gives its number in the kaṭapayādi numeral system, mentioned in the previous section. For example the formal name using the kaṭapayādi for the Śankarabharanam rāga is "Dhīra Śankarabharanam", where "dhīra" in the kaṭapayādi system denotes 29. However, it must be said that the complete list and exact order of the meḷakarta rāgas are rarely emphasized in musical training, and are present largely as background reference.

Rhythm and numbers. The rhythmic basis (*tāla*) of Karnāṭak music is similarly provided by an alphabet consisting of finger tapping, and hand clapping and waving gestures. A vocalist almost always keeps rhythm using these gestures even while performing. In the most familiar tāla system of Karnāṭak music, there are seven basic combinations of these gestures. These coupled with five forms of the tapping gesture gives a system of 35 tālas, analogous to the system of the meḷakarta rāgas. However, unlike the meḷakarta rāgas, this system has little correspondence with the actual rhythmic structure used in most compositions. Only 3 of the 35 tālas are commonly used, and two other commonly used tālas do not find a place in the table of the 35 tālas.

The striking beauty and complexity of rhythm in Indian classical music derives from exploiting combinatorial possibilities in rhythm in another sense. The player of a percussion instrument like the *tabla* in Hindustāni music and the *mṛdaṅgam* in Karnāṭak music acquires, over time, a large stock of rhythmic phrases, which can be combined in creative ways to fit into the structure of a tāla. A striking aspect of both Hindustāni and Karnāṭak music is that complex rhythm patterns are both spoken and played on the instrument. The spoken form, called "*solkaṭṭu*" (literally "bundle of words") in Karnāṭak music consists of sets of syllables, each of which corresponds to and has a sound similar to a stroke played on the drum. Thus one may find a complex and intricate rhythm piece, several minutes long, first verbally recited in full, and then played exactly on a tabla or mṛdaṅga? Even when a percussionist trains, both forms are learned: "Throughout my training, I learned literally everything in two forms, spoken and played" (Nelson, 2008, p. 3).

The rhythm player in Indian classical music plays both solo and accompanies a vocalist or instrumentalist. It is in solo performance (often fitted into a vocal or instrumental concert) that the percussionist displays his (rarely, her) full repertoire and skill. Rhythm pieces are built up from complex phrases and sentences, which in turn are built up from a set of basic phrases and the use of rests or pauses. The basic rhythm phrases are easily recognizable to most people familiar with Indian music. For example, common four syllable phrases in Karnāṭak music are "ta ka di mi" and "ta ka jo ṇu"; a five syllable phrase might be "ta di ki ṭa tom"; a seven syllable phrase might be built as a combination of four and three—"ta ka di mi ta ki ṭa" or as a combination of two and five—"ta ka ta di ki ṭa tom".

The tāla structure provides the basic framework in which phrases are set and played. For example, the most commonly used tāla, the *ādi tāla*, consists of eight beats per cycle. Each beat is typically split into pulses, which may follow binary splits—2, 4 or 8 syllables per beat, or may follow splits based on three—3, 6, 12 syllables per beat. The percussionist designs a piece stringing together stock phrases and rests to cover several cycles of the tāla, creating contrasts, tensions and resolutions. The player often improvises on the fly while playing out a designed piece. The design and improvisation are called *kaṇakku* (literally "calculation"). Since the pulses, beats and cycles of the tāla must synchronize at crucial points during a piece, calculation and arithmetic are fundamental to the percussionists design and performance. Examples of simple and complex rhythm pieces for solo playing can be found in Nelson's *Solkaṭṭu Manual* (2008) and also in the solkaṭṭu recordings available on the web.

The classical dance traditions in India, which are also a live and vibrant aspect of the culture, share the tāla system and structure of percussion music. The syllables and spoken phrases are also a basic part of classical dance. Rhythmic compositions are often spoken out, much like in percussion music, before being performed as dance. The syllables used are similar with slight variations. Besides the classical traditions of music and dance, there are many vibrant traditions of folk music and dance spread across different regions in India. There is no doubt that the classical and the folk traditions influenced each other over the centuries. Hence it is possible that some of the aspects discussed above have corresponding features in the folk traditions. It is more than likely that research on these aspects will reveal interesting connections with numbers and mathematics.

Sanskrit prosody. The oldest extant text in Sanskrit is the *Ṛg Veda* from the second millennium BC. The four vedas, of which the *Ṛg Veda* is the oldest, are composed in specific metrical forms and have been preserved largely through an oral tradition centered around sacred ritual. The earliest authoritative discussion of these metrical forms is the work on prosody by Piṅgala from the mid-third century BC (Sridharan, 2005). The vedic metrical forms are classified on the basis of a count of the number of syllables. One of the most widely used metrical forms from the later vedic to classical periods, is the *anuṣṭubh*, a verse composed in four lines (*pādas*), each of which contains eight syllables. For example, the opening lines in anuṣṭubh verse of the Bhagavad Gita are

g g g g l g g g
dharmakshetre kurukshetre
l l g g l g l g
samaveta yuyutsavah

In the lines quoted above, each syllable is marked following the rules of Sanskrit prosody with a "*l*" or a "*g*", which corresponds to a light or a heavy syllable (*l* = *laghu* – literally "light" meaning short, *g* = *guru* – literally "heavy" meaning long). All Sanskrit poetry has the structure of the light and heavy syllables. Since there are no accents in the Sanskrit language, the meter is determined by the structure of the light and heavy syllables. The anuṣṭubh form has the number of syllables in a pāda fixed at eight, but the number of time units or "morae" is not fixed. Hence the duration taken to speak different lines of the anuṣṭubh stanza may be different. Many of the classical metrical forms have a fixed number of morae instead of fixed syllabic length, where the light syllables have a value of one mora and the heavy syllables a value of two morae.

A basic question that arises with regard to a metrical form is how many different possibilities there are of a given form. How many different combinations of laghu and guru syllables are possible when the form has a fixed syllabic length of n syllables? It is easy to see that this is 2^n since there are two possibilities (l and g) for each syllable. Similarly one can ask how many possibilities exist if the line has a fixed moraic length. As far as is known, the first text to deal with such problems is possibly Piṇgala's *Chandaḥ-sūtra*, which deals with enumerating metrical forms of a given syllabic length. Piṇgala's date is uncertain but it is possible that he lived around the time of Pāṇini in the third century BC. There was a connected (if not continuous) tradition of mathematical work on the problems related to prosody and music, that reached a mature form in the work of Nārāyaṇa Paṇḍiṭa in the fourteenth century CE.

There are several aspects of this tradition that are of potential interest to the mathematics education community. The first is that the mathematics associated with these combinatorial enumeration problems is interesting even from a contemporary perspective, and hence unexpectedly deep. At the same time, large parts of it are accessible without a knowledge of advanced mathematics, and there are several connections with what is learned in school or in early university education. The second is that numbers in the context of these problems primarily represent not quantity but serial (ordinal) position. That the mathematics associated with such representations can be interesting is a fresh and different perspective that may enrich students' experience of numbers. Finally, the methods used to solve these problems rely on uniquely representing positive integers in a variety of ways, which are vast and interesting extensions of the familiar representations of numbers in base ten or other bases. In the subsequent sections, we explore the mathematical aspects of this tradition.

THE FOUR PROBLEMS RELATED TO COMBINATORICS OF METRICAL FORMS

One of the basic questions that arise in the context of a poetic or musical form is what possibilities there are of a given type. Consider a syllable string consisting of exactly three syllables, each of which may be light or heavy. What are the syllable forms that are possible? This is the first problem discussed by Piṇgala. Piṇgala arrives at the fact that there are 2^n possibilities for a string of length n, by first enumerating the forms in a systematic manner. The systematic enumeration of forms is called "*prastāra*". Piṇgala discusses six problems associated with such forms, of which we focus on the following four problems in this chapter.

1. *Prastāra*: What are the combinations of light and heavy syllables that are possible for a given length of syllables? How do we enu-

merate all possibilities in order? What is the rule that allows one to carry out this enumeration?

2. *Saṇkhyā*: How many combinations are possible for a given syllabic length?
3. *Uddiṣṭa*: Given a string in the enumeration, how can one obtain the exact number of this string in the enumeration sequence?
4. *Naṣṭa* (converse of Uddiṣṭa): Given a number in the enumeration sequence, how can one obtain the string corresponding to this number?

We also find in Piṇgala a treatment of the *Lagakriya* problem, that is, to find the number of metres of a given length with a specified number of gurus (or equivalently, laghus). This problem, which we shall not discuss in this chapter, gives rise to the construction of what is now known as the Pascal's triangle (Sridharan, 2005).

First, we discuss the problem of enumeration or generating the prastāra for syllable strings of length n. To simplify the exposition we use the letters "*a*" and "*b*" to stand for heavy (*g*) and light (*l*) syllables respectively. Also, we have adopted a left to right convention because of the familiarity of dictionary order, which is the reverse of the convention adopted by Piṇgala. Table 79.2 gives the complete set of two, three and four letter "words" made from the letters "*a*" and "*b*". Notice that the lists are in dictionary order.

We see that each prastāra or enumeration can be obtained from the previous one by a recursive rule. To get the list of two letter words, we first prefix "*a*" to all the one letter words to get half of the two letter words, then prefix "*b*" and get the remaining half. Similarly to get the list of three letter words, we prefix "*a*" to the two letter words to obtain four of the three letter words, and prefix "*b*" to obtain the remaining four. The recursive rule actually follows from the fact that the order of enumeration is exactly the dictionary order. The rule allows us to generate the entire list from the previous one. However, it is not local enough to allow us, given a line in a particular prastāra, to generate the next line. For example, one may ask, which string comes just after "*bab*" in the prastāra of

Table 79.2. "Words" From a Two Letter Alphabet

One Letter Words	*Two Letter Words*	*Three Letter Words*	
a	*aa*	*aaa*	*baa*
b	*ab*	*aab*	*bab*
	ba	*aba*	*bba*
	bb	*abb*	*bbb*

three letter words? Piṇgala, as interpreted by later commentators, gives a rule to solve this problem, which amounts to the following. Going from right to left, change the first "*a*" that you encounter to a "*b*", replace all the letters to its right with a string of "*a*"s, and leave the rest of the string unchanged. The rule gives "*bba*" as the string immediately following "*bab*." It can be checked if this rule applies to all the lines in the prastāras in Table 79.2. (Note the similarity with the procedure for adding one when the numbers are expressed in the binary system with "*a*" standing for 0 and "*b*" for 1.)

We can see that the length of each prastāra is double that of the previous one, arriving at the fact that the length of a prastāra for a string of n syllables is 2^n. This is the saṅkhyā problem. We consider next the problem of uddiṣṭa by considering the following example: what is the exact position of the string "*bba*" in the prastāra for three letter words? The following line of reasoning allows us to construct a rule to solve this problem.

First, we assign numbers in the sequence starting from "0" instead of "1". Thus "*aaa*" occupies the zeroth position in the sequence of three letter words.

Since the first letter of the word *bba* is *b*, it cannot occur in the first four words of the prastāra. Its position *within* the last four is exactly the same as the position of *ba* in the prastāra for two letter words. In other words, the position of *bba* is $4+x$, where x is the position of *ba* in the two letter sequence. This gives us a recursive rule, since the position of *ba* in the two letter sequence is $2+y$, where y is the position of *a* in the one letter sequence, which is in fact zero. Thus we arrive at the position of *bba* as $4+2+0=6$. By substituting "1" for "*b*" and "0" for "*a*", we realize that "*bba*" is actually the binary representation of the number 6: "110", and 6 can be obtained by adding the powers of 2 with the digits as coefficients: $1\times2^2 + 1\times2^1 + 0\times2^0$. If we wish to enumerate the sequence in the natural manner from 1 to 8, then we need to increment this number by one. The position of *bba* in the sequence is then 7.

The problem of naṣṭa, which is the converse of the uddiṣṭa problem, is to obtain the string from the number that gives its position in the sequence, *given the total number of syllables*. The rule can be explained by taking the same example as above, and asking what is the string in the 7th position (using the natural numbering from 1 to 8) in the prastāra of three letter words? We arrive at this by the following rule: if the number is odd, add 1 and halve the number, write "*a*". If the number is even, halve the number and write "*b*". For the next step write the letter to the left of the previously obtained letters. So for the first step, we add $7+1=8$, halve 8 to obtain 4, and write "*a*". For the next step, 4 is even, so we halve 4 to get 2, and write "*b*" to the left of "a". Next, since 2 is even, we halve it to obtain 1 and write "*b*" to the left. Since we have obtained three letters we stop. If the string length is more than three, we continue the process till the correct number of syllables are obtained. (Note that once we obtain "1", the syllables for all the subsequent steps will be "a"s.) Reading the letters obtained, we get "*bba*".

We can also apply this rule to a number larger than 8, say, 9. Since 9 is odd, add 1, halve 10 to obtain 5 and write "*a*". In the next step, since 5 is odd, add 1 and halve to obtain 3, write "*a*" to the left. In the next step, add 1 and halve to obtain 2, write "*a*" to the left. In the next step, halve to obtain 1 and write "*b*" to the left . In this way, we obtain the string "*baaa*" as the 9th in the prastāra for four letter words. Note that in every case, we can also obtain the string by decrementing the number by 1, writing its binary representation, and substituting "*a*" for "0" and "*b*" for "1".

How does the procedure described above for solving the naṣṭa procedure work? We can understand this by considering another way to obtain the prastāra for three letter words from that of the two letter words, which is the following. Take the first word in the two letter prastāra, namely, "*aa*". We alternately append an "*a*" and a "*b*" to the *right* to get the first two words of the three letter prastāra. That is, we get "*aaa*" and "*aab*". Similarly to get the next two words we alternately append an "*a*" and a "*b*" to the *right* of the second word in the two letter prastāra, namely "*ab*". Thus we get "*aba*" and "*abb*". We can verify that these four words are the first four words in the three letter sequence. A little thought reveals that this is just another way of preserving the dictionary order as one moves from two to three letters. In general, we can get the $k+1$ letter sequence from the k letter sequence, by taking in order each row in the k letter sequence and generating two rows for the $k+1$ letter sequence by appending first an "*a*" and then a "*b*" to the right.

Hence from each word in the k letter sequence, we get two words in the $k+1$ letter sequence. More precisely, from the nth word in the k letter sequence, we get the $(2n{-}1)^{\text{th}}$ word – which always ends in an "*a*" – and the $(2n)^{\text{th}}$ word—which always ends with a "*b*"—in the $k+1$ letter sequence.

Let us now try to understand the naṣṭa procedure for three letter words. We know that if the row number is odd, it will be of the form $2n$-1, and the string corresponding to this row number will always end with an "*a*". Now we add one and divide by 2, which gives us n. By recursion, the next step is to find the string corresponding to the row number n, in the prastāra for two letter words. If the original row number is even, it is of the form $2n$, and the string corresponding to this row number will always end with an "*b*". Now we divide by 2, which gives us n. By recursion, the next step is to find the string corresponding to the row number n, in the prastāra for two letter words.

From the solutions to the uddiṣṭa and naṭṭa problems, we can set up a one to one correspondence between numbers and strings, which Indian mathematicians were clearly aware of. From the recognition of such correspondence, it is a major leap to ask the question in what sense each word in the prastāra of three letter words in Table 79.2 represents the numbers 1 to 8. The answer clearly is that the words are just binary representations of numbers (0 to 7 in the standard binary representation instead of 1 to 8). It is not clear if Piṇgala or other mathematicians actually saw the strings as we see them now, that is, as representations for numbers. However, one wonders what really lay behind the interest in the uddiṣṭa and naṣṭa problems, which do not have any apparent practical significance.

To summarize, we note that Piṇgala provides rules for (i) obtaining the prastāra for syllabic meters, that is, meters of a fixed syllabic length in terms of strings of a binary alphabet (ii) obtaining the total number of such strings (meters) (iii) obtaining the sequence number of a given string and (iv) obtaining the string from its number in the sequence. A rule that generates a unique sequence of strings (prastāra) allows one to formulate the uddiṣṭa and naṣṭa problems. The rules obtained as solutions to the problems of uddiṣṭa and naṣṭa set up a one-to-one correspondence between strings and numbers, allowing the possibility of interpreting the strings as representations of numbers. In the next section, we consider these four problems in the context of a different type of verse—verses with fixed moraic length.

Mātrāvṛttas and Fibonacci Numbers

The metrical forms that we considered in the previous sections were those with fixed syllabic length, where the syllables could be either light or heavy. In this section we consider metrical forms called mātrāvṛttas, where the number of morae, or time units (*mātrās*) is fixed. Here too, syllables may be light (*l*) or heavy (*g*), with *l* having a value of one time unit and *g* a value of two time units. Thus a meter of the form *llg* would be 4 units long; so would a meter of the form *gg* or *llll*. The four problems discussed in the previous section (prastāra, saṇkhyā, uddiṣṭa and naṣṭa) can be posed with regard to the mātrāvṛttas. The solutions to these problems for the mātrāvṛttas are found in the work of Indian mathematicians beginning with Virahāṅka in the seventh century CE. Table 79.3 presents the prastāras for mātrāvṛttas of lengths 1 to 6.

From Table 79.3, we can see how a recursive rule allows us to generate the prastāra for a given length from the previous two prastāras. For example, to generate the prastāra of length 4, append a "*g*" at the end of each string in the prastāra of length 2, and an "*l*" at the end of each string in

Table 79.3. Prastāras for Mātrāvṛttas of Different Moraic Lengths

Total Length = 1	*Total Length = 2*	*Total Length = 3*	*Total Length = 4*	*Total Length = 5*	*Total Length = 6*	
l	*g*	*lg*	*gg*	*lgg*	*ggg*	*lggl*
	ll	*gl*	*llg*	*glg*	*llgg*	*glgl*
		lll	*lgl*	*lllg*	*lglg*	*lllgl*
			gll	*ggl*	*gllg*	*ggll*
			llll	*llgl*	*llllg*	*llgll*
				lgll		*lglll*
				glll		*gllll*
				lllll		*llllll*

the prastāra of length 3. Thus the number of strings in the prastāra of length 4 is the number of strings in the prastāra of length 3 plus the number of strings in the prastāra of length 2. Using the notation S_n for the number of strings in the prastāra of length *n*, we can write

$$S_4 = S_3 + S_2$$

Generalizing, we have, $S_n = S_{n-1} + S_{n-2}$

This is exactly the recursive relation for the so-called Fibonacci numbers. Virahāṅka's text may well be the first to write down the recurrence relation for the Fibonacci numbers, although it may have been known earlier. The recurrence relation gives the solution to the saṅkhyā problem of finding the number of strings in the prastāra of length *n*. The discussion of the mathematics associated with such prastāras was a continuing tradition. Later mathematicians, to name a few, such as Halāyudha, Kedārabhaṭṭa and Hemacandra, discussed these problems. Like in the case of the varṅa prastāras, the solutions to the problems of Uddishta and naṣṭa, are also discussed for the mātrāvṛttas. The mathematical rationale underlying the solutions is the fact that any positive integer is either a Fibonacci number or can be expressed uniquely as a sum of non-consecutive Fibonacci numbers (Sridharan, 2006). This can be easily checked by the following argument. Take any positive integer N_0. If N_0 is a Fibonacci number, we stop since the unique sum is the number itself. If N_0 is not a Fibonacci number, then there is a largest Fibonacci number S_n such that $S_n < N_0$. Now consider the number $N_1 = N_0 - S_n$. Since $S_{n+1} > N_0$, we have $S_n + S_{n-1} > N_0$ or $S_{n-1} > N_1$. We repeat the process for N_1. The construction ensures that the process will yield a unique S_n at each step and will terminate since the sequence is strictly decreasing. The number N_0 is the sum of all the Fibonacci numbers obtained in this manner. The process also ensures that we do not obtain consecutive Fibonacci numbers at any stage. Thus every positive integer can be expressed uniquely as a sum of non-consecutive Fibonacci numbers.

The naṣṭa problem can now be solved using a simple algorithm. Suppose we want to find the string corresponding to the sequence number 7 in the prastāra for mātrā length of six (see Table 79.3). We note that the prastāra for a mātrā length of six units has 13 strings. We first subtract 7 from 13, which is the total number of strings. The result 6, is to be expressed as a sum of Fibonacci numbers, which we obtain as 5 + 1. Now we write down all the Fibonacci numbers from 1 to 13 and write down an "*l*" or a "*g*" below each of them using the following rule. For all Fibonacci numbers that appear in the expression, we write down a "*g*" below this number and skip the next Fibonacci number (put a "dash" below it). Below all the remaining Fibonacci numbers write down an "*l*". So for 6=5+1, we write a "*g*" below 1 and put a dash below 2. We also write a "*g*" below 5 and put a dash below 8. We write *l* below the remaining Fibonacci numbers. As seen below, we obtain the string as: *glgl*.

1	2	3	5	8	13
g	–	*l*	*g*	–	*l*

We verify from Table 79.3 that this is the seventh string in the prastāra for 6 mātrās. We consider one more example: what is the string that is number 4 in the prastāra for 6 mātrās? First we subtract 4 from 13, this gives 9. Next we express 9 as a sum of Fibonacci numbers. We obtain 9 = 8 + 1. Now we apply the rule and write "*l*" and "*g*" below each of the Fibonacci numbers from 1 to 13 in the following manner:

1	2	3	5	8	13
g	–	*l*	*l*	*g*	–

We obtain the string "*gllg*" as the fourth string in the prastāra, which can be verified from Table 79.3. We leave it to the reader to verify the algorithm in other cases. As can be seen, the algorithm depends on the fact that each number can be expressed as a sum of Fibonacci numbers uniquely, where there are no consecutive Fibonacci numbers. The uddiṣṭa problem needs one to proceed in the converse direction. The reader is refered to the article by Sridharan (2006) for an exposition of the uddiṣṭa rule.

It is well known that Fibonacci numbers occur widely in many natural contexts. The mātrā prastāra provides a context for grasping the recurrence relation among Fibonacci numbers that is accessible. The idea that Fibonacci numbers form a "base" in which all positive integers can be expressed may be a surprising and interesting fact for many students. One can use a string of zeros and ones to represent a number in this "base", where the position of the ones indicate the Fibonacci numbers which appear in the sum. There will, of course, be no consecutive ones in this number representation.

EXTENSIONS AND OTHER PRASTĀRAS

For the two prastāras discussed earlier (Tables 79.2 and 79.3), we write down the recursive relation and note the similarity in the two relations.

$S_n = S_{n-1} + S_{n-1} = 2 \times S_{n-1}$ (varṇa prastāras, Table 79.1)

$S_n = S_{n-1} + S_{n-2}$ (mātrā prastāras, Table 79.2)

The recursive relations suggest different kinds of extensions. One possible extension is of the form

$S_n = S_{n-1} + S_{n-1} + S_{n-1} = 3 \times S_{n-1}$

This gives rise to the ternary sequence or powers of 3: 3^0, 3^1, 3^2... One can obtain unique representations of positive integers using powers of three, which would correspond to the canonical base 3 representation. Such prastāras are discussed in the work of Nārāyaṇa Paṇḍiṭa in the fourteenth century. In fact, Nārāyaṇa Paṇḍiṭa discussed such relations in their general form (i.e., corresponding to base n representation) (Singh, 2001). However, we do not discuss these any further in this chapter.

An extension of the recursive relation for mātrā prastāras, that is, the Fibonacci relation, could be

$S_n = S_{n-1} + S_{n-2} + S_{n-3}$

The numbers obtained through this recursive relation and the associated mathematics were again discussed by Nārāyaṇa Paṇḍiṭa. Here too, he analysed the most general form of this relation ($S_n = S_{n-1} + S_{n-2} + ... + S_{n-q}$), where q is an arbitrary number less than n. Oddly enough, another recurrence relation was analysed by the musicologist Śārṇgadeva before Nārāyaṇa Paṇḍiṭa, who studied the problem for its connection not to prosody, but to rhythm or tāla patterns. In prosody, we considered time units with values of 1 and 2, the laghu and the guru respectively. In the context of tāla patterns, Śārṇgadeva considers four time units: *druta*, laghu, guru and *pluta*. The druta is half the duration of a laghu and a pluta is thrice the duration of a laghu. Re-adjusting to whole number values, we get the following values for the four units: druta – 1, laghu – 2, guru – 4, and pluta – 6. The question that may be asked is what combinations are possible for a sequence of units that has a total duration of say, seven time units. We note that the complete list of such combinations can be derived in the following manner:

- Append a pluta (P) at the end of all strings of duration 1 unit.
- Append a guru (G) at the end of all strings of duration 3 units
- Append a laghu (L) at the end of all strings of duration 5 units
- Append a druta (D) at the end of all strings of duration 6 units

The tāla sequences obtained by applying the algorithm are shown in Table 79.4. The recursive relation for the prastāra of 7 units can be written therefore as $S_7 = S_6 + S_5 + S_3 + S_1$. Generalizing, we get

$$S_n = S_{n-1} + S_{n-2} + S_{n-4} + S_{n-6}$$

The numbers S_n obtained using this recursive relation have been called Śārṅgadeva numbers in analogy with the Fibonacci numbers (Sridharan, Sridharan & Srinivas, 2010). The above recursive relation allows one to solve the saṅkhyā problem, namely, to find the number of strings for a given total duration. Śārṅgadeva also provides solutions to the naṣṭa and uddiṣṭa problems. As one may guess, these depend on the fact that every positive integer can be uniquely expressed as a sum of Śārṅgadeva numbers. We do not discuss the mathematical aspects of the Śārṅgadeva number representation in this chapter and the interested reader is referred to Sridharan, Sridharan and Srinivas (2010). As mentioned, Nārāyaṇa Paṇḍita in his *Gaṇitakaumudī* of 1356 CE discusses general recurrence relations of this from a purely mathematical point of view unconnected to applications in prosody or music. Nārāyaṇa Paṇḍita's work brings this tradition to its culmination.

Table 79.4. Tāla Combinations of a Total Duration of 7 Units

1	DP	9	GDL	17	DDGD	25	DLLDD
2	DLG	10	LLDL	18	GLD	26	LDLDD
3	LDG	11	DDLDL	19	LLLD	27	DDDLDD
4	DDDG	12	DLDDL	20	DDLLD	28	GDDD
5	DGL	13	LDDDL	21	DLDLD	29	LLDDD
6	DLLL	14	DDDDDL	22	LDDLD	30	DDLDDD
7	LDLL	15	PD	23	DDDDLD	31	DLDDDD
8	DDDLL	16	LGD	24	DGDD	32	LDDDDD
						33	DDDDDDD

The prastāras discussed by Śārṅgadeva which are related to rhythm forms are called tāla prastāras. Śārṅgadeva also considers tāna prastāras, or combinations of musical notes. An example of tāna prastāras considered by Śārṅgadeva is the enumeration of all phrases containing the svaras S, R, G, M (the first four notes of the seven-note musical scale of Indian classical music discussed earlier), where each svara occurs only once. Śārṅgadeva describes a rule for constructing the rows of the prastāra, the number of rows being given by 4 factorial (4!). The prastāra is shown in Table 79.5.

Śārṅgadeva discusses the saṅkhyā, naṣṭa and uddiṣṭa problems for the tāna prastāras. The solution to the latter two problems are based on fact that any positive integer m less than or equal to n! can be uniquely represented as follows:

$$m = d_0 0! + d_1 1! + d_2 2! + \ldots + d_{n-1}(n-1)!$$

Where d_i are integers such that $d_0 = 1$ and each d_i lies between 0 and i both inclusive. This is a variant of the general form for the factorial representation of integers (Sridharan, Sridharan, & Srinivas, 2010).

CONCLUDING REMARKS

In the preceding sections, we have discussed the generation of string sequences or prastāras in the context of prosody and music. We considered four kinds of prastāras. Two of these were discussed in greater detail: the varna prastāras, which are the combinations for verses having a fixed length of syllables, and the mātrā prastāras, which are the combinations where the moraic length is fixed. Two more prastāras were discussed briefly, those associated with rhythm forms (tāla prastāras) and those associated with combinations of notes (tāna prastāras). For each of these prastāras, four problems can be considered:

- Prastāra: the rule for generating the prastāra itself,

Table 79.5. Tāna Prastāra for the First Four Notes

1. S R G M	7. S R M G	13. S G M R	19. R G M S
2. R S G M	8. R S M G	14. G S M R	20. G R M S
3. S G R M	9. S M R G	15. S M G R	21. R M G S
4. G S R M	10. M S R G	16. M S G R	22. M R G S
5. R G S M	11. R M S G	17. G M S R	23. G M R S
6. G R S M	12. M R S G	18. M G S R	24. M G R S

- Saṅkhyā: the total number of combinations or strings in the prastāra
- Uddiṣṭa: obtaining the sequence number of a given string
- Naṣṭa: recovering the string when the sequence number is given.

As discussed, the last two problems are of particular importance, since their solution involves decomposing any given positive integer uniquely into numbers of a particular form. In the case of the varna prastāras, the numbers were powers of two. For the mātrā prastāras, these were Fibonacci numbers. For the tāla and the tāna prastāras, these were the Śārṅgadeva and the factorial numbers respectively. All of these lead to different kinds of unique representations for the positive integers. The idea that the base 10 system of number representation is only one among many different kinds of representations is a powerful idea that is made accessible by the consideration of the combinatorial problems such as those discussed by Indian mathematicians. The fact that these problems are associated with cultural forms—music, dance and prosody—that are still a living part of our experience can bring these domains closer to mathematics. The historical perspective on Indian mathematical traditions suggests the mathematics "embedded" in these cultural forms did not remain merely implicit, but were explored explicitly by mathematicians, and led to the development of a productive tradition of combinatorial problems within mathematics. We believe that the discussion of the mathematics associated with such problems holds promise in mathematics education and in the popularization of mathematical ideas. The details of how connections can be made between school mathematics and historical traditions, such as the one that we have discussed in this chapter, requires both further research and more work with learners of mathematics.

REFERENCES

Note: For a more exhaustive set of references, see Sridharan, Sridharan and Srinivas (2010).

Emch, G. G., Sridharan, R. & Srinivas, M. D. (Eds.) (2005). *Contributions to the History of Indian Mathematics*, Delhi: Hindustan Book Agency.

Krishnaswamy, A. (2003). On the twelve basic intervals in South Indian classical music. Paper presented at the 115th Audio Engineering Society Convention. Retrieved from http://ccrma.stanford.edu/%7Earvindh/cmt/aes11503.pdf

Nelson, D. (2008). *Solkaṭṭu Manual: An Introduction to the Rhythmic Language of South Indian Music.* Middletown, CT: Wesleyan University Press.

Plofker, K. (2009). *Mathematics in India*. Princeton, NJ: Princeton University Press.

Seshadri, C. S. (Ed.) (2010). *Studies in the History of Indian Mathematics*, Delhi: Hindustan Book Agency.
Singh, P. (2001). The *Gaṇitakaumudī* of Nārāyaṇa Paṇḍita, Chapter XIII, (English translation with notes), *Ganita Bharati, 23,* 18-82.
Sridharan, R. (2005) Sanskrit Prosody, Piṅgala Sutras and Binary Arithmetic. In G. G. Emch, R. Sridharan and M. D. Srinivas (Eds.) *Contributions to the History of Indian Mathematics* (pp. 33-62), Delhi: Hindustan Book Agency.
Sridharan, R. (2006). Pratyayas for Mātrāvrttas and Fibonacci Numbers. *Mathematics Teacher, 42*, 120-137.
Sridharan, Raja, Sridharan, R. & Srinivas, M. D. (2010) Combinatorial Methods in Indian Music: Pratyayas in Sangitaratnakara of Sarngadeva. In Seshadri, C.S. (Ed.) *Studies in the History of Indian Mathematics* (pp. 55-112), Delhi: Hindustan Book Agency.

Lightning Source UK Ltd.
Milton Keynes UK
UKHW022259281020
372363UK00004B/12